Research Design
and Statistical Analysis

Research Design and Statistical Analysis

Jerome L. Myers
Arnold D. Well
University of Massachusetts

HarperCollins*Publishers*

Sponsoring Editor: Laura Pearson
Project Editor: David Nickol
Design Supervisor: Heather A. Ziegler
Cover Design: Wanda Lubelska Design
Production: Willie Lane/Sunaina Sehwani
Compositor: House of Equations, Inc.
Printer and Binder: R. R. Donnelley & Sons Company

Research Design and Statistical Analysis

Library of Congress Cataloging-in-Publication Data

Myers, Jerome L.
    Research design and statistical analysis / Jerome L. Myers, Arnold
D. Well.
        p.   cm.
    Includes bibliographical references and index.
    ISBN 0–673–46414–8
    1. Experimental design.   2. Mathematical statistics.   I. Well, A.
(Arnold)   II. Title.
QA279.M933 1991   90–21469
001.4′34—dc20   CIP

91 92 93 94   9 8 7 6 5 4 3 2 1

Contents in Brief

Contents in Detail

CHAPTER 13   ANALYSIS OF COVARIANCE   435

CHAPTER 14   MORE ABOUT CORRELATION   465

Preface

In writing this book, we had two overriding goals. The first was to provide a textbook from which graduate and advanced undergraduate students could readily learn. Over the years we have experimented with various organizations of the content and have concluded that bottom-up is better than top-down learning. In view of this, most chapters begin with an informal intuitive discussion of key concepts to be encountered, followed by a simple numerical example. At that point, having given the student a foundation on which to build, we provide a more formal justification of the calculations and an extensive discussion of relevant assumptions. The same principle underlies the ordering of the chapters. Although it is tempting to begin with a development of the general linear model and then treat topics such as analysis of variance as special cases, such a treatment runs the risk of quickly burying the student in a sea of regression coefficients and matrix manipulations. We treat the special case of analysis of variance first. Most students are already familiar with the calculations through an introductory statistics course or undergraduate research experience. Even if they have never seen an F ratio, the concept of variances is familiar, as is the idea of pooling group variances to get an estimate of error variance (in the t test). Ultimately, the general regression framework is presented. Then, the by now familiar analyses of variance, covariance, and trend are reconsidered as special cases. We feel that learning statistics involves many passes; that idea is embodied in our text, with each successive pass at a topic being more general.

Our second goal was to provide a sourcebook for researchers. One implication of this is an emphasis on concepts and assumptions; to put it another way, formulas are not enough. No text can present every design and analysis researchers will encounter in their own research or in their readings of the research literature. In view of this, we build a conceptual foundation that should permit the reader to generalize to new situations and to comprehend the advice of statistical consultants and the content of articles on statistical methods. We do this by emphasizing such basic concepts as sampling distributions, expected mean squares, design efficiency, and statistical models. We pay close attention to the assumptions that are made, the consequences of their violation, the detection of those violations, and alternative methods in the face of severe violations. Our concern for alternatives to standard analyses has led us to integrate several nonparametric procedures into relevant design chapters rather than to throw them together in a single last chapter as is so often the case. Our approach permits us to explicitly compare the pros and cons of alternative data analysis procedures within the research context to which they apply.

Our concern that this book serve the researcher has also influenced its coverage. In our roles as consultants to our colleagues and their students, we are frequently reminded that research is not just experimental. For the most part, standard textbooks have ignored the needs of many researchers, who observe the values of independent variables rather than manipulate them. Such needs are clearly present in social and clinical psychology, where sampled social and personality measures are taken as predictors of behavior. Even in traditionally experimental areas, such as cognitive psychology, variables are often sampled. For example, the effects of word frequency and length upon measures of text reading are often of interest. Such observational studies require a knowledge of correlation and regression analysis. Too often, ignorant of anything other than analysis of variance, researchers arbitrarily dichotomize the independent variable scale and compare the "high" and "low" conditions. The result is a loss of power and information. Our book provides extensive coverage of these research situations and the proper analyses.

Another implication of the goal of providing a useful reference book for researchers is an emphasis on the use of statistical computer packages. Analysis of real data is almost never done without a computer, nor should it be. Even in the simplest research design, readily available packages permit much finer-grained analyses than would be attempted with a hand-held calculator. Throughout the book, we discuss several mainframe and microcomputer packages and provide examples of their output. The book also includes an appendix with control information for each output in the preceding chapters and a more general introduction to the use of SYSTAT, BMDP, and $SPSS^X$.

Many individuals have influenced our thinking about, and teaching of, statistics. Discussions with our colleague Alexander Pollatsek have been invaluable, as has been the feedback of our teaching assistants over the years. Most recently, these have included Mary Bryden-Miller, Joseph V. DiCecco, Patricia Collins, Katie Franklin, Jill Greenwald, Randall Hansen, Pam Hardiman, Susan Lima, Laurel Long, Robert F. Lorch, Jr., Edward J. O'Brien, David Palmer, and Jill Shimabukuro. We would also like to thank the students in our statistics courses who encouraged us in this effort and made useful suggestions about earlier drafts of the book.

We would like to express our gratitude to Richard Welna, formerly of Scott, Foresman, who encouraged us in the early stages of this work, and to David Nickol, our project editor at HarperCollins, for his assistance in bringing it to completion. Special thanks are due James Dykes, Jr., and Keith Stanovich for helpful reviews. We also wish to thank the American Statistical Association, the Biometric Society, and the Biometrika Trustees for their permission to reproduce statistical tables.

Most of all, we thank our wives, Nancy and Susan, for their patience and encouragement.

Jerome L. Myers
Arnold D. Well

Chapter 1

Preliminary Considerations

1.1 INTRODUCTION

We undertake research studies to determine whether, and to what extent, variables of interest are related. We may wish to determine if students taught by one method read better than those taught by another, if there is a relationship between cigarette smoking and blood pressure, or if anxiety is reduced more by therapy A than by therapy B. Often, our goal is not merely to find out whether two variables are related, but whether changes in one of the variables cause changes in the other. In such cases, the variable presumed to be doing the causing (here, teaching method, cigarette smoking, or type of therapy) is called the *independent variable*, and the variable that is studied to determine the effects of the independent variable is called the *dependent variable*.

In this first chapter, we discuss different kinds of independent variables and introduce ideas about research designs and statistical analyses that will be developed throughout the rest of the book.

1.2 MANIPULATED AND OBSERVED INDEPENDENT VARIABLES

In experiments, the independent variable is *manipulated*. This means that the researcher is in control of the process that assigns subjects to different levels of the independent variable. In correlational studies, the independent variable is *observed*. That is, the researcher measures the performance of subjects who happen to have different values on the independent variable. The advantage of the experiment is the ease of making causal inferences.

Suppose we were interested in comparing the effectiveness of two methods of teaching reading. Manipulating the independent variable (here, teaching method) would involve assigning each of the students selected to participate in the study to one of the two teaching methods. The goal would be to make this assignment in a way that minimized systematic differences between the instructional groups on *irrelevant variables*, factors that influence performance but are not of current interest. Irrelevant variables like the age and intelligence of the child and the degree of parental encouragement could have large effects on reading performance, and, if uncontrolled, could make it difficult to interpret any data we collected from the two instructional groups.

Suppose we did not manipulate the independent variable but instead merely observed it. This would involve seeking out groups of students already being taught by the different teaching methods and measuring their reading performance. If we did this, some irrelevant variables might vary systematically across instructional groups. For example, students taught by method A might tend to have better teachers and more encouraging parents than students taught by method B. If, after a period of instruction, the students taught by method A performed better, it would be difficult to sort out how much (if any) of the superior performance was due to the teaching methods per se, and how much was due to the better teachers and more encouraging parents. We describe this situation by saying that teaching method is *confounded* with teacher quality and parental attitude.

One way of controlling the effects of irrelevant variables is to assign students to teaching method by using *randomization*; that is, to employ some procedure that gives each student an equal chance of being assigned to each teaching method. For example, assignment could be made by tossing a coin, consulting a table of random digits, or drawing numbers from a hat. Randomization does not ensure that the groups of students are perfectly matched on the irrelevant variables. It does guarantee that there are no *systematic* differences. Between-group differences in irrelevant variables are limited to "chance" factors associated with random assignment, and if the experiment was repeated many times, neither teaching method would have an advantage due to these factors. When a procedure like randomization is used to prevent systematic differences due to irrelevant variables, it seems fair to conclude that significant group differences in the dependent variable are caused by differences in the independent variable.

Although probably nobody would choose to evaluate teaching method without controlling the most important irrelevant variables in some fashion, there are many situations in which it is difficult or impossible to manipulate the independent variable of interest. It is not possible to assign human subjects to different levels of cigarette smoking, age, or type of brain damage. When, for example, incidence of heart disease is tabulated for groups of individuals with different levels of cigarette consumption, these groups also differ in other ways, so that it is more difficult to make causal inferences than if an experiment had been performed. The finding that there is a correlation between cigarette smoking and heart disease does not, by itself, establish that cigarette smoking causes heart disease. Instead, it is possible that certain genetic or environmental factors (e.g., those that lead to high levels of stress) cause the cigarette smoking as well as the heart disease. It is possible to collect data on all the important variables and develop causal models that can be tested against the obtained data. However, such analyses are more complicated and inferences are much less direct than those that follow from performing an experiment.

1.3 ERROR VARIABILITY

We indicated in the previous section that when irrelevant variables are allowed to vary systematically across levels of the independent variable, it is difficult to determine whether the independent variable is responsible for any group differences that might be found. However, even when a procedure like randomization has been used to eliminate systematic bias, the presence of irrelevant variables will result in *error variability*, variability among scores that cannot be attributed to the effects of the independent variable.

There will always be some error variability, even among scores that have been obtained under the same experimental treatment. Scores can be thought of as consisting of two components: a *treatment component* determined by the independent variable, and an *error component* that comes from irrelevant variables. Error variability may occur because scores come from different individuals who differ in age, intelligence, and motivation, or because they come from the same individuals at different times and vary because of changes in such variables as attentiveness, practice, and fatigue. Error variability does not come only from individual differences among subjects: differences in background noise, room temperature, and the way in which the instructions are read may also contribute to the error component.

The presence of large amounts of error variability will tend to obscure the effects of the independent variable. In our example, if each teaching method was equally effective and there was no error variability, every score in the two instructional groups would have exactly the same value. However, there will always be error variability. Within each instructional group, differences in such variables as age, intelligence, background noise, and attentiveness will lead to differences in the scores. Because the values of the irrelevant variables will not be exactly the same in each group, the error variability will also lead to between-group differences. Therefore, if we found that two groups differed in reading performance, the differences could be due, at least in part, to error variability. Much of the remainder of the book will deal with principles and techniques of inferential statistics that have been developed to help us decide whether differences observed at different levels of the independent variable are "real" or merely due to error variability.

One of the major tasks for the researcher is to minimize the effects of error variability. In collecting data, many steps can be taken to reduce the size of the error component in the scores. For example, care might be taken to ensure that subjects in each condition receive exactly the same instructions and that intervals between trials are exactly the same for all subjects. If only one level of an irrelevant variable is present, it cannot give any advantage to any one level of the independent variable nor contribute to the variability among the scores. Each research study will have its own potential sources of error variability, but by careful analysis of each situation the researcher can eliminate or minimize many of them.

The researcher can also minimize the effects of error variability by choosing an efficient research design. We noted earlier that although randomization prevents systematic differences between groups due to the irrelevant variables, there will still be random differences. Instead of using a completely randomized design, we could include some of the irrelevant variables as additional independent variables. For example, we could divide the subject pool into three levels of intelligence (low, medium, and high) and randomly assign students from each level of intelligence to the two teaching methods, giving six combinations of teaching method and intelligence. The advantage of this design is that much of the error variability in the reading scores that is due to differences in intelligence can be removed if the appropriate statistical analysis is used. The design that includes intelligence as an independent variable is said to be more *efficient* than the design that employs completely random assignment of subjects to teaching method, because the effects of error variability are reduced. For some independent variables (teaching method is not one of them) even greater efficiency can be achieved by testing the same subject at each level of the independent variable. This *repeated-measures design* will be discussed briefly in Section 1.5 and in detail in Chapter 8.

Usually, the greater efficiency that comes with more complicated designs has a cost: more information is required so that some kind of matching can be performed, or the appropriate statistical analysis is based on more stringent assumptions about the nature of the data. A major theme of this book is that there are many possible designs and many considerations in the choice of a design. One would like to have as efficient a design as possible, consistent with the resources and subjects that are available and the assumptions that must be made about the data. Ideally, the researcher should be aware of the pros and cons of the different designs and the trade-offs that must be considered in making the best choice.

1.4 USING MORE THAN ONE INDEPENDENT VARIABLE IN A DESIGN

A *factorial design* is one in which the effects of two or more independent variables (also called *factors*) are studied simultaneously. Including several factors in the same design allows us to study both their individual and their joint effects.

To be more concrete, consider an experiment designed to assess the effects of fatigue and background noise on test performance. Suppose we chose two levels of fatigue (high and low) and two levels of noise (high and low). We could conduct an experiment by randomly assigning subjects to the four conditions defined by combining the levels of the two factors. If performance varied across the levels of one of the factors by more than could reasonably be expected due to error variability, we would say that the factor had a significant *main effect*. In the current example, we might expect to find significant main effects for both the noise and fatigue factors because high levels of both variables should have negative effects on performance.

It is also possible that there will be joint effects of fatigue and background noise over and above their main effects. Such joint effects are referred to as *interactions*; we say there is an interaction between two factors if the effects of one of the factors are not the same at each level of the other. Even if high levels of noise had negative effects on performance when averaged over levels of fatigue, it is possible that the effects of noise would be different for the two fatigue conditions. Because high levels of noise could raise arousal level, the negative effects of noise might be smaller in the high-fatigue than in the low-fatigue condition. If there is a difference in performance between the high- and low-noise conditions, averaged over fatigue conditions, there is a main effect of noise. If this difference is not the same for both the high- and low-fatigue conditions, there is an interaction between fatigue and noise.

We could test whether there were effects of fatigue and noise by manipulating each variable by itself in separate experiments. However, if we did so, we could not test whether the factors interacted. Also, even if our primary interest was in the main effect of one of the factors, we might choose to use a factorial design in order to obtain greater efficiency. In Section 1.3 we considered a factorial design in which an irrelevant variable (intelligence) was included as a factor, even though we were not specifically interested in the effects of intelligence; including intelligence as a factor in the design allowed us to test for the effects of teaching method with greater efficiency.

Although the example we just considered employed only two factors, in principle any number of factors may be used. We can test the effects of each factor included in the design as well as the joint effects of every pair of factors. Higher-order interactions involving more than two factors can also be tested.

1.5 WITHIN-SUBJECTS AND BETWEEN-SUBJECTS FACTORS

If a score is obtained for each subject at each level of an independent variable, the independent variable is called a *within-subjects* factor. If each subject is tested at only a single level of the independent variable, it is a *between-subjects* factor. When subjects are tested at every level, the different treatment conditions will automatically be matched on certain irrelevant variables such as intelligence. As we indicated, this can lead to greater efficiency. However, the use of within-subjects factors may lead to biases due to order effects.

After a subject is tested in one treatment condition, performance in a second condition may tend to be better because of practice effects, or worse because of fatigue. For this reason, when within-subjects factors are employed, the order in which the different treatments are presented is usually randomized so that each treatment has an equal chance of being presented in each position in the sequence of treatments (see Chapter 8 for further discussion of this type of design). The randomization of treatment conditions might be constrained to ensure that each treatment appears equally often in each ordinal position. If order was systematically manipulated, it would be possible to remove variability due to temporal effects as well as variability due to individual differences.

Certain independent variables, such as gender, are inherently between-subjects factors, and it would not be possible to use others, such as teaching method, as within-subjects factors because of strong carryover effects. If subjects are taught to read by one method, it is not possible to remove this knowledge so that they may subsequently be taught by another method. However, in many studies, the investigator may choose whether a particular independent variable is to be a between- or within-subjects factor, and the use of within-subjects factors frequently results in greater efficiency.

1.6 CATEGORICAL AND CONTINUOUS INDEPENDENT VARIABLES

Values taken on by *quantitative* variables differ in amount: drug dosage may be 20, 40, or 60 milligrams, and study time in a memory experiment may be 5, 7, or 10 minutes. In contrast, values taken on by *qualitative* variables differ inherently in type: eye color may be blue, brown, or green, a teaching method may or may not involve computer-assisted instruction, and party affiliation may be Democrat or Republic.

The distinction between *categorical* and *continuous* variables is somewhat different. Categorical variables usually have relatively few levels, and cases or subjects are considered to belong to one of these levels because they possess, or have been assigned, the characteristic that defines the category associated with the level. In contrast, continuous variables are quantitative and can take on many values. An inherently qualitative variable must be treated as categorical; however, variables that are inherently quantitative may be treated as either categorical or continuous variables. If we were interested in studying the effects of smoking, we would treat cigarette consumption as a continuous variable if each subject was assigned a value corresponding to the average number of cigarettes he or she smoked per day. On the other hand, we would treat cigarette consumption as a categorical variable if we classified subjects as belonging to one of four categories: nonsmoker, light, moderate, and heavy (corresponding, say, to 0, 1–9, 10–20, and more than 20 cigarettes per day, respectively). It should be emphasized that assignment to a level of a categorical variable is an all-or-none thing. In the present example, subjects would be considered to be equally good heavy smokers whether they smoked 21 or 60 cigarettes a day.

In experiments, manipulated independent variables are always treated as categorical variables, whether they are inherently quantitative or qualitative. When quantitative independent variables such as drug dosage, intelligence test score, height, number of siblings, or cigarettes smoked per day are observed, they usually take on many values. The researcher has the option of preserving these values for subsequent analyses or of classifying them into smaller numbers of categories (e.g., instead of receiving a numerical IQ score, a subject might be classified as below average, average, or above average), thereby creating a categorical variable. Some investigators prefer to transform quantitative variables into categorical ones, because they believe that this somehow simplifies the situation or because they are familiar with statistical analyses (such as analysis of variance) that are appropriate for designs with categorical independent variables. We shall argue later that this procedure is often not appropriate, because it ignores valuable information and because the "simplification" achieved is often illusory.

1.7 ANALYSIS OF VARIANCE AND REGRESSION ANALYSIS

Statistical tests are procedures for determining whether differences among treatment conditions are likely to be "real" (i.e., due in part to the individual or joint effects of the independent variables) or whether they could easily be due solely to error variability.

In a set of procedures referred to as analysis of variance (ANOVA), tests are based on the partitioning of variability in the dependent variable. Components of variability associated with the main effects of the factors and the joint effects of combinations of factors are obtained and compared with the component that reflects error variability. For example, for a design with a single independent variable, the variability in the dependent variable would be partitioned into two components: one associated with within-group score differences (this provides an estimate of the error variability), and a second component associated with between-group differences that might be in part due to error variability and in part due to the effects (if any) of the independent variable. The decision about whether the independent variable has statistically *significant* effects is based on whether the between-group differences are large compared with what might reasonably be expected if the independent variable had no effect so that between-group differences would be due only to error variability.

The use of ANOVA is appropriate in certain "simple" situations—specifically, when all the independent variables are categorical and when the variability in the dependent variable can be neatly partitioned into nonoverlapping components associated with the effects of the independent variables and with error variability. We cannot explain this last requirement clearly until we have introduced a good deal of additional material, but essentially it is that all the independent variables must be uncorrelated with one another. This will generally be the case in experiments in which the same number of scores are obtained from each treatment group, but not otherwise. The advantage of ANOVA is that it is conceptually simple. What at one time was an additional advantage of ANOVA, but has been made less relevant by the availability of computers and statistical software packages, is that the computations are quite simple and can be performed on a hand calculator. Because of its conceptual and computational simplicity and because it can be used to analyze the data from most experiments, ANOVA has traditionally been used extensively in fields such as psychology.

In contrast, multiple regression analysis (MRA) is a more general method of analyzing changes in the dependent variable that are associated with changes in the independent variables. Multiple regression can be used to analyze designs employing categorical, continuous, and combinations of continuous and categorical independent variables. In addition, MRA imposes no restriction that the independent variables must be uncorrelated (although, as we shall see, interpretation is more complicated—for logical as well as for statistical reasons—when highly correlated independent variables are considered).

Although ANOVA can perhaps best be thought of as a special case of regression analysis, until recently ANOVA and MRA were regarded by many researchers as two separate enterprises. Because ANOVA is generally adequate to handle the analysis of data from true experiments, many students have been taught about ANOVA but not about MRA. This limited background leaves them unequipped to deal appropriately with many issues that arise in correlational studies and unable to understand the issues associated with analyzing data from experiments in which not all treatment conditions have equal numbers of scores. Researchers with very limited statistical repertoires often find themselves in the position of having to phrase all of their research questions in terms of one of the few techniques they have mastered. For example, inherently continuous independent variables may be treated as categorical variables so that ANOVA may be used, even when other types of analyses may be more appropriate. For many research areas in psychology, it is no longer enough to know the procedures of ANOVA.

In this book, we provide introductions to both analysis of variance and multiple regression. We believe that learning about both ANOVA and MRA not only provides the researcher with the flexibility needed to ask many research questions, but also results in better understanding of both ANOVA and MRA than if each was presented without the other. Our strategy will be to provide a thorough introduction to ANOVA before introducing MRA, because ANOVA is conceptually simpler and is typically used in simpler situations.

1.8 CONCLUDING REMARKS

When a research study is conducted, the statistical analysis should be planned in detail before the data are collected. The alternative is a post hoc search for an analysis that is both sensitive to the effects of the independent variables and consistent with the design and the nature of the dependent variable, and such an analysis may not exist.

We have introduced many ideas about research designs and statistical analyses in this preliminary chapter. We do not expect the reader to understand many of these ideas fully until they are explained in detail later in the book. However, if nothing else, the reader should have learned that in designing a research study many considerations enter into the choice of the design and the statistical analysis: the researcher must consider design efficiency as well as the power of possible statistical analyses and the kinds of assumptions about the data that must be made in order to use the statistical analyses.

In this book, we attempt to provide the reader with the background necessary to make these decisions. No text can present every design and analysis that researchers will encounter in their own work or in the research literature. We will, however, consider many common designs, and attempt to build a conceptual structure that permits the reader to generalize to new situations and to comprehend both the advice of statistical consultants and articles on statistical methods. We do this by emphasizing such basic concepts as design efficiency, statistical models, sampling distributions, and expected mean squares; by paying close attention to the assumptions that are made and the consequences of violations of these assumptions; and by considering some alternative methods that may be used when the assumptions are severely violated.

1.9 EXERCISES

The following terms provide a useful review of some of the concepts in the chapter. Define, describe, or identify each of them:

independent variable	repeated-measures design
dependent variable	within-subjects factor
manipulated independent variable	between-subjects factor
observed independent variable	quantitative variable
irrelevant variable	qualitative variable
randomization	categorical variable
error variability	continuous variable
factorial design	

Chapter **2**

Samples and Populations

2.1 INTRODUCTION

Chapters 2 and 3 present several topics that are essential to understanding the remainder of this book. In Chapter 2, we distinguish among sample, population, and sampling distributions. We discuss some ways of displaying the data in samples in order to obtain a sense of the population distributions. We define several statistics that describe characteristics of samples, population parameters that these statistics estimate, and certain criteria that are important in deciding what statistics to calculate. In Chapter 3, we discuss four distributions that play a central role in statistical inference and the relations between these distributions. We briefly consider some applications of these distributions to drawing inferences about population parameters. We recognize that most of our readers have been exposed to some of these topics in an introductory statistics course. However, our coverage should serve to review some topics, and to expand on others. Furthermore, some topics will be new to many students; we suspect that this is true of our discussions of ways of plotting data, criteria for estimation, and linear combinations.

A research example is useful to provide a sense of the ideas we will treat in this chapter. Suppose that a state board of education wants to compare the effectiveness of computer-aided instruction and traditional methods upon arithmetic learning in third graders. The board might decide to conduct a pilot study in which 40 children are taught by one of two instructional methods: computer-assisted instruction (the experimental method), or a more standard method currently being used in the schools (the control method). It is important to understand that the board is not interested solely, or even primarily, in whether the use of computers will lead to better arithmetic performance in one study. The purpose of the research is to assess the effect of the two methods of instruction upon all the test scores that might have been obtained. The third graders in the study can be viewed as a sample from the entire population of third graders in which the board is interested. This presumably includes not only all current third graders but third graders who might be taught by the same methods in the future. Inferences about the scores in this hypothetical population will be based on the sample of scores available from the study.

Suppose the difference between the average experimental and control group scores is 4.2. What exactly can be concluded? These results do not conclusively establish that computer-aided arithmetic instruction is better. Performance may be affected by many factors other than the method of instruction, factors that are beyond the researcher's control. The two groups cannot be perfectly equated for ability, previous experience, motivation, and a host of other relevant factors. Even if instruction was exactly as effective in the control condition as in the experimental condition, these "chance" factors would almost certainly result in one of the two groups performing somewhat better than the other. If the study were repeated with another 40 students selected in the same way, we would expect a different set of group means.

Despite this uncertainty, we can draw inferences about the effect of the two instructional methods upon arithmetic performance in the population of third graders. Finding that average performance is better in the experimental condition is evidence that computer-assisted instruction is better in general, and the bigger the performance advantage for the experimental condition, the stronger this evidence is. The strength of this evidence will also depend upon the extent and nature of the variability in our sample. If the 20 scores within each condition are similar, we can be reasonably confident that a replication of the experiment will produce similar results; little variability within a sample suggests little variability within the population from which the sample was drawn and, therefore, little variability across samples. On the other hand, if the scores in each group vary greatly, or if the experimental condition's advantage is due to one or two scores, we would expect considerable variability across samples and, therefore, we should have less faith that the experimental method produces an advantage in the population from which we drew our sample.

One implication of these comments is that when we make inferences about a population, we must consider three types of distributions. The first of these is the hypothetical population distribution. The second is the observed distribution of sample scores. As we indicated, the observed sample distribution is the basis for our inferences about the population distribution. We are never able to observe the population of scores; the best we can do is draw samples from it and use sample statistics to draw conclusions about population parameters. The statistics are measures calculated from the data, such as the sample mean, whereas the parameters are hypothetical entities that are not observable, such as the population mean. We will provide definitions of several population parameters in this chapter.

A third kind of distribution is essential to the process by which we use sample statistics to draw inferences about population parameters. This is the distribution of the statistic of interest across samples; this *sampling distribution*, like the population distribution, is not observable. As an example, assume that the statistic we calculate from the data is the difference between the experimental and control group means. Now assume that we rerun the experiment with a new sample of third graders. If we repeat this sampling process many times, we will have many different values of the statistic of interest

—in this instance, the difference between the two group means. We can conceive of a distribution of these many values of the difference in group means. The properties of this sampling distribution are very important to us. If the sampling distribution of a statistic has little variability, the statistic will not vary greatly across samples and, consequently, we can have high confidence in any inference drawn on the basis of a single sample. In this chapter, we will present some of the properties of sampling distributions of some commonly used statistics, and we will consider how sample, population, and sampling distributions are related. In Chapter 3, we will review several theoretical distributions that are important in statistical inference, and concepts that underlie the inferential process.

Readers may find it helpful to read Appendixes A and B in conjunction with Chapters 2 and 3. They review the algebras of summation and expectation, respectively. Although the definitions, rules, and derivations presented in those appendixes are basic to the entire text, the first applications appear in these chapters.

2.2 SAMPLE DISTRIBUTIONS: DISPLAYING THE DATA

Many researchers' first response to a data set is to calculate averages and then perform some statistical test without ever looking at the data. However, the average of a data set tells us only one thing about that set and about the population from which it was drawn. Knowledge of variability and shape of the sample distribution is also important; such knowledge may reveal something about the processes that gave rise to the data and should have a bearing on the inferential procedures we subsequently employ. Do scores collected under one condition vary more than those collected under another? This may provide insight into the effects of these conditions and serve as a warning to reconsider using statistical tests that assume the two sampled populations have equal variance. Are there outliers—a few scores that are very far from the bulk of the data? Is the distribution skewed (asymmetric)? If it isn't, are there more scores in the tails than we might expect if we had sampled from a normally distributed population of scores? Outliers, skew, heavy tails all tell us something about the population from which we have sampled, and serve as signals that we may be able to draw more valid inferences by transforming our data in some way, or by using procedures other than old standbys such as the t and F tests.

We strongly recommend looking carefully at the data, and in this section we consider several ways of plotting data and various statistics that provide information about the shape of the distribution. We will also note alternatives to the mean and standard deviation that are frequently less sensitive to the presence of outliers.

2.2.1 Stem-and-Leaf Plots

The two panels of Figure 2.1 present one way to display data that provides considerably more information than the histograms typically found in introductory statistics texts. Stem-and-leaf plots provide useful information about the location (central tendency), variability, and shape of the distribution, and also identify outlying data points.

(a) Sample from normal population

```
        STEM AND LEAF PLOT OF VARIABLE: NORMAL     , N = 50
MINIMUM IS:      133.000
LOWER HINGE IS:      394.000
MEDIAN IS:      514.000
UPPER HINGE IS:      618.000
MAXIMUM IS:      956.000


            1   3
            1   5
            2
            2   7
            3   12344
            3 H 78999
            4   01122344
            4   568
            5 M 114
            5   56667789
            6 H 0112
            6   67789
            7   2
            7   59
            8   03
        ***OUTSIDE VALUES***
            9   5
```

(b) Sample from exponential population

```
        STEAM AND LEAF PLOT OF VARIABLE: EXPON     , N = 50
MINIMUM IS:      302.000
LOWER HINGE IS:      349.000
MEDIAN IS:      428.000
UPPER HINGE IS:      602.000
MAXIMUM IS:      1340.000


            3 H 0001113344444
            3   56679
            4 M 0012222334
            4   5799
            5
            5   66899
            6 H 000224
            6   56
            7
            7   8
            8   3
        ***OUTSIDE VALUES***
           11   1
           12   7
           13   4
```

Figure 2.1 Stem-and-leaf plots of samples from a normally distributed (panel a) and an exponentially distributed (panel b) population.

Each of the two plots in Figure 2.1 provides a summary of 50 scores randomly sampled from a different population. The particular program used to present this information is the GRAPH module of SYSTAT, but similar results can be obtained from SPSS[X] and other statistical packages.[1] The program first prints some important numbers. As you might guess, the minimum and maximum are the lowest and highest scores. The median is the value of the score at the 50th percentile of the sample distribution; that is, the score defining the middle of the distribution. To obtain the value of the median, first rank the scores. Then find the score that has a *depth* of

$$d_M = \frac{N + 1}{2} \qquad (2.1)$$

For our samples, in which $N = 50$, $d_M = 25.5$; the median is the score halfway between the 25th and 26th scores when the scores are ranked from smallest to largest. If we had only 49 scores, the median would have a depth of $(49 + 1)/2$, or 25; its value would be that of the 25th score. The median is useful because it is a *resistant* statistic; unlike the sample mean, it is not affected by extreme scores. Furthermore, in one sense the median typifies the data better than any other value; the average absolute distance of scores from the median is less than their average absolute distance from any other value.

The lower and upper hinges (H_L and H_U) are closely related to the 25th and 75th percentiles; these are the values that cut off the upper and lower 25% of the scores. We define the depth of the lower hinge (d_{LH}) as

$$d_{LH} = \frac{[d_M] + 1}{2} \qquad (2.2)$$

where the brackets around d_M indicate that we are discarding the fractional part; therefore, because the depth of the median is 25.5, $[d_M] = 25$. Applying Equation 2.2, we see that the lower hinge has a depth of $(25 + 1)/2$, or 13. Therefore, the lower hinge is the 13th score when the scores are ranked from lowest to highest. The upper hinge is the 13th score when we start from the highest-ranked score.

What can we learn from the hinges? First, the distance between the upper and lower hinges, $H_U - H_L$, is a measure of spread, one more resistant to outliers than is the standard deviation. In our discussion of box plots (Section 2.2.2) we will see a nice graphic display of these distances, enabling us to compare the spreads for different samples very easily. Second, $H_U - H_L$ can be used to provide a useful definition of outliers; we will show how this is done when we discuss the distributions below the summary statistics in Figure 2.1.

Third, the ratio of the distance between the hinges to the standard deviation is a useful descriptor of the sample and sheds some light on the nature of the sampled population. In a normally distributed population, the middle half of the area of the normal curve falls between scores that are 1.349 standard deviation units apart (see Appendix Table D.2); that is,

---

[1] A brief introduction to SYSTAT, BMDP, and SPSS[X] is presented in Appendix E.

$$\frac{H_U - H_L}{\sigma} = 1.349$$

Therefore, a sample from a normal population should yield a ratio close to 1.349. For the sample in panel a, the standard deviation is 170.737, and therefore the standardized ratio is

$$\frac{H_U - H_L}{\hat{\sigma}} = \frac{618 - 394}{170.737} = 1.312$$

For the sample in panel b, the ratio is

$$\frac{602 - 349}{229.447} = 1.103$$

These numbers are one indication that the samples come from different populations, and that the first sample is more likely to have been drawn from a normally distributed population than is the second one. The ratios are at best weak indicators of the nature of the population, but they do provide some information that is not available from any measure of variability, or location.

We can provide a still better sense of the population distribution by plotting the distribution of the sample. The single value in the left column of the stem-and-leaf plot is called the *stem*; in this case, multiply it by 100. Each *leaf* (the values to the right of the stem) is multiplied by 10 and added to the stem. Thus, for the distribution in panel b, the first row represents three values of 300, three more of 310, two more of 330, and five of 340. Note that some accuracy is lost; for example, 310 actually could be any value from 310 to 319. The two H's and the M indicate the groups containing the hinge and median values. The plot also marks "outside values," or outliers. We will consider outliers shortly, but first let's look at the shapes of the sampled distributions.

Both samples were drawn from populations with mean (μ) 500 and standard deviation (σ) 200. The sample distribution in panel a of Figure 2.1 is roughly symmetric with a peak approximately in the middle. In fact, these scores were drawn from a normally distributed population, described by the classic bell-shaped curve. The scores in panel b were drawn from an exponentially distributed population; this distribution is shaped much like that of the sample in panel b with a pileup of low scores and a straggling tail (skew) to the right.

Now let's consider the outliers. A score Y is defined as an outlier if

$$Y > H_U + (1.5)(H_U - H_L) \quad \text{or} \quad Y < H_L - (1.5)(H_U - H_L)$$

For example, in panel b of Figure 2.1, an outlier is any value greater than

$$602 + (1.5)(602 - 349) = 981.5$$

There are several possible reasons for outliers. First, they can occur by chance. The data plotted in panel a were a random sample from a normal population, and the one outlier is a chance result; according to an equation presented by Hoaglin, Mosteller, and Tukey (1983, p. 40), we would expect outliers in no more than about .015 of samples of size 50 from normal populations. A second reason for outliers is the nature of the sampled population. As we noted, panel b contains data drawn from an exponential population. Outliers often occur in samples from such populations. However, samples from symmetric populations with heavy tails (a heavier incidence of extreme scores than in a normally distributed population) also often have several outliers. Inferential errors may be quite frequent when standard tests such as the t and F are performed on data drawn from such skewed or heavy-tailed distributions. In these cases, tests based on the ranks of the scores may yield more valid inferences. Blair and Higgins (1980, 1985) present considerable evidence on this point (see also our discussion of these topics in Chapters 3 and 4). In short, outliers are an indication that inferential procedures based on ranks may yield more valid results than the methods most typically used.

Outliers may also be caused by clerical errors, and the numerical value of an outlier should always be checked to make sure that it is not the result of such an error. Outliers may also indicate subjects who are different from the other members of the sample and whose data therefore should be put aside when doing further analyses. For example, in studies of text memory, we have occasionally found that the same individuals who had extremely low recall scores answered very few comprehension questions correctly following each text and also had very fast reading times. Our impression is that these individuals were not carefully attending to the text, and therefore their memory data are suspect.

In summary, the stem-and-leaf plot of the sample distribution, together with information about location (the median), spread (the distance between the hinges), and outliers, provides a description of important aspects of the data. More details about the construction and interpretation of such plots may be found in Tukey (1977) or Velleman and Hoaglin (1981).

2.2.2 Box Plots

Figure 2.2 presents a different way to summarize a data set. Again, the graphs have been plotted using SYSTAT, but, again, such plots can be obtained using any one of several programs. The left and right boundaries of the box, marked by +'s, correspond to the hinges. The plus inside the box marks the location of the median. Lines are drawn from the hinges to the furthest point that is not an outlier. Outliers are indicated by asterisks. The parentheses containing the median indicate a confidence interval (see Section 3.3.5 for a discussion of this concept); in general, narrower intervals imply that the sample median is more likely to be close to the population median, and nonoverlapping intervals for two different box plots implies a difference in the medians of their respective populations.

The first plot is for the data set drawn from a normal distribution ($\mu = 500$, $\sigma = 200$) and plotted earlier in panel a of Figure 2.1. The sample distribution is roughly symmetric about its median, and the position of the outlier is clearly marked. The second plot is for a data set that we didn't include in Figure 2.1. The population distribution looks very much like the normal except that there are somewhat more scores in the tail than there would be for a normal distribution. This "mixed-normal" distribution was created by sampling with probability of .9 from the same normal distribution represented by the first box plot and with probability of .1 from a second nor-

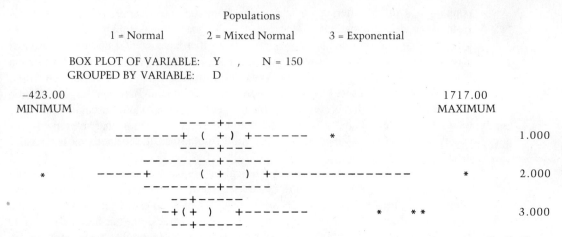

Figure 2.2 Box plots of samples from a normally distributed population (1), a mixture of two normally distributed populations (2), and an exponentially distributed population.

mal distribution with the same mean but a standard deviation of 600 rather than 200. This distribution, like the normal, is symmetric but has more values in the tail; such a distribution is often referred to as "heavy-" or "long-tailed."

The two box plots provide a very nice comparison of the two distributions. We can quickly see that the population medians are very similar (in fact, they are both 500), but that the second sample represents a more variable population. This last conclusion is based on the distance between the hinges, the distance of the outliers from the hinges, and the distance between the two parentheses (width of the confidence interval for the median). Such differences in variability are often of interest in themselves. For example, the two plots might represent performances on a cognitive task following administration of a placebo and a drug, respectively; the plot suggests that the drug aids some subjects but impairs the performance of others. We might wish to investigate further the relation between subject characteristics and effect of drug.

The third box plot is for the sample of 50 scores drawn from the exponential distribution (panel b of Figure 2.1). The skew is very evident in the markedly longer right tail, the three outliers, and the off-center position of the median. In summary, the box plot provides a ready means of comparing locations, spreads, and shapes of sample distributions. The location and spread measures used here have the advantage over the mean and variance of not being sensitive to outliers; the median and interhinge distance are resistant statistics. Even if inferential procedures using means and variances are planned, the plots of Figure 2.2 help to obtain a sense of the data, and determine whether the assumptions underlying various planned inferential procedures are met by the population distribution.

2.2.3 Normal Probability Plots

Because so many commonly used statistical procedures are based on the assumption that the data were sampled from a normally distributed population, it is helpful to have several ways of looking at possible violations of this assumption. Skewness and heavy tails in stem-and-leaf and box plots indicate nonnormality. However, a more direct measure is available in such computing packages as BMDP, SPSS^X, and SYSTAT. Basically, these programs rank order the scores and then plot their expected z values[2] (assuming normality) against the actual scores. Figure 2.3 presents such plots for the three samples of 50 scores that we previously graphed using box plots. Plus signs indicate individual points whereas numbers represent several scores whose points are too close to distinguish by individual plusses. Panel a presents the data from a normally distributed population; the plot is reasonably described by a straight line, as it generally will be for samples from normally distributed populations. Panel b contains the plot for the sample drawn from two different normally distributed populations. The heavy tails evident in the box plot of Figure 2.2 are now reflected in an early and late horizontal component of the curve; symmetric heavy-tailed distributions usually will be seen in S-shaped functions like this one. Panel c is based on the sample taken from the exponential distribution. The pileup of scores in the left tail of the distribution seen in Figures 2.1 and 2.2 is reflected by a very steep positive slope, whereas the long right-hand tail seen in Figures 2.1 and 2.2 is now reflected in the tendency of the plot to level off as Y increases.

2.3 SAMPLE DISTRIBUTIONS: SOME BASIC STATISTICS

In the preceding section, we described the median and interhinge distance as resistant measures of location and variability. Readers are probably more familiar with the mean and standard deviation. These statistics are the measures of location and variability encountered most frequently in the literature, and they have an important role in many statistical procedures. The purpose of this section is to briefly review the definitions and properties of these statistics. The arithmetic mean provides a measure of the location, or central tendency, of the sample distribution; the variance (or its square root, the standard deviation) indicates how representative of the individual scores the mean is. We will also consider the covariance (or the closely related correlation coefficient), which provides a measure of the relation between two distributions of scores. Be aware, however, that other indices of location, variability, and relatedness sometimes have advantages over the mean, variance, and correlation, and that inferential procedures based upon these other indices may yield more valid inferences under some conditions. We will develop these points further as this chapter progresses.

Table 2.1 presents several statistics, the symbol for each, the formula that defines it, an equivalent formula useful for computations, and a summary of several properties of these statistics. These properties will be referred to at various points in this book to justify certain results, and readers should be familiar with them. The review of the algebra of summation in Appendix A includes derivations of the computing formulas and proofs of the properties listed in Table 2.1.

---

[2] For those unfamiliar with the term, a z score is the deviation of a score from the average score divided by the standard deviation of the scores.

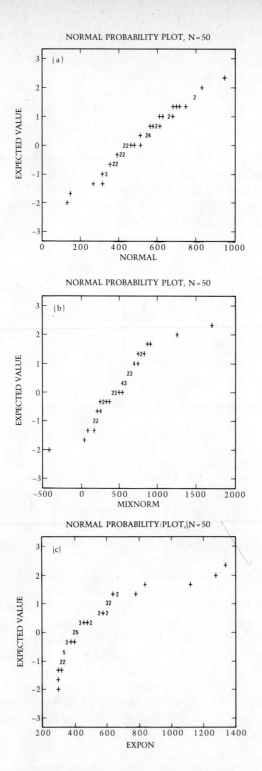

Figure 2.3 Normal probability plots for samples drawn from normal (a), mixed-normal (b), and exponential (c) populations.

TABLE 2.1 SOME IMPORTANT SAMPLE STATISTICS

Statistic	Symbol	Defining formula (n = number of scores)
Mean	$\bar{Y}$	$\sum_i Y_i/n$
Sum of squares	SS	$\sum_i (Y_i - \bar{Y})^2$
Variance	$\hat{\sigma}^2$	$SS/(n-1)$
Standard deviation	$\hat{\sigma}$	$\sqrt{\hat{\sigma}^2}$
Sum of products	SP	$\sum_i (Y_i - \bar{Y})(X_i - \bar{X})$
Covariance	$\hat{\sigma}_{XY}$	$SP/(n-1)$
Correlation coefficient	r_{XY}	$\hat{\sigma}_{XY}/\hat{\sigma}_X\hat{\sigma}_Y = SP/\sqrt{SS_X SS_Y}$

Computing Formulas for SS and SP

$$SS = \Sigma Y^2 - \frac{(\Sigma Y)^2}{n} \qquad SP = \Sigma XY - \frac{(\Sigma X)(\Sigma Y)}{n}$$

Properties of the Mean
1. The sum of deviations of all scores about their mean is zero; that is, $\Sigma(Y - \bar{Y}) = 0$.
2. Adding a constant, k, to all scores is equivalent to adding a constant to the mean; that is, $\Sigma(Y + k)/n = \bar{Y} + k$, where $\bar{Y}$ is the original mean.
3. Multiplying all scores by a constant, k, is equivalent to multiplying the mean by the constant; that is, $\Sigma kY/n = k\bar{Y}$.
4. The mean is that value such that the sum of squared deviations about it is a minimum; that is, $\Sigma(Y - \bar{Y})^2$ is a minimum.

Properties of the Variance and Standard Deviation
1. Adding a constant to all scores leaves the variance and standard deviation unchanged.
2. Multiplying all scores by a constant, k, is equivalent to multiplying the original variance by k^2, and the original standard deviation by k.

A few additional comments are in order. Although the definition of the mean presented in Table 2.1 is familiar, even students of statistics have trouble with a slight variation. This difficulty is illustrated in a study by Pollatsek, Lima, and Well (1981), who posed the following problem to college students who had a course in introductory statistics: "A student spent two semesters at College A and had a 3.2 GPA. The same student attended College B for three semesters and earned a 3.8 GPA. What was the student's overall GPA?" Only 38% of the students responded correctly. Most added the two values and divided by 2, a direct, but incorrect, application of the formula in Table 2.1. In this case, the two observations should be weighted differently to reflect the fact that one is based on two semesters and the other on three. This is done by multiplying each different value of Y by its probability $[p(y)]$ and adding these products. In the example, $\bar{Y} = (3.2)((2/5) + (3.8)(3/5) = 3.56$. The formula for this *weighted mean* is

$$\bar{Y} = \sum_y yp(y) \tag{2.3}$$

The index of summation, y, indicates that the operation is to be carried out for all distinct values of Y. As a result, the probabilities of the values of Y must sum to 1; that is,

$$\sum_y p(y) = 1$$

The usual formula for the mean (Table 2.1) is a special case of Equation 2.3, in which each of n scores gets a weight of $1/n$.

Several points about the variance also should be noted. First, we will often refer to its numerator, the *sum of squares* (SS; short for sum of squared deviations) in this book. Second, readers may have previously seen the sum of squares divided by n, rather than by $n - 1$ as it is in Table 2.1. When the variance is used solely to describe the variability in the data, the average squared deviation of scores about the mean is a reasonable measure, and n is the proper denominator. However, when the sample variance is used as an estimate of a population variance, which is the case in statistical tests and other inferential procedures, $n - 1$ is the proper denominator. We will discuss the reason for this in Section 2.7, which deals with estimation of population parameters. You should also note that statistical packages such as $SPSS^X$, BMDP, SAS, and SYSTAT divide by $n - 1$ when calculating the standard deviation. Finally, we should point out that the symbol $\hat{\sigma}^2$ is meant to indicate an estimate of the population variance; the caret (^) means "estimate of," and σ^2 is the common symbol for the population variance.

The covariance and correlation both reflect the direction and magnitude of the relation between two measures, X and Y. A major difference between them is that the magnitude of the covariance depends upon the measurement scales used. For example, the covariance of a set of heights and weights would increase by a factor of 12 if height were measured in inches instead of feet. Dividing by the product of the standard deviations yields a statistic, r, that is independent of the measurement scales. The reason for this is that a change in the measurement scale will change the covariance and standard deviation by the same factor, leaving the ratio (r) unchanged. We will say more about correlation and related concepts in Chapters 12–16.

2.4 POPULATION DISTRIBUTIONS

The assumptions we make about how the population of scores is distributed is a critical part of the inferential process. Therefore, properties of such theoretical distributions are considered in this section. We begin by discussing discretely distributed and continuously distributed variables. After defining these distributions, we define population means, standard deviations, and covariances. These definitions rest upon the concept of an expected value and, therefore, readers may find it helpful to review Appendix B.

2.4.1 Discrete Random Variables

In this section we develop one particular discrete distribution, the binomial. The reason for this is not that the binomial is more useful than other discrete distributions (though it is useful in many statistical applications) but that it provides a relatively simple context within which to develop the idea of a theoretical population distribution and to introduce a few basic ideas about statistical inference. We begin with an example: suppose each of several children takes a test consisting of four multiple-choice problems. Let Y symbolize the score obtained on the test. The variable Y is referred to as a *random variable* because there are certain probabilities with which Y takes on values within its possible range of values. It is a *discrete* random variable because there are values within that range that cannot occur. In this example, Y can take on any of the five integer values from 0 to 4.

Consider the possible sequences of correct (C) and erroneous (E) answers a child may give to the four problems. As Table 2.2 demonstrates, there are only 2^4, or 16, possible sequences. Each of these is associated with a value of Y, the number of correct responses in four trials. Given a *statistical model*, a set of assumptions about how responses are generated, we can assign a probability to each value of Y; we denote this probability by $p(Y = y)$, or more simply, $p(y)$. The complete set of values of $p(y)$

TABLE 2.2 POSSIBLE SEQUENCES OF CORRECT (C) AND ERROR (E) RESPONSES

Sequence	Number correct (Y)	$p(y)$
<E, E, E, E>	0	1/16 = .0625
<E, E, E, C>	1	
<E, E, C, E>	1	
<E, C, E, E,>	1	4/16 = .25
<C, E, E, E>	1	
<E, E, C, C>	2	
<E, C, E, C>	2	
<E, C, C, E>	2	
<C, E, E, C>	2	6/16 = .3725
<C, E, C, E>	2	
<C, C, E, E>	2	
<C, C, C, E>	3	
<C, C, E, C>	3	
<C, E, C, C>	3	4/16 = .25
<E, C, C, C>	3	
<C, C, C, C>	4	1/16 = .0625
		$\Sigma p(y) = 1$

constitutes a *probability distribution*. Different sets of assumptions will lead to different distributions. In the current situation, for the purpose of illustration, we have made two assumptions:

1. Each child's response to each problem is a guess that has equal probability of being correct or in error. It is as if each child drew a response from a box containing many slips of paper, half labeled "correct" and half labeled "error" and all slips had an equal opportunity to be chosen.
2. The responses are *independent* of each other; that is, the probability of a correct response on any trial does not depend on the outcomes of any other trials. In terms of sampling from the box of correct and error responses, we assume that the slip removed to determine a response to one problem is replaced in the box, and that the box is then shuffled before the next draw.

Given the preceding two assumptions, the 16 sequences in Table 2.2 are equally likely, and the $p(y)$ values are readily obtained. Note that, because Y must take on some value, the sum of these values of $p(y)$ equals 1.

Again, we emphasize that different assumptions lead to different probability distributions. If there were four alternatives for each multiple choice problem rather than two, assumption 1 would have to be modified to state that the probability of a correct response on each problem was .25 (as though 25% of the slips of paper in the box were now marked "correct" and 75% marked "error"). All the sequences of responses and their corresponding values of Y listed in Table 2.2 would still be possible, although the probabilities associated with them would change. As you might guess, the probability of getting three or four problems correct would now be much less than indicated in Table 2.2. If the children were not guessing and we could assume that the probability of a correct response was, for example, .8, still another probability distribution would be indicated.

These distributions are all members of a general class in which (1) there are n trials, each of which can have one of two outcomes, say A and $\tilde{A}$ (not A); (2) the outcome probabilities, π and $1 - \pi$ (probabilities of A and $\tilde{A}$, respectively), are constant over trials; and (3) the probabilities on any trial do not depend upon the outcomes of other trials. Then the theoretical distribution of y can be generated by the *binomial probability function*:

$$p(y;n,\pi) = \frac{n!}{y!(n - y)!} \pi^y(1 - \pi)^{n - y} \tag{2.4}$$

Keep in mind that the probability distribution specified by Equation 2.4 is a theoretical probability distribution which was generated on the basis of a statistical model. This model of the population distribution, together with our sampled scores, is the basis for various kinds of inferences about the population. For example, if our assumptions are valid, and if enough data are collected, the proportions of observed scores that take on the values 0, 1, 2, 3, and 4 should closely match the theoretical probabilities in Table 2.2. Large differences between the observed and theoretical values of $p(y)$ would lead us to conclude that the model is wrong in some respect. Perhaps the trials are not independent, or perhaps the binomial function is a correct characterization of the population distribution, but the children are not choosing between the two alternatives with equal probability.

Even if the model is valid, how closely we can expect the observed proportions to match the theoretical probabilities depends on the size of the sample; the larger the sample, the more likely it is that the observed proportions will be close to the theoretical probabilities. The theoretical probabilities may be viewed as the proportions of an infinitely large set of scores having particular values of Y, assuming the statistical model is correct.

We have used the binomial distribution to illustrate this introduction to discrete probability distributions, but several points hold not only for all discrete distributions but for the continuous distributions we will consider next. First, our inferences rest on a statistical model, a set of assumptions about the population of scores. Second, even when our assumptions are correct, for small samples the sample distribution may not resemble the population distribution. Thus, the inferential process cannot be error-free. Third, inferences are less likely to be in error when we base them on more data because the sample distribution is then more likely to approximate the population distribution.

2.4.2 Continuous Random Variables

Variables that can take on any value within their range are called *continuous random variables*. A common example in psychological research is response time, which theoretically takes on any value from zero to infinity. Of course, observed response times usually fall between some boundaries, such as 200 milliseconds and 10 seconds. Even within such boundaries, continuity is more theoretical than real because the best laboratory timing devices rarely record in units smaller than thousandths of a second. Not all the possible values of a continuous random variable can be recorded. Nevertheless, the concept is important because many of the statistics on which inferences are based are continuous variables.

One important distinction between discrete and continuous distributions is that $p(y)$, the probability that Y takes on a particular value, is meaningful only with reference to discrete distributions. Therefore, we need some other way of characterizing continuous distributions. Before we consider how to do this, we will discuss why it is not meaningful to refer to the probability of particular values of Y when Y is a continuous variable. We may achieve some insight if we consider a relatively crude clock capable of registering response times within a tenth of a second. Times longer than .95 but shorter than 1.05 seconds will be registered as 1 second. Now substitute a more accurate clock capable of measuring to the nearest hundredth of a second. Only response times in the interval from .995 to 1.005 seconds now will be registered as 1 second. Of course, there will be fewer times between .995 and 1.005 seconds than between .95 and 1.05 seconds; the probability of registering 1 second is lower with the more accurate clock. Extending the argument, it should become apparent that the probability of a response time of exactly 1 second duration is essentially zero.

The issue may be understood better by considering the continuous distribution in Figure 2.4. The area segment between y_1 and y_2 represents the proportion of observations that fall between these two values. If y_1 and y_2 are placed closer together, the

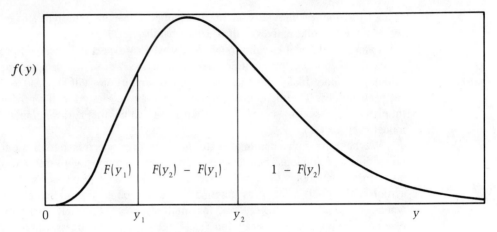

Figure 2.4 Example of a continuous distribution.

probability of a score between these two values becomes smaller. Theoretically, we could make that probability as close to zero as we desired by just reducing the separation between y_1 and y_2. Therefore, it makes no sense to speak of the probability of some exact value y when Y is a continuous random variable. To distinguish among continuous random variables having different distributions, we need some function of Y other than its probability to represent the distribution. One function we can define is $F(y)$, the cumulative probability function, the area under the curve to the left of some point y. For example, $F(y_1)$ is just the probability of getting a value less than y_1; in terms of Figure 2.4, it's the area to the left of y_1. The area in the segment between y_1 and y_2 is $F(y_2) - F(y_1)$. This is the probability that y takes on a value between y_1 and y_2. This probability forms the basis for another function often used to characterize a distribution: let

$$f(y) = \frac{F(y_2) - F(y_1)}{y_2 - y_1}$$

when $y_2 - y_1$ is very small. This ratio is called the *probability density function*. We can view $f(y)$ as the height of the curve at the value y. To see why this is so, realize that we are dividing an area (between y_1 and y_2) by its width ($y_2 - y_1$) and that area/width = height. Even though the area gets increasingly close to zero as the width gets smaller, the ratio approaches a constant value that ordinarily is greater than zero. The probability density $f(y)$ takes on different values for different values of y. If we can find the formula for $f(y)$, we can calculate the probability density for any y. In other words, if we had a formula for $f(y)$ for the distribution of Figure 2.4, we could plug values of y into it and plot the distribution in that figure.

In contrast to $f(y)$, $F(y)$ *is* a probability, a proportion of the curve up to the point y. It will be of primary interest in most inferential procedures; significance tests are based on the area in the tails of some theoretical distribution. Nevertheless, $f(y)$ is basic because the cumulative distribution is derived from it.

How is the cumulative distribution derived from the probability density distribution? If the density function is discrete, we could just add the probabilities of each value of y. We can do something like that with continuous distributions. Think of dividing the area under a continuous distribution into very narrow rectangles, each of which has height $f(y)$ and width dy. Thus $f(y) \times dy$ is the height times the width, or the area, associated with a particular value of y; it is the continuous analog of $p(y)$. If we want the value of $F(1)$, we can add the areas of the rectangles that lie between $y = 0$ and $y = 1$. The value of the sum approaches a constant as the rectangles become very narrow. That limiting value of the sum is called the integral. Mathematicians would indicate the operation of obtaining the area between 0 and y_1 by

$$F(y_1) = \int_0^{y_1} f(y) \, dy$$

and the area between y_1 and y_2 by

$$F(y_2) - F(y_1) = \int_0^{y_2} f(y) \, dy - \int_0^{y_1} f(y) \, dy = \int_{y_1}^{y_2} f(y) \, dy$$

One other point about density and cumulative distributions should be noted. When we considered discrete random variables, we saw that the sum of all the $p(y)$ had to equal 1; the random variable had to take on one of its possible values. Similarly, the integral of a continuous random variable over the entire range of values has to be 1, or $f(y)$ does not characterize a true density function. For example, if Y is a random variable whose upper bound is 100, $F(100)$ must equal 1.

To summarize, for continuous probability distributions, probability is represented by an area under the curve, not by the height of the curve. The height of the curve at the point where Y equals some specific value y is not $p(y)$, the probability that Y equals y, but $f(y)$, the probability density function at that point. This function represents the limiting ratio obtained when the probability of a very small interval around y is divided by the width of that interval. The density function has a different formula for different distributions and provides a way of characterizing a continuous distribution just as Equation 2.4 characterizes the binomial distribution. The probability of falling between any two y values corresponds to the area under the curve between those two values (see Figure 2.4) and can be obtained from the $f(y)$ by a mathematical operation called integration. It is these areas (particularly those in the tails of continuous distributions) that we typically must consider in inferential procedures.

2.4.3 Population Parameters

Table 2.3 presents the population parameters that are estimated by the statistics presented in Table 2.1. Each parameter is defined in terms of expected values. It may help to compare Tables 2.1 and 2.3. In particular, note that the formulas for statistics and parameters are very similar. One difference is that the sample mean $\bar{Y}$ is replaced by the population parameter μ in defining variances and covariances. A second difference is that we sum each of n values and divide by n (or $n - 1$) to obtain sample statistics, whereas we take the expectation of the variable of interest to obtain the population parameter. For example, the sample variance is defined as

TABLE 2.3   SOME IMPORTANT POPULATION PARAMETERS

Parameter	Symbol	Defining formula
Mean	μ	$E(Y)$
Variance	σ^2	$E(Y - \mu)^2$
Standard deviation	σ	$\sqrt{\sigma^2}$
Covariance	σ_{XY}	$E(X - \mu_X)(Y - \mu_Y)$
Correlation coefficient	ρ_{XY}	$\sigma_{XY}/\sigma_X\sigma_Y$

Notes
1. If Y is a discrete random variable,

$$E(Y) = \Sigma yp(y) \quad \text{and} \quad E(Y - \mu)^2 = \Sigma(y - \mu)^2 p(y)$$

and if Y is a continuous random variable,

$$E(Y) = \int yf(y) \, dy \quad \text{and} \quad E(Y - \mu)^2 = \int(y - \mu)^2 f(y) \, dy$$

Expected values (E) are discussed further in Appendix B.
2. Alternative forms of the variance and covariance, similar to the computing formulas of Table 2.1, are

$$\sigma^2 = E(Y^2) - \mu^2$$

and

$$\sigma_{XY} = E(XY) - \mu_X\mu_Y$$

$$\hat{\sigma}^2 = \frac{\Sigma(Y - \bar{Y})^2}{n - 1}$$

whereas the population variance is

$$\sigma^2 = E(Y - \mu)^2$$

The properties listed for the corresponding statistics in Table 2.1 hold for the parameters of Table 2.3 as well. For example, multiplication of all values of a random variable by a constant results in the variance being multiplied by the constant squared; that is, $\sigma_{kY}^2 = k^2\sigma_Y^2$.

2.5 LINEAR COMBINATIONS

So far in this chapter, we have presented formulas for the means and variances of distributions of individual scores and for covariances and correlations of pairs of scores. But suppose the random variable of interest is itself a combination of variables. An instructor might give three tests and want to calculate the average total score for the class, or the standard deviation of total scores. The variable in this case is

$$T_i = Y_{i1} + Y_{i2} + Y_{i3}$$

where T_i represents the sum of scores for the ith student in the class. Another possibility is that the third test is a final covering the entire course and therefore is given twice as much weight as the first two tests. In that case, the total would not be the sum of the three scores but rather

$$T_i = (1)(Y_{i1}) + (1)(Y_{i2}) + (2)(Y_{i3})$$

There are many other such possible combinations of scores that may be of interest. For example, consider a study in which pairs of students are matched for their ability to learn arithmetic. One member of each pair is assigned to an experimental condition, the other to a control. A difference score for the ith pair of subjects (D_i) is the experimental score (Y_{iE}) minus the control score (Y_{iC}), or

$$D_i = (1)(Y_{iE}) + (-1)(Y_{iC})$$

The mean itself is a combination of scores. To get an average of a set of n scores, we would ordinarily add all n scores and divide by n. But note that this operation can also be represented by

$$\bar{Y}_. = \frac{1}{n}(Y_1) + \frac{1}{n}(Y_2) + \cdots + \frac{1}{n}(Y_n)$$

All of the above are examples of *linear combinations* and can be written in the general form

$$L = w_1 Y_1 + \cdots + w_j Y_j + \cdots + \quad (2.5)$$
$$= \Sigma w_j Y_j$$

L is referred to as a linear combination because the Y_j are not raised to a power other than 1. The weights (w_j) can be any numbers; as the preceding examples illustrate, they need not be equal to each other, or integers, or even positive numbers.

Many statistics of interest to researchers are linear combinations. In order to draw inferences about the population parameters estimated by these statistics, we need to know something about means and standard deviations of linear combinations. Therefore, we consider these matters next.

2.5.1 Means of Linear Combinations

Table 2.4 presents three test scores for each of eight students, as well as the means, variances, and covariances for all tests. A final grade, L, is obtained by weighting the first two tests by .25, and the third test by .5. For example, applying Equation 2.5, the final score for the first student is

$$L_1 = (.25)(83) + (.25)(76) + (.5)(93) = 86.25$$

To obtain the mean final grade, $\bar{L}_.$, we added the eight values of L_i and divided by 8. Note that we get the same result if we just weight the mean for tests 1 and 2 by .25 and the mean of test 3 by .5. That is,

$$\bar{L}_. = (.25)(76.125) + (.25)(77.125) + (.5)(82.375) = 79.5$$

In general, if we have t scores for each of n subjects,

$$\bar{L}_. = \frac{1}{n} \sum_{i=1}^{n} L_i$$

$$= \frac{1}{n} \sum_{i=1}^{n} \left(\sum_{j=1}^{t} w_j Y_{ij} \right)$$

We may rearrange terms to get

$$\bar{L}_. = \sum_{j=1}^{t} w_j \left(\frac{1}{n} \sum_{i=1}^{n} Y_{ij} \right)$$

$$= \sum_{j=1}^{t} w_j \bar{Y}_{.j} \tag{2.6}$$

The quantity $\bar{Y}_{.j}$ is the average score on the jth test; the "dot" subscript is designed to remind us that we have obtained this mean by averaging over individuals. Equation 2.6 states that the average of a set of linear combinations is a linear combination of the average scores.

The same general rule can be shown to hold for a population of values of L. The mean of that population can be represented as the linear combination of the means of the variables being combined:

$$E(L) = \Sigma w_j \mu_j \tag{2.7}$$

where μ_j is the expected value of the population consisting of the scores on the jth variable.

2.5.2 Variances of Linear Combinations

Not surprisingly, the variance of a set of values of L is related to the variances of the variables that are being combined. Less obviously, it is also related to the covariances of each possible pair of variables. We will shortly examine these relationships using the variances and covariances of Table 2.4. However, before doing so, we first will consider the variances of sums and differences. These are somewhat simpler than the linear combination in Table 2.4 and therefore provide us with a good starting point for this presentation.

Table 2.5 presents a set of X and Y scores, the sum of the two scores, and the difference between them. Note that the sum of scores for each individual is a linear combination in which both weights are $+1$ and the difference is a linear combination in which the first score is weighted $+1$ and the second -1. The table also presents the variance for each variable ($\hat{\sigma}_X^2$, $\hat{\sigma}_Y^2$), the variance of the sums, the variance of the differences, and the covariance ($\hat{\sigma}_{XY}$) of the two variables.

Note that the following relation holds between the variance of the sum and the variances and covariance of the two measures:

$$\hat{\sigma}_{X+Y}^2 = \hat{\sigma}_X^2 + \hat{\sigma}_Y^2 + 2\hat{\sigma}_{XY} = \hat{\sigma}_X^2 + \hat{\sigma}_Y^2 + 2r_{XY}\hat{\sigma}_X\hat{\sigma}_Y \tag{2.8}$$

TABLE 2.4   STATISTICS BASED ON A LINEAR COMBINATION $[L = .25(Y_1 + Y_2) + .5Y_3]$

	Data			
	Y_1	Y_2	Y_3	Final (L)
	83	76	93	86.25
	71	68	80	74.75
	90	84	86	86.50
	61	70	65	65.25
	77	83	82	81.00
	67	74	92	81.25
	70	68	75	72.00
	90	94	86	89.00
Column means =	76.125	77.125	82.375	79.50

The variances and covariances are

$$
\begin{array}{c} \\ Y_1 \\ Y_2 \\ Y_3 \end{array}
\begin{array}{ccc} Y_1 & Y_2 & Y_3 \\ \left[\begin{array}{ccc} 115.554 & 80.839 & 57.946 \\ 80.839 & 84.982 & 36.375 \\ 57.946 & 36.375 & 84.839 \end{array}\right] \end{array}
$$

Note The diagonal elements are the variances and the off-diagonal elements are the covariances. The matrix is symmetric because the covariance of Y_j and $Y_{j'}$, is the same as the covariance of $Y_{j'}$ and Y_j.

The variance of the difference scores has a similar form:

$$\hat{\sigma}_{X+Y}^2 = \hat{\sigma}_X^2 + \hat{\sigma}_Y^2 - 2\hat{\sigma}_{XY} = \hat{\sigma}_X^2 + \hat{\sigma}_Y^2 - 2r_{XY}\hat{\sigma}_X\hat{\sigma}_Y \qquad (2.9)$$

The only difference between Equations 2.8 and 2.9 is in the sign of the covariance term.

These equations will be very important in constructing significance tests and confidence intervals for linear combinations. It may help to understand where they come from. Suppose we have sum of two scores, X and Y, for each of n people. In order to obtain the variance of these sums, we first subtract the average of all n sums from each sum; for the ith individual, we have

$$(Y_i + X_i) - (\bar{Y} + \bar{X})$$

which may be rewritten

$$(Y_i - \bar{Y}) + (X_i - \bar{X})$$

Next, we must square these n quantities. From elementary algebra, the square of quantities having the general form $a + b$ is

$$(a + b)^2 = a^2 + b^2 + 2ab$$

TABLE 2.5   VARIANCES OF SUMS AND DIFFERENCES

	X	Y	X + Y	X − Y
	3	5	8	−2
	12	10	22	2
	8	7	15	1
	10	11	21	−1
	6	8	14	−2
	3	7	10	−4
Sum =	42	48	90	−6
Mean =	7	8	15	−1
Variance =	13.6	4.8	32.0	4.8
		$\hat{\sigma}_{XY} = 6.8$		

The relationship between the variances of the rightmost two columns and $\hat{\sigma}_X^2$, $\hat{\sigma}_Y^2$, and $\hat{\sigma}_{XY}$ is easily shown:

$$\hat{\sigma}_{X+Y}^2 = \hat{\sigma}_X^2 + \hat{\sigma}_Y^2 + 2\hat{\sigma}_{XY} = 13.6 + 4.8 + (2)(6.8) = 32.0$$

$$\hat{\sigma}_{X-Y}^2 = \hat{\sigma}_X^2 + \hat{\sigma}_Y^2 - 2\hat{\sigma}_{XY} = 13.6 + 4.8 - (2)(6.8) = 4.8$$

Therefore, the squared deviation of a sum from the average sum is

$$[(X_i + Y_i) - (\bar{X} + \bar{Y})]^2 = [(X_i - \bar{X}) + (Y_i - \bar{Y})]^2$$
$$= (X_i - \bar{X})^2 + (Y_i - \bar{Y})^2 + 2(X_i - \bar{X})(Y_i - \bar{Y})$$

The squared deviation terms to the right of the equal sign are the basis for the variances of the two variables, and the product term is the basis for the covariance. Summing each of these terms for the n subjects and dividing by $n - 1$, we obtain Equation 2.8.

Now consider a population with two scores for each individual. These scores can be summed or subtracted from each other. The variances of such populations of sums and difference scores is expressed in a form parallel to Equations 2.8 and 2.9:

$$\sigma_{X+Y}^2 = \sigma_X^2 + \sigma_Y^2 + 2\rho_{XY}\sigma_X\sigma_Y \tag{2.10}$$

$$\sigma_{X-Y}^2 = \sigma_X^2 + \sigma_Y^2 - 2\rho_{XY}\sigma_X\sigma_Y \tag{2.11}$$

These expressions will prove very important in the development of many inferential procedures. Frequently, the two measures being combined will be assumed to be independently distributed. In Appendix B, we prove that when two variables are independently distributed, their covariance (and correlation) must be zero. Therefore, the covariance terms in Equations 2.10 and 2.11 will be zero when X and Y are independently distributed. Assuming independence, the variances of sums and differences will be the same: both will be the sums of the variances for the two variables.

So far we have considered a special case of the variance of linear combinations: only two variables have been considered, and they have either been added or subtracted. We can generalize the equations to deal with any number of variables and with weights other than 1 and -1. For example, reconsidering Table 2.4, we can express the relation between the variance of the final grades ($\hat{\sigma}_L^2$) and the variances for each of the three tests ($\hat{\sigma}_j^2$). The variance of the rightmost column, obtained in the usual way, is 67.429. We can obtain the same result by calculating

$$\hat{\sigma}_L^2 = w_1^2\hat{\sigma}_1^2 + w_2^2\hat{\sigma}_2^2 + w_3^2\hat{\sigma}_3^2$$
$$+ 2[w_1 w_2 r_{12}\hat{\sigma}_1\hat{\sigma}_2 + w_1 w_3 r_{13}\hat{\sigma}_1\hat{\sigma}_3 + w_2 w_3 r_{23}\hat{\sigma}_2\hat{\sigma}_3]$$

where $\hat{\sigma}_j^2$ is the variance of the jth measure, w_j is the weight by which that measure is multiplied, and $r_{jj'}$ is the correlation of the jth and j'th variables; recall that $r_{jj'}\hat{\sigma}_j\hat{\sigma}_{j'}$ is the covariance of Y_j and $Y_{j'}$. The reason we multiply the covariances by 2 is because the covariance of Y_j and $Y_{j'}$ is the same as that for $Y_{j'}$ and Y_j. Substituting the values of the variances and covariances in Table 2.4 into the preceding equation, we have

$$\hat{\sigma}_L^2 = .25^2(115.554 + 84.982) + .5^2(84.839)$$
$$+ 2[(.25)(.25)(80.839) + (.25)(.5)(57.946 + 36.375)] = 67.429$$

The general form of the equation for the sample variance of a linear combination is

$$\hat{\sigma}_L^2 = \sum_j w_j^2\hat{\sigma}_j^2 + \sum\sum_{j \neq j'} w_j w_{j'} r_{jj'}\hat{\sigma}_j\hat{\sigma}_{j'} \tag{2.12}$$

If there are only two independent measures with weights $+1$, or $+1$ and -1, Equation 2.12 reduces to Equation 2.8 or 2.9. Equation 2.12 is derived in the appendix at the end of the chapter.

2.6 SAMPLING DISTRIBUTIONS

The central problem in using the statistics of a single study to draw inferences about population parameters is that the values of the statistics are not identical to those of the parameters they estimate. We need some idea of how much in error our estimates are, and some way of taking this error into account when testing hypotheses and constructing confidence intervals. In this section, we develop a conceptual framework for dealing with sampling variability.

2.6.1 What Is a Sampling Distribution?

The concept of a sampling distribution is implicit in statistical inference. For example, consider the following marketing study. Fifty individuals are sampled from some well-defined population and asked to rate a new brand of breakfast cereal. The ratings range from 1 ("strongly dislike") to 11 ("strongly like") with 6 as the neutral point. We might wish to test whether the mean of the sampled population is different from 6. The mean of the sampled ratings is 8.6. If the sample mean changed little from one sample to another, this value would provide strong evidence against the hypothesis that $\mu = 6$. On the other hand, if the sample mean was quite variable over samples, then a sample value of 8.6 could well have occurred even when the population mean was 6.

The critical point is that it is useful to picture many random replications of the 50-subject sampling experiment with each replication giving rise to a value of $\bar{Y}$. This hypothetical probability distribution of $\bar{Y}$ is called the *sampling distribution of the mean* for samples of size 50. As we can see from the preceding example, knowing the properties of this sampling distribution enables us to assess inferences made on the basis of a single sampled value of $\bar{Y}$. Every statistic has a sampling distribution, because each time a new sample is drawn from a population the sample statistic is based on a new set of values. For now, we will focus on the mean and variance of the sampling distribution of $\bar{Y}$. These two properties of the sampling distribution of the mean will prove useful to know when studying subsequent developments.

2.6.2 Some Properties of the Sampling Distribution of the Mean

We never can observe the sampling distribution of a statistic because we never take a large number of samples from the same population. Fortunately, if we make certain valid assumptions about the population, we can infer the properties of the sampling distribution without actually drawing even one sample. This point may be clearer if we consider an example. Assume that the integers from 1 to 6 are each written on a separate slip of paper; these are then placed in a box. The box is thoroughly shaken to provide an equal opportunity for each value to be drawn. Once a slip of paper is drawn, its value (Y) is noted, the slip is replaced, the slips are reshuffled, and a slip is again drawn. On any trial, this process is equally likely to yield any integer from 1 to 6, and the trial outcome will be independent of that for any other trial. The population distribution for a very large series of trials will look like that drawn in the top panel of Figure 2.5. Using the equations for population means and variances presented in Table 2.3, verify the numerical values presented in Figure 2.5 for the population mean and variance.

Now assume an experiment in which we have two boxes, each of which contains the integers from 1 to 6. A trial consists of drawing one slip from each box. The mean of the two values is calculated. The addition and multiplication rules for probability enable us to calculate the sampling distribution of this mean. For example, the lowest possible mean, $\bar{Y} = 1$, would occur if both Y_1 and Y_2 were 1. Assuming that (1) we always replace the slips before drawing a new sample, and (2) the boxes are thoroughly shuffled before each draw, the probability that the mean is 1 is $1/6 \times 1/6$, or $1/36$. A mean of 1.5 will occur if $Y_1 = 1$ and $Y_2 = 2$, or $Y_1 = 2$ and $Y_2 = 1$; this probability is $1/36 + 1/36$. The middle panel contains the sampling distribution of the mean of this two-draw experiment. Similarly, the bottom panel contains the sampling distribution of the mean for an experiment in which the mean is based on a trial in which slips of paper are drawn from three boxes. Calculate some of the ordinate values for the bottom two panels to verify that you understand how these sampling distributions are generated.

Several points follow from Figure 2.5. First, we can generate sampling distributions and calculate their parameters (such as the mean and variance) without actually

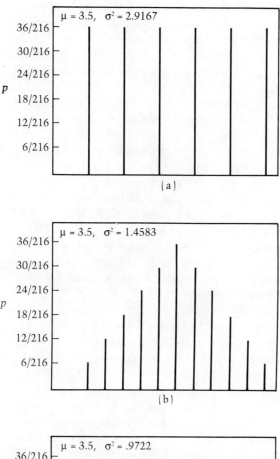

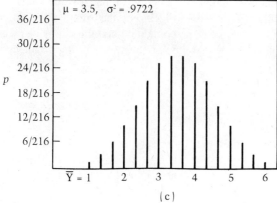

Figure 2.5 A population distribution (a) and sampling distribution of the mean for $n = 2$ (b) and $n = 3$ (c).

carrying out the sampling process. This fact enables us to draw inferences based on long-run probabilities. Of course the correctness of these inferences will depend upon the validity of the assumptions upon which the derivation of the distribution was based. If the draws were not independent, or if the outcomes were not equally likely, the actual probabilities of various values of the mean would not equal the theoretical ones presented in Figure 2.5.

The second point to recognize is that the sampling distributions in the bottom two panels look quite different from the population distribution in the top panel. In the population distribution, the six outcomes are equally likely. In the sampling distributions, values near the mean, 3.5, are more likely than extreme values. Furthermore, the sampling distribution appears even less flat as n is increased from 2 to 3. The sampling distributions have a peak because there are more combinations of outcomes of draws that give rise to the middle values than to the extreme values. For example, when $n = 2$, there is only one way to get a mean of 1 ($Y_1 = Y_2 = 1$). A mean of 3.5 can occur in any of six ways (<1,6>, <6,1>, <2,5>, <5,2>, <3,4>, <4,3>). Even when the population distribution is extremely skewed, as the sample size is increased, the sampling distribution of the mean will become symmetric.

The third aspect to focus on in Figure 2.5 is the mean and variance. For example, the expected value of the sampling distribution for $n = 2$ is 3.5. The expected value of the sampling distribution of the mean is the same as that of the population. The variance, however, changes with n. The value when $n = 2$ (middle panel) is exactly half of the population variance (top panel), and the variance of the sampling distribution of the mean when $n = 3$ (bottom panel) is exactly one third that of the population variance. The variance of the sampling distribution of the mean is further decreased by using still larger samples. The important point is that the average of a large set of sample means will be the same as the population mean and that the sample-to-sample variability of the sample mean will be less when n is large. Therefore, a single sample mean is more likely to be close to the population mean being estimated when the sample is large than when it is small. This makes sense; the larger the sample, the more likely it is to resemble other samples from the same population and the closer its mean will be to that of those other samples. The variability of the mean is important enough to warrant further consideration.

2.6.3 The Standard Error of the Mean

The most commonly used measure of the sampling variability of a statistic is its *standard error* (SE). The SE of the mean is obtained by taking the square root of the variance of the sampling distribution of the mean. To derive a general formula, we first recall that the mean is a linear combination:

$$\bar{Y} = \left[\frac{1}{n}\right] Y_1 + \left[\frac{1}{n}\right] Y_2 + \cdots + \left[\frac{1}{n}\right] Y_i + \cdots + \left[\frac{1}{n}\right] Y_n$$

If the Y_i are independently distributed, the covariance terms in Equation 2.12 equal zero and

$$\sigma_{\bar{Y}}^2 = \left[\frac{1}{n}\right]^2 \sigma_1^2 + \cdots + \left[\frac{1}{n}\right]^2 \sigma_i^2 + \cdots + \left[\frac{1}{n}\right]^2 \sigma_n^2$$

But what is the meaning of σ_i^2? Assume that many samples of n scores have been drawn at random from the same population. Then Y_i is the ith score observed in each sample, and σ_i^2 is the variance of all those scores in the ith position in their respective samples. There is no reason why the variance of the scores observed first within their samples should be different from the variance of those observed second, or third. In other words, the n values of σ_i^2 should equal each other. Furthermore, if there are infinitely many samples, the variance of the scores in any position should equal the population variance. Therefore,

$$\sigma_{\bar{Y}}^2 = n \left[\frac{1}{n}\right]^2 \sigma^2 = \frac{\sigma^2}{n}$$

This is exactly what we observed when we computed the variances of the distributions in Figure 2.5. Taking square roots, we now have a formula for the SE of the mean of a sample of n independently distributed observations:

$$\sigma_{\bar{Y}} = \frac{\sigma}{\sqrt{n}} \tag{2.13}$$

Think of this as an index of the error in predicting the population mean on the basis of a single sample mean. That error decreases as n increases; we argued before that this made good sense. The error also decreases as the population variance decreases. That too makes sense: the less variable the population of scores is, the more similar samples drawn from that population will be and, therefore, the less the sample means will vary.

2.6.4 Differences between Means

A very common research situation involves the comparison of two conditions such as an experimental and control treatment. Two groups of subjects are viewed as random samples from their respective populations. One question is whether the population means, μ_1 and μ_2, are different. Our response to this question will be based on the difference between the group means, $\bar{Y}_1 - \bar{Y}_2$. Suppose $\bar{Y}_1 - \bar{Y}_2 = 5$. Is this a large enough difference to lead us to conclude that μ_1 and μ_2 are not equal? Or could the difference of five points be due to chance variation? To answer this we need a measure of chance variation. More precisely, just as we previously needed the SE of the mean to characterize the variability among sample means, we now need a measure of the sample-to-sample variability of the difference between two means.

Assume that the two groups in the experiment under consideration have n_1 and n_2 subjects, respectively. Suppose the experiment were run infinitely often, each time with two randomly sampled groups of scores of sizes n_1 and n_2. Each replication yields two group means and therefore a difference, $\bar{Y}_{.1} - \bar{Y}_{.2}$. We can conceive of a sampling distribution of the statistic $\bar{Y}_{.1} - \bar{Y}_{.2}$. The mean of that sampling distribution will be

$$E(\bar{Y}_{.1} - \bar{Y}_{.2}) = E(\bar{Y}_{.1}) - E(\bar{Y}_{.2}) = \mu_1 - \mu_2$$

To arrive at an expression for the variance of the sampling distribution, we first note that the difference between the means $(\bar{D})$ can be written as a linear combination:

$$\bar{D} = (1)\bar{Y}_{.1} + (-1)\bar{Y}_{.2}$$

Applying Equation 2.11 for the variance of a difference, and setting the covariance term to zero because we have two independent groups of scores, we have

$$\sigma_{\bar{D}}^2 = \sigma_{\bar{Y}_1}^2 + \sigma_{\bar{Y}_2}^2$$

But, from Equation 2.13, we know that the variance of the mean is the population variance divided by the sample size. Therefore,

$$\sigma_{\bar{D}}^2 = \frac{\sigma_1^2}{n_1} + \frac{\sigma_2^2}{n_2}$$

where σ_j^2 is the variance of scores in the population from which group j was sampled. We can now obtain the standard deviation of the sampling distribution of the difference between the means:

$$\sigma_{\bar{Y}_1 - \bar{Y}_2} = \sqrt{\frac{\sigma_1^2}{n_1} + \frac{\sigma_2^2}{n_2}} \tag{2.14}$$

Several points about the preceding development should be kept in mind. First, Equation 2.14 rests upon the assumption that the two groups of scores are independently distributed. Often, this will not be true. For example, if the same subjects were tested under two different conditions, the two scores would be correlated and a covariance term would have to be incorporated into Equation 2.14. The general form of this is Equation 2.12. Second, always sum the variances of the mean prior to taking the square root. We find students occasionally "improving" upon Equation 2.14 by taking the square roots of each right-hand term and summing these. If the difference in the two procedures isn't obvious, compare $\sqrt{3^2 + 4^2}$ with $\sqrt{3^2} + \sqrt{4^2}$.

The expected values and standard deviations of sampling distributions play an important role in statistical inference. However, in most cases, we do not know the values of these population parameters but must obtain numerical estimates from sampled data. In the next section, we will consider what makes a good estimator of a population parameter. The criteria presented are in terms of the sampling distribution of that estimator.

2.7 ESTIMATING POPULATION PARAMETERS

An infinite number of possible estimators of any single population parameter exists. The population mean might, for example, be estimated by the sample mean, the sample median, or even the first score drawn from the sample. The choice of an estimator may seem intuitively obvious. Why not just estimate the population mean by the sample mean, the population variance by the sample variance, and so on? The answer is that the "obvious" estimator may not be a very good estimator. For example, suppose we wanted to estimate the value of the largest score in a population; call this parameter G.

Intuitively, we might use g, the largest score in the sample, to estimate G. However, g will be less than G in most samples. Why use an estimator which, on the average, will give a value that is systematically too small?

Sometimes intuition suggests competing choices for estimators. Suppose a sample is taken from a symmetrically distributed population of scores. Then, the population mean and median are identical. In that case, intuition leaves two choices for an estimator. Do we take the sample mean as the estimate? the sample median? Does it matter?

Clearly, we need something more than intuition to guide us in estimating (and testing hypotheses about) population parameters. Which quantity best estimates a particular population parameter can be decided by establishing criteria for good estimators and then examining how closely various estimators meet these criteria. The criteria that are generally agreed on are based on knowledge of the sampling distribution of the estimator. We will consider three important criteria for selecting estimates in turn.

2.7.1 Unbiasedness

Suppose we wish to estimate some population parameter θ (theta); θ might be a mean, a variance, or any other quantity of interest. A statistic, $\hat{\theta}$, is calculated from a sampled set of n scores. Our first criterion is that the mean of the sampling distribution of $\hat{\theta}$ should be equal to θ:

$$E(\hat{\theta}) = \theta \tag{2.15}$$

Although any one value of the estimator will not equal the parameter value, *averaging over many samples*, the estimate should be correct. Estimators conforming to Equation 2.15 are called unbiased estimates.

One example of biased estimation is the use of the largest score in a sample (g) to estimate the largest score in a population (G). We noted earlier that $E(g) < G$. A second example is S^2 as an estimator of σ^2 where $S^2 = \Sigma(Y_i - \bar{Y})^2/n$. In Appendix B, we show that $E(S^2) < \sigma^2$. More precisely,

$$E(S^2) = \left[\frac{n-1}{n} \right] \sigma^2$$

Multiplying both sides by $\dfrac{n}{(n-1)}$ gives

$$E \left(\frac{n}{n-1} \right) S^2 = \sigma^2$$

but

$$\left(\frac{n}{n-1} \right) S^2 = \frac{\Sigma(Y - \bar{Y})^2}{n-1} = \hat{\sigma}^2$$

Because it is unbiased, $\hat{\sigma}^2$ is preferred to S^2 when inferences about the population are involved.

The sample mean is an unbiased estimator of the population mean; as is demonstrated in Appendix B, $E(\bar{Y}) = \mu$. Bear in mind, however, that if the distribution is symmetric many other statistics are also unbiased estimators of μ; the median is one of these. Evidently, there are other considerations besides bias in choosing an estimator.

2.7.2 Consistency

Again, let $\hat{\theta}$ be some estimator of θ. It is a consistent estimator of θ if its value is more likely to be within some specified distance from θ as n increases. A familiar example of a consistent estimator is the sample mean; because $\sigma_{\bar{Y}}^2 = \sigma^2/n$, it is evident that the sampling variability of $\bar{Y}$ about μ declines as n increases.

Consistency is an important property of an estimator. Inferences based on consistent estimators are more likely to be correct as sample size increases. Nevertheless, even consistency combined with unbiasedness is not a sufficient basis for selecting between possible estimators of a parameter. A third criterion will be considered next.

2.7.3 Efficiency

Assume that a sample of size n has been drawn from a symmetric population. In that case, the sample mean and median are both unbiased estimators of the population mean because the population mean and median have the same value in any symmetric distribution. Furthermore, both the sample mean and median are consistent estimates of μ. They do differ in one respect, however. For any sample size n, the sampling distributions of the median and mean will differ in their variances. The estimator whose sampling distribution has the smaller variance should be preferred because any single estimate will tend to be closer to the parameter being estimated. Assume, for example, that many large samples are drawn from a normally distributed population. A mean and median are computed for each sample, and sums of squared deviations of the estimates about μ are then calculated, one for the sample means and another for the medians. For large samples, the variance of the sample means will be approximately 64% of the variance of the sample medians. This is expressed by saying that the *relative efficiency* of the median to the mean (as estimators of the mean of a normally distributed population) is .64. Conversely, the relative efficiency of the mean to the median is 1/.64 or 157%.

In general, assume a population parameter θ that can be estimated by either of two statistics, $\hat{\theta}_1$ or $\hat{\theta}_2$. The relative efficiency (R.E.) of $\hat{\theta}_1$ to $\hat{\theta}_2$ is

$$\text{R.E.} = \frac{E(\hat{\theta}_2 - \theta)^2}{E(\hat{\theta}_1 - \theta)^2} \tag{2.16}$$

Thus, relative efficiency is the ratio of two averages of squared deviations of estimates about the same population parameter. Note that this is a measure of the efficiency of the estimator in the denominator relative to that in the numerator.

2.7.4 Which Estimator?

Most of the estimation and hypothesis testing procedures presented in this and similar books, and in published journal articles, make use of the sample mean and the unbiased variance estimate $\hat{\sigma}^2$. If the population from which the data are drawn has a normal distribution, these statistics will be efficient relative to their competitors. Consequently, estimates based upon them are more likely to be close to the true value of the parameter being estimated, and hypothesis tests are more likely to lead to correct inferences. But what if the population distribution is not normal? We will address this question by considering the relative efficiencies of several estimators of μ for different population distributions. Our hope is that the results to be presented will encourage more thought about the choice of estimators and the statistical procedures based upon them.

To examine the efficiencies of various estimators, a computer was used to draw 2000 random samples of size 20 from a normally distributed population that had $\mu = 0$ and $\sigma = 1$. Three statistics were calculated for each sample. These were the mean ($\bar{Y}$), the median ($\tilde{Y}$), and the 10% trimmed mean ($\bar{Y}_{.10}$); this last statistic is obtained by rank ordering the scores in the sample and discarding the highest and lowest 10%, the top and bottom two scores for $n = 20$. The variances of the 2000 values of these three statistics are presented in the first column of Table 2.6. The column also contains the efficiencies of $\tilde{Y}$ and $\bar{Y}_{.10}$ relative to $\bar{Y}$; these are obtained by taking ratios of the variances. It should be clear that when the population of scores is normal, the mean is the more efficient statistic and therefore the better estimator of the population mean.

The situation is quite different if we make one change. Suppose 19 of the 20 subjects in each sample were drawn from the population with $\mu = 0$ and $\sigma = 1$; however, one subject is drawn from a population with $\mu = 0$ and $\sigma = 3$. This second population looks much like the first except that extreme scores are more likely. Think of it as comprising those rare individuals who come to the study hung over from the previous night's party. Such subjects might contribute to the variance, increasing the proportion of very small and very large scores. Variances of the three statistics and efficiencies relative to the mean are presented in the second column of Table 2.6. The interesting result here is that the variances of the sampling distributions of both the trimmed mean and the median are markedly less than that of the mean.

Contrary to popular mythology and intuition, the sample mean is not always the best estimator of the population mean. This happens because the sampling variance of the mean is increased much more than that of the trimmed mean or median by the inclusion of a deviant score in some samples. We say "some samples" because the score from the high-variance distribution is most likely to come from its middle; only

TABLE 2.6   VARIANCES AND RELATIVE EFFICIENCIES[a] OF THREE ESTIMATES OF A POPULATION MEAN

Statistic	Normal distribution		Mixed-normal distribution	
	Variance	Relative efficiency	Variance	Relative efficiency
$\bar{Y}$	.0513	1.0000	.2602	1.0000
$\tilde{Y}$	.0733	.6999	.0790	3.2927
$\bar{Y}_{.10}$	.0539	.9518	.0614	4.2378

[a] Relative efficiency for each statistic is its sampling variance divided into that of the sample mean $\bar{Y}$.

some samples will include a very small or very large value. Our sampling procedure is probably quite representative of what happens in many studies. The result of the occasional inclusion of these deviant scores is that we have less confidence in our inferences about population parameters. In many cases, the researcher might be well advised to use inferential procedures that do not rest upon the sample mean. Several nonparametric, or distribution-free, procedures will be presented in this book; these procedures will be particularly useful in fairly simple designs, but less so in more complex designs involving several independent variables. Another possible approach implicit in the results presented in Table 2.6 is to trim data from the tails of sample distributions. This involves adjusting estimates of population variances. Hogg, Fisher, and Randles (1975) described a t test based on trimming, and compared it with several nonparametric tests (as well as with the standard t test) for distributions exhibiting various degreees of tail weight and skew. That article also suggests ways of estimating tail weight and skew, and of using these estimates to select the best hypothesis testing procedure.

To sum up the developments of this section, unbiasedness, consistency, and efficiency are desirable properties in the statistics we use in drawing inferences. The prevalent use of inferential procedures based on $\bar{Y}$ and $\hat{\sigma}$ reflect the fact that these statistics are known to have these properties under many conditions. However, there will be situations in which the data are so distributed that other statistics will be more efficient. The researcher should be aware of this and, when such situations arise, consider alternative approaches to inference.

2.8 CONCLUDING REMARKS

This chapter focused on two broad concerns, ways to obtain a description of the data and the relation between the data and the population from which it was sampled. With respect to the description of data, it must be emphasized that we begin to understand a data set by looking at the data. We would be embarassed by such a seemingly obvious utterance if we weren't convinced that most researchers don't look at data; they look at means for various conditions, and they look at significance values associated with a test statistic. Researchers should take advantage of such devices as stem-and-leaf and box plots, now readily available in most computer packages, including several designed for microcomputers. Alternatives to means and standard deviations, such as medians and distance between hinges, should also be examined. Obtaining a complete description of data is important for several reasons. First, variability, shape, and outliers can tell something about the process underlying the data. Second, such information provides a better basis for choosing inferential procedures. A nice example is our demonstration that when the population has heavier tails than the normal, the sample median is actually a better estimator of the population mean than the sample mean is. Similarly, as we will demonstrate at several points in this text, tests based on ranks are often superior, sometimes very much so, to tests based on the mean and standard deviation when the data have been sampled from skewed or heavy-tailed distributions.

With respect to the relation between sample and population, the concept of a sampling distribution is crucial. Inferential statements are meaningful only in terms of such distributions. When we say that the means of two conditions are significantly different, we are making a statement about the sampling distribution of the difference between the two means. When we use a sample statistic to estimate a population parameter, calculations of the standard error of the sampling distribution provide a sense of how good an estimate we have, and considerations of properties of estimators (such as bias, consistency, and relative efficiency) are implicit in our use of this estimator in preference to other alternatives. The material we have presented on these matters is basic to understanding the rest of this book, and we urge readers to make sure they have a firm grasp of these ideas.

2.9 EXERCISES

2.1 The following terms provide a useful review of many concepts in the chapter. Define, describe, or identify each of them:

statistic	skew
parameter	weighted mean
box plot	sum of squares
stem-and-leaf plot	random variable
light- (or short-) tailed distribution	expected value
heavy- (or long-) tailed distribution	probability density function
normal probability plot	linear combination
depth of a statistic	unbiased estimator
resistant statistic	relative efficiency
hinge	consistent estimator
outlier (in terms of interhinge	trimmed mean
distance)	standard error of the mean

2.2 We have scores for 16 individuals on two measures of problem-solving ability:

$$X = 40\ 22\ 28\ 17\ 26\ 52\ 34\ 30\ 32\ \ 2\ 25\ 42\ 46\ 32\ 45\ 38$$

$$Y = 21\ 40\ 34\ 34\ 16\ 37\ 21\ 38\ 32\ 11\ 34\ 38\ 26\ 27\ 33\ 47$$

Find (a) the mean of the X scores; (b) the median of the X scores; (c) $(\Sigma Y_i)^2$; (d) ΣY_i^2; (e) $\hat{\sigma}_{XY}$; (f) $\hat{\sigma}_X^2$; (g) r_{XY}; (h) the upper and lower hinges for the Y scores.

2.3 It is known that in a particular large school, the mean IQ is 100. For some reason, you wish to select a random sample of five students. The first student you select has an IQ of 150. Given the above information, answer the following questions and justify your answers.

(a) What is your best estimate of the mean IQ of the next four students you select?

(b) What is your best estimate of the mean IQ of all five students in the sample?

(c) Do either of your answers to (a) and (b) change if the sample size is increased to 10? If so, what is the nature of the change?

2.4 Following are several sets of scores in ranked order. For each data set, is there any indication that it does not come from a normal distribution? Explain, citing whatever statistics you wish to make your point.

(a) X = 10 16 50 50 50 55 55 55 57 61 61 62 63 72 73 75 83 85 107 114

(b) X = 15 25 26 37 37 39 45 45 48 49 49 52 53 61 61 63 68 70 72 76

(c) X = 9 9 10 12 14 14 15 16 16 16 17 18 24 28 31 32 32 35 47 59

2.5 Following are summary statistics and stem-and-leaf plots, computed by SYSTAT, for two data sets. The X and Y values are scores on a standardized reading test for two groups of poor readers after a three-month instructional period; different instructional methods (X and Y) were used with the two groups of subjects. One of the researchers argues that it doesn't matter which method is used, whereas the second researcher argues that there is an advantage for method Y. What results would you present in support of each researcher's position? State and defend your own position. If you were going to do a statistical test on the data, are there any aspects of their distributions that are important to consider?

```
                STEM AND LEAF PLOT OF VARIABLE: X, N = 50
MINIMUM IS:        300.000
LOWER HINGE IS:       305.000
MEDIAN IS:         315.000
UPPER HINGE IS:       326.000
MAXIMUM IS:        370.000


              30    011111223344
              30 H 556689
              31    011222
              31 M 5577899
              32    000034
              32 H 66
              33    012
              33    568
              34    003
              34    5
         ***OUTSIDE VALUES***
              37    0


                STEAM AND LEAF PLOT OF VARIABLE: Y, N = 50
MINIMUM IS:      288.000
LOWER HINGE IS:     300.000
MEDIAN IS:       307.000
UPPER HINGE IS:     324.000
MAXIMUM IS:      453.000


              28    8899
              29    01144
              29    799
              30 H 000123344
              30 M 555689
              31    1112244
              31    77
              32 H 34
```

```
   32   566
   33   0013
   33   7
***OUTSIDE VALUES***
   37   77
   40   8
   45   3
```

2.6 An instructor grades students on a scale that runs from 0 to 3. The results for the class are as follows:

Score	Proportion of class
3	.4
2	.3
1	.2
0	.1

Calculate (a) the mean and (b) the variance for the class. (c) Is it necessary to know the class size to do the calculations? Explain.

2.7 Consider the hypothetical population that corresponds to a random variable Y, where Y takes on each of the values 2, 4, 6, and 8 with probability .25.

(a) What are the values of $E(Y)$ and $\text{Var}(Y)$?

(b) Samples consisting of two scores are drawn with replacement from this population, and the mean of each sample, $\bar{Y}$, is obtained. The sampling distribution of $\bar{Y}$ is as follows. Graph this distribution.

$$\bar{Y} = \quad 2 \qquad 3 \qquad 4 \qquad 5 \qquad 6 \qquad 7 \qquad 8$$
$$p(\bar{Y}) = .0625 \quad .125 \quad .1875 \quad .250 \quad .1875 \quad .125 \quad .0625$$

(Note: The $p(\bar{Y})$ are obtained by enumerating the different ways a value of $\bar{Y}$ can occur and then multiplying $.25^2$ by the number of ways. For example, $p(\bar{Y} = 5) = p(Y_1 = 2 \text{ and } Y_2 = 8) + p(Y_1 = 8 \text{ and } Y_2 = 2) + p(Y_1 = 4$ and $Y_2 = 6) + p(Y_1 = 6 \text{ and } Y_2 = 4) = (4)(.25^2) = .250$; the subscripts refer to the first and second scores drawn in a sample.)

(c) Find $E(\bar{Y})$ and $\text{Var}(\bar{Y})$ for the distribution in (b). How do these values relate to your answer to part (a)?

2.8 We have a population in which $p(X=0) = .8$ and $p(X = 1) = .2$.

(a) Calculate the population mean and variance $[E(X)$ and $\text{Var}(X)]$.

(b) Assume we draw samples of size 3. If we define the outcome of the experiment as a value of Y where $Y = \Sigma X$, there are four possible outcomes. Complete the following table (S^2 is the sum of squares divided by n, whereas $\hat{\sigma}^2$ is the sum of squares divided by $n - 1$):

Y	$p(Y)$			$\bar{X}$	S_X^2	$\hat{\sigma}_X^2$
0	$.8^3$		$= .512$			
1	$(3)(.8^2)(.2)$		$= .384$			
2	$(3)(.8)(.2^2)$		$= .096$			
3	$.2^3$		$= .008$			

In calculating the sample statistics, be careful. The population probabilities don't enter in (why not?). For example, if $Y=2$,

$$\bar{X} = \frac{\Sigma X}{n} = \frac{Y}{n} = \frac{(2)(1) + (1)(0)}{3} = \frac{2}{3}$$

and

$$S_X^2 = \frac{\Sigma X^2 - n\bar{X}^2}{n} = \frac{(2)(1^2) + (1)(0^2)}{3} - \left(\frac{2}{3}\right)^2 = \frac{2}{9}$$

(c) Using the entries in the above table, find $E(Y)$, $E(\bar{X})$, $E(S_X^2)$, and $E(\hat{\sigma}_X^2)$. How do $E(Y)$ and $E(\bar{X})$ compare with the value of $E(X)$ obtained in part (a)? How do $E(S^2)$ and $E(\hat{\sigma}^2)$ compare with the value of σ^2 obtained in part (a)? What do these results say about which sample statistics are biased or unbiased estimators?

2.9   There are three exams in a course and the statistics are

Test	$\bar{Y}$	$\hat{\sigma}_Y^2$
1	72	169
2	84	121
3	62	144

Also, the correlations are $r_{12} = .7$, $r_{13} = .5$, $r_{23} = .6$.

The instructor is concerned about the drop in performance between the first two tests and the third one. In order to determine whether this drop is statistically significant, the instructor calculates some statistics. First define

$$D = \frac{1}{2}(Y_1 + Y_2) - Y_3$$

the difference between the average of the first two exams and the score on the third. Now find the average value of D ($\bar{D}$) and its standard deviation ($\hat{\sigma}_{\bar{D}}$).

2.10 Each subject in a population is tested on two trials in an experiment. The scores on the two trials will be referred to as X_1 and X_2, respectively, and are either 1 or 0, depending on whether the trial result was a success or a failure. The joint distribution of X_1 and X_2 is

X_1

		1	0
	1	.4	.2
X_2	0	.3	.1

For example, .4 of the population succeeded on both trials; that is, $P(X_1 = 1$ and $X_2 = 1) = .4$.

(a) Find the mean of the population of scores for trial 1 [$E(X_1)$].

(b) Find the variance of X_1 (σ_1^2).

(c) Calculate the possible values of $X_1 - X_2$. Using the joint probabilities presented above, find the mean and variance of the sampling distribution of these difference scores.

(d) Find the covariance of X_1 and X_2 (σ_{12}).

2.11 In a study of the effects of motivation on signal detection performance, each of 20 subjects is tested under four levels of payoff for correct detections; call these m_1 through m_4. One approach to statistical tests of hypotheses about the effects of these factors is based on a knowledge of the variances and covariances for the four conditions. These can be summarized in a variance-covariance matrix:

$$
\begin{array}{c c}
& \begin{array}{cccc} m_1 & m_2 & m_3 & m_4 \end{array} \\
\begin{array}{c} m_1 \\ m_2 \\ m_3 \\ m_4 \end{array} &
\left[\begin{array}{cccc}
25 & 16 & -12 & 28 \\
16 & 36 & -18 & 21 \\
-12 & -18 & 16 & 16 \\
28 & 21 & 16 & 49
\end{array} \right]
\end{array}
$$

The entries on the main diagonal are the variances of the 20 scores for these conditions. The off-diagonal entries are covariances. These variances and covariances, together with knowledge of the means for the four conditions, permit us to test many hypotheses about the effects of motivation and to estimate the sizes of these effects. Let us consider some possibilities.

(a) We may be interested in comparing the effectiveness of the two lowest payoff levels, m_1 and m_2. That is, we wish to estimate, or test hypotheses about, $E(Y_1 - Y_2)$. This requires us to calculate the variance of the difference between the scores in the two conditions, $\text{Var}(Y_1 - Y_2)$. Calculate this quantity.

(b) Suppose we want to compare the average score for the lowest two levels with that for the highest two levels; that is, we are interested in $E[(1/2)(Y_1 + Y_2) - (1/2)(Y_3 + Y_4)]$. What is the variance of this difference score?

(c) Because the condition m_1 involves no payoff, we decide to compare it with the average of the three nonzero payoff conditions. Therefore, calculate the variance of the 20 values of $Y_1 - (1/3)(Y_2 + Y_3 + Y_4)$.

2.12 (a) Suppose we wish to compare performances under two experimental conditions, A_1 and A_2. We might have two different groups of n subjects each. We can calculate

$$\bar{D} = \bar{Y}_{.1} - \bar{Y}_{.2}$$

Write an expression for $\text{Var}(\bar{D})$, the variance of the sampling distribution of $\bar{D}$, in terms of σ_1^2 and σ_2^2. These are the variances of two populations of scores, one obtained under A_1 and the other under A_2.

(b) Suppose we run the above experiment by having only one sample of n subjects, each of whom is tested under both A_1 and A_2. Then we calculate $D_i = Y_{i1} - Y_{i2}$ for each subject, giving us a set of n difference scores. (i) Write an expression for the variance of a population of such difference scores in terms of σ_1^2, σ_2^2, and ρ_{12}; ρ (rho) is the correlation of the two scores in the population. (ii) Convert the variance of D into an expression for $\text{Var}(\bar{D})$, the variance of the sampling distribution of $\bar{D}$ using this "repeated-measures" design. (iii) Comparing the result in (b)(ii) with that in (a), which design do you think is more likely to lead to correct inferences? Why?

2.13 We have a population of values of Y whose distribution is

$$Y = 1 \quad 2 \quad 3$$
$$p(Y) = .4 \quad .2 \quad .4$$

(a) Find the mean and variance of the population.

(b) We draw samples of size 3 from the population. The sampling distribution of the mean is presented below. To illustrate how the probabilities were calculated, consider $p(\bar{Y} = 2)$. A sample of three 2's, $<2, 2, 2>$, has probability $.2^3$. A mean of 2 would also be observed if the sample contained 1, 2, and 3 in any order. The probability of such a sample is $(.4)(.2)(.4)$, and there are six possible orders. Combining these results gives $P(\bar{Y} = 2) = .2^3 + (6)(.4^2)(.2) = .20$. The sampling distribution of the mean is

$$\bar{Y} = 1 \quad 4/3 \quad 5/3 \quad 2 \quad 7/3 \quad 8/3 \quad 3$$
$$p(\bar{Y}) = .064 \quad .096 \quad .240 \quad .200 \quad .240 \quad .096 \quad .064$$

Note that the probabilities sum to 1 as they should. Now calculate the mean and variance of this sampling distribution. How do they compare with your answers to part (a)? Plot the sampling distribution of the mean, and comment on its shape relative to that of the population distribution.

(c) Suppose that samples of size 25 were drawn from the population. What would the mean and variance of the sampling distribution of the mean be? What do you think the sampling distribution would look like? (Don't try to calculate it; reason from your previous answer and our discussion of the sampling distribution of the mean.)

(d) Next we consider the sampling distribution of $\hat{\sigma}^2$, still using samples of size 3. The probabilities of the different sample variances were obtained by considering which samples could give rise to a particular value of $\hat{\sigma}^2$. For example, the probability that the sample variance is zero is the probability of all 1's, or all 2's, or all 3's; this is $.4^3 + .2^3 + .4^3$, or .136. The sampling distribution is

$$\hat{\sigma}^2 = 0 \quad 1/3 \quad 4/3 \quad 1$$
$$p(\hat{\sigma}^2) = .136 \quad .288 \quad .384 \quad .192$$

Calculate $E(\hat{\sigma}^2)$ and verify that it equals σ^2. What are the values of S^2? What is the relation of $E(S^2)$ to σ^2?

2.14 Consider a population of subjects each of whom is tested on each of four trials. The mean score for the population of subjects is 70 for every trial; the variance of each of the four populations is 100.

(a) If a mean of the four scores is calculated for every subject in the population, will the mean of its sampling distribution be 70, less than 70, or more than 70? Explain briefly.

(b) Will the variance of the sampling distribution of the mean be 100/4, less, or greater? Explain briefly.

2.15 In this problem, we consider why it is important to weight two means when combining them to get a single estimate of a population mean. In the first year of a state program for teaching illiterates to read, 20 individuals were enrolled. The mean score on an evaluation test for these 20 people was $\bar{X}_1 = 52$. Deciding the program was successful, the state appropriated more funds the following year, enrolling 60 students. The mean test score for these 60 was 56. In assessing the two sets of results, it was decided to combine the two means to obtain a single estimate of performance for the sampled population. Let μ be the mean of the population from which the two samples were drawn,

(a) Let $\bar{U} = (1/2)(\bar{X}_1 + \bar{X}_2)$ and $\bar{W} = (20/80)(\bar{X}_1) + (60/80)(\bar{X}_2)$; $\bar{U}$ and $\bar{W}$ stand for "unweighted" and "weighted," respectively. Express $E(\bar{U})$ and $E(\bar{W})$ in terms of μ. Comment on whether either is biased.

(b) Noting that $\bar{U}$ and $\bar{W}$ are both linear combinations of X_1 and X_2, express $\sigma_{\bar{U}}^2$ and $\sigma_{\bar{W}}^2$ as functions of σ^2, the population variance. (To obtain a general result, replace 20 and 60 by n_1 and n_2 and express $\sigma_{\bar{U}}^2 - \sigma_{\bar{W}}^2$ as a function of σ^2, n_1, and n_2.)

(c) On the basis of your answers to (a) and (b), which estimate of μ is better? Why?

2.16 A population of voters consists of equal numbers of conservatives and liberals. Furthermore, .9 of the liberals prefer the Democratic candidate in the upcoming election, whereas only .3 of the conservatives prefer the Democratic candidate.

(a) What is the probability of sampling an individual from the entire population who prefers the Democratic candidate?

(b) In Appendix B, we prove that the variance of a proportion p is $p(1 - p)/n$. With this in mind, what is the variance of the sampling distribution of P, the proportion of Democratic voters in a sample of 50 individuals who are randomly selected from the population of voters?

(c) Suppose you are a pollster who knows that the population is equally divided between liberals and conservatives, but you don't know what the proportion of Democratic voters is. You sample 50 individuals with the constraint that 25 are liberals and 25 are conservatives; this is referred to as stratified sampling.

    (i) From the information presented at the start of this problem, what is the variance of the sampling distribution of P_L, the proportion of Democratic voters in a sample of 25 liberals?

    (ii) What is the variance of the sampling distribution of P_C, the proportion of Democratic voters in a sample of 25 conservatives?

    (iii) The proportion of Democrats in the stratified sample is $P = (1/2)(P_L + P_C)$. What is the variance of the sampling distribution of P when stratification is employed?

    (iv) In view of your answers to (b) and (c)(iii), is it better to stratify or not? Explain your answer.

APPENDIX: The Mean and Variance of a Linear Combination

Consider n subjects with m scores for each. If the jth score for the ith subject is denoted by Y_{ij} and is multiplied by a weight w_j, we can calculate a linear combination of the m scores:

$$L_i = \sum_j^m w_j Y_{ij} \tag{2.17}$$

The mean of the n values of L_i is

$$\bar{L}_. = \frac{1}{n}\sum_i^n L_i = \frac{1}{n}\sum_i^n \sum_j^m w_j Y_{ij}$$

Rearranging terms, we have

$$\bar{L}_. = \sum_j w_j \frac{1}{n}\sum_i Y_{ij} = \sum_j w_j \bar{Y}_{.j} \tag{2.18}$$

The variance of the L_i is, by definition,

$$\hat{\sigma}_L^2 = \frac{1}{n-1}\sum_i^n (L_i - \bar{L}_.)^2 \tag{2.19}$$

To derive the relationship between this variance and the variances and covariances of the Y_{ij}, we begin by substituting the expressions in Equations 2.17 and 2.18 into Equation 2.19:

$$\hat{\sigma}_L^2 = \frac{1}{n-1}\sum_i \left[\sum_j w_j Y_{ij} - \sum_j w_j \bar{Y}_{.j}\right]^2$$

$$= \frac{1}{n-1}\sum_i \left[\sum_j w_j (Y_{ij} - \bar{Y}_{.j})\right]^2$$

Expanding the squared expression, we have

$$\hat{\sigma}_L^2 = \frac{1}{n-1}\sum_i \left[\sum_j w_j^2 (Y_{ij} - \bar{Y}_{.j})^2 + \sum\sum_{j \neq j'} w_j w_{j'} (Y_{ij} - \bar{Y}_{.j})(Y_{ij'} - \bar{Y}_{.j'})\right]$$

Rearranging terms so that we first sum with respect to i, we have

$$\hat{\sigma}_L^2 = \frac{1}{n-1}\left[\sum_j w_j^2 \sum_i (Y_{ij} - \bar{Y}_{.j})^2 + \sum\sum_{j \neq j'} w_j w_{j'} \sum_i (Y_{ij} - \bar{Y}_{.j})(Y_{ij'} - \bar{Y}_{.j'})\right]$$

The sum of squared deviations divided by $n-1$ is a variance, and the sum of cross products of deviations divided by $n-1$ is a covariance; therefore,

$$\hat{\sigma}_L^2 = \sum_j w_j \hat{\sigma}_j^2 + \sum\sum_{j \neq j'} w_j w_{j'} r_{jj'} \hat{\sigma}_j \hat{\sigma}_{j'}$$

Chapter 3

Some Important Distributions

3.1 INTRODUCTION

Four theoretical distributions—the normal, the t, the chi square (χ^2), and the F—play an important role in the vast majority of data analyses reported in the research literature. In this chapter, we attempt to provide a sense of these distributions—what they look like, their applications, the assumptions involved in applying them, and the relations among them. To some extent, this material reviews content typically taught in an introductory statistics course. For example, most readers will have had some previous exposure to at least the normal and t distributions. Nevertheless, we believe that our treatment of these distributions will provide both a beneficial review and an important supplement to previous exposures.

The presentation in this chapter is far from exhaustive. Entire textbooks have been written about applications of both the chi-square and F statistics; indeed much of this book will discuss the applications of the F statistic to data analysis. What we have tried to do is focus on those topics that we consider to be basic to future developments in this book. This means an emphasis on assumptions involved in applying these statistical distributions in data analysis. It also means a review of basic ideas in hypothesis testing and of other aspects of inference, such as interval estimation.

3.2 THE NORMAL DISTRIBUTION

In Chapter 2 we discussed general properties of theoretical distributions. In this and the next few sections, we will consider some specific continuous distributions that play a prominent role in statistical inference. We begin with the normal distribution because the validity of many statistical procedures rests upon the assumption that the distribution of the population of scores is normal. Furthermore, the normal distribution provides a relatively simple context within which to review inferential procedures such as hypothesis testing and interval estimation. We will do just that in Section 3.3. We will then consider three distributions that are derived from the normal: the t, the chi square, and the F.

The normal distribution is characterized by its density function

$$f(y) = \frac{1}{\sigma\sqrt{2\pi}} e^{-(y-\mu)^2/2\sigma^2} \tag{3.1}$$

where μ and σ are the mean and standard deviation of the population, and π and e are mathematical constants. The random variable Y can take on any value between $-\infty$ and $+\infty$, and the curve is symmetric about the mean.

Infinitely many normal distributions are possible, one for each combination of mean and variance. However, inferences based on these normal distributions are aided by the fact that all of the possible normal distributions are related to a single distribution. This *standardized normal distribution* is obtained by subtracting the distribution mean from each score and dividing the difference by the distribution standard deviation:

$$z = \frac{Y - \mu}{\sigma} \tag{3.2}$$

The variable z is often referred to as a *standardized normal deviate*. If the Y's are normally distributed, the corresponding z's also will be. However, the mean of the z distribution will be 0 and its standard deviation will be 1. This is proved in Appendix B.

Standardization directly provides us with information about the relative position of an individual score and is very helpful. For example, assume a normally distributed population of scores with $\mu = 500$ and $\sigma = 15$. A value Y of 525 would correspond to a z score of 1.67; $z = (525 - 500)/15 = 1.67$. Turning to Appendix Table D.2, we find that $F(z) = .9525$ when z is 1.67. $F(z)$ is the proportion of standardized scores less than z in a normally distributed population of such scores. In this example, we may conclude that the score of 525 exceeds .9525 of the population. Of course, this conclusion is valid only if our values of μ and σ are correct and Y is normally distributed.

Equation 3.2 defined a z score as $(Y - \mu)/\sigma$. In fact, this is just a special case of a general formula for a z score. Instead of Y, we could have any observed quantity; examples would be the sample mean, the difference between two sample means, or some other statistic. Call this V for *observed variable*. To transform V into a z score, subtract its expected value from it (just as μ was subtracted from Y in Equation 3.2). Then divide the difference by σ_V, the standard error (SE) of the sampling distribution of V. Thus, a general formula for z is

$$z = \frac{V - E(V)}{\sigma_V} \tag{3.3}$$

where σ_V is the standard error of V. Let us now consider the case in which V is the sample mean $\bar{Y}$, and use the z score of Equation 3.3 to review some general ideas about inferential processes.

3.3 INFERENCES BASED ON THE NORMAL DISTRIBUTION

Broadly speaking, there are two kinds of inferences with which researchers are concerned. Most frequently, we ask whether a parameter has a specified value. For example, is the mean of the population 100? Is the difference between the means of the experimental and control population zero? The process by which we address such questions is often referred to as hypothesis, or significance, testing. The second kind of question involves asking what the value of the parameter is. In Section 2.7, we discussed some aspects of this process of estimation. In this section, we consider one way of indicating the reliability of our estimate. We do this by constructing bounds, called confidence intervals, on the estimate. We will use the z distribution to illustrate both hypothesis testing and interval estimation, but we will emphasize certain general principles that apply to hypothesis tests and confidence intervals based on any theoretical distribution.

3.3.1 Testing a Hypothesis about a Population Mean

Suppose that the average score on a standardized test of reading for first graders is 68, and the standard deviation of the population is known to be 39. Further suppose that a new program for teaching reading is tried with a sample of 169 students. At the end of a year of instruction, a standardized test of reading skills is administered to this sample. If the average score in this sample is higher than the present norm of 68, we must decide whether this is just a chance effect or whether the new method of instruction has produced a reliable improvement in reading performance. These two possibilities are formalized as the null and alternative hypotheses, which are designated H_0 and H_1, respectively.

Letting μ represent the mean of the population of first graders taught by the new method, we can restate our two hypotheses as

$$H_0: \mu \leqslant 68 \quad \text{and} \quad H_1: \mu > 68$$

Once these two hypotheses have been formulated, we need a test statistic whose value will enable us to decide between them. Recall the general form of the z statistic (Equation 3.3):

$$z = \frac{V - E(V)}{\sigma_V}$$

In the present example, we replace V by the sample mean, $E(V)$ by the mean specified by H_0 (μ_{hyp}), and σ_V by the standard error of the mean. Consequently,

$$z = \frac{\bar{Y} - \mu_{hyp}}{\sigma/\sqrt{n}} \tag{3.4}$$

Substituting the values from our example into Equation 3.4, we have

$$z = \frac{\bar{Y} - 68}{39/\sqrt{169}}$$

The parameter μ_{hyp} is the value of the mean expected if the null hypothesis is true. Our null hypothesis actually specifies a range of values, but we take μ_{hyp} to be 68 because values of z larger than 68 are evidence against H_0, and if we can reject H_0 for $\mu_{hyp} = 68$, we can reject it for any value smaller than 68.

The next step is to determine those values of z that, if obtained in the study, would lead to rejection of H_0 in favor of H_1. Such values constitute a *rejection region*. This is a set of possible values of z that are consistent with H_1 and very improbable if H_0 is assumed to be true. Indeed, these values are so unlikely if H_0 is assumed that their occurrence leads us to reject H_0. An arbitrarily chosen value, α (alpha), defines exactly how unlikely "so unlikely" is. Traditionally, researchers have set α at .05. We want very strong evidence against H_0 before we reject it.

We will now establish a rejection region for our study. Turning to Table D.2 we look for $F(z) = .95$; this means that .05 (our α level) of the normal curve lies above the corresponding value of z. That value, the critical value, is 1.645. Therefore, if the value of z calculated from our data exceeds 1.645, we will reject H_0.

To summarize the steps so far:

1. State the null and alternative hypotheses.
2. Decide on a test statistic; in the present example, this is the z defined by Equation 3.4.
3. Decide on a value of α and establish a rejection region.

Only one other step is necessary:

4. Run the experiment and calculate the value of the test statistic. If it lies within the rejection region, reject H_0; otherwise, do not reject it.

In the example, assume that the mean of the 169 scores is 74. Then, substituting into Equation 3.4, we have

$$z = \frac{74 - 68}{3} = 2.0$$

Because the value of z exceeds the criterion value 1.645, we conclude that the new program for teaching reading is superior to the traditional program. More formally, we reject the null hypothesis that the mean of the population of reading scores obtained under the new program is equal to or less than the old mean of 68 in favor of the alternative that it is higher.

The second step in the testing process—the choice of a test statistic—deserves further comment. The use of the z test implies two assumptions. The first is that the population of scores is normally distributed. In fact, as we shall see in Section 3.4, with samples this large the mean (and therefore z) will tend to have a normal sampling distribution even if the population is not normally distributed. The second assumption— that the scores are independently distributed—is more critical. Recall that the expression for the standard error of the mean was derived under the assumption of independence. Suppose this assumption is incorrect; perhaps some students in our sample copied from each other. Then the variance of the mean should include positive covariance terms (see Equation 2.12) and $\sigma/\sqrt{n}$ underestimates the true standard error. The consequence would be that the z statistic would be spuriously inflated, and we would reject H_0 with probability greater than .05.

3.3.2 One- and Two-Tailed Tests

The test described for the study in the preceding section is called *one-tailed* or *directional*. Because we were interested in whether the training method improved entrance test scores, the rejection region consisted of only the largest values of z. Any disadvantage due to training was not of interest. In the context of this research example, that makes sense. On the other hand, one can conceive of situations in which normative data, or a theory, dictate a specific value of μ_{hyp}, and we wish to detect any departure from that hypothesized value. Suppose we wanted to know whether reading scores in a city differ from the national norms. Let μ_{hyp} be the national average score on a standardized reading test, and let Y be the mean of children in a particular city. Then the null hypothesis is

$$H_0: \mu = \mu_{hyp}$$

and the alternative hypothesis is

$$H_1: \mu \neq \mu_{hyp}$$

In this case, we say that the significance test is *two-tailed* or *nondirectional.* If the population of scores is normally and independently distributed and we know the population variance, we again can use the z test. Accordingly, we turn to Appendix Table D.2. Because our test is two-tailed, we need two criterion values; call these z_L and z_U (for lower and upper). Assuming that $\alpha = .05$, these values should be chosen so that .025 (half of α) of the distribution lies below z_L and .025 lies above z_U. From Table D.2, we find that .975 $(1 - .025)$ of the z distribution is below 1.96. Therefore, $z_U = 1.96$ and, because the distribution is symmetric about zero, $z_L = -1.96$. Our decision rule is: reject if $z < -1.96$ or $z > 1.96$. An equivalent rule is to reject whenever the absolute value, $|z|$, is greater than 1.96.

3.3.3 Type 1 and Type 2 Errors

When a hypothesis is tested, two types of errors are possible. If the null hypothesis is true, rejection is a *Type 1 error.* The probability of such an error is α, and it is determined by the experimenter in the way we have illustrated. Suppose the null hypothesis is false. Failure to reject a false null hypothesis is a *Type 2 error;* its probability is β (beta). The probability of rejecting a false null hypothesis is called the *power* of the test. The sum of power and β is 1.

The following table may help to clarify the meanings of α, β, and power:

	Decision	
	Reject	Fail to reject
H_0 true	$P(\text{Type 1 error}) = \alpha$	$1 - \alpha$
H_0 false	Power $= 1 - \beta$	$P(\text{Type 2 error}) = \beta$

The rows represent two mutually exclusive events: H_0 is either true or false. Given either of these two events, the researcher may make one of two mutually exclusive decisions: reject or do not reject H_0. The probability in each cell is a *conditional probability,* $P(\text{decision} | \text{event})$. We read this as "the probability of the decision given the event." For example, α is $P(\text{reject} | H_0 \text{ true})$, "the probability of rejection given that

H_0 is true." Similarly, power is $P(\text{reject} | H_0 \text{ false})$. Given an event, one of the two decisions must be made, and therefore the conditional probabilities in each row sum to 1.

We choose the value of α, and in this way we control the Type 1 error rate. For example, if we set $\alpha = .05$, we implicitly acknowledge that, *if the null hypothesis is true*, the probability is .05 that we will obtain a value of our test statistic in the rejection region. Therefore, the Type 1 error rate is .05. We cannot choose the value of β, or power, in quite the same way. However, we can take power into consideration in planning an experiment. To do so, we must understand how power varies as a function of several factors—how false the null hypothesis is, the population variance, the α selected, the sample size, and whether the alternative hypothesis is one- or two-tailed. We next consider how this is done.

3.3.4 The Power of a Statistical Test

Again consider our example of the sample of 169 reading scores. Recall that $\mu_{\text{hyp}} = 68$ and σ is known to be 39. Assuming a directional test with $\alpha = .05$, we reject H_0 if $z > 1.645$. Now suppose we decide that it is important to reject the null hypothesis if the new instructional method yields an improvement of seven points; smaller effects are of little practical importance because they would not warrant the expense of instituting the new method throughout the school system. To summarize, we wish to test $H_0: \mu \leqslant 68$ against the specific alternative hypothesis

$$H_A: \mu \geqslant 75$$

The question is: what power does our test of H_0 have if this alternative is true?

For any statistical test, answering this requires the following steps:

1. The null and alternative (H_0 and H_1) hypotheses must be specified as well as α. These determine our rejection region—the values of the test statistic that result in rejecting H_0 in favor of H_1. In the present example, we have specified a one-tailed alternative and we will set $\alpha = .05$.

2. A specific alternative hypothesis must be established. This may be a departure from the null hypothesis that we expect on the basis of pilot data, or one that is of theoretical or practical importance. In the present example, H_A was specified as $\mu = 75$. Note the distinction between H_1, which establishes a range of values of the parameter inconsistent with the null hypothesis, and H_A, which establishes a specific alternative. We can only calculate the power of the test against a specific alternative.

3. Calculate an effect size—a measure of the distance between the parameter values specified by H_0 and H_A. In the present example, we calculate the distance between μ_A (the mean under H_A, 75 in our example) and μ_{hyp} (68) in z units:

$$z_A = \frac{\mu_A - \mu_{\text{hyp}}}{\sigma/\sqrt{n}} = \frac{7}{3} = 2.333$$

4. For many common tests, tables or charts are available that enable us to read the value of power once we know the effect size and other information, such as α

and n. Cohen (1977) has published such tables, and in Chapter 4 we will consider charts for calculating the power of the F test. However we arrive at it, the power value is the probability of sampling a value in the rejection region when H_A is true. In the case of the z test, power may be calculated using the normal probability table, D.2. Figure 3.1 shows the normal distribution of z scores, assuming H_A to be true; note that the mean is at 2.333. We have indicated the area corresponding to the power of the test; this is that area under the curve that is to the right of $z = 1.645$, the cutoff established when we selected our α level. We can find the size of this area by shifting the distribution of Figure 3.1 to the left so that its mean is at zero. If we did that, the critical z score would be $1.645 - 2.333$, or $-.688$. We then turn to the standardized normal distribution of Table D.2 and look for $F(.688)$, the area to the left of $z = .688$. Because the normal distribution is symmetric, this is also the area to the right of $-.688$. This area, the power of our test, is .75.

In general, the power of a test depends upon several factors. First, power increases as α increases, because the increase in α requires an increase in the range of values included in the rejection region. If α were .10, the critical value in Figure 3.1 would shift from 1.645 to 1.28, increasing the rejection region and, consequently, the area above it. Second, power is affected by the nature of H_0 and H_1. If our statement of H_0 were two-tailed, with α still at .05, the decision rule would have been to reject H_0 if $z < -1.96$ or $z > 1.96$. Then there would be two critical values in Figure 3.1, -1.96 and 1.96. Power would correspond to the areas to the right of 1.96 and to the left of -1.96 under the H_A distribution. As we can see in Figure 3.1, the probability of $z < -1.96$ if H_A is true is essentially zero, and the probability that $z > 1.96$ is less

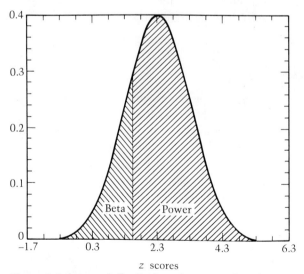

Figure 3.1 A normal distribution of z scores when the alternative to the null hypotheses (H_A) is true. The mean is 2.33, and the rejection region consists of all values greater than 1.645. The text explains the basis for these values.

than the probability that $z > 1.645$. Therefore, the one-tailed test is more powerful against the specific alternative, $\mu_A = 2.333$. On the other hand, this one-tailed test has virtually no power against specific alternatives of the form $\mu < \mu_{hyp}$, whereas the two-tailed test has the same probability of rejecting H_0 against these alternatives as against those of the form $\mu > \mu_{hyp}$.

Two other factors affecting power are the population variance and the sample size. In the case of the z test, reduced variance and larger n yield smaller standard errors of the mean. As the SE decreases, the sample mean is more likely to be close to the true parameter value. Therefore, if H_0 is false, a smaller SE increases the probability of getting values of Y consistent with the alternative hypothesis; power is increased. We can influence variability by our choice of measures and experimental design, as well as by controlling extraneous factors that might contribute to chance variability. How large an n we need will depend on the other factors noted above and the power we want, as well as the smallest size effect we want to be able to reject with that power. In Chapter 4, we will show how we can use that information together with plots of power functions to determine sample sizes to use in conjunction with the analysis of variance. Cohen (1977) provides tables for several different tests that enable us to decide on the sample size.

The study of power functions has other implications for researchers. Power functions for different statistical tests of the same H_0 can be compared. Assuming that the choice among tests is not dictated by some other factor (such as validity of assumptions, ease of calculations, or availability of tables), the test with the higher power function should be chosen. We also should consider the effects of violations of assumptions upon the power of various statistical tests. For example, the t test is more powerful than other tests that can be used to compare two experimental conditions when the population is normally distributed. However, when the population of scores is not normally distributed, other tests may achieve markedly more power (Blair and Higgins 1980, 1985).

3.3.5 Confidence Intervals for μ

Although researchers seem to concentrate their data analysis efforts on tests of hypotheses, those efforts often might be better addressed to estimating the value of the parameter. In order to have a sense of the reliability of such estimates, we calculate a confidence interval, a pair of numbers that bound the parameter being estimated. As in the preceding sections, we will use the standardized normal distribution and the parameter μ to illustrate the principles involved. However, many of the points we will make apply to confidence intervals for other parameters, and are valid when other theoretical distributions provide the basis for constructing the interval.

Once again consider our new reading program. Recall that we have a sample of 169 subjects with $\bar{Y} = 74$ drawn from a population whose standard deviation is 39. Assume that we want to calculate a confidence interval for the mean of a population of students taught in the new reading program. To understand the process by which we do this, imagine drawing many samples of 169 scores from the population and computing a z for each sample. Assuming the scores are independently sampled from a normal population, Table D2 tells us that .95 of the sampled z values will lie between -1.96 and 1.96. That is,

$$P\left[-1.96 \leqslant \frac{\bar{Y} - \mu}{\sigma_{\bar{Y}}} \leqslant 1.96\right] = .95$$

To obtain the bounds on μ, consider each inequality separately. First,

$$-1.96 \leqslant \frac{\bar{Y} - \mu}{\sigma_{\bar{Y}}}$$

Solving for μ, we have the upper bound

$$\mu \leqslant \bar{Y} + 1.96\sigma_{\bar{Y}}$$

Next, the inequality

$$\frac{\bar{Y} - \mu}{\sigma_{\bar{Y}}} \leqslant 1.96$$

gives the lower bound on μ:

$$\bar{Y} - 1.96\sigma_{\bar{Y}} \leqslant \mu$$

Putting it all together, we have

$$P(\bar{Y} - 1.96\sigma_{\bar{Y}} \leqslant \mu \leqslant \bar{Y} + 1.96\sigma_{\bar{Y}}) = .95$$

What does this equation mean? Assume we randomly draw many samples of size 169 from a normally distributed population. In each case the sample mean is computted. That value, as well as those of n and σ, are inserted into the preceding equation yielding a set of limits for each sample. These limits will vary from sample to sample because each sample yields a different value of $\bar{Y}$. The equation tells us that we can expect .95 of the sets of limits to enclose the true value of μ.

That's fine if an infinitely large number of samples is collected. However, we ran one study that resulted in a single sample of 169 reading scores. We substitute the value of $\bar{Y}$ from this study into the equation, getting the confidence interval

$$CI = 74 \pm (1.96)\left[\frac{39}{13}\right] = 74 \pm 5.88 = 68.12, 79.88$$

We have .95 confidence that μ lies between 68.12 and 79.88. That does *not* mean that "the probability is .95 that μ lies between 68.12 and 79.88"; μ is either in this interval or it isn't. However, because we know that, in the long run, .95 of the samples will yield intervals enclosing the true parameter value, our confidence is .95 that the one interval calculated for this data set contains μ.

The confidence interval not only provides us with bounds on our estimate of a parameter but also tests any null hypothesis about that parameter. For the data set in our example, a two-tailed test will reject H_0: $\mu = \mu_{hyp}$ at the .05 significance level if, and only if, μ_{hyp} lies outside the .95 confidence interval; that is, if μ_{hyp} is less than 68.12 or greater than 79.88. For example, suppose we wish to test H_0: $\mu = 68$. Because 68 is below the lower bound of the .95 confidence interval, we can infer that a two-tailed test with $\alpha = .05$ would reject H_0. We can also evaluate the outcome of tests of any other null hypothesis just by noting whether the values specified by H_0 (that is, μ_{hyp}) falls within the confidence interval.

The relation between confidence intervals and significance tests may be understood by considering the usual decision rule for a two-tailed test: reject H_0 if

$$\frac{\bar{Y} - \mu_{\text{hyp}}}{\sigma_{\bar{Y}}} > z_{.025} \qquad \text{or} \qquad \frac{\bar{Y} - \mu_{\text{hyp}}}{\sigma_{\bar{Y}}} < -z_{.025}$$

where $z_{.025}$ is the value of z exceeded by .025 of the distribution. Some algebra will show that this is equivalent to the rule: reject if

$$\mu_{\text{hyp}} < \bar{Y} - z_{.025}\sigma_{\bar{Y}} \qquad \text{or} \qquad \mu_{\text{hyp}} > \bar{Y} + z_{.025}\sigma_{\bar{Y}}$$

Thus, the null hypothesis will be rejected at the .05 level (two-tailed) whenever the hypothesized value of the population mean is less than the lower bound or more than the upper bound of a .95 confidence interval.

We can also use the confidence interval to carry out one-tailed tests of significance. For example, suppose we want to test $H_0: \mu \leqslant 68$ against the directional alternative $H_1: \mu > 68$. Because the lower bound of our .95 confidence interval is 68.12, we have .975 confidence that the true population mean is greater than 68.12. Therefore, $H_0: \mu \leqslant 68$ is very unlikely to be true. In fact, we can reject this null hypothesis at the .025 level of significance.

The confidence interval permits tests of all possible null hypotheses simultaneously, and the interval width provides a sense of the test's power. The narrower the interval, the wider the range of values of μ that will be rejected. The relation between interval width and power can be understood by considering a general formula for the confidence interval:

$$P\left[\bar{Y} - \frac{z_{\alpha/2}\sigma}{\sqrt{n}} \leqslant \mu \leqslant \bar{Y} + \frac{z_{\alpha/2}\sigma}{\sqrt{n}}\right] = 1 - \alpha \qquad (3.5)$$

where $\pm z_{\alpha/2}$ are the critical values of z required for a two-tailed test of significance at the α level. Note that the interval width decreases (both bounds approach $\bar{Y}$) as n increases or σ decreases; more data or less variability yields more precise estimates of the parameter μ. Both of these factors also affect the power of the z test; indeed, anything that decreases the interval width will increase power against a specific null hypothesis.

One other factor influences the width of the confidence interval; the level of confidence chosen. Note that higher confidence implies smaller α, which in turn requires a large value of $z_{\alpha/2}$; as a result, the interval width increases. As we saw in Section 3.3.4, smaller α levels also result in a loss of power. There is a trade-off here; the only way we can be assured of both a narrow interval (or high power) and high confidence (or low type 1 error rate) is to collect lots of data. Nobody gets something for nothing.

The equation under consideration can be viewed as a special case of a general formula for the confidence interval based on the z distribution. Consider the expected value of any sampling distribution; call this E. This parameter is estimated by some statistic V. Then, the $1 - \alpha$ confidence interval for E is calculated as

$$\text{CI} = V \pm z_{\alpha/2}\sigma_V \qquad (3.6)$$

If V is replaced by $\bar{Y}$ and σ_V by $\sigma/\sqrt{n}$, we have the previously presented bounds for μ. If V is the difference between the means of two conditions, and σ_V is the standard error of that difference (see Section 2.6.4), we can calculate a confidence interval for the difference between two population means. We will do exactly this, using the *t* instead of the *z* distribution. The normal distribution has been useful for illustrating several ideas about inference, but the *t* distribution will prove more appropriate for most inferences about means because the *t* test does not require that we know the population standard deviation. Before we elaborate on this point and review the use of the *t* statistic, we will introduce one other idea that is important in interpreting the results of many statistical procedures.

3.4 THE CENTRAL LIMIT THEOREM

In Section 2.6.2, we considered the sampling distribution of the means of samples drawn from a population in which all values are equally likely. As the sample size increased, the sampling distribution of the mean showed a pronounced peak (see Figure 2.5). If the mean had been based on a large enough sample, the sampling distribution would have looked very much like a normal distribution. Even if the population of scores is skewed, the sampling distribution of the mean is still approximately normal if *n* is large enough. Usually, *n*'s of 40 or more suffice even when the samples are drawn from quite skewed populations. Sample sizes as small as 15 may result in approximately normal distributions of sample means if the population of scores has a symmetric distribution.

Figure 2.5 illustrates the operation of the *central limit theorem:* the mean of a sample of *n* independently drawn observations will have a sampling distribution that approaches normality as *n* increases regardless of the shape of the population distribution from which the scores are drawn. In general, the central limit theorem applies to any linear combination of a set of scores.

The central limit theorem also plays an important role in many statistical tests and in constructing confidence intervals. Suppose the statistic that estimates some population parameter can be viewed as a linear combination of scores. Then the sampling distribution of the statistic will be approximately normal if *n* is large, and the standardized normal table will provide the basis for drawing inferences about the parameter of interest.

Even tests that are not directly based on the standardized normal distribution often rest on the assumption that the data have been sampled from a normally distributed population; we will see this when the *t*, chi-square, and *F* tests are considered. Because of the central limit theorem, statistics of interest such as group means may be normally distributed (if *n* is large enough) even if the individual scores are not. Therefore the distribution of the test statistic will often approximate its theoretical distribution.

3.5 THE t DISTRIBUTION

3.5.1 The t Statistic

Recall that the denominator of the *z* score is the standard error (SE) of the observed quantity that is in its numerator. We usually do not know the value of this parameter but use an estimate based on the data in its place. If *n* is large, say more than 40, the estimate of the SE will vary only slightly across samples and the normal distribution of Table D.2 will adequately approximate the sampling distribution of the test statistic. In many experiments, n is not large; consequently, the estimate of the SE of the mean will vary over samples. In such cases, the sampling distribution of $(\bar{Y} - \mu_{hyp})/(\hat{\sigma}/\sqrt{n})$ will not be normal. This distribution, the t, has a more prominent peak than the standardized normal. Also, more of the t distribution's area is in its tails. As a result, when n is small and σ is estimated, inferences based on the normal distribution may be in error.

The t statistic has the general definition

$$t = \frac{V - E(V)}{\hat{\sigma}_V} \tag{3.7}$$

where, as before, V refers to an observed variable that estimates a population parameter $E(V)$, and $\hat{\sigma}_V$ is the sample statistic that estimates the standard error of the sampling distribution of V. Note that the t statistic is identical to the z statistic except that the denominator of Equation 3.7 is an *estimate* of the standard error of V. As in the case of z, it is assumed that scores are independently and normally distributed. If $E(V)$ is a population mean μ, then V is $\bar{Y}$ and $\hat{\sigma}_V$ is the sample standard deviation $\hat{\sigma}$ divided by $\sqrt{n}$. There is a family of t distributions whose members more closely approximate the normal distribution as sample size increases. To be more precise, the shape of the t distribution depends on something called *degrees of freedom* or, more frequently, *df*. The concept of degrees of freedom is closely related to sample size but is not quite the same thing. Because the concept is basic not only to using the t statistic but to many other statistics employed by researchers, it merits some discussion.

3.5.2 Degrees of Freedom (df) and the t Distribution

The df associated with any quantity is the number of independent observations upon which that quantity is based. The meaning of "independent observations" is best illustrated by an example. Suppose that we are asked to choose 10 numbers that sum to 50. We can freely choose any 9 values but the 10th must be 50 minus the sum of the first 9. In this case, there are 10 scores but because only 9 can be chosen independently, there are only 9 df. There is a restriction due to the fact that the numbers must sum to 50, and that costs us a "degree of freedom." The same situation occurs when we calculate a sample standard deviation. Recall that we must subtract each score from the mean; however, as we noted earlier, the sum of deviations of all scores about their mean must be zero. That is,

$$\Sigma(Y - \bar{Y}) = 0$$

Rewriting this last result gives

$$\Sigma Y = n\bar{Y}$$

If the sample mean is 5 and n is 10, we have the original example in which the sum of 10 scores must equal 50. Therefore, the sample standard deviation is based on $10 - 1$, or 9, df.

At this point, it looks like df is always just $n - 1$. That's true if the statistic of interest involves only one restriction. But suppose we draw two samples from some population; one sample is of size n_1 and the other of size n_2. We want to estimate the population variance, but we have two estimates. As will be seen shortly, these can be averaged; however, the point now is that there are two restrictions if two sample variances are computed: the sum of the n_1 scores in the first sample must equal $n_1 \bar{Y}_1$, and the sum of the n_2 scores in the second sample must equal $n_2 \bar{Y}_2$. There are $n_1 - 1$ df associated with the variance for the first sample and $n_2 - 1$ associated with that for the second sample. The df associated with a statistic involving some combination of these two variances will be df $= (n_1 - 1) + (n_2 - 1) = n_1 + n_2 - 2$. The message is that df is not necessarily the number of scores minus 1. Rather,

df = number of independent observations

  = total number of observations minus number of restrictions on those observations

In the two-sample example, there are $n_1 + n_2$ observations and two restrictions caused by taking deviations about each of the sample means.

Turn now to Table D.3 in the appendix. Each row of the table corresponds to a different value of df. The columns are proportions of the sampling distribution of t exceeding some cutoff; note that each column is headed by levels of significance for one-tailed tests (P) and two-tailed tests $(2P)$. Find the column corresponding to $P = .025$ (and $2P = .05$) and the row for df $= 9$. The critical value in the cell is 2.262. This value is exceeded by .025 of the sampling distribution of t when there are 9 df; .05 of the distribution is greater than 2.262 or less than -2.262. Now look down the same column to the row labeled infinity. The critical value in that cell is 1.96. This means that the probability is .025 of exceeding 1.96 and the probability of $t > 1.96$ or $t < -1.96$ is .05. This is exactly the critical value in Table D.2, the normal probability table. In general, the critical value of t decreases as df increases, rapidly approaching the critical value in Table D.2 for the normal distribution. The reason for this is that our estimate of σ exhibits less sampling variability as n increases. In short, as n increases, $\hat{\sigma}$ more closely approximates σ, and therefore the distribution of t more closely approximates that of a normally distributed z score.

3.6 INFERENCES ABOUT MEANS USING THE t DISTRIBUTION: THE ONE-SAMPLE CASE

The t test is an important tool in data analysis. We often wish to test hypotheses or construct confidence intervals for parameters that are linear combinations. Although standard errors of the relevant statistic are not usually known, estimates can be derived from Equation 2.12 and substituted into Equation 3.7. Of most frequent interest is the special case of the difference between means obtained in two different conditions. At some stage of the data analysis, researchers usually carry out such tests even when the research involves several factors, with several levels of each. In Section 3.7 we will review such comparisons of two independent means. In this section, we review the use of the t test in the somewhat simpler situation in which there is a single set of scores.

3.6.1 Testing H_0: $\mu = \mu_{hyp}$

In Section 3.3.1, we considered a research study in which reading scores were obtained from a sample of students taught by a new instructional method. The question was whether this method resulted in improved performance compared with the average for the population taught by methods in use up to that time. That average was 68, so we tested H_0: $\mu \leqslant 68$ against the H_1: $\mu > 68$. We assumed that we knew the population standard deviation and tested the null hypothesis, using the standardized normal probability (z) distribution. Let's reconsider that research example. This time assume that 25 students are taught by the new instructional method. Further assume that the data are examined using normal probability, stem-and-leaf, and box plots; these analyses provide no reason to doubt that the data were drawn from a normally distributed population. In contrast to the developments in Section 3.3, we do not assume that we know σ and use the estimate calculated from our data. That estimate $\hat{\sigma}$ equals 28.12, and the sample mean $\bar{Y}$ equals 78.2. We choose $\alpha = .05$ and note that our test is one-tailed. We then turn to Appendix Table D.3; the entry in the row corresponding to 24 ($n - 1$) degrees of freedom and the column for $P = .05$ is 1.711. Therefore, we reject H_0 if $t > 1.711$. Equation 3.7 provides the basis for a formula for this *single-sample t test*:

$$t_{n-1} = \frac{\bar{Y} - \mu_{hyp}}{\hat{\sigma}/\sqrt{n}} \tag{3.8}$$

The subscript $n-1$ indicates the df. Inserting the values from our data into this formula, we have

$$t_{24} = \frac{78.2 - 68}{28.12/5} = 1.814$$

The value of t exceeds the criterion value of 1.711; therefore, it is reasonable to conclude that the new program for teaching reading is superior to the traditional program. More formally, we reject the null hypothesis that the mean of the population of reading scores obtained under the new program is equal to the old mean of 68 in favor of the alternative that it is higher.

3.6.2 A Confidence Interval for μ

As we noted, the confidence interval permits us to test the null hypothesis while getting a sense of what the parameter value actually is. Furthermore, the width of the interval provides an index of the power of our test. For these reasons, we will calculate a confidence interval for the mean of the population of reading scores obtained under the new reading program. The form of this interval is the same as that when the z statistic was used (Equation 2.23) except that z is replaced by t and σ by $\hat{\sigma}$ in our calculations:

$$P\left[\bar{Y} - (t_{n-1,\alpha/2})\frac{\hat{\sigma}}{\sqrt{n}} \leqslant \mu \leqslant \bar{Y} + (t_{n-1,\alpha/2})\frac{\hat{\sigma}}{\sqrt{n}}\right] = 1 - \alpha \tag{3.9}$$

Assume that we want a .90 interval. Then, substituting into the preceding equation, we have

$$P(\bar{Y} - 1.711\hat{\sigma}_{\bar{Y}} \leqslant \mu \leqslant \bar{Y} + 1.711\hat{\sigma}_{\bar{Y}}) = .90$$

and the .90 confidence interval is

$$CI = 78.2 \pm (1.711) \left[\frac{28.12}{5} \right] = 78.2 \pm 9.623$$

$$= 68.58, \; 87.82$$

We have .90 confidence that the mean of a population of scores obtained with the new method of teaching reading lies between 68.58 and 87.82. It is also important to understand that the confidence interval provides a test of any null hypothesis. For the data set in our example, any null hypothesis specifying a value of μ less than 68.58 or greater than 87.82 will be rejected by a two-tailed test with $\alpha = .10$. For comparison purposes, consider a .95 confidence interval; the critical t value would be 2.064 (rather than 1.711), and the CI would be bounded by 66.59 and 89.81. Note that a wider range of values of μ_{hyp} will now be accepted, indicating a loss of power when we use a .05 rather than a .10 value of α.

3.6.3 The *t* Test Using Difference Scores

The significance tests and confidence intervals of Equations 3.8 and 3.9 can be applied directly to differences among means if the means are correlated. This occurs when the same subject is tested under both treatments (*repeated-measures design*), or two members of a matched pair are randomly assigned to the two treatments being compared (*matched-subjects design;* matching is on the basis of some measure related to Y). In either of these designs, there are only n independent measures—the differences that can be calculated for each subject, or pair. Consider a population of such difference scores. The null hypothesis that the two treatments are equally effective is now $H_0: \mu_D = 0$; μ_D is the mean of the population of difference scores. The significance test is exactly the test reviewed in Section 3.6.1 for the one-sample case. Instead of a group of n scores, there is a group of n difference scores. Accordingly, we again can calculate a t statistic distributed on $n - 1$ df; the numerator involves the mean difference rather than the mean score, and the denominator is the standard deviation of difference scores divided by n, the number of difference scores.

3.6.4 The Normality Assumption

The probability values of Table D.3 were derived under the assumption that the values of Y were randomly sampled from a normally distributed population. This assumption is probably more often false than true. For example, distributions of rating data tend to be symmetric, but extreme ratings occur infrequently; such distributions are often referred to as "short-tailed" because there are fewer scores in the tail than would be expected if the distribution were normal. Often, distributions are symmetric but "long-tailed"; there are more extreme scores than in the normal distribution. Neither of these violations of the normality assumption appears to have much effect upon Type 1 error rates. If many samples are independently drawn from such populations and a t test computed for each sample, true null hypotheses will be rejected in 4% to 6% of the samples when α is .05. This is close enough to .05 to be attributable to sampling error.

Problems arise when samples come from a very skewed distribution. We have drawn 1000 samples of sizes 10 and 40 from the right-skewed exponential distribution described in Section 2.2. The population had a mean of zero, and we tested the null hypothesis that μ equaled zero at the .05 level. When the alternative was $H_1 : \mu > 0$, there were far too few rejections. The Type 1 error rates for $n = 10$ and $n = 40$ were .007 and .020, respectively. When the alternative hypothesis was $H_1 : \mu < 0$, there were many more than .05 rejections. For $n = 10$ and 40, the obtained Type 1 error rates were .155 and .098.

Although our data are frequently skewed, they are rarely as skewed as in the exponential condition; those Type 1 error rates represent a worst case. Even there, tests of H_0 against a two-tailed alternative did not fare as badly as tests against a one-tailed alternative; when n was 40, the sum of the two error rates was .118 (.020 + .098) as opposed to a theoretical proportion of .10. More importantly, the situation is clearly better when $n = 40$ than when $n = 10$. If we used still larger samples, the empirical error rates would eventually stabilize at .05 even in the exponential condition; this is a consequence of the central limit theorem. Larger data sets not only yield increased power but also tend to offset clear violations of the normality assumption.

Previous experiments and theoretical considerations will often give the researcher reason to expect skewed data. Response times, number of responses in some fixed time, and proportion correct are among those measures that frequently have skewed distributions. Severe skewness often will be obvious in a plot of the frequency distribution of the data. Other signals are that the sample mean and median are far apart and that the median is far from halfway between the 25th and 75th percentiles; stem-and-leaf and box plots are very useful in assessing these potential problems. If there is evidence of skewness, every effort should be made to collect larger samples. If small samples are drawn from very skewed populations, there are few remedies. One possibility is to transform the data by performing some operation on all the scores and then applying the t test. For example, the logarithm of exponentially distributed scores will tend to be more nearly normally distributed than will the original scores. We will have more to say about transformations in Chapter 4.

Departures from normality may also affect the power of the t test. More precisely, if the distribution of scores is long-tailed, the standard error of the mean may be large and, consequently, power may be low. Blair and Higgins (1985) have shown that under such conditions a test based on the ranks of the scores (or difference scores), the Wilcoxon signed rank (WSR) test, is often more powerful than the t test. The t has a small (usually less than .06) power advantage when n is less than 10, or when the distribution of Y is either normal or short-tailed. For heavy-tailed distributions with larger n, the advantage of the WSR is often quite large. In view of this, any doubts the researcher has usually should be resolved in favor of using the WSR test. There are two important qualifications about this advice, however. First, the Type 1 error rate of the t test will be less affected by departures from symmetry. Second, the power of the WSR procedure is weakened by the presence of tied scores. Nevertheless, the WSR test has a clear place in our arsenal of analytic procedures. Chapter 8 will present a description of the test and a more detailed discussion (also, see Siegel and Castellan 1988, Lehmann 1975).

3.7 INFERENCES ABOUT MEANS USING THE t DISTRIBUTION: THE TWO-SAMPLE CASE

In Section 3.6.3, we considered the t test for data obtained when scores under two conditions were correlated. The correlation might stem from matching subjects on some measure related to the dependent variable or from testing the same subject under both conditions. An alternative to such correlated-scores designs is an *independent-groups* design. In this section we review the formulas for this design that are used to test whether the means of two populations differ and to establish confidence intervals for the difference in the two means. These formulas require an expression for the standard error of the statistic estimating that difference. Therefore, before presenting tests and confidence intervals based on the t distribution, we will develop a formula for $\hat{\sigma}_{\bar{Y}_1 - \bar{Y}_2}$.

3.7.1 The Standard Error of the Difference between Two Independent Means

Assume the existence of two populations of scores differing only with respect to the treatment administered. In comparing two methods of teaching arithmetic to third-grade children, we might have an experimental and a control population. We draw samples of size n_1 from the first population and of size n_2 from the second. Two sample means are calculated, and $\bar{Y}_2$ is subtracted from $\bar{Y}_1$. Suppose that we repeated this sampling and computational process many times. We then would have a sampling distribution of the difference $\bar{Y}_1 - \bar{Y}_2$. In Section 2.6.4, we showed that the variance of that sampling distribution (Equation 2.14) is

$$\sigma^2_{\bar{Y}_1 - \bar{Y}_2} = \frac{\sigma_1^2}{n_1} + \frac{\sigma_2^2}{n_2}$$

assuming that the scores are independently distributed. Let us impose one other assumption: the effect of the difference in treatments is independent of the value of the score. It is as if the population of experimental scores was created by merely adding a constant (which could be zero) to every score in the control population. Under this assumption, the two populations may differ in their expected values but not in their shapes. Therefore, $\sigma_1^2 = \sigma_2^2 = \sigma^2$; this is known as the *assumption of homogeneity of variance*. The implication is that the variance of the difference of means can be rewritten as

$$\sigma^2_{\bar{Y}_1 - \bar{Y}_2} = \frac{\sigma^2}{n_1} + \frac{\sigma^2}{n_2} = \sigma^2 \left[\frac{1}{n_1} + \frac{1}{n_2} \right]$$

We have a single population variance (σ^2) and two possible estimates of it, the variances of the two groups sampled in our experiment. To obtain the best single estimate, we need to average the two group variances. Because the estimate based on the larger group is more likely to be close to the true variance (our variance estimates are consistent statistics), the best estimate is a weighted average of the two group variances. This is often referred to as the *pooled variance estimate*, or $\hat{\sigma}^2_{pooled}$, and is calculated as

$$\hat{\sigma}^2_{pooled} = \left(\frac{n_1 - 1}{n_1 + n_2 - 2}\right)\hat{\sigma}^2_1 + \left(\frac{n_2 - 1}{n_1 + n_2 - 2}\right)\hat{\sigma}^2_2 \qquad (3.10)$$

$$= \frac{SS_1 + SS_2}{df_1 + df_2}$$

where $SS_j = \Sigma_i(Y_{ij} - \bar{Y}_{.j})^2$. Note that the weight on each group variance is obtained by dividing the df for that group by the sum of the df for the two groups $[(n_1 - 1) + (n_2 - 1)]$. The df rather than the n's are used in these weights because this gives an unbiased estimate of σ^2.

We can now state the expression for the estimate of the SE of the sampling distribution of the difference of two independent means:

$$\hat{\sigma}_{\bar{Y}_1 - \bar{Y}_2} = \sqrt{\hat{\sigma}^2_{pooled}\left(\frac{1}{n_1} + \frac{1}{n_2}\right)} \qquad (3.11)$$

3.7.2 Testing Hypotheses about $\mu_1 - \mu_2$

Assume that samples of sizes n_1 and n_2 are randomly drawn from two independent normally distributed populations of scores. The populations have means μ_1 and μ_2, respectively, and the sample means are $\bar{Y}_1$ and $\bar{Y}_2$. Both populations are assumed to have the same variance σ^2. We wish to test the null hypothesis that the difference between the population means has some specified value; call this $E(\bar{D})$, the difference expected under the null hypothesis. In most research, $E(\bar{D})$ will be zero, but, occasionally, prior knowledge or theory dictates a different value. The null hypothesis can be tested by the following t statistic:

$$t_{n_1 + n_2 - 2} = \frac{\bar{D} - E(\bar{D})}{\hat{\sigma}_{\bar{D}}} \qquad (3.12)$$

$$= \frac{(\bar{Y}_1 - \bar{Y}_2) - (\mu_1 - \mu_2)}{\hat{\sigma}_{\bar{Y}_1 - \bar{Y}_2}}$$

The formula for the denominator is given by Equation 3.11. The subscript $n_1 + n_2 - 2$ is the number of df associated with the pooled variance estimate and serves to remind us that this t statistic will be distributed on that number of df.

Table 3.1 presents an example of the application of Equation 3.12 to the scores of 10 experimental (E) and 8 control (C) subjects in an experiment. Alpha is set equal to .05, and the null hypothesis, H_0: $\mu_E - \mu_C = 0$, is tested against the two-tailed alternative, H_1: $\mu_E - \mu_C \neq 0$. The df for this two-sample t test are 16, and therefore the critical value of t is 2.12. Because the alternative hypothesis is two-tailed, the null

TABLE 3.1   AN EXAMPLE OF THE TWO-SAMPLE t TEST

Group	Data										$\bar{Y}_{.j}$	$\hat{\sigma}_j^2$
E	77	80	88	83	87	86	69	68	62	68	76.80	88.622
C	68	64	55	73	54	78	56	62			64.75	90.250

$$H_0: \mu_1 - \mu_2 = 0 \quad H_1: \mu_1 - \mu_2 \neq 0 \quad \alpha = .05 \quad t_{16} = 2.12$$

The pooled estimate of the population variance is

$$\hat{\sigma}_{pooled}^2 = \frac{(9)(88.622) + (7)(90.25)}{16} = 89.343$$

The standard error of the difference in the means is

$$\hat{\sigma}_{\bar{Y}_1 - \bar{Y}_2} = \sqrt{(89.343)[1/10 + 1/8]} = 4.483$$

Therefore,

$$t_{16} = \frac{76.8 - 64.75}{4.483} = 2.688$$

which exceeds the criterion t (2.12) for a two-tailed alternative; H_0 is rejected.

A .95 confidence interval for $\mu_1 - \mu_2$ is

$$CI = (76.8 - 64.75) \pm (2.12)(4.483)$$

$$= 2.546, \ 21.554$$

hypothesis will be rejected if $t < -2.12$ or $t > 2.12$. The t statistic calculated in Table 3.1 exceeds 2.12, and we conclude that the mean score of a population of experimental subjects exceeds that of a similar population run under the control condition.

3.7.3 A Confidence Interval for $\mu_1 - \mu_2$

If the confidence level is $1 - \alpha$, then the bounds on a parameter, $E(V)$, are

$$CI = V \pm t_{\alpha/2} \hat{\sigma}_V \qquad (3.13)$$

Equation 3.9, which presented the bounds on the mean of a population, is a special case in which $V = \bar{Y}$ and $E(V) = \mu$. When we want bounds on the difference between two population means, $E(V) = \mu_1 - \mu_2$, the bounds are

$$CI = (\bar{Y}_1 - \bar{Y}_2) \pm t_{\alpha/2} \hat{\sigma}_{\bar{Y}_1 - \bar{Y}_2} \qquad (3.14)$$

where $\pm t_{\alpha/2}$ are the critical values for a two-tailed t test at the α significance level. Calculations are illustrated at the bottom of Table 3.1 for the data of that table.

3.7.4 Correlated Scores or Independent Groups?

We have now presented two different experimental designs for comparing the means obtained with two different treatments. For example, consider a comparison of two methods of teaching third graders arithmetic. In a matched-pairs design, the 40 children are rank-ordered on the basis of some measure of ability, then divided into 20 ordered pairs. The members of each pair are then randomly assigned to the two instructional methods. In an independent-groups design, assignment to instructional methods is completely random except for the restriction that there are 20 children in each condition. What are the advantages and disadvantages of the two methods?

The independent-groups design has two advantages. First, it involves more df in the t test. Because each group provides an independent estimate of the population variance, each based on n scores, the t is distributed on $2(n-1)$ df. The matched-pairs design involves only $n-1$ df because there is a single set of n difference scores. Looking at Table D3, it is evident that the critical value of t becomes smaller as df increases. In fact, the power of the t test increases, and the confidence interval is narrower for larger df. A second advantage of the independent-groups design is that it does not require an additional measure for matching subjects. Sometimes, such a measure can be difficult to obtain.

Why, then, should we go to the trouble of matching subjects? The answer to this lies in Equation 2.11 for the variance of difference scores. The variance of the sampling distribution of the difference of two means may be written as

$$\sigma_{\bar{D}}^2 = \sigma_{\bar{Y}_1}^2 + \sigma_{\bar{Y}_2}^2 - 2\rho_{\bar{Y}_1\bar{Y}_2}\sigma_{\bar{Y}_1}\sigma_{\bar{Y}_2}$$

Because $\sigma_{\bar{Y}}^2 = \sigma^2/n$, and assuming homogeneous variances, the equation can be re-written as

$$\sigma_{\bar{D}}^2 = \left[\frac{1}{n}\right](2\sigma^2 - 2\rho\sigma^2) = \left[\frac{1}{n}\right](2\sigma^2)(1-\rho)$$

where ρ is the correlation between scores in the two conditions. That correlation should be zero for the independent-groups design but greater than zero in both matched-pairs and repeated-measures designs. As a result, $\sigma_{\bar{D}}^2$ will be smaller in correlated-scores designs (σ_{cor}^2) than in independent-groups designs (σ_{ind}^2). In fact, $\sigma_{cor}^2/\sigma_{ind}^2$ will equal $1-\rho$. Therefore, correlated-scores designs will have smaller denominators and, consequently, larger t ratios than independent-group designs. There will be some trade-off because the t statistic for the correlated-scores design is distributed on fewer df than that for the independent-groups design. Nevertheless, if the same subjects are tested under both conditions, or if subjects are matched on the basis of a measure that is related to the dependent variable in the experiment, ρ usually will be large enough to more than offset the loss in df. Ordinarily, the t test will be more powerful and the confidence interval narrower for the correlated-scores than for the independent-groups design. The repeated-measures design is used extensively in behavioral research, largely for the reasons just presented. Note, however, that not all variables can be manipulated in this way. It would not make sense to use both methods of arithmetic instruction on the same subjects. Furthermore, the researcher should be aware of the possibility of "contrast effects"; some experimental treatments (e.g., one amount of reward) have a very different effect when the same subject has been exposed to other treatments (e.g., other amounts of reward) than when the subject has experienced only that treatment.

3.7.5 The Assumption of Normality

When using the t distribution as a basis for inferences about two population means, we assume that the two populations of scores are independently and normally distributed, and have the same variance. As we noted earlier, violation of the independence assumption will often result in an inflated Type 1 error rate. Nonnormality is less of a problem, at least in large samples. From our discussion of the central limit theorem, we know that the sampling distribution of the difference of means approaches normality as the combined sample size $(n_1 + n_2)$ increases. As a consequence, the actual Type 1 error rates associated with the test statistic will closely approximate the values in Table D3 for the t distribution if $n_1 + n_2$ is moderately large. "Moderately large" may be as small as 20 if $n_1 = n_2$ and if the two populations have symmetric distributions, or even if they are skewed but have the same direction and degree of skewness. Our rather liberal attitude with respect to skewness may seem surprising in view of the fact that the one-sample t required quite large n's to achieve honest Type 1 error rates when the parent population was skewed. However, here we are concerned with the sampling distribution of the difference between independent means. If two populations are skewed in the same direction, and if the samples are equal in size, then the differences in sample means are as likely to be positive as negative, and the sampling distribution of those differences will tend to be symmetric. For most situations the researcher will encounter, combined sample sizes of 40 should be sufficient to guarantee an honest Type 1 error rate.

Nonnormality of the sampled populations does affect the power of the two-sample t test. When this condition exists because the populations are skewed or have outliers, the sampling distribution of $\overline{Y}_1 - \overline{Y}_2$ will tend to be long-tailed. In such cases, Wilcoxon's rank sum test, or the equivalent Mann-Whitney U test, often will be much more powerful than the t test, and it is only slightly less powerful even when the population has a normal or a short-tailed distribution (Blair and Higgins 1980). The Wilcoxon test is available in most statistical packages and is described in all nonparametric textbooks. It is a special case of another nonparametric test, the Kruskal-Wallis H test. We will present this procedure in Chapter 4.

3.7.6 The Assumption of Homogeneity of Variance

The denominator of the equation for the two-sample t test is based on the pool of two variance estimates (Equation 3.10); the underlying assumption is that the two group variances estimate the same population variance. If this is not true—if the population variances are *heterogeneous*—then the sampling distribution of the t statistic of Equation 3.12 may not have a true t distribution. Table 3.2 gives some sense of what may happen in this case. We drew 2000 pairs of samples of various sizes from two normal populations with identical means but different variances. Proportions of rejections for α's equal to .01 and .05 are presented.

Several points about the results should be noted. First, if the two sample sizes are equal, there is little distortion in Type 1 error rate unless n is very small and the ratio of the variances is quite large. Even then, in terms of percent distortion, the effect is most marked when α is at .01. Second, when n's are unequal, whether the Type 1 error

TABLE 3.2 EMPIRICAL TYPE 1 ERROR RATES FOR THE t TEST AS A FUNCTION OF POPULATION VARIANCES AND SAMPLE SIZES

n_1	n_2	σ_1^2/σ_2^2	$\alpha = .05$	$\alpha = .01$
5	5	4	.060	.014
5	5	16	.061	.019
5	5	100	.066	.024
15	15	4	.056	.010
15	15	16	.054	.015
15	15	100	.059	.017
5	10	4	.095	.031
5	10	.25	.020	.003
10	15	4	.073	.023
10	15	.25	.040	.006
10	20	4	.091	.031
10	20	.25	.021	.003
20	30	4	.067	.027
20	30	.25	.037	.004

rate is inflated or deflated depends upon the direction of the relation between sample size and population variance. The reason for this can be understood by considering the fact that the denominator of the t is based on a weighted average of two variance estimates; the weights are proportions of degrees of freedom. Therefore, when the larger group is drawn from the population with the larger variance, the larger variance estimate receives more weight than the smaller estimate. The denominator of the t test tends to be large and the t small; the rejection rate is less than it should be. Conversely, when sample size and population variance are negatively correlated, the smaller variance estimate gets the larger weight; the denominator of the t statistic tends to be small and the t large; the rejection rate is inflated.

Unequal sample sizes should be avoided when possible. However, we recognize that there will be many cases in which sample sizes will differ, often markedly. For example, the response rate to questionnaires may be quite different for two populations such as male and female, or college- and non-college-educated. We do not advocate discarding data from the larger sample; this would increase sampling variability for statistics computed from that sample. Instead, we recommend an alternative to the standard t test. One such alternative is a t that does not use the pooled estimate of the population variance. The denominator of this statistic would be that of the z test for two independent groups with variance estimates instead of known population variances. We define

$$t' = \frac{(\bar{Y}_1 - \bar{Y}_2) - (\mu_1 - \mu_2)}{\sqrt{\hat{\sigma}_1^2/n_1 + \hat{\sigma}_2^2/n_2}} \tag{3.15}$$

This statistic is sometimes referred to as Welch's (1938) t. If the scores have been drawn from normally distributed populations, t' is distributed approximately as t but not with the usual df. The df are

$$df' = \frac{(\hat{\sigma}_1^2/n_1 + \hat{\sigma}_2^2/n_2)^2}{\hat{\sigma}_1^4/[n_1^2(n_1 - 1)] + \hat{\sigma}_2^4/[n_2^2(n_2 - 1)]} \tag{3.16}$$

We round the df to the nearest integer when evaluating the value of t' or when obtaining confidence intervals.

SYSTAT, SPSS^X, and BMDP's programs for the t tests yield values of t' as well as the usual t values. In addition, when the statement ROBUST is included in the TEST paragraph of BMDP3D, trimmed means are calculated and trimmed versions of both the standard and Welch t are calculated. The standard errors and df are adjusted to take such trimming into account. Yuen (1974) and Yuen and Dixon (1973) demonstrate that these trimmed t tests are more powerful than the untrimmed alternatives when the population distribution is long-tailed. However, the approach appears to be appropriate only with symmetric distributions (the same number of scores are trimmed from each tail of the sample distribution), and we would advise careful examination of the sample distribution before using these trimmed tests. Whichever test is chosen, *the decision must be made before conducting the test.* In other words, the choice of a test should be dictated by the variance ratios and the shape of the distribution, and not on the basis of which test rejects H_0 with the smallest estimated α.

3.8 THE CHI-SQUARE DISTRIBUTION

The chi-square (χ^2) distribution is important both because of its role in data analyses and its relation to other distributions, such as the normal, t, and F. With respect to data analysis, χ^2 can be used to test hypotheses about the variances of populations and to construct confidence intervals for those variances. More often, the χ^2 distribution is used to test either goodness of fit of a theoretical distribution to frequency data or the independence of two categorical variables. We do not present these applications of χ^2 in this book. Inferences about variances are discussed in several more extensive presentations of χ^2 (e.g., Hays 1988; Hildebrand 1986) and both goodness-of-fit and independence tests are presented in almost every introductory statistics text. In this section, we will concentrate on defining the χ^2 statistic and its relation to several other distributions that play a prominent role in this book.

3.8.1 The Chi-Square Statistic

Assume that n scores are independently and randomly sampled from a normally distributed population of scores with mean μ and standard deviation σ. Each score is transformed into a z score, which is then squared; the squared z scores are then summed. This sum will be referred to as χ^2; that is,

$$\chi^2 = \sum_{i=1}^{n} \frac{(Y_i - \mu)^2}{\sigma^2} \tag{3.17}$$

If many such random samples of size n are drawn from a normal population, and χ^2 is calculated for each sample, the sampling distribution will have a characteristic probability density that will depend upon degrees of freedom. The quantity in Equation 3.17 has n df, one for each of the n values of Y entering into the χ^2 statistic. Note that, because the values of μ and σ are parameter values and not estimates, no df are lost from the total of n.

An important property of the χ^2 distribution is that the sum of independently distributed χ^2's also has a χ^2 distribution. Its df is the sum of the df's for the component χ^2's. For example, a quantity obtained by summing a chi-square statistic on 2 df (χ^2_2) and one on 3 df (χ^2_3) will have a χ^2 distribution on 5 df (χ^2_5). This is referred to as the *additivity property* for independent χ^2 statistics.

Appendix Table D4 presents critical values of χ^2 for various numbers of df. Because a chi-square variable on 1 df is simply a squared z score, the first row of Table D4 is related to Table D2 for the standardized normal distribution. For example, approximately 95% of a normal population lies between $z = -1.96$ and $z = 1.96$. Therefore, approximately .95 of the χ^2_1 distribution should be between zero and $(1.96)^2$, or 3.841; this in turn implies that .05 of the distribution will exceed 3.841. This is confirmed in the first row of Table D4. As df increase, larger values are required at any α level.

Equation 3.17 defines χ^2 in terms of deviations of scores about the population mean. To establish the relation of the t and F statistics to χ^2, we first must note that the ratio of the numerator of a sample variance to its population variance also has a χ^2 distribution; however, this statistic is distributed on $n - 1$, rather than n, df. That is,

$$\sum_{i=1}^{n} \frac{(Y_i - \bar{Y}.)^2}{\sigma^2} = \frac{(n-1)\hat{\sigma}^2}{\sigma^2} = \chi^2_{n-1} \tag{3.18}$$

Carefully note the difference between Equations 3.17 and 3.18; the latter involves deviations from the sample, rather than the population, mean. Because the sum of the deviations of the n scores about their mean must equal zero, the sampling distribution of the statistic in Equation 3.18 has $n - 1$ df associated with it. We can show this more precisely by considering the identity

$$(Y_i - \mu) = (Y_i - \bar{Y}) + (\bar{Y} - \mu)$$

Squaring both sides and summing over the n values, we have

$$\Sigma(Y_i - \mu)^2 = \Sigma(Y_i - \bar{Y}.)^2 + n(\bar{Y}. - \mu)^2 + 2(\bar{Y}. - \mu)\Sigma(Y_i - \bar{Y}.)$$

Note the application of the summation rules of Appendix A. Because $\Sigma(Y_i - \bar{Y}.)=0$, after dividing both sides of the preceding equation by the variance of the sampled population we get

$$\frac{\Sigma(Y - \mu)^2}{\sigma^2} = \frac{\Sigma(Y - \bar{Y})^2}{\sigma^2} + \frac{n(\bar{Y} - \mu)^2}{\sigma^2}$$

The rightmost term may be rewritten as $(\bar{Y} - \mu)^2/(\sigma^2/n)$. In this form, it should be evident that it is a squared z score and therefore is distributed as χ^2 on 1 df, provided the sampled population is normal and σ^2 is the variance of that population. We also know that the term on the left of the equals sign is a χ^2 variable on n df. Using the additivity property, we have

$$\frac{\Sigma(Y - \bar{Y})^2}{\sigma^2} = \chi_n^2 - \chi_1^2 = \chi_{n-1}^2$$

Thus, the statistic defined in Equation 3.18 is distributed on $n - 1$ df.

3.8.2 Relation of the t to the χ^2 Distribution

The single-sample t statistic used to test a null hypotheses about a population mean (Equation 3.8) can be rewritten as

$$\frac{\bar{Y} - \mu}{\left[\dfrac{1}{(n-1)n}\Sigma(Y - \bar{Y})^2\right]^{1/2}}$$

To demonstrate the relation of t_{n-1} to χ_{n-1}^2, we square the preceding expression and divide numerator and denominator by the population variance σ^2. Rearranging terms, we can write

$$t_{n-1}^2 = \frac{(\bar{Y} - \mu)^2/(\sigma^2/n)}{[\Sigma(Y - \bar{Y})^2/\sigma^2]/(n - 1)}$$

The numerator is a squared z score; therefore, if the population is normally distributed, the numerator is a chi-square variable on 1 df. The denominator is just the quantity of Equation 3.18 divided by $n - 1$ df. Therefore, assuming the sampled population is independently and normally distributed, we have

$$t_{n-1} = \frac{\sqrt{\chi_1^2}}{\sqrt{\chi_{n-1}^2/(n - 1)}} \tag{3.19}$$

The t statistic for comparing the means of two independently sampled groups is also related to χ^2. Recall that the formula for this t is

$$t_{n_1+n_2-2}^2 = \frac{[(\bar{Y}_1 - \bar{Y}_2) - (\mu_1 - \mu_2)]^2}{\hat{\sigma}_{pooled}^2[(1/n_1) + (1/n_2)]}$$

Dividing numerator and denominator by $\sigma^2[1/n_1) + (1/n_2)]$ gives

$$t_{n_1+n_2-2}^2 = \frac{[(\bar{Y}_1 - \bar{Y}_2) - (\mu_1 - \mu_2)]^2/\sigma^2[(1/n_1) + (1/n_2)]}{\hat{\sigma}_{pooled}^2/\sigma^2}$$

The numerator of this last expression is a squared z score or, equivalently, a χ^2 variable on 1 df. The denominator is

$$\frac{\hat{\sigma}_{pooled}^2}{\sigma^2} = \frac{1}{\sigma^2}\left[\frac{(n_1 - 1)\hat{\sigma}_1^2 + (n_2 - 1)\hat{\sigma}_2^2}{n_1 + n_2 - 2}\right]$$

Substituting for $\hat{\sigma}^2_{pooled}$ in terms of Equation 3.10 and again rearranging terms, we have

$$\frac{\hat{\sigma}^2_{pooled}}{\sigma^2} = \frac{df_1\hat{\sigma}^2_1/\sigma^2 + df_2\hat{\sigma}^2_2/\sigma^2}{df_1 + df_2}$$

The quantities df_j and $\hat{\sigma}^2_j$ are the df and variance estimate for the jth group. If the sampled scores have been drawn from an independently and normally distributed population, and if $\hat{\sigma}^2_1$ and $\hat{\sigma}^2_2$ are both unbiased estimates of the same population variance (in other words, if the population variances are homogeneous), then the last expression can be rewritten as a function of χ^2 variables. The denominator is then

$$\frac{\chi^2_{df_1} + \chi^2_{df_2}}{df_1 + df_2} = \frac{\chi^2_{df_1 + df_2}}{df_1 + df_2}$$

Thus the numerator of the squared t statistic is a χ^2 variable on 1 df, and the denominator is a χ^2 variable on $n_1 + n_2 - 2$ df, divided by those df. But note that this is true only if the conditions of independence, normality, and homogeneity of variance are met. If these conditions are violated, the two-sample t statistic will not necessarily be a ratio of χ^2 variables, and therefore will not be distributed as t.

To summarize the developments so far, the χ^2 statistic on f df is the sum of f, independently and normally distributed, squared z scores. A t statistic on f df is the square root of a ratio in which the numerator is a χ^2 variable (on 1 df), and the denominator is a χ^2 variable (on f df) that is divided by f.

3.9 THE *F* DISTRIBUTION

The F statistic is probably the most frequently cited statistic in the research literature in fields ranging from agriculture to zoology. To get some feeling for why this might be so, consider the common case in which we obtain a mean score under several experimental conditions. Asking whether these means differ significantly is asking whether the variance of these means is greater than chance variance. Because the F statistic is a ratio of two sample variances, it is ideally suited for answering this question. If the variance of sample means is very large relative to a sample variance that reflects only chance effects, we can reasonably conclude that the experimental conditions really do differ in their effect. The F also is employed to investigate other questions, and we will see many of these applications in this book.

To understand the F distribution, we begin with a simple sampling experiment. Assume the existence of two normally distributed populations with variances σ^2_1 and σ^2_2, respectively. Further assume that we randomly sample n_1 scores from the first population and n_2 scores from the second population; the variance estimates calculated from the sample data are $\hat{\sigma}^2_1$ and $\hat{\sigma}^2_2$. Because we have assumed random sampling from normally distributed populations of scores, it follows from Equation 3.18 that

$$\frac{\hat{\sigma}^2_1}{\sigma^2_1} = \frac{\chi^2_{n-1}}{n_1 - 1} \quad \text{and} \quad \frac{\hat{\sigma}^2_2}{\sigma^2_2} = \frac{\chi^2_{n-2}}{n_2 - 1}$$

We next calculate a ratio based on these quantities:

$$F = \frac{\hat{\sigma}_1^2/\sigma_1^2}{\hat{\sigma}_2^2/\sigma_2^2} = \frac{\chi_{df_1}^2/df_1}{\chi_{df_2}^2/df_2} \tag{3.20}$$

The sampling procedure is repeated many times with the constraint that the sample sizes are n_1 and n_2. The resulting distribution of ratios has $(n_1 - 1)$ df associated with the numerator and $(n_2 - 1)$ df associated with the denominator. We speak of the ratio as being distributed as F on $(n_1 - 1)$ and $(n_2 - 1)$ df.

Note that if the two groups of scores are drawn from populations with the same variance, Equation 3.19 becomes

$$F = \frac{\hat{\sigma}_1^2}{\hat{\sigma}_2^2} \tag{3.21}$$

The population variances are identical and therefore drop from the equation.

This last result is particularly important in many inferential procedures: the ratio of two variance estimates based on samples drawn independently from the same normal population (or from two populations with identical variances) has the F distribution on $(n_1 - 1)$ and $(n_2 - 1)$ df. This ratio is actually a ratio of χ^2 variables divided by their df. We should also note that in the special case where the numerator has only 1 df, $F = t^2$. This can be seen by comparing Equation 3.20 with the square of Equation 3.19.

As with χ^2 and t, there is really a family of F distributions, one for every possible pair of values of df_1 and df_2. For any distribution in this family, $E(F) = df_2/(df_2 - 2)$. As the denominator df get larger, the mean of the sampling distribution of F approaches 1. Accordingly, very large or very small ratios of sample variances serve as a signal that the two samples are not drawn from populations with the same variance.

F distributions are not symmetric about their mean but are typically skewed to the right. This can be seen in Figure 3.2, which presents two F distributions. Note that the values of F must be positive because F is a ratio of sums of squared deviations. Also note that the value of F exceeded by .05 of the distribution is smaller when there are 20 denominator df than when there are 5. In general, a smaller critical value of F is required in significance tests when there are more df in either the numerator or the denominator. Therefore, as either df_1 or df_2 increases, so does the power of the F test.

Critical values of the F statistic are presented in Appendix Table D5. To illustrate their use, let us consider an example. Suppose we have an experimental (E) group of 13 subjects and a control (C) group of 9 subjects. Perhaps these are two groups of students taught by different instructional methods. Further suppose that we have reason to believe that scores will be more variable in the C condition. The null hypothesis in this study is tested against a one-tailed alternative. We have $H_0: \sigma_C^2 \leqslant \sigma_E^2$, and the alternative is $H_1: \sigma_C^2 > \sigma_E^2$.

The F ratio provides a way of testing the null hypothesis. If the population variances are the same, the ratio of sample variances should not be significantly different from 1. The observed ratio, $\hat{\sigma}_C^2/\hat{\sigma}_E^2$, is 37.62/14.43, or 2.61. That seems to suggest that H_1 is correct; however, as with previous hypothesis tests, we need to set α and deter-

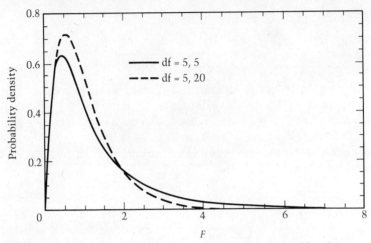

Figure 3.2 *F* distributions for two combinations of degrees of freedom. When $df_1 = df_2 = 5$ (solid curve), the value of *F* exceeded by .05 of the distribution is 6.39; when $df_1 = 5$ and $df_2 = 20$ (dashed curve) the comparable value is 2.90.

mine the appropriate critical value. Let us set α at .05. To find the critical value, we turn to Appendix Table D5.

The *F* ratio in this study is distributed on 8 and 12 df because the numerator variance is based on 9 scores and the denominator variance on 13. Find the critical *F* values in the column for 8 df and the block for 12 df. The block contains several rows, each corresponding to a different value of α. Finding the row labeled .05, we see that the value in the cell corresponding to that row and the column for 8 numerator df is 2.85. Five percent of the *F* distribution on 8 and 12 df lies to the right of this value. We will sometimes refer to this critical value of *F* as F_{α,df_1,df_2}; for example, $F_{.05,8,12} = 2.85$. The first subscript after the α value always refers to the numerator df, the second to the denominator df.

In this example, because the ratio of sample variances does not exceed the critical value, we cannot reject H_0. This does not mean that the population variances are identical. In fact, the population variances may be very different; this experiment was a poor one involving small samples and, consequently, had little power to reject H_0 even if it were false.[1]

---

[1] The preceding example was designed to illustrate the use of the *F* distribution table. In practice, there are better approaches to determining whether the variances of two groups of scores are significantly different (e.g., Levene 1960). If the populations are not normal, the variance ratio will not be distributed as *F* and matters become worse as the sample sizes increase because the sample distributions become more like the (nonnormal) populations. However, as we shall see in Chapter 4, the effects of nonnormality are quite different in the analysis of variance. In this application of the *F* statistic, the numerator is a variance of means, not of individual scores. Under the central limit theorem, the sampling distribution of the means will approach normality as *n* increases.

3.10 CONCLUDING REMARKS

Inferential procedures based on the normal, chi-square, t, and F distributions play a major role in the remainder of this book. This reflects the prominence of these procedures in the research literature, a prominence resulting from ease of calculation, applicability to many situations of interest, and a reliance on statistics that are good estimators of parameters under many conditions. Perhaps most importantly, *when their underlying assumptions are met*, the tests based on these distributions are uniformly most powerful tests; no alternative will have greater power to reject the null hypothesis being tested. Bear in mind, however, that, in the face of violations of assumptions, other procedures may provide more valid inferences. These may be tests based on other distributions (such as alternatives to the t test based on ranked data) or modifications of the usual test statistic (such as t' for heterogeneous variances). The implication is that it is important to attempt to assess whether the sampled populations of scores conform to the assumptions underlying the analysis. Methods described for plotting data in Chapter 2 will help in this effort.

Regardless of which statistical procedure we use, it is easy to fall into the trap of too literally translating statistical results into scientific conclusions. Certainly the significance or nonsignificance of a test statistic should be a big factor in drawing conclusions about treatment effects, particularly when decisions about such things as α, β, and n have been made before the experiment. But assuming nonsignificance, how sure can we be that an important effect does not exist when the probability of the test statistic falls .1% above α, or when we find that the observed variability is greater than the estimate used to decide $n?$ What should we conclude about results that are barely significant or fall just short of significance when the assumptions underlying the test are not met by our data? Should we draw the same inference about two nonsignificant results when the qualitative trends in one set of data are consistent with expectations based on available data and theory, yet no recognizable pattern exists in the second data set? How should published results of others be interpreted, when it is apparent that most experimenters give no thought to β in planning their experiment, that few even preselect $\alpha?$

Under such circumstances, the test statistic can be at best a rough indicator of population effects rather than a sharp inferential tool. There are no simple answers, but we reject any one-to-one relation between the significance or nonsignificance of a test statistic and the existence or nonexistence of treatment effects, or the tenability or nontenability of a theory under investigation. In drawing inferences, scientists have the responsibility of adding to the test statistic a priori expectations, and their knowledge of the literature and of the particular experimental conditions (is there, for example, reason to suspect that some variable whose effects are not analyzable obscured the effects of independent variables?), and of the size and direction of effects; they must then subjectively weight these factors. When the results of the data analysis (regardless of the type), conflict with those factors not built into the test, then the experimenter should reserve judgment. The ultimate criterion of the credibility of experimental conclusions is whether or not these conclusions are supported by subsequent replications of the experiment or by differently designed investigations of the hypotheses in question.

EXERCISES

3.1 The following terms provide a useful review of many concepts in the chapter. Define, describe, or identify each of them:

α	t distribution
β	chi-square distribution
Type 1 error	F distribution
Type 2 error	degrees of freedom
one- and two-tailed tests	single-sample t test
rejection region	repeated-measures design
power	matched-subjects design
null hypothesis	homogeneity of variance
alternative hypothesis	pooled variance estimate
confidence interval	correlated-scores design
central limit theorem	independent-groups design
normal distribution	Welch's t statistic

3.2 A standard IQ test yields scores that are normally distributed with $\mu = 100$ and $\sigma = 15$. Let Y be a randomly selected score on this test.

(a) What is $P(Y > 130)$? $P(85 < Y < 145)$? $P(Y > 70)$? $P(70 < Y < 80)$?

(b) What scores define the middle 80% of the distribution?

(c) What is the 75th percentile (scores such that 75% of the scores are below it)?

(d) If 1000 students are tested, approximately how many will have scores between 90 and 110?

(e) What is the probability that the score of a randomly selected student will be greater than 115?

(f) What is the probability that the mean IQ of a group of 10 randomly selected students is greater than 115?

3.3 A test of logical reasoning is developed such that for women $\mu = 200$ and $\sigma = 60$, and for men $\mu = 170$ and $\sigma = 50$. For both men and women, scores are approximately normally distributed.

(a) What is the probability that a randomly selected woman will have a score greater than 170?

(b) What is the probability that the mean of a sample of nine randomly selected women will be greater than 170?

(c) What is the probability that a randomly selected man will have a higher score than a randomly selected woman?

(d) What is the probability that the mean of a random sample of 20 men will be greater than the mean of a random sample of 20 women?

3.4 Assume that X and Y are independently and normally distributed measures. The relevant population parameters are

	X	Y
μ	30	20
σ	20	16

(a) What is the probability of sampling an individual whose X score is (i) less than 25? (ii) greater than 60? (iii) between 15 and 40?

(b) We define a new random variable: $W = X + Y$. (i) What is the probability of sampling a W score greater than 35 from the population? That is, what is $P(W > 35)$? (ii) What is $P(40 < W < 62)$? (iii) What is $P(X > \mu_Y)$?

(c) An individual's X score is at the 85th percentile for X scores (i.e., exceeds .85 of the population of X scores); her Y score is at the 30th percentile for the population of Y scores. What is the percentile rank of her W score?

3.5 Assume that we have a population of scores uniformly distributed between 0 and 1. This means that $f(y)$ is a line bounded by 0 and 1 with a slope of zero and that $F(y)$ (the probability of sampling a score less than y) is y. For example, the probability of sampling a score less than .8 is .8. The mean and standard deviation of this uniform population are .5 and $1/\sqrt{12}$.

(a) What is the probability of sampling a score less than .6?

(b) What is the probability that in a sample of two scores, both will be less than .6? Express your answer as a probability raised to a power.

(c) What is the probability that in a sample of 20 scores, all 20 will be less than .6?

(d) Suppose we draw a sample of 20 scores from this uniformly distributed population and calculate its mean. We repeat this random sampling process many times and plot the sampling distribution of the mean. Describe the shape of that sampling distribution. What is its mean and variance?

(e) On the basis of your answer to part (d), what is the probability that the mean of a sample of 20 scores drawn from the uniformly distributed population is less than .6?

(f) Briefly state the justification for your approach to part (e). Why isn't a similar approach appropriate in answering part (c)?

3.6 In this problem, we will use the normal probability distribution to test a hypothesis about a proportion.

Well-established national norms shows that .40 of the individuals who have been institutionalized and then released after treatment for a particular mental disorder are reinstitutionalized within five years. A new drug that has been developed to prevent reoccurrences of the disorder is given to an experimental sample of 64 patients. We wish to determine whether the probability of failure (i.e., reinstitutionalization) with the new drug is less than .4.

(a) Let π = the probability of failure. State the null and alternative hypotheses.

(b) Let p be the proportion of failures in the sample of 64 patients. The mean of the sampling distribution of p is π and its variance is $\pi(1 - \pi)/n$. In the actual study, only 12 of the 50 subjects were reinstitutionalized after a five-year period. Using Equation 3.3 and the information provided, test the null hypothesis.

(c) In part (b), we used the normal probability (z) table to test an hypothesis about a population probability. What assumption about the sampling distribution of p is implied by this procedure? What is the justification for that assumption? Do you think the assumption would be as justified if the sample had only 10 people in it? or if the value of π specified by the null hypothesis were .1? Explain.

3.7 Given the following data from a three-group experiment:

	Treatment group 1	Treatment group 2	Control group (group 3)
$\bar{Y}$	29	25	19
σ	10	12	14
n	15	15	20

All tests referred to in the following questions should be performed using $\alpha = .05$.

(a) Test whether the mean of treatment group 1 is significantly different from that of the control group. State H_0 and H_1 before doing the test.

(b) Find the 95% confidence interval for $\mu_1 - \mu_3$.

(c) Test whether the mean of the two treatment groups differs significantly from the control.

3.8 Data are collected from 10 subjects as they go through four trials of a learning experiment. The data are as follows:

Subject	Trial 1	Trial 2	Trial 3	Trial 4
1	10	10	14	16
2	17	13	19	17
3	21	24	19	19
4	11	13	15	17
5	14	16	15	19
6	13	15	14	17
7	9	14	11	10
8	16	17	14	21
9	14	14	18	13
10	19	17	21	20

For all tests, assume $\alpha = .05$, two-tailed.

(a) Is there a significant change in performance from trial 1 to trial 2? State H_0, H_1, and the rejection region for the significance test and then perform the test.

(b) Find the 95% confidence interval for $\mu_1 - \mu_2$.

(c) Does the average score on the first two trials differ significantly from that of the average on the last two trials? State H_0, H_1, and the rejection region; hypotheses should be stated in terms of a linear combination of the four population trial means. Carry out the test and state your conclusion.

3.9 A sample of nine 30-day-old protein-deficient infants are given a motor skills test. The mean for a normal population is 60. The data are

$$40\ 69\ 75\ 42\ 38\ 47\ 37\ 52\ 31$$

(a) Find a .95 confidence interval for the mean of the protein-deficient population. Does the mean of a population of protein-deficient children differ significantly from that of a normal population?

(b) Using the value of $\hat{\sigma}$ calculated in part (a), estimate the sample size needed to have a confidence interval width of only 40 points.

(c) After three months the scores of the nine children are

$$42\ 62\ 71\ 40\ 41\ 40\ 35\ 45\ 28$$

Calculate difference scores and test whether there has been any change from the first test to the second. State H_0 and H_1, and the rejection region and your conclusion assuming $\alpha = .05$. How would your answer change if we wished to test whether there was an improvement from the first test to the second?

(d) Calculate the covariance between the two sets of scores. Then use this to calculate $\hat{\sigma}_{\bar{D}}$ and compare the result to the direct calculation of $\hat{\sigma}_{\bar{D}}$ in part (c).

3.10 An arithmetic skills test is given to 8- and 10-year-old boys and girls. There are 10 children in each of the four cells of this research design. The means and standard deviations are given below:

		8 years	10 years
Boys	$\bar{Y} =$	58	72
	$\hat{\sigma} =$	2.7	2.1
Girls	$\bar{Y} =$	53	60
	$\hat{\sigma} =$	2.9	2.2

(a) Do 10-year-old girls score significantly higher than 8-year-old girls? (i) State H_0, H_1, and the rejection region. (ii) Carry out the test and briefly state your conclusion.

(b) (i) Calculate a .90 confidence interval for the difference in means for 8- and 10-year-old girls ($\mu_{10} - \mu_8$). (ii) Suppose you had not done the significance test in part (a); explain precisely but briefly exactly how you could test the H_0 in part (a) from the information in the CI.

(c) There is considerable data showing that boys do better than girls on tests such as this arithmetic test. An interesting question is whether this advantage increases with age. In other words, is the difference between boys and girls greater at age 10 than at age 8? (i) State H_0 and H_1 in terms of a linear combination of the four population means. (ii) Carry out a t test of your null hypothesis, briefly reporting the conclusion.

3.11 We have sampled two independent groups of scores from three different pairs of populations. For each of these three "experiments," computer output from SYSTAT follows; these are means, standard deviations, and stem-and-leaf plots. For each experiment, state briefly which test you would choose from among the t, t', and Wilcoxon ranked sum. Briefly, but precisely, justify your choice in terms of Type 1 and Type 2 error rates and the statistics presented below. In some cases, as in the real world, there may not be a single best, or even adequate, choice. If that's the case, make some decision, justifying it on the basis of the data.

```
THE FOLLOWING RESULTS ARE FOR:     THE FOLLOWING RESULTS ARE FOR:
               EXPT      =   1.000              EXPT      =    1.000
               GROUP     =   1.000              GROUP     =    2.000
TOTAL OBSERVATIONS:     25          TOTAL OBSERVATIONS:       15


                          Y                                    Y
N OF CASES                    25    N OF CASES                    15
MEAN                     101.840    MEAN                      98.667
STANDARD DEV               7.755    STANDARD DEV              16.233
-----------------------------------------------------------------
```

```
THE FOLLOWING RESULTS ARE FOR:     THE FOLLOWING RESULTS ARE FOR:
               EXPT      =   1.000              EXPT      =    1.000
               GROUP     =   1.000              GROUP     =    2.000


STEM AND LEAF PLOT OF VARIABLE: Y, N = 25  STEM AND LEAF PLOT OF VARIABLE: Y, N = 15

MINIMUM IS:        89.000      MINIMUM IS:        68.000
LOWER HINGE IS:       96.000    LOWER HINGE IS:       87.000
MEDIAN IS:        102.000      MEDIAN IS:        99.000
UPPER HINGE IS:      106.000    UPPER HINGE IS:      110.500
MAXIMUM IS:       119.000      MAXIMUM IS:       130.000


           8   9                        6   8
           9   0223                     7   9
           9 H 5699                     8 H 6779
          10 M 0112224                  9 M 79
          10 H 5568                     10   18
          11   001                      11 H 0126
          11   59                       12
                                        13   0
-----------------------------------------------------------------
```

```
THE FOLLOWING RESULTS ARE FOR:     THE FOLLOWING RESULTS ARE FOR:
               EXPT      =   2.000              EXPT      =    2.000
               GROUP     =   1.000              GROUP     =    2.000

TOTAL OBSERVATIONS:     35          TOTAL OBSERVATIONS:       25


                          Y                                    Y
N OF CASES                    35    N OF CASES                    25
MEAN                     103.829    MEAN                     108.640
STANDARD DEV              11.094    STANDARD DEV              11.532
-----------------------------------------------------------------
```

```
THE FOLLOWING RESULTS ARE FOR:      THE FOLLOWING RESULTS ARE FOR:
            EXPT      =     2.000             EXPT      =     2.000
            GROUP     =     1.000             GROUP     =     2.000


STEM AND LEAF PLOT OF VARIABLE: Y, N = 35   STEM AND LEAF PLOT OF VARIABLE: Y, N = 25

MINIMUM IS:         94.000          MINIMUM IS:        100.000
LOWER HINGE IS:        95.000       LOWER HINGE IS:        102.000
MEDIAN IS:          99.000          MEDIAN IS:         104.000
UPPER HINGE IS:       110.000       UPPER HINGE IS:        112.000
MAXIMUM IS:        140.000          MAXIMUM IS:        154.000

            9    44444                       10    001
            9 M  5555566678889              10 H  22223333
           10    114                         10 M  4455
           10    56699                       10    7
           11 H  11114                       10    89
           11                                11
           12    12                          11 H  22
           12    9                           11    4
        ***OUTSIDE VALUES***                 11    67
           14    0                        ***OUTSIDE VALUES***
                                             12    8
                                             15    4
-----------------------------------------------------------------------

THE FOLLOWING RESULTS ARE FOR:      THE FOLLOWING RESULTS ARE FOR:
            EXPT      =     3.000             EXPT      =     3.000
            GROUP     =     1.000             GROUP     =     2.000


TOTAL OBSERVATIONS:     35           TOTAL OBSERVATIONS:     35


                        Y                                   Y
N OF CASES             35            N OF CASES             35
MEAN                91.371           MEAN               101.086
STANDARD DEV        21.004           STANDARD DEV        17.394
-----------------------------------------------------------------------

THE FOLLOWING RESULTS ARE FOR:      THE FOLLOWING RESULTS ARE FOR:
            EXPT      =     3.000             EXPT      =     3.000
            GROUP     =     1.000             GROUP     =     2.000


STEM AND LEAF PLOT OF VARIABLE: Y, N = 35    STEM AND LEAF PLOT OF VARIABLE: Y, N = 35

MINIMUM IS:         42.000          MINIMUM IS:         52.000
LOWER HINGE IS:        83.000       LOWER HINGE IS:        93.500
MEDIAN IS:          91.000          MEDIAN IS:          99.000
UPPER HINGE IS:       100.000       UPPER HINGE IS:       109.500
MAXIMUM IS:        165.000          MAXIMUM IS:        145.000

            4    2                           5    2
        ***OUTSIDE VALUES***            ***OUTSIDE VALUES***
            5    8                           7    77
            6    2                           8    1
            6    66                          8    569
            7    3                           9 H  2344
            7    6                           9 M  5677889
```

```
 8 H 1334              10    1113
 8   567               10 H 6678
 9 M 0001244           11    14
 9   6679              11    589
10 H 00                12    3
10   568               12    6
11   123           ***OUTSIDE VALUES***
11   7                 13    4
***OUTSIDE VALUES***   14    5
16    5
```

3.12 In the first of two experiments, three groups of subjects are required to solve problems under varying levels of environmental stress (noise: low, medium, and high). The experimenter has hypothesized an inverted U-shaped function: she believes performance should be best under medium stress and about equal for the high- and low-stress conditions. The necessary information for the three groups is presented below.

	Low	Medium	High
n	15	18	21
$\bar{Y}$	67.333	68.611	66.048
$\hat{\sigma}$	6.102	6.137	6.128

To test her theory, the experimenter carries out two statistical tests. In each case, state H_0, H_1, and the rejection region, and carry out the test, reporting your conclusion.

(a) According to the theory, the average performance of low and high populations should not differ from each other.

(b) According to the theory, the average of a medium population should be higher than the average of the combined low and high populations.

3.13 The experiment described in the previous problem was rerun using a different design in which each subject was tested at all three levels of stress. There were 16 subjects with three scores for each subject: one for problems under low stress, one for those done under medium stress, and one for those done under high stress. Presented below are the means for each level and the matrix of variances and covariances; the diagonal entries are the variances, and the off-diagonal entries are the covariances.

	Low	Medium	High
Mean	69.563	75.375	68.375

	Variance-Covariance matrix		
	Low	Medium	High
Low	57.329		
Medium	50.575	95.583	
High	28.375	39.383	28.650

(a) For these data and this design, test whether the average performances under low and high stresses differ significantly from each other. State all the usual statables, including your conclusion.

(b) Test whether the average under medium stress is significantly higher than the average of the combined low and high conditions.

3.14 At the start of 1983, the Western Mass Widget Works began a new procedure designed to increased productivity. The number of widgets produced per hour for the same 15 workers are presented for 1982 and 1983:

| 1982 | 20 26 17 13 24 19 18 25 25 14 17 23 20 19 22 |
| 1983 | 17 25 16 15 26 22 22 24 31 21 25 31 29 28 32 |

(a) Find the .95 CI for the change in productivity from 1982 to 1983.

(b) Prior to installation of the new procedures, there had been a debate over the results. Some thought production would go up, and others thought it would go down. State the null and alternative hypotheses and, with α at .05, come to a conclusion.

(c) Hard-Nosed Harrigan, V.P. in charge of data analysis, is interested only in whether the new methods have improved productivity. Again state H_0 and H_1 and carry out the significance test.

3.15 Several researchers have compared laboratory reading (subjects knew they would be tested for recall) with natural reading (subjects read material without knowing they would be tested). In one such study, everyone was tested on the same materials on each of two different days. Free-recall percentages (correct responses) were

Lab: Day 1: 45 60 42 57 63 38 36 51
        Day 2: 43 28 18 40 37 23 16 18

Natural: Day 1: 64 51 44 48 49 55 32 31
            Day 2: 21 38 19 16 24 27 22 35

(a) For each group, find the .95 confidence interval for the population mean of the change in recall over the two days.

(b) We wish to compare the two groups on day 1. State the null and alternative hypotheses. Can we reject H_0 at the .05 level?

(c) From part (a), we have a change score for each subject. We wish to test whether the amount of change is the same for the two populations of readers. State the null and alternative hypotheses; briefly justify the latter. Do the test at the .05 level.

3.16 Equation 3.18 states that $[(n-1)\hat{\sigma}^2]/\sigma^2$ is distributed as a chi-square variable if each score in the sample is randomly drawn from an independently and normally distributed population of scores. Let us see what this implies.

(a) Suppose we draw many samples of size n from a normally distributed population. We calculate the ratio, $[(n-1)\hat{\sigma}^2]/\sigma^2$, for each sample. If $n = 6$, (i) what is the probability that this ratio is less than 9.216? (ii) What is the probability that the ratio lies between 1.145 and 7.289?

(b) The population sampled in part (a) has a variance of 10. Still assuming $n = 6$, in what proportion of samples will $\hat{\sigma}^2$ be less than 8.702?

3.17 Based on a review of large amounts of data, it is well established that the variance of a population of ratings of the quality of a particular wine is 12.64. A new method for training raters is established in hopes of reducing the variance. In a sample of 10 judges trained under the new method, $\hat{\sigma}^2$ is 3.51.

(a) Would you conclude that the new method has effectively reduced the variance of ratings? Explain your reasoning.

(b) Suppose the population of scores was not normally distributed. Why is this a problem for the approach you took in part (a)? Would it be less of a problem if your sample size were larger? Explain.

3.18 We have two samples of readers, 5 boys and 11 girls. We form a ratio of the variances of the two samples, $\hat{\sigma}_B^2 / \hat{\sigma}_G^2$; call this F in accord with Equation 3.20.

(a) If many samples of sizes 5 and 11 are drawn, (i) what is the proportion of F values greater than 2.61 that we should expect? (ii) less than 4.47?

(b) What assumptions are implied in your approach to answering part (a)?

3.19 We draw a sample of 9 boys and 13 girls and calculate the variances of their scores on a reading test. Assuming $\alpha = .05$, how large must $\hat{\sigma}_B^2 / \hat{\sigma}_G^2$ be for you to conclude that the variance is greater in the population of boys' scores?

Chapter **4**

Between-Subjects Designs: One Factor

4.1 INTRODUCTION

In this chapter, we will consider a very simple research design in which the performances of several groups of subjects are compared. We will develop a framework within which we can compare the means of the populations represented by the groups in the study. For example, suppose we had some performance measure from individuals suffering from different mental disorders. Then the factor of interest would be mental disorder, and the levels of this factor might be schizophrenic, manic-depressive, and obsessive-compulsive. Another example might involve obtaining ratings of some public policy from members of different religions. Here the factor of interest is religion, and the levels would include Protestant, Catholic, Jewish, and unaffiliated. Strictly speaking, both of these are examples of observational studies because the subjects are selected from different populations rather than assigned to different treatments. In true experiments that involve random assignment of subjects to levels of an independent variable, the independent variable is said to be manipulated and the design is often referred to as *completely randomized*. Different subjects might be tested after different delays following a study period; here the factor is delay and the levels might be 1, 3, 5, or 7 days. The independent variable need not be quantitative; a pharmaceutical company might be interested in assessing the performance of subjects who have been randomly assigned to different drugs designed to alleviate hypertension.

Whether the levels of the independent variable are observed or manipulated, the data analysis has much the same form and the underlying assumptions are the same. What characterizes the designs of this and the following chapter is that each subject yields a single measure. These designs are often referred to as *between-subjects* designs; all the variability in the data is due to differences between subjects. If the same subject were tested under several conditions, some of the variability would be *within subjects;* there would be variability within each subject's set of scores.

Between-subjects designs have the advantage of simplicity. The underlying statistical model involves fewer assumptions than is the case in designs in which each subject yields scores on several trials or under several conditions. Each additional assumption underlying the derivation of the test statistic is one more assumption that can be violated, undermining the validity of the statistical inference. The between-subjects design also has the advantage of computational simplicity relative to other designs. This is less important in this era of electronic calculators and fast computers than it was in the past. Nevertheless, it is nice to be able to obtain and check results quickly.

The chief disadvantage of the between-subjects design is its relative inefficiency. Because individuals differ on so many dimensions, chance variability will tend to be great, often obscuring real effects and reducing the power of the statistical test. Matching subjects on the basis of some measure other than the dependent variable will often lead to a reduction in this chance variability. Within-subjects designs in which a single subject is exposed to all levels of the independent variable will also tend to be more efficient than most between-subjects designs. We will have more to say about the issue of chance variability in different designs when we consider alternatives to the one-factor design.

To help you learn the material in this book, we shall first present only limited aspects of a topic, then gradually develop additional concepts and computations. In this chapter, the following limitations have been placed on the presentation:

1. We consider only the subset of between-subjects designs that involve a single independent variable; these are one-factor designs.
2. We consider only independent variables whose levels are fixed. That is, we define the population of levels as consisting only of those that have been selected for the experiment.
3. We consider only tests of the general null hypothesis that $\mu_1 = \mu_2 = \cdots = \mu_j = \cdots = \mu_a$, where μ_j is the mean of a population of scores of individuals tested under A_j, the jth level of the independent variable A. We assume that there are a levels of A represented in the study.

In addition to tests of the null hypothesis about the population means, we shall consider estimates of population parameters. For the most part, the chapter will focus on an approach known as *analysis of variance,* or ANOVA, and the F distribution described in Chapter 3 will be prominently featured. However, we also will consider nonparametric alternatives.

4.2 THE *F* TEST OF THE NULL HYPOTHESIS

4.2.1 Some Basic Ideas

We begin our discussion of the analysis of variance with a simple example. The upper panel of Table 4.1 contains average speed of traversing a runway, in feet per second, for four groups of eight rats. All rats were allowed access to a sucrose solution for 20 seconds at the end of each of five runs. The groups differed in percentage of sucrose in the solution offered.

Before the different sucrose solutions were applied, the rats might have been thought of as a sample from a single, infinitely large, population. Assume that each individual in this parent population is randomly assigned to one of the four treatments (solutions) applied in the experiment. There are now four very large *treatment popula-*

TABLE 4.1   DATA AND CALCULATIONS FOR FOUR GROUPS OF RATS IN A RUNWAY STUDY

The data and summary statistics

		Percentage of sucrose in water		
8	16	32	64	
1.4	3.2	6.2	5.8	
2.0	6.8	3.1	6.6	
3.2	5.0	3.2	6.5	
1.4	2.5	4.0	5.9	
2.3	6.1	4.5	5.9	
4.0	4.8	6.4	3.0	
5.0	4.6	4.4	5.9	
4.7	4.2	4.1	5.6	
$T_{.j} = 24.0$	37.2	35.9	45.2	$T_{..} = 142.30$
$\bar{Y}_{.j} = 3.00$	4.65	4.49	5.65	$\bar{Y}_{..} = 4.45$
$\Sigma_i Y_{ij}^2 = 86.54$	186.78	171.67	264.24	
$\hat{\sigma}_j^2 = 2.08$	1.97	1.51	1.27	

Mean squares and F ratio based on above data

$$MS_A = 8 \sum_{j=1}^{4} \frac{(\bar{Y}_{.j} - \bar{Y}_{..})^2}{3}$$

$$= \frac{8}{3}[(3 - 4.45)^2 + (4.65 - 4.45)^2 + (4.49 - 4.45)^2 + (5.65 - 4.45)^2]$$

$$= 9.56$$

$$MS_{S/A} = \frac{\Sigma \hat{\sigma}_j^2}{4} = \frac{2.08 + 1.97 + 1.51 + 1.27}{4} = 1.71$$

$$F = MS_A/MS_{S/A} = 5.59$$

tions, systematically differing from each other in the percentage of sucrose in the solution awarded following each run. Assuming random assignment of the 32 rats to the four treatments, we may view each set of eight scores as a random sample from a population of scores obtained under the corresponding treatment. Usually, the first question of interest is whether there are any differences among the means of the four populations of scores. More precisely, we wish to test

$$H_0: \mu_1 = \mu_2 = \mu_3 = \mu_4$$

The alternative hypothesis is that there is at least one inequality among the means of the treatment populations.

Suppose the null hypothesis is true; the four sucrose solutions don't differ in their effects upon running speeds. Even if this were so, the four group means will differ from each other. By chance, the eight rats in one group may be stronger, swifter, better motivated, or healthier than those in another group. Also, there may be chance variations in other factors that affect running speeds, such as the times at which the animals are tested. The test of the null hypothesis is really a test of whether the $\bar{Y}_{.j}$ differ more than would be expected on the basis of these chance factors. If they do, then something more than chance variation is involved. That "something more" is presumably the effect of the independent variable (percentage of sucrose in our example), and the null hypothesis should be rejected.

In analysis of variance, we attempt to determine whether more than chance variability is involved by calculating two independent measures of variability. One of these, MS_A (the *mean square for A*), is based upon the spread among the group means and therefore reflects both chance variability and—if H_0 is false—the effects of the independent variable. The second measure of variability, $MS_{S/A}$ (*mean square for subjects within levels of A*), reflects only chance variability. If MS_A is much larger than $MS_{S/A}$, we may decide that the spread among the group means is too large to have resulted from chance variability and, therefore, conclude that H_0 is false. Let us define these two measures of variability more precisely and calculate them for the data of Table 4.1.

4.2.2 The Mean Squares

Assume that we have four populations of scores, one corresponding to each of the four sucrose solutions. We may view Y_{ij}, the ith score in the jth group of scores in the observed data set, as being the sum of two components: its treatment population mean μ_j, and some deviation from that mean due to chance. This deviation is often referred to as the *error* component of the score; we will denote it by ϵ_{ij} (lowercase Greek epsilon). Then

$$Y_{ij} = \mu_j + (Y_{ij} - \mu_j) = \mu_j + \epsilon_{ij} \tag{4.1}$$

The variance of the jth treatment population is the variance of the ϵ_{ij} because μ_j is a constant component of all the scores in that population. We assume that this error variance is the same for all the populations, and we label it σ_e^2.

If the treatment population means are also equal, we can view our four groups of observed scores as four independent random samples of eight scores from the same parent population. In that case, the variance of the four group means provides an estimate of $\sigma_{\bar{Y}}^2$, the variance of the sampling distribution of the mean. That is,

$$\sum_{j=1}^{4} \frac{(\bar{Y}_{.j} - \bar{Y}_{..})^2}{3} \triangleq \sigma_{\bar{Y}}^2$$

The symbol "$\triangleq$" is read "is an estimate of." The left side is divided by 3 because this is 1 less than the number of means; we want an unbiased estimate of the variance of the sampling distribution of the means. (From now on, sample variances will be unbiased variance estimates, their denominators will be df, not numbers of observations.)

Recall from Section 2.6.2 that the variance of the sampling distribution of the mean equals the variance of the population of scores divided by n, the number of scores upon which the mean is based. More succinctly,

$$\sigma_{\bar{Y}}^2 = \frac{\sigma_e^2}{n}$$

Therefore,

$$\sum_{j=1}^{4} \frac{(\bar{Y}_{.j} - \bar{Y}_{..})^2}{3} = \frac{\hat{\sigma}_e^2}{8}$$

We can solve for $\hat{\sigma}_e^2$ in the preceding equation by multiplying both sides of the equation by n (8 in this example). In other words, *if the null hypothesis is true* (that is, the group means vary only by chance), n times the variance of the group means gives an unbiased estimate of the variance due to error, or chance factors, in this design. This estimate of error variance is often referred to as the *between-groups mean square* and is denoted MS_A; the subscript labels the independent variable. In general,

$$MS_A = n \sum_{j=1}^{a} \frac{(\bar{Y}_{.j} - \bar{Y}_{..})^2}{a - 1} \tag{4.2}$$

where we have a groups of n scores. Table 4.1 contains the calculation for our numerical example. Again, we emphasize that this quantity estimates the variance of the error components in any treatment population, providing that (1) H_0 is true and (2) the population variances are homogeneous.

Another estimate of error is available from the four group variances. The variance of each group of eight scores provides an estimate of σ_e^2, and the four group estimates can be averaged to provide a single "within-groups" estimate of σ_e^2. Of course, this assumes homogeneity of variance; averaging four different sample variances implies they are all estimates of the same population variance. The result of averaging the group variances is called the *within-group mean square* and is denoted by $MS_{S/A}$; as we noted earlier, S/A is read "subjects within levels of A." The general formula for a levels of A is

$$MS_{S/A} = \frac{1}{a} \sum_{j=1}^{a} \hat{\sigma}_j^2$$

$$= \frac{1}{a} \sum_{j}^{a} \sum_{i}^{n} \frac{(Y_{ij} - \bar{Y}_{.j})^2}{n - 1}$$

$$= \frac{1}{a(n-1)} \sum_{j}^{a} \sum_{i}^{n} (Y_{ij} - \bar{Y}_{.j})^2 \tag{4.3}$$

This quantity has also been calculated in Table 4.1.

4.2.3 The F Test

If the null hypothesis is true, MS_A and $MS_{S/A}$ both estimate the error variance within a treatment population. Therefore, their ratio should be about 1. Usually the ratio will be a little more or a little less than 1; it would be surprising if two independent estimates of the same population variance were identical.

Suppose the null hypothesis is false. For example, suppose that running speed does increase as a function of sucrose percentage. Then the means of the four groups of scores in Table 4.1 will differ not only because the rats in the different groups differ by chance but also because they are subjected to different sucrose solutions. In other words, when H_0 is false, MS_A, which is n times the variance of the group means, reflects not only chance variability but also variability due to the independent variable. However, the within-group variance should not be affected by the independent variable because all subjects in a group receive the same treatment. Therefore, when H_0 is false the ratio $MS_A/MS_{S/A}$ should be greater than 1. That ratio has been labeled F in Table 4.1 and is equal to 5.59. An important question is, how large must the ratio be before we can conclude that more than chance variability is contributing to the numerator? Before answering that, let us review the logic of the F test.

Under the assumptions of the null hypothesis, homogeneity of variance, and independently distributed scores, MS_A and $MS_{S/A}$ are two independent estimates of the population error variance σ_e^2. From Chapter 3, we know that if we add the assumption that the populations of scores are normally distributed, the ratio of two independent estimates of the same population variance has an F distribution. Therefore, under these assumptions, the ratio $MS_A/MS_{S/A}$ is distributed as F. Because the numerator is an estimate of the variance of a population means, it has $a - 1$ df. The denominator has $a(n - 1)$ df because the variance estimate for each group is based on $n - 1$ df and a group variances are averaged. In the example of Table 4.1, the df are 3 and 28, respectively. Suppose $\alpha = .01$; then we find from Table D5 that an F of 4.57 is required for significance. Becuase our calculated value exceeds this, we can reject H_0 at the .01 level. We conclude that the means of the four populations of rats receiving these sucrose concentrations are not identical.

4.3 THE ANALYSIS OF VARIANCE

The mean squares defined in the preceding section may be viewed as ratios of sums of squared deviations to degrees of freedom. These sums of squared deviations ordinarily are referred to as "sums of squares" (SS). In this section we show that SS_A and $SS_{S/A}$ account for the total variability in the data of the one-factor design; then computing formulas for these quantities will be presented. We will see that these formulas lead to the same F ratio presented before. This partitioning of the total variability into several sources will prove helpful when we encounter designs involving more than one factor.

The developments in this section involve two indices of summation: i indicates a value from 1 to n within each group, and j indicates a value from 1 to a, where a is the number of groups. Appendix A provides an explanation of the use of this notation, using several examples.

4.3.1 Partitioning the Total Variability

Consider the following identity, which states that the deviation of a score (Y_{ij}) from the grand mean ($\bar{Y}_{..}$) consists of two components: (1) the deviation of a score from the mean of its own treatment group ($\bar{Y}_{.j}$), and (2) the deviation of the group mean from the grand mean:

$$Y_{ij} - \bar{Y}_{..} = (Y_{ij} - \bar{Y}_{.j}) + (\bar{Y}_{.j} - \bar{Y}_{..})$$

Squaring both sides yields

$$(Y_{ij} - \bar{Y}_{..})^2 = (Y_{ij} - \bar{Y}_{.j})^2 + (\bar{Y}_{.j} - \bar{Y}_{..})^2 + 2(Y_{ij} - \bar{Y}_{.j})(\bar{Y}_{.j} - \bar{Y}_{..})$$

If we sum over i and j for both sides of this equation, using the rules for summation presented in Appendix A, we obtain

$$\sum_{j}^{a}\sum_{i}^{n}(Y_{ij} - \bar{Y}_{..})^2 = \sum_{j}^{a}\sum_{i}^{n}(Y_{ij} - \bar{Y}_{.j})^2 + n\sum_{j}^{a}(\bar{Y}_{.j} - \bar{Y}_{..})^2 \qquad (4.4)$$

$$+ 2\sum_{j}^{a}\sum_{i}^{n}(Y_{ij} - \bar{Y}_{.j})(\bar{Y}_{.j} - \bar{Y}_{..})$$

Rearranging the cross-product term, we have

$$\sum_{j}\sum_{i}(Y_{ij} - \bar{Y}_{.j})(\bar{Y}_{.j} - \bar{Y}_{..}) = \sum_{j}(\bar{Y}_{.j} - \bar{Y}_{..})\sum_{i}(Y_{ij} - \bar{Y}_{.j})$$

$$= \sum_{j}(\bar{Y}_{.j} - \bar{Y}_{..})(0) = 0$$

The last result stems from the fact that the sum of deviations of all observations about their average is zero (Appendix A provides a proof).

Consider Equation 4.4 again, ignoring the cross-product term, which has been proved equal to zero. The term $\sum_i\sum_j(Y_{ij} - \bar{Y}_{..})^2$ is the numerator of the variance of all an scores about the grand mean and henceforth will be referred to as the *total sum of squares* (SS_{tot}). The term $\sum_i\sum_j(Y_{ij} - \bar{Y}_{.j})^2$ is the *within-groups sum of squares* ($SS_{S/A}$). It is the sum (or "pool") of the numerators of the within-group variances. The term $n\sum_j(\bar{Y}_{.j} - \bar{Y}_{..})^2$ is n times the numerator of the variance of the a group means about the grand mean. This is usually referred to as the *between-groups sum of squares* (SS_A). The relation among these terms is

$$SS_{tot} = SS_{S/A} + SS_A \qquad (4.5)$$

The degrees of freedom enter into a similar relation. SS_{tot} is associated with $an - 1$ df because it represents the variability of an scores about the grand mean. We have previously noted that $df_A = a - 1$ and $df_{S/A} = a(n - 1)$. Now note that

$$df_{tot} = df_{S/A} + df_A$$

The df are important not only because they are necessary components of the F ratio and determiners of the F distribution, but these quantities also provide a check on the partitioning of the total variability in more complex designs, where some term in the analysis may be overlooked or the variance wrongly analyzed in some other way. The check assumes that we have the correct df for each term in the analysis; in that case they must sum to the total number of scores minus 1.

4.3.2 Summarizing the Analysis of Variance

Table 4.2 summarizes much of what we have presented so far. The numerical results are for the data set of Table 4.1. The leftmost column contains the *sources of variance* (SV), which follow from our partitioning of the total variability. Moving to the right, we next find the df. The rationale for these formulas and values was presented earlier.

TABLE 4.2   ANALYSIS OF VARIANCE OF THE DATA OF TABLE 4.1

SV	df	SS	MS	F
Total	$an - 1 = 31$	$\sum\limits_{j}^{a} \sum\limits_{i}^{n} (Y_{ij} - \bar{Y}_{..})^2 = 76.44$		
A (between groups)	$a - 1 = 3$	$n\sum\limits_{j}^{n} (\bar{Y}_{.j} - \bar{Y}_{..})^2 = 28.67$	$SS_A/df_A = 9.56$	$MS_A/MS_{S/A} = 5.59^a$
S/A (within groups)	$a(n - 1) = 28$	$\sum\limits_{j}^{a} \sum\limits_{i}^{n} (Y_{ij} - \bar{Y}_{.j})^2 = 47.77$	$SS_{S/A}/df_{S/A} = 1.71$	

[a] $p < .01$

The formulas for sums of squares are the numerators of the expressions in Equations 4.2 and 4.3. These define numerators of variances and are the way we should think about the sums of squares. Nevertheless, there are equivalent expressions that are easier to compute. Although most of your calculations with real data sets will be done on computers using professionally written analysis programs, computing formulas may prove useful for small data sets analyzed using electronic hand-held calculators. Therefore, Table 4.3 presents such formulas and their application to the data set of Table 4.1. These formulas yield the same results as in Table 4.2 and thus provide a check on our work.

Note the use of the T notation in Table 4.3. $T_{..}$ is the grand total of all an scores; the two dots indicate summation with respect to individuals and groups. Similarly, $T_{.j}$ is the total of the n scores in the jth group; here the replacement of the usual i subscript by a dot indicates summation over individuals within the group. The computing formulas can be derived from the definitional formulas of Table 4.2 using the rules of summation presented in Appendix A. However, the df provide a basis for immediately writing out the computing formula for any term in analyses for any experimental design. The details of that approach are presented in Appendix 4.1. Once we have numerical values for the SS and df, the MS and the F ratio can be calculated. As we noted earlier, the F is distributed on 3 and 28 df in our example, and is significant at the .01 level.

In handling actual experimental data, we would not be finished with the analysis at this point. Many interesting questions are still unanswered. Do all four means differ significantly from one another? Or are the 16% and 32% solutions essentially equivalent, as a quick look at the data suggests? If the treatment means are plotted as a function of concentration, what type of equation best describes the relation? We will introduce the conceptual and computational machinery needed to deal with these questions in Chapters 6 and 7.

4.3.3 Unequal Group Sizes

The presentation so far has been restricted to the case where n is equal for all groups. We now remove this restriction and present the formulas and calculations for the case in which n_j, the number of subjects in the jth treatment group, varies over levels of j.

TABLE 4.3   COMPUTING FORMULAS FOR SUMS OF SQUARES APPLIED TO THE DATA OF
                TABLE 4.1

Define the *correction term*

$$C = \frac{(\text{sum of all scores in the data matrix})^2}{\text{total number of scores}} = \frac{T_{..}^2}{an}$$

$$= \frac{(24 + 37.2 + 35.9 + 45.2)^2}{32} = 632.79$$

$$SS_{tot} = \sum_j^a \sum_i^n Y_{ij}^2 - C$$

$$= 86.54 + 186.78 + 171.67 + 264.24 - 632.79$$

$$= 709.23 - 632.79$$

$$= 76.44$$

$$SS_A = \frac{\sum_j T_{.j}^2}{n} - C$$

$$= \frac{(24.0)^2 + (37.2)^2 + (35.9)^2 + (45.2)^2}{8} - 632.79$$

$$= 661.46 - 632.79$$

$$= 28.67$$

$$SS_{S/A} = SS_{tot} - SS_A = 76.44 - 28.67 = 47.77$$

Table 4.4 contains comprehension scores for children taught to read by one of three different methods of instruction. Some children were absent from school for several instructional sessions, so their comprehension scores have not been included in the analysis. Table 4.5 contains the revised formulas for sums of squares, their computing formula equivalents, the calculations, and the summary of the analysis of variance.

The SS_{tot} is defined exactly as in a design in which all n's are the same; it is the numerator of the variance of the N scores ($N = \sum_j n_j$) about the grand mean of the data set. To obtain SS_A we subtract each group mean from the grand mean, square this difference, and weight it by the group frequency (n_j). Note that if the n_j are equal, this formula is exactly the same as the one in Table 4.2. The $SS_{S/A}$ again is calculated as $SS_{tot} - SS_A$. We can think of the $MS_{S/A}$ as an average of the a group variances just as we did in the equal n case. The only difference is that we give more weight to those group variances that are based on more observations. We can express this weighted average as

TABLE 4.4 DATA AND SUMMARY STATISTICS FOR A COMPARISON OF THREE METHODS OF TEACHING READING

Method							$T_{.j}$	n_j	$\bar{Y}_{.j}$	$\Sigma_i Y_{ij}^2$	$\hat{\sigma}_j$
1	83	64	94	55	85	57					
	77	76	63	77			731	10	73.100	54,923	12.853
2	85	63	74	74	68	86					
	59	68					577	8	72.125	42,271	9.672
3	88	96	95	67	89	81					
	79	87	75	63	70	89	979	12	81.583	81,181	10.917
Totals							2,287	30	76.233	178,375	

TABLE 4.5 ANALYSIS OF VARIANCE OF THE DATA OF TABLE 4.4

Sum of squares calculations

$$SS_{tot} = \sum_j \sum_i (Y_{ij} - \bar{Y}_{..})^2 = \sum_j \sum_i Y_{ij}^2 - C$$

$$= 178,375 - \frac{2287^2}{30} = 4029.367$$

$$SS_A = \sum_j n_j (\bar{Y}_{.j} - \bar{Y}_{..})^2 = \sum_j \frac{T_{.j}^2}{n_j} - C$$

$$\frac{731^2}{10} + \frac{577^2}{8} + \frac{979^2}{12} - \frac{2287^2}{30} = 576.675$$

$$SS_{S/A} = SS_{tot} - SS_A = 3452.692$$

The ANOVA table

SV	df	SS	MS	F	
Total	29	4,029.367			
A	2	576.675	288.388	2.255	$(p = .124)$
S/A	27	3,452.692	127.877		

$$MS_{S/A} = \frac{n_1 - 1}{N - a}\hat{\sigma}_1^2 + \cdots + \frac{n_j - 1}{N - a}\hat{\sigma}_j^2 + \cdots + \frac{n_a - 1}{N - a}\hat{\sigma}_a^2$$

where $\hat{\sigma}_j^2$ is the variance of the n_j scores in group A_j about the group mean, and its weight, $(n_j - 1)/(N - a)$, is the ratio of its df to the summed df's of all the group variances. The intuition underlying this averaging process is that we are assuming that the a population variances are the same and we want an estimate of this common variance. However, we have a such estimates, one for each group. It makes sense to average these, but, if the n's are not equal, the group variances based on larger n's will be better estimates of the population variance and therefore are given more weight in the averaging process. Weighting is by df rather than n's because this yields an unbiased estimator of σ_e^2.

4.4 THE MODEL FOR THE ONE-FACTOR DESIGN

In Section 4.2, we presented a somewhat informal statement of assumptions and justification of the ratio of mean squares as an F ratio. In this section, we will take a closer look at these assumptions and see how they lead to an important concept, that of *expected mean squares*.

4.4.1 The Structural Model

Reconsider Equation 4.1:

$$Y_{ij} = \mu_j + \epsilon_{ij}$$

Equation 4.1 is unchanged if we add and subtract the constant μ to the right side, resulting in

$$Y_{ij} = \mu + (\mu_j - \mu) + \epsilon_{ij}$$

where $\mu = \Sigma \mu_j / a$. This may be rewritten as

$$Y_{ij} = \mu + \alpha_j + \epsilon_{ij} \tag{4.6}$$

Thus the model underlying our data analysis asserts that the score of the ith individual in the jth group is the sum of the following three components:

1. *The parent population mean μ.* This quantity may be viewed as the average of the treatment population means (the μ_j) and is a constant component of all scores in the data matrix.
2. *The effect of treatment A_j, α_j.* This is a constant component of all scores obtained under A_j but may vary over treatments (levels of j). Throughout this chapter we assume that the a levels of A exhaust the population of levels of A. Therefore, as a result, $\Sigma_j \alpha_j = 0$; the sum of all deviations of values about their average, $\Sigma(\mu_j - \mu)$, is zero. The null hypothesis asserts that the a values of α_j are all zero.[1]
3. *The error ϵ_{ij}.* This is the deviation of the ith score in group j from μ_j and reflects uncontrolled, or chance, variability. It is the only source of variation within the jth group, and if the null hypothesis is true, the only source of variation in the data matrix.

---

[1] Researchers often refer to the effect of an independent variable as in "A had a significant effect on performance." A somewhat verbose, but precise, translation of such a statement is that one or more of the levels of A had an effect; that is, at least one of the α_j is not zero.

Equation 4.6 is not sufficient for deriving parameter estimates and significance tests. In addition, the following assumptions about the distribution of ϵ_{ij} are required:

1. The ϵ_{ij} are independently distributed. This means that the probability of sampling some value of ϵ_{ij} does not depend on other values of ϵ_{ij} in the sample. An important consequence of this is that the ϵ_{ij} are uncorrelated.
2. The distribution of the ϵ_{ij} is normal, with mean zero, in each of the a treatment populations.
3. The distribution of the ϵ_{ij} has variance σ_e^2 in each of the a treatment populations; that is, $\sigma_1^2 = \cdots = \sigma_j^2 = \sigma_e^2$. This is often referred to as the assumption of *homogeneity of variance*.

Although we never have access to the populations of scores, plots of the data will provide information about the validity of our assumptions. Section 2.2.1 described some procedures that will prove helpful. In later sections of this chapter, we'll discuss what problems may arise when our assumptions are violated, and how the data analysis might be changed in response to these problems. For now, we assume that Equation 4.6 is valid and that the populations of error components are independently and normally distributed with equal variances. This permits us to derive a more formal justification of the F test presented in this chapter.

4.4.2 Expected Mean Squares

In Section 4.2 we argued that the ratio of mean squares was a reasonable test of the null hypothesis of equality of the treatment population means. The idea is that if H_0 is true, MS_A and $MS_{S/A}$ are both estimates of σ_e^2, the variance of the treatment populations. Because they both estimate the same quantity, their average values over many random replications of the experiment should be about the same size. If H_0 is false, however, MS_A also reflects the separation among the treatment population means in addition to chance variability. In that case, on the average, MS_A will tend to be larger than $MS_{S/A}$. In this section we will provide a somewhat more rigorous version of this argument, one based on the average values of MS_A and $MS_{S/A}$ over many replications of the experiment.

Suppose we draw a samples of n scores from their respective treatment populations, and calculate the two mean squares. Now suppose that we draw another a samples of n scores, and again calculate MS_A and $MS_{S/A}$. We could repeat this sampling experiment many times and arrive at two sampling distributions: one for MS_A and another for $MS_{S/A}$. Given the intuitive arguments of Section 4.2, the average value of MS_A will reflect both error variance and treatment effects, whereas the average value of $MS_{S/A}$ will reflect only error variance. These averages of the sampling distributions of the two mean squares are the expected values of the mean squares or the *expected mean squares* (EMS). They play a very important role both in understanding the analysis of variance and in deciding a number of practical issues. To cite just one application, in more complex designs there will be many possible sources of variance; knowledge of the EMS will be needed in order to construct F tests correctly.

Given the structural model of Equation 4.6, and assuming that the ϵ_{ij} are independently distributed with variance σ_e^2, we can prove that the EMS are

$$E(\text{MS}_A) = \sigma_e^2 + n\sum_j \frac{(\mu_j - \mu)^2}{a - 1} \tag{4.7}$$

and

$$E(\text{MS}_{S/A}) = \sigma_e^2 \tag{4.8}$$

To better understand what this means, let's return to our sampling experiment. Suppose we draw four samples of five scores each from their respective treatment populations, calculate the mean squares, and repeat this procedure many times. Further suppose the variance is 1000 for each of the four treatment populations and that the means of the populations are 48, 55, 67, and 82. Then

$$\sigma_e^2 = 1000$$

$$\mu = \frac{1}{4}(48 + 55 + 67 + 82) = 63$$

and

$$n\sum_j \frac{(\mu_j - \mu)^2}{a - 1} = \frac{5[(-15^2) + (-8^2) + (4^2) + (19^2)]}{3} = 1110$$

According to Equation 4.7, the average value of MS_A will be 2110 over many random replications of the experiment; the average for $\text{MS}_{S/A}$ will be 1000.

Look again at Equations 4.7 and 4.8. Note that if the null hypothesis is true (i.e., the μ_j are all equal), both expectations equal σ_e^2; in any experiment, the two mean squares will not be identical, but they rarely should be very different. On the other hand, if the μ_j differ, MS_A has a larger expected value than does $\text{MS}_{S/A}$. The F ratio generally will be greater than 1. From Equation 4.7, it appears that the size of the F ratio will be a function of n and of how far apart the μ_j are. In short, we can expect greater power when we run more subjects and when we deal with larger effects.

In the design of this chapter, there are only two mean squares and it therefore requires no great insight to decide that $\text{MS}_{S/A}$ is the *error term,* or denominator of the F test, against which MS_A is to be tested. Life will become a little more complicated in other designs, beginning with Chapter 8. At that point we will provide you with rules for generating EMS for each SV. The rule for choosing an error term against which to test a particular SV of interest will always be the following:

> Choose an error term such that its EMS will equal that for the term being tested if the null hypothesis is true.

The approach taken in proving Equation 4.7 is to express SS_A in terms of the structural model of Equation 4.6 and then to find its expectation, assuming independence of observations and homogeneity of variance. We have done this in some detail in Appendix 4.2. We suspect that you will take our word for the result; however, going through such proofs can be very helpful, both because it forces you to think about such useful concepts as sampling distributions and expected values and because it provides a better insight into the role various assumptions play in the analysis of variance. Before working through the proof in Appendix 4.2, you may find it helpful to review the material on rules of summation and the algebra of expectations in Appendixes A and B, respectively.

If Equations 4.7 and 4.8 are valid, the ratio of mean squares is sensitive to violations of the null hypothesis. If, in addition, the treatment populations are normally distributed, the ratio of mean squares will be distributed as F on $a - 1$ and $a(n - 1)$ df, and Table D5 will provide a valid assessment of Type 1 error rates. Note that the normality assumption was not invoked in deriving the EMS; it is necessary, however, if the ratio of mean squares is to have an F distribution.

4.5 ASSUMPTIONS UNDERLYING THE F TEST

Although the critical values of F in Appendix Table D5 are derived from the assumptions presented previously, it does not follow that violations of the assumptions necessarily invalidate the F test. For example, in view of our discussion of the central limit theorem and the t test (in Chapter 3), we might guess that the ratio of mean squares will be distributed approximately as F even when the populations are not normal. In this section, we look at the role of assumptions more closely. We ask what the consequences of violations of assumptions are and, in those cases in which there are undesirable consequences, what alternatives to the standard analysis exist.

4.5.1 Validity of the Structural Model

It is important to bear in mind that the analysis of variance for the one-factor design begins with the assumption of Equation 4.6. The data analyst takes the position that only one factor systematically influences the data and the residual variability ($MS_{S/A}$) represents random error. It happens quite frequently that this model is not valid for the research design. For example, suppose half of the subjects at each level of A are run by one experimenter and half by another. The investigator may view the experimenter variable as irrelevant and not consider it within the statistical analysis. However, in many studies, subjects will respond differently to different experimenters. If the experimenter is a meaningful source of variability, the $MS_{S/A}$ represents both error variance and differences among experimenters. But the variance due to experimenters will not contribute to MS_A because all experimenters are running subjects at all levels of A. This situation violates the principle that the numerator and denominator of the F ratio should have the same expectation when H_0 is true. In this situation, the denominator has a larger expectation because the "irrelevant" variable makes a contribution. The result is a loss of power, which can be considerable if the irrelevant variable has a large effect. We say that the F test is *negatively biased* in this case.

As a general rule, the researcher should formulate a complete structural model, one that incorporates all systematically varied factors, even those thought to be irrelevant or uninteresting. In the example given, this would mean treating the study as one involving two factors, A and *experimenter,* and carrying out the analysis presented in the next chapter. If there is strong evidence that *experimenter* contributes only chance variability, the researcher may assume the model of this chapter; the data may be analyzed as if it came from a one-factor design. What constitutes "strong evidence?" On the basis of sampling studies and experience, we would reanalyze the data ignoring factors like *experimenter* if such factors were not significant at the .25 level in the initial data analysis. The advantage of the one-factor ANOVA (if appropriate) is that its error mean square will have a few more df than that for the two-factor analysis; power increases with both numerator and denominator df.

4.5.2 The Independence Assumption

When only one observation is obtained from each subject, and subjects are randomly assigned to treatments or randomly sampled from distinct populations, the assumption that the scores are independently distributed should be met. There are exceptions, however, that are often unrecognized by researchers. For example, suppose we wished to compare attitudes on some topic for males and females; before being tested, subjects participate in three-person discussions of the relevant topic. The scores of individuals who were part of the same discussion group will tend to be positively correlated. If this failure of the independence assumption is ignored (and it often is; see Anderson and Ager 1978 for a review), there will be a positive bias—an inflation of Type 1 error rate—in an *F* test of the sex effect. Why this is so, and the nature of the proper analysis, will be explained in Chapter 10. We should also be aware that the assumption of independence is routinely violated in the designs of Chapters 8 and 9 in which a subject is tested under several treatment levels. The consequences of this, and the remedies, are discussed in those chapters.

4.5.3 The Normality Assumption

The Type 1 error probability associated with the *F* test is little affected by sampling from nonnormal populations unless the samples are quite small and the departure from normality extremely marked. This reflects the role played by the central limit theorem; the distribution of means and their differences will tend to be normal as *n* increases even when the distribution of the parent populations is not. The effects of nonnormality upon the distribution of the *F* statistic has been demonstrated by mathematical proofs (Scheffé 1959) and computer sampling studies (Donaldson 1968, Lindquist 1953, pp. 78–90). This generally rosy view of the impact of nonnormality should be qualified by noting that substantial errors can occur in estimating confidence intervals for variance components (such as σ_e^2) if the treatment populations are not normally distributed.

A particularly flagrant, but not uncommon, violation of the normality assumption occurs when the independent variable is discretely distributed, as it is whenever rating data or response frequencies are analyzed. Lunney (1970) has run computer studies in which the dependent measure had only two values; group sizes were 3 through 31, the number of groups ranged from 2 through 5, and the probability of a success on a single trial varied from .1 to .9. Except for the most extreme *p* values, 20 error df sufficed to yield a close approximation to theoretical α levels; 40 error df were needed when *p* was .1 or .9. These findings have been corroborated and extended by Hsu and Feldt (1969) and Bevan, Denton, and Myers (1974). These investigators varied the number of rating points and the shape of the distribution. Even with $a = 2$, $n = 4$ (six error df), and only three scale points, theoretical α levels were closely approximated in all but the most skewed distributions. In most cases, the empirical α was no more deviant from the theoretical value than the value due to sampling error obtained when sampling from a continuous normal population.

Thus far, we have limited our discussion to the effects of nonnormality upon Type 1 error rate. In Section 3.7, we pointed out that a nonparametric procedure sometimes provides more power against false null hypotheses than a parametric test when the normality assumption is violated. In the next section, we will consider a nonparametric alternative to the F test. We will first describe the test, and then note some conditions under which it may be more powerful than the F test.

4.5.4 The Kruskal-Wallis *H* Test

This requires rank-ordering all the scores and computing the sum of ranks for each group. We will start with a brief justification of the formula, then consider a reanalysis of the data of Table 4.4 using the H test, and finally compare the power of the F and H tests for a few conditions.

Let $N = \Sigma_j n_j$, the total number of scores; n_j is the number of scores in group j. If all N scores are rank-ordered, ignoring treatment level, the mean of all N ranks is $(N + 1)/2$ and the variance is $N(N + 1)/12$. Let $\bar{R}_{.j}$ represent the mean of the ranks for the jth treatment group. Then a quantity very much like the SS_A may be calculated for the group means:

$$\Sigma n_j (\bar{R}_{.j} - \bar{R}_{..})^2 = \Sigma n_j \left[\bar{R}_{.j} - \frac{N + 1}{2} \right]^2$$

The statistic H is the preceding quantity divided by the theoretical variance, $N(N + 1)/12$. If the treatment distributions are the same (that is, H_0 is true), we can view the mean ranks as having been drawn from the same sampling distribution. Furthermore, if the n_j are reasonably large, this sampling distribution will be approximately normal. Then H is the ratio of a sum of squared deviations of approximately normally distributed variables divided by their theoretical variance. In other words, H is an approximately chi-square distributed variable; the appropriate df are $a - 1$.

A little algebra translates the expression on the right side of the preceding equation into a more convenient form for computing:

$$H = \left[\frac{12}{N(N + 1)} \right] \left[\Sigma \frac{T_{.j}^2}{n_{.j}} \right] - 3(N + 1) \tag{4.9}$$

where $T_{.j}$ is the sum of the ranks for group A_j. Table 4.6 presents a reanalysis of the data of Table 4.4. The 30 scores have been assigned ranks. In case of ties, the median rank is assigned. For example, three subjects had a score of 63; because there were only three lower scores, these subjects are tied for the fourth, fifth, and sixth places and are given the median rank of 5. As we indicated above, the ranks have been summed for each condition and the results substituted into Equation 4.9. The null hypothesis is rejected if H exceeds the critical value for χ^2 on $a - 1$ df. For this example, assuming $\alpha = .05$, we need a value of H of 5.991 or greater to reject H_0. As can be seen in Table 4.6, the result of the analysis is not significant. The p value is only slightly less than that obtained in the F test.

TABLE 4.6   KRUSKAL-WALLIS *H* TEST APPLIED TO THE DATA OF TABLE 4.4

Data

Method of instruction					
Method 1		Method 2		Method 3	
Score	Rank	Score	Rank	Score	Rank
83	20	85	21.5	88	25
64	7	63	5	96	30
94	28	74	12.5	95	29
55	1	74	12.5	67	8
85	21.5	68	9.5	89	26.5
57	2	86	23	81	19
77	16.5	59	3	79	18
76	15	68	9.5	87	24
63	5			75	14
77	16.5			63	5
				70	11
				89	26.5

$T_{.j} =$ 132.5   96.5   236 $T_{..} = 465$

Note: If the rankings are correct, $\Sigma T_{.j}$ $(T_{..})$ must equal $N(N + 1)/2$, or $(30)(31)/2 = 465$ in this example. If $T_{..}$ did not equal 465, we would check the rankings.

Calculations

$$H = \left[\frac{12}{N(N + 1)} \right] \left[\Sigma \frac{T_{.j}^2}{n_{.j}} \right] - 3(N + 1)$$

$$= \left[\frac{12}{(30)(31)} \right] \left[\frac{132.5^2}{10} + \frac{96.5^2}{8} + \frac{236^2}{12} \right] - (3)(31)$$

$$= (.0129)(7560.990) - 93 = 4.56$$

Assuming the chi-square distribution on 2 df, the *p* value associated with this result is .102.

In Chapter 3, we cited published evidence that tests based on ranks can be more powerful than parametric tests when the data have been drawn from populations that are skewed, or have more values in their tails than is true of the normal distribution ("heavy-tailed" distributions). Table 4.7 illustrates this point using the results of some computerized sampling experiments we have run. In order to explain the table, we ask you to consider three normally distributed populations, all with $\mu = 0$ and $\sigma^2 = 1$. Assume that we randomly draw three groups of five scores, one group from each population. Both the *F* and the *H* statistic are calculated and evaluated for significance

TABLE 4.7. EMPIRICAL REJECTION RATES FOR THE F AND KRUSKAL-WALLIS H TESTS
($a = 3$)

		Distance between adjacent population means					
		0		.4		.8	
n	Test =	F	H	F	H	F	H
	Normal	.054	.049	.167	.144	.504	.441
5	Mixed-normal	.055	.055	.119	.122	.363	.357
	Exponential	.048	.047	.197	.243	.578	.616
	Normal	.048	.036	.308	.285	.876	.852
10	Mixed-normal	.038	.040	.205	.234	.644	.720
	Exponential	.035	.047	.365	.528	.859	.936

against a .05 critical value. We repeat this experiment 1000 times. The leftmost values in the first row labeled "Normal" tell us that the F was significant in 54 of the 1000 samples and the H in 49 of them. In other words, the empirical Type 1 error rate is very close to the theoretical value of .05. If we had run the experiment many more times, we would have obtained rejection rates even closer to .05.

The next two rows present the outcomes when the populations are either heavy-tailed or severely skewed. In the mixed-normal condition, the probability was .9 of drawing a score from this same normal distribution and .1 of drawing a score from a normal distribution with mean 0 and standard deviation 3; this essentially created a distribution with occasional outliers. The exponential distribution is extremely skewed with $\mu = 1$ and $\sigma = 1$. Despite the violation of the normality assumption, the empirical Type 1 error rates for both the F and H tests are quite close to the theoretical value of .05.

We next separated the three distributions by .4 or .8. For $n = 5$, the F test has greater power in the normal case, the H test does better in the exponential case, and there is little to choose between the two tests when mixed-normal distributions are sampled. For $n = 10$, the H test has more power in both the exponential and mixed-normal conditions. In general, the F will have a slight advantage for normal and short-tailed distributions, and for most distributions when n is very small; the H will do relatively better when distributions are skewed or heavy-tailed, and its advantage in those conditions will become more pronounced as sample size increases.

Using the information in these simulations implies that we have a reliable sense of the shape of the population distribution. Unfortunately, the observed data distribution may be misleading unless n's are much larger than those usually used by researchers. Nevertheless, it makes sense to plot the distributions. Several computer packages make available normal probability plots, histograms, box plots, and stem-and-leaf plots; we find this last plot particularly useful for examining the distribution's shape. The larger the n, and the more consistent the data pattern across conditions, the more reliable will be our sense of the population distribution and the better informed will be our choice of statistical test. To the degree that the data conform to prior expectations about the distribution shape, such decisions will be strengthened. For example, the distribution of rating data usually will be light-tailed (few scores at the end points of the scale); if a stem-and-leaf plot is consistent with this, the *F* test is probably more powerful than its nonparametric alternatives. Reaction times tend to be skewed; with this measure, a skewed distribution of the obtained data would indicate using the Kruskal-Wallis *H* test because of its power advantage over the *F* with such data. Examine the research literature and previous data sets to try to get a sense of what your data may look like. Theoretical models of the data, when available, may also provide useful clues to the shape of the distribution.

The choice between the *F* and *H* tests should reflect considerations in addition to relative power. First, the *H* test is sensitive to differences in the shapes of treatment population distributions. This means that heterogeneous population variances will affect the test in much the way that the *F* test is affected; the nature of these effects upon *F* will be considered in the next section. In fairness, we should note that the distortion in Type 1 error rates under heterogeneous variances is probably somewhat less for the *H* test than for the *F* test (Tomarken and Serlin 1986).

The second consideration is that transformation of scores into ranks may distort relative distances among conditions, thereby making the *H* test inappropriate for analyzing the joint effects of several independent variables. For example, suppose we believe that one method of instruction has a large advantage over another with low-ability students but only a small advantage with high-ability students. Even if this is true of the population means on the original data scale, it is quite possible that it will not be reflected in the rankings. Therefore, a test of these relative differences based upon rank transformations of the original scores could provide misleading results.

4.5.5 The Homogeneity of Variance Assumption

Heterogeneous variances result in inflated Type 1 error rates as long as the *a* groups are of the same size. The inflation is usually less than .02 at the .05 level, and less than .005 at the .01 level, provided the ratio of the largest to smallest variance is no more than 4 to 1. Even larger ratios may not be a problem, but this will depend upon sample size, the number of groups, and the shape of the population distribution. The results of computer simulations employing these factors are discussed in articles by Clinch and Keselman (1982) and Tomarken and Serlin (1986).

These simulation studies clearly demonstrate that heterogeneous variances are a problem when group sizes are not equal. This is also true of the two-sample *t* test. It helps to consider expected mean squares to understand why heterogeneous variances are more troublesome when *n*'s are unequal. For simplicity, assume the null hypothesis is true. Then, using the approach of Appendix 4.2, it can be shown that

$$E(\text{MS}_A) = E\left[\frac{1}{a-1} \sum_j^a n_j (\bar{Y}_{.j} - \bar{Y}_{..})^2 \right] \tag{4.10}$$

$$= \frac{1}{a-1} \sum_j^a \left[\frac{N - n_j}{N} \right] \sigma_j^2$$

and

$$E(\text{MS}_{S/A}) = \sum \left[\frac{1}{N-a} \sum_j^a \sum_i^{n_j} (Y_{ij} - \bar{Y}_{.j})^2 \right] \tag{4.11}$$

$$= \frac{1}{N-a} \sum_j^a (n_j - 1)\sigma_j^2$$

where $N = \Sigma_j n_j$, $\bar{Y}_{..} = \Sigma_j \Sigma_i Y_{ij}/N$, and σ_j^2 is the variance of the jth treatment population. If the n_j all equal n, then $N = an$, and the expressions both reduce to the same quantity:

$$E(\text{MS}_A) = E(\text{MS}_{S/A}) = \sum \frac{\sigma_j^2}{a}$$

This equality of expectations doesn't mean that the ratio of mean squares will be distributed as F when variances are heterogeneous; we have already noted that there will be inflation of Type 1 error rate. Nevertheless, the equality does constrain the degree to which the distribution of mean squares will vary from the theoretical F distribution.

Now consider the more general case in which the n_j are not equal. Turning to Equation 4.10, note that the larger n_j is, the smaller the contribution of the jth population variance to $E(\text{MS}_A)$ because $N - n_j$ decreases as n_j increases. However, the opposite is true for $E(\text{MS}_{S/A})$; the larger n_j is, the greater the contribution of σ_j^2 to the denominator because $n_j - 1$ increases as n_j does. Therefore, when large variances are paired with large group sizes, they will increase the denominator relatively more than the numerator. Thus, when the large groups have the larger variances, the true Type 1 error rate will be below the theoretical value and power against false null hypotheses is likely to be reduced; this is negative bias. On the other hand, when larger variances are paired with small group sizes, they tend to contribute relatively more to the numerator of the F test than to its denominator. In this case, the bias is positive; the Type 1 error rate is inflated.

These distortions of error rate can be quite marked. Consider the case in which population variances and group sizes are inversely paired. When the nominal α is .05, the probability of rejecting a true null hypothesis has been found to be anywhere from .14 (Tomarken and Serlin 1986) to .22 (Clinch and Keselman 1982). When the pairing was direct, Type 1 error rates often dropped to the .02 level. There are essentially two ways to deal with these problems. The first is to find an alternative to the usual F test. The second is to find a transformation of the data that will yield homogeneity of variance on the new data scale. We will consider each of these approaches.

4.5.6 The Welch (1947) and Brown-Forsythe (1974) Tests

Both of these tests provide approximate F statistics that do not require the homoegeneity-of-variance assumption. The error degrees of freedom in both tests incorporate information about the group variances. Neither test is superior under all conditions, and so we will present the formulas for both. After that, we will discuss the merits of the two procedures relative to each other and to the standard F test.

The formulas for the Welch and Brown-Forsythe statistics (F_W and F^*, respectively) are presented in Table 4.8. The table also contains several results generated by the computer package, BMDP7D, based upon an analysis of the unequal-n data of Table 4.4. The F_W and F^* results are not very different from that obtained with the usual F test because the variances are very homogeneous in this data set. The small F value for Levene's test for variances confirms this. In that test absolute deviations of scores about their group means are calculated; an ANOVA is then performed on these deviation scores.

Clinch and Keselman (1982) and Tomarken and Serlin (1986) have compared the standard F test with F_W and F^*. When the homogeneity-of-variance assumption is met, all three procedures produce honest Type 1 error rates. The F has slightly more power when H_0 is false than do the two alternatives, but the difference in power is typically less than .04 when α is .05, and even less when α is .01. On the other hand, both F_W and F^* yield more honest Type 1 error rates when variances are heterogeneous. They also are much more powerful than the F test when variances and group sizes are directly paired. In summary, the advantage of the standard F over the two alternatives is slight when variances are homogeneous, but the alternative procedures are definitely to be preferred when variances are heterogeneous. Therefore, if there is any sign that the variances are heterogeneous, F_W or F^* should be used.

The choice between F_W and F^* depends upon a number of factors. These include group size, the degree of heterogeneity of variance, the spacing of the treatment population means, and the shape of the parent populations. The following guidelines, based on a review of several simulation studies, should be useful:

1. When the ratio of largest to smallest group variance is 6:1 or more, the average group size is at least 10, and samples are drawn from normally distributed populations, F_W will yield Type 1 error rates close to the nominal values, whereas those for F^* will be slightly inflated (about a .06 false rejection rate at the .05 level). More importantly, for most patterns of spacing of the means, F_W will have significantly more power than F^*. However, F^* is more powerful for one pattern of means: if all group means except one are approximately equal, and if the deviant mean is paired with the largest group variance, F^* would be the test to use (Tomarken and Serlin 1986).

2. If variance heterogeneity is moderate and the treatment populations are normally distributed, there is little basis for preference. However, if the populations are skewed, F_W has an inflated Type 1 error rate (Clinch and Keselman 1982).

3. For small group sizes ($n < 6$), F^* had Type 1 error rates closer to the nominal value and power equal to or better than that of F_W (Brown and Forsythe 1974; Dijkstra and Werter 1981).

Some cautions are in order. First, it should be evident that the relative merits of the Welch and Brown-Forsythe tests depend upon a complex combination of factors. Researchers with heterogeneous variances would be well advised to turn to the articles cited and study the patterns of error rates and power. Second, decide between tests on the basis of the parameters of your study—n's, variances, spacing of means, shape of the distribution. Try to resist the temptation to compute all three F's, and choose that

TABLE 4.8   THE WELCH (F_w) AND BROWN-FORSYTHE ($F*$) FORMULAS AND BMDP7D'S ANALYSIS OF THE DATA OF TABLE 4.4

The Brown-Forsythe test

$$F* = \frac{\Sigma_j n_j (\bar{Y}_{.j} - \bar{Y}_{..})^2}{\Sigma_j c_j \hat{\sigma}_j^2}$$

where $N = \Sigma n_j$, $\bar{Y}_{..} = \Sigma n_j \bar{Y}_{.j}$, and $c_j = 1 - (n_j/N)$

$$df_1 = a - 1$$

and

$$df_2 = \frac{(\Sigma_j c_j \hat{\sigma}_j)^2}{\Sigma_j [c_j^2 \hat{\sigma}_j^2/(n_j - 1)]}$$

The Welch test

$$F_W = \frac{A}{B}$$

where $A = \dfrac{1}{a-1} \Sigma w_j (\bar{Y}_{.j} - \bar{Y}_{..})^2$

$$B = 1 + \left[\frac{2(a-2)}{a^2-1}\right] \Sigma \frac{[1 - (w_j/u)]^2}{n_j - 1}$$

and $w_j = n_j/\hat{\sigma}_j^2$, $u = \Sigma w_j$, $\bar{Y}_{..} = \Sigma w_j \bar{Y}_{.j}/u$

$$df_1 = a - 1$$

$$\frac{1}{df_2} = \left[\frac{3}{a^2 - 1}\right] \Sigma \frac{[1 - (w_j/u)]^2}{n_j - 1}$$

BMDP7D Output

MEAN	73.100	72.125	81.583
STD.DEV.	12.853	9.672	10.917
S. E. M.	4.065	3.420	3.151
MAXIMUM	94.000	86.000	96.000
MINIMUM	55.000	59.000	63.000
CASES INCL.	10	8	12

ANALYSIS OF VARIANCE TABLE FOR MEANS

SOURCE	SUM OF SQUARES	DF	MEAN SQUARE	F VALUE	TAIL PROBABILITY
METHOD	576.6750	2	288.3375	2.25	0.1243
ERROR	3452.6917	27	127.8775		

EQUALITY OF MEANS TESTS; VARIANCES ARE NOT ASSUMED TO BE EQUAL

WELCH		2,	17	2.37	0.1239
BROWN-FORSYTHE		2,	25	2.30	0.1206

LEVENE'S TEST FOR VARIANCES		2,	27	0.66	0.5250

one procedure to report that yields the most significant result. That temptation can be almost overwhelming in the face of output from a computer package such as BMDP7D, which contains all three F's and their associated p values. The problem is that such an approach may yield an F that reflects not a true effect of your independent variable but an inflated Type 1 error rate.

One other point should be considered. A number of tests of heterogeneity of variance have been proposed. The idea is that if the test result is significant, alternatives to the standard F test should be used in assessing differences among means. We do not recommend the use of these preliminary tests. Several are sensitive to departures from normality and may yield significant results for that reason. As Games, Keselman, and Clinch (1979) have noted, no single test combines an honest Type 1 error rate with good power over a variety of population distributions. We would rather see researchers place their efforts in plotting their data, examining the distributions, and noting discrepancies among the group variances. Ratios of largest to smallest variance of 4:1 or more when n's are equal, and even 2:1 when n's are unequal, signal possible problems with the usual F test. As we have noted, even when all assumptions are met, the performance of F is only slightly better than that of F_W or F^*. Therefore, when in doubt about the validity of the homogeneity of variance assumption, it makes sense to choose the appropriate alternative procedure.

4.5.7 Transformations of the Data

One other response to violations of assumptions underlying the F test has been to transform the data, for example by raising all scores to some power. Transformations have been used (1) to transform skewed distributions into more nearly normal distributions; (2) to reduce heterogeneity of variance; and (3) to remedy a condition known as nonadditivity in designs in which each subject is tested on several trials or under several treatment levels. We will delay discussion of this third purpose until Chapter 8. For now, we note that a transformation that best achieves one purpose may not be equally suitable for other purposes, although it is true that transformations that equate variances do tend to yield more normally distributed scores.

Our focus here will be on transformations designed to achieve homogeneous variances. Such transformations can be derived if the relation between μ_j and σ_j^2 is known. One example of this occurs when the data are proportions, for example percent correct. In that case, the scores in each treatment population are binomially distributed and the variance can be represented as $\sigma_j^2 = k\mu_j(1 - \mu_j)$, where k is a constant. The appropriate transformation here is

$$Y_{ij}' = \text{arc sin}\sqrt{Y_{ij}}$$

For example, if $Y = .5$, its square root is .707, and the angle whose sine is .707 (the arc sin) is 45°; Y' is 45. Tables of the arc sin transformation may be found in Snedecor and Cochran (1967) and Fisher and Yates (1963). In general, the simplest way to perform this, or any of the most often encountered transformations, is to use a statistical computing package such as BMDP, SSPS, SAS, or SYSTAT. Usually, a single statement added to the control statements will do the job.

A fairly general approach to deciding which transformations should be used is to plot $\log(\hat{\sigma}_j)$ against $\log(\bar{Y}_{.j})$. If this relation is approximately described by a straight line, find its slope (the amount of change in $\hat{\sigma}$ for each unit of change in $\bar{Y}_{.j}$). The appropriate transformation would be

$$Y'_{ij} = Y_{ij}^{1-\text{slope}}$$

If $1 - \text{slope} = 0$, take the log of each score instead of raising it to a power. The square root transformation is often encountered; here the slope is .5, and so each score is raised to the power .5. This situation is not unusual when the data are in the form of frequency counts. A slope of 1, requiring a log transformation, often occurs when data are markedly skewed; reaction times may be amenable to this transformation. It is also applicable when the scores themselves are standard deviations.

Suppose the data clearly establish the relation between treatment group means and standard deviations. Then the appropriate transformation will tend to yield more powerful tests of the null hypothesis (Dunlap and Levine 1982) but on a different scale from the original one. That's a very big "but." If relative distances between means are of interest, analyses of transformed data run into interpretative problems. For example, suppose we wished to determine whether the difference in performance under two instructional methods is the same for low- and high-ability subjects. Even if the differences were the same at the two ability levels on the original data scale, the transformation well might produce a different result on the new scale. Conversely, one method might have more of an advantage for low- than for high-ability subjects on the original data scale, but the result of the comparison might be quite different on the new scale.

A related problem occurs in testing hypotheses about specific functional relations between the independent and dependent variables. For example, a linear relation on the original data scale will be nonlinear on the transformed scale. Unless we can translate our original linearity hypothesis into a hypothesis about the functional relation on the new scale, any gain due to variance stabilization will be more than offset by the ambiguity of the results of the analysis.

This is not to say that transformations will always cloud the interpretation of the data. Carefully chosen transformations may clarify interpretation. Suppose, for example, we have a theory that predicts that Y will be a power function of some independent variable X; that is, $Y = aX^b$. Then plotting log Y as a function of log X is quite helpful. If the theory is correct, the relation will be linear; the exponent b is now the slope of the line and a is its intercept. It is much easier to test the validity of the straight-line relation and to estimate its parameters than to do the same for the original power function. But note that this transformation was dictated by our theoretical interests, not a desire to remedy some violation of assumptions underlying the F test based on the original scale. Unfortunately, transformations to stabilize variances often will conflict with theoretical interests.

For the designs of this chapter, the Welch and Brown-Forsythe tests provide quite reasonable control of Type 1 error rates and good power in the face of nonnormality and heterogeneous variances. They do so using the original data. In view of this, transformations to achieve variance homogeneity or normality should be used only when theory dictates the transformation, or when the transformed scale is readily interpretable and allows for a simpler or more powerful analysis of the data.

4.6 ASSESSING THE IMPORTANCE OF VARIABLES

So far, our discussion of the analysis of variance has been concerned only with tests of significance—with the question, do the different treatment levels have different effects? There is another question that is at least as important: how much do the means of the treatment populations differ? In fact, we could argue that this question of the size of the effects is more important than the question of whether the effects are significant; with enough data, even effects of little theoretical or practical import can yield significant results. Furthermore, only by estimating the magnitude of the effects can we assess the relative importance of several variables. One approach to this goal is to estimate $\Sigma_j(\mu_j - \mu)^2/(a - 1)$, the component of variance in Equation 4.7 that corresponds to the variability of the μ_j. We will begin by using Equations 4.7 and 4.8 to derive an estimate of this quantity, and then we will develop a measure of the relative contribution of the independent variable A to the total population variance. The approach will extend to designs that are considerably more complicated than the simple one-factor design of this chapter.

4.6.1 Estimating Variance Components

The developments of this section will be easier to follow if we restate Equations 4.7 and 4.8 here:

$$E(\text{MS}_A) = \sigma_e^2 + n\theta_A^2 \tag{4.7}$$

$$E(\text{MS}_{S/A}) = \sigma_e^2 \tag{4.8}$$

The term θ_A^2 is shorthand for $\Sigma_j(\mu_j - \mu)^2/(a - 1)$; we use a lowercase Greek theta (θ) rather than sigma (σ) as a reminder that A is a fixed-effect variable. We will have more to say about the distinction between random and fixed effects in Chapter 8. For now, the point is that θ_A^2 represents the variance of the set of a treatment population means estimated by the group means, whereas σ_e^2 represents the variance of a population of scores from which the observed n scores have been sampled.

Suppose we want to estimate the variance of the four treatment population means for the data set of Table 4.1. From our knowledge of *EMS*, we know that MS_A estimates $\sigma_e^2 + n\theta_A^2$ and $\text{MS}_{S/A}$ estimates σ_e^2. This, together with some simple algebra, yields

$$\frac{\text{MS}_A - \text{MS}_{S/A}}{n} = \hat{\theta}_A^2$$

where the diacritic "ˆ" denotes "estimate of." In general, *for any design*, to estimate a component of variance, we should

1. Calculate the difference between the *MS* that would be in the numerator of an *F* test of that component and its error term, or denominator.
2. Divide this difference by the one or more coefficients by which the component of interest is multiplied.

Applying the foregoing procedure to the data of Table 4.1, we conclude that

$$\hat{\theta}_A^2 = \frac{9.56 - 1.71}{8} = .98$$

If H_0 is true, $MS_{S/A}$ may, by chance, be greater than MS_A and the estimate of θ_A^2 will be negative. Such a result is meaningless since variances cannot be negative. We conclude that our best estimate is zero; the negative estimate is assumed to be a chance result.

What do we gain from such estimation? We get a direct measure of how important our independent variable is. Neither the F ratio nor its level of significance provide this, since both these quantities are influenced by n and error variance. This point should become clear if we suppose that the results presented in Table 4.1 are based on 40 subjects in each group instead of eight, and that MS_A is now 10.51. The revised F is now 6.15, but the estimate of θ_A^2 is .22, one-fourth its previous value. The larger F is testimony to the greater amount of data collected rather than to increased effectiveness of our experimental manipulation. The F ratio does not tell us whether an effect is large or small but only (with some probability) whether it is zero or not. Nor can F ratios be compared to provide knowledge of the relative effectiveness of two variables—or the same variable in two different experiments—unless the df are identical for the two variables and they are tested against the same error variance. Estimates of the magnitudes of effects are prerequisite, moreover, to establishing quantitative behavioral laws.

4.6.2 Estimating Variance Proportions

While the absolute value of the variance due to an independent variable is informative, its size relative to the size of the variances that other sources sponsor is even more so. The values of the F ratio and of $\hat{\theta}_A^2$ may show that A is an important source of variance, but the estimated variance due to A relative to the variance estimated for the population as a whole will reveal to us whether there are other important sources of variance. We use ω^2 to designate this relative magnitude of the effect of a variable. The computation of ω^2 depends on the design, the assumed structural equation that relates data to population parameters, and whether effects are random or fixed.

Assuming Equation (4.6), we have

$$Y_{ij} - \mu = \alpha_j + \epsilon_{ij}$$

Squaring both sides and taking expectations over the set of treatment populations, we have

$$E(Y_{ij} - \mu)^2 = \frac{\Sigma \alpha_j^2}{a} + E(\epsilon_{ij})^2$$

The cross-product term vanishes because of the presumed independence of α and ϵ. The last equation can be rewritten

$$\sigma_Y^2 = \left(\frac{a-1}{a} \right) \theta_A^2 + \sigma_e^2 = \delta_A^2 + \sigma_e^2$$

Note that whereas $\theta_A^2[= \Sigma\alpha_j^2)/(a-1)]$ is not truly the variance of the α_j, $\delta_A^2[= (\Sigma\alpha_j^2)/a]$ is.

We define

$$\omega_A^2 = \frac{\delta_A^2}{\sigma_Y^2} \tag{4.12}$$

the proportion of the total population variance attributable to the effects of A.

From the *EMS* for this design, point estimates are readily obtained:

$$\hat{\delta}_A^2 = \left[\frac{a-1}{a}\right]\hat{\theta}_A^2 \tag{4.13}$$

$$= \left[\frac{a-1}{a}\right]\left[\frac{MS_A - MS_{S/A}}{n}\right]$$

$$\hat{\sigma}_e^2 = MS_{S/A}$$

Substituting Equation 4.13 into Equation 4.12 and replacing population parameters by estimates, we get:[2]

$$\hat{\omega}_A^2 = \frac{[(a-1)/a](1/n)(MS_A - MS_{S/A})}{[(a-1)/a](1/n)(MS_A - MS_{S/A}) + MS_{S/A}} \tag{4.14}$$

Multiplying numerator and denominator by an and dividing by $MS_{S/A}$ yields

$$\hat{\omega}_A^2 = \frac{(a-1)(F_A - 1)}{(a-1)(F_A - 1) + na} \tag{4.15}$$

where

$$F_A = \frac{MS_A}{MS_{S/A}}$$

For the data of Table 4.1, we have

$$\hat{\omega}_A^2 = \frac{(3)(4.59)}{(3)(4.59) + 32} = .30$$

We estimate that A accounts for 30% of the variance in the set of treatment populations. While this is not trivial, 70% of the variance is due to factors for which we have not accounted.

4.7 POWER OF THE F TEST

In most studies we are interested in rejecting the null hypothesis. Therefore, it is important that our statistical procedure has a high probability of doing just that if the independent variable has effects, and particularly if those effects are not trivially small.

---

[2]If we hold ω_A^2 to be $E(\hat{\omega}^2)$, Equation 4.14 provides an approximation since $E(\hat{\delta}_A^2)/E(\hat{\sigma}_e^2) \neq E(\hat{\omega}^2)$; that is, the ratio of expected values does not generally equal the expected value of a ratio. The approximation is reasonably accurate, however, and much simpler than the correct expression.

The ability to calculate the power of a test is important to the researcher at two points: first, in the planning of the experiment power calculations can play an important role. Second, in the evaluation of the adequacy of an experiment after collecting the data, power calculations can be helpful. We will first consider how the power of the F test is evaluated when the data have been collected. Then we will show how the researcher can decide on a sample size before running the experiment in order to achieve a desired level of power.

4.7.1 Assessing the Power of the F Test

There are several situations in which it is important to have a sense of the power of the data analysis. In one case, the experimenter predicts an effect of the independent variable, and there is a clear trend among the observed means, one that supports the experimenter's prior hypothesis. Unhappily, the result is not statistically significant. Either there really is no effect in the sampled population and the pattern of observed means is due to chance, or there is an effect but the experiment had too little power to detect it. In a second case, the experimenter predicts no effect and does not obtain a significant result. In this case, the experimenter would like to claim that power to detect real effects was high, and therefore the failure to achieve significance was because the effect did not exist or was trivially small. Finally, in a third situation, the experimenter predicts no effect but gets a significant result. Here, the experimenter would like to claim that the effect, although possibly real, was not large enough to be of theoretical or practical interest.

In all of these situations, estimates of θ_A^2 and ω^2 will be important in discussing the results because they help provide a sense of the absolute and relative size of the variance of the treatment population means. Calculations of power supplement these statistics. If the F test did not have a significant result, but power was low to detect effects of the magnitude estimated from the data, then the experiment should be rerun with larger n's or with a different design or procedure aimed at reducing error variance. Whether or not the result was significant, if ω^2 was very small and the power to detect the effects estimated from the data was very high, we may be dealing with an effect of little theoretical or practical interest. If the effect is of interest, the power calculations provide us with a sense of the sample size needed in further investigations of this independent variable. In view of these considerations, we need to be able to calculate the power of the F test.

The power of the F test (and of other tests; see the discussion in Chapter 3) is in part a function of the experimental design, but for the time being we will assume that we are running the design of this chapter; we have a conditions with n subjects in each condition. Then the power of the F test will depend upon the following factors:

1. The significance level α. As we reduce the rejection region, say from .05 to .01, we lower the probability of rejecting false, as well as true, null hypotheses. In other words, reduced Type 1 error rate is accompanied by reduced power.
2. The values of a and n. More precisely, increases in either numerator or denominator df yield increased power. Ordinarily, the value of a is determined by the goals of the experiment; n is usually more arbitrarily selected, although constrained by practical concerns such as time, effort, and cost.

3. The error variance σ_e^2. The less noise in our data, the easier it will be to detect treatment effects. Therefore, power increases with decreases in error variance.
4. The size of the effects to be detected. In the case of the F test, we will have more power to reject the null hypothesis of equal treatment population means as the variance of those means increases.

To determine the power of our test, we need to state the numerical values of these four factors and then to find some way of relating power to them. Assuming we have run the experiment, we know df_1 and df_2 and have selected a level of significance. We can use $MS_{S/A}$ as an estimate of σ_e^2. Ordinarily, we take $\hat{\delta}_A^2$ as our estimate of the size of the variance of effects to be detected; $\hat{\delta}_A^2$ was defined in Equation 4.13. Given this information, we construct the following index of power:

$$\phi = \sqrt{\frac{n\Sigma(\mu_j - \mu)^2/(df_2 + 1)}{\sigma_e^2}} = \sqrt{\frac{n\delta_A^2}{\sigma_e^2}} \tag{4.16}$$

We do not know ϕ, but we can estimate it. $MS_{S/A}$ estimates σ_e^2, and the estimate of δ_A^2 is provided by Equation 4.13. Inserting these estimates into Equation 4.16, we arrive at a value of $\hat{\phi}$, which we then use as an entry to the charts in Appendix D6. Here, power is graphed as a function of ϕ for several combinations of df and α values. To illustrate the use of this table, consider a study in which productivity measures are obtained from four groups of 10 employees who vary in experience. The group means and variances are

	A_1	A_2	A_3	A_4
$\bar{Y}_{.j}$	15.25	18.50	20.25	20.00
$\hat{\sigma}_j^2$	18.14	13.43	20.28	23.71

From these values, we calculate $MS_A = 42.33$ and $MS_{S/A} = 18.89$; the F ratio equals 2.24. With df $= 3$ and 28, and setting $\alpha = .05$, we require $F = 2.95$ to reject H_0. Despite an apparent trend in the data, we cannot conclude that performance improves with experience in the population of workers.

A reasonable question is whether this F test had sufficient power to detect effects of experience. One way to approach this is to use the data we collected to assess power. Substitution of the mean squares calculated before into Equation 4.13 yields $\hat{\delta}_A^2 = 2.564$. Substituting this value and the values of $MS_{S/A}$ and n into Equation 4.16, we have

$$\hat{\phi} = \sqrt{\frac{(8)(2.564)}{18.89}} = 1.042$$

Now turn to the chart for which $df_1 = 3$. Find 1.04 (essentially 1) on the abscissa (X axis) scale for $\alpha = .05$ and consider only the set of curves labeled $\alpha = .05$. The curve for $df_2 = 30$ is close enough to the actual value of 28. Find the point on the ordinate (Y axis) that intersects with $\phi = 1$ at the curve for $df_2 = 30$. That point on the ordinate is roughly .32. That is the power value we are seeking.

It is easy to become so focused on the process that we forget what the result means. What the estimated power of .32 means is the following. Given the sample size we ran, and assuming the $MS_{S/A}$ is a reasonable estimate of σ_e^2, we have power of about .32 to detect effects of experience of the order of magnitude estimated from our data. Put somewhat differently, if the treatment population means are about as different as our sample means suggest, we still have a .68 $(1 - .32)$ probability of making a Type 2 error. What this suggests is that our experiment isn't very informative. It may be that experience has little effect upon productivity, but it also may be that there is an important effect that we have failed to detect because we had too little power.

4.7.2 Selecting Sample Size

Suppose we decide to rerun the study of productivity with a new sample of subjects. We would like to have better power to reject the hypothesis of no effect of experience, assuming that the actual variance of the μ_j is estimated by the data summarized above. How many subjects should we run in this new experiment? We might decide to double the n, to use 16 subjects in each of the four conditions. Substituting into Equation 4.16, we have

$$\hat{\phi} = \sqrt{\frac{(16)(2.564)}{18.89}} = 1.47$$

Again turning to Appendix D6, with df_2 now equal 60, we find power to be about .65. This is considerably better than we had with eight subjects. However, we might ask what it would take to obtain .95 power. By substituting various values of n into $\sqrt{n}\,(\sqrt{2.564/18.89})$ and referring the resulting value of $\hat{\phi}$ to the power chart, we can answer this. When $n = 33$, $\hat{\phi} = 2.12$, $df_2 = 128$ (essentially ∞), and power = .95. This may be more subjects than are available. If this is the case, we should run the largest number that is practical. The power calculations are still important, however, because they provide one measure of the effectiveness of our research plan.[3]

4.8 CONCLUDING REMARKS

The one factor design is not often directly applied in psychological research. Many studies involve repeated measurements on subjects, calling for parceling out subject effects in addition to treatment effects. Even studies in which there are no repeated measurements usually involve more than one treatment variable. Nevertheless, notation, derivations, computations, null hypothesis testing, and the model underlying the use of the F test are all extremely important, and will be involved in our further work. If you master them, you will more easily understand the complicated designs and analyses of subsequent chapters.

---

[3]In this example, in which we are planning a replication, we should consider the original data set as well. Using the procedures of the next chapter, we can test whether the pattern of differences among the means is replicated. If it is, we would pool the data from the two experiments, providing a more powerful test than is available from either data set alone.

EXERCISES

4.1 The following terms provide a useful review of many concepts in the chapter. Define, describe, or identify each of them:

completely randomized design	consequences of nonnormality
between-subjects design	Kruskal-Wallis H test
within-subjects design	assumption of homogeneity of
parent population	variance
treatment population	consequences of heterogeneity of
source of variance	variance
mean squares	Welch test (F_W)
structural model	Brown-Forsythe test (F^*)
error component of a score	data transformation
expected mean square	ω^2
assumptions underlying ANOVA	factors influencing the power of the
negatively biased F test	F test

4.2 Carry out the analysis of variance (ANOVA) on the following data sets:

(a)

A_1	A_2	A_3
28	38	64
23	39	73
21	57	61
38	36	48
38	38	72
49	48	52
28	52	54
33	40	60
34	39	54
29	45	60

(b)

A_1	A_2	A_3	A_4	A_5
24	48	42	96	73
07	91	82	67	81
46	63	75	88	33
45	69	76	24	44
97	26		92	94
	22		83	77
	45			60
				89
				25

4.3 I have a data set with three groups of five scores each. Because the scores involve decimal values, I multiply every score by 100.

(a) How will the mean squares change (relative to an analysis on the original data set)?

(b) Should the F ratio change?

(c) In general, what happens to a variance when every score is multiplied by a constant?

(d) Suppose we just added a constant, say 10, to all 15 scores. How would that effect the mean squares and F ratio?

(e) Suppose we added 5 to all scores in the first group, 10 to all scores in group 2, and 15 to all scores in group 3? Should MS_A change? $MS_{S/A}$? Explain briefly. Would the result have been any different if we subtracted 5 from each of the group 1 scores, left the group 2 scores unchanged, and added 5 to all the group 3 scores? Explain.

4.4 In deriving the EMS (see Appendix 4.2), we assume that $\Sigma_j \alpha_j = 0$. We do not assume, however, that $\Sigma_j \Sigma_i \epsilon_{ij} = 0$ but rather that $E(\Sigma\Sigma\epsilon_{ij}) = 0$. What is the basis for this distinction between α_j and ϵ_{ij}? Would the situation change if the levels of the treatment variable A were randomly sampled from a population of levels? If the population of subjects were exhaustively sampled? Explain.

4.5 Following are summary statistics from a three-group experiment. Present the analysis of variance table when (a) $n_1 = n_2 = n_3 = 10$ and (b) $n_1 = 6$, $n_2 = 8$, and $n_3 = 10$.

	A_1	A_2	A_3
$T_{\cdot j} =$	30	45	70
$\hat{\sigma}_j^2 =$	3.2	4.1	5.7

4.6 A sample of humanities majors are divided into three groups of 10 each in a study of statistics learning. One group receives training on relevant concepts *before* reading the text, a second receives the training *after* reading the text, and a third is a no-training *control*. Summary statistics on a test are

	Before	After	Control
$\bar{Y}_{\cdot j} =$	24	16	13
$\hat{\sigma}_j^2 =$	72	62	76

(a) Do an ANOVA based on these statistics and report whether the F is significant at the .01 level. State the rejection region and df. What would a significant result tell you?

(b) The Before group seems to perform differently from the other two groups. We would like to test whether the mean of the Before population is better than the average of the other two populations combined. In other words, we wish to test the following null hypothesis:

$$H_0: \psi = \frac{1}{2}(\mu_A + \mu_C) - \mu_B = 0$$

where ψ is a linear contrast. The best estimate of this contrast is

$$\hat{\psi} = \frac{1}{2}(\bar{Y}_A + \bar{Y}_C) - \bar{Y}_B$$

(i) What is the estimate of the variance of the sampling distribution of $\hat{\psi}$ (assume homogeneity of variance)? (ii) Do a t test of the null hypothesis. (Note: This problem parallels several in Chapters 2 and 3 where the relevant information was first presented.)

4.7 For the following data set, calculate the squared t value using the square of the usual t formula. Now calculate the F statistic using the raw-score formulas for ANOVA, as illustrated in Table 4.5. F should equal t^2. After comparing the two sets of results, rewrite the formula for the t statistic as a function of MS_A and $MS_{S/A}$.

The data are

$$A_1 \quad 27\ 18\ 16\ 33\ 24$$
$$A_2 \quad 23\ 33\ 26\ 19\ 38$$

4.8 An experimenter drew four independent random samples each of size 5 from one normally distributed population. The variance of the four sample means about their grand mean is 84. He also has a sample of 15 independent observations that he believes are from the same population from which the original four samples were taken. The variance of the 15 scores is 384. Present a statistical test (and do it) to determine whether the 15 scores do come from the same population as the four samples of 5. In two or three sentences, state what this problem has to do with the ANOVA for a one-factor design. What is the major difference between this problem and that application?

4.9 The following problem is best done using a computer package. We recommend BMDP7D, which provides an analysis of variance for one-factor designs, Levene's test of homogeneity of variance, the Brown-Forsythe test, and Welch's test, as well as histograms for each group showing the distribution of the scores, plots of group standard deviations (σ) as a function of group means, and log(σ) as a function of log(mean). Following is a copy of the control information for BMDP7D and the data set. The PLOT command is optional but gives the s.d.-mean functions. HIST asks for the histograms for each group and gives us a feeling for the shape of the distribution.

The control information is

```
/PROBLEM    TITLE IS 'BMDP7D ANOVA OF EX4DATA'.
/INPUT      VARIABLES ARE 2.
            FORMAT IS FREE.
/VARIABLE   NAMES ARE A, Y.
/GROUP      CODES(1) ARE 1,2,3.
            NAMES(1) ARE A1,A2,A3.
/HIST       GROUPING IS A.
/PRINT      PLOT.
/END
```

The data are

A_1	133	1	49	18	3									
A_2	1	98	50	185	173	176	16	7	381	3				
A_3	142	174	22	256	2	28	50	262	12	44	216	26	276	48	360

(a) Is there any indication of heterogeneity of variance in the data. Explain.

(b) Calculate F, F^*, and F_W (these are provided by BMDP7D). Do the relative significance values of the three F's makes sense in terms of our discussion of F^* and F_W in this chapter? Explain.

(c) If you were going to transform the data, what transformation would you use? Why? [Note: Look at the plot of $\log(\hat{\sigma}_j)$ against $\log(\bar{Y}_{.j})$; BMDP7D prints the regression coefficient for the plot, and that can be used to decide on the transformation as described in Section 4.5.7.]

(d) To redo the analysis using transformed data and BMDP, edit your control file by inserting a transformation paragraph just before the HIST paragraph. For example, if you wanted to raise each score to the $-.25$ power, the instruction is

    /TRAN Y=Y**(-.25).   ("**" is "raise to the power of")

If you want the log of each score, the instruction is

    /TRAN Y = LN(Y).   (Don't forget the period.)

Other possible transformations are presented in the BMDP manual. Do the appropriate one and then comment on any changes in the results, including the different tests and plots. More specifically, are the variances more homogeneous? (If not, you'd better try a different transformation.) What about the relations among the three F tests now? Is the ordinary F test more or less significant than before? Does this make sense in terms of the discussion of this chapter?

4.10 Consider the following data based on three different experimental conditions; there are 15 subjects in each condition of the experiment.

A_1	2	10	10	11	14	18	25	26	30	37	40	45	53	58	88
A_2	14	15	18	18	21	24	31	39	43	46	48	57	60	61	91
A_3	22	29	31	34	35	39	42	50	53	54	59	67	74	87	92

Is there any indication that the populations are not normally distributed? Carry out both F and H tests and comment on the outcomes.

4.11 A pilot study compares a new drug for alleviating pain with a placebo. Ratings by 20 patients result in the following ANOVA:

SV	df	MS	F
A	1	60	4.0
S/A	18	15	

(a) Estimate θ^2 and ϕ for this data set. If we set α at .05, what power do we have to reject H_0, assuming the true value of ϕ is as large as we estimated?

(b) Assuming the estimates arrived at above, what n is needed in a new experiment to have power $= .5$?

(c) Suppose that $n = 10$ and the drug population mean is six points higher than that for the placebo. What power would an F test have to reject the null hypothesis?

4.12 We have four groups of data:

A_1	A_2	A_3	A_4
27	17	29	31
35	19	33	39
40	12	36	47
19	25	23	27
26	20	22	32

(a) Calculate group means and variances and do the ANOVA. Are the results significant at the .05 level?

(b) Suppose we wanted to reject the null hypothesis when $\omega^2 = .4$. Both ω^2 and ϕ^2 are functions of θ^2 and σ_e^2. A little algebra shows that

$$\phi^2 = \frac{\omega^2 n}{1 - \omega^2}$$

With α at .05, what power did our experiment have to detect an ω^2 of .4?

4.13 (a) Calculate $\hat{\theta}_A^2$, $\hat{\sigma}_e^2$, and $\hat{\omega}_A^2$ for this three-group experiment ($n = 10$):

SV	df	MS
A	2	80
S/A	27	5

Suppose $n = 5$ and we had

SV	df	MS
A	2	42.5
S/A	12	5

Now repeat the calculations. Comparing the two results, what is the effect of n upon the F ratio? Upon the other three quantities?

(b) If $F = 1$, what is the value of $\hat{\omega}_A^2$? Briefly explain.

(c) An alternative measure of the importance of a variable is

$$\hat{\eta}^2 = SS_A/(SS_A + SS_{S/A}).$$

(i) What does SS_A estimate (i.e., what is its expectation)?

    (ii) What does $SS_{S/A}$ estimate? (Because you know the expectation of the mean squares, the expectation of the sums of squares is easily obtained.)

    (iii) What does $\hat{\eta}_A^2$ estimate?

    (iv) Compute the value of $\hat{\eta}_A^2$ for each of the two ANOVA tables. What is the effect of n in this case?

    (v) Suppose that $F = 1$. What is the value of $\hat{\eta}_A^2$ for a groups of n subjects each?

(d) In view of the foregoing answers, which measure ($\hat{\omega}^2$ or $\hat{\eta}^2$) do you prefer and why?

4.14 Given the following ANOVA table:

SV	df	MS
A	2	80
S/A	27	40

(a) Assuming n is equal for each group, what is n?

(b) Is there a significant treatment effect if $\alpha = .05$?

(c) Find the values of $\hat{\theta}_A^2$, $\hat{\delta}_A^2$, $\hat{\omega}_A^2$, and $\hat{\eta}_A^2$ ($\hat{\eta}^2$ is defined in Problem 13). Which of these is the best measure of the importance of the treatment? Why?

(d) If the error variance and treatment effect are as estimated from the table, what is the power of the F test for $\alpha = .01$?

(e) How many subjects would we need in each group to have a power of .80 if $\alpha = .01$?

4.15 Suppose we have a theory of learning and memory that claims (1) the probability of learning a previously unlearned item when it is presented is .2, no matter how many times it has been presented before; and (2) once the item has been learned, it will be retained perfectly. An experiment to test the theory is run. Three groups of subjects are each presented with a list of 50 items to learn. Group 1 is presented with the list once, group 2 is presented with the list two times, and group 3 is presented with the list three times. The obtained ANOVA table is

SV	df	SS	MS	F
A	2	151.67	75.83	2.54
S/A	12	358.80	29.50	

(a) Can we reject the null hypothesis $\mu_1 = \mu_2 = \mu_3$ if $\alpha = .05$?

(b) According to the theory, the expected number correct for condition 1 is 10 items; that is, $.2 \times 50$. The expected number for condition 2 is 18 items; a subject either learns an item on the first presentation ($.2 \times 50 = 10$) or doesn't learn it on the first trial but does on the second ($.8 \times .2 \times 50 = 8$). Similarly, the expected number for the third condition is 24.4. If the theory is correct and if $\alpha = .05$, what is the power of the ANOVA to reject the null hypothesis of no differences among the three means? Has the experiment provided us with information about whether our theory is correct?

4.16 In Appendix 4.2, we derived $E(MS_A)$. Following the approach presented there, derive $E(M_{S/A})$. Be sure to make explicit the assumptions involved at each step.

APPENDIX 4.1: Computing Formulas for Sums of Squares (SS) Based on Degrees of Freedom (df)

Computing formulas can be derived from the definitional formulas of Table 4.2 using the rules of summation presented in Appendix A. The df provide a basis for a simpler approach, which applies to any term and to analyses for any design. Consider the $SS_{S/A}$ first:

1. Expand the df expression.

$$df_{S/A} = a(n-1) = an - a$$

2. Each df in the expanded term corresponds to a squared quantity. Therefore, we now have

$$\sum_{j=1}^{a} \sum_{i=1}^{n} Y_{ij}^2 - \sum_{j=1}^{a} T_{.j}^2$$

Note that each squared quantity must have subscripts that correspond to the indices of summation immediately to its left. Any indices that are not subscripted are replaced by dots, indicating that summation has taken place over that index before squaring. For example, $T_{.j}^2$ indicates that the n scores at A_j have been summed and that total was then squared.

3. Divide each squared quantity by the number of values that were added before the total was squared. Because Y_{ij} is a single score, Y_{ij}^2 is divided by 1. $T_{.j}^2$ is divided by n, however. Therefore,

$$SS_{S/A} = \sum_{j}^{a} \sum_{i}^{n} Y_{ij}^2 - \sum_{j}^{a} \frac{T_{.j}^2}{n}$$

This formula is equivalent to those in Tables 4.2 and 4.3. The expressions for SS_A and SS_{tot} in Table 4.3 may be justified in much the same way, beginning with the df.

APPENDIX 4.2: Deriving Expected Mean Squares

In order to derive $E(MS_A)$ for the one-factor design, we first must define $\bar{Y}_{.j}$ and $\bar{Y}_{..}$ in terms of the parameters of Equation 4.6. The group mean is

$$\bar{Y}_{.j} = \frac{\Sigma_i Y_{ij}}{n} = \frac{\Sigma_i (\mu + \alpha_j + \epsilon_{ij})}{n}$$

$$= \frac{n\mu + n\alpha_j + \Sigma_i \epsilon_{ij}}{n}$$

$$= \mu + \alpha_j + \bar{\epsilon}_{.j} \tag{4.17}$$

Also, the grand mean is

$$\bar{Y}_{..} = \frac{\Sigma_j \Sigma_i Y_{ij}}{an} = \frac{\Sigma_j \Sigma_i (\mu + \alpha_j + \epsilon_{ij})}{an}$$

$$= \frac{an\mu + n\Sigma_j \alpha_j + \Sigma_j \Sigma_i \epsilon_{ij}}{an}$$

Note that $\Sigma_j \alpha_j = 0$ because the sum of deviations of any set of observations (the μ_j) about their average (μ) equals zero. Therefore, the last equation becomes

$$\bar{Y}_{..} = \mu + \bar{\epsilon}_{..} \tag{4.18}$$

We may now consider $E(SS_A)$. Substituting Equations 4.17 and 4.18 into the definitional formula for SS_A, and subtracting, we have

$$E(SS_A) = E\left[n\sum_j (\bar{Y}_{.j} - \bar{Y}_{..})^2 \right] = E\left[n\sum_j (\alpha_j + \bar{\epsilon}_{.j} - \bar{\epsilon}_{..})^2 \right] \tag{4.19}$$

Expanding the rightmost term, we obtain

$$E(SS_A) = nE\left[\sum_j \alpha_j^2 + \sum_j \bar{\epsilon}_{.j}^2 + \sum_j \bar{\epsilon}_{..}^2 + 2\sum_j \alpha_j \bar{\epsilon}_{.j} - 2\sum_j \alpha_j \bar{\epsilon}_{..} - 2\sum_j \bar{\epsilon}_{.j} \bar{\epsilon}_{..} \right] \tag{4.20}$$

Because the expectation of a sum equals the sum of the expectations, the expectation of each of the six terms to the right of the equals sign can be evaluated separately. Consider the leftmost of these six terms. Because α_j is a constant component of every score in the jth treatment population (see Equation 4.6), its value doesn't change over replications of the experiment. Therefore,

$$nE\left[\sum_j \alpha_j^2 \right] = n\sum_j \alpha_j^2$$

Turn now to the next term in Equation 4.20, $nE(\Sigma\bar{\epsilon}_{.j}^2)$. $E(\bar{\epsilon}_{.j}^2)$ is an average squared deviation; that is, it is the variance of the sampling distribution of the mean error component. In Section 2.5.2, we saw that the variance of the sampling distribution of a mean based on n independent observations was σ^2/n. Therefore,

$$E(\bar{\epsilon}_{.j}^2) = \sigma_{\bar{e}_j}^2 = \frac{\sigma_{e_j}^2}{n}$$

and

$$nE\left[\sum_j \bar{\epsilon}_{.j}^2 \right] = n\sum_j E(\bar{\epsilon}_{.j}^2) = n\sum_j \frac{\sigma_{e_j}^2}{n}$$

Because homogeneity of variance is assumed, this becomes

$$nE\left[\sum_j \bar{\epsilon}_{.j}^2 \right] = nE\left[\sum_j \frac{\sigma_e^2}{n} \right] = a\sigma_e^2$$

Similarly,

$$nE\left[\sum_j \bar{\epsilon}_{..}^2 \right] = n\sum_j E(\bar{\epsilon}_{..}^2) = na\left(\frac{\sigma_e^2}{na} \right) = \sigma_e^2$$

The next term, $-2nE(\Sigma_j \bar{\epsilon}_{.j}\bar{\epsilon}_{..})$, simplifies if we note that $\Sigma_j \bar{\epsilon}_{.j} = a\bar{\epsilon}_{..}$. We can write

$$-2nE\left[\sum_j \bar{\epsilon}_{.j}\bar{\epsilon}_{..}\right] = -2nE(a\bar{\epsilon}_{..}^2) = -2na\left(\frac{\sigma_e^2}{na}\right) = -2\sigma_e^2$$

The last two terms in Equation 4.20 are equal to zero because they reflect the covariance of α_j with the average error components in a data set. For example,

$$nE\left[\sum_j \alpha_j \bar{\epsilon}_{.j}\right] = n\sum_j E(\alpha_j \bar{\epsilon}_{.j})$$

but $E(\alpha_j \epsilon_{.j})$ is a covariance (see Section 2.4.3); because α_j is a constant component of all scores in the jth treatment population, it will not covary with the average error component in the group $(\bar{\epsilon}_{.j})$ or in the entire data set $(\bar{\epsilon}_{..})$.

Collecting terms, we have

$$E(\text{SS}_A) = n\sum_j \alpha_j^2 + a\sigma_e^2 + \sigma_e^2 - 2\sigma_e^2 + 0 + 0 = (a-1)\sigma_e^2 + n\sum_j \alpha_j^2$$

Dividing by the df, $a - 1$, in order to convert to a mean square, we have the result of Equation 4.7.

We will not prove Equation 4.8 $[E(\text{MS}_{S/A})]$ at this point. Instead, we have included it as a problem at the end of this chapter. In doing that problem, note the steps we have gone through:

1. Write the equation for the sums of squares in terms of population parameters.
2. Expand the squared quantities, placing the expectation operator in front of each resulting term in the expansion.
3. Recognize that expectations of squared average deviations are variances of sampling distributions of means, and use this to convert those expectations into population variances.
4. Carefully note where the assumptions of independence and homogeneity of variance are being used.

Chapter 5

Between-Subjects Designs: Several Factors

5.1 INTRODUCTION

In this chapter, we extend the between-subjects designs of Chapter 3 to include more than one factor. For example, consider an experiment in which texts are presented one word at a time to a subject at a rate of either 300, 450, or 600 words per minute (wpm). The texts are either intact or the order of sentences is scrambled. Thus we have a 2 × 3 design: two text conditions at each of three rates. Each of 48 subjects is assigned at random to one of the resulting six cells of the design with the restriction that each cell contains eight subjects.

Including several factors within the same experiment allows us to consider several issues. In the example just described, the researchers believe that recall is better for intact than for scrambled texts, and that recall declines as presentation rate increases. The major issue, however, is whether the advantage of intact over scrambled text is the same at all presentation rates. We might expect the difference between recall of intact and scrambled texts to decrease as rate increases. At higher rates of presentation, readers may have more difficulty integrating sentences with each other; therefore it should matter less whether the sentences could be integrated (intact condition) or not (scrambled condition). When the differences between mean performances for one variable (e.g., text) change with the level of the other variable (e.g., rate), we say that the two variables interact. As our example suggests, the question of *interaction* may be of as much or more interest as the effect of each variable alone.

In the next section, we will develop the analysis for two-factor designs, illustrating the approach with a data set based on the experiment described above. In subsequent sections we will present a more formal development of the model underlying the analysis and then extend the approach to designs involving more than two factors.

5.2 TWO-FACTOR DESIGNS: THE ANALYSIS OF VARIANCE

5.2.1 The Data

Table 5.1 presents the data for 48 subjects run in the text recall experiment described in the preceding section. The scores are percentages of idea units recalled. Cell means and variances are also presented. The variances are reasonably homogeneous, the ratio of largest to smallest being less than 3:1. Averaging over the 24 subjects in the three rate conditions shows that recall is better for intact than for scrambled texts. Next, averaging over both texts shows that recall gets worse as the presentation rate increases. Of course, these effects may be due to chance variability; there are different people in the different conditions and their scores will differ even if the variables have no effect. We will do some tests shortly to examine whether the results are statistically significant, but we first continue our look at the data.

TABLE 5.1   DATA AND SUMMARY STATISTICS FOR A TWO-FACTOR EXPERIMENT

Text	Rate		
	300	450	600
Intact	72	49	40
	63	71	49
	57	63	36
	52	48	50
	69	68	54
	75	65	46
	68	52	46
	74	63	26
cell mean =	66.250	59.875	43.375 $\bar{Y}_{intact} = 56.5$
cell variance =	68.492	79.549	81.415
Scrambled	65	56	41
	45	55	42
	53	49	57
	53	52	39
	51	35	36
	58	57	52
	53	45	52
	57	49	48
cell mean =	54.375	49.750	45.875 $\bar{Y}_{scramb} = 50.0$
cell variance =	33.977	52.214	55.267

$$\bar{Y}_{300} = 60.3125 \quad \bar{Y}_{450} = 54.8125 \quad \bar{Y}_{600} = 44.6250$$
$$\text{Grand mean} = 53.25$$

The next question is whether the two independent variables interact. Figure 5.1 provides a plot of the six cell means that helps us consider this question. There is an apparent interaction that can be described in either of two ways. We can say that the advantage of intact over scrambled text decreases as rate increases; indeed, there is a small reversal at 600 wpm. Another way of describing the interaction is to say that the intact text yields a greater decrease in performance as rate increases than does the scrambled text. However we describe these six means, it appears that the effects of one variable change as a function of the level of the other variable. This suggests that there is an interaction between text and rate. We soon will consider this more closely.

5.2.2 The Analysis

Just as we did in Chapter 4, we want to partition the total variability in the data set. This will be done by dividing the deviation of a score from the grand mean into components related to the effects of the factors and their interactions, and the error component. These deviations will form the basis for various sums of squares to be used in our F tests. Some notation will be helpful in doing this. In general, we will refer to the scores as Y_{ijk}, where i indicates the position in a cell, j indicates the type of text (1 = intact and 2 = scrambled), and k indicates the rate (1 = 300 wpm, 2 = 450 wpm, and 3 = 600 wpm). For example, Y_{312} is the third score in the cell defined by the intact text read at a rate of 450 wpm. The six cell means are designated by $\bar{Y}_{jk}$; for example, $\bar{Y}_{23}$ is the mean of the eight scores in the scrambled text, 600-wpm condition (i.e., 45.875). The text means are designated by $\bar{Y}_{.j.}$; the rate means by $\bar{Y}_{..k}$. Thus $\bar{Y}_{.1.} = 56.5$ and $\bar{Y}_{..1} = 60.3125$. The dot subscripts indicate that we have summed over the indices in those positions to arrive at the mean. Following this logic, $\bar{Y}_{...}$ denotes the grand mean of all 48 scores.

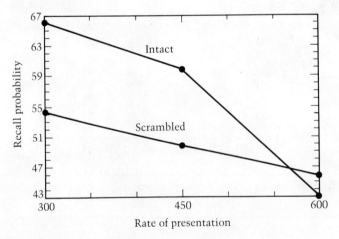

Figure 5.1 A plot of the cell means for the data of Table 5.1.

The deviation of a score from the grand mean can be viewed as consisting of two parts, the deviation of the score from its cell mean and the deviation of the cell mean from the grand mean. That is,

$$Y_{ijk} - \bar{Y}_{...} = (Y_{ijk} - \bar{Y}_{.jk}) + (\bar{Y}_{.jk} - \bar{Y}_{...})$$

The deviation of a score from the cell mean reflects chance, or error, variability; a subject's score differs from those of other people in the same condition, and therefore from the average for that condition, because of individual differences or random variation within the experimental conditions. The deviation of the cell mean from the grand mean may be decomposed further into components that reflect the questions of interest to the researcher. In our example, the cell deviation can be partitioned into a text component, a rate component, and a third component reflecting the interaction of text and rate. We will refer to this cell deviation as the *cell effect;* the text and rate components of this effect will be referred to as *main effects,* and the interaction component will be referred to as the *interaction* effect. On occasion we will refer to some variable such as text as having a significant effect (or a significant main effect). More precisely, we mean that the population effects at the levels of the variable are not all zero or, equivalently, that the population means at these levels are not all the same. The relation of the cell effect to the main and interaction effects is

$$\bar{Y}_{.jk} - \bar{Y}_{...} = (\bar{Y}_{.j.} - \bar{Y}_{...}) + (\bar{Y}_{..k} - \bar{Y}_{...}) + (\bar{Y}_{.jk} - \bar{Y}_{.j.} - \bar{Y}_{..k} + \bar{Y}_{...})$$

cell effect = main effect of text + main effect of rate + interaction

The above equation gives us more insight into the interaction effect; subtracting the two main effect terms from both sides, we see that the interaction is what remains of the cell effect after the main effects have been subtracted.

With the data of Table 5.1, Table 5.2 translates the preceding equation into numbers. Note that the sum of each column is zero. Furthermore, the sum of the interaction effects for either of the text conditions, or for any of the rate conditions, is zero. This is because the sum of deviations of quantities about their average must be equal to zero, as we prove in Appendix A.

At the bottom of Table 5.2, the effects in each column have been squared and summed, and each squared effect has been multiplied by the number of subjects upon which it is based. The term based on the leftmost column has been labeled SS_{cells} to reflect that it is the sum of the squared deviations of cell means about the grand mean. As the equation at the bottom of the table indicates, this variability of the cell means is partitioned into several components, each of which forms the basis for a test of a different null hypothesis. Make sure that you understand how we have calculated the numerical value of each of these SS terms and verify that the SS_{cells} is the sum of the SS values for text, rate, and text × rate.

Table 5.3 presents the analysis of variance (ANOVA) of the data of Table 5.1. Generic formulas for df and SS have been presented, assuming a texts and b rates; numerical values are then presented next to these formulas. The sources of variance (SV) reflect, first, the partitioning of the total variability into a between-cell (SS_{cells}) and a within-cell ($SS_{S/TR}$) component and, second, the further partitioning of the between-cells component into main and interaction effects as suggested by Table 5.2.

TABLE 5.2   DEVIATIONS OF CELL MEANS FOR DATA OF TABLE 5.1

Text	Rate	$(\bar{Y}_{.jk} - \bar{Y}_{...}) = (\bar{Y}_{.j.} - \bar{Y}_{...}) + (\bar{Y}_{..k} - \bar{Y}_{...}) + (\bar{Y}_{.jk} - \bar{Y}_{.j.} - \bar{Y}_{..k} + \bar{Y}_{...})$			
	300	13.000	3.25	7.0625	2.6875
Intact	450	6.625	3.25	1.5625	1.8125
	600	−9.875	3.25	−8.6250	−4.5000
	300	1.125	−3.25	7.0625	−2.6875
Scrambled	450	−3.500	−3.25	1.5625	−1.8125
	600	−7.375	−3.25	−8.6250	4.5000

$$8 \sum_j \sum_k (\bar{Y}_{.jk} - \bar{Y}_{...})^2 = (3)(8)\sum_j (\bar{Y}_{.j.} - \bar{Y}_{...})^2 + (2)(8)\sum_k (\bar{Y}_{..k} - \bar{Y}_{...})^2 + 8 \sum_j \sum_k (\bar{Y}_{.jk} - \bar{Y}_{.j.} - \bar{Y}_{..k} + \bar{Y}_{...})^2$$

$$3026.5 = \quad\quad 507 \quad\quad + \quad\quad 2027.025 \quad\quad + \quad\quad 492.475$$

$$\text{SS}_{\text{cells}} = \quad\quad \text{SS}_{\text{text}} \quad\quad + \quad\quad \text{SS}_{\text{rate}} \quad\quad + \quad\quad \text{SS}_{\text{text} \times \text{rate}}$$

TABLE 5.3   THE ANALYSIS OF VARIANCE OF THE DATA OF TABLE 5.1

SV	df	SS	MS	F
Total	$abn - 1 = 47$	$\sum_i^n \sum_j^a \sum_k^b (Y_{ijk} - \bar{Y}_{...})^2 = 5623$		
Between cells	$ab - 1 = 5$	$n\sum_j^a \sum_k^b (\bar{Y}_{.jk} - \bar{Y}_{...})^2 = 3026.500$		
Text (T)	$a - 1 = 1$	$nb\sum_j^a (\bar{Y}_{.j.} - \bar{Y}_{...})^2 = 507.000$	$\dfrac{\text{SS}_T}{\text{df}_T} = 507.000$	$\dfrac{\text{MS}_T}{\text{MS}_{S/TR}} = 8.201^a$
Rate (R)	$b - 1 = 2$	$na\sum_k^b (\bar{Y}_{..k} - \bar{Y}_{...})^2 = 2027.025$	$\dfrac{\text{SS}_R}{\text{df}_R} = 1013.687$	$\dfrac{\text{MS}_R}{\text{MS}_{S/TR}} = 16.397^a$
$T \times R$	$(a-1)(b-1) = 2$	$\text{SS}_{\text{cells}} - \text{SS}_T - \text{SS}_R = 492.475$	$\dfrac{\text{SS}_{T \times R}}{\text{df}_{T \times R}} = 246.063$	$\dfrac{\text{MS}_{T \times R}}{\text{MS}_{S/TR}} = 3.980^b$
S/TR	$ab(n-1) = 42$	$\text{SS}_{\text{tot}} - \text{SS}_{\text{cells}} = 2596.500$	$\dfrac{\text{SS}_{S/TR}}{\text{df}_{S/TR}} = 61.821$	

[a] $p < .01$

[b] $p < .05$

Note: For the example in Table 5.1, $n = 8$, $a = 2$, and $b = 3$.

These SV then dictate the values of the df. The variability among cell means is based on 5 df because deviations are calculated for six cell means about the grand mean of the 48 scores. The between-cell variability is not of interest in itself because it has several possible sources. The six cell means may differ because they represent different texts, different rates of presentation, or different combinations of text and rate. There is 1 df for the text SV because the mean of the 24 scores from the intact condition is compared with the mean of the 24 scores from the scrambled condition. Similarly, there are 2 df for the rate SV because the variability of 3 means is being calculated. Calculating the SS_{TR} involves taking deviations of cell means about the grand mean and removing that portion of the variability due to text and rate; therefore, the 5 df for the cells SV is reduced by 1 (text) and 2 (rate), leaving 2 df for text $\times$ rate. In general, if there are a levels of a variable A and b levels of B, the interaction df are

$$(ab - 1) - (a - 1) - (b - 1) = (a - 1)(b - 1)$$

reflecting the adjustment of cell variability for the variability due to A and B. In practice, we can generate the df for any interaction just by multiplying the df for the interacting main variables.

The $df_{S/AB}$ may be thought of as the difference between df_{tot} and df_{cells}; in our example, this is $47 - 5$, or 42. More generally, if there are ab cells with n scores in each,

$$df_{S/AB} = (abn - 1) - (ab - 1) = ab(n - 1)$$

We can also view these df as the result of pooling the df for variability within each cell $[ab \times (n - 1)]$.

The formulas for the SS for text, rate, and text $\times$ rate are essentially instructions to operate on the effects in Table 5.2. These formulas were presented in Table 5.2 for the text memory experiment, and the numerical values in Table 5.3 were calculated using them. Note that, just as the three right-hand deviation scores in Table 5.2 summed to $\bar{Y}_{.jk} - \bar{Y}_{...}$, SS_T, SS_R, and SS_{TR} sum to SS_{cells}. Computing formulas for each of these sums of squares are presented in Table 5.4 and applied to the data of Table 5.1, letting $a = 2$ (text), $b = 3$ (rate), and $n = 8$. The raw-score formulas are based on the relation of SS to df as was done in Section 4.3.

Returning to Table 5.3, we find that as in the one-factor design, the MS are ratios of SS to df. All three F ratios are formed by using the $MS_{S/TR}$ in the denominator. For example, the F ratio for the text SV is $587/61.821 = 8.201$. The reason for this choice of denominator follows from the discussion of expected mean squares (EMS) in Chapter 4. Under the assumptions presented here, both MS_T and $MS_{S/TR}$ estimate σ_e^2 if the means of the scrambled and intact text populations do not differ. Therefore, forming a ratio of these two MS follows the rule (see Chapter 4) that the numerator and denominator MS of an F ratio must have the same expectation when the null hypothesis represented by the numerator is true. The same rationale also justifies testing MS_R and MS_{TR} against $MS_{S/TR}$.

As the experimenters hypothesized, mean recall for intact texts is significantly better than for scrambled texts. To understand what this means, consider six populations differing with respect to type of text and rate of presentation. The F test of text

TABLE 5.4   DEFINING AND COMPUTING FORMULAS FOR SUMS OF SQUARES
OF TABLE 5.3

$$C = \frac{T_{...}^2}{nab}$$

$$= \frac{(2556)^2}{48} = 136,107$$

$$SS_{tot} = \sum_i \sum_j \sum_k (Y_{ijk} - \bar{Y}_{...})^2 = \sum_i \sum_j \sum_k Y_{ijk}^2 - C$$

$$= 141,730 - 136,107 = 5623$$

$$SS_{cells} = n \sum_j \sum_k (\bar{Y}_{.jk} - \bar{Y}_{...})^2 = \frac{1}{n} \sum_j \sum_k T_{.jk}^2 - C$$

$$= \frac{530^2 + 479^2 + \cdots + 367^2}{8} - 136,107 = 3026.5$$

$$SS_T = nb \sum_j (\bar{Y}_{.j.} - \bar{Y}_{...})^2 = \frac{1}{nb} \sum_j T_{.j.}^2 - C$$

$$= \frac{(530 + 479 + 347)^2 + (435 + 398 + 367)^2}{24} - 136,107 = 507$$

$$SS_R = na \sum_k (\bar{Y}_{..k} - \bar{Y}_{...})^2 = \frac{1}{na} \sum_k T_{..k}^2 - C$$

$$= \frac{(530 + 435)^2 + (479 + 398)^2 + (347 + 367)^2}{16} - 136,107$$

$$= 2027.025$$

$$SS_{TR} = n \sum_j \sum_k (\bar{Y}_{.jk} - \bar{Y}_{.j.} - \bar{Y}_{..k} + \bar{Y}_{...})^2 = SS_{cells} - SS_T - SS_R$$

$$= 3026.5 - 507 - 2027.025 = 492.475$$

$$SS_{S/TR} = \sum_i \sum_j \sum_k (Y_{ijk} - \bar{Y}_{.jk})^2 = SS_{tot} - SS_{cells}$$

$$= 5623 - 3026.5 = 2596.5$$

effects addresses the null hypothesis that the average of the three populations having the intact text does not differ from the average for the three populations having the scrambled text. In terms of Table 5.1, it is the *marginal* means, $\bar{Y}_{intact}$ and $\bar{Y}_{scrambled}$ (56.5 and 51.0), that differ significantly. At the risk of belaboring the point, this F test tells us nothing about differences among the means of populations represented by individual cells.

The experimenters also predicted that performance would deteriorate significantly as rate of presentation increases, and this prediction was also verified by the F test. In this case, the interpretation is that the population means based on each of the three rates differ significantly; that is, the null hypothesis

$$H_0: \mu_{300} = \mu_{450} = \mu_{600}$$

is rejected. Note that the three means being compared are each an average over intact and scrambled populations. Furthermore, a significant result only tells us that some difference exists within the set of three population means, but does not specify which difference(s) is (are) responsible for the result.

Turning now to the interaction of text and rate, we find that the F in Table 5.3 is significant at the .05 level; the advantage of intact over scrambled text decreases significantly as rate increases. Another way of saying this is that performance for intact texts declines more rapidly than for scrambled text as presentation rate increases. Let us more closely consider the nature of interaction.

5.2.3 More About Interaction

Look again at Table 5.3. Consistent with the partitioning of the deviation of the cell mean from the grand mean (Table 5.2), SS_{TR} can be obtained by subtracting SS_T and SS_R from SS_{cells}. In other words, the interaction SS represents the variability among cell means that still remains when variability due to the main effects of the two factors has been removed. Although this definition of interaction is correct, it is not very satisfying. Ordinarily, we would like to make a statement about an interaction in terms of the tabulation or plot of the original cell means. One way to develop this sort of interpretation is to consider what the data would look like if there was no interaction. When there is no interaction, the effects of the factors in the design are *additive;* each cell mean is obtained by adding the main effects to the grand mean. If the interaction components in Table 5.2 were zero, the cell means would be the sum of the grand mean and the text and rate components. For example, the mean for the intact, 300-wpm cell would be $53.25 + 3.25 + 7.0625 = 63.5625$. Ignoring the interaction column in Table 5.2, the six cell means become

		Rate		
		300	450	600
Text	Intact	63.5625	58.0625	47.8750
	Scrambled	57.0625	51.5625	41.3750
	Difference	6.5	6.5	6.5

If we were to plot two functions, one for each text condition, they would be parallel, with the two points at each rate separated by 6.5. Of course, no sets of means obtained from real data will ever be exactly parallel; however, *interaction is a significant departure from parallelism*. With respect to our example of the memory experiment, the significant F test of interaction indicates that the convergence of the intact- and scrambled-text curves in Figure 5.1 is due to more than just chance variability.

Look again at the table, in which the interaction component makes no contribution. Note that the difference between the 300- and 450-rate means is 5.50 for both text conditions. Also the difference between the 450- and 600-rate means is 10.1875 in both rows. The point is that, when there is no interaction, the difference between any two row means will be the same in all columns or, equivalently, the difference between any two column means will be the same in all rows.

In the original data of Table 5.1, the pattern of differences among cell means is very different. We have

		Rate		
		300	450	600
Text	Intact	66.250	59.875	43.375
	Scrambled	54.375	49.750	45.875
	Difference	11.875	10.125	−2.500

With the original data, the difference between the two rows varies markedly over columns. Verify that column differences also vary over rows. The extent to which such differences vary provides a sense of the interaction. Of course, we still need a statistical test to ensure that the variation in such differences is significant.

5.2.4 Simple Effects

An adequate interpretation of the data requires joint consideration of the effects of the treatment variables and their interactions. In the example developed in this chapter, the test of the text main effects revealed that, *when averaged over the three rates,* recall is significantly better for integrated than for scrambled texts. If the interaction had not been significant, we would conclude that, in the population, the advantage of the intact over the scrambled text is the same for each of the rates as it is for the marginal means. But the interaction was significant, and the status of differences between text conditions at each rate is less clear. For example, nothing in the ANOVA of Table 5.3 tells us whether the advantage of the scrambled text at the 600 rate is significant or just due to chance. The comparison of row means within a particular column (or column means within a particular row) involves the concept of simple effects. To illustrate such simple effects in the recall study, consider the following table:

		Rate			
		300	450	600	$\bar{Y}_{.j.} - \bar{Y}_{...}$
Text	Intact	5.9375	5.0625	−1.2500	3.25
	Scrambled	−5.9375	−5.0625	1.2500	−3.25

The column at the far right contains the main effects of the intact and scrambled texts. These were calculated by subtracting the intact- and scrambled-text means (56.5 and 50.0) from the grand mean (53.25). The significant F test of the text SV indicates that these values differ significantly from zero; in other words, the marginal text means differ significantly from the grand mean and, therefore, from each other. The entries in the other three columns are the simple effects. These six values were obtained by subtracting each cell mean from the column (rate) mean; thus, for the means in Table 5.1, the simple effect of the intact text at the 450 rate is $59.875 - 54.8125$. Note that the main effect in each row is the average of the three simple effects in that row. Therefore, if all interaction effects were zero, all simple effects of a variable would equal the main effect of that variable.

We could also look at the simple effects of rate for a particular text. For example, although it appears that recall scores are decreasing with increased rate for the scrambled text, the differences between the rates are not large; they could be due to chance. To see if this is the case, we begin by subtracting from each cell mean its row mean (i.e., we find $\bar{Y}_{.jk} - \bar{Y}_{.j.}$). The resulting simple effects of rate within each text condition are tabled below:

		Rate		
		300	450	600
Text	Intact	9.750	3.345	-13.125
	Scrambled	4.375	$-.250$	-4.125
	$\bar{Y}_{..k} - \bar{Y}_{...}$	7.0625	1.5625	-8.6250

For example, the simple effect of the 300 rate at the intact-text condition is $66.25 - 56.5$. Again note that the main effects of a variable are averages of that variable's simple effects; the main effects in the very last row are an average of the two values above them. The above table gives us another way of looking at the data. As in Figure 5.1, it is evident that the rate effects are much smaller for the scrambled text than for the intact text. The effects in the table make it somewhat more evident that the 450-rate condition yields better recall than the average for the intact text (the sign of the simple effect is positive) but worse recall than the average for the scrambled text (here, the sign is negative).

We have carefully defined simple effects and attempted to clarify their relation to main and interaction effects because these simple effects should become the focus of the analysis if the interaction is significant. The question that remains is how to test for significance of the simple effects. Suppose we want to determine whether rate does have significant effects in the scrambled text condition: do the three rate means differ significantly when the text is scrambled? The numerator for this F test is the sum of squares for rate within the scrambled text condition. It is obtained by viewing the design as involving one factor, rate, with three levels and eight subjects at each level; the data for the intact text are ignored. Now the computing formulas of Chapter 4 can be applied. Sum the squared simple effects of rate in the scrambled-text row of the preceding table and multiply by n (8):

$$SS_{Rate/Scrambled} = n\sum_{k}^{3}(\bar{Y}_{.2k} - \bar{Y}_{.2.})^2$$
$$= 8[(4.35)^2 + (-.25)^2 + (-4.125)^2]$$
$$= 288$$

There are two possible denominators against which to test terms like this. The MS_{error} might be based only on the variances for the three cells involved in the comparison. In the example above, this error term would be $(1/3)(33.977 + 542.215 + 55.112)$; it is distributed on $3(n - 1)$, or 21, df. Alternatively, the error mean square for the original ANOVA based on six groups might be used. This has the advantage of having 42 df, and consequently yields a more powerful F test. If the six variances are reasonably homogeneous, there is no reason not to pool them and gain error df. On the other hand, if the three cells not involved in the comparison have variances clearly different from the three scrambled-text cells, the error term should be based only on the scrambled-text cells. To decide the proper error term for our example, we set α at .25, then ran BMDP7D, treating the design as one in which there is a single variable with six levels. Levene's F ratio (see Section 4.5.6) is less than 1, indicating that the six population variances are very similar, if not identical. We therefore tested $MS_{Rate/Scrambled}$ against $MS_{S/TR}$. The resulting F ratio was $(288/2)(61.821) = 2.35$, which is not significant at the .05 level with 2 and 42 df. Despite the trend in Figure 5.1, we do not have sufficient evidence to conclude that presentation rate influences recall when the text is scrambled.

Before deciding to use the overall error term based on all six cells, we required strong evidence of homogeneity of variance. This seems to contradict our more liberal attitude in Chapter 4. There is an important distinction, however. Assume we wish to compare K means. If our error term is based only on the corresponding K variances, and if the K sample sizes are equal, we can tolerate considerable heterogeneity of variance. This is because the expectations of numerator and denominator mean squares will be equal under the null hypothesis as long as the n's are constant. In contrast, suppose we wish to compare K means, and our error term is a pool of not only the corresponding K variances but other variances as well. If these other variances are very different from the K variances, considerable bias can be introduced into the F test.

When the interaction is significant, tests of simple effects can shed light on the reason for that interaction. Even in the absence of a significant interaction, a case can be made that differences predicted prior to data collection should be examined. Bear in mind, however, that testing several simple effects, or testing the largest effects observed, increases the risk of obtaining significant results by chance. This problem will be discussed more fully in Chapter 6, and procedures for guarding against such inflation of Type 1 error rate will be introduced there.

5.3 TWO-FACTOR DESIGNS: THE GENERAL CASE

5.3.1 Layout of the Design

The design in our example is often referred to as 2 × 3 ("two by three"). In general, there are two independent variables, A and B, with a levels of A, b levels of B, and n

scores in each of the ab cells. The first subscript for each score, i in the general case, indexes the score within each cell and varies from 1 to n. The second subscript, j, indexes the level of A and varies from 1 to a. The third subscript, k, indexes the level of B and varies from 1 to b.

The ab groups of n subjects are considered to be random samples, one from each of ab populations. These populations could be viewed as having been drawn from one infinitely large parent population prior to the imposition of the ab experimental conditions. Thus, if we conclude that the ab population means differ, we can attribute this to the factors in our study rather than to any variables other than those factors. If neither main nor interaction effects are significant, we cannot reject the hypothesis that the ab treatment populations are identical.

5.3.2 A Structural Model

The analysis of the text memory data in Section 5.2 implies a theory about the relation of scores to population parameters. Such a theory is sometimes referred to as a structural model because it specifies the structure (components) of a score in terms of population effects. The structural model is a statement about what factors, or combinations of factors, may account for the variability in the data set. In order to understand that model, we need to define certain population parameters. Table 5.5 defines the population means, effects based upon these means, and the quantities that estimate these parameters. The parameters provide a basis for thinking about why the scores in our data set may vary.

The EMS for the SV encountered in Table 5.3 also are presented in Table 5.5, and show how the population parameters may contribute to the variability in the data. Note that σ_e^2 contributes to each EMS. The added term is a coefficient times a θ^2 term reflecting the SV itself. The coefficient is the number of scores upon which each of the relevant means are based. For example, the A SV reflects the variability among a means that are each based on bn scores, whereas the AB SV involves the variability of ab means that are each based on n scores. $MS_{S/AB}$ is the appropriate error term for all F tests because, if H_0 is true, the numerator and denominator MS will have the same expectation, σ_e^2.

Scores obtained under the same combination of A and B may differ simply because they were obtained from different individuals or because of other chance factors such as variation in the time of day. These individual differences and errors of measurement contribute to the error component defined in Table 5.5. Deciding whether the variability among the means is due to anything other than chance variability is what the analysis is about. These possible other contributors to variability are the three effects listed in Table 5.5: α, β, and $\alpha\beta$. For example, consider the factor A; if it doesn't matter what level of A is administered to subjects, the $\mu_{j.}$ would be identical and, therefore, the α_j would all equal zero. The F test of the A source of variance will evaluate the variance of the $\overline{Y}_{.j.}$; the issue is whether the variance of those means is about what one would expect on the basis of chance variability alone, or whether the variance of the $\overline{Y}_{.j.}$ is so large as to suggest that the population means, the $\mu_{j.}$, vary. An important point to keep in mind is that this test won't tell us about whether the factor A has an effect at any particular level of B. Each mean being compared is based on all bn scores at its level of A.

TABLE 5.5 POPULATION PARAMETERS AND ESTIMATES, AND EXPECTED MEAN SQUARES (EMS) FOR A TWO-FACTOR DESIGN

The model

$$Y_{ijk} = \mu + \alpha_j + \beta_k + (\alpha\beta)_{jk} + \epsilon_{ijk}$$

where $Y_{ijk} = i$th score at jth level of A and kth level of B

Population means	Estimate
$\mu_{jk} = $ mean of population of scores at A_j and B_k	$\bar{Y}_{jk} = \dfrac{T_{.jk}}{n}$
$\mu_{j.} = \dfrac{1}{b}\sum_k \mu_{jk}$	$\bar{Y}_{.j.} = \dfrac{T_{.j.}}{bn}$
$\mu_{.k} = \dfrac{1}{a}\sum_j \mu_{jk}$	$\bar{Y}_{..k} = \dfrac{T_{..k}}{an}$
$\mu_{..} = \dfrac{1}{ab}\sum_j \sum_k \mu_{jk}$	$\bar{Y}_{...} = \dfrac{T_{...}}{abn}$

Population effects

$\alpha_j = \mu_{j.} - \mu$, *main effect* of treatment A_j
$\beta_k = \mu_{.k} - \mu$, *main effect* of treatment B_k
$(\alpha\beta)_{jk} = (\mu_{jk} - \mu) - \alpha_j - \beta_k$
$\quad = \mu_{jk} - \mu_{j.} - \mu_{.k} + \mu$, *interaction effect* of A_j and B_k
$\epsilon_{ijk} = Y_{ijk} - [\mu + \alpha_j + \beta_k + (\alpha\beta)_{jk}]$
$\quad = Y_{ijk} - \mu_{jk}$, *error component*

Expected mean squares

SV	EMS	Definitions
A	$\sigma_e^2 + nb\theta_A^2$	$\theta_A^2 = \dfrac{\sum_j (\mu_{j.} - \mu_{..})^2}{a-1} = \dfrac{\sum_j \alpha_j^2}{a-1}$
B	$\sigma_e^2 + na\theta_B^2$	$\theta_B^2 = \dfrac{\sum_k (\mu_{.k} - \mu_{..})^2}{b-1} = \dfrac{\sum_k \beta_k^2}{b-1}$
AB	$\sigma_e^2 + n\theta_{AB}^2$	$\theta_{AB}^2 = \dfrac{\sum_j \sum_k (\mu_{jk} - \mu_{j.} - \mu_{.k} + \mu_{..})^2}{(a-1)(b-1)} = \dfrac{\sum_j \sum_k (\alpha\beta)_{jk}^2}{(a-1)(b-1)}$
S/AB	σ_e^2	

In summary, variability in the data has four possible sources:

1. *The error component ϵ_{ijk}.* We assume that the errors are independently and normally distributed with mean zero and variance σ_e^2 within each treatment population defined by a combination of levels of A and B.

2. *The main effect of treatment A_j, or α_j.* The factor A is assumed to have fixed effects; that is, the a levels have been arbitrarily selected and are viewed as representing the population of levels. Then $\Sigma_j \alpha_j = 0$. We shall test the null hypothesis that

$$H_0: \mu_{1.} = \mu_{2.} = \cdots = \mu_{j.} = \cdots = \mu_{a.}$$

If all the means are identical, the α_j must all be zero. Therefore, an equivalent statement of the null hypothesis is

$$H_0: \alpha_1 = \alpha_2 = \cdots = \alpha_j = \cdots = \alpha_a = 0$$

3. *The main effect of treatment B_k, or β_k.* This too is a fixed-effect variable and so $\Sigma_k \beta_k = 0$. We shall test

$$H_0: \mu_{.1} = \mu_{.2} = \cdots = \mu_{.k} = \cdots = \mu_{.b}$$

or, equivalently,

$$H_0: \beta_1 = \beta_2 = \cdots = \beta_k = \cdots = \beta_b = 0$$

4. *The interaction effect of A_j and B_k, or $(\alpha\beta)_{jk}$.* Because both A and B have fixed effects, $\Sigma_k (\alpha\beta)_{jk} = \Sigma_j (\alpha\beta)_{jk} = 0$. The relevant null hypothesis is

$$H_0: (\alpha\beta)_{11} = (\alpha\beta)_{12} = \cdots = (\alpha\beta)_{jk} = \cdots = (\alpha\beta)_{ab} = 0$$

The relations among these population parameters and the data may be summarized by the following *structural model:*

$$Y_{ijk} = \mu + \alpha_j + \beta_k + (\alpha\beta)_{jk} + \epsilon_{ijk} \tag{5.1}$$

Given this model and following the method of deriving EMS in Chapter 4, the EMS formulas in Table 5.5 can be obtained.

5.4 THREE-FACTOR DESIGNS: THE ANALYSIS OF VARIANCE

In the previous section, subjects read either intact or scrambled text, at presentation rates of either 300, 450, or 600 words per minute; recall was then tested. Consider a variation of this experiment in which the researcher included a third factor; on the basis of a prior test, subjects were classified as either good or poor readers. Table 5.6 presents the mean and variance of each cell of this three-factor experiment. Figure 5.2 presents a plot of the 12 cell means. In both reading groups, recall is worse at high presentation rates than at low, and worse for scrambled than for intact texts. However, the means for the poor readers are lower than those for the good readers. The plot of

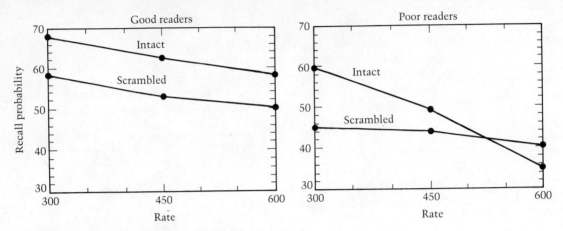

Figure 5.2 A plot of the cell means of Table 5.6.

the means suggests one other difference between good and poor readers: whereas the two lines (for the two texts) are roughly parallel for the good readers, they converge, eventually crossing, as rate increases for the poor readers. Another way of saying this is that for poor readers, but not for good readers, the advantage of intact over scrambled text decreases as rate increases.

The issue is whether these trends in the data represent real population effects or just chance variability. To determine this, we again perform an analysis of variance. As we did with the two-factor data, we begin by partitioning the total variability among the 12 cell means into several components of variability. Some notation will be helpful in presenting this partitioning. In general, we have three independent variables: *A*, *B*, and *C* and the relevant indices are

$$i = 1, 2, \ldots, n \qquad j = 1, 2, \ldots, a$$
$$k = 1, 2, \ldots, b \qquad m = 1, 2, \ldots, c$$

In our example, $n = 6$ (subjects in each cell), $a = 2$ (texts), $b = 3$ (rates), and $c = 2$ (good and poor readers). The deviation of any score from the grand mean can be viewed as

$$Y_{ijkm} - \bar{Y}_{\ldots} = (Y_{ijkm} - \bar{Y}_{.jkm}) + (\bar{Y}_{.jkm} - \bar{Y}_{\ldots})$$

We can further partition the deviation of the cell mean from the grand mean into several components:

$$\bar{Y}_{.jkm} - \bar{Y}_{\ldots} = (\bar{Y}_{.j..} - \bar{Y}_{\ldots}) + (\bar{Y}_{..k.} - \bar{Y}_{\ldots}) + (\bar{Y}_{...m} - \bar{Y}_{\ldots}) \qquad (5.2)$$
$$+ (\bar{Y}_{.jk.} - \bar{Y}_{.j..} - \bar{Y}_{..k.} + \bar{Y}_{\ldots}) + (\bar{Y}_{.j.m} - \bar{Y}_{.j..} - \bar{Y}_{...m} + \bar{Y}_{\ldots}) +$$
$$(\bar{Y}_{..km} - \bar{Y}_{..k.} - \bar{Y}_{...m} + \bar{Y}_{\ldots}) + (\bar{Y}_{.jkm} + \bar{Y}_{.j..} + \bar{Y}_{..k.} + \bar{Y}_{...m} - \bar{Y}_{.jk.} - \bar{Y}_{.j.m} - \bar{Y}_{..km} - \bar{Y}_{\ldots})$$

The first three terms on the right side of the equality represent main effects of *A*, *B*, and *C*, and the next three represent first-order interaction effects (*AB*, *AC*, and *BC*). These types of effects were all encountered in our earlier discussion of two-factor

TABLE 5.6   CELL MEANS (AND VARIANCES) FOR A THREE-FACTOR EXPERIMENT, $n = 6$

Readers	Text	Rate			Mean
		300	450	600	
	Intact	67.833 (19.767)	62.667 (33.466)	58.333 (18.622)	62.944
Good					
	Scrambled	58.500 (53.905)	53.000 (30.803)	50.167 (50.566)	53.889
	Mean	63.167	57.833	54.250	58.417
	Intact	60.167 (22.963)	49.500 (50.297)	35.000 (21.197)	48.222
Poor					
	Scrambled	45.333 (44.524)	44.000 (39.200)	40.667 (26.266)	43.333
	Mean	52.750	46.750	37.833	45.778

Text	Rate			Mean
	300	450	600	
Intact	64.000	56.083	46.667	55.583
Scrambled	51.917	48.500	45.417	48.611
Mean	57.958	52.292	46.042	52.097

designs. The last term is new; it represents the deviation of the cell mean from the grand mean adjusted for main and first-order interaction effects. That may be clearer if we rewrite this term in the equivalent form:

$$(\bar{Y}_{.jkm} - \bar{Y}_{....}) - (\bar{Y}_{.j..} - \bar{Y}_{....}) - (\bar{Y}_{..k.} - \bar{Y}_{....}) - (\bar{Y}_{...m} - \bar{Y}_{....})$$
$$- (\bar{Y}_{.jk.} - \bar{Y}_{.j..} - \bar{Y}_{..k.} + \bar{Y}_{....}) - (\bar{Y}_{.j.m} - \bar{Y}_{.j..} - \bar{Y}_{...m} + \bar{Y}_{....})$$
$$- (\bar{Y}_{..km} - \bar{Y}_{..k.} - \bar{Y}_{...m} + \bar{Y}_{....})$$

We refer to this term as the second-order (or three-way) interaction effect of A, B, and C.

The above partitioning of the deviation of a score from the grand mean forms the basis for the partitioning of sum of squares and degrees of freedom in Table 5.7. The only new SV is ABC; its df follow from the preceding definition:

$$\text{df}_{ABC} = (abc - 1) - (a - 1)(b - 1) - (a - 1)(c - 1) - (b - 1)(c - 1)$$
$$- (a - 1) - (b - 1) - (c - 1)$$
$$= (a - 1)(b - 1)(c - 1)$$

TABLE 5.7 PARTITIONING THE TOTAL DEGREES OF FREEDOM AND SUMS OF SQUARES IN A THREE-FACTOR DESIGN

SV	df	SS
Total	$abcn - 1$	$\sum_i \sum_j \sum_k \sum_m (Y_{ijkm} - \bar{Y}_{....})^2$
Between cells	$abc - 1$	$n \sum_j \sum_k \sum_m (\bar{Y}_{.jkm} - \bar{Y}_{....})^2$
A	$a - 1$	$nbc \sum_j (\bar{Y}_{.j..} - \bar{Y}_{....})^2$
B	$b - 1$	$nac \sum_k (\bar{Y}_{..k.} - \bar{Y}_{....})^2$
C	$c - 1$	$nab \sum_m (\bar{Y}_{...m} - \bar{Y}_{....})^2$
AB	$(a - 1)(b - 1)$	$nc \sum_j \sum_k (\bar{Y}_{.jk.} - \bar{Y}_{.j..} - \bar{Y}_{..k.} + \bar{Y}_{....})^2$
AC	$(a - 1)(c - 1)$	$nb \sum_j \sum_m (\bar{Y}_{.j.m} - \bar{Y}_{.j..} - \bar{Y}_{...m} + \bar{Y}_{....})^2$
BC	$(b - 1)(c - 1)$	$na \sum_k \sum_m (\bar{Y}_{..km} - \bar{Y}_{..k.} - \bar{Y}_{...m} + \bar{Y}_{....})^2$
ABC	$(a - 1)(b - 1)(c - 1)$	$n \sum_j \sum_k \sum_m (\bar{Y}_{.jkm} + \bar{Y}_{.j..} + \bar{Y}_{..k.} + \bar{Y}_{...m} - \bar{Y}_{.jk.} - \bar{Y}_{.j.m} - \bar{Y}_{..km} - \bar{Y}_{....})^2$
S/ABC (within cells)	$abc(n - 1)$	$(n - 1) \sum \sum \sum \hat{\sigma}^2_{jkm} = SS_{tot} - SS_{B.\,cells}$

Note: $\hat{\sigma}^2_{jkm}$ is the variance of the n scores in the cell defined by A_j, B_k, and C_m.

Table 5.8 presents the numerical results of the ANOVA, as well as the computing formulas for the SS. As always, the MS are ratios of SS to df. All of the F ratios have been constructed by dividing the MS for the effect of interest by $MS_{S/ABC}$. Note that this error term is an average of the 12 cell variances in Table 5.6. The justification for the F ratios lies in the EMS column of Table 5.8; proper F ratios are formed if numerator and denominator MS have the same expectation when H_0 is true. In the designs of this and the preceding chapter, the numerator MS estimates σ_e^2 under the null hypothesis. The structural model underlying these estimates is presented in Table 5.9, together with definitions of parameters and the variance components contained in the EMS expressions.

With the exception of the AC and BC interactions, all sources in Table 5.8 are significant at least at the .05 level. Our focus now must be on understanding these effects. The main effects are straightforward; good readers remember better than poor readers, intact text is better recalled than scrambled text, and recall decreases with increases in presentation rate. The interactions merit somewhat more discussion.

TABLE 5.8 ANOVA OF THE DATA OF TABLE 5.6 AND COMPUTING FORMULAS FOR SUMS OF SQUARES

(a) The ANOVA

SV	df	SS	MS	F^a	EMS
Total	71	8357.374			
Between cells	11	6299.486			
Text (A)	1	875.014	875.014	25.512^b	$\sigma_e^2 + nbc\theta_A^2$
Rate (B)	2	1705.444	683.556	24.862^b	$\sigma_e^2 + nac\theta_B^2$
Reader (C)	1	2875.347	2.875.347	83.834^b	$\sigma_e^2 + nab\theta_C^2$
AB	2	355.444	177.772	5.182^b	$\sigma_e^2 + nc\theta_{AB}^2$
AC	1	78.125	78.125	2.278	$\sigma_e^2 + nb\theta_{AC}^2$
BC	2	129.778	64.889	1.892	$\sigma_e^2 + na\theta_{BC}^2$
ABC	2	280.333	140.167	4.087^c	$\sigma_e^2 + n\theta_{ABC}^2$
S/ABC	60	2057.888	34.298		σ_e^2

^{a} The error term for all F tests is $\mathrm{MS}_{S/ABC}$.
$^b p < .01.$
$^c p < .05.$

(b) Computations for sums of squares

$$C = \frac{T_{....}^2}{abcn} = 195{,}416.681$$

$$\mathrm{SS}_{tot} = \Sigma\Sigma\Sigma\Sigma Y_{ijkm}^2 - C = 203{,}954.055 - C = 8357.374$$

$$\mathrm{SS}_{B\,cells} = \frac{\Sigma\Sigma\Sigma T_{.jkm}^2}{n} - C = 213{,}023.986 - C = 6299.486$$

$$\mathrm{SS}_A = \frac{\Sigma T_{.j..}^2}{nbc} - C = 196{,}291.694 - C = 875.014$$

$$\mathrm{SS}_B = \frac{\Sigma T_{..k.}^2}{nac} - C = 197{,}122.125 - C = 1705.444$$

$$\mathrm{SS}_C = \frac{\Sigma T_{...m}^2}{nab} - C = 198{,}292.023 - C = 2875.347$$

$$\mathrm{SS}_{AB} = \frac{\Sigma\Sigma T_{.jk.}^2}{nc} - C - \mathrm{SS}_A - \mathrm{SS}_B = 198{,}352.583 - C - \mathrm{SS}_A - \mathrm{SS}_B$$
$$= 355.444$$

$$\mathrm{SS}_{AC} = \frac{\Sigma\Sigma T_{.j.m}^2}{nb} - C - \mathrm{SS}_A - \mathrm{SS}_C = 199{,}245.167 - C - \mathrm{SS}_A - \mathrm{SS}_C$$
$$= 78.125$$

$$\mathrm{SS}_{BC} = \frac{\Sigma\Sigma T_{..km}^2}{na} - C - \mathrm{SS}_B - \mathrm{SS}_C = 200{,}127.250 - C - \mathrm{SS}_B - \mathrm{SS}_C$$
$$= 129.778$$

$$\mathrm{SS}_{ABC} = \mathrm{SS}_{B\,cells} - \mathrm{SS}_A - \mathrm{SS}_B - \mathrm{SS}_C - \mathrm{SS}_{AB} - \mathrm{SS}_{AC} - \mathrm{SS}_{BC} = 280.333$$

TABLE 5.9   STRUCTURAL MODEL AND PARAMETER DEFINITIONS FOR THE THREE-FACTOR DESIGN

The model

$$Y_{ijkm} = \mu + \alpha_j + \beta_k + \gamma_m + (\alpha\beta)_{jk} + (\alpha\gamma)_{jm} + (\beta\gamma)_{km} + (\alpha\beta\gamma)_{jkm} + \epsilon_{ijkm}$$

$$\text{where } \alpha_j = \mu_{j..} - \mu_{...}, \; A \text{ main effect}$$

$$\beta_k = \mu_{.k.} - \mu_{...}, \; B \text{ main effect}$$

$$\gamma_m = \mu_{..m} - \mu_{...}, \; C \text{ main effect}$$

$$(\alpha\beta)_{jk} = \mu_{jk.} - \mu_{j..} - \mu_{.k.} + \mu_{...}, \; AB \text{ interaction effect}$$

$$(\alpha\gamma)_{jm} = \mu_{j.m} - \mu_{j..} - \mu_{..m} + \mu_{...}, \; AC \text{ interaction effect}$$

$$(\beta\gamma)_{km} = \mu_{.km} - \mu_{.k.} - \mu_{..m} + \mu_{...}, \; BC \text{ interaction effect}$$

$$(\alpha\beta\gamma)_{jkm} = \mu_{jkm} + \mu_{j..} + \mu_{.k.} + \mu_{..m} - \mu_{jk.} - \mu_{j.m} - \mu_{.km} - \mu_{...},$$

$$ABC \text{ interaction effect}$$

$$\text{and } \epsilon_{ijkm} = Y_{ijkm} - \mu_{jkm}, \text{ the error component}$$

Definitions of θ^2 terms in EMS

$$\theta_A^2 = \frac{\sum\limits_j (\mu_{j..} - \mu_{...})^2}{a - 1} = \frac{\sum\limits_j \alpha_j^2}{a - 1}$$

$$\theta_B^2 = \frac{\sum\limits_k (\mu_{.k.} - \mu_{...})^2}{b - 1} = \frac{\sum\limits_k \beta_k^2}{b - 1}$$

$$\theta_C^2 = \frac{\sum\limits_m (\mu_{..m} - \mu_{...})^2}{c - 1} = \frac{\sum\limits_m \gamma_m^2}{c - 1}$$

$$\theta_{AB}^2 = \frac{\sum\limits_j \sum\limits_k (\mu_{jk.} - \mu_{j..} - \mu_{..k} + \mu_{...})^2}{(a - 1)(b - 1)} = \frac{\sum\limits_j \sum\limits_k (\alpha\beta)_{jk}^2}{(a - 1)(b - 1)}$$

$$\theta_{AC}^2 = \frac{\sum\limits_j \sum\limits_m (\mu_{j.m} - \mu_{j..} - \mu_{..m} + \mu_{...})^2}{(a - 1)(c - 1)} = \frac{\sum\limits_j \sum\limits_m (\alpha\gamma)_{jm}^2}{(a - 1)(c - 1)}$$

$$\theta_{BC}^2 = \frac{\sum\limits_k \sum\limits_m (\mu_{.km} - \mu_{.k.} - \mu_{..m} + \mu_{...})^2}{(b - 1)(c -)} = \frac{\sum\limits_k \sum\limits_m (\beta\gamma)_{km}^2}{(b - 1)(c - 1)}$$

$$\theta_{ABC}^2 = \frac{\sum\limits_j \sum\limits_k \sum\limits_k (\mu_{jkm} + \mu_{j..} + \mu_{.k.} + \mu_{..m} - \mu_{jk.} - \mu_{j.m} - \mu_{.km} - \mu_{...})^2}{(a - 1)(b - 1)(c - 1)}$$

$$= \frac{\sum\limits_j \sum\limits_k \sum\limits_m (\alpha\beta\gamma)_{jkm}^2}{(a - 1)(b - 1)(c - 1)}$$

5.5 INTERACTION EFFECTS IN THE THREE-FACTOR DESIGN

The interpretation of first-order interaction effects is like that made in two-factor designs. We need only average over the values of the third factor. To illustrate, let us look more closely at the significant text × rate (*AB*) interaction. Averaging over the type of reader reduces Table 5.6 to the following two-factor table:

		Rate		
		300	450	600
Text	Intact	64.000	56.083	46.667
	Scrambled	51.917	48.500	45.417

Averaging over both good and poor readers, it is clear that the advantage of intact over scrambled text decreases as rate increases. Another way of looking at the data is in terms of the effects of rate for each text condition: the performance decrement with increased rate is more pronounced for the intact text. You can construct similar tables of means to examine any two-factor interaction.

Why is the second-order (three-factor) interaction significant? The answer may be found in Figure 5.2. The two lines for the good readers are virtually parallel, whereas those for the poor readers converge and cross at the fastest rate. There is a text × rate × reader–type interaction because the simple text × rate interaction effects are quite different for the two groups of readers. It is important to realize that, although we have chosen to speak in terms of differences in the AB simple interactions at different levels of C, a three-factor interaction means that the simple interaction effects of any two variables vary as a function of the level of the third variable.

Researchers often understand this last statement to mean that whenever the plot of the AB combinations looks different at different levels of C, the three-factor interaction is likely to be significant. It is important to understand that this is not correct. Saying that the simple interaction effects of AB are the same at all levels of C is not the same as saying that the pattern of the ab means is the same at all levels of C. The following set of means should help us to understand this point:

	C_1		C_2	
	B_1	B_2	B_1	B_2
A_1	15	7	9	5
A_2	5	13	11	23

Figure 5.3 presents a plot of the eight cell means under consideration. If these were population means, would you think there is a second-order interaction? The pattern of means looks different at C_1 than at C_2, and we find that our students usually believe that an ABC interaction is present. In fact, if we compute SS_{ABC}, we find it is exactly zero, so there cannot be an $A \times B \times C$ interaction. We can come to the same conclusion without actually calculating SS_{ABC}. We can calculate the simple AB effects for each cell ($\mu_{jkm} - \mu_{j.m} - \mu_{.km} + \mu_{..m}$); they are identical at C_1 and C_2. There is no ABC interaction because the difference between the slopes of the A_2 and A_1 lines is the same at C_1 as at C_2.

In describing main effects, we compare marginal means. With two-factor interactions, we examine whether a set of functions are parallel. As the example of Figure 5.3

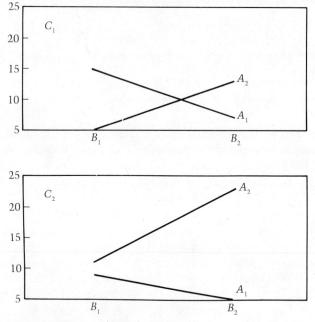

Figure 5.3 Means from a 2^3 experiment.

indicates, three-factor interactions are sometimes more difficult to understand and describe. However, in 2^3 designs there is a measure of interaction that proves useful in thinking about the issue. Begin by defining the simple AB interaction at C_1 as a difference between differences:

$$I_{AB/C_1} = (15 - 7) - (5 - 13) = 16$$

We would get the same result if we calculated the column differences first: $(15 - 5) - (7 - 13) = 16$. In fact, I_{AB} is just the difference between the positive and negative diagonal sums: $(15 + 13) - (7 + 5) = 16$. Taking the diagonal differences at C_2, we obtain

$$I_{AB/C_2} = (9 + 23) - (11 + 5) = 16$$

Whenever the simple interaction of A and B has the same value of the interaction measure at both C_1 and C_2, the SS_{ABC} will be zero. Of course, the same conclusion follows if we compare the AC interactions at the two levels of B, or the BC interactions at the two levels of A. Try it and see for yourself.

One implication of this example is that the pattern of means can be deceptive. Some patterns will provide clear signals of three-factor interaction. If there is no AB interaction at one or more levels of C but a simple interaction at least at one other level of C, then there should be an ABC interaction. That was the situation in the example of the text memory experiment. Also, if the lines in one panel of Figure 5.2 converge, whereas those in the other panel diverge, an ABC interaction is indicated. If the two AB plots are the same (or displaced by a constant amount), except for one point, there is reason to expect a three-factor interaction. Other patterns, such as that in Figure 5.3, are ambiguous, however. Often, it is unclear whether our predictions about the pattern of means require a significant second-order interaction. In such cases, the best we can do is examine the results of the ANOVA. A significant second-order interaction implies that the magnitude of the first-order interactions of any two variables depends on the level of the third variable. In this case it makes sense to calculate the interaction measure, *I*, for the various possible simple two-factor interactions at each level of the third factor and use this to provide a sense of the cause of the significant three-factor interaction. If the second-order interaction is not significant, focus on the significant first-order interactions, if there are any.

5.6 MORE THAN THREE INDEPENDENT VARIABLES

The analysis and interpretation of data for between-subjects designs involving more than three factors are in all respects straightforward generalizations of the material presented for the simpler cases. Each variable and each possible combination of variables are potential contributors to the total variability, and so is the variability among subjects within each combination of all the variables. As might be guessed, the df for any higher-order interaction are a product of the df for the variables entering into the interaction. For example, an *ABCD* interaction would have $(a - 1) \times (b - 1) \times (c - 1) \times (d - 1)$ df. Although computations of SS can be developed by following rules presented for simpler designs, it would seem a masochistic exercise to carry out such calculations in this era of high-speed computing packages. Perhaps more to the point, once data are typed and saved in a computer file, the researcher can carry out many supplementary analyses with relatively little effort. Such analyses would include plots that present information about the distribution of scores, tests of simple effects, and other comparisons that will be described in the next chapter.

5.7 POWER CALCULATIONS IN FACTORIAL DESIGNS

In Chapter 4, we presented a formula for estimating ϕ (Greek phi), an index of power. Together with values of df_1 and df_2, ϕ enables us to use the power charts of Appendix D.6. In this section, we will provide a more general formula for ϕ to be used in any factorial design, and we will illustrate power calculations, using the design of Table 5.6 and the results of the ANOVA in Table 5.8. The general form of the power index, ϕ, is

$$\phi = \frac{\sqrt{(\text{coefficient}) \times (\text{sum of squared effects})/(df_{SV} + 1)}}{\sqrt{\text{error variance}}}$$

The meaning of the above terms will be clearer if we consider a specific design. We will provide formulas for the power to test effects in a three-factor experiment, using the test of *AB* as an example.

The approach depends upon the EMS of Table 5.8 and the definitions of variance components in Table 5.9. For example, to test the *A* effects, the sum of the squared effects is $\Sigma_j \alpha_j^2 = (a - 1)\theta_A^2$. The coefficient is bcn, the number of scores at a level of A, and df_{SV} is $a - 1$. Therefore,

$$\hat{\phi}_A = \frac{\sqrt{bcn\Sigma_j \hat{\alpha}_j^2/a}}{\sqrt{MS_{S/ABC}}}$$

Similarly, we have power indices for interaction effects:

$$\hat{\phi}_{AB} = \frac{\sqrt{cn\Sigma_j \Sigma_k (\widehat{\alpha\beta})_{jk}^2/[(a - 1)(b - 1) + 1]}}{\sqrt{MS_{S/ABC}}}$$

and

$$\hat{\phi}_{ABC} = \frac{\sqrt{n\Sigma_j \Sigma_k \Sigma_m (\widehat{\alpha\beta\gamma})_{jkm}^2/[(a - 1)(b - 1)(c - 1) + 1]}}{\sqrt{MS_{S/ABC}}}$$

Numerical estimates of the quantities in the above expressions can be obtained from the EMS in Table 5.8.

In Table 5.8, the F ratio for the text $\times$ rate (AB) interaction was significant at the .01 level. Suppose, however, that the MS_{AB} had been smaller than the value in Table 5.8; say it had a value of 104.389. In that case,

$$F = \frac{104.389}{34.298} = 3.044$$

and the F would have a value slightly less than that required for significance at the .05 level. If we had reason to believe that the test of AB lacked power, we might be encouraged to run the experiment again. At the very least, a write-up of the results might qualify the lack of significance by noting the lack of power. To calculate power, we first need an estimate of the sum of squared AB interaction effects. Note that in a three-factor design

$$E(MS_{AB}) = \sigma_e^2 + \frac{nc\Sigma_j \Sigma_k (\alpha\beta)_{jk}^2}{(a - 1)(b - 1)}$$

and

$$E(MS_{S/ABC}) = \sigma_e^2$$

Therefore, the desired estimate is

$$nc\Sigma \Sigma (\widehat{\alpha\beta})_{jk}^2 = (a - 1)(b - 1)(MS_{AB} - MS_{S/ABC})$$

and the estimate of $\hat{\phi}_{AB}$ is

$$\hat{\phi}_{AB} = \frac{\sqrt{(a - 1)(b - 1)(MS_{AB} - MS_{S/ABC})/[(a - 1)(b - 1) + 1]}}{\sqrt{MS_{S/ABC}}}$$

The necessary values are $(a - 1)(b - 1) = 2$, $MS_{AB} = 104.389$, and $MS_{S/AB} = 34.298$, and

$$\hat{\phi}_{AB} = \frac{\sqrt{46.727}}{\sqrt{34.298}} = 1.1672$$

Turning to the chart in Appendix D.6 for $df_1 = 2$, and finding the curve for $df_2 = 60$, the power to detect an effect of the size estimated by our data is about .40 (with $\alpha = .05$). It is clear that our F test lacked power; the Type 2 error rate is almost .60. On the other hand, it is also evident that the AB interaction is not very large, relative to chance variability. To see why this is so, assume that we doubled the n, running 144 subjects instead of 72. Then $\hat{\phi}$ would be multiplied by $\sqrt{2}$ and would be 1.65. The df_2 would be 132; use the infinity curve in the power chart. This doubling of sample size does increase power, but only to about .72.

Power calculations for other SV follow the approach just illustrated. Researchers should carry out such calculations more frequently than they do. They give a sense of how much data are required to have a reasonable chance of rejecting false null hypotheses. In many cases, power against certain specified false hypotheses will be low even with large n, suggesting that the investigator might be better advised to focus on reducing error variance. One approach to doing that will be discussed in Section 5.10.

5.8 POOLING IN FACTORIAL DESIGNS

Once again consider the experiment in which recall is studied as a function of type of text (intact or scrambled), presentation rate (300, 450, and 600 wpm), and reading ability (good or poor). Further assume that there are three subjects in each cell of the design and that the results of an ANOVA are as shown in the upper part of Table 5.10. Given the relatively small number of error df and the nonsignificance of terms involving reading ability (C), the researchers may decide not to include C and its interactions among the SV, recognizing that not including these terms will result in the dfs associated with them being added to the error df. This will increase the power of the F tests for the A and B main effects and the $A \times B$ interaction. The result is the ANOVA in the middle part of Table 5.10. This analysis is based on the assumption that population effects due to sources involving C are negligible.

This neglect of certain SV occurs fairly commonly, often with less motivation than we have provided. Sometimes a variable is ignored merely because its effects are not of interest; examples might include position of the correct response in a discrimination task, order of presentation of tests, and set of stimulus materials to be learned. In referring to the middle part of Table 5.10, we speak of *pooling* because the error SS is now a pool of $SS_{S/ABC}$ and all other SS involving C. Note that SS_{tot}, SS_A, SS_B, and $SS_{A \times B}$ are the same in both analyses; $SS_{S/AB}$ in the middle panel is the sum of the remaining terms from the panel above. A potential advantage of pooling stems from the fact that $df_{S/AB}$ is also a pool of df equal to the sum of all df for terms involving C; therefore tests of A, B, and AB may have more power. It is also possible that the new error MS will be smaller than the original one if the SV that are neglected have little variability associated with them. This situation would also result in greater power when effects are tested against the pooled error term.

The analysis ignoring C implies a structural model in which main and interaction effects involving C are zero. If this is the case, then all mean squares for SV involving C have expectations equal to σ_e^2, and it is reasonable to pool them with $MS_{S/ABC}$. However, if there is variance due to C or its interactions, the pooled error term ($MS_{S/AB}$)

TABLE 5.10 THREE ANOVAS ILLUSTRATING POOLING OF MEAN SQUARES

Complete ANOVA

SV	df	SS	MS	F
Text (A)	1	137.556	137.556	4.51[a]
Rate (B)	2	221.430	110.715	3.63[a]
Reader (C)	1	56.120	56.120	1.84
AB	2	204.352	102.176	3.35
AC	1	35.377	35.377	1.16
BC	2	65.266	32.633	1.07
ABC	2	55.502	27.751	.91
S/ABC	24	732.000	30.500	
Total	35	1507.603		

All C terms pooled

SV	df	SS	MS	F
Text (A)	1	137.556	137.556	4.37[a]
Rate (B)	2	221.430	110.715	3.52[a]
AB	2	204.352	102.176	3.25
S/AB	30	944.265	31.476	
Total	35	1507.603		

C interactions pooled

SV	df	SS	MS	F
Text (A)	1	137.556	137.556	4.49[a]
Rate (B)	2	221.430	110.715	3.62[a]
Reader (C)	1	56.120	56.120	1.83
AB	2	204.352	102.176	3.34[a]
Error	29	888.145	30.626	
Total	35	1507.603		

[a] $p < .05$.

may result in a negatively biased F test of A, B, and AB; power could be lost rather than gained by the pooling procedure. To see why this is so, consider any two mean squares, MS_1 and MS_2. To pool them, add their SS and divide by the summed df:

$$MS_{pool} = \frac{SS_1 + SS_2}{df_1 + df_2} = \frac{df_1}{df_1 + df_2}MS_1 + \frac{df_2}{df_1 + df_2}MS_2$$

The pooled MS is a weighted average of the terms being pooled. That means that its expectation is an average of the expectations of the pooled terms. If, for example, $E(MS_C) = \sigma_e^2 + 18\theta_C^2$, then variance due to C will contribute to the pooled error term but not to the numerator of the F test of A, B, or AB. In that case, assuming H_0 true, $E(MS_A)/E(MS_{S/ABC}) < 1$; the F test is negatively biased.

In summary, pooling can be advantageous when all pooled terms are functions only of σ_e^2, particularly when the original error term has relatively few df associated with it. On the other hand, if some components of the pool reflect sources of variance other than error, the ratio of mean squares is not a proper F ratio. In view of these possible benefits and problems, the question of when, if ever, to pool has no simple answer. We recommend that pooling any MS with the error MS under the complete model be done only when (a) there is a priori reason to believe that the pooled terms will reflect only chance variability; and (b) a test of the term considered for pooling against the original error MS is not significant at the .25 level. In the example of Table 5.10, none of the interaction terms involving C were significant by this criterion when tested against $MS_{S/ABC}$, but the C SV was. Therefore, we would analyze the data by assuming a structural model having A, B, AB, and C sources, but no interaction terms involving C. The result of that analysis is in the bottom part of Table 5.10. Note that the AB interaction term, which was not significant in the two preceding analyses, now is. This is because there are more error df (and therefore more power) than in the first analysis based on the complete structural model, and the error MS is smaller than in the second analysis (because that error term included variability due to C).

5.9 UNEQUAL CELL FREQUENCIES

We have proceeded as though equal-sized experimental groups have been sampled from equal-sized treatment populations. When this is not the case, problems may arise. We noted in Chapter 4 that unequal n's exaggerate the consequences of heterogeneity of variance. Furthermore, when n's are not equal, different analyses from those presented so far may be required. The choice of analysis is not always simple, and the interpretation is sometimes complicated. The most straightforward case is that in which population sizes are unequal but proportional, and sample sizes reflect those proportions. We will consider this situation first.

5.9.1 Proportional Population and Sample Sizes

Assume a study of how political party preference and previous political participation affect attitudes toward current political issues. Further assume that large-scale sampling studies have reliably established that Democrats, Republicans, and Independents exist in the population of interest in the ratio 4:3:3, and that two-thirds of each of these three subpopulations voted in the last election. To reflect these population sizes, we might run the following numbers of subjects in the study:

	Democrats	Republicans	Independents
Voted	24	18	18
Did not vote	12	9	9

Note that the ratio of row frequencies is constant over columns (24:12, 18:9, and 18:9 all equal 2:1) and the ratio of column frequencies is constant over rows (24:18:18 and 12:9:9 both equal 4:3:3); this constancy defines proportionality of cell frequencies.

Formulas for SS_A, SS_B, and SS_{AB} for this proportional-n case are presented in Table 5.11. They are quite similar to those for the equal-n case except that the n's now vary, as their subscripts indicate. This means that multiplication or division by n must precede summation. One other similarity to the equal-n case is that the error mean square, $MS_{S/AB}$, is again an average of all the within-cell variances. However, it is a weighted average. Just as in the t test for two independent groups, each cell variance is weighted by the df associated with the variance for that cell, divided by the $df_{S/AB}$.

The null hypothesis tested in the proportional-n case is a statement about weighted cell means. To illustrate, in the 2×3 voting study cited above, the null hypothesis for the main effect of political affiliation (columns) is

$$H_0: \frac{n_{11}\mu_{11} + n_{21}\mu_{21}}{n_{.1}} = \frac{n_{12}\mu_{12} + n_{22}\mu_{22}}{n_{.2}} = \frac{n_{13}\mu_{13} + n_{23}\mu_{23}}{n_{.3}}$$

The interaction null hypothesis is, as usual, $\mu_{jk} - \mu_{j.} - \mu_{.k} + \mu_{..} = 0$ for all combinations of j and k; the row, column, and grand means are weighted averages of the μ_{jk}.

TABLE 5.11   FORMULAS FOR SUMS OF SQUARES FOR A TWO-FACTOR DESIGN WITH PROPORTIONAL CELL FREQUENCIES

$$SS_A = \sum_j n_{j.}(\bar{Y}_{.j.} - \bar{Y}_{...})^2 = \sum_j \frac{T_{.j.}^2}{n_{j.}} - C$$

$$SS_B = \sum_k n_{.k}(\bar{Y}_{..k} - \bar{Y}_{...})^2 = \sum_k \frac{T_{..k}^2}{n_{.k}} - C$$

$$SS_{AB} = \sum_j \sum_k n_{jk}(\bar{Y}_{.jk} - \bar{Y}_{.j.} - \bar{Y}_{..k} + \bar{Y}_{...})^2$$

$$= \sum_j \sum_k \frac{T_{.jk}^2}{n_{jk}} - C - SS_A - SS_B$$

$$SS_{tot} = \sum_i \sum_j \sum_k (Y_{ijk} - \bar{Y}_{...})^2 = \sum_i \sum_j \sum_k Y_{ijk}^2 - C$$

$$SS_{S/AB} = (n_{..} - ab)MS_{S/AB} = \sum_j \sum_k (n_{jk} - 1)\hat{\sigma}_{jk}^2$$

$$= SS_{tot} - SS_A - SS_B - SS_{AB}$$

$$C = T_{...}^2/n_{..}$$

where $n_{j.}$ = number of scores at A_j
$\quad n_{.k}$ = number of scores at B_k
$\quad n_{jk}$ = number of scores at A_j and B_k
$\quad n_{..}$ = total number of scores

Several computer packages are available that provide calculations equivalent to those performed by the formulas in Table 5.11; among the easiest to use for this purpose are SSPS[X] ANOVA and SAS ANOVA. Other programs, such as BMDP2V and SYSTAT's MGLH weight the cell means equally. Still other programs require a sequence of tests or a control statement indicating that proportional weighting is desired; BMDP4V includes a WEIGHTS statement allowing the user to define the appropriate weighting. In short, different statistical programs do different things when cell frequencies are not equal (and there is even variation among programs within the same family such as SSPS, BMDP, or SAS). A careful reading of the manual is a necessity.

5.9.2 Disproportionate Cell Frequencies

Table 5.12 illustrates the problem that can arise when cell frequencies are not proportional. Looking only at the cell means, we find no evidence of an AB interaction; both $\bar{Y}_{.11} - \bar{Y}_{.21}$ and $\bar{Y}_{.12} - \bar{Y}_{.22}$ equal 15. According to the discussion of Section 5.2.4, in the absence of interaction we should expect main effects to equal simple effects; $\bar{Y}_{.1.} - \bar{Y}_{.2.}$, should also equal 15. The obtained difference, however, is 18. We obtain a result that appears even stranger if we do an analysis of variance. We find that $SS_A = 1620$, $SS_B = 980$, and $SS_{AB} = (6200 - 4500) - 1620 - 980 = -900$. The minus sign is neither a computational nor a typographical error. Nevertheless, from the development of analysis of variance presented thus far, the result is utter nonsense. From the cell means, the value of SS_{AB} should be zero; a negative value bears no interpretation within our framework.

The reason for these strange computational results may become clearer if we consider an extreme case of disproportionate cell frequency. Suppose the ns were

	B_1	B_2
A_1	0	8
A_2	8	0

TABLE 5.12   EXAMPLE INVOLVING DISPROPORTIONATE CELL FREQUENCIES

		B_1	B_2	Totals	$\bar{Y}_{.j.}$
A_1	$T_{.1k}$	40	200	240	
	n_{1k}	2	8	10	
	$\bar{Y}_{.1k}$	20	25		24
A_2	$T_{.2k}$	40	20	60	
	n_{2k}	8	2	10	
	$\bar{Y}_{.2k}$	5	10		6
	$T_{..k}$	80	220		
	$n_{.k}$	10	10	$\bar{Y}_{...} = 15$	
	$\bar{Y}_{..k}$	8	22		

Now, SS_A and SS_B are identical; both are based solely on the difference between the A_2B_1 and A_1B_2 means and thus the A and B main effects are perfectly correlated. If we now enter two scores in the other two cells, as in Table 5.12, the correlation of effects is no longer perfect, although it will still be large. The magnitude of both SS_A and SS_B will depend primarily (though no longer entirely) on the difference between the A_2B_1 and A_1B_2 means.

Another way to picture the problem is to consider the partitioning of $SS_{\text{B cells}}$, the variability among the cell means. We have

$$\sum_j \sum_k n_{jk}(\bar{Y}_{.jk} - \bar{Y}_{...})^2 = \sum_j \sum_k n_{jk}(\bar{Y}_{.j.} - \bar{Y}_{...})^2 + \sum_j \sum_k n_{jk}(\bar{Y}_{..k} - \bar{Y}_{...})^2$$

$$+ \sum_j \sum_k n_{jk}(\bar{Y}_{.jk} - \bar{Y}_{.j.} - \bar{Y}_{..k} + \bar{Y}_{...})^2$$

$$+ 2\sum_j \sum_k n_{jk}(\bar{Y}_{.j.} - \bar{Y}_{...})(\bar{Y}_{..k} - \bar{Y}_{...})$$

$$+ 2\sum_j \sum_k n_{jk}(\bar{Y}_{.j.} - \bar{Y}_{...})(\bar{Y}_{.jk} - \bar{Y}_{.j.} - \bar{Y}_{..k} + \bar{Y}_{...})$$

$$+ 2\sum_j \sum_k n_{jk}(\bar{Y}_{..k} - \bar{Y}_{...})(\bar{Y}_{.jk} - \bar{Y}_{.j.} - \bar{Y}_{..k} + \bar{Y}_{...})$$

If the n_{jk} are proportional or equal, the last three terms are zero. Otherwise, they can be either positive or negative. The reason $SS_{\text{B cells}} - SS_A - SS_B$ is negative in our example is that it includes the three cross-product terms, which have a relatively large negative total in this data set. The important point is that the cross-product terms are numerators of correlation coefficients between the A and B effects, the A and AB effects, and the B and AB effects, respectively. If these correlations are not zero, then the three sum-of-squares terms are not independently distributed and the usual F test computations cannot independently assess the significance of main and interaction effects.

Figure 5.4 contains a graphic representation of the situation when cell frequencies are disproportionate. The square represents SS_{tot}. The circles represent SS_A, SS_B, and SS_{AB}. When cell frequencies are equal or proportional, these three circles do not overlap, the overlap represents the cross-product terms in Equation 5.2.[1] When effects have overlapping variability, the sums of squares are referred to as *nonorthogonal*. This poses a difficult problem in data analysis and interpretation. For example, we could calculate SS_A in the usual way. If so, it consists of the areas *t, u, v,* and *w*. Or we could adjust the SS_A for the contribution of the other main effect B. In that case, we have an adjusted SS_A consisting of the areas *t* and *w*. Or we could adjust SS_A for the contributions of B and AB, in which case the adjusted SS_A would consist only of *t*. Similar options are available with respect to the other SV. There has been considerable debate over the proper method of adjustment. Our view is that the proper analysis depends

---

[1] The overlapping areas in Figure 5.4 represent covariation among A, B, and AB. Such covariation can be positive or negative. Therefore, the variability due to any one source after adjustment for either or both of the other sources may be smaller (as Figure 5.4 implies) or larger than the original unadjusted variability that is represented by the entire circle.

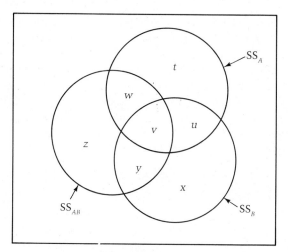

Figure 5.4 Partitioning of variability in a two-factor experiment.

upon the researcher's assumptions about the weight to be given to the levels of the independent variables in the study. We will consider two common cases.

Chance Variation in Cell Frequencies

The variables in our studies are either manipulated (e.g., stimulus intensity, type of therapy) or have levels that existed prior to the study (e.g., sex, type of mental disorder, political affiliation). When all variables are manipulated, we usually view the data in our cells as random samples from equal-sized treatment populations. In such cases, it makes sense to give all levels equal weight; in other words, we plan on equal cell frequencies. Despite our intent, however, chance variations in the n_{jk} can, and often do, occur. Subjects fail to show or become ill, or equipment malfunctions. Sometimes the subjects can be replaced but often this is difficult. In a study of effects of different therapies carried out with individuals for an entire year, replacement of a patient who stopped attending therapy sessions after seven months would be impractical.

It is important to emphasize that equal weighting of the cell means implies *chance* variation in the loss of data. Suppose certain therapy methods were more aversive than others and resulted in a larger loss of patients. Then the data loss is related to the experimental condition, and equal weighting would not be appropriate.

Weighting the means equally is equivalent to adjusting each main and interaction effect for its relation to the other effects (Carlson and Timm 1974). In terms of Figure 5.4, the adjusted SS_A, SS_B, and SS_{AB} correspond to the areas t, x, and z, respectively. BMDP2V and SYSTAT use only this adjustment method. It is also the default method (others can be specified) in SPSS[X] MANOVA. The Type IV MS in SAS GLM is also the result of this method.

There are some statisticians who oppose this adjustment method. Appelbaum and Cramer (1974, Cramer and Appelbaum 1980) have argued that main effects should not be adjusted for interaction effects when the usual test of interaction has a nonsignificant result. The logic is that such tests of main effects will have more power when small or no interaction variances are ignored in the analysis. The trouble with this approach is that weak tests of the interaction may cause us to fail to adjust for substantial interaction variance, thus increasing the Type 1 error rate in tests of the main effect. Until well-justified guidelines exist (other than the traditional .05 significance level) for nonsignificance of the interaction, we prefer to adjust all SS for overlap with the other SS in those cases in which inequality of cell frequencies can be attributed to chance losses of data.

Systematic Variation in Cell Frequencies

Consider a study in which outcome of therapy is rated; the independent variables are method of therapy and severity of symptoms. The number of patients completing therapy may vary as a function of the levels of both of these variables. Thus, it is improper to view the treatment populations as equal in size and improper to weight the cell means equally. We may have reliable prior information about the relative sizes of the populations; these can be used to weight the cell means in an analysis of variance. More likely, we will assume that the observed frequencies reflect the relative population sizes and weight the means accordingly. BMDP4V accomplishes this using the control statement /WEIGHT BETWEEN ARE SIZES. BMDP4V allows the data analyst to select other weights as well. This can also be accomplished in other packages by using control statements that define contrasts among cell means. Such contrasts will be discussed further in Chapter 6.

Sometimes when the levels of the independent variables involve prior groupings (e.g., disease, sex, race), a priori analysis of the relations among those variables may dictate the adjustments to be made. For example, suppose we hypothesize that an individual's race (R) influences his or her educational level (E), which in turn influences the dependent variable, an attitude measure. Under this assumption, neither race nor its interaction with educational level should contribute significantly to the variability in the data if the effect of educational level were removed. In terms of Figure 5.4, this implies that SS_E corresponds to its entire circle (u, v, x, and y), the adjusted SS_R corresponds to the areas t and w, and the adjusted SS_{ER} corresponds to the area z. Such adjustment sequences can be carried out by many computer packages. This can be done with SPSS[X] MANOVA by specifying the order of adjustment in the DESIGN control statement. In our example, we would use the statement DESIGN = E, R, R BY E. Reversing the order of R and E in the statement would result in an unadjusted SS_R and in SS_E adjusted for R. Similarly, the TYPE I MS in SAS GLM reflects the order in the MODEL statement.

We have omitted mention of several approaches to unequal n's which have been used by many researchers. In an effort to equally weight cell means, researchers have randomly dropped scores to equate cell frequencies, estimated missing scores for the same reason, and used the "method of unweighted means" (Horst and Edwards 1982) to analyze the data. This last method is computationally simple but gives results equivalent to adjusting each SS for its overlap with all others only when there are exactly two levels of all variables. Dropping scores results in a loss of df and consequently of power. The reliability of missing score estimates will depend upon the number of scores available in the cell. In any event, the availability of many computer packages designed to do unequal-n analyses makes such shortcuts and approximations superfluous. (It should also be noted that any program for regression analysis can accomplish any series of adjustments; Chapter 16 will discuss this further.) It is critical, however, to recognize that all unequal-n programs do not do the same thing. Think through your assumptions about your independent variables and their consequences for the data analysis. Then read the manual for the computer package you wish to use to make sure that you are using the right program or the right option within that program.

5.10 INTRODUCING FACTORS TO REDUCE ERROR VARIANCE: THE TREATMENTS × BLOCKS DESIGN

Suppose we are interested in testing the effects of instructions (A) upon reading skill scores (Y), and subjects are randomly assigned to the levels of A. The variability among subjects within instructional groups—the error term against which instructional effects are tested—is generally large due to individual differences in factors that affect performance. Even subjects treated alike will differ in their scores because of differences in such attributes as experience, motivation, and ability.

As an alternative to the one-factor design, we could divide the subjects into b blocks on the basis of a concomitant variable, X, which is believed to be highly correlated with Y. For example, suppose we had four instructional conditions ($a = 4$) whose effects on reading skill we wished to compare. Then X might be a measure of reading readiness obtained prior to the start of the school year. Assume that we have 80 subjects available for this study. We might rank these 80 children in terms of their reading readiness score. We could then divide them into five blocks of 16 children; block 1 (B_1) would consist of the 16 children with the highest reading readiness scores, B_2 would consist of the 16 with the next highest X values, and so on. Then we would randomly assign each of the 16 children in B_1 to one of the four instructional conditions (level of A) with the constraint that there would be exactly four children in each instructional condition. We would do the same thing for each of the other reading readiness blocks. The result is that we have a two-factor design in which $a = 4$, $b = 5$, and $n = 4$. This two-factor (treatments × blocks) design usually will have less error variance than the one-factor design based on the same total number of subjects because the four subjects within each $A \times B$ cell will be less variable in reading readiness than the 20 subjects within each level of A would have been if we had used a one-factor design.

Turn now to Table 5.13, in which df and EMS have been presented for both the one factor design and its treatment × blocks counterpart. Note that in order to compare the two designs, we assume the same number of subjects (abn; bn subjects at each level of A) in both of them. We have also used different subscripts for the population error variances for the two designs. If X and Y are correlated, $\sigma_{t \times b}^2$ should be less than σ_{cr}^2, as we argued above. This means that $MS_{S/AB}$ should be less than $MS_{S/A}$. It also means that MS_A will be smaller in the blocking than in the one-factor design. This is because the different levels of A in the blocking design are roughly equivalent in read-

TABLE 5.13 EXPECTED MEAN SQUARES FOR A ONE-FACTOR AND A BLOCKING DESIGN

Design	SV	df	EMS
One-factor	A	$a - 1$	$\sigma_{cr}^2 + bn\theta_A^2$
	S/A	$a(bn - 1)$	σ_{cr}^2
Treatments $\times$ blocks	A	$a - 1$	$\sigma_{t \times b}^2 + bn\theta_A^2$
	B	$b - 1$	$\sigma_{t \times b}^2 + an\theta_B^2$
	AB	$(a - 1)(b - 1)$	$\sigma_{t \times b}^2 + n\theta_{AB}^2$
	S/AB	$ab(n - 1)$	$\sigma_{t \times b}^2$

ing readiness. However, even though both numerator and denominator mean squares are reduced by blocking, the ratio will be increased if θ_A^2 is greater than zero. Subtracting a constant from numerator and denominator of a ratio greater than 1 increases that ratio.

Table 5.13 provides some sense of the conditions under which blocking will be advantageous. If θ_B^2 or θ_{AB}^2 are large, then σ_{cr}^2 will be much larger than $\sigma_{t \times b}^2$ because the block variability represents some of the error variance in the one-factor design. How much σ_{cr}^2 is reduced by blocking depends in part on how many blocks are used. As b increases, the range of X values within a block should decrease, and the less variability in X, the less variability there should be in Y. We would expect less error variance in reading skill scores if reading readiness scores varied very little within a block than if they varied considerably. The effectiveness of blocking in reducing error variance in Y also depends on how highly X and Y are correlated. Dividing subjects into blocks on the basis of the lengths of their big toes presumably would not be helpful in reducing error variance for reading skill scores.

The df column in Table 5.13 suggests that the reduction in error variance due to blocking could be offset by a loss in error df. There are $a(b - 1)$ more df associated with $MS_{S/A}$ than with $MS_{S/AB}$. Therefore, it is possible that blocking could yield a smaller error term for the F test of A, but a less powerful test because of the smaller number of error df. However, if the total number of observations is large, both error df will be large, and the difference between the two values will not have as much of an influence on power as will the difference in error variances. Also, the larger the correlation between X and Y, the less important will be the difference in error df; as we noted above, the higher the correlation (ρ) between X and Y, the greater the reduction in error variance due to blocking.

Blocking is a particularly useful approach to research design when measures related to a planned dependent variable are available. To use the blocking design most effectively, the researcher should consult an article by Feldt (1958), in which he addresses the question of how many blocks should be used. If the total N is held constant, an increase in the number of blocks yields decreased within-cell variance; however, it also results in a decrease in error df. Therefore, there is an optimal number of blocks that increases with total N and the correlation of X and Y.

Feldt has tabled this optimal value for various combinations of a, N, and ρ, and also has compared the blocking design with other approaches to decreasing error variance.

5.11 CONCLUDING REMARKS

The between-subjects factorial designs of this and the preceding chapter have several advantages. Assuming equal cell frequencies, the analysis of the data is much simpler than for most other designs. For any given number of scores, the error df will be larger than for any comparable design. The requirements of the underlying model are more easily met by these between-subjects designs than by other designs, and violations of assumptions are less likely to affect the distribution of the F ratio. The between-subjects designs share one major deficit. Because the within-cell variance, which is the basis for the error term for all F tests, is a function of individual differences, the efficiency of the design is low compared to that of other designs. That is, other designs, which allow the removal of individual difference variability from the error variance, generally will yield smaller error terms and therefore more precise estimates of population effects and more powerful tests of hypotheses. The between-subjects designs are most useful whenever subjects are relatively homogeneous for the variable being measured; whenever a large n is available, compensating somewhat for the variability of measurements; or whenever it is feasible to block on the basis of a concomitant measure related to the dependent variable. Also, we should be aware that there are many experiments in which it is impossible to do anything but assign different subjects to different levels of the variable. This is self-evident when the independent variable is personality type or training technique. It may also be true when much time is needed to obtain a measure from the subject, and it is therefore preferable to obtain only one measure from each subject.

EXERCISES

5.1 The following terms provide a useful review of some concepts in the chapter. Define, describe, or identify each of them:

    main effect
    interaction effect (first-order, second-order)
    simple effect

    pooling mean squares
    disproportionate cell frequencies
    treatment $\times$ blocks design

5.2 **(a)** Plot the marginal and cell means for the data set that follows. Without doing an ANOVA, indicate which of the effects appear to be significant.

	A_1	A_2	A_3	A_4
	14	22	31	18
B_1	12	34	33	21
	26	24	43	19
	42	46	20	41
B_2	15	18	25	30
	27	17	15	44

(b) Carry out an ANOVA on the above data and present the results in a table.

5.3 We have the following set of cell means and variances ($n = 5$):

	Means				Variances		
	A_1	A_2	A_3		A_1	A_2	A_3
B_1	32	22	24	B_1	85	75	80
B_2	28	18	20	B_2	90	80	70
B_3	24	14	16	B_3	77	87	76

(a) Plot the cell and marginal means and indicate which (if any) effects appear to be significant.

(b) Estimate α_1, α_2, α_3, β_1, β_2, β_3, $(\alpha\beta)_{11}$, and $(\alpha\beta)_{23}$.

(c) Calculate SS_A, SS_B, and SS_{AB}.

(d) What is $MS_{S/AB}$ and $SS_{S/AB}$?

(e) Which effects are significant if $\alpha = .05$?

(f) Estimate θ_A^2, θ_B^2, and θ_{AB}^2.

5.4 We have the following cell means and variances based on ten scores per cell:

	Means				Variances		
	A_1	A_2	A_3		A_1	A_2	A_3
B_1	36	43	53		11	20	22
B_2	33	36	44		7	9	15

(a) Carry out the ANOVA and present the tabled results.

(b) Test the simple effects of B at A_3.

(c) Test the simple effects of A at B_2.

(d) Briefly justify your choice of error terms for parts (b) and (c).

5.5 In order to conserve space, most psychology journals do not publish ANOVA tables except where the analysis is complicated and there are many significant sources of variance. Thus, usually all we can expect to find is a table of means and a report of the obtained values of F (though some journals now publish the error MS's). At times, however, we wish that the researcher had published other analyses on the data. Often we can perform these analyses for ourselves, even though the researcher has not provided us with the raw data from the study. Suppose we have been given the following table of cell means from a between-subjects design with two factors:

	A_1	A_2	A_3
B_1	11	12	10
B_2	3	10	14

and we have been told only that the AB interaction is significant, $F = 3.93$, $P < .05$.

(a) If there are four subjects in each cell, reconstruct the entire ANOVA summary table, including the SS's, df's, MS's, and the F's for each source of variance.

(b) Is the A main effect significant at the $\alpha = .05$ level? What are your best estimates for δ_A^2 and δ_{AB}^2? Note that analogous to the discussion of variance components in one-factor designs (Section 4.6.2), δ_{AB}^2 is the true variance of the $(\alpha\beta)_{jk}$ components; that is, $\Sigma\Sigma(\alpha\hat\beta)_{jk}^2/ab$, and can be estimated from
$$\hat\delta_{ab}^2 = [(a-1)(b-1)/ab]\hat\theta_{AB}^2.$$

(c) If the effects are as estimated in the present study, what is the power for the test of (i) the A main effect? (ii) the AB interaction?

(d) How many subjects per cell would be required to yield a power of .8 ($\alpha = .05$) for the test of the A main effect?

5.6 Consider a study in which each S is presented with a set of digits. He is then presented with a probe digit and must respond "yes" or "no"; the probe was or wasn't in the list. We hypothesize that observed reaction time is $RT = t_e + t_c + t_r$, where t_e = time to encode (read) the probe digit, t_c = time to compare it with all the members of the memorized list, and t_r is time to say "yes" or "no." Note that only t_c should change with the number of digits in memory.

Suppose we have a two-factor design. One variable is L, list length, the number of digits memorized. The other is Q, quality of the probe. It may be clear and easy to read or fuzzy and tough to read. We think that this should only influence the time it takes to encode the stimulus. Let $t_r = 300$. Make up values of t_e and t_c for high- and low-quality stimulus and $L = 2, 3$, and 4. The numbers should be consistent with the theory. Then plot your "observed" RT. If you analyze the variance, what terms should be significant? Which not?

5.7 Consider each of the following sets of hypotheses. Which SV should be significant? Plot a data set consistent with the theory.

(a) In a bar press experiment, we believe that

$$Y = K \times D \times P$$

where Y = bar pressing rate (our dependent variable)

$\quad\quad\quad D$ = hours of deprivation

$\quad\quad\quad K$ = constant

$\quad\quad\quad P$ = number of practice trials

(b) In impression formation studies, we give subjects some information on the attractiveness (A) and intelligence (I) of an individual and then ask them to rate the individual. We believe that R (rating) $= (A + I)/2$.

(c) We have depressives and nondepressives (D), who are given either a success or a failure experience (outcome O) and are led to believe the task involves luck or effort (attribution A). We hypothesize that

(i) Expectancies about success in future tasks are always higher for non-depressives than for depressives.

(ii) Under effort instructions, success leads to higher expectancies than failure does; this difference is larger for depressives than for non-depressives.

(iii) Under luck instructions, success raises expectancies by the same amount for both depressives and nondepressives. Furthermore, the average difference in expectancies between success and failure is lower than it was under effort instructions.

5.8 Patients in a mental hospital are divided into experimental groups on the basis of their socioeconomic level (three levels) and the kind of treatment they receive (two levels, psychotherapy and behavior therapy). The investigator predicts that (1) psychotherapy will be less effective than behavior therapy, and (2) psychotherapy will be more effective the higher the socioeconomic level of the patient, but that this will not be true for behavior therapy. In fact, no main effect of socioeconomic level is predicted.

State which SV's (sources of variance) should be significant if the predictions hold. Then set up a matrix of cell means consistent with the predictions (assuming errorless data).

5.9 We have the following data for a three-factor design:

		C_1				C_2	
	A_1	A_2	A_3		A_1	A_2	A_3
B_1	47	28	27	B_1	33	24	33
B_2	23	14	23	B_2	29	10	9

(a) Without carrying out an ANOVA, which of the following appear to be large?
  (i) the main effects?
  (ii) the first-order interactions?
  (iii) the ABC interaction?
  (iv) the simple AB interaction at C_1?
  (v) the simple effects of B at $A_1 C_1$?

(b) Assuming the following variance estimates, and $n = 5$, carry out the ANOVA.

		C_1				C_2	
	A_1	A_2	A_3		A_1	A_2	A_3
B_1	421	412	382	B_1	350	364	312
B_2	224	118	265	B_2	124	208	116

(c) Test the simple effects of A at $B_1 C_1$. Justify your choice of error term.

5.10 In a study designed to examine changes in attention with age, 180 children are required to sort decks of cards into two piles according to the value of a relevant dimension contained on the card. For example, if the relevant dimension is *shape*, the child may be asked to sort all the cards in a deck that contain a circle into one pile and all the cards that contain a square into another pile. To investigate how well a child is able to focus on a single dimension, the amount of *irrelevant* information present on the card is varied across different experimental conditions; there is either no irrelevant information (I_1), one irrelevant dimension such as color (I_2), or two irrelevant dimensions. To summarize the design, there are three age levels (three, five, and seven years), three levels of irrelevant information, and both male and female subjects; there are 10 subjects in each of these 18 cells. The dependent variable is the average (mean) time required to sort a deck of cards.

(a) Write down the ANOVA table for this design (SV's, df's, EMS's). Remember that you know the number of levels of each variable, so the df's and the coefficients of the EMS terms should be expressed as numbers, not letters.

(b) The following hypotheses are made:
   (i)   Older children are generally faster at doing the task.
   (ii)   Irrelevant information interferes with performance, and there is more interference for younger children than for older children.
   (iii) The tendency for younger children to be more influenced by the irrelevant information than older children is more pronounced for boys than for girls.

If the hypotheses are correct, what terms would you expect to be significant?

5.11 Four- and six-year-old subjects (age, A) are asked to sort a set of pictures by their color or by their size (type of sort, T). They are then tested for recall, cued either by a color or the word "big" or "small" (cue, C). The researcher hypothesizes that (1) six-year-olds will recall better under all conditions; (2) sorting by color will yield better recall than sorting by size; (3) but more so for younger subjects; (4) performance will be better when the cue and sort match (e.g., sort by size and cued by size) than when they don't, for both ages. While not every SV is predictable, many are on the basis of these hypotheses. Construct a set of means consistent with the hypotheses and plot them. Which SV should be significant?

5.12 Ninety-six children are subjects in a study of perceptual discrimination. Half of the children are six years old, and half are nine years old. Half the subjects are tested with two-dimensional objects and half with three-dimensional objects. Half are required to discriminate on the basis of shape, half on the basis of color. Thus there are eight groups differing in age (A: 6 or 9), dimensions (D: 2 or 3), and relevant cue (C: shape or color). We have the following hypotheses:

(a) Nine-year-olds will make fewer errors than six-year-olds.

(b) On the average, it will be easier to discriminate three-dimensional objects than two-dimensional objects.

(c) The difference between two- and three-dimensional objects will be more marked for six-year-olds than for nine-year-olds.

(d) The difference between two- and three-dimensional objects will hold for shape but not for color.

What SV's should be significant if the data conform to the hypotheses?

5.13 Consider several three-factor designs with independent variables A, B, and C. In each case you will be given information about some effects and asked what you can conclude about other effects. In some cases, the correct answer may be that no conclusion can be drawn.

(a) There are no A or B main effects, but there are nonzero simple effects of B at A_2. What can we conclude about

   (i)   the AB interaction?
   (ii) the simple A effects at different levels of B?
   (iii) the C main effect?

(b) There is no A main effect and no AB or BC interactions. There is a C main effect, an AC interaction, and nonzero simple B effects at A_1. What can we conclude about

   (i)   the B main effect?
   (ii) the simple A effects at each level of C?
   (iii) the ABC interaction?

(c) There is a BC interaction but no B main effect or ABC interaction. What can we conclude about

   (i)   the simple BC interactions at each level of A?
   (ii) the simple B effects at each level of C?
   (iii) the overall AB interaction?

(d) There is no A main effect or AB interaction. There is a B main effect, nonzero simple effects of A at C_1, and a simple AB interaction at C_1. What can we conclude about

   (i)   the AC interaction?
   (ii) the simple effects of C at A_1?
   (iii) the ABC interaction?
   (iv) the BC interaction?

5.14 A small sample of citizens were asked to rate their attitude toward a presidential candidate. These subjects were from four regions (R) of the country (west, midwest, northeast, and south) and were either Republicans, Democrats, or Independents (political preference, P). There were six subjects in each cell of the design. The results of an ANOVA were

SV	df	MS	F
R	3	750	2.50
P	2	900	3.00
RP	6	447	1.49
S/RP	60	300	

(a) Place limits on the p values for each of the three terms to be tested.

(b) Does P or R contribute a larger proportion of the estimated total variance in the population? Calculate values of δ^2 for P and R to support your answer.

(c) Suppose we rerun the study with 20 subjects in each of the cells. Given the above data as the basis for estimating the population interaction variance, and the error variance, what power would we now have?

5.15 We have the following ANOVA table:

SV	df	MS	F
A	3	56.8	2.84
B	2	40.0	2.00
AB	6	40.0	2.00
S/AB	36	20.0	

(a) Estimate ω^2 for A, B, and AB. Note that analogous to the discussion of variance proportions in one-factor designs (Section 4.6.2; also see Exercise 5.5), $\omega_A^2 = \delta_A^2/\sigma_Y^2$ and $\omega_{AB}^2 = \delta_{AB}^2/\sigma_Y^2$ where $\sigma_Y^2 = \delta_A^2 + \delta_B^2 + \delta_{AB}^2 + \sigma_e^2$.

(b) If the effects and error variance are as indicated in the table, what power would we have for A, B, and AB (at $\alpha = .01$) if the study were redone with $n = 8$?

5.16 (a) Estimate the proportion of population variance (ω^2) that A, B, and AB contributes for the following table.

SV	df	MS
A	2	512
B	4	512
AB	8	152
S/AB	75	62

(b) Assuming that the AB effects observed in this study are close to the true population values, what power did this experiment have to reject the null hypothesis of no interaction? Assume that $\alpha = .05$.

5.17 A is the variable of interest in the following data set; however, there are two levels of an incidental variable E (experimenter), and the design is balanced; that is, there are five subjects in each of the six cells. The data are

A	E	Scores				
1	1	35	28	29	23	21
2	1	18	27	30	21	27
3	1	14	16	25	15	17
1	2	21	24	26	18	13
2	2	17	19	14	21	25
2	3	16	24	23	10	17

(a) In many cases, the E variable is ignored because it is not of interest. Assuming E has no effect, state the structural model and present an ANOVA table, including EMS, consistent with that model.

(b) Assuming E has an effect, present the structural model and the ANOVA table.

(c) Which do you believe to be the more appropriate model. Why?

(d) If you are in doubt about the choice of models, which would you choose? Why?

5.18 Consider the following cell means and variances ($n = 5$):

	Means				Variance estimates		
	A_1	A_2	A_3		A_1	A_2	A_3
B_1	36	26	24	B_1	75	79	84
B_2	25	15	23	B_2	61	82	85
B_3	23	19	34	B_3	90	71	77

(a) If the cell means were based on errorless data, which effects would be significant?

(b) Generate the entire ANOVA table that would be obtained if the data were analyzed as a two-factor design with the structural model

$$Y_{ijk} = \mu + \alpha_j + \beta_k + (\alpha\beta)_{jk} + \epsilon_{ijk}$$

(c) What effects would be significant at $\alpha = .05$?

(d) Generate the ANOVA table that would be obtained if the same data were analyzed as a one-factor design according to the model

$$Y_{ij} = \mu + \alpha_j + \epsilon_{ij}$$

Is the A main effect significant if $\alpha = .05$?

(e) Find ω_A^2 for the one- and two-factor analyses.

(f) If the effects are really as indicated in the data, what is the power for the A main effect at $\alpha = .05$ for each design? (You need only get a rough estimate, i.e., within .10 of the actual value.)

5.19 Ten study groups consisting of two males and two females are taught statistics under one method of instruction and ten more groups of two males and two females are taught by another method. Performance on an exam is the dependent measure in an ANOVA. The ANOVA table is

SV	df	MS	EMS
M	1	3627	$\sigma_e^2 + 4\sigma_{G/M}^2 + 40\theta_M^2$
X	1	128	$\sigma_e^2 + 2\sigma_{GX/M}^2 + 40\theta_X^2$
MX	1	123	$\sigma_e^2 + 2\sigma_{GX/M}^2 + 20\theta_{MX}^2$
G/M	18	420	$\sigma_e^2 + 4\sigma_{G/M}^2$
GX/M	18	32	$\sigma_e^2 + 2\sigma_{GX/M}^2$
Residual	40	30	σ_e^2

M stands for method, *X* for sex, and *G* for groups. Although we have not yet discussed this design, the above table and the discussion of pooling in Section 5.8 provide all the information you need to answer the following questions.

(a) Compute the appropriate *F* tests of *M*, *X*, and *MX* and report whether each is significant at the .05 level.

The investigator decides she is justified in reanalyzing her data using the following SV:

SV	df	MS	F	EMS
M	1	3627	?	?
X	1	128	?	?
MX	1	123	?	?
G/M	18	420		?
Pooled residual	58	?		?

(b) What is the value of $MS_{pooled\ residual}$?

(c) What are the *F* ratios now?

(d) What has the investigator assumed in doing this new ANOVA? Answer this by filling in the EMS. All you need do is write the appropriate subscripts (e.g., *e*, *M*, *G/M*, etc.); don't worry about the coefficients.

(e) What is the justification for this assumption in the original data analysis? What is the potential advantage of the reanalysis if the assumption is correct? What is the potential disadvantage if the assumption is incorrect?

5.20 Following is a data set with unequal *n*'s. The cell totals (T_{jk}) and the *n*'s (n_{jk}) are given (T_{jk}/n_{jk}):

	B_1	B_2
A_1	20/2	40/4
A_2	16/8	4/2

(a) Calculate SS_{cells}. This follows the procedure illustrated for SS_A in Table 4.5.

(b) In the same way you did part (*a*), calculate SS_A and SS_B. Then subtract these from SS_{cells} to get SS_{AB}. Do you see any problem with this procedure? Explain.

(c) Calculate $\hat{\alpha}_1$ and $\hat{\alpha}_2$. This requires finding the marginal (row) means for A_1 and A_2 and subtracting them from the grand mean. If you have done this correctly, $\Sigma_j n_{.j} \hat{\alpha}_j = 0$ ($n_{.j}$ is the total *n* for row *j*).

(d) Subtract $\hat{\alpha}_j$ from each cell mean in row *j*. Look at the table of means that results. Using these adjusted means, what are SS_B and SS_{AB}? How does this result compare with your answer in part (b)?

5.21 Consider the following table of cell means. We will adjust these means for the effects of *A* under different assumptions about cell frequencies and, by observing what happens to the column means, infer the consequences of equal, proportional, and disproportional cell frequencies for ANOVA. The cell means are

	B_1	B_2	B_3
A_1	12	8	22
A_2	8	6	13
A_3	1	3	16

(a) Assume that the cell frequencies are all the same. (i) Calculate the row and column means. (ii) Calculate estimates of the row effects (the $\hat{\alpha}_j$). (iii) Subtract each value of $\hat{\alpha}_j$ from the three cell means in the corresponding row. How do the column means of this adjusted (for A effects) matrix compare with those in part (i)?

(b) Assume the original cell means presented above. This time, however, assume the following cell frequencies (n_{jk}):

	B_1	B_2	B_3
A_1	4	8	12
A_2	3	6	9
A_3	1	2	3

Redo (i)–(iii) of part (a). Be careful in calculating the row, column, and grand means; the values being averaged must be weighted by their corresponding frequency. For example, the mean of the first row is [(4)(12) + (8)(8) + (12)(22)]/24. Similarly, the grand mean is the weighted average of the 12 cell means (or of the three row or column means).

(c) Again assume the original set of means. This time, the n_{jk} are

	B_1	B_2	B_3
A_1	5	10	4
A_2	5	5	10
A_3	10	5	5

Again do (i)–(iii).

(d) Review your answers to this problem and draw a conclusion about the effects of equal, proportional, and disproportional cell frequencies about the partitioning of variability in ANOVA.

Chapter 6

Contrasts Among Means

6.1 INTRODUCTION

In Chapters 4 and 5, we discussed how ANOVAs provide information about main effects and interactions. However, ANOVAs test very general hypotheses about population means. A significant main effect or interaction tells us that "something is going on" in the data but not necessarily what it is. For example, consider a study of the effects of incentive on learning in preschool children. There are four groups of 16 subjects differing with respect to incentive. Three groups of children begin with a stack of chips. In condition G (gain) the child receives an additional chip for each correct response made during a 20-trial sequence; in condition L (loss), the child loses a chip for each incorrect response; in condition GL, the child gains for a correct response and loses for an error. The fourth group is a control (CON, neither gain nor loss); these children are given no initial stack of chips and are merely instructed to see how often they can get the experimenter to say "right." The dependent variable is the number of errors in a block of 20 trials. The cell means and ANOVA table are given in Table 6.1. The significant F ratio allows us to conclude that the four treatment effects are not all equal; however, such an "omnibus F test" does not allow us to pinpoint where the differences are. For example, we cannot conclude that the mean of the control population differs from that of any of the three incentive populations, or from the mean of the three incentive populations combined. But it is just such differences between group means, and between means of subsets of groups, that usually are of most interest to us. This chapter will present significance tests and confidence intervals to enable us to draw inferences about such comparisons among means.

Many such comparisons can be constructed and tested. Therefore, it is important to keep in mind that the more tests that are performed, the greater the probability of a Type 1 error. To understand why this is so, think of tossing a coin that has a .95 probability of coming up heads and a .05 probability of coming up tails. Although the probability of a tail is .05 on any one toss, the probability of at least one tail in the set of tosses becomes very large if we toss the coin many times. In the same way, even if all a population means are identical and the probability of a Type 1 error is small for any single significance test, the probability of obtaining at least one significant result by chance becomes large when we perform many significance tests on the set of means. What to do about this is a major concern of this chapter.

**TABLE 6.1   MEANS AND ANOVA TABLE FOR AN
EXAMPLE OF A ONE-FACTOR DESIGN
($n = 16$)**

		Incentive (A)		
	Control (CON)	Gain (G)	Loss (L)	Gain or Loss (GL)
$\bar{Y}_{.j}$	8.8	4.2	3.4	2.5

SV	df	SS	MS	F
A	3	377.4	125.8	8.39
S/A	60	900	15	

In much of this chapter, we will use the example of Table 6.1 to illustrate various sorts of comparisons among means that might be of interest to a researcher. We will present statistics to test whether these comparisons involve significant differences. These will be t and F statistics, but the formulas will look somewhat different from other t and F statistics we have seen before. More important, these statistics will not be evaluated against the usual criteria, the values in Tables D.3 and D.5. Because of the problem noted—the risk of increased Type 1 error rate when several significance tests are performed—criteria other than the ordinary t and F critical values have been developed by statisticians. After presenting the test statistics, we will consider these alternative criteria for their evaluation, noting when each is appropriate. Finally, we will extend the developments of this chapter to multifactor designs.

6.2 CONTRASTS IN A ONE-FACTOR DESIGN

6.2.1 Examples of Contrasts

Let us reconsider Table 6.1 The main effect of incentive is highly significant, so that the null hypothesis $\mu_{CON} = \mu_G = \mu_L = \mu_{GL}$ can be rejected. There are a large number of specific null hypotheses that might be of interest. Most commonly, researchers are interested in *pairwise comparisons;* these involve pitting one condition against another. With four levels of incentive, six such comparisons are possible, each requiring a test of a null hypothesis of the form $H_0: \mu_j - \mu_{j'} = 0$. Each of these null hypotheses can be expressed in terms of a linear combination of the population means. For example,

$$H_{01}: (1)\mu_{CON} + (-1)\mu_G + (0)\mu_L + (0)\mu_{GL} = 0 \tag{6.1}$$

$$H_{02}: (1)\mu_{CON} + (0)\mu_G + (-1)\mu_L + (0)\mu_{GL} = 0$$

Other potentially interesting null hypotheses that can be expressed as linear combinations involve the averages of sets of treatment population means. Two of many possible examples are

$$H_{03}: \mu_{\text{CON}} - \frac{\mu_G + \mu_L + \mu_{GL}}{3} = 0 \qquad (6.2)$$

and

$$H_{04}: \frac{\mu_G + \mu_L}{2} - \frac{\mu_{GL} + \mu_{\text{CON}}}{2} = 0$$

These two hypotheses address the questions of whether the average of the three incentive population means differs from the mean of the control population and whether the average of the G and L population means differs from the average of the GL and CON population means.

We define a *contrast* or *comparison* as a linear combination of population means,

$$\psi = \sum_{j=1}^{a} w_j \mu_j \qquad (6.3)$$

with the constraint that the sum of the weights should add to zero (i.e., $\Sigma_j w_j = 0$). Some authors use the two terms interchangeably; others use the term *comparison* to refer to *pairwise comparisons,* linear combinations that involve only two means. Because, according to our definition, any four numbers that sum to zero may serve as contrast weights, there are an infinite number of potential contrasts among means in the current example. The vast majority of these are of no interest. For example, we are probably not interested in testing the null hypothesis

$$\frac{-1}{16}\mu_{\text{CON}} + \frac{-15}{16}\mu_G + \frac{37}{58}\mu_L + \frac{21}{58}\mu_{GL} = 0$$

Even if we confine ourselves to pairwise comparisons and to comparisons between simple averages of population means, the number of possible contrasts increases rapidly as a gets larger. For the current example, in which $a = 4$, there are six possible pairwise comparisons, 12 contrasts in which a single mean is compared with the average of two others, 4 contrasts in which a single mean is compared with the average of three others, and 3 contrasts in which the average of two means is compared with the average of the other two.

6.2.2 The *t* Statistic for Testing Contrasts

In this section, we will show how the t statistic of Chapter 3 can be used to test hypotheses about contrasts. However, in order to guard against increased Type 1 error rate caused by carrying out many tests on the means, we will not evaluate the value of t against the usual criterion of Table D.3. Instead, our criteria will be more conservative and will depend upon the number of tests performed as well as several other factors that we will discuss later in this chapter. In this section we focus on the nature of the calculated t statistic. We then will relate this approach to the partitioning of SS_A into components representing the various contrasts. Following this we will consider various criteria for evaluating t tests of contrasts.

We know from Chapter 3 that for any population in which the scores are independently and normally distributed, $(V - E)/\hat{\sigma}_V$ is distributed as t, where V (for variable) is a normally distributed sample statistic that estimates the population parameter E and

$\hat{\sigma}_V$ estimates the standard error of V. When testing contrasts, we replace the parameter E by the contrast $\Sigma_j w_j \mu_j$; we will represent such contrasts by the Greek letter ψ (psi). In what follows we will present formulas for $\hat{\psi}$ and $\hat{\sigma}_{\hat{\psi}}$, and justify their substitution into the generic formula for t presented earlier.

The intuitive formula for $\hat{\psi}$ is the contrast of the sample means, $\hat{\psi} = \Sigma w_j \bar{Y}_{.j}$. It is also the best estimate of ψ. The reason for this is that, as we discussed in Chapter 2, the sample mean is an unbiased, consistent, and efficient estimator of the population mean. Therefore, the best estimate of a weighted average of treatment population means is the weighted average of the sample means. For example, our best estimates of the contrasts presented in H_{01} and H_{03} (Equations 6.1 and 6.2) are

$$\hat{\psi}_1 = \bar{Y}_{\text{CON}} - \bar{Y}_G = 8.8 - 4.2 = 4.6$$

and

$$\hat{\psi}_3 = \bar{Y}_{\text{CON}} - \frac{1}{3}(\bar{Y}_G + \bar{Y}_L + \bar{Y}_{GL}) = 5.43$$

The statistic $\hat{\psi}$ must not only estimate ψ, but it must also be normally distributed if the ratio we construct is to have the t distribution. This condition will be met if the scores in each of the a treatment populations are normally distributed. As we noted in Chapter 3, linear combinations of normally distributed variables are themselves normally distributed. The statistic $\hat{\psi}$ is a linear combination of group means, which in turn are linear combinations of individual scores.

Finally, we need to derive an expression for $\hat{\sigma}_{\hat{\psi}}$, the estimated standard error of $\hat{\psi}$. Because $\hat{\psi}$ is a linear combination of the group means, the expression for the variance of the sampling distribution of $\hat{\psi}$ follows the general form of the variance of a linear combination given in Equation 2.12. Therefore,

$$\sigma_{\hat{\psi}}^2 = \sum w_j^2 [\text{Var}(\bar{Y}_{.j})]$$

Because the variance of the sampling distribution of a mean is the variance of the parent population divided by the sample size, the last expression may be rewritten as

$$\sigma_{\hat{\psi}}^2 = \sum \frac{w_j^2 \sigma_j^2}{n_j}$$

where σ_j^2 is the variance of the jth treatment population. The assumption of homogeneity of variance yields further simplification; because $\sigma_1^2 = \cdots = \sigma_j^2 = \cdots = \sigma_a^2 = \sigma_e^2$, we can rewrite the last equation as

$$\sigma_{\hat{\psi}}^2 = \sigma_e^2 \sum_j \frac{w_j^2}{n_j}$$

From our discussion in Chapter 4, we know that the best estimate of the population variance, σ_e^2, is $\text{MS}_{S/A}$. Recall from Chapter 4 that, when the group n's are equal, this is obtained by summing the a group variances and dividing by a. If the group n's are not equal, the weighted average of the group variances is obtained. In general, the best estimate of $\sigma_{\hat{\psi}}^2$ is

$$\hat{\sigma}_{\hat{\psi}}^2 = \hat{\sigma}_e^2 \sum_j \frac{w_j^2}{n_j} = \text{MS}_{S/A} \sum_j \frac{w_j^2}{n_j} \tag{6.4}$$

We now have formulas for both $\hat{\psi}$ and the estimate of its standard error. Under the usual assumptions of independence, normality, and homogeneity of variance, and assuming H_0 ($\psi = 0$) is true, their ratio is distributed as t on $\Sigma n_j - a$ df. That is,

$$t = \frac{\hat{\psi}}{\hat{\sigma}_{\hat{\psi}}} = \frac{\Sigma w_j \bar{Y}_{.j}}{\sqrt{MS_{S/A} \Sigma w_j^2 / n_j}} \qquad (6.5)$$

We can test any contrast of means by choosing the proper weights and calculating the quantity defined in Equation 6.5. The null hypothesis can be tested against either directional or nondirectional alternatives. Procedures for evaluating the significance of this statistic will be discussed in Section 6.3.

To illustrate the calculations, we will construct the t statistic to be used in testing ψ_1 and ψ_3 (see Equations 6.1 and 6.2) for the means of Table 6.1. The t statistic required to test the H_0: $\psi_1 = 0$ is calculated by substitution into Equation 6.3:

$$t = \frac{\Sigma w_j \bar{Y}_{.j}}{\sqrt{MS_{S/A} \Sigma w_j^2 / n_j}} = \frac{\bar{Y}_{CON} - \bar{Y}_G}{\sqrt{MS_{S/A}[(1)^2/16 + (-1)^2/16]}}$$

$$= \frac{8.8 - 4.2}{\sqrt{15[1/16 + 1/16]}} = 3.359$$

The t for the contrast ψ_3 is

$$t = \frac{\bar{Y}_{CON} - (1/3)(\bar{Y}_G + \bar{Y}_L + \bar{Y}_{GL})}{\sqrt{MS_{S/A}[(1/144) + (1/144) + (1/144) + (1/16)]}}$$

$$= \frac{8.8 - 3.37}{\sqrt{15(12/144)}} = 4.860$$

We prefer to use unit contrast weights (i.e., 1 and -1) and sets of fractional weights such as $<1/2, 1/2, -1/2, -1/2>$ or $<1/3, 1/3, 1/3, -1>$ in which the positive and negative weights sum to $+1$ and -1, to emphasize that we are usually interested in comparing means or averages of sets of means. However, for testing hypotheses of the form $\psi = 0$, the result would be exactly the same if all weights were multiplied by any constant (e.g., if we used $<1, 1, 1, -3>$ or $<2, 2, 2, -6>$ instead of $<1/3, 1/3, 1/3, -1>$). Some people prefer to work with integer weights because this may make the calculations a bit simpler. In addition, if we wish to specify a contrast in a statistical package, we can generally use integer or decimal weights but not fractional ones. The reader should verify that the value of t remains at 4.86 if we use the weights $<1, 1, 1, -3>$. However, it should be noted that multiplying the original weights by 3 makes the value of the contrast and the width of any confidence interval for ψ three times as large.

6.2.3 The Sum of Squares Associated with a Contrast

The t statistic of Equation 6.5 provides one approach to testing hypotheses about contrasts. An alternative, but equivalent, test is based on partitioning the sum of squares for the A source of variance. We develop this approach now in order to emphasize the continuity between the analysis of variance and tests of contrasts, and to enhance our understanding of contrasts. To begin with, recall that SS_A is a function of the variability among the a group means. If this variability is significant, we conclude that the μ_j differ, but exactly where that difference lies is not clear. The idea in this section is that a hypothesis that can be represented as a contrast among the four group means can be tested by obtaining a part of SS_A and testing it against $MS_{S/A}$. The F ratio formed in this way is just the square of the t in Equation 6.5 and will lead to the same conclusions provided the criterion for significance is the square of that used for the t test. Intuitively, if the contrast represents a major reason for the variance of the four group means, its sum of squares will represent a large fraction of SS_A. We will use $SS_{\hat{\psi}}$ to label this sum of squares. A formula will be derived directly from Equation 6.5.

From Chapter 3 we know that $t^2_{df} = F_{1,df}$. Therefore, we square the rightmost expression in Equation 6.5 and, dividing numerator and denominator by $\Sigma_j w_j^2/n_j$, we have

$$t^2 = F = \frac{(\Sigma_j w_j \bar{Y}_{.j})^2/(\Sigma w_j^2/n_j)}{MS_{S/A}} \tag{6.6}$$

Because the numerator is the numerator of an F ratio, it is a mean square and is labeled $MS_{\hat{\psi}}$. Because this mean square is distributed on 1 df, the mean square and the sum of squares are the same: $MS_{\hat{\psi}} = SS_{\hat{\psi}}$. Therefore,

$$SS_{\hat{\psi}} = \frac{(\Sigma_j w_j \bar{Y}_{.j})^2}{\Sigma w_j^2/n_j} \tag{6.7}$$

We next want to show that $SS_{\hat{\psi}}$ is a fraction of SS_A, and that, because the size of $SS_{\hat{\psi}}$ depends upon the choice of contrast weights, it enables us to locate the source of the variability more precisely among the a means. We will consider a set of three questions based on the incentive study whose results are summarized in Table 6.1. Each question implies a specific null hypothesis, a set of weights for the treatment means, and a sum of squares calculation. The questions, null hypotheses, and weights are:

1. Does the mean of the control population differ from the mean of the three combined incentive populations?

$$H_{01}: \mu_{CON} - \frac{1}{3}(\mu_G + \mu_L + \mu_{GL}) = 0$$

2. Does the mean of the *GL* population differ from that of the other two combined incentive populations?

$$H_{02}: \frac{1}{2}(\mu_G + \mu_L) - \mu_{GL} = 0$$

3. Do the *G* and *L* population means differ?

$$H_{03}: \mu_G - \mu_L = 0$$

Using Equation 6.7 and the means given in Table 6.1 (note that $n = 16$), we can calculate a sum of squares for each contrast. For H_{01}, we have

$$SS_{\hat{\psi}_1} = \frac{[(1)(8.8) + (-1/3)(4.2) + (-1/3)(3.40) + (-1/3)(2.5)]^2}{(1/16)[(1^2) + (-1/3^2) + (-1/3^2) + (-1/3^2)]}$$

$$= 354.253$$

Dividing this by $MS_{S/A}$ (see Table 6.1), we have

$$F = \frac{354.253}{15} = 23.617$$

which is the square of 4.86, the value of the t calculated for this contrast in Section 6.2.2.

Similarly, we can show that $SS_{\psi_2} = 18.027$ and $SS_{\psi_3} = 5.12$. If we add the three SS terms we have calculated, we have $354.253 + 18.027 + 5.12 = 377.4$. It is no accident that this sum is the value of SS_A. We chose our three questions, and therefore the corresponding weights, so that this would be the case. Think of SS_A as a pie containing variability instead of apples. We can choose three, or $a - 1$ in general, contrasts in such a way that the pie is divided into nonoverlapping pieces. In our example, SS_{ψ_1} turned out to be the biggest piece; it accounted for 94% of the variability (SS_A) among the four group means.

To continue the analogy of SS_A to a pie, it can be divided into three pieces in many different ways. In general, it is possible to find infinitely many sets of $a - 1$ contrasts such that the members of each set account for all the df and all the variability in SS_A. For example, in our discussion of trend analysis in the next chapter, we will introduce hypotheses corresponding to the following set of three weights:

1. $-3,\ -1,\ \ 1,\ \ \ 3$
2. $-1,\ \ \ 1,\ \ 1,\ -1$
3. $-1,\ \ \ 3,\ -3,\ \ \ 1$

If you use these three sets of weights and the means in Table 6.1 to calculate three SS values, you will find that these contrast sums of squares again add to 377.4 (SS_A).

Will every set of $a - 1$ contrasts result in SS_{ψ} that add to SS_A? Not at all; in each of the preceding examples, the three contrasts had a particular property that resulted in their accounting for different portions of the variability, for nonoverlapping pieces of the pie. When this is the case, the contrasts are said to be *orthogonal*. Before defining this property more precisely, let's consider two examples in which orthogonality doesn't occur. First, suppose we wanted to test two null hypotheses:

$$H_{01}: \mu_{CON} - \mu_G = 0$$

and

$$H_{02}: \mu_{CON} - \frac{1}{3}(\mu_G + \mu_L + \mu_{GL}) = 0$$

Now if μ_G differs from μ_{CON}, there's a good chance that the average of all three incentive conditions differs from μ_{CON} because one component of that average differs from μ_{CON}. In other words, there's a positive relation between the values of the two contrasts. This lack of independence between the two contrasts is called *nonorthogonality*. In this example, it shows itself when we calculate the contrast sums of squares: $SS_{\psi_1} = 169.28$ and $SS_{\psi_2} = 354.253$. Their sum, 523.533, is greater than SS_A—the two pieces of the pie must overlap.

Our second example of nonorthogonality is even more extreme. Consider the following contrasts:

$$\psi_1 = (-1)\mu_{CON} + (1)\mu_G$$

$$\psi_2 = (-1)\mu_{CON} + (1)\mu_L$$

$$\psi_3 = (1)\mu_G + (-1)\mu_L$$

We can see that $\psi_3 = \psi_1 - \psi_2$, so that if we know the values of two of these contrasts we must also know the value of the third. Therefore, one of the three contrasts is totally redundant. The three sums of squares values total to less than SS_A in this example.

We can't always tell whether two contrasts are orthogonal or not by adding the sums of squares associated with them. However, there is a rule for distinguishing between orthogonal and nonorthogonal contrasts: two contrasts, $\psi_p = \Sigma_j w_{jp}\mu_j$ and $\psi_q = \Sigma_j w_{jq}\mu_j$, are orthogonal if their weights are uncorrelated.[1] If there are n scores at each level of A, the criterion for orthogonality can be stated as

$$\sum_j w_{jp}w_{jq} = 0 \tag{6.8}$$

Equation 6.8 follows from the fact that the correlation between the two sets of weights is zero if the covariance is zero. Because the sum, and therefore the mean, of each set of weights equals zero by definition, the covariance of the pth and qth set of weights is

$$\frac{1}{a}\sum_j (w_{jp} - \bar{w}_{.p})(w_{jq} - \bar{w}_{.q}) = \frac{1}{a}\sum_j w_{jp}w_{jq}$$

If the groups vary in size, the criterion is

$$\sum_j \frac{w_{jp}w_{jq}}{n_j} = 0 \tag{6.9}$$

As an example of nonorthogonality, consider H_{01} and H_{03} of Equations 6.1 and 6.2. We have

$$\sum_j w_{j1}w_{j3} = (1)\left[\frac{1}{3}\right] + (0)\left[\frac{1}{3}\right] + (0)\left[\frac{1}{3}\right] + (-1)(-1) = \frac{4}{3}$$

However, other pairs of contrasts are orthogonal. For example, if the n's are equal and the weights are $<1, -1, 0, 0>$ and $<-1/2, -1/2, 1/2, 1/2>$,

$$(1)\left[-\frac{1}{2}\right] + (-1)\left[-\frac{1}{2}\right] + (0)\left[\frac{1}{2}\right] + (0)\left[\frac{1}{2}\right] = 0$$

Two points about orthogonality deserve emphasis. First, whether or not two contrasts are orthogonal depends on the contrast weights, not on the values of the obtained group means. One way of thinking about this is that orthogonality depends on what questions are addressed by the contrasts, not on what the answers turn out to be. The

---

[1] If the sample means are independently and normally distributed, orthogonality implies statistical independence of the contrasts.

second point is that we choose to test particular contrasts because they are of substantive interest, whether or not they are orthogonal to one another. For example, researchers commonly test pairwise comparisons; these are not orthogonal, but are often of interest and should be tested when they are.

This last comment brings us back to one of the central issues of this chapter, the problem of controlling the inflation of Type 1 error rate when many contrasts are tested. With the preceding discussion of test statistics and of the relation between sums of squares and contrasts, we are finally ready to tackle the error rate problem.

6.3 THE PROPER UNIT FOR THE CONTROL OF TYPE 1 ERROR

As we argued in Section 6.1, when the number of hypothesis tests increases, so does the probability of Type 1 error. Therefore, if the probability of Type 1 error for a test is set without regard to how many tests might be conducted, the error rate for the entire collection of tests might rise to totally unacceptable levels. In this section, we argue that when testing hypotheses about contrasts, the proper unit for control of Type 1 error should not be the individual contrast, but a set of contrasts called a *family*. We consider ways of defining what is meant by a family of contrasts and procedures that set a limit on the probability of Type 1 error for the family.

We begin by distinguishing between the *error rate per contrast* (EC)—the probability that a single contrast results in a Type 1 error—and the *error rate per family* (EF)—the probability that a set, or family, of contrasts will have at least one Type 1 error. For a family of K independent tests,

$$EF = p(\text{at least one Type 1 error for the family})$$

$$= 1 - p(\text{no Type 1 error for the family})$$

$$= 1 - [p(\text{no Type 1 error on one test})]^K$$

The probability that a single test does not yield a Type 1 error is

$$p(\text{no Type 1 error}) = 1 - p(\text{Type 1 error})$$

$$= 1 - EC$$

Therefore,

$$EF = 1 - (1 - EC)^K \tag{6.10}$$

If a family consists of six independent tests that are each conducted at EC = .05, EF is $1 - (1 - EC)^6 = .265$, considerably more than .05. If the six tests are not independent, EF is less than .265, although its exact value is difficult to calculate (e.g., see Sidak 1967). In any event, the bigger the family, the more EF exceeds EC.

One extreme view we could take with respect to the discrepancy between EC and EF is simply to ignore EF. We could choose a value for EC without considering the total number of contrasts that might be made (say EC = .05), and test each contrast using this value. The obvious problem with this approach is that even if there were no true treatment effects, we would have a good chance of finding "significant" results if we performed a large enough number of tests, because the EF would then be large. Publishing such findings could result in wasted effort spent investigating and attempting to replicate effects that did not exist. In view of this, we want to choose ECs for the members of a family of contrasts that will keep the probability of a Type 1 error for the family within acceptable bounds. The EC selected will depend on the size of the family; to keep the EF constant, we must decrease EC as family size increases. Therefore, before deciding what values of EC to use, we must decide exactly what we mean by the term *family of contrasts*.

An investigator working in a research area over a period of years might perform hundreds of experiments and test thousands of hypotheses. If we considered these thousands of tests to form a single family and set EF equal to .05, ECs for the individual tests would be infinitesimally small. Although this ultraconservative approach would result in an extremely low Type 1 error rate, the Type 2 error rate would soar to unacceptably high levels. In other words, the investigator could be confident that significant results reflected real effects, but could miss finding many real effects. Because power decreases when Type 1 error is reduced, the definition of family must be based on a compromise between the concerns about Type 1 and Type 2 errors.

Two candidates for what should be considered to be a family have received serious support:

1. All the contrasts made in a single experiment
2. All the contrasts associated with a single source of variance in a single experiment

According to the first suggestion, all the contrasts performed in a three-factor design would constitute a single family. According to the second, there would be seven families, one for each main effect and interaction.

Defining the family as the set of contrasts associated with a single source of variance results in a reasonable compromise between Type 1 and Type 2 errors. Another reason for defining the family in this way is that investigators usually precede tests of contrasts by overall F tests. It will be easier to relate the results of contrast tests to the overall test if they are based on the same unit, a single source of variance.

Another factor in determining the size of the family of tests is whether we have planned some specific set of tests before viewing the data or whether we decide to test certain differences among means that appear large when the data are viewed. We consider this issue next.

6.4 PLANNED VERSUS POST HOC CONTRASTS

Consider two scenarios:

1. We have the four types of incentives introduced in the example of Table 6.1. We find that the largest difference for any pair of means is between CON and GL and that this difference is significant at the .05 level.
2. Before conducting the experiment, we predict on the basis of theory or previous research that CON will differ from GL. After the data are collected, we find that this pairwise comparison is significant at the .05 level.

We should have much more confidence that there is a real difference between CON and GL in the second scenario than in the first. If the null hypothesis is true and we choose (without looking at the data) a single contrast to be tested at $\alpha = .05$, the probability of a Type 1 error is .05. This is not true if we examine the data and then choose the largest pairwise difference, CON against GL, for testing. This is equivalent to carrying out six (the number of possible pairwise comparisons) significance tests; the probability of at least one Type 1 error is more than .05, the probability for a single test. The two scenarios just described form the basis for the distinction between planned (or a priori) and post hoc (or a posteriori) contrasts.

By planned contrasts, we mean those chosen before the data have been seen. Typically, a family of planned contrasts is fairly small, and the EF can be held to any desired level by making use of a result that follows from Equation 6.10. Recall that

$$EF \leqslant 1 - (1 - EC)^K$$

with the equality holding when the contrasts are independent. Expansion of $(1 - EC)^K$ enables us to show that, for small values of EC (and .05 is small), $(1 - EC)^K$ is approximately equal to $1 - K(EC)$. Therefore, EF is approximately K times as large as EC.

In general, if K tests are conducted with error rates $EC_1, EC_2, \ldots, EC_K$,

$$EF \leqslant \sum_i EC_i \tag{6.11}$$

where EC_i is the probability of a Type 1 error for the ith contrast. The relationship expressed in Equation 6.11 is known as the *Bonferroni inequality,* and it is the basis for a procedure for testing planned contrasts. From the Bonferroni inequality we see that if each of the K contrasts that make up the family is tested at $EC = EF/K$, the probability of a Type 1 error for the family cannot exceed EF. If, for example, the family contains four planned contrasts, EF will not be larger than .05 if each member contrast is tested at $EC = .0125$.

Planned contrasts may be conducted whether or not the overall F tests are significant. In fact, there is no logical reason why tests of planned contrasts cannot be conducted instead of the standard analysis of variance. However, investigators who conduct tests of families of planned contrasts usually test the overall hypotheses as well, and for reasons that are discussed later in the chapter we recommend this practice.

When we talk of post hoc contrasts, we refer to situations in which detailed analyses are conducted following the standard overall tests of ANOVA. The ANOVA has indicated that "something is going on" in the data, and specific contrasts suggested by the data are tested to determine whether they reflect effects large enough to be taken seriously. The effective size of a family of post hoc contrasts is determined not by the number of contrasts actually tested but by those that conceivably might have been tested, had the data suggested it was worth doing so.

In summary, a family of contrasts is associated with a single source of variance. The size of the family is determined by the number of contrasts that might be tested. For a family of planned contrasts, the decision about how many contrasts to test is made on logical or theoretical grounds or on the basis of previous research. A family of post hoc contrasts, on the other hand, might be considered to consist of every conceivable contrast that might be tested, or might be restricted to a well-defined subset such as the set of all possible pairwise contrasts.

6.5 CONTROLLING TYPE 1 ERROR FOR FAMILIES OF PLANNED CONTRASTS

6.5.1 The Bonferroni *t* Procedure (sometimes referred to as the Dunn procedure)

As mentioned in the previous section, EF can be held to any desired level through use of the Bonferroni inequality. If each of the K members of a family are tested at EF/K, the probability of a Type 1 error for the entire family will not exceed EF. Suppose that for the incentive experiment of Table 6.1 we decided, before seeing the data, that the following three contrasts of means were of interest:

$$\psi_1 = \mu_{CON} - \frac{\mu_G + \mu_L + \mu_{GL}}{3} \tag{6.12}$$

$$\psi_2 = \mu_G - \mu_{GL}$$

$$\psi_3 = \mu_L - \mu_{GL}$$

Using the Bonferroni procedure, we can set the EF at .05 for the family of three contrasts by setting the EC at .05/3, or .017. One way to do this is to use a statistical software package; such packages print out the exact significance levels for each test. If we conduct the test using a calculator, we need to find the critical t value that cuts off .017/2 = .0085 in each tail (assuming a two-tailed test); however, such values are not available in the usual table of the t distribution. Therefore, we have constructed Table D.8, which presents critical values of t as a function of the number of contrasts (K) tested. From this table, we find that when EF = .05, df_{error} = 60, and K = 3, the critical t value is 2.463. An alternative approach uses an approximation that expresses t in terms of the normal standard deviate, z, for which extensive tables are readily available. One such approximation is

$$t_\alpha = z_\alpha + \frac{z_\alpha^3 + z_\alpha}{4(df_{error} - 2)} \tag{6.13}$$

We want a value of z that cuts off a total of .017 in both tails of the normal distribution, or .0085 in each tail. By linear interpolation, we obtain z_α = 2.39. Substituting into Equation 6.13, we have t = 2.459, which is quite close to the value we obtained from Table D.8. We will use 2.46 as the critical value for testing the contrasts in Equation 6.12.

In carrying out these tests, we will conclude that the population contrast is not zero if the t we calculate (using Equation 6.5 or the equivalent) is greater than 2.46 or less than −2.46. In Section 6.2.2, we calculated the t for the first contrast as t = 4.86. Therefore, the first null hypothesis can be rejected. The t values for the other two contrasts are 1.242 and .657. Both are considerably smaller than 2.51, and therefore these contrasts do not differ significantly from zero.

A few authors prefer that the members of a family of planned contrasts be chosen so that they are mutually orthogonal, because using $EC = EF/K$ is conservative when the contrasts are not orthogonal and because of the concern that answers to questions asked of the data should provide nonredundant information. Although we should remember that information obtained by testing nonorthogonal contrasts is, to some extent, overlapping, we believe that the primary criterion for deciding which contrasts to test is whether the questions they ask are interesting, not whether the contrasts are orthogonal.

An additional point that should be mentioned is that ECs need not be equal for all members of a family. If one member of a family is deemed to be more interesting than the other two, we may be willing to have a higher Type 1 error rate for testing it in order to achieve greater power for the test of that contrast. If we wish to hold EF to .05, we can conduct the more interesting test at $EC = .025$ and split the remaining .025 equally between the other two tests. If we use such a strategy, it is crucial to understand that the decision to have a higher EC for one contrast than for others to be tested must be made on a priori grounds. Testing some contrast at a higher α level than other contrasts on the basis of the observed pattern of means is cheating, pure and simple.

Confidence intervals can also be found for members of a family of contrasts. The probability is at least $1 - EF$ that for *all K* members of a family, ψ is contained within the limits

$$\hat{\psi} \pm t_{EF}\hat{\sigma}_{\hat{\psi}} \tag{6.14}$$

To obtain the .95 confidence intervals for the three members of the family in Equation 6.12, we need the critical value of t and the values of $\hat{\sigma}_{\hat{\psi}}$, the standard error of the contrast. For $EF = .05$ (two-tailed), we found the necessary value of t to be 2.46. The expression for the standard error was presented in Equation 6.4:

$$\hat{\sigma}_{\hat{\psi}} = \sqrt{MS_{S/A}\Sigma\frac{w_j^2}{n_j}}$$

Therefore, substituting into Equation 6.14, we obtain the interval for ψ:

$$(8.8 - 3.37) \pm 2.46\sqrt{15\left[\frac{4}{3}\right]\left[\frac{1}{16}\right]} = 2.68, 8.18$$

The intervals for the other two contrasts in the family extend from .33 to 3.07 and $-.47$ to 2.27.

When confidence intervals are based upon the EF, as in the preceding example, they are interpreted somewhat differently than when they are based on the EC. In order to see the distinction, assume that we run the experiment of Table 6.1 many times. In each replication we calculate the confidence interval for the three contrasts used in the previous example. The critical value of t will be 2.46, the value based on a family of three contrasts. We expect that in .95 of the replications, each of the three intervals will contain the population contrast value it estimates. In other words, the probability is .95 that the confidence intervals for all members of the family of contrasts will contain their corresponding ψ value. Because of this property, intervals based on EF are referred to as *simultaneous confidence intervals*. If we had based the confidence interval on the EC, and used $t = 1.98$, the critical value of t ordinarily used for $\alpha = .05$, we would expect that .95 of all intervals constructed (as opposed to .95 of all families of three intervals) would contain the true value of ψ. Note that if a given interval has a .95 probability of containing the population parameter, the probability that a family of three such intervals all contain their respective parameters will be less than .95.

Although we have discussed the use of the Bonferroni inequality with multiple t tests in this section, it should be emphasized that the Bonferroni procedure can be used with any set of statistical tests, whether parametric or nonparametric. Also, several modifications of the Bonferroni procedure have been proposed (Sidak 1967, Holland and Copenhaver 1988). These provide slightly more power at a cost of increased computational complexity.

Finally, we note again that most statistical packages allow convenient tests of contrasts and provide significance levels that can be compared with the ECs based on the Bonferroni inequality.

6.5.2 Dunnett's Test for Comparing Treatment Groups with a Control Group

There are occasions on which the contrasts of interest are between each of the $a - 1$ treatment groups and a control group. We could, of course, use the Bonferroni t for each of the contrasts. However, because no two of the set of $a - 1$ contrasts are orthogonal, the Bonferroni approach would be somewhat conservative. Dunnett (1955) has developed a test that takes account of the nonorthogonality in this special case. We calculate the t statistic of Equation 6.5 with weights of 1 and -1 for the two means involved in the comparison. Critical values for evaluating this statistic are found in Table D.9. We must find $d_{\mathrm{EF},K,\mathrm{df}_e}$, where K is the total number of means including that of the control group, and df_e are the degrees of freedom associated with the MS_e, the error mean square. For a one-factor design, $K = a$, $\mathrm{df}_e = \Sigma n_j - a$, and $\mathrm{MS}_e = \mathrm{MS}_{S/A}$. Table D.9 is two-tailed. For a one-tailed test with $EF = .05$, you would look up $\alpha = .10$. Assuming a two-tailed test with EF = .05 in the example of Table 6.1 ($K = 4$, $\mathrm{df}_e = 60$), the critical d value would be 2.43. For comparison, consider the Bonferroni procedure; exactly the same test statistic is calculated as in the Dunnett test. Because each of the incentive conditions is contrasted with the control, we enter Table D.8 with $K = 3$ and find the criterion value to be 2.46. The Dunnett criterion is generally smaller than the Bonferroni, so that the Dunnett test is slightly more powerful and yields slightly narrower simultaneous confidence intervals.

6.6 CONTROLLING TYPE 1 ERROR FOR POST HOC CONTRASTS

6.6.1 The General Case

Often, observed patterns in the data suggest the presence of effects that had not been anticipated and chosen for a priori testing. When the corresponding null hypotheses are tested to determine whether these effects are worth taking seriously, we are obligated to be quite conservative in choosing EC. We must remember that in testing contrasts "after the fact" we are, in effect, investigating the family of all possible contrasts.

Scheffé (1959) has developed a procedure for controlling Type 1 error when all possible contrasts associated with a source of variance belong to the family. According to the Scheffé procedure, if the source of variance with which the contrast is associated has f degrees of freedom, the value of t obtained for the contrast is evaluated against

$$S = \pm\sqrt{fF_{\mathrm{EF},f,\mathrm{df}_e}} \qquad (6.15)$$

where df_e refers to the degrees of freedom associated with the error term. For example, in performing post hoc contrasts on the means of Table 6.1, $f = a - 1 = 3$, $\mathrm{df}_e = 60$, and $F_{.05,3,60} = 2.76$. Therefore, to test a contrast while holding the EF at .05, the criterion value of the calculated t statistic is $\sqrt{(3)(2.76)}$, or 2.88; the contrast is significant if $t > 2.88$ or $t < -2.88$.

Confidence intervals can also be obtained. The probability is $1 - \mathrm{EF}$ that the values of all possible contrasts in a family are simultaneously contained by intervals of the form

$$\hat{\psi} \pm \hat{\sigma}_{\hat{\psi}}\sqrt{fF_{\mathrm{EF},f,\mathrm{df}_e}} \qquad (6.16)$$

In the one-factor example of Table 6.1, the standard error of the contrast ($\hat{\sigma}_{\hat{\psi}}$) is $\sqrt{\mathrm{MS}_{S/A}\Sigma w_j^2/n_j}$. To obtain a familywise confidence interval for ψ_1 of Equation 6.12, recall that $\mathrm{MS}_{S/A} = 15$, $\Sigma w_j^2 = 4/3$, and $n = 16$. Therefore, the interval is

$$(8.8 - 3.37) \pm \sqrt{15\left[\frac{4}{3}\right]\left[\frac{1}{16}\right]}(2.88) = 2.210,\ 8.650$$

Note that this interval is slightly wider than that obtained earlier with the Bonferroni procedure. This is the price we pay for selecting contrasts to evaluate on the basis of looking at the data.

Experimenters who have used both the standard ANOVA tests and the Scheffé procedure have sometimes been perplexed to find significant main effects or interactions but no significant contrasts. The source of this apparent contradiction is the fact that the overall F test has exactly the same power as the *maximum possible contrast* tested by the Scheffé procedure. However, the maximum possible contrast may be of little interest and may not have been computed [it could be something like $(11/37)\mu_1 + (26/37)\mu_2 - (17/45)\mu_3 - (28/45)\mu_4$]. Despite the fact that a significant overall F test indicates that at least one contrast is significant by the Scheffé criterion, there is no guarantee that any obvious or interesting contrast will be significant.

Scheffé, recognizing the conservative nature of his procedure, has suggested that EF be set to .10 when using it. This may strike some as heresy, but EF and EC are different concepts, so it may not be necessary to demand that traditional EC levels of significance be applied to EF. Even if EF is set to .10, EC will generally be quite low.

As with the Bonferroni and Dunnett procedures, there is no logical necessity that the Scheffé tests of contrasts be preceded by a significant overall F. On the other hand, even the maximum possible contrast will not be significant unless the overall F test is. There seems little point in spending energy on a series of Scheffé post hoc tests unless first determining, by the overall F test, whether there is any possibility of finding a significant contrast. In many cases, the researcher will achieve more power by designating every contrast of possible interest a priori and using the Bonferroni instead of the Scheffé procedure. Although the power of the Bonferroni tests decreases as the size of the family increases, the number of tests can be quite large before the critical value exceeds that of the Scheffé procedure. Again consider the example of Table 6.1. No matter how many contrasts are tested, the criterion value of S is 2.88 when EF = .05. For comparison, turn to Table D.8, which presents the critical values for the Bonferroni test. With df = 60 and EF = .05, the critical value is less than 2.88 as long as less than nine contrasts are planned. If $K = 9$, the critical values for the two procedures are about the same. For $K > 9$, the value of S will be smaller and the S test therefore will be more powerful and give rise to narrower confidence intervals. For more detailed comparisons of the two procedures, articles by Dunn (1961), Perlmutter and Myers (1973), and Keselman (1974) should be consulted.

6.6.2 Pairwise Contrasts

Because the Scheffé procedure can be thought of as controlling EF for the family of all possible contrasts, it is very conservative. If our interest is confined to pairwise contrasts, more powerful post hoc procedures are available.

If the factor A has a levels, there are $a(a - 1)/2$ possible pairwise contrasts among the means of A. For the example of Table 6.1, assuming the two-tailed EF is .05, the Bonferroni criterion for the six possible pairwise comparisons would be 2.73. The Scheffé criterion would be 2.88. A smaller critical value for this situation is obtained from a procedure developed by Tukey (1953). The significance tests and confidence intervals that are part of Tukey's procedure are based on the sampling distribution of q, the Studentized range statistic. This statistic is identified as the range of a set of observations (largest $-$ smallest) divided by an estimate of the standard deviation of the population from which the sample of observations has been drawn. It is assumed that the observations are drawn independently from a normally distributed population. For testing pairwise contrasts on the means of A, the a observations are the treatment group means. The assumption that they are independently sampled from the same normal distribution implies that the null hypothesis is true and that there is homogeneity of variance. Our best estimate of the standard error of the mean is $\hat{\sigma}_{\bar{Y}} = \sqrt{\hat{\sigma}^2/n}$, where n is the number of scores going into each mean in the contrast. For the one-factor design,

$$\hat{\sigma}_{\bar{Y}} = \sqrt{\frac{MS_{S/A}}{n}} \qquad (6.17)$$

For any pairwise contrast, $\bar{Y}_{.j} - \bar{Y}_{.j'}$ (where $\bar{Y}_{.j}$ is the larger of the two means), if the null hypothesis is true and the assumptions met, the quantity

$$q = \frac{\bar{Y}_{.j} - \bar{Y}_{.j'}}{\sqrt{MS_{S/A}/n}} \qquad (6.18)$$

will be distributed as q_{a,df_e}; critical values of q are presented in Table D.10. By comparison, the t statistic we have calculated for pairwise contrasts in the procedures discussed so far is

$$t = \frac{\bar{Y}_{.j} - \bar{Y}_{.j'}}{\sqrt{\left(\dfrac{2}{n}\right) MS_{S/A}}} \qquad (6.19)$$

Comparing Equations 6.18 and 6.19, we see that $t = q/\sqrt{2}$. Therefore, to use the Tukey procedure to control EF for the entire family of pairwise contrasts, compare the calculated value of t to $q_{EF,a,df_e}/\sqrt{2}$. Entering Table D.10 with $a = 4$ and EF $= .05$ (note that the rejection region for the q distribution is all in the upper tail), we find the critical value of q is 3.74. We compare the calculated t value for any pairwise contrast with $3.74/\sqrt{2}$, or 2.64. For comparison purposes, note that for $K = 6$, $df_e = 60$, and EF $= .05$, the critical Bonferroni value is 2.73. This is larger than the value of 2.64 obtained when the Tukey procedure is used; the latter will yield narrower confidence intervals and more power when all pairwise contrasts are tested. This advantage becomes somewhat more pronounced as the number of levels of A increases.

In the preceding discussion, we calculated t and compared it to $q/\sqrt{2}$ to emphasize that all our procedures involve the same test statistic; only the criterion changes. However, there is an alternative, but equivalent, way to run the Tukey test that is more efficient for evaluating all pairwise differences. The only restriction is that this method requires equal group sizes.[2] The steps are

1. Rank-order all a means.
2. Define a critical difference:

$$d_{crit} = q_{a,df_e,EF}\sqrt{\frac{MS_{S/A}}{n}}$$

It follows from Equation 6.18 that if the calculated value of q exceeds the critical value of q, the difference between the means must exceed d_{crit}.

3. Beginning with the largest observed difference and proceeding in order of size of the difference, compare each in turn with d_{crit}. Any observed differences that exceed d_{crit} are significant. Stop testing when a difference is encountered that is not significant.

Tukey's approach also provides simultaneous confidence intervals for pairwise contrasts. The probability is $1 - $ EF that the values of all possible pairwise contrasts are contained within intervals of the form

$$\hat{\psi} \pm q_{EF,a,df_e}\hat{\sigma}_{\bar{Y}} \qquad (6.20)$$

where $\hat{\sigma}_{\bar{Y}}$ is defined by Equation 6.17. In the example of Table 6.1, the .95 Tukey confidence interval for the difference between the control and gain groups would be

$$(8.8 - 4.2) \pm 3.74\sqrt{\frac{15}{16}} = .979, \ 8.221$$

---

[2] Using the Tukey test in the manner just described does not require the n's to be equal.

For comparison purposes, note that the Bonferroni interval (assuming $K = 6$) is

$$(8.8 - 4.2) \pm 2.727 \sqrt{15 \left[\frac{2}{16}\right]} = .866, 8.334$$

which is slightly wider than the Tukey interval. If all pairwise comparisons are to be made, the Tukey test has greater power and its confidence intervals are narrower.

We have proceeded in this section as if all group sizes were the same. If the group sizes in a comparison are unequal, say $n_j \neq n_{j'}$, the $2/n$ in Equation 6.19 is replaced by $(1/n_j) + (1/n_{j'})$; this more general form of the t statistic is also tested against $q/\sqrt{2}$. This form of the Tukey test is often referred to as the Tukey-Kramer test.

6.7 SUMMARY OF TESTING PROCEDURES

To summarize the developments of the preceding two sections, we have discussed a number of procedures for controlling familywise error rate. Each procedure can be thought of as employing exactly the same test statistic,

$$t = \frac{\Sigma_j w_j \bar{Y}_{.j}}{\sqrt{MS_{S/A} \Sigma_j w_j^2 / n_j}}$$

but with different criteria for significance. The criteria and the conditions under which each applies are summarized in Table 6.2.

Bear in mind that $t^2 = F$. In the case of the Scheffé procedure, or when a computer package prints out F ratios for tests of contrasts, it may be more convenient to use the F statistic. Then the criterion should just be the square of the criterion stated in Table 6.2. Also, note that the foregoing equation for t, and the criteria for its evaluation, are

TABLE 6.2   CRITERIA ASSOCIATED WITH DIFFERENT PROCEDURES FOR CONTROLLING TYPE 1 ERROR

Procedure	Criterion	When applied
Bonferroni t	$t_{EF/K}$	Testing K planned contrasts
Dunnett	$d_{EF,a}$	Comparing treatment groups with a control group
Scheffé	$\sqrt{fF_{EF,f,df_e}}$	Testing any post hoc contrast
Tukey/Tukey-Kramer	$q_{EF,a}/\sqrt{2}$	Testing any pairwise contrast among a means

All criteria are applied to the same test statistic:

$$t = \frac{\Sigma_j w_j \bar{Y}_{.j}}{\sqrt{MS_e \Sigma_j w_j^2 / n_j}}$$

Note: MS_e and df_e are the error mean square and its degrees of freedom. For the Scheffé criterion, f is the degrees of freedom associated with the omnibus test of the main or interaction sources of variance.

easily generalized for use in multifactor designs. We will consider such applications later in this chapter.

6.8 TESTS WHEN THERE IS HETEROGENEITY OF VARIANCE

When the assumption of homogeneity of variance has been violated, we calculate the test statistic differently but use the criteria listed in Table 6.2. For pairwise comparisons, the new test statistic, t^*, is just the square root of the Brown-Forsythe statistic, F^*. This was presented in Section 4.5.6 as an omnibus F test for situations in which variances were heterogeneous. A more general form of t^*, sometimes referred to as Welch's test, is applicable to any set of contrast weights. The formula is presented in Table 6.3 together with a numerical example. Note that the df_e are a function of the variances (as was the case in using F^*). To use this statistic in conjunction with any of the criteria of Table 6.2, round the df_e to the nearest integer and use this value to enter the appropriate table of critical values.

This approach has been shown to provide tests that are quite robust and powerful for all cases that have been investigated (e.g., Brown and Forsythe 1974b, Keselman, Games, and Rogan 1979, Kohr and Games 1977, Tamhane 1979).

TABLE 6.3 AN EXAMPLE OF THE CALCULATION OF t^* TO TEST CONTRASTS WHEN VARIANCES ARE HETEROGENEOUS

	A_1	A_2	A_3	A_4
$\bar{Y}_{.j}$	20	22	24	26
$\hat{\sigma}_j^2$	4	7	17	16
n_j	9	12	8	10

H_0: $(-1/2)(\mu_1 + \mu_2) + (1)\mu_3 = 0$

Using the Welch procedure, we get

$$t^* = \frac{\Sigma w_j \bar{Y}_{.j}}{\sqrt{\Sigma w_j^2 \hat{\sigma}_j^2 / n_j}}$$

$$= \frac{-(1/2)(20 + 22) + 24}{\sqrt{(-1/2)^2(4/9) + (-1/2)^2(7/12) + (1)^2(17/8)}}$$

$$= \frac{3}{\sqrt{2.3819}} = 1.94$$

The df associated with this t^* statistic are

$$df = \frac{(\Sigma w_j^2 \hat{\sigma}_j^2 / n_j)^2}{\Sigma w_j^4 \hat{\sigma}_j^4 / n_j^2 (n_j - 1)}$$

$$= \frac{2.3819^2}{(-1/2)^4 4^2/(81)(8) + (-1/2)^4 7^2/(144)(11) + (1)^4 17^2/(64)(7)} = 8.75$$

6.9 OTHER PROCEDURES THAT MIGHT BE ENCOUNTERED

In a recent review of pairwise multiple comparison procedures, Jaccard, Becker, and Wood (1984) reported an informal survey of the psychological literature in which they found that many investigators use tests that do not adequately control Type 1 error rate. These procedures include the use of multiple t tests without any regard for EF, the Newman-Keuls test, the Duncan multiple-range test, and the Fisher least significant difference (LSD) test. We describe the last three briefly because they are still commonly encountered.

A characteristic of the Tukey procedure is that the more levels of A there are, the bigger $|\bar{Y}_{.j} - \bar{Y}_{.j'}|$ has to be in order for the difference to be significant; q_a increases with a. Moreover, the same criterion is used to test all pairwise differences. The Newman-Keuls procedure is a sequential test in which if there are a means, q_a is used to test the most extreme pairwise difference, just as in the Tukey test. However, if the most extreme difference is significant, a more liberal criterion, q_{a-1}, is used for the next most extreme differences, and so on. The advantage is that all pairwise differences except the largest one are tested with more power than would be the case with the Tukey procedure. However, the disadvantage is that Type 1 error is no longer adequately controlled if there are more than three levels of A.

The Duncan test is another sequential procedure. The EC is allowed to increase as the number of means tested increases; the rationale is that designs with more levels are more likely to contain real differences. For a test on a ordered means, the probability of a Type 1 error is set to $1 - (1 - \alpha)^{a-1}$. This can result in a very high Type 1 error rate. For example, if $\alpha = .05$, the Type 1 error probability for a test on 10 ordered means is .40.

The Fisher least significance difference test is a two-stage procedure. No tests of contrasts are conducted unless the overall null hypothesis is rejected by a standard ANOVA. However, if the overall null hypothesis can be rejected, all pairwise differences of interest may be tested by using the standard t, without regard for how many pairwise differences there are. In other words, once the significant ANOVA test has indicated that "something is going on," the familywise error rate is not controlled.

6.10 CONTRASTS IN MULTIFACTOR DESIGNS

Although the developments of this chapter have been in terms of the one-factor design exemplified in Table 6.1, the calculations and procedures for controlling EF are readily extended to the factorial designs of Chapter 5. We begin by introducing an example of a two-factor design. Table 6.4 presents cell means and an ANOVA for a study in which subjects were sampled from three clinical populations and assigned to one of four types of therapies. There are six subjects in each of the 12 cells. The $A \times B$ interaction is significant at $\alpha = .05$, as are the main effects of A and B. Therefore, we can reject the null hypotheses:

TABLE 6.4   MEANS AND ANOVA TABLE FOR AN EXAMPLE OF A TWO-FACTOR DESIGN

(a) Group means

		Clinical population (B)			
		Depressives (DP)	Schizophrenics (SC)	Bipolars (BI)	$\bar{Y}_{j\cdot}$
	Control (CON)	2	3	−2	1.0
Therapy	Analytic therapy (AT)	8	7	12	9.0
type (A)	Behavior therapy (BT)	10	8	12	10.0
	Cognitive therapy (CT)	12	4	8	8.0
	$\bar{Y}_{\cdot\cdot k}$	8.0	5.5	7.5	
			$\bar{Y}_{\cdot\cdot\cdot} = 7.0$		

(b) ANOVA ($n = 6$, $MS_{S/AB} = 10.00$)

SV	df	SS	MS	F
A	3	900	300	30.0
B	2	84	42	4.2
AB	6	324	54	5.4
S/AB	60	600	10	

$$\alpha_{CON} = \alpha_{AT} = \alpha_{BT} = \alpha_{CT} = 0$$

$$\beta_{DP} = \beta_{SC} = \beta_{BI} = 0$$

$$(\alpha\beta)_{CON,DEP} = \cdots = (\alpha\beta)_{CT,BI} = 0$$

Now we wish to extend the procedures developed so far in this chapter to delineate possible sources of these omnibus effects.

6.10.1 Further Analyses of Main Effects

Suppose that we wish to test whether the mean of the populations of control subjects differs from the mean of all the therapy populations. In other words, we wish to test

$$H_0: \psi = \frac{1}{3}(\mu_{AT} + \mu_{BT} + \mu_{CT}) - \mu_{CON} = 0 \qquad (6.21)$$

This contrast is only one of many possible contrasts on the row (therapy, A) or column (clinical population, B) means. In general, to test such contrasts, we use the t statistic of Equation 6.5, replacing $MS_{S/A}$ by $MS_{S/AB}$, and replacing n_j by bn if we are contrasting the A conditions, or by an if we are contrasting the B conditions. Thus, to test the null hypothesis presented in Equation 6.21, we note that there are 6×3, or 18 (bn), scores in each therapy condition, and

$$\hat{\psi} = \frac{1}{3}(9 + 10 + 8) - 1 = 8$$

$$MS_{S/AB} = 10$$

$$\sum_j w_j^2 = 3(\frac{1}{3})^2 + (-1)^2 = \frac{4}{3}$$

Therefore,

$$t = \frac{\hat{\psi}}{\sqrt{MS_{S/AB}\Sigma_j w_j^2/bn}} = \frac{8}{\sqrt{10(4/3)/18}} = 9.30$$

Evaluation of the significance of this or any other t based on the marginal (row or column) means follows the guidelines of Table 6.2. For example, suppose this test is only one of five tests planned on the A marginal means. Then we will enter Table D.8 with $K = 5$ and $df_e = 60$. Or, assuming EF = .05, each t would be evaluated in the standard table of t values (D.3) against the critical value required at the .01 (EF/5) level. On the other hand, if we were interested only in all possible pairwise comparisons among the four therapy conditions, we would use the Tukey procedure, entering the table of q (Studentized range) values with $a = 4$ and $df_e = 60$.

One other point should be noted. Authors and computer packages often present values of F instead of t when reporting tests of contrasts. Because a contrast has 1 df, $F = t^2$, or we could calculate F directly as a sum of squares divided by $MS_{S/AB}$. We will refer to the numerator sum of squares as $SS_{p(A)}$, the sum of squares for the pth contrast among the $\bar{Y}_{.j.}$. This sum of squares is calculated by inserting the appropriate weights and means into Equation 6.7 and replacing n_j by bn, the number of scores upon which each mean is based. When the null hypothesis of Equation 6.21 is tested, the result is

$$SS_{p(A)} = \frac{[(1/3)(9 + 10 + 8) - (1)]^2}{[(3)(1/3)^2 + (-1)^2]/(6)(3)} = 864$$

Dividing by $MS_{S/AB}$, we obtain $F = 864/10 = 86.4$, the square of the t calculated earlier.

Although the calculations have been illustrated by contrasting the therapy (row) means, we could also test contrasts on the clinical population (column) means. As an exercise, the reader should verify that contrasting the mean of the BI populations against that for the combined SC and DP populations yields $SS_{q(B)} = 9$ and $t = .949$.

6.10.2 Further Analyses of Interaction Effects

Given the significant interaction of Table 6.4, we can only conclude that the effects of therapy differ across populations (or, equivalently, that the differences among the clinical population means depend upon the method of therapy used). The pattern of cell means suggests some possible reasons for the interaction. In particular, the contrast between CON and the three therapies, stated in Equation 6.21, appears to vary over clinical populations. We can see this by calculating an estimate of the population contrast for each clinical population:

$$\hat{\psi}_{p(A)/B_1} = \frac{1}{3}(8 + 10 + 12) - 2 = 8$$

$$\hat{\psi}_{p(A)/B_2} = \frac{1}{3}(7 + 8 + 4) - 3 = 3.33 \tag{6.22}$$

$$\hat{\psi}_{p(A)/B_3} = \frac{1}{3}(12 + 12 + 8) - (-2) = 12.67$$

where $p(A)/B_k$ refers to a particular contrast among the a means at the kth level of B. In this example, the average of the three contrasts is

$$\hat{\psi}_{p(A)} = \frac{8 + 3.33 + 12.67}{3} = 8 \tag{6.23}$$

The same result is obtained if we weight the row means of Table 6.4:

$$\hat{\psi}_{p(A)} = \frac{1}{3}(9 + 10 + 8) - (1)(1) = 8$$

If we look at the three contrast effects calculated in Equation 6.22, it appears that they vary over the levels of B; that is, the difference between the CON mean and the mean for the three therapies seems to depend upon the clinical population. This suggests that we test whether the variance of the three effects we have calculated is significant. In general, the issue is whether there is a significant $p(A) \times B$ source of variance. We could also contrast the three populations under each therapy and ask whether those contrasts effects vary as a function of therapy. We then would test $q(B) \times A$, where $q(B)$ refers to the particular contrast among the levels of B. The calculations needed to test $p(A) \times B$ [or $q(B) \times A$] are considered next.

Is $\psi_{p(A)}$ the Same at Each Level of B?

This question implies the null hypothesis

$$H_0: \psi_{p(A)} \text{ is the same at each level of } B \tag{6.24}$$

or

$$H_0: p(A) \times B = 0$$

or

$$H_0: \sum_j w_{jp}\mu_{j1} = \sum_j w_{jp}\mu_{j2} = \sum_j w_{jp}\mu_{j3}$$

Note that w_{jp} is the weight associated with A_j for the pth contrast.

The null hypothesis of Equation 6.24 can be tested by doing the following calculation $F = MS_{p(A) \times B}/MS_{S/AB}$. Because we are testing the variance of b contrast effects (for example, the variance of the three contrast effects in Equation 6.21), the numerator of this F ratio is on $b - 1$ df. Another way of thinking about the numerator df is that $p(A)$ has 1 df and B has $b - 1$ df; therefore $p(A) \times B$ has $1 \times (b - 1) = b - 1$ df and $MS_{p(A) \times B} = SS_{p(A) \times B}/(b - 1)$. The definition of $SS_{p(A) \times B}$ is somewhat similar to that of the sums of squares for a main effect. Recall that for a one-factor design with n scores in each group, SS_A is defined as n times the sum of the squared deviations of group means about the grand mean. When we consider the variability of contrast effects instead of group means, the sum of squares is n times the sum of squared deviations of the contrast effects about their average, divided by the sum of squared weights. That is,

$$SS_{p(A) \times B} = \frac{n\Sigma_k(\hat{\psi}_{p(A)/B_k} - \hat{\psi}_{p(A)})^2}{\Sigma_j w_{jp}^2} \tag{6.25}$$

From Equations 6.22 and 6.23, we find that $\hat{\psi}_{p(A)/B_k} = 8$, 3.33, and 12.67, and their mean, $\hat{\psi}_{p(A)}, = 8$. Therefore, for this example,

$$SS_{p(A) \times B} = \frac{6[(8 - 8)^2 + (3.33 - 8)^2 + (12.67 - 8)^2]}{(3)(1/3)^2 + (-1)^2} = 196$$

A computing form of Equation 6.25 can be expressed as

$$SS_{p(A) \times B} = \sum_k SS_{p(A)/B_k} - SS_{p(A)} \tag{6.26}$$

$SS_{p(A) \times B_k}$ is calculated by substitution of the appropriate values into Equation 6.7. For example, consider the contrast between CON and the mean effect of the three therapies $[\psi_{p(A)}]$ in the DP population (B_1) of Table 6.4:

$$SS_{p(A)/B_1} = \frac{[(1/3)(8 + 10 + 12) - 2]^2}{[(3)(1/3)^2 + (-1)^2]/6} = 288$$

Verify that $SS_{p(A)/B_2} = 50$ and $SS_{p(A)/B_3} = 722$. From Section 6.10.1, $SS_{p(A)} = 864$ and substituting into Equation 6.26,

$$SS_{p(A) \times B} = (288 + 50 + 722) - 864 = 196$$

Then $MS_{p(A) \times B} = 196/2$, and the F ratio is $98/10 = 9.8$. This ratio provides a test of whether the contrast between the CON and the average of the three therapy treatments varies significantly as a function of the clinical population being treated. The F would be evaluated on 2 and 60 df.[3]

Exactly the same reasoning can be employed to determine whether a contrast on the means of B is the same at each level of A. The null hypothesis, H_0: $q(B) \times A = 0$, can be tested by obtaining

$$SS_{q(B) \times A} = \sum_j SS_{q(B)/A_j} - SS_{q(B)}$$

---

[3] If the test was decided upon before viewing the data, and was one of a set of K planned tests, F should be evaluated with $\alpha = EF/K$. If the test was decided upon after examining the data, there is no simple way of controlling EF. The Scheffé procedure applies to single df terms, but we have $b - 1$ df.

The test statistic on $a - 1$ and $ab(n - 1)$ df is

$$F = \frac{SS_{q(B) \times A}/(a - 1)}{MS_{S/AB}}$$

Do $p(A)$ and $q(B)$ Interact?

Look again at Equation 6.22. One reason for the variation among the three values of $\hat{\psi}_{p(A)/B_k}$ is that the difference between the control and the average of the three therapy means seems to be much greater for the BI population (B_3) than for either of the other two clinical populations. This suggests a further analysis of the SS_{AB} in which we contrast the $p(A)$ value at B_3 with those at B_1 and B_2. As an illustration of the possibilities, we will consider whether the $p(A)$ value at B_3 differs from the average of the $p(A)$ values for B_1 and B_2. Is 12.67 significantly different from $(1/2)(8 + 3.33)$? We will refer to this contrast among contrast effects as $\psi_{p(A) \times q(B)}$. It is distributed on 1 df because it involves the contrast between a single observed value (essentially a difference among differences) and its expected value (usually zero). Therefore, we can either use the t test of Equation 6.5 or the SS_{ψ} formulation of Equation 6.7. The only real issue is how the means are to be weighted. We can best answer that by considering the null hypothesis more closely:

$$H_0: \psi_{p(A) \times q(B)} = 0$$

In our example,

$$\psi_{p(A)} = \frac{1}{3}(\mu_{AT} + \mu_{BT} + \mu_{CT}) - \mu_{CON}$$

and

$$\psi_{q(B)} = \mu_{BI} - \frac{1}{2}(\mu_{DP} + \mu_{SC})$$

In testing the null hypothesis, we assign a weight, $w_{jp,kq}$, to each of the ab cell means. For example, the mean for the cell defined by AT and BI receives a weight equal to 1/3 (the AT weight) times 1 (the BI weight). Similarly, the mean for the cell defined by CT and DP receives a weight equal to $(1/3)(-1/2)$, or $-1/6$. The full set of weights is presented in Table 6.5. Now we can define the effect of interest as

$$\psi_{p(A) \times q(B)} = \sum_j \sum_k w_{jk,pq}\mu_{jk} \tag{6.27}$$

where $w_{jk,pq}$ is the product of w_{jp} and w_{kq}. The null hypothesis can be restated as

$$H_0: \psi_{p(A) \times q(B)} = \sum_j \sum_k w_{jk,pq}\mu_{jk} = 0$$

Except for the fact that the contrast weights apply to the cell means rather than the marginal means, testing $p(A) \times p(B)$ is like testing any other contrast. The null hypothesis can be tested using

$$t = \frac{\hat{\psi}_{p(A) \times q(B)}}{\hat{\sigma}_{\hat{\psi}}} = \frac{\sum_j \sum_k w_{jk,pq}\bar{Y}_{.jk}}{\sqrt{MS_{S/AB}\sum_j \sum_k (w_{jk,pq})^2/n_{jk}}} \tag{6.28}$$

TABLE 6.5   CELL WEIGHTS FOR TESTING THE NULL HYPOTHESIS, $p(A) \times q(B) = 0$)

		B			Weights for $p(A)$
		DP	SC	BI	
	CON	$(-1)(-1/2)$ $= 1/2$	$(-1)(-1/2)$ $= 1/2$	$(-1)(1)$ $= -1$	-1
	AT	$(1/3)(-1/2)$ $= -1/6$	$(1/3)(-1/2)$ $= -1/6$	$(1/3)(1)$ $= 1/3$	$1/3$
A	BT	$(1/3)(-1/2)$ $= -1/6$	$(1/3)(-1/2)$ $= -1/6$	$(1/3)(1)$ $= 1/3$	$1/3$
	CT	$(1/3)(-1/2)$ $= -1/6$	$(1/3)(-1/2)$ $= -1/6$	$(1/3)(1)$ $= 1/3$	$1/3$
	Weights for $q(B)$	$-1/2$	$-1/2$	1	

In the example under consideration, substitution into Equation 6.5 gives

$$t = \frac{(1/2)(2) + (-1/6)(8) + \cdots + (1/3)(12) + (1/3)(8)}{\sqrt{10[(1/2)^2 + (-1/6)^2 + \cdots + (1/3)^2 + (1/3)^2]/6}}$$

$$= \frac{7}{\sqrt{(10)(2)/6}} = 3.83$$

Using Equation 6.7, we can show that the sum of squares is 147. Then the F ratio is $147/10 = 14.7$, the square of 3.83. Of course, the df_e for either the t or F test is $ab(n - 1)$, or 60, as it has been in the preceding analyses. Whether t or F is tested, EF should be controlled. For planned contrasts, the Bonferroni procedure is used just as in previous applications. If the contrast is tested as the result of "data snooping," the Scheffé procedure should be used. For this case, the t is evaluated against

$$S = \sqrt{(a - 1)(b - 1)F_c}$$

where F_c is the critical F value for $(a - 1)(b - 1)$ and $ab(n - 1)$ df, and the designated EF.

In closing this section, we note that $SS_{p(A) \times q(B)}$ is one component of the overall SS_{AB}. It accounts for a part of the interaction variability. Other parts could be calculated by finding sets of weights orthogonal to the weights in Table 6.5. In fact, we can partition SS_{AB} into $(a - 1)(b - 1)$ single df terms of the form $SS_{p(A) \times q(B)}$. As we noted in our earlier discussion of partitioning SS_A into orthogonal components, an infinite number of partitionings are possible.

6.10.3 Testing Contrasts When There Are More than Two Factors

The computations of test statistics and the procedures for evaluating them when there are three or more factors are straightforward extensions of the development thus far.

The t test for the one-factor design, Equation 6.5, can be written in more general form for any between-subjects design as

$$t = \frac{\hat{\psi}}{\sqrt{MS_e(\text{sum of squared weights})/N_M}}$$

where $\hat{\psi}$ is the contrast of interest, MS_e is the error term for the ANOVA, and N_M is the number of scores on which each of the weighted means in the contrast is based. Furthermore, the error df will be those ordinarily associated with MS_e. For example, in a design involving the factors A, B, C, and D, a contrast of the marginal means for the levels of C would mean that $N_M = abdn$. MS_e would be $MS_{S/ABCD}$, and df_e would be $abcd(n-1)$.

Similarly, the sum of squares for any multifactor design is just a modification of Equation 6.7. We would have

$$SS_{\hat{\psi}} = \frac{\hat{\psi}^2}{(\text{sum of squared weights})/N_M}$$

This would then be divided by MS_e to form an F ratio.

6.11 CONCLUDING REMARKS

This chapter has had two major foci: computations of t and F statistics for testing contrasts, and procedures using these statistics to control familywise error rates. With respect to the computations, we have included them at least as much for conceptual as for computational purposes. In most cases the calculations can be done by computer. Packages like SAS (PROC GLM) and SPSS^X (ONEWAY AND MANOVA) have CONTRAST commands. Any contrast can be tested by using the CONTRAST, AMATRIX, and CMATRIX commands in SYSTAT. The VALUE (of the weights) and the TYPE statement in BMDP4V together permit contrasts to be tested.

With respect to controlling error rate, our recommendations for between-subjects designs were summarized in Table 6.2. Those recommendations and the comments that follow should be considered guidelines, not rules. There are no firm rules either for defining a family of contrasts or for controlling error rate for the family. We balance concerns about control of Type 1 and Type 2 error by defining the family more or less broadly and choosing more or less conservative procedures.

In multifactor designs, we recommend that the family consist of the contrasts that are associated with a single source of variance. The number of contrasts in each family is determined by whether we have planned the contrasts in advance or whether the choice about which contrasts to test is determined by looking through the data to see which effects seem to be present. In the latter case, the family consists of all the contrasts that might possibly be tested. When relatively few contrasts are planned, the Bonferroni procedure should be used to control EF except when only contrasts between treatment groups and a control group are of interest. The Bonferroni procedure is conceptually very simple and can be used with any test statistic.

When decisions about what contrasts to test are made after viewing the data, the Scheffé procedure can be used to limit EF. However, because in large designs the majority of possible contrasts would never be tested no matter what the data looked like, the Scheffé tests are very conservative. If we can decide in advance how many contrasts may possibly be of interest and use the Bonferroni procedure, each contrast might be tested with more power, even if there are a large number of contrasts.

If the set of all pairwise contrasts are of interest, the Tukey procedure can be used. All of the procedures discussed can be thought of as using the same test statistic, but different criteria for significance. When there is concern about violations of the homogeneity of variance assumption, in each case the test statistic can be replaced by the t^* statistic and the different criteria for significance evaluated by using the degrees of freedom described in Section 6.8.

Our next set of comments deal with those situations in which different kinds of tests are conducted on the same source of variance. First, we note that the overall ANOVA F test and the Scheffé procedure are consistent. This means that if the overall test is conducted at $\alpha = .05$ and EF $= .05$ for any contrasts tested by the Scheffé procedure, the probability of a Type 1 for the combination of both procedures remains at .05. Rejection of the overall hypothesis is not necessarily followed by a significant contrast because the power of the F is equal to that of the "maximum contrast," which may not be obvious or interesting and therefore may not be tested.

Second, the Tukey test is a test of the omnibus null hypothesis. That is, if the largest pairwise difference is significant, we can reject the hypothesis of equality of the a values of μ_j. Therefore, there is no need to conduct an ANOVA before the Tukey procedure. In fact, there will be occasions in which the omnibus F and the Tukey test will yield different conclusions. (For a discussion of the conditions under which this will happen and when each test is more powerful, see Myers 1979.) Nevertheless, investigators usually do carry out the ANOVA before applying the Tukey test. Because the ANOVA carries additional information such as ω^2 values, we do not object to this procedure.

Third, investigators usually combine tests of small sets of planned contrasts with overall F tests. Our theories are rarely, if ever, developed to the point that we are willing to forgo checking whether effects other than those we anticipated are present in the data. Although performing an overall F test at $\alpha = .05$ in addition to testing a small family of planned contrasts using EF $= .05$ will increase the probability of Type 1 error, this seems to be a worthwhile price to pay for the additional information provided by the F. If there is concern that the error probability is too high, α and EF can both be made smaller.

Note the distinction between testing all possible pairwise contrasts and testing selected planned contrasts. A Tukey test of the largest difference between means is a test of the overall null hypothesis. A test of some planned contrast does not have the same status. Although a significant result in this case implies rejection of the overall null hypothesis, nonsignificance has no implications beyond the means involved in the contrast being tested.

Finally, we note that although we have considered tests of population means exclusively in this chapter, the ideas we have introduced apply generally to any set of statistical tests. For example, the Bonferroni inequality can be applied directly to any procedure, and Delucchi (1983) has discussed analogs of the Scheffé procedure for contrasts on the proportions in contingency tables.

EXERCISES

6.1 The following terms provide a useful review of many concepts in the chapter. Define, describe, or identify each of them:

omnibus F test

contrast

pairwise comparison

$\sigma_{\hat{\psi}}^2$

$SS_{\hat{\psi}}$

error rate per comparison

error rate per family

Bonferroni inequality

Bonferroni procedure

simultaneous confidence interval

Dunnett's test

Scheffé's test

Tukey/Kramer test

6.2 Ninety children, varying in age ($A_1 = 5$, $A_2 = 7$, and $A_3 = 9$), are taught one of three mnemonic methods (methods for memorizing; B_1, B_2, and B_3). All subjects are then shown a series of objects and their recall is scored. Thus we have nine groups of 10 subjects each. The cell means and variances are

	Means				Variances		
	A_1	A_2	A_3		A_1	A_2	A_3
B_1	44	58	78	B_1	75	79	84
B_2	56	66	83	B_2	61	82	85
B_3	52	70	79	B_3	90	71	77

(a) B_2 and B_3 both involve the use of imagery, whereas B_1 involves repeating the object names. Therefore a contrast of the B_1 mean against the average of the B_2 and B_3 means is of interest. (i) State H_0. (ii) Calculate the value of t for a test of this null hypothesis.

(b) We could also have calculated an F test for the hypothesis in part (a); find the SS associated with the contrast.

(c) There is some reason to believe that the relative effectiveness of imagery decreases with increasing age. The cell means appear to bear this out; the size of the contrast tested in parts (a) and (b) appears to decrease as A increases. State the relevant null hypothesis and calculate the F statistic that tests this hypothesis.

6.3 Reconsider the data of Table 5.1.

(a) Use a contrast to test whether recall for the 450-wpm rate condition differs significantly from the average recall for the other two rate conditions.

(b) Test whether the contrast indicated in part (a) is significantly different for the intact and the scrambled-text conditions.

6.4   Each cell in the following table contains a mean based on 10 scores:

	A_1	A_2	A_3
B_1	20	10	6
B_2	6	10	8

(a) Find the sums of squares accounted for by each of the following contrasts:

$$\psi_1 = \mu_1. - \frac{\mu_2. + \mu_3.}{2} \quad \text{and} \quad \psi_2 = \mu_2. - \mu_3.$$

(b) Are the two contrasts orthogonal?

(c) Find SS_A and compare it with the sum of the sums of squares found in part (a).

(d) We wish to determine whether either of the above contrasts varies as a function of B. Find the SS terms associated with each of the relevant significance tests. Add these terms and compare them with SS_{AB}.

6.5   Main and interaction effects on 1 df can be viewed as contrasts of cell means (or cell totals). For example, if we have three levels of A and two levels of B, SS_B can be calculated as SS_ψ by assigning weights of 1 to the three cells at B_1 and weights of -1 to the three cells at B_2. The advantage of this is that it provides a very quick way to calculate many commonly encountered quantities.

Following are the means for an experiment described in Chapter 5; we have collapsed over the three presentation rates, so each mean is based on 18 scores.

	Text	
Reader	Scrambled	Intact
Good	62.944	53.889
Poor	48.222	43.333

(a) To obtain SS_{Reader}, what weights should be assigned to the four cells? Calculate SS_{Reader} using these weights in the SS_ψ equation and check your result against that in Table 5.8. Do the same for (b) SS_{Text} and (c) $SS_{R \times T}$.

6.6   In the design presented below, each cell total is based on 10 scores.

	C_1		C_2	
	B_1	B_2	B_1	B_2
A_1	34	23	62	45
A_2	28	58	37	41

A rereading of the discussion of interaction in Section 5.5 may prove helpful in doing this problem.

(a) What weights would be assigned to each cell in order to calculate the SS_B as a contrast? Carry out the calculations.

(b) Assign the contrast weights needed to calculate SS_{AC}. Do the calculations.

(c) Assign the contrast weights needed to calculate SS_{ABC}. Do the calculations.

(d) For the following design, what weights should be assigned to the cell totals to use the contrast formula to calculate SS_{BC}? Do the calculations.

	C_1		C_2	
	B_1	B_2	B_1	B_2
A_1	36	19	22	38
A_2	21	34	31	40
A_3	27	45	42	46

6.7 We have five treatment conditions in a problem-solving study, each with $n = 20$. Two groups, $F1$ and $F2$, are given instructions designed to facilitate problem solving. The third group is a control group given neutral instructions. The fourth and fifth groups, $I1$ and $I2$, are given instructions designed to interfere with problem solving. The data are

	F1	F2	C	I1	I2
$\bar{Y}_{.j}$	14.6	14.9	13.8	11.8	11.7
$\hat{\sigma}_j^2$	3	4	5	4	4

Test each of the following hypotheses at $\alpha = .05$.

(a) The average of the facilitation group population means is greater than the mean of the control population.

(b) The average of the interference population means is different from the mean of the control population.

(c) The difference between the average of the facilitation means and the control mean is not the same as the difference between the control mean and the average of the interference means.

6.8 Productivity measures are obtained for psychology Ph.D.'s graduating in certain research areas and from a variety of universities. There are three variables in the study:

Research area	Clinical/Social/Cognitive	(A_1, A_2, and A_3)
Institution type	Public/Private	(B_1 and B_2)
Geographical area	East/Midwest/South/West	(C_1, C_2, C_3, and C_4)

State the weights for the cell means needed for tests of each of the following hypotheses:

(a) Public and private university Ph.D.'s. are equally productive.

(b) Midwestern Ph.D.'s are more productive than nonmidwestern Ph.D.'s.

(c) The regional difference referred to in (b) is less marked for private than for public universities.

(d) The regional difference is more marked for clinicians than for nonclinicians.

6.9 This problem may further your understanding of orthogonality. We have the following group means, each based on 10 scores:

A_1	A_2	A_3
24	16	14

(a) Calculate SS_A.

(b) Calculate the sum of squares for each of the following contrasts:

 (i) $\psi_1 = \mu_1 - \mu_2$

 (ii) $\psi_2 = (1/2)(\mu_1 + \mu_2) - \mu_3$

 (iii) $\psi_3 = \mu_1 - \mu_3$

What should be true of the relation between SS_A and the sums of squares for ψ_1 and ψ_2? Why?

(c) We can remove the effect associated with ψ_1 from the data by setting the means at A_1 and A_2 equal to their average. The adjusted means are

A_1	A_2	A_3
20	20	14

Redo part (b)(ii) and (iii). Are either of the sums of squares different from those calculated for the original (unadjusted) means? Explain.

(d) Suppose the group sizes were not equal; the n_j's are 8, 10, and 12, respectively. Returning to the original means, calculate SS_A. Then calculate the sums of squares for ψ_1 and ψ_2. Does the relation between SS_A and the sums of squares for ψ_1 and ψ_2, noted in part (b), still hold? Explain.

(e) As we noted in this chapter, two contrasts, ψ_p and ψ_q, are orthogonal if $\Sigma_j w_{jp} w_{jq}/n_j = 0$. This condition will be met if we redefine ψ_2 as

$$\psi_{2'} = n_1 \mu_1 + n_2 \mu_2 - (n_1 + n_2)\mu_3$$

Calculate the SS for this contrast on the original means. Add it to the SS for ψ_1 and compare the resulting sum to SS_A.

6.10 In a study of conflict in parachutists, GSR measures were obtained for five different groups of five subjects who differed with respect to when the measures were taken: two weeks before the jump (BJ-2), one week before (BJ-1), on the day of the jump prior to jumping (DJ-P), and on the day of the jump after jumping (DJ-A). There was also a control group of normal (nonjumping) cowards (C). The MS_{error} for the ANOVA $= 4.0$ and the means were

	BJ-2	BJ-1	DJ-A	DJ-P	C
Mean	5	5	7	9	2

(a) Suppose the investigator had planned to compare each of the four experimental groups with the control (C). Assuming a two-tailed test with EF $= .05$, carry out the test of DJ-P against C, using (i) the Bonferroni procedure, (ii) the Dunnett procedure.

(b) Suppose that the experimenter tested all possible pairwise comparisons. Reevaluate your conclusion about DJ-P against C, using the appropriate procedure for controlling the EF at .05.

(c) Briefly comment on the relative power of these three procedures, justifying your conclusion by citing relevant information in your preceding answers. In one or two sentences, explain why these situations give rise to the differences in power that you indicate.

(d) Calculate the confidence intervals obtained with each of the three procedures and relate the results to your answer to part (c).

6.11 A sample of humanities majors are divided into three groups of 10 each in a study of statistics learning. One group receives training on relevant concepts *before* reading the text, a second receives the training *after* reading the text, and a third is a no-training *control*. Summary statistics on a test are

	Before	After	Control
$\bar{Y}_{.j} =$	24	16	13
$\hat{\sigma}_j^2 =$	72	62	76

(a) Do an ANOVA based on these statistics and report whether the F is significant at the .01 level. State the rejection region and df. What would a significant result tell you?

(b) The *before* group seems to perform differently from the other two groups. We would like to test whether the mean of the *before* population is higher than the average of the other two populations combined. In other words, we wish to test the following null hypothesis:

$$H_0: \psi = \frac{1}{2}(\mu_A + \mu_C) - \mu_B = 0$$

where ψ is a linear contrast. The best estimate of this contrast is

$$\hat{\psi} = \frac{1}{2}(\bar{Y}_A + \bar{Y}_C) - \bar{Y}_B$$

(i) What is the estimate of the variance of the sampling distribution of $\hat{\psi}$ (assume homogeneity of variance)?

(ii) Calculate the t statistic appropriate for testing H_0.

(c) Evaluate the test statistic you just calculated, assuming (i) the test had been planned before viewing the data, (ii) the test was a result of viewing the data.

6.12 We have five group means, each based on 10 scores, with $MS_{S/A} = 4.0$. The means are

A_1	A_2	A_3	A_4	A_5
8.6	9.5	9.2	8.0	10.4

(a) We plan five contrasts with EF = .05. Test the contrast of A_5 against the average of the other four groups. State the criterion required for significance and whether H_0 can be rejected.

(b) Suppose we had decided on the contrast in part (a) after inspecting the data. Now what is the result of your significance test? Be sure to show your criterion statistic.

(c) Find the confidence intervals corresponding to the tests in parts (a) and (b). Explain the difference in widths.

(d) Suppose we did all possible pairwise tests. Actually do A_1 against A_2. What is the criterion statistic? What conclusion do you reach about H_0?

(e) Suppose the only contrast we planned pitted the average of A_1 and A_2 against the average of the remaining three groups. Do the calculations and report the results, showing the criterion statistic.

6.13 In an attitude change study, four groups of subjects are presented with persuasive messages about a topic. Two groups read the messages, a positive message for one group and a negative message for the other. Two other groups receive the messages by viewing a videotape. A fifth, control, group receives no message. Each group has its attitude assessed by a questionnaire for which larger scores mean a more positive attitude. There are seven subjects in each group, and $MS_{Error} = MS_{S/A} = 20$. The group means are

A_1	A_2	A_3	A_4	A_5
Video/positive	Video/negative	Read/positive	Read/negative	Control
71	42	63	47	52

(a) Determine which experimental conditions differ significantly from the control, using the Dunnett test with EF = .05.

(b) Test the hypothesis that the difference between the positive and negative messages is the same whether they are read or presented by videotape. Assume this is a single planned contrast.

(c) By how much would two group means have to differ before they would be considered to be significantly different by the Tukey test with EF = .05?

Chapter 7

Trend Analysis

7.1 INTRODUCTION

In Chapter 6 procedures for comparing the means of treatment groups were discussed. Our primary concern was to answer questions that are most appropriate when the independent variable is qualitative. We focused on pairwise differences among means and on contrasts of the averages of two subsets of groups. Although such contrasts can be tested even when the independent variable is quantitative, it is usually more informative to consider the overall trend in the treatment group means than to make specific comparisons among group means. We might want to determine whether there is a trend for means to increase as the level of the independent variable increases, or whether the function relating the means and the level of the independent variable is significantly curved.

An example may make clearer what kinds of questions this chapter will address. Consider an experimental study of generalization. A mild shock is presented in the presence of a rectangle of light that is 11 inches high and 1 inch wide. Subjects are then randomly divided into five groups of 10, each of which is tested in the presence of a rectangle of light, but with no shock. The independent variable is the height of the rectangle of light on these test trials; the five groups see a rectangle whose height is either 7, 9, 11, 13, or 15 inches high. An average galvanic skin response (GSR) measure is obtained for each subject. The experimenters' hypothesis is that two processes are at work in this experiment. First, they believe that the magnitude of conditioned responses should vary directly with the magnitude of the test stimulus; this implies that GSR scores should increase as a function of the height of the rectangle of light. Second, the experimenters expect a generalization effect. There should be a trend for GSR scores to be higher the closer the test stimulus is to the training stimulus. The result of this generalization process would be a symmetric inverted U-shaped curve. If only this process were operating, GSR would be highest in response to the 11-inch test stimulus and lowest for the 7- and 15-inch test stimuli.

The five group means, each of which is based on 10 scores, are plotted in Figure 7.1. If the experimenters' theory is correct, this function is a sum of two effects, an effect that increases with stimulus height and an effect that increases and then decreases as height increases, with a peak when the height is 11. Therefore, in this example, we

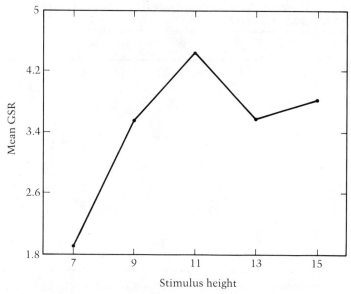

Figure 7.1 Mean GSR plotted as a function of stimulus height.

want to test each of the two kinds of effects independently. Trend analysis permits us to answer the following questions: (1) If we fit a straight line to the five means, will its slope be significantly different from zero? This addresses the hypothesis that there is an increase in the population means with increasing test stimulus height. (2) If we fit a symmetric inverted U-shaped curve to the five group means, will the points on this function vary significantly? This addresses the hypothesis that there is a generalization effect in the sampled population of GSR scores. There are other questions we could ask about these data. For example, note the slight upturn in the rightmost data point. If this upturn is due to more than chance variability, it is not accounted for by the straight line representing stimulus height effects and the inverted U-shaped curve representing generalization effects. There is the possibility that some process that produces an S-shaped function is also at work.

7.2 TESTING LINEAR TREND

7.2.1 The Equation for a Straight Line

We begin our discussion of trend analysis by considering linear trend because this provides a simple context within which to develop the basic ideas. Our example will be the GSR experiment discussed in the previous section. The group means plotted in Figure 7.1 are presented in Table 7.1 together with some other statistics that we will soon consider. We can draw a straight line to summarize the trend of the five means. That line is described by the equation

$$\hat{\bar{Y}}_{.j} = b_0 + b_1 X_j \tag{7.1}$$

TABLE 7.1 OBSERVED AND PREDICTED GROUP MEANS, AND CALCULATIONS, FOR A ONE-FACTOR EXPERIMENT

						Average
Stimulus height (X)	7	9	11	13	15	11
Observed mean GSR ($\bar{Y}_{.j}$)	1.910	3.560	4.440	3.580	3.830	3.464
Predicted mean GSR ($\hat{\bar{Y}}_{.j}$)	2.692	3.078	3.464	3.850	4.236	3.464
$\hat{\bar{Y}}_{.j} - \bar{Y}_{..}$	−.772	−.386	0	.386	.772	0
Variance ($\hat{\sigma}_j^2$)	2.218	2.563	1.964	2.881	1.659	2.257

$$b_1 = \frac{\Sigma(X_j - \bar{X})(\hat{\bar{Y}}_{.j} - \bar{Y}_{..})}{\Sigma(X_j - \bar{X})^2} = \frac{(-4)(-1.554) + \cdots + (4)(.366)}{(-4)^2 + \cdots + (4)^2} = \frac{7.720}{40} = .193$$

Note 1: We obtain the predicted means from Equation 7.5. For example,

$$\hat{\bar{Y}}_{.1} = \bar{Y}_{..} + b_1(X_1 - \bar{X}) = 3.464 + (.193)(-4) = 2.692$$

Note 2: The above statistics can be inserted into Equations 7.6 and 7.7 to test H_0: $\beta_1 = 0$. From Equation 7.6 we have

$$SS_{lin} = n\Sigma(\hat{\bar{Y}}_{.j} - \bar{Y}_{..})^2$$
$$= 10[(-.772)^2 + (-.386)^2 + \cdots + (.386)^2] = 14.900$$

From Equation 7.7 we have

$$F = \frac{14.9}{2.257} = 6.602$$

where X_j is the stimulus length for the jth group (7, 9, 11, 13, and 15) and $\hat{\bar{Y}}_{.j}$ is the mean value for the jth group predicted by Equation 7.1 if we have numerical values of b_0 and b_1. The constant b_1 is referred to as the slope of the line or the linear regression coefficient; it is the amount $\hat{\bar{Y}}_{.j}$ increases with each unit of increase in X_j. The constant b_0 is called the intercept (or Y-intercept) and is the value we expect for $\bar{Y}_{.j}$ if X_j is zero.

We can also construct a straight-line equation for the population means:

$$\mu_j = \beta_0 + \beta_1 X_j \tag{7.2}$$

where β_0 and β_1 are the parameters estimated by b_0 and b_1. When we test for linear trend, we test the null hypothesis

$$H_0: \beta_1 = 0$$

In words, the null hypothesis is that the straight line that best fits the treatment population means is flat.

Several points about this last statement must be understood. First, the μ_j may vary even if the above hypothesis is true—that is, even if there is no linear trend. To see this, suppose that the μ_j fall on a perfectly symmetric inverted U-shaped function with its peak at $X = 11$. For example, the population means might be 1.5, 2, 3, 2, 1.5. These means exhibit no linear trend; a best-fitting straight line would have a slope (β_1) of zero. Nevertheless, there is variability among the five means.

The second point to consider is that the straight lines described by Equations 7.1 and 7.2 are lines of best fit. There are many straight lines we can draw to describe a set of means. We choose our values of b_0 and b_1 to minimize $\Sigma(\hat{Y}_j - \bar{Y}_{.j})^2$, the sum of squared distances between $\hat{Y}_j$ and $\bar{Y}_{.j}$, the predicted and observed values of the group means. This is usually referred to as a *least-squares criterion* and it is what we mean by "best fit."

The third point to note is that rejection of the null hypothesis of no linear trend doesn't allow us to conclude that the population means are well fit by a straight line but only that the best-fitting straight line has a slope other than zero. In other words, the best-fitting straight line is not necessarily a good fit. Still, inferences about linear trend are important. In the conditioning experiment, a significant linear trend would support the hypothesis that the magnitude of a conditioned response tends to increase as the magnitude of the stimulus increases.

7.2.2 Variability of the Predicted Means

One way of testing linear trend follows directly from the preceding discussion of the equation for a line of best fit. This approach involves the following steps:

1. Obtain values of b_0 and b_1.
2. Substitute these values into Equation 7.1 to calculate the predicted group means, the $\hat{Y}_j$.
3. Calculate the sums of squared deviations of these predicted group means about their average, $\bar{Y}_{..}$.
4. Test this sum of squares against the usual within-groups error term.

There is a computationally simpler way to calculate the sum of squares which we will illustrate later. However, the present approach is useful to consider because it emphasizes the nature of the sum of squares we calculate; namely, that it represents the variability of a set of means predicted by using Equation 7.1.

In order to carry out these four steps, we need the least-squares formulas for calculating b_0 and b_1. These are

$$b_1 = \frac{\Sigma(\bar{Y}_{.j} - \bar{Y}_{..})(X_j - \bar{X}_.)}{\Sigma(X_j - \bar{X}_.)^2} \tag{7.3}$$

$$b_0 = \bar{Y}_{..} - b_1\bar{X}_. \tag{7.4}$$

Given these equations we can rewrite Equation 7.1:

$$\hat{Y}_j = b_0 + b_1 X_j = (\bar{Y}_{..} - b_1\bar{X}_.) + b_1 X_j \tag{7.5}$$

$$= \bar{Y}_{..} + b_1(X_j - \bar{X}_.)$$

The regression coefficient b_1 has been calculated in Table 7.1. Using this value and Equation 7.5, we have calculated $b_1(X_j - \bar{X})$, the deviations of predicted group means from the grand mean. If the null hypothesis of no linear trend is true, $\beta_1 = 0$; therefore, we would expect b_1 to be small and the values of $\hat{\bar{Y}}_j$ to be close to $\bar{Y}_{..}$, the grand mean. Large variability of the predicted group means about the grand mean supports rejection of the null hypothesis of no linear trend. This variability of the means predicted by the best-fitting straight line is labeled SS_{lin} ("sum of squares for linearity") and is defined as

$$SS_{lin} = n\sum(\hat{\bar{Y}}_j - \bar{Y}_{..})^2 \tag{7.6}$$

The F test of the null hypothesis that $\beta_1 = 0$ is

$$F_{1,a(n-1)} = \frac{SS_{lin}}{MS_{S/A}} \tag{7.7}$$

The SS_{lin} is calculated very much like the SS_A in Chapter 4. Therefore, it may be surprising to find that it is distributed on only 1 degree of freedom. After all, we have five squared deviations. However, the SS_{lin} only reflects the deviation of b_1 from zero, and is therefore distributed on a single degree of freedom.

   Equations 7.6 and 7.7 have been applied to the data of Table 7.1, and the results are tabled there. The F ratio on 1 and 45 df is quite significant. We therefore reject the null hypothesis of no linear trend. It appears that the μ_j increase in magnitude as X, the height of the rectangle of light, increases. However, this test tells us nothing about the shape of the function that describes the population means. To draw inferences about whether the function is curved, we must consider functions somewhat more complicated than straight lines. In Section 7.3, we do just that. However, we first will present a much simpler way to calcuate SS_{lin}.

7.2.3 SS_{lin} as a Single df Contrast

From Equation 7.5 we know that

$$\hat{\bar{Y}}_j - \bar{Y}_{..} = b_1(X_j - \bar{X})$$

Substituting the right-hand side into Equation 7.6, further substituting for b_1 from Equation 7.3, and simplifying, we can rewrite Equation 7.6 as

$$SS_{lin} = \frac{[\sum(X_j - \bar{X})\bar{Y}_j]^2}{\sum(X_j - \bar{X})^2/n} \tag{7.8}$$

The detailed steps transforming Equation 7.6 into 7.8 are presented in Appendix 7.1. Equation 7.8 has also been applied to the generalization data in Table 7.8; the result ($SS_{lin} = 14.9$) is the same as with the earlier application of Equation 7.6.

   The important point to recognize is that Equation 7.8 is identical to Equation 6.7 for SS_ψ; the identity is complete if we merely replace $X_j - \bar{X}$ by w_j. The advantage of recognizing the SS_{lin} is the sum of squares for a single df contrast is that we do not have to calculate b_1 and use it to calculate the predicted group means. Furthermore, we soon will show that tests of other trends can also be viewed as tests of single df contrasts. The only difference is that the $X_j - \bar{X}$ in Equation 7.8 are replaced by other weights.

The linear weights for the generalization example were the deviations of the stimulus lengths about their average: $-4, -2, 0, 2,$ and 4. From Chapter 6, we know that SS_{lin} is unchanged if we multiply or divide all the weights by a constant; this is because the squared constant appears in both numerator and denominator of Equation 7.8. Therefore, we can get the same value of SS_{lin} if we divide $X_j - \bar{X}$ by 2; the new weights are

$$\xi_{j1} = -2, -1, 0, 1, 2$$

The ξ_1 (Greek xi; the subscript 1 refers to linearity) are weights that can be used in Equation 7.8 in place of $X_j - \bar{X}$ to test the linearity hypothesis whenever (1) the values of the independent variable are equally spaced, and (2) each mean is based on the same number of scores. If the X_j are equally spaced, $X_j - \bar{X}$ will differ from ξ_{j1} by a constant multiplier and, as already noted, SS_{lin} will not be affected. The reason for the requirement of equal n is that we want $\Sigma n_j \xi_{j1} = 0$. For the general case in which spacing or n's are unequal, refer to Chapters 12 and 15 on bivariate and multiple regression.

Turn now to Appendix Table D.7, labeled "Coefficients of Orthogonal Polynomials." Find the block of coefficients for $a = 5$ (five levels of the independent variable) and look at the first row, the linear coefficients. These are the ξ_{j1} listed earlier. The table also lists linear coefficients for other values of a—that is, for experiments in which there are more or fewer levels of the independent variable. For each row of linear coefficients, (1) $\Sigma\xi_{j1} = 0$, and (2) provided the values of X (the independent variable) are equally spaced, the linear coefficients are a straight-line function of X. From now on, when testing whether the slope of the best-fitting straight line differs significantly from zero, we can use Equation 7.8 but with the linear coefficients of Table D.7 replacing $X_j - \bar{X}$, provided the values of X are equally spaced and the n_j are all equal.

As you may have guessed by now, the coefficients in the rows labeled quadratic, cubic, and so on, enable us to test other hypotheses about the shape of the function that best describes the treatment population means. We turn now to discuss just these hypotheses and how they are tested.

7.3 TESTING NONLINEAR TRENDS

7.3.1 A General Test

One question we might wish to ask is whether the group means in Table 7.1 depart significantly from the best-fitting straight line. The null hypothesis is that the population means fall on a straight line; there is no curvature. The experimenters expect generalization, which implies that this null hypothesis should be false; the population means should deviate from a straight line. A general test of the null hypothesis of no curvature follows from the recognition that the variability among the group means, SS_A, can be partitioned into two components. The first of these is the SS_{lin}, which we discussed in the preceding section. Recall that this reflects the difference between the best-fitting straight line and a line with slope of zero. The second component of SS_A is SS_{nonlin} ("sum of squares for nonlinearity"), which reflects the departure of the observed group means from the best-fitting straight line. This partitioning of SS_A follows from the identity

$$(\bar{Y}_{.j} - \bar{Y}_{..}) = (\bar{Y}_{.j} - \hat{\bar{Y}}_{.j}) + (\hat{\bar{Y}}_{.j} - \bar{Y}_{..})$$

Squaring both sides and summing over all subjects and groups, we have

$$n\Sigma(\bar{Y}_{.j} - \bar{Y}_{..})^2 = n\Sigma(\bar{Y}_{.j} - \hat{\bar{Y}}_{.j})^2 + n\Sigma(\hat{\bar{Y}}_{.j} - \bar{Y}_{..})^2$$

$$SS_A \quad = \quad SS_{nonlin} \quad + \quad SS_{lin}$$

In the preceding section, we pointed out that SS_{lin} was distributed on 1 df. Therefore, SS_{nonlin} must be distributed on $a - 2$ df. A more direct rationale follows from observing that we have a data points but lose 2 df for estimating β_0 and β_1.

To calculate SS_{nonlin}, it is simplest to just subtract SS_{lin} from SS_A. Therefore, the F test of the null hypothesis that the population means do not depart from a straight line is

$$F = \frac{(SS_A - SS_{lin})/(a - 2)}{MS_{S/A}} \tag{7.9}$$

Applying Equation 7.9 to the data of Table 7.1, we have

$$SS_{nonlin} = SS_A - SS_{lin} = 35.242 - 14.9 = 20.342$$

and

$$F_{3,45} = \frac{20.342/3}{2.257} = 3.004$$

which is significant at the .05 level. We conclude that the function relating the μ_j and X_j is not a straight line.

Let us summarize what we have learned so far from the analysis of trend in the generalization example. First, the test of linearity reveals that there is a trend for the treatment population means to increase as X increases; the linear regression coefficient is significantly greater than zero. Second, the test of nonlinearity reveals that the straight line by itself is not sufficient to account for the variation in the population means. Consistent with the idea of stimulus generalization, the best-fitting function appears to be curved.

The SS_{nonlin} can be further partitioned into $a - 2$ components each distributed on 1 df. We next consider how these components are calculated and what they represent.

7.3.2 Orthogonal Polynomials

In the example of the generalization experiment (Table 7.1), we have so far established that there are both linear and nonlinear components of the population function. In many analyses, tests of these two components will be enough. However, more precise theories motivate more precise statistical tests. For example, in the generalization experiment the theory specifies two independent processes that combine to generate the treatment means. The absolute magnitude of the stimulus is thought to produce a linear effect; for each increment of one unit in X, the μ_j should increase by some constant amount. Distance of the test stimulus from the training stimulus results in a quadratic effect; if only this generalization effect were present, the μ_j would be a symmetric inverted U-shaped function of X. Note that this statement of the theory is more specific than just stating that there will be deviations from the best-fitting straight line. The theory requires a very specific kind of nonlinearity. It says that the group means are adequately described by a *second-order polynomial function* of the form.

$$\bar{Y}_{.j} = b_0 + b_1 X_j + b_2 X_j^2 \tag{7.10}$$

This function is also called a *quadratic function*, and b_2 is often referred to as the *quadratic coefficient*.

Equation 7.10 is a special case of the general *polynomial function of order $a - 1$*:

$$\bar{Y}_{.j} = b_0 + b_1 X_j + b_2 X_j^2 + \cdots + b_p X_j^p + \cdots + b_{a-1} X_j^{a-1} \tag{7.11}$$

Note the restriction that if there are a points, the order of the polynomial is at most $a - 1$ (it can be less because b_{a-1}, b_{a-2}, . . . can be zero). For example, if we have only two data points we can draw a line between them, thus establishing a linear function of the form

$$Y = b_0 + b_1 X$$

However, we do not have enough data to determine more than two coefficients, and therefore two data points restrict us to estimating a first-order polynomial function.

In the generalization example, there are five group means, and so the data could conceivably be fit by a function having cubic (X^3) and quartic (X^4) terms. Our theory, however, holds that only the linear and quadratic components are necessary to account for the variation among the treatment population means. Because the SS_{lin} was significant, we have already demonstrated the presence of a linear component. Now we would like to determine whether, in accord with the theory, the only significant nonlinear component is the quadratic. In order to construct independent tests of the quadratic, cubic, and quartic components of the function for our five group means, we make use of the orthogonal polynomial coefficients of Appendix Table D.7.[1] Turning to the table, again focus on the block for which $a = 5$. Several points hold for the four rows of coefficients:

1. The plot of the coefficients in a given row is closely related to the component we wish to test. In Figure 7.2, we have plotted each row as a function of X. Note that the linear coefficients, the ξ_{j1}, lie on a straight line. The quadratic coefficients, the ξ_{j2}, lie on a symmetric U-shaped function; multiplication by -1 would give us the inverted U hypothesized for the generalization experi-

---

[1] Appendix 7.2 shows how these coefficients are derived and provides a better sense of their relation to the polynomial terms of Equation 7.11. The derivations can be used for cases in which the n's are not equal and/or the values of X are not equally spaced. However, it is far simpler in such cases to use standard regression analysis programs as discussed in Chapters 12 and 15.

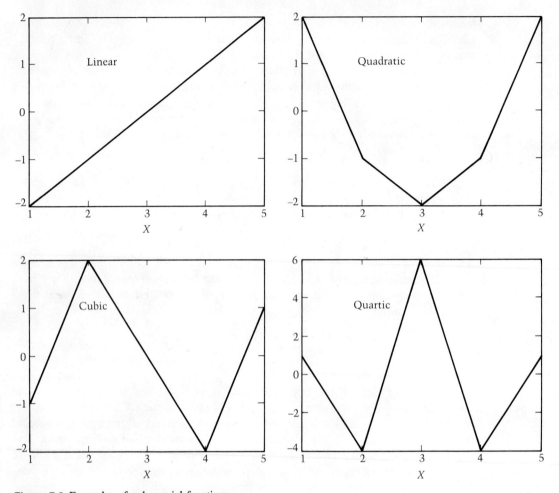

Figure 7.2 Examples of polynomial functions.

ment. As we shall soon show, sums of squares and values of test statistics are not affected by reversing the sign of the coefficient or, indeed, by multiplying all coefficients by any constant.

2. As with the linear coefficients, and all the sets of contrast weights encountered in Chapter 6, the coefficients sum to zero. That is,

$$\sum_j \xi_{jp} = 0$$

where ξ_{jp} is the jth value in the pth row.

3. All pairs of rows are orthogonal by the definition provided in Chapter 6. Recall that a necessary condition for two sets of weights, w_{jp} and w_{jq}, to be orthogonal is that $\sum_j w_{jp} w_{jq}$ must be zero. This requirement is met for the six pairs of rows here. For example,

$$\sum_j \xi_{j1} \xi_{j2} = (-2)(2) + (-1)(-1) + (0)(-2) + (1)(-1) + (2)(-2) = 0$$

Using the ξ_{jp}, we can rewrite Equation 7.11 as a sum of orthogonal polynomial components. The advantage of this form of the polynomial function over Equation 7.11 is that the orthogonal components can each be tested independently. The general form of this relation between the group means and the ξ_{jp} is

$$\bar{Y}_{.j} = b_0' + b_1'\xi_{j1} + b_2'\xi_{j2} + \cdots + b_p'\xi_{jp} + \cdots + b_{a-1}'\xi_{j(a-1)} \tag{7.12}$$

where

$$b_0' = \bar{Y}_{..} \tag{7.13}$$

and, for $p > 0$,

$$b_p' = \frac{\sum_j \xi_{jp} \bar{Y}_{.j}}{\sum_j \xi_{jp}^2} \tag{7.14}$$

The orthogonality of the components, as well as their relation to the group means, is demonstrated in Table 7.2. We have calculated the terms in Equation 7.12 for each value of j. The first row contains the linear coefficient, b_1', followed by $b_1'\xi_{11}$,

TABLE 7.2   ORTHOGONAL COMPONENTS AND ANOVA OF THE FUNCTION PLOTTED IN FIGURE 7.1

(a) Orthogonal components

Component	b_p'	$b_p'\xi_{1p}$	$b_p'\xi_{2p}$	$b_p'\xi_{3p}$	$b_p'\xi_{4p}$	$b_p'\xi_{5p}$	$n\Sigma(b_p'\xi_{jp})^2$
Linear	.3860	−.7720	−.3860	0	.3860	.7720	14.900
Quadratic	−.3243	−.6486	.3243	.6486	.3243	−.6486	14.723
Cubic	.1880	−.1880	.3760	0	−.3760	.1880	3.534
Quartic	.0546	.0546	−.2183	.3274	−.2183	.0546	2.085
$b_0'(\bar{Y}_{..})$			3.464	3.464	3.464	3.464	3.464
Sum		1.91	3.56	4.44	3.58	3.83	35.242

Note: The sum of each column labeled $b_p'\xi_{jp}$ is $\bar{Y}_{.j}$, and the sum of the rightmost column is SS_A.

(b) ANOVA

SV	df	SS	MS	F
A	4	35.242	8.811	3.904[a]
lin(A)	1	14.900	14.900	6.602[b]
quad(A)	1	14.723	14.723	6.523[b]
cubic(A)	1	3.534	3.534	1.566
quart(A)	1	2.085	2.085	.924
S/A	45	101.565	2.257	

[a] $p < .01$

[b] $p < .05$

$b_1' \xi_{21}, \ldots, b_1' \xi_{51}$. The last entry in the first row, $SS_{p(A)}$, is n times the sum of the squared values of $b_1' \xi_{j1}$. In general,

$$SS_{p(A)} = n\sum_j (b_p' \xi_{jp})^2$$

and in our experiment

$$SS_{\text{lin}} = 10[(-.7720)^2 + \cdots + (.7720)^2] = 14.900$$

the result we obtained in Section 7.2.

The next three rows are similar except that the regression coefficients (b_p') and the ξ values they multiply are either quadratic, cubic, or quartic, depending on the row. The row following the quartic row contains the value of b_0' $(\bar{Y}_{..})$ in each of the five columns corresponding to the five experimental groups. Adding the mean to the components above it yields the values in the last row. Those column sums are the group means, as Equation 7.12 states. That the four polynomial terms are orthogonal components of the group means is demonstrated in two ways. First, we can apply the usual cross products test:

$$\sum_j (b_p' \xi_{jp})(b_q' \xi_{jq}) = b_p' b_q' \sum_j \xi_{jp}\xi_{jq} = 0$$

For example,

$$(-.7720)(-.6486) + (-.3860)(.3243) + \cdots + (.7720)(-.6486) = 0$$

This orthogonality does not hold for the terms in the original polynomial function, Equation 7.11; to obtain independent tests of the components, we needed to rewrite that equation in the form of Equation 7.12.

The rightmost column of Table 7.2 contains both a second piece of evidence that the polynomial components of the group means are orthogonal and the basis for our significance tests. The entries in that column are the $SS_{p(A)}$; i.e., SS_{lin}, SS_{quad}, SS_{cubic}, and SS_{quart}. The bottom entry in the column is the sum of these four single df terms; it is also the value of SS_A. In short, the sum of squares calculations for the polynomial terms correspond to a partitioning of SS_A into $a - 1$ nonoverlapping pieces, each of which is distributed on a single degree of freedom. Given the values in the rightmost column of Table 7.2, we form F ratios, dividing the $SS_{p(A)}$ by $MS_{S/A}$. This has been done in the table, and we find that only the linear and quadratic components are significant. We reject the null hypotheses

$$H_0: \beta_1' = 0 \quad \text{and} \quad H_0: \beta_2' = 0$$

and we fail to reject the null hypotheses

$$H_0: \beta_3' = 0 \quad \text{and} \quad H_0: \beta_4' = 0$$

This is consistent with our original theory about the results of the experiment; the best-fitting straight line has a slope greater than zero, indicating that response magnitude increases with the magnitude of the test stimulus, and there is a quadratic component reflecting stimulus generalization. Our significance test results lead us to conclude that these are the only two processes at work; linear and quadratic components appear adequate to describe the variation among the group means.

The grand mean of the 50 scores and the linear and quadratic components of Table 7.2 are plotted in the upper panel of Figure 7.3; they have been summed in the bottom panel to provide our best estimate of the shape of the population function. If we summed the cubic and quartic components as well, the points representing the observed group means would fall on the resulting curve. However, our significance tests make it clear that these two components do not reflect real trends in the population. Therefore, the deviations of the five group means about the function plotted in the bottom panel are attributed to chance variability.

As we demonstrated in calculating SS_{lin} in Section 7.2, we do not need to calculate the b'_p to obtain the sum of squares for any orthogonal component. In general, the sum of squares for the pth component is

$$SS_{p(A)} = \frac{n(\Sigma_j \xi_{jp} \bar{Y}_{.j})^2}{\Sigma_j \xi_{jp}^2} \tag{7.15}$$

$$= \frac{(\Sigma_j \xi_{jp} T_{.j})^2}{n\Sigma_j \xi_{jp}^2}$$

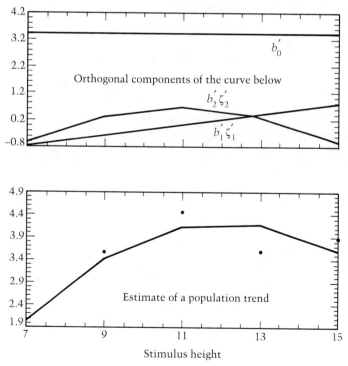

Figure 7.3 Plots of significant trend components and their sum.

An example of the calculations using the first form of Equation 7.15 is

$$SS_{quad(A)} = \frac{10[(2)(1.91)+(-1)(3.56)+(-2)(4.44)+(-1)(3.58)+(2)(3.83)]^2}{2^2+(-1)^2+(-2)^2+(-1)^2+2^2}$$

$$= \frac{(10)(-4.54)^2}{14} = 14.723$$

which is the result presented in the $SS_{p(A)}$ column of Table 7.3. Because this sum of squares is distributed on 1 df, we form our F ratio by dividing $SS_{quad(A)}$ by $MS_{S/A}$; this F is distributed on 1 and 45 df in our example. The reader should use Equation 7.14 and the weights in Appendix Table D.7 to verify our results for the cubic and quartic components.

7.3.3 Strategies in Testing Trend

We generally prefer to carry out the minimum number of significance tests required to evaluate our theory. This is because we use the Bonferroni procedure of Chapter 6 to control for the familywise error rate associated with these tests of polynomial components; the fewer tests we perform, the more power each test has. Therefore, although we have presented F tests for each of the four possible polynomial components in our example, in practice we would perform only three tests. Because our theory predicted linear and quadratic effects, and no polynomial contribution beyond these, we would test the linear and quadratic components, as indicated above. In addition, in order to determine whether the hypothesized components are sufficient to account for the pattern of the a means, we would also test the residual A variability; the numerator sum of squares for this test is calculated simply as

$$SS_{A\,res} = SS_A - SS_{lin} - SS_{quad}$$

and $MS_{A\,res} = SS_{A\,res}/df_{A\,res}$, where the df are $(a-1)-2$ in our example. If $MS_{A\,res}$ is significant when tested against $MS_{S/A}$, we have evidence that one or more of the polynomial components included in $SS_{A\,res}$ (the cubic and quartic in our example) are contributing to the A effects. This significant result may be followed by tests of the components of the residual term if the outcome of such tests will help us reformulate our theory. In practice, however, cubic and higher-order terms usually are difficult to interpret.

7.4 MULTIFACTOR DESIGNS

Consider an extension of our experiment on stimulus generalization. Assume that we sample our subjects from three populations: subjects diagnosed as severely schizophrenic (SS), subjects diagnosed as mildly schizophrenic (MS), and control subjects (C) who are not part of a clinical population. We have 30 subjects from each population, and six are assigned to each of the five test stimulus conditions described in the Introduction to this chapter. In general, we have a levels of A, b levels of B, and n subjects in each of the ab cells. In this example, A will denote clinical population ($a = 3$), B will denote stimulus height ($b = 5$), and $n = 6$. The cell means for this design are presented in Table 7.3, as is the analysis of variance. Note that the average of the 15 cell variances, $MS_{S/AB}$, is the error term for the F tests in the table. It will also be the error term for the tests of polynomial coefficients that we will present in this section.

Given a design of this sort, there are a number of questions we can ask. We can analyze the main effects of stimulus height into their trend components just as we did in the one-factor design. Averaging over the three clinical populations, are there linear and quadratic trends? Of somewhat more interest is the analysis of interaction effects. Personality theorists have hypothesized that schizophrenics tend to discriminate less among stimuli than do normals. This hypothesis has two possible consequences: (1) As the severity of schizophrenia increases, response magnitude should show a less pronounced increase with increases in stimulus magnitude. If so, the linear coefficient of the generalization gradient $[\text{lin}(B)]$ should be lowest in the severely schizophrenic and highest in the controls. (2) As the severity of schizophrenia increases, there should be more generalization; the flattest generalization gradient should be found in the severely schizophrenic population. This implies that the quadratic coefficients of the generalization function $[\text{quad}(B)]$ will vary across clinical populations and be closest to zero for the severely schizophrenic population. In terms of Equation 7.12, we expect β_1' and β_2' to vary across the three populations with their highest values for the controls and

TABLE 7.3 GROUP MEANS FOR A TWO-FACTOR EXPERIMENT AND THE ANOVA

(a) The means

Clinical population (A)	Height of test stimulus (B)					
	7	9	11	13	15	$\bar{Y}_{.j.}$
Controls	1.24	3.96	4.44	4.48	3.08	3.44
Mildly schizophrenic	3.33	3.99	4.63	3.75	4.05	3.95
Severely schizophrenic	3.99	4.32	3.82	4.84	4.13	4.22
$\bar{Y}_{..k} =$	2.853	4.090	4.297	4.357	3.753	$\bar{Y}_{...} = 3.87$

Note: $MS_{S/AB}$ is the average of the 15 cell variances and equals 1.167. Also, each cell mean is based on six scores.

(b) The ANOVA

SV	df	SS	MS	F
A	2	9.414	4.707	4.033[a]
B	4	27.261	6.815	5.840[b]
AB	8	25.751	3.219	2.758[a]
S/AB	75	87.525	1.167	

[a] $p < .05$
[b] $p < .01$

lowest for the severely schizophrenic. We also expect that the average function for the two schizophrenic populations will have smaller linear and quadratic components than that for the control population.

Other questions about the interaction are indicated if both factors are quantitative. For example, A might be the number of training trials, and B the height of the test stimulus. Now we might ask questions about the polynomial function that relates the linear or quadratic component of the stimulus height curve to the number of training trials. For example, does β_2' increase as a linear function of number of training trials?

Throughout the rest of this chapter, the calculations will be exactly those presented in Chapter 6. The only difference is that weights such as w_{jp} and w_{kq} are now replaced by values of ξ obtained from Appendix Table D.7. The questions we will ask in this section are somewhat different from those considered in Chapter 6; the emphasis is on evaluating and comparing trends rather than differences among sets of means. Nevertheless, you may find it helpful to review the material on multifactor designs in the preceding chapter, particularly the calculations involved in the analysis of interaction.

7.4.1 The Analysis of Main Effects

The questions here are about the slope and shape of the average curve for the three clinical groups. The relevant data are the means for each of the test stimuli based on the 18 subjects tested with that stimulus. More generally, we are interested in the qth polynomial component of B, and each of the b means is based on an scores. The variability due to this component is

$$\text{SS}_{q(B)} = \frac{an(\Sigma_k \xi_{kq} \bar{Y}_{..k})^2}{\Sigma_k \xi_{kq}^2} \tag{7.16}$$

or

$$\text{SS}_{q(B)} = \frac{(\Sigma_k \xi_{kq} T_{..k})^2}{an\Sigma_k \xi_{kq}^2} \tag{7.16'}$$

Equation 7.16 is identical to Equation 7.15 for the one-factor design except that the $\bar{Y}_{..k}$ (or $T_{..k}$) are obtained by averaging (or summing) over the a levels of A (the three clinical populations in our example), and, therefore, n is replaced by an in the equations. Applying Equation 7.16 to the data of Table 7.3, we have calculated the orthogonal polynomial components of SS_B. The calculations are presented in Table 7.4. The results of significance tests will be presented in a complete ANOVA table when we have analyzed the interaction variability.

7.4.2 Analysis of Interaction Effects

As we have noted, in normal populations there is a tendency for GSR to increase as the magnitude of the test stimulus increases, but this is combined with a generalization gradient, an inverted U-shaped curve. We hypothesized that such linear and quadratic trends should be less pronounced in schizophrenic populations, who generally are less

TABLE 7.4 ORTHOGONAL POLYNOMIAL COMPONENTS OF SS_B OF TABLE 7.3

$$SS_{lin(B)} = \frac{an(\Sigma_k \xi_{k1} \bar{Y}_{..k})^2}{\Sigma_k \xi_{k1}^2}$$

$$= \frac{(3)(6)[(-2)(2.853) + (-1)(4.090) + (0)(4.297) + (1)(4.357) + (2)(3.753)]^2}{(-2)^2 + (-1)^2 + 0^2 + 1^2 + 2^2}$$

$$= \frac{76.880}{10} = 7.668$$

$$SS_{quad(B)} = \frac{an(\Sigma_k \xi_{k2} \bar{Y}_{..k})^2}{\Sigma_k \xi_{k2}^2}$$

$$= \frac{(3)(6)[(2)(2.853) + (-1)(4.090) + (-2)(4.297) + (-1)(4.357) + (2)(3.753)]^2}{2^2 + (-1)^2 + (-2)^2 + (-1)^2 + 2^2}$$

$$= \frac{263.581}{14} = 18.827$$

$$SS_{cubic(B)} = \frac{an(\Sigma_k \xi_{k3} \bar{Y}_{..k})^2}{\Sigma_k \xi_{k3}^2}$$

$$= \frac{(3)(6)[(-1)(2.853) + (2)(4.090) + (0)(4.297) + (-2)(4.357) + (1)(3.753)]^2}{(-1)^2 + 2^2 + 0^2 + (-2)^2 + 1^2}$$

$$= \frac{2.420}{10} = .242$$

$$SS_{quart(B)} = \frac{an(\Sigma_k \xi_{k4} \bar{Y}_{..k})^2}{\Sigma_k \xi_{k4}^2}$$

$$= \frac{(3)(6)[(1)(2.853) + (-4)(4.090) + (6)(4.297) + (-4)(4.357) + (1)(3.753)]^2}{1^2 + (-4)^2 + 6^2 + (-4)^2 + 1^2}$$

$$= \frac{35.280}{70} = .504$$

able to discriminate among stimuli. This implies that the $A \times B$ (clinical population $\times$ stimulus height) interaction should be significant as, in fact, it was (see the ANOVA in Table 7.3). But the significant interaction merely tells us that the curves for the three clinical populations are not parallel. Although this is necessary to support the theoretical propositions stated, it is not sufficient. We predicted something more specific than a lack of parallelism; we predicted that the nonparallelism of curves for the three populations would be due to variation in the linear and quadratic components. Panel a of Table 7.5 presents the regression coefficients (b'_{jq}) for the three clinical groups. They were calculated using Equation 7.14; for example, the cubic regression coefficient for the controls (A_1) is

$$b'_{13} = \frac{\Sigma_k \xi_{k1} \bar{Y}_{.1k}}{\Sigma_k \xi_{k1}^2} = \frac{(-1)(1.24) + \cdots + (1)(3.08)}{(-1)^2 + \cdots + (1)^2} = \frac{.80}{10} = .08$$

The notation b'_{13} refers to the third (cubic) component for the first level of A.

The numerical estimates of the population regression coefficients in Table 7.5 are consistent with our predictions for the three types of subjects. The linear and quadratic regression coefficients are larger for the control subjects than for either schizophrenic group, suggesting that the controls discriminate better among the test stimuli. The cubic and quartic coefficients appear to be small for all three groups. Of course, we still have to carry out statistical tests. The differences we have noted may be due to chance variability, and the apparently small cubic and quartic coefficients might differ significantly from zero when tested against a measure of error variance.

Testing the Equality of Regression Coefficients

Suppose we wish to compare the quadratic components of the curves for the three clinical populations; that is, we wish to test the null hypothesis that

$$H_0: \beta'_{12} = \beta'_{22} = \beta'_{32}$$

We would calculate $SS_{A \times \text{quad}(B)}$, a measure of the variability in the quadratic component of B as a function of the level of A. The general equation for such terms is identical to Equation 6.26 for qualitative contrasts. Restated here, it is

$$SS_{A \times \text{quad}(B)} = \sum_j SS_{q(B)/A_j} - SS_{q(B)} \qquad (7.17)$$

where $SS_{q(B)/A_j}$ is the sum of squares for the qth component of B, calculated only for the A_j data. The calculations are illustrated for $SS_{A \times \text{quad}(B)}$ in panel b of Table 7.5. Because we are comparing three (in general, a) quadratic coefficients, the sum of squares is distributed on 2 (or $a - 1$) df. Therefore, the $MS_{A \times \text{quad}(B)}$ is 15.747/2, or 7.874.

The meaning of $MS_{A \times \text{quad}(B)}$ may be clearer if we consider another way to calculate it. Taking the values of the three quadratic coefficients from panel a of Table 7.5, calculate their variance as you would for any three numbers:

$$\frac{[-.62 - (-.2733)]^2 + [-.16 - (-.2733)]^2 + [-.04 - (-.2733)]^2}{2} = .0937$$

This variance of the b'_{j2} is multiplied by 6 (in general, n) and by 14 (the sum of the squared ξ_{k2}); the result is $84 \times .0937 = 7.874$, the same result we obtained using Equation 7.17. The point of viewing the calculations in this way is that it makes very explicit that $MS_{A \times \text{quad}(B)}$ is a function of the variance of the a quadratic regression coefficients, and thus directly bears on the null hypothesis that the β'_{j2} are the same at all levels of A.

The calculations for other polynomial components are identical to those for the quadratic in panel b of Table 7.5 except that the ξ values are changed. For example, to calculate the term $SS_{A \times \text{cubic}(B)}$, the coefficients would be $-1, 2, 0, -2$, and 1, and $\Sigma_k \xi_{k3}^2 = 10$. The results of the calculations of the polynomial components of the main and interaction sums of squares are presented in a general ANOVA table in panel c of Table 7.5. The higher GSR scores for the schizophrenic groups appear to represent a

TABLE 7.5   REGRESSION COEFFICIENTS, SAMPLE CALCULATIONS, AND ANOVA OF THE DATA OF TABLE 7.3

(a) Regression coefficients for each level of A

Clinical population (A)	Regression coefficients (b_p')			
	Linear (b_1')	Quadratic (b_2')	Cubic (b_3')	Quartic (b_4')
Controls (A_1)	.42	−.62	.08	−.04
Mildly schizophrenic (A_2)	.12	−.16	.12	.06
Severely schizophrenic (A_3)	.08	−.04	−.09	−.08

(b) Calculation of $SS_{A \times quad(B)}$

$$SS_{quad(B)/A_1} = \frac{n(\Sigma_k \xi_{k2} \overline{Y}_{.1k})^2}{\Sigma_k \xi_{k2}^2}$$

Then

$$SS_{quad(B)/A_1} = \frac{6[(2)(1.24) + (-1)(3.96) + (-2)(4.44) + (-1)(4.48) + (2)(3.08)]^2}{2^2 + (-1)^2 + (-2)^2 + (-1)^2 + 2^2}$$

$$= \frac{452.054}{14} = 32.290$$

Replacing the means in this ratio by those at A_2 and A_3, we get

$$SS_{quad(B)/A_2} = \frac{30.106}{14} = 2.150$$

$$SS_{quad(B)/A_3} = \frac{1.882}{14} = .134$$

Summing the three values of $SS_{quad(B)/A_j}$ and subtracting $SS_{quad(B)}$ (see Table 7.4), we have

$$SS_{A \times quad(B)} = (32.290 + 2.150 + .134) - 18.827 = 15.747$$

(c) ANOVA of the data of Table 7.3

SV	df	SS	MS	F
A	2	9.414	4.707	4.033[a]
B	4	27.261	6.815	5.840[b]
lin(B)	1	7.688	7.688	6.588[a]
quad(B)	1	18.827	18.827	16.130[b]
cubic(B)	1	.242	.242	.207
quart(B)	1	.504	.504	.432
AB	8	25.751	3.219	2.758[a]
$A \times$ lin(B)	2	4.144	2.072	1.776
$A \times$ quad(B)	2	15.747	7.874	6.747[b]
$A \times$ cubic(B)	2	1.492	.746	.639
$A \times$ quart(B)	2	4.368	2.180	1.872
S/AB	75	87.525	1.167	

[a] $p < .05$

[b] $p < .01$

reliable population effect; the significant F test of A attests to this. The significant effect of test stimulus height (B) seems largely due to a linear and quadratic trend. Not only are $\text{lin}(B)$ and $\text{quad}(B)$ the only significant components of the variability among the five means, but they account for 97% of the B sum of squares. We obtain this figure by dividing the sum of $SS_{\text{lin}(B)}$ and $SS_{\text{quad}(B)}$ by SS_B. Given the negligible residual variability, we ordinarily would not further analyze the residual into a cubic and quartic component; we do so in this case only to provide readers with a chance to check their understanding of the calculations.

The partitioning of SS_{AB} provides some additional insights. The $A \times \text{lin}(B)$ term is not significant. Taken together with the significant linear component of the B main effect, we conclude that all three populations exhibit increases in GSR with increasing size of the test stimulus but that the rate of increase does not differ among the populations. On the other hand, the quadratic components of the three group curves $[A \times \text{quad}(B)]$ do differ significantly; this accounts for about 61% of the interaction variability in the data set $(15.747/25.751 = .612)$. The reason for this can be found in the original data set (Table 7.3) and the values of the quadratic regression coefficients (panel a of Table 7.5). The curve appears flatter and, accordingly, the quadratic coefficients are closer to zero in the schizophrenic groups than in the control group. In other words, the schizophrenics generalize more to stimuli further from the training stimulus.

Contrasting Regression Coefficients

In the example of this section, we might want to contrast the average of the quadratic regression coefficients for the two schizophrenic groups with that for the controls. The formula for the sum of squares for contrasts of this sort is

$$SS_{p(A) \times q(B)} = \frac{n(\Sigma_j \Sigma_k w_{jp} \xi_{kq} \bar{Y}_{.jk})^2}{\Sigma_j \Sigma_k (w_{jp} \xi_{kp})^2} \tag{7.18}$$

or

$$SS_{p(A) \times q(B)} = \frac{(\Sigma_j \Sigma_k w_{jp} \xi_{kq} T_{.jk})^2}{n \Sigma_j \Sigma_k (w_{jp} \xi_{kq})^2} \tag{7.18'}$$

To contrast the average of the quadratic coefficients for the two schizophrenic groups with that for the control group, we specify a weight for each of the three clinical groups. These are

$$w_{j1} = 2, \ -1, \text{ and } -1$$

for the C, SS, and MS conditions, respectively. The subscript 1 indicates that this is the first of two contrasts we will consider. Multiplying each value of w_{11} by each value of ξ_{k2}, we obtain the products for the control group:

$$w_{11}\xi_{k2} = (2)(2), \ (2)(-1), \ (2)(-2), \ (2)(-1), \ (2)(2)$$

$$= 4, \quad -2, \quad -4, \quad -2, \quad 4$$

and for each of the schizophrenic groups:

$$w_{21}\xi_{k2} = (-1)(2), \ (-1)(-1), \ (-1)(-2), \ (-1)(-1), \ (-1)(2)$$

$$= -2, \quad 1, \quad 2, \quad 1, \quad -2$$

Applying Equation 7.18, we have

$$SS_{p(A)\times quad(B)} = \frac{6[(4)(1.24)+(-2)(3.96)+ \cdots +(1)(4.84)+(-2)(4.13)]^2}{4^2+(-2)^2+ \cdots +1^2+(-2)^2}$$

$$= \frac{1271.962}{84} = 15.142$$

This sum of squares is distributed on 1 df and can be tested against $MS_{S/AB}$; we obtain

$$F_{1,75} = \frac{15.142}{1.167} = 12.975$$

Clearly, the average of the quadratic coefficients for the two schizophrenic groups is significantly less than the quadratic coefficient for the control group.

At this point we might ask whether the quadratic coefficients differ significantly for the two schizophrenic groups. Do severely schizophrenic patients generalize more (have a less negative value of β_2') than mildly schizophrenic patients? The appropriate weights for this contrast are

$$w_{j2} = 0, \ 1, \ \text{and} \ -1$$

for the C, MS, and SS conditions, respectively. The reader should verify that the sum of squares for this contrast of quadratic coefficients is .605, which is not significant. Thus, we conclude that the two schizophrenic conditions differ from the normal with respect to quadratic curvature but do not differ from each other.

These two contrasts of quadratic coefficients are othogonal, as can be seen by summing the cross products of the w_{j1} and the w_{j2}. Because $SS_{A\times quad(B)}$ is distributed on 2 df, orthogonality implies that the sums of squares based on w_{j1} and w_{j2} should sum to $SS_{A\times quad(B)}$. This is the case; $15.142 + .605 = 15.747$.

To summarize the results of the analysis of interaction, the significant test of $A \times quad(B)$, coupled with nonsignificant tests of the interaction of A with the other polynomial components (see panel c of Table 7.5), shows that the AB interaction variability is largely attributable to differences among the three clinical populations in the quadratic curvature of the generalization functions. The two tests based on the further breakdown of $SS_{A\times quad(B)}$ into orthogonal components, in turn, reflect a pronounced difference between the schizophrenics and the controls but little difference between the two schizophrenic populations.

Analysis of Interaction When A and B Are Both Quantitative

Suppose that, in the experiment on generalization, A is also a quantitative variable. For example, A is the number of pairings of shock and training stimulus (the 11-inch-high rectangle of light), and B is once more the height of the test stimulus. The numbers of pairings are equally spaced: 5, 10, and 15. In such an experiment, we have two sets of

ξ's: ξ_{jp} for the levels of A, and ξ_{kq} for the levels of B. We can calculate sums of squares for such terms as $\text{lin}(A) \times \text{lin}(B)$ and $\text{quad}(A) \times \text{cub}(B)$. The calculations are the same as for any single df term of the form $\text{SS}_{p(A) \times q(B)}$. The interpretation is a little different from that presented for the case in which A represents a qualitative variable, such as clinical population.

Panel a of Figure 7.4 plots the means for an experiment in which we have three equally spaced levels of A (number of training trials) and five equally spaced levels of B (size of test stimulus). The means (again based on GSR measures) are exactly the same ones we have used throughout this section, but A now represents 5, 10, and 15 training trials. We calculate a quadratic coefficient for each of the three curves in this panel and plot these three values as a function of level of A in panel b. It is evident that

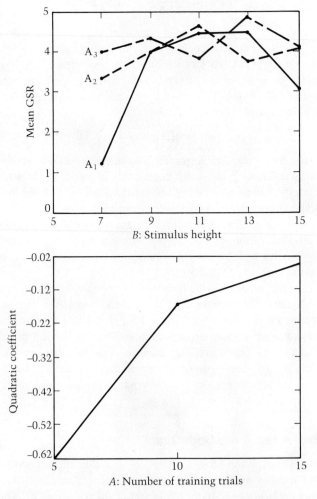

Figure 7.4 Cell means and quadratic coefficients for the data of Table 7.3.

b_2' increases (becomes less negative) as the number of training trials increases; this corresponds to the fact that the generalization functions in panel a are flatter (more generalization) for more training trials. Because the data are the same as before, $SS_{A \times quad(B)}$ still equals 15.747 and is still significant. This tells us that the three values plotted in panel b do vary significantly as a function of number of training trials.

Once again, we partition $SS_{A \times quad(B)}$ into two single df components. We obtain $SS_{lin(A) \times quad(B)}$ by replacing the w_{jp} of Equation 7.18 by the values -1, 0, and 1, the linear values of ξ when there are only three levels. Finding the products of the linear weights for A and the quadratic weights for B, we have for A_1:

$$(-1)(2), \ (-1)(-1), \ (-1)(-2), \ (-1)(-1), \ (-1)(2)$$

or

$$-2, \quad 1, \quad 2, \quad 1, \quad -2$$

The products are all zero at A_2, and at A_3 they are -1 times the A_1 values. Substituting these values into Equation 7.18 gives us

$$SS_{lin(A) \times quad(B)} = \frac{6[(-2)(1.24) + (1)(3.96) + \cdots + (-1)(4.84) + (2)(4.13)]^2}{(-2)^2 + (1)^2 + \cdots + (-1)^2 + (2)^2}$$

$$= \frac{395.6064}{28} = 14.129$$

When divided by $MS_{S/AB}$ this result is significant. What it tells us is that the line that best fits the three points in panel b of Figure 7.4 has a slope significantly different from zero. In terms of the original means in panel a, the apparent flattening of the generalization functions with increased numbers of training trials is a significant trend.

Now substitute the quadratic weights for $a = 3$ (Appendix Table D.7) into Equation 7.18; these weights are 1, -2, and 1, respectively. Then,

$$SS_{quad(A) \times quad(B)} = \frac{135.946}{84} = 1.618$$

Note that $SS_{lin(A) \times quad(B)} + SS_{quad(A) \times quad(B)} = 15.747$, the value of $SS_{A \times quad(B)}$. This reflects the fact that our two sums of squares components are orthogonal. Most of the $A \times quad(B)$ variability is due to the linear variation in quad(B). In terms of Figure 7.4, the significant value of $SS_{lin(A) \times quad(B)}$ tells us that the best-fitting straight line for the three points in panel b has a slope different from zero, and the nonsignificant value of $SS_{quad(A) \times quad(B)}$ indicates that the three quadratic regression coefficients do not depart significantly from a straight line. It appears that the quadratic coefficient of the generalization function increases (becomes less negative) linearly as the number of training trials increases.

7.5 CONCLUDING REMARKS

Trend analysis is a powerful tool for analyzing functional relations among variables. However, it is important to keep in mind that it is dangerous to draw conclusions about levels of the manipulated variables that were not included in the experiment. Reconsider the example just discussed. Our conclusion that there is a linear increase in the quadratic coefficient of the generalization function holds only for the values of A in our experiment. We can only guess as to what the function would look like if we had included values other than 5, 10, and 15 training trials. Extrapolation, drawing conclusions beyond the range of values used, can be misleading. It is unlikely that with more than 15 training trials the linear trend will continue. This would imply that with enough training trials the quadratic coefficient of the generalization function would become positive (that is, that the inverted U will become a U), which is rather unlikely. Even interpolation, conclusions about effects of values not in the experiment but within the range used, can be misleading. Remember that for fixed-effect variables, *statistical* inferences can be made only for the levels included in the experiment. Other inferences are based on our knowledge of the situation. In general, the more values of the quantitative independent variable that are present in the experiment, the more likely interpolation is to be valid.

Another concern is the routine application of trend analysis whenever one or more independent variables are quantitative. Any set of a data points can be fitted by a polynomial of order $a - 1$, but if the population function is not polynomial (a sine curve, for example), the polynomial analysis will be misleading. It is also dangerous to identify statistical components freely with psychological processes. It is one thing to postulate a cubic component of A, to test for it, and to find it significant, thus substantiating the theory. It is another matter to assign psychological meaning to a significant component that has not been postulated on a priori grounds. An unexpected significant component would be of interest and should alert the experimenter to the possible need for revising the behavioral hypotheses. Since calculating several polynomial F tests will increase the overall Type 1 error rate, however, significant results established on an a posteriori basis should require subsequent experimental validation before they are drawn into the body of scientific conclusions. Furthermore, when several trend tests are carried out on a set of data, inflation of the error rate for the family of tests can be a problem. For planned tests, the approach developed in Section 6.5 applies. For post hoc trend tests, the familywise error rate can be controlled by setting the significance level for the individual tests equal to the familywise error rate divided by the total number of orthogonal tests. With these caveats in mind, trend analysis can be a powerful tool for establishing the true shapes of data functions. As such, these methods of analyses should go hand in hand with developing precise quantitative behavioral theories.

APPENDIX 7.1: Calculating SS_{lin}

The text (Equation 7.6) defines SS_{lin} as

$$SS_{lin} = n\sum (\hat{\bar{Y}}_{.j} - \bar{Y}_{..})^2$$

Thus it is a measure of the variability of the predicted group means (predicted by a straight line that meets the least-squares criterion) about their average. Note that the average of the predicted group means is the grand mean of the data. We will demonstrate this and derive a computing formula for SS_{lin}. We begin with Equation 7.5, which defined the straight-line relationship between $\bar{Y}_{.j}$ and X_j:

$$\hat{\bar{Y}}_j = \bar{Y}_{..} + b_1(X_j - \bar{X}_.)$$

The average of the predicted group means is

$$\frac{\Sigma\hat{\bar{Y}}_j}{a} = \bar{Y}_{..} + b_1\Sigma\frac{(X_j - \bar{X}_.)}{a}$$

Because the sum of deviations of values about their average must be zero, $\Sigma(X_j - \bar{X}_.) = 0$, and the average of the predicted group means is just the grand mean of the data.

To obtain a computing formula for SS_{lin}, we substitute Equation 7.5 into Equation 7.6. The result is

$$SS_{lin} = n\Sigma[\bar{Y}_{..} + b_1(X_j - \bar{X}_.) - \bar{Y}_{..}]^2$$
$$= n\Sigma[b_1(X_j - \bar{X}_.)]^2$$
$$= nb_1^2\Sigma(X_j - \bar{X}_.)^2$$

Equation 7.4 provides an expression for b_1 that we substitute into the preceding equation. This gives us

$$SS_{lin} = n\left[\frac{[\Sigma(\bar{Y}_{.j} - \bar{Y}_{..})(X_j - \bar{X}_.)]^2}{[\Sigma(X_j - \bar{X}_.)^2]^2}\right]\Sigma(X_j - \bar{X}_.)^2$$

$$= \frac{n[\Sigma(\bar{Y}_{.j} - \bar{Y}_{..})(X_j - \bar{X}_.)]^2}{\Sigma(X_j - \bar{X}_.)^2}$$

Note that $\Sigma(\bar{Y}_{.j} - \bar{Y}_{..})(X_j - \bar{X}_.) = \Sigma\bar{Y}_{.j}(X_j - \bar{X}_.) - \Sigma\bar{Y}_{..}(X_j - \bar{X}_.)$. But $\Sigma\bar{Y}_{..}(X_j - \bar{X}_.) = \bar{Y}_{..}\Sigma(X_j - \bar{X}_.) = (\bar{Y}_{..})(0)$.
Therefore,

$$SS_{lin} = \frac{n[\Sigma\bar{Y}_{.j}(X_j - \bar{X}_.)]^2}{\Sigma(X_j - \bar{X}_.)^2}$$

APPENDIX 7.2: Finding Numerical Values for Orthogonal Polynomial Coefficients

Table D.7 presents values of ξ' for various values of a under the assumptions that there are equal numbers of measures for each level of A and that the levels of A are equally spaced. The derivational technique frequently gives noninteger values of ξ. These values were multiplied by λ to give the integers ξ' of Table D.7; remember that multiplying a set of weights by a constant λ will not change the sum of squares based on those weights.

For several reasons, it is worthwhile to show how the tabled polynomial coefficients are derived. We shall probably be more comfortable with the analyses if we can sense where the coefficients come from. More important, there may be occasions in which n's are not equal, or more frequently, in which the levels of A are not equally spaced. Suppose we have a study of delay of reward with four delay intervals. The values of X, delay magnitude, and numbers of subjects are

$$
\begin{array}{ccccc}
X_j & 0 & 1 & 3 & 4 \\
n_j & 4 & 5 & 5 & 4
\end{array}
$$

The linear coefficients ξ_{j1} are of the form

$$\xi_{j1} = a_1 + X_j$$

Substituting the actual delay intervals, we have

$$\xi_{11} = a_1 + 0$$

$$\xi_{21} = a_1 + 1$$

$$\xi_{31} = a_1 + 3$$

$$\xi_{41} = a_1 + 4$$

We now impose the restriction $\sum_j n_j \xi_{jp} = 0$. Then

$$4(a_1) + 5(a_1 + 1) + 5(a_1 + 3) + 4(a_1 + 4) = 0$$

and on solving this, we have

$$18a_1 + 36 = 0$$

$$a_1 = -2$$

Then

$$\xi_{11} = -2$$

$$\xi_{21} = -2 + 1 = -1$$

$$\xi_{31} = -2 + 3 = 1$$

$$\xi_{41} = -2 + 4 = 2$$

The quadratic coefficients are of the form

$$\xi_{j2} = a_2 + b_2 X_j + X_j^2$$

Then

$$\xi_{12} = a_2 + (0)(b_2) + 0^2$$

$$\xi_{22} = a_2 + b_2 + 1$$

$$\xi_{32} = a_2 + 3b_2 + 9$$

$$\xi_{42} = a_2 + 4b_2 + 16$$

We have two unknowns, a_2 and b_2. Because of the orthogonality requirement, we also have two simultaneous equations:

$$\sum_j n_j \xi_{j2} = 0$$

$$\sum_j n_j \xi_{j1} \xi_{j2} = 0$$

The second equation represents the orthogonality requirement. Applying these restrictions yields

$$(4)(a_2) + (5)(a_2 + b_2 + 1) + (5)(a_2 + 3b_2 + 9) + (4)(a_2 + 4b_2 + 16) = 0$$

$$(4)(-2)(a_2) + (5)(-1)(a_2 + b_2 + 1) + (5)(1)(a_2 + 3b_2 + 9)$$
$$+ (4)(2)(a_2 + 4b_2 + 16) = 0$$

Simplifying, we have

$$18a_2 + 36b_2 + 114 = 0$$

$$42b_2 + 168 = 0$$

Then

$$b_2 = \frac{-168}{42} = -4$$

and

$$18a_2 + (36)(-4) + 114 = 0$$

$$a_2 = \frac{5}{3}$$

Substituting into the original expressions for ξ_{j2} gives

$$\xi_{12} = \frac{5}{3}$$

$$\xi_{22} = \frac{5}{3} + (-4) + 1 = -\frac{4}{3}$$

$$\xi_{32} = \frac{5}{3} + (3)(-4) + 9 = -\frac{4}{3}$$

$$\xi_{42} = \frac{5}{3} + (4)(-4) + 16 = \frac{5}{3}$$

Let $\lambda = 3$. Then we have integer values

$$\xi'_{12} = 5$$

$$\xi'_{22} = -4$$

$$\xi'_{32} = -4$$

$$\xi'_{42} = 5$$

The cubic coefficients will involve three unknowns: a_3, b_3, c_3. We also have three simultaneous equations:

$$\sum n_j \xi_{j3} = 0$$

$$\sum n_j \xi_{j1} \xi_{j3} = 0$$

$$\sum n_j \xi_{j2} \xi_{j3} = 0$$

We leave the solution as an exercise.

Usually, the values of X are equally spaced or can be transformed to a scale on which they are equally spaced; if $A = 1, 2, 4, 8$, for example, then $\log_2 X = 0, 1, 2, 3$. When spacing is equal, and observations at each value of X occur equally frequently, Table D.7 immediately provides the desired coefficients.

EXERCISES

7.1 The following terms provide a useful review of some concepts in the chapter. Define, describe, or identify each of them:

equation for a straight line
linear regression coefficient
intercept
significant linear trend
significant nonlinear trend

polynomial equation
orthogonal polynomials
significant quad(A) $\times$ B term
significant quad(A) $\times$ lin(B) term

7.2 Four groups of eight subjects are each tested on a problem; time to solve is the dependent variable. The independent variable is the number of previous practice problems. The group means are

Problems:	1	2	3	4
Mean time:	6.49	4.82	4.25	3.80

The average within-group variance is 1.42.

(a) Calculate the value of b_1, the least-squares linear regression coefficient.

(b) Using the result in part (a), calculate the predicted mean for each of the four groups.

(c) Use Equation 7.6 to calculate SS_{lin}. Redo the calculations using the single df formula for a contrast. Carry out the significance test. What would a significant F ratio tell us about the results of this experiment?

(d) A somewhat different test of SS_{lin} has been proposed in several statistics textbooks. The procedure is to test SS_{lin} against $MS_{residual}$; $SS_{residual} = SS_{tot} - SS_{lin}$, $df_{residual} = (an - 1) - 1$, and $MS_{residual} = SS_{residual}/df_{residual}$. What advantage might this test have over the one described in this chapter? What is the possible danger?

(e) Calculate SS_A and subtract SS_{lin}. Are the data adequately described by a straight line? Explain.

7.3 Errors in a memory task are recorded for male and female subjects of ages 5, 6, 7, and 8. There are five subjects in each of the eight cells, the average within-cell variance is 8.75, and the cell means are

	Age			
	5	6	7	8
Male	16.3	7.2	6.5	7.4
Female	12.1	7.3	6.7	6.2

(a) Combining males and females, is there a linear trend in the decline of errors with age? Carry out the appropriate test.

(b) The large decrease in errors between ages 5 to 6, together with the apparent leveling off with further age increases, suggests a curvilinear component of the age function. Carry out the statistical test of nonlinearity.

(c) There is some indication that the decrease in errors with age is less pronounced for females. Carry out the appropriate trend test to address this issue.

(d) The mean of the Sex × lin(Age) term may be clearer if we redo part (c) in the following way. Find the equation for the best-fitting straight line for the males, and calculate the predicted mean errors at each of the four ages. Do the same for the females. At this point you have a table of means much like the preceding one except that those means were observed and these are predicted. Calculate the Sex × Age interaction sum of squares using the predicted means (remember that $n = 5$) and compare it to $SS_{\text{Sex} \times \text{lin(Age)}}$.

7.4 There are three training methods (T) and four equally spaced amounts of practice (P) with five subjects in each cell. The total number of errors in each cell is

	P_1	P_2	P_3	P_4
T_1	39	28	26	27
T_2	42	31	28	27
T_3	28	27	29	26

(a) $SS_{S/TP} = 87.84$. Present a complete ANOVA table including tests of the polynomial components of P and $T \times P$.

(b) We wish to know if the linear trend for T_1 differs significantly from the average linear trend for T_2 and T_3. (i) State the null hypothesis in terms of the regression coefficients for the three training populations; the notation may be β_{11}, β_{12}, and β_{13}, where the first subscript indicates linearity and the second the level of T. (ii) Carry out the significance test.

7.5 In a study of visual processing, subjects are required to scan a row of D digits on a screen and report if all of a set of T target digits are in the row. For example, on one trial the subject might be required to check for two targets, the numbers 3 and 7. The subject responds "yes" if both are present among the D digits scanned, and "no" otherwise. A model that views the subject as a serial processor predicts that response time on negative trials (RT_{neg}) will be

$$RT_{\text{neg}} = k_1 + k_2 DT$$

where k_1 and k_2 are constants. What terms, including trend components of main and interaction effects, should be significant if the theory is correct?

7.6 In trying to evaluate displays to be used in various detection tasks, a researcher ran the following experiment. Subjects viewed a screen on which targets appeared for 1 second; the subject had to report the location of the target as quickly and accurately as possible. One independent variable is the level of contrast (C) of the target to its background. A second is the structure of the screen, which is divided into one to five vertical areas (A). The researcher hypothesized that a certain amount of segmentation of the screen should aid location of the target; for example, dividing the screen in half should make it easier to find the target than when there is a single area to be searched. However, there should be some point at which further segmentation of the screen results in deterioration of performance because the subject will become confused and repeatedly search the same area. The shape of the function of A should also depend upon the level of C; the optimal value of A should be higher if contrast is higher. The cell means (proportion detected), based on 10 subjects each, are

		\multicolumn{5}{c}{Number of areas}				
		A_1	A_2	A_3	A_4	A_5
	C_1	42	53	60	68	52
Contrast level	C_2	37	44	51	40	36
	C_3	29	41	36	33	32

The average of the cell variances is 2.084. Perform a trend analysis and present your conclusions about the slopes and shapes of the average of the three functions, and about the differences among the functions.

7.7   In a study of memory, 12 groups of 10 subjects are run. They are all required to tell whether a comparison tone is the same as a standard presented several seconds earlier or different from it. The groups differ in D, duration of the interstimulus interval; $D = 1$, 2, or 3 seconds. They also differ in I, the stimulus presented during the interstimulus interval: blank, noise, tone 15 hertz above standard, or tone 30 hertz above standard. It is hypothesized that

   H_1: At all levels of I, memory scores will drop between $D = 1$ and 2, and drop somewhat less between $D = 2$ and 3.

   H_2: The rate of decay will be greater with $I =$ noise, next greatest when $I =$ tone $+$ 30, next greatest when $I =$ tone $+$ 15, and least for the blank interval.

   (a) Draw a set of curves consistent with these hypotheses.

   (b) What trend components of main and interaction effects would you expect to be significant, if these hypotheses are correct?

7.8   A large-scale study of programmed instruction is carried out with three variables: method (linear program, branching program, material is just read); ability level of students (low, average, high); and instruction time per day (30, 45, or 60 minutes). Consider the ability levels to be equally spaced. Several hypotheses are:

   H_1: Programmed instruction is superior to nonprogrammed instruction.

   H_2: Performance improves as instruction time increases but the improvement is smaller between 45 and 60 minutes than between 30 and 45 minutes.

   H_3: The improvement in performance with higher ability level is less marked for the programmed instruction conditions than for the reading control.

   H_4: The better the reader the less improvement with increased instructional time.

   (a) Plot a set of means consistent with these hypotheses.

   (b) What significance tests would you carry out to test this set of hypotheses?

7.9   Given the following ANOVA results, plot a set of means consistent with the pattern of significant results (marked by an asterisk *):

SV	df
A	2
*B**	3
lin(*B*)*	1
quad(*B*)	1
cubic(*B*)	1
*AB**	6
A × lin(*B*)	2
A × quad(*B*)*	2
A × cubic(*B*)	2

7.10   Assume a significant *AB* interaction. Further assume that there is a significant lin(*A*) × quad(*B*) component. Plot a set of means consistent with this, assuming three levels of *A* and four levels of *B*.

Chapter **8**

Repeated-Measures Designs

8.1 INTRODUCTION

In Chapter 4, we considered an experimental design in which percentage of sucrose was the independent variable and mean running time in a block of trials was the dependent variable. In this *between-subjects* design, each of the 32 rats in the experiment was randomly assigned to one of four sucrose solutions. Instead of the between-subjects design, we might have used a *within-subjects,* or *repeated-measures,* design in which each subject is tested at all levels of the factor of interest. For example, we could test each rat with all four of the sucrose solutions, with the order of solutions randomized over four days independently for each rat. Table 8.1 presents a data set for eight rats based on such an experiment, together with the ANOVA. Each row of the data set represents a different subject and each column represents a percentage of sucrose. In general, we have n subjects, each tested under a treatment levels. A single score is denoted by Y_{ij}, the score for the ith subject under the jth level of the treatment variable A.

In one sense there is nothing new about this design. Its layout is essentially that of a two-factor ($S \times A$) design with one score in each of the an cells of the design. Recognizing this makes the calculations of sums of squares and mean squares easy; as we can see in Table 8.1, they are nearly the same as in the two-factor design of Chapter 5. Of course, there is one difference: because there is only one score in each cell, we cannot calculate a within-cell error term. Instead, we use MS_{SA} as the denominator for the F test of A. We will shortly present the structural model and other assumptions underlying the data analysis, and use these to generate expected mean squares, thus justifying the use of MS_{SA} as our error term. Before then, however, we will discuss some ways in which the design differs from the between-subjects designs considered so far.

One obvious advantage of the repeated-measures design is that we need fewer subjects than in the between-subjects design. This is important when our subjects are members of a population that is limited in size, as in some clinical populations, or when

TABLE 8.1 DATA SET AND ANOVA FOR A SUBJECTS × TREATMENTS DESIGN

Data set

Percentage of sucrose in water

Subject	8	16	32	64	$T_{i.}$
1	1.4	3.2	3.2	3.0	10.8
2	2.0	2.5	3.1	5.8	13.4
3	1.4	4.2	4.1	5.6	15.3
4	2.3	4.6	4.0	5.9	16.8
5	4.7	4.8	4.4	5.9	19.8
6	3.2	5.0	6.2	5.9	20.3
7	4.0	6.8	4.5	6.5	21.8
8	5.0	6.1	6.4	6.6	24.1
$T_{.j}$	24.0	37.2	35.9	45.2	$T_{..} = 142.3$

ANOVA

SV	df	SS	MS	F
S	$n-1=7$	$a\sum_i (\bar{Y}_{i.} - \bar{Y}_{..})^2 = 35.387$	$\dfrac{SS_S}{n-1} = 5.055$	$\dfrac{MS_S}{MS_{SA}} = 8.568$
A	$a-1=3$	$n\sum_j (\bar{Y}_{.j} - \bar{Y}_{..})^2 = 28.671$	$\dfrac{SS_A}{a-1} = 9.557$	$\dfrac{MS_A}{MS_{SA}} = 16.198$
SA	$(n-1)(a-1)=21$	$\sum_i \sum_j (Y_{ij} - \bar{Y}_{i.} - \bar{Y}_{.j} + \bar{Y}_{..})^2$ $= 12.382$	$\dfrac{SS_{SA}}{(n-1)(a-1)} = .590$	
Total	$an-1=31$	$\sum_i \sum_j (Y_{ij} - \bar{Y}_{..})^2 = 76.440$		

Calculations

$$T_{i.} = \sum_j Y_{ij} \quad T_{.j} = \sum_i Y_{ij} \quad T_{..} = \sum_i \sum_j Y_{ij}$$

$$C = \frac{T_{..}^2}{na} = \frac{142.3^2}{32} = 632.790$$

$$SS_S = \frac{\Sigma_i T_{i.}^2}{a} - C = \frac{10.8^2 + \cdots + 24.1^2}{4} - 632.790 = 35.387$$

$$SS_A = \frac{\Sigma_j T_{.j}^2}{n} - C = \frac{24.0^2 + \cdots + 45.2^2}{8} - 632.790 = 28.671$$

$$SS_{SA} = SS_{tot} - SS_S - SS_A = 76.440 - 35.387 - 28.671 = 12.382$$

subjects are difficult to recruit, as when the task is very boring or dangerous, or when subjects are expensive animals such as monkeys. Even without these constraints on subject availability, the repeated-measures design may prove more practical than a between-subjects design. For example, if it takes very little time to obtain a score from a subject, then it is more efficient to run one subject under several conditions than to run several subjects, each under a different condition.

Practicality is an important factor in the choice of a design but efficiency—the size of the error variance—is even more important. Therefore, to understand the potential advantages of repeated-measures designs, we should consider the nature of the error variance. Recall that the structural model for the one-factor between-subjects design is

$$Y_{ij} = \mu + \alpha_j + \epsilon_{ij}$$

We might think of ϵ_{ij} in a between-subjects design as the sum of two components: (1) an *individual differences* component that occurs because subjects who differ from one another in ability, training, and other personal characteristics would respond differently even if tested at exactly the same moment of time under exactly the same conditions and (2) a *measurement error* component that occurs because even the same individual would respond differently when tested on different occasions because of fluctuations of attention, changes in the physical environment, chance variability in the stimulus, and a host of other factors. Because the individual differences component is usually quite variable, σ_e^2 (and $MS_{S/A}$, our best estimate of it) tends to be large in between-subjects designs. Other designs achieve greater efficiency than between-subjects designs by removing some or all of the individual differences component from ϵ.

In the treatment × blocks design discussed in Chapter 5, greater efficiency was achieved by introducing a blocking factor that allowed some of the individual differences variability to be removed from the total variability. In repeated-measures designs, the idea of blocking is taken a step further; a blocking factor, subjects, is introduced which has n levels, one for each subject. As we can see in the ANOVA of Table 8.1, we now are able to remove variability due to individual differences by subtracting a sum of squares for subjects from the total sum of squares. Both MS_A and the error term, MS_{SA}, still reflect measurement error. However, because of the elimination of variability due to individual differences, the error variance in the repeated-measures design will be much smaller than that in the comparable between-subjects design. As a result, when H_0 is false, F ratios will be larger and treatment effects will be more easily detected.

Repeated-measures designs make efficient use of subjects, both in the sense of using fewer subjects than between-subjects designs, and in the sense of having less error variance. However, not all independent variables lend themselves to such designs. For example, subject variables such as sex, intelligence, and clinical category must be treated as between-subjects factors. Except under rather unusual circumstances, a given subject cannot be expected to contribute one score as a male and a second score as a female. Also, for some types of independent variable, once subjects are run at one level, it does not make sense to run them at a second level. For example, in an experiment designed to compare the effectiveness of different methods of teaching mathematics, knowledge achieved by being exposed to one of the methods cannot be miraculously expunged so that it can be relearned using a second method.

Although the between-subjects designs we considered in Chapters 4 and 5 are inefficient, they are relatively simple. Scores in different groups can be considered to be independent and the within-cell variance can be used as the error term for testing any effect. The repeated-measures designs we introduce in the present chapter are more efficient, but we pay for the increased efficiency with some additional complexity. When we treat A as a within-subjects variable, certain considerations arise that were not encountered with between-subjects designs. For one thing, because each subject contributes scores to different levels of a within-subjects variable, we expect these scores to be correlated, not independent of one another, and this will have implications for the validity of the F test calculated in Table 8.1. For another, we cannot use within-cell variance as the error term to test all effects. In a pure repeated-measures design, there is no within-cell variance; because subjects is a factor in the design, each cell contains a single score. There will usually be an appropriate error term for each test of interest, but what we use as the error term will depend on which source of variance is tested. These inferential issues—implications of correlated scores and choice of error term—will be developed in subsequent sections.

8.2 THE ADDITIVE MODEL FOR THE $S \times A$ DESIGN

We first develop and discuss a very simple structural model for the design of Table 8.1. In this *additive model*, subject (S) and treatment (A) effects are assumed to add (together with an error component) to account for the deviation of a single score, Y_{ij}, from the grand mean, μ. We will then consider the "nonadditive" model in which an $S \times A$ interaction effect is added to the main effects to account for the structure of the data. Although the nonadditive model provides a more realistic account of most data sets, it raises certain inferential problems. Therefore, we begin our discussion of repeated measures designs with the additive model.

8.2.1 The Structural Equation

Consider a group of n subjects, each of whom is tested once under each level of the treatment variable A. The order in which the subject participates in the different treatment conditions is randomly determined, and the randomization is carried out independently for each subject. We assume that the n subjects are a random sample from an infinite population of subjects. We view Y_{ij}, the score of the ith subject under treatment A_j, as being composed of a "true score," μ_{ij}, and "measurement error," ϵ_{ij}; that is,

$$Y_{ij} = \mu_{ij} + \epsilon_{ij} \tag{8.1}$$

Next we want to express the true score μ_{ij} in terms of the population grand mean and effects of subjects and treatments. To do this we have to define some averages of the true scores. In turn, this requires us to distinguish between *fixed-effect* and *random-effect* variables. A treatment variable A is said to have fixed effects when we assume that its levels have been arbitrarily selected. Therefore, to find the average true score for subject i, we average over the a levels of A selected for our experiment:

$$\mu_i = \sum_{j=1}^{a} \frac{\mu_{ij}}{a} \tag{8.2}$$

The variable S is said to be a random-effect variable because the n subjects in the experiment are a random sample from a population of subjects. Therefore, the average of the population of true scores at any level of A is an expected value of the infinite set of true scores under A_j. Accordingly,

$$\mu_j = E(\mu_{ij}) \tag{8.3}$$

We define the population mean based on all subjects and the a levels of A as

$$\mu = \sum_{j=1}^{a} \frac{\mu_j}{a} = E(\mu_i) \tag{8.4}$$

Equations 8.2–8.4 allow us to define subject and treatment effects. The subject effect, denoted by η (Greek eta), is

$$\eta_i = \mu_i - \mu \tag{8.5}$$

The effect of A_j is

$$\alpha_j = \mu_j - \mu \tag{8.6}$$

We assume that the deviation of the true score μ_{ij} from the grand mean μ is due to the subject and treatment components defined in Equations 8.5 and 8.6. Therefore, we can now rewrite Equation 8.1 in terms of η_i and α_j:

$$Y_{ij} = \mu_{ij} + \epsilon_{ij} = \mu + \eta_i + \alpha_j + \epsilon_{ij} \tag{8.7}$$

This equation is the structural model under the assumption of additivity; the term *additivity* reflects the adding of η_i and α_j in the model and the assumption that there are no interaction effects. In the nonadditive model, to be considered next, we will add a term to Equation 8.7 to reflect an $S \times A$ interaction, and the consequences of that additional assumption will be developed.

To complete our statement of the additive model, we state certain conditions on α_j, η_i, and ϵ_{ij}:

1. The α_j, η_i, and ϵ_{ij} are distributed independently of each other.
2. If A is a fixed-effect variable (so that the entire population of levels of A is considered to be represented in the design), then

$$\sum \alpha_j = \sum (\mu_j - \mu) = 0$$

because the sum of all deviations of treatment means about their mean must be 0. A measure of variability of the treatment effects is given by

$$\delta_A^2 = \sum \frac{(\mu_j - \mu)^2}{a}$$

or

$$\theta_A^2 = \sum \frac{(\mu_j - \mu)^2}{a - 1}$$

If there is no variability in the μ_j's, there is no A main effect.

3. Because we have assumed that the subjects in the experiment are a random sample from an infinite population, the n values of η_i sampled in the experiment are unlikely to sum to zero; that is,

$$\sum_{i=1}^{n} \eta_i \neq 0$$

However, the average value of all such effects for the population of subjects will be zero; that is,

$$E(\eta_i) = 0$$

We assume that the population of η_i values is distributed independently and normally with variance

$$\text{Var}(\eta_i) = E(\eta_i^2) = \sigma_S^2$$

4. The error component ϵ_{ij} is assumed to be distributed independently and normally with

$$E(\epsilon_{ij}) = 0$$

and

$$\text{Var}(\epsilon_{ij}) = E(\epsilon_{ij}^2) = \sigma_e^2$$

The assumption that the errors are independent implies that

$$\text{Cov}(\epsilon_{ij}, \epsilon_{i'j'}) = E(\epsilon_{ij}\epsilon_{i'j'}) = 0 \quad \text{for } i \neq i' \text{ or } j \neq j'$$

With these definitions and assumptions before us, we may now take another look at the F tests of Table 8.1.

8.2.2 Expected Mean Squares and Design Efficiency

On the basis of Equation 8.7 and the assumptions about the α_j, η_i, and ϵ_{ij} stated, the expected mean squares can be derived. We have

SV	EMS
S	$\sigma_e^2 + a\sigma_S^2$
A	$\sigma_e^2 + n\theta_A^2$
$S \times A$	σ_e^2

It should be evident that, under the additive model, MS_{SA} is the appropriate error term for testing null hypotheses about both S and A. In either test, if the null hypothesis is true, the numerator and denominator of the F ratio have the same expected value.

The ratio of expected mean squares for the test of A looks very much like that for the test of A in the one-factor between-subjects design. There is a difference, however.

As we noted in the introduction to this chapter, the error variance in the between-subjects design has contributions from both individual differences and measurement error. In fact, if we let σ_m^2 represent just the variance of the errors of measurement, then the error variance in the between-subjects design may be rewritten as

$$E(\text{MS}_{S/A}) = \sigma_m^2 + \sigma_S^2$$

On the other hand, for the within-subjects design, variability due to subjects has been removed as a source of variance separate from the error term; therefore, the error variance in this design is

$$E(\text{MS}_{SA}) = \sigma_m^2$$

If H_0 is false (that is, $\theta_A^2 > 0$), then $E(\text{MS}_A)/E(\text{MS}_{SA})$ is greater than $E(\text{MS}_A)/E(\text{MS}_{S/A})$ because, for the within-subjects design,

$$\frac{E(\text{MS}_A)}{E(\text{MS}_{SA})} = \frac{\sigma_m^2 + n\theta_A^2}{\sigma_m^2} = 1 + \frac{n\theta_A^2}{\sigma_m^2}$$

and for the between-subjects design,

$$\frac{E(\text{MS}_A)}{E(\text{MS}_{S/A})} = \frac{\sigma_m^2 + \sigma_S^2 + n\theta_A^2}{\sigma_m^2 + \sigma_S^2} = 1 + \frac{n\theta_A^2}{\sigma_m^2 + \sigma_S^2}$$

Unless $\theta_A^2 = 0$ or there is no variance due to individual differences (an unlikely occurrence),

$$\frac{n\theta_A^2}{\sigma_m^2} > \frac{n\theta_A^2}{\sigma_m^2 + \sigma_S^2}$$

Therefore, the within-subjects design should yield a more powerful test of the effects of A than the between-subjects design, and the advantage will increase as σ_S^2 increases. Detecting significant effects is like trying to detect a signal in the presence of a noisy background; the "noise" for the within-subjects design is just due to errors of measurement, but is increased by the presence of individual differences in the between-subjects design. The smaller error variance for the within-subjects design not only yields larger F ratios but also more reliable estimates of population parameters, such as μ_j, $\mu_j - \mu_{j'}$, or θ_A^2.

8.3 THE NONADDITIVE MODEL FOR THE $S \times A$ DESIGN

In many experiments, the effects of treatments may vary over subjects. For example, the effects of rate of presentation of text material upon comprehension may depend upon such individual factors as reading ability, familiarity with the topic, current state of alertness, and motivation to perform well in the experiment. In short, the additive model of Equation 8.7 may not provide a valid description of the structure of data. In this section we consider a more general, nonadditive, model that includes a term to represent $S \times A$ interaction effects.

8.3.1 The Structural Equation

For the nonadditive model, we add an interaction component to the additive model given in Equation 8.7. That equation becomes

$$Y_{ij} = \mu + \eta_i + \alpha_j + (\eta\alpha)_{ij} + \epsilon_{ij} \tag{8.8}$$

Assumptions about the distribution of the terms that were in Equation 8.7 are unchanged. The interaction effect associated with the ijth cell is defined as

$$(\eta\alpha)_{ij} = (\mu_{ij} - \mu) - \eta_i - \alpha_j \tag{8.9}$$

$$= (\mu_{ij} - \mu) - (\mu_i - \mu) - (\mu_j - \mu)$$

$$= \mu_{ij} - \mu_i - \mu_j + \mu$$

At each level of A_j, $(\eta\alpha)_{ij}$ is assumed to be normally and independently distributed. The mean of each of these a distributions of interaction effects, $E(\eta\alpha)_{ij}$, is 0, and the variance is $E(\eta\alpha)_{ij}^2$, which we will refer to as σ_{SA}^2.

For each subject in the population, the mean interaction effect is also zero; in this case, $\Sigma_j(\eta\alpha)_{ij}/a = 0$. Notice that we expressed the mean of the population of interaction effects at A_j as an expected value, whereas we expressed the mean of the interaction effects for a subject as an arithmetic mean. The distinction lies in the fact that S is a random-effect variable and A is a fixed-effect variable. When S is random and A is fixed, the average of the interaction components for any subject is zero because all the levels of A that have been selected go into the average. However, the averages of the interaction components for each of the levels of A will generally not be zero. Although $E(\eta\alpha)_{ij}$, the average of the interaction components for the hypothetical population of subjects associated with A_j, is zero, $(1/n)\Sigma_i(\eta\alpha)_{ij}$, the average of the interaction components at A_j for the n subjects actually sampled to be in the experiment, is not. This distinction is important because it indicates that the $S \times A$ interaction should contribute to the variability among the levels of A but not to the variability among the levels of S. As we shall see, the expected mean square for A contains a component due to the interaction of S and A, but the expected mean square term for S does not.

Table 8.2 presents the true scores for a rather small population of four subjects. Note that the treatment population (column) means are identical; the null hypothesis is true. Also note that the average interaction effect in each column and in each row is zero; this is an algebraic result of the definition of an interaction effect. Now let's run an "experiment" by randomly sampling two of the four subjects, say S_1 and S_2. Ordinarily, we'd have more subjects in our experiment, but this should be enough to make our point. We also have set all ϵ_{ij} equal to zero; that is, we have made the observed scores (Y_{ij}) in our experiment equal to the true scores so that σ_e^2 will not be a factor in our discussion. Take the "data" from the first two subjects and calculate the $\bar{Y}_{.j}$; they are 9, 9.5, and 5.5. Note that although the μ_j are identical, the $\bar{Y}_{.j}$ are not. That's not because of error of measurement, because we have none in this artificial data set. It is entirely due to the fact that the two average *sampled* interaction effects are different at the three levels of A. To demonstrate this, we have subtracted each $(\eta\alpha)_{ij}$ from the corresponding Y_{ij}. For example, Y_{11} is $8 - (-1) = 9$ and $Y_{21} = 10 - 3 = 7$; this population of scores obtained by removing the effects of nonadditivity is also presented in Table 8.2. Because there are no interaction effects (and no error components), the treatment means in the data set, like those in the population, are identical. The point of all this is that interaction among subjects and treatments will contribute to the variability among the $\overline{Y}_{.j}$, and therefore to MS_A. Which treatment means will be most raised or lowered relative to their population values will depend upon the set of interaction effects that have been sampled from each treatment population.

One other point follows from our two-subject experiment. Notice that the subject means, $\overline{Y}_{i.}$, are unchanged if we subtract the interaction effects from the scores in Table 8.2; therefore MS_S is not affected by the pattern of sampled interactions. The reason for this is that A is a fixed-effect variable. When we get the mean for a subject, we average over all the $(\eta\alpha)_{ij}$ in a row; that average will always be zero.

This artificial example was intended to provide some intuition about the nature of the expected mean squares. We next present a more precise statement of these expectations and consider certain implications.

8.3.2 Expected Mean Squares

Table 8.3 presents the sources of variance (SV), degrees of freedom (df), and expected mean squares (EMS) under the additive and nonadditive models. Definitions of the components of variance are also presented. Although we have not repeated the calculations of Table 8.1, it should be noted that these do not differ for the two models. In

TABLE 8.2   DATA FOR A "POPULATION" OF FOUR SUBJECTS

With interaction effects present

	A_1		A_2		A_3			
	Y_{i1}	$(\eta\alpha)_{i1}$	Y_{i2}	$(\eta\alpha)_{i2}$	Y_{i3}	$(\eta\alpha)_{i3}$	μ_i	η_i
S_1	8	-1	10	1	9	0	9	$-.5$
S_2	10	3	9	2	2	-5	7	-2.5
S_3	11	-1	12	0	13	1	12	2.5
S_4	9	-1	7	-3	14	4	10	.5
μ_j	9.5		9.5		9.5		$\mu = 9.5$	

Without interaction effects present

	Y_{i1}	Y_{i2}	Y_{i3}	μ_i	η_i
S_1	$8 - (-1) = 9$	$10 - 1 = 9$	$9 - 0 = 9$	9	$-.5$
S_2	$10 - 3 = 7$	$9 - 2 = 7$	$2 - (-5) = 7$	7	-2.5
S_3	$11 - (-1) = 12$	$12 - 0 = 12$	$13 - 1 = 12$	12	2.5
S_4	$9 - (-1) = 10$	$7 - (-3) = 10$	$14 - 4 = 10$	10	.5
μ_j	9.5	9.5	9.5		

TABLE 8.3 EXPECTED MEAN SQUARES FOR AN $S \times A$ DESIGN

Model definition

The additive model: $Y_{ij} = \mu + \eta_i + \alpha_j + \epsilon_{ij}$

The nonadditive model: $Y_{ij} = \mu + \eta_i + \alpha_j + (\eta\alpha)_{ij} + \epsilon_{ij}$

where $\eta_i = \mu_i - \mu$, $\quad \alpha_j = \mu_j - \mu$, $\quad (\eta\alpha)_{ij} = \mu_{ij} - \mu_i - \mu_j + \mu$

$$\mu_{ij} = E(Y_{ij}), \quad \mu_j = E(\mu_{ij}), \quad \mu_i = \sum_j \frac{\mu_{ij}}{a}, \quad \mu = E(\mu_i) = \sum_j \frac{\mu_j}{a}$$

$$\sum_j \alpha_j = \sum_j (\eta\alpha)_{ij} = 0 \quad \text{and} \quad E[(\eta\alpha)_{ij}] = E(\epsilon_{ij}) = 0$$

$$E(\epsilon_{ij}^2) = \sigma_e^2, \quad E(\eta_i^2) = \sigma_S^2, \quad E[(\eta\alpha)_{ij}^2] = \sigma_{SA}^2, \quad \text{and} \quad \sum_j \frac{\alpha_j^2}{a-1} = \theta_A^2$$

SV	df	Additive EMS	Nonadditive EMS	F
S	$n-1$	$\sigma_e^2 + a\sigma_S^2$	$\sigma_e^2 + a\sigma_S^2$	$\dfrac{\text{MS}_S}{\text{MS}_{SA}}$
A	$a-1$	$\sigma_e^2 + n\theta_A^2$	$\sigma_e^2 + \sigma_{SA}^2 + n\theta_A^2$	$\dfrac{\text{MS}_A}{\text{MS}_{SA}}$
SA	$(n-1)(a-1)$	σ_e^2	$\sigma_e^2 + \sigma_{SA}^2$	

both cases, the A source of variance would be tested against MS_{SA}. However, the two models will have different consequences for our interpretation of the F test, and for our ability to perform other inferences, such as tests of the subject term, estimates of ω^2, and estimates of missing scores. We will have more to say about all these topics shortly.

Not surprisingly, the EMS for the SA source of variance under the nonadditive model now contains contributions from both σ_e^2 and σ_{SA}^2. Under that model, σ_{SA}^2 also contributes to the EMS for A but not to that for S. This follows from the numerical example presented in Table 8.2 and our discussion of its consequences. We saw there that the average of the n interaction effects at each level of A ordinarily will vary; therefore, the MS_A may reflect variability among those average interaction effects (σ_{SA}^2). In contrast, the average of the a interaction effects for each subject is zero; therefore the MS_S will not reflect σ_{SA}^2.

The expected mean squares in Table 8.3 indicate that if we have nonadditivity, our test of the subject effect is negatively biased. This is because $E(\text{MS}_S)/E(\text{MS}_{SA}) < 1$ when the null hypothesis is true. From Table 8.3,

$$\frac{E(\text{MS}_S)}{E(\text{MS}_{SA})} = \frac{\sigma_e^2 + a\sigma_S^2}{\sigma_e^2 + \sigma_{SA}^2}$$

If H_0 is true (i.e., $\sigma_S^2 = 0$, the ratio is clearly less than 1. This may result in too many Type 2 errors when the S effect is tested. However, this is not a particularly important consequence of nonadditivity because we are not usually interested in testing the S effect. Even if we are, the subject variability is almost always large enough to be detected by the most conservative test. On the other hand, there are times when estimates of σ_S^2 are desirable, and no exact estimate is available if there is $S \times A$ variance in the population.

Another implication of the expected mean squares for the nonadditive model is that the $S \times A$ interaction reduces the efficiency of the design. Looking at $E(\text{MS}_A)/E(\text{MS}_{SA})$ in the two cases, we find that the ratio for the nonadditive model is smaller than for the additive model. That is,

$$1 + \frac{n\theta_A^2}{\sigma_e^2 + \sigma_{SA}^2} \leqslant 1 + \frac{n\theta_A^2}{\sigma_e^2}$$

Thus, if the null hypothesis is false, the $S \times A$ interaction effects will tend to reduce the power of the F test. It is important to understand, however, that such interaction variance will almost always be less than the individual difference variance associated with between-subjects designs. Therefore, even when the data do not conform to the additive model, the repeated-measurements design will yield more powerful tests of the null hypothesis than will the between-subjects design with the same number of observations.

The expected mean squares in Table 8.3 do not suggest bias in the test of the independent variable, A. Nevertheless, there is a potential problem, which we consider next.

8.4 TESTING THE EFFECTS OF THE INDEPENDENT VARIABLE, *A*

As we have frequently stated, a necessary condition for a mean square to be a proper error term to test some null hypothesis is the following: If H_0 is true, then the expectations of the numerator and denominator mean squares should be identical. As is evident in Table 8.3, this requirement clearly is met by the expected mean squares in both the additive and nonadditive cases. However, meeting this requirement is not sufficient for $\text{MS}_A/\text{MS}_{SA}$ to have an F distribution. Rouanet and Lepine (1970) and Huynh and Feldt (1970) have shown that an assumption called *sphericity* (or *circularity*) must also be met; when that assumption is violated, Type 1 error rates will be inflated, sometimes severely. In what follows, we will try to provide a clear sense of the sphericity assumption, its consequences, and procedures to use when the assumption is violated.

8.4.1 Sphericity (Homogeneity of Variances of Difference Scores)

The concept of sphericity is illustrated in panel a of Table 8.4. We have data for five subjects at three levels of A. We also have calculated all possible difference scores for each subject: $d_{i,12} = Y_{i1} - Y_{i2}$, $d_{i,13} = Y_{i1} - Y_{i3}$, and $d_{i,23} = Y_{i2} - Y_{i3}$. The assump-

TABLE 8.4   DATA EXHIBITING SPHERICITY (SET A) AND NONSPHERICITY (SET B)

(a) Data set A (exhibits sphericity)

	A_1	A_2	A_3	$Y_{i3} - Y_{i2}$	$Y_{i2} - Y_{i1}$	$Y_{i3} - Y_{i1}$
S_1	21.050	7.214	26.812	19.598	−13.836	5.760
S_2	6.915	29.599	16.366	−13.233	22.684	9.451
S_3	3.890	21.000	41.053	20.053	17.110	37.163
S_4	11.975	12.401	18.896	6.495	.426	6.921
S_5	31.169	34.786	31.872	−2.914	3.617	.703
Mean	15.000	21.000	27.000	6.000	6.000	12.000
$\hat{\sigma}^2$	124.000	132.000	100.000	208.000	208.000	208.000

Note: $MS_{SA} = 104 = (1/2)\sigma_d^2$

(b) Data set B (exhibits nonsphericity)

	A_1	A_2	A_3	$Y_{i3} - Y_{i2}$	$Y_{i2} - Y_{i1}$	$Y_{i3} - Y_{i1}$
S_1	1.7	3.9	6.0	2.1	2.2	4.3
S_2	4.4	6.5	14.5	8.0	2.1	10.1
S_3	7.8	13.3	18.6	5.3	5.5	10.8
S_4	6.6	9.4	14.5	5.1	2.8	7.9
S_5	9.1	15.2	23.5	8.3	6.1	14.4
Mean	5.92	9.66	15.42	5.76	3.74	9.50
$\hat{\sigma}^2$	8.557	21.793	41.457	6.378	3.653	13.914

Note: $MS_{SA} = 3.991 = (1/2)(6.378 + 3.653 + 13.914)/3$

tion of sphericity states that the three populations of difference scores have identical variances. In general, if we have a treatment levels, there will be $(1/2)(a)(a-1)$ possible populations of difference scores, and it is assumed that all have the same variance, σ_d^2. Note that the three sample values of σ_d^2 are identical in the example in panel a; the sphericity assumption is met by these data.[1] The data in panel b exhibit considerably more heterogeneity of variance of difference scores (nonsphericity); if such heterogeneity exists in the population, Type 1 error rates will be increased. We will shortly consider what to do about this, but first we note some conditions under which the sphericity assumption will hold.

Additivity is a sufficient, but not necessary, condition for the sphericity assumption to hold. Additivity is sufficient because if there are no $S \times A$ interaction effects, the difference between the scores under any two treatments, A_j and $A_{j'}$, does not vary over

---

[1] We will never observe perfect sphericity in a real data set, but by eliminating error from these "data" we will be able to make some points more clearly.

subjects. But this means that the variance of the difference scores based on A_j and $A_{j'}$ is zero, and this is true for all j and j'. Thus, additivity implies constant (zero) variance of difference scores. However, it is evident from data set A of Table 8.4 that additivity is not necessary for sphericity to hold. Although there is an interaction between A and S, the variances of difference scores are identical.

Another condition that is sufficient, but not necessary, for sphericity is called *compound symmetry*. This requires homogeneity of the population treatment variances and homogeneity of the population covariances; that is, we assume

$$\sigma_1^2 = \cdots = \sigma_j^2 = \cdots = \sigma_a^2$$

and

$$\rho_{12}\sigma_1\sigma_2 = \cdots = \rho_{jj'}\sigma_j\sigma_{j'} = \cdots$$

where ρ_{12} is the population correlation between the scores at A_1 and A_2 and $\rho_{12}\sigma_1\sigma_2$ is the covariance. Recall from Chapter 2 that if $d_{12} = Y_1 - Y_2$, then

$$\sigma_d^2 = \sigma_1^2 + \sigma_2^2 - 2\rho_{12}\sigma_1\sigma_2 \tag{8.10}$$

If compound symmetry holds, all variances of difference scores will involve the exact same variances and covariances of scores, and therefore they must be identical. Although sphericity is a consequence of compound symmetry, we can see from data set A of Table 8.4 that sphericity can occur without compound symmetry. Although the variances of difference scores are identical, the variances of the individual scores are not the same, nor are the covariances for pairs of treatments.

Nonsphericity, or heterogeneity of variance of difference scores, is analogous in both form and consequences to heterogeneity of variance in the between-subjects designs of Chapters 4 and 5. In Chapter 4, the error term $MS_{S/A}$, was the average of the group variances. We showed that if the null hypothesis is true, and if there are n scores at each treatment level, then $E(MS_A) = E(MS_{S/A})$ even if the group variances are very different from each other. However, even though this heterogeneity of variance doesn't affect the ratio of expected mean squares, it does affect the sampling distribution of the ratio of mean squares. More precisely, when the null hypothesis is true, heterogeneity of variance inflates the probability of sampling large F values.

We have a similar situation in the repeated-measures design. As can be seen in panels a and b of Table 8.4, the error term MS_{SA} is one-half the average of the three values of $\hat{\sigma}_d^2$; this relation between the variances of the difference scores and MS_{SA} will hold for any number of levels of A. If these variances are very different, the Type 1 error rate will be inflated, as is the case when group variances differ in the between-subjects design. Mauchly (1940) derived a test of the null hypothesis that the variances of difference scores are homogeneous, but, if the population distributions are not normal, the Mauchly test tends to yield significant results even when sphericity holds. Therefore, we recommend against using this test to determine whether there is a problem in the data set. Rogan, Keselman, and Mendoza (1979) present some simulation results relevant to this point, together with a good discussion of the general topic of the analysis of repeated measurements.

There are three data analysis strategies that protect the researcher against inflation of Type 1 error rates due to nonsphericity. These are (1) the *multivariate analysis of variance,* or MANOVA; (2) the *univariate F test with ϵ-adjusted degrees of freedom;* and (3) *tests of planned contrasts.* The first two involve calculating the covariances of scores. The reason for this lies in the relation, expressed in Equation 8.10, between the variances and covariances of scores and the variance of the difference scores. Let us consider each of the three approaches in turn.

8.4.2 Multivariate Analysis of Variance (MANOVA)

In MANOVA, an *F* statistic is calculated, but the calculations involve covariance terms and the *F* is distributed on $a - 1$ and $n - (a - 1)$ df rather than on $a - 1$ and $(n - 1)(a - 1)$ df. The assumption of sphericity is not required. Comparisons with the ϵ-adjustment approach to be considered next have revealed no consistent power advantage for either approach over all conditions (O'Brien and Kaiser 1985); both also provide an honest Type 1 error rate. Because the ϵ-adjusted approach builds on the univariate *F* test developed thus far in this book, and is therefore conceptually simpler, we will focus on it. Readers interested in learning about MANOVA, as well as related multivariate analyses, will find many textbooks on multivariate statistics (e.g., Dillon and Goldstein 1984, Harris 1985, Morrison 1976). A nonmathematical introduction that focuses on how to use the SPSS[X] MANOVA program is given by O'Brien and Kaiser (1985).

8.4.3 The ϵ-Adjusted *F* Test

This procedure stems from a paper by Box (1954), in which he showed that the statistic MS_A/MS_{SA} is distributed as *F* even when the assumption of sphericity is violated. However, with violations of sphericity, the degrees of freedom associated with the *F* are not $a - 1$ and $(a - 1)(n - 1)$ but $(a - 1)\epsilon$ and $(a - 1)(n - 1)\epsilon$. The adjustment ϵ is a function of the degree of nonsphericity; as we shall see, its calculations involve the variances and covariances of scores at each treatment level. If the variances of difference scores are homogeneous—that is, if there is sphericity—ϵ will be 1 and the usual *F* test can be used. Under conditions in which the assumption is severely violated, ϵ approaches a lower bound of $1/(a - 1)$. In this case, *F* would be distributed on 1 and $n - 1$ df. In short, as nonsphericity increases, the accompanying inflation of Type 1 error rate is compensated for by requiring a larger *F* (because df are less) for significance.

Calculations of $\hat{\epsilon}$, the estimate of ϵ, require calculating covariances for all pairs of levels of *A,* as well as variances for each level of *A.* Fortunately, two common statistical computer programs, BMDP2V and SPSS[X] MANOVA, do these calculations. Part of the BMDP2V output for the data from data set B of Table 8.4 is presented in Table 8.5. The format of the ANOVA table requires some comment. The first SV listed corresponds to T^2/an (the correction term, *C*), and the null hypothesis tested is that the average true score is equal to 0. The error term for testing this null hypothesis is what we have been calling the *S* source of variance. The error term for testing the *A* effect is the MS_{SA}. The *p* level for the conventional degrees of freedom is presented in the "tail prob" column. The next column presents the *p* value for the df adjusted by the ϵ estimate first presented by Box (1954) and subsequently extended to the designs of Chapter

TABLE 8.5 OUTPUT FROM BMDP2V FOR DATA SET B OF TABLE 8.4

	SOURCE	SUM OF SQUARES	DEGREES OF FREEDOM	MEAN SQUARE	F	TAIL PROB.	GREENHOUSE-GEISSER PROB.	HUYNH-FELDT PROB.
1	MEAN	1601.66667	1	1601.66667	25.09	.0074		
	ERROR	255.30000	4	63.82500				
2	A	229.02533	2	114.51267	28.69	.0002	.0025	.0009
	ERROR	31.92800	8	3.99100				

ERROR TERM   EPSILON FACTORS FOR DEGREES OF FREEDOM ADJUSTMENT

	GREENHOUSE-GEISSER	HUYNH-FELDT
2	.6284	.7808

9 by Greenhouse and Geisser (1959). The p value in the far right column is based on degrees of freedom adjusted by $\tilde{\epsilon}$, a second estimator of ϵ (Huynh and Feldt 1976). The value of $\tilde{\epsilon}$ is always at least as large as that of $\hat{\epsilon}$. Therefore, the $\tilde{\epsilon}$ adjustment will generally result in a more powerful F test of A; however, the Type 1 error rate appears to be close to the nominal level. For these reasons, we recommend the Huynh-Feldt adjustment.

Researchers who do not have access to the BMDP2V or SPSS[X] MANOVA programs still can save considerable labor in calculating $\hat{\epsilon}$ (or $\tilde{\epsilon}$) by using a statistical program that calculates the variance at each level of A and the covariance for each pair of levels of A. SYSTAT's CORR (correlation) module will do just that when given the data and the command COVARIANCE. The output is a *variance-covariance matrix,* **S,** whose diagonal elements are the values of $\hat{\sigma}_j^2$ and whose off-diagonal elements are the covariances of Y_j and $Y_{j'}$. These values have been calculated for data set B of Table 8.4, and **S** is presented in Table 8.6 together with equations for calculating $\hat{\epsilon}$ and $\tilde{\epsilon}$. The elements of **S** have been inserted into these equations to illustrate the calculations.

In the rare case in which the researcher has neither a program for calculating $\hat{\epsilon}$ directly, such as BMDP2V, nor a program for calculating **S,** the calculations can be tedious and time-consuming. Greenhouse and Geisser have suggested a three-step approach to significance testing that frequently will avoid the necessity of carrying out such calculations:

1. First test the F ratio in question with the conventional degrees of freedom [i.e., $a - 1$, and $(n - 1)(a - 1)$]. If the F test is not significant using these df, it certainly won't be if the df are reduced. The null hypothesis cannot be rejected.

2. If the F test using conventional df is significant, perform the conservative F test using 1 and $n - 1$ df [i.e., the test that assumes ϵ takes on its lowest possible value of $1/(a - 1)$]. If this conservative F test is significant, the null hypothesis can be rejected without further testing.

3. If the conventional F test is significant but the conservative F test is not, the correction factor ϵ must be estimated.

TABLE 8.6   CALCULATION OF $\hat{\epsilon}$ AND $\tilde{\epsilon}$ USING DATA SET B OF TABLE 8.4

The variance-covariance matrix is

$$\mathbf{S} = \begin{bmatrix} 8.557 & 13.349 & 18.050 \\ 13.349 & 21.793 & 28.436 \\ 18.050 & 28.436 & 41.457 \end{bmatrix}$$

We define the following statistics based on **S**:

$\bar{S}_{..}$ = the mean of all a^2 (9) elements of the matrix = 21.275

$\bar{S}_{jj}$ = the mean of the a (3) variances (main diagonal elements)

$$= \frac{8.557 + 21.793 + 41.457}{3} = 23.936$$

$\bar{S}_j$ = the mean of the jth row (or column) where

$\bar{S}_1 = 13.319$, $\bar{S}_2 = 21.193$, and $\bar{S}_3 = 29.314$

$\sum_j \sum_{j'} S_{jj'}^2$ = the sum of the squares of the nine elements

$$= 4892.049$$

$\sum_j S_j^2$ = the sum of the squared row (or column) means

$$= 13.319^2 + 21.193^2 + 29.314^2$$
$$= 1485.846$$

The Greenhouse-Geisser adjustment of the df is

$$\hat{\epsilon} = \frac{a^2(\bar{S}_{jj} - \bar{S}_{..})^2}{(a-1)(\sum_j \sum_{j'} S_{jj'}^2 - 2a\sum_j \bar{S}_j^2 + a^2\bar{S}_{..}^2)}$$

$$= \frac{9(23.936 - 21.275)^2}{2[4892.049 - (2)(3)(1485.846) + 9(21.275^2)]} = .628$$

and the Huynh-Feldt adjustment is

$$\tilde{\epsilon} = \frac{n(a-1)\hat{\epsilon} - 2}{(a-1)[n-1-(a-1)\hat{\epsilon}]}$$

$$= \frac{(5)(2)(.628) - 2}{2[5 - 1 - (2)(.628)]} = .781$$

Given the availability of computers and the fact that estimates of ϵ are automatically calculated by some statistical packages, we recommend that the df correction be used. Notice, however, that the correction does not apply if A only has two levels. In that case, we test $H_0: \mu_2 - \mu_1 = 0$. There is only one variance of difference scores, and therefore the sphericity assumption is irrelevant. We can generalize this point to any single df contrast of means. For example, we might have three means and wish to test $H_0: \psi = \mu_3 - (1/2)(\mu_1 + \mu_2) = 0$. We have a single contrast score for each subject, $C_i = Y_{i3} - (1/2)(Y_{i1} + Y_{i2})$, and therefore only one variance of contrast scores. Our best estimate of the population contrast ψ is the mean of the contrast scores (i.e., $\hat{\psi} = \bar{C}$). The estimated standard error of $\hat{\psi}$ is $\hat{\sigma}_{\hat{\psi}} = \hat{\sigma}_C/\sqrt{n}$. If we test $\hat{\psi}$ against $\hat{\sigma}_{\hat{\psi}}$, sphericity is not an issue; it is as though we performed a test on the mean of any set of n numbers. Because such tests of contrasts are not affected by nonsphericity, and because such tests are of interest in their own right, we will consider them more closely. In what follows, we assume that readers are familiar with the material in Chapter 6.

8.4.4 Testing Single Degree-of-Freedom Contrasts in Within-Subjects Designs

In view of the preceding comments, we suggest that the researcher plan contrasts prior to viewing the data and test these against their respective variances. For example, using data set B of Table 8.4, we might plan to test the null hypotheses that the contrasts ψ_1 and ψ_2 are zero, where these contrasts are defined as

$$\psi_1 = \mu_2 - \mu_1 \quad \text{and} \quad \psi_2 = \mu_3 - \frac{1}{2}(\mu_1 + \mu_2)$$

Table 8.7 presents values of these contrasts for each of the five subjects in Table 8.4. Two equivalent tests of each null hypothesis, t and F, are illustrated in the table. The denominators for both the t and F tests are based on the variance of the contrast being tested; the denominator of the t is $\hat{\sigma}_{\hat{\psi}} = \hat{\sigma}_C/\sqrt{n}$, and that of the F is $\text{MS}_{S \times \hat{\psi}}$, which equals $\hat{\sigma}_{\hat{\psi}}^2/(\Sigma_j w_j^2/n)$ or $\hat{\sigma}_C^2/\Sigma w_j^2$. An alternative testing procedure is to use $\sqrt{\text{MS}_{SA}\Sigma_j w_j^2/n}$ for the denominator of the t statistic or MS_{SA} for the denominator of the F statistic. We recommend $\text{MS}_{S \times \hat{\psi}}$ rather than MS_{SA} because if there is nonsphericity in the data set, the values of $\text{MS}_{S \times \hat{\psi}}$ may vary greatly as a function of ψ. Consequently, MS_{SA} will be too large for some contrasts and too small for others.[2]

Ordinarily, we use statistical packages to test within-subjects contrasts directly. For example, we can use the CMATRIX command in SYSTAT or the CONTRAST = SPECIAL command in SPSS[X] followed by one or more rows of contrast weights. An alternate approach that is less direct but provides a better understanding of the nature of tests of within-subjects contrasts is illustrated in Table 8.7. This approach first creates a new variable that represents the value of a contrast for each subject and then tests hypotheses about this contrast variable. It is also easily implemented by using statistical packages. For example, BMDP3D will create variables corresponding to our two contrasts if we use

```
/TRANSFORM    CON1 = Y1-Y2
              CON2 = .5*(Y1+Y2)-Y3.
```

---

[2] If sphericity holds for a data set, MS_{SA} would result in a more powerful test than $\text{MS}_{S \times \hat{\psi}}$ because the associated df are $(n-1)(a-1)$ rather than $n-1$. However, we can never *know* the variances of difference scores in the population and, if the sphericity assumption is violated, test statistics based on MS_{SA} will not be properly distributed. Therefore, we (and every computer package of which we know) use $\text{MS}_{S \times \hat{\psi}}$ to test hypotheses about ψ.

TABLE 8.7 t AND F TESTS OF TWO CONTRASTS BASED ON DATA SET B OF TABLE 8.4

	$C_{1i} = Y_{i2} - Y_{i1}$	$C_{2i} = Y_{i3} - (1/2)(Y_{i1} + Y_{i2})$
S_1	2.2	3.20
S_2	2.1	9.05
S_3	5.5	8.05
S_4	2.8	6.50
S_5	6.1	11.35
$\hat{\psi} = \bar{C}$	3.74	7.63
$\hat{\sigma}_C^2$	3.6530	9.2332
$\hat{\sigma}_{\hat{\psi}}^2 = \hat{\sigma}_C^2/n$	3.653/5	9.2332/5

$$t = \frac{\hat{\psi}}{\hat{\sigma}_{\hat{\psi}}}$$

$$F = \frac{\hat{\psi}^2/(\Sigma_j w_j^2/n)}{MS_{S \times \hat{\psi}}}$$

$$MS_{S \times \hat{\psi}} = \frac{\hat{\sigma}_{\hat{\psi}}^2}{\Sigma_j w_j^2/n}$$

$$t = \frac{3.74}{\sqrt{3.653/5}} = 4.376 \qquad t = \frac{7.63}{\sqrt{9.2332/5}} = 5.615$$

$$F = \frac{3.74^2/(2/5)}{(3.653/5)/(2/5)} = 19.145 \qquad F = \frac{7.63^2/(1.5/5)}{(9.2332)/5)/(1.5/5)} = 31.526$$

Note: $F = t^2$ for both contrasts.

With a few additional instructions, BMDP3D will perform one-sample t tests on each of the contrast variables.

Assuming these tests of contrasts are planned before the data have been looked at, the Bonferroni test procedure described in Chapter 6 provides a way of controlling the error rate for the family of planned contrasts. Even when only pairwise comparisons are of interest, the Bonferroni procedure appears to meet the goals of controlling familywise error rate and maximizing power better than various alternatives, including modifications of the Tukey test designed to counter the problem of sphericity (Maxwell 1980). The problem with using procedures such as the Scheffé and Tukey is that they involve MS_{SA}. Unless the sphericity assumption holds, MS_{SA} will generally not have the correct value for any given comparison of interest. Boik (1981) showed that even extremely small departures from sphericity can create serious distortions in the width of confidence intervals formed by using the Scheffé procedure in repeated-measures designs, and Maxwell (1980) and Mitzel and Games (1981) have shown that the Type 1 error rate tends to be inflated.

Although procedures that compensate for nonsphericity have been proposed (such as adjusting the degrees of freedom in the Scheffé procedure by $\hat{\epsilon}$ or $\tilde{\epsilon}$) and found to maintain the appropriate error rate per family, the Scheffé confidence interval will be too long or too short for almost all comparisons. Using a series of Monte Carlo simulations, Maxwell, Delaney, and Sternitzke (1983) found that although several modifications of the Scheffé procedure did not adequately control Type 1 error, it was controlled at a satisfactory level by the Bonferroni procedure.

The message seems to be clear. Because of its conceptual simplicity, flexibility, and ability to control error, the Bonferroni procedure should be used when families of contrasts are tested in repeated-measures designs.

8.5 ESTIMATING VARIANCE COMPONENTS AND ω^2, THE PROPORTION OF THE POPULATION VARIANCE ACCOUNTED FOR

As we pointed out in Chapter 4, it is sometimes of interest to evaluate the actual magnitude of variance components such as θ_A^2 or σ_S^2. Such estimates also can be combined to calculate $\hat{\omega}^2$, an estimate of the proportion of total population variance due to some component. If the additive model provides a good description of the population of scores, the estimates are readily obtained. In the first part of this section, we will present estimates assuming the additive model. We will then note the implications of nonadditivity. Finally, we will consider a very useful application of these developments, the estimation of the reliability of a measuring instrument such as a set of test items or of judges.

8.5.1 Estimation Assuming the Additive Model

We will use the data of Table 8.1 to illustrate the estimation of θ_A^2, σ_S^2, and the related values of ω^2. The mean squares and expected mean squares based on that design and data set were presented in Table 8.3. We first estimate σ_S^2. Note that, under the additive model,

$$E(\text{MS}_S) - E(\text{MS}_{SA}) = (\sigma_e^2 + a\sigma_S^2) - \sigma_e^2 = a\sigma_S^2$$

Therefore, for the $S \times A$ design,

$$\hat{\sigma}_S^2 = \frac{\text{MS}_S - \text{MS}_{SA}}{a} \tag{8.11}$$

For the data of Table 8.1, we have

$$\hat{\sigma}_S^2 = \frac{5.055 - .590}{4} = 1.116$$

Similarly,

$$\hat{\theta}_A^2 = \frac{\text{MS}_A - \text{MS}_{SA}}{n} \tag{8.12}$$

which, for the data of Table 8.1, is

$$\hat{\theta}_A^2 = \frac{9.557 - .590}{8} = 1.121$$

Recall that $\theta_A^2 = \Sigma_j(\mu_j - \mu)^2/(a - 1)$, which, strictly speaking, is not a variance. If we want an estimate of the variance component of A—that is, of $\delta_A^2 = \Sigma_j(\mu_j - \mu)^2/a$ —we calculate

$$\hat{\delta}_A^2 = \left[\frac{a - 1}{a}\right]\hat{\theta}_A^2 \tag{8.13}$$

For the data set we are considering,

$$\hat{\delta}_A^2 = \frac{3}{4}(1.281) = .841$$

We now have all the pieces we need to estimate ω_A^2, the proportion of the total population variance due to the independent variable, A (sucrose percentage in our example). From the additive model (Equation 8.7), it follows that

$$Y - \mu = \alpha + \eta + \epsilon$$

Because the three terms on the right side are assumed to be independently distributed, the variance of their sum is the sum of their variances; that is, the variance of the population of Y values is

$$\sigma_Y^2 = \delta_A^2 + \sigma_S^2 + \sigma_e^2$$

Then our estimate of ω_A^2 is

$$\hat{\omega}_A^2 = \frac{\hat{\delta}_A^2}{\hat{\sigma}_Y^2} = \frac{\hat{\delta}_A^2}{\hat{\delta}_A^2 + \hat{\sigma}_S^2 + \hat{\sigma}_e^2} \tag{8.14}$$

Equations 8.11 and 8.13 provide estimates of σ_S^2 and δ_A^2, and MS_{SA} estimates σ_e^2. After making these substitutions in the preceding equation, we multiply numerator and denominator by an and collect terms, which gives us

$$\hat{\omega}_A^2 = \frac{(a - 1)(MS_A - MS_{SA})}{(a - 1)(n - 1)MS_{SA} + (a - 1)MS_A + nMS_S} \tag{8.15}$$

From the ANOVA in Table 8.1, we have

$$\hat{\omega}_A^2 = \frac{(3)(9.557 - .590)}{(3)(7)(.590) + (3)(9.557) + (8)(5.055)} = .330$$

The treatment variable is not only significant but contributes much of the variance of the scores in the sampled population. We can also assess ω_S^2 by simply changing the numerator in Equation 8.15 to $n(MS_S - MS_{SA})$. That is,

$$\hat{\omega}_S^2 = \frac{n(MS_S - MS_{SA})}{(a - 1)(n - 1)MS_{SA} + (a - 1)MS_A + nMS_S} \tag{8.16}$$

Substituting from Table 8.1, we get

$$\hat{\omega}_S^2 = \frac{(8)(5.055 - .590)}{(3)(7)(.590) + (3)(9.557) + (8)(5.055)} = .438$$

As would be expected, individual differences contribute even more of the total population variance than does A.

8.5.2 Estimation Assuming the Nonadditive Model

If treatments and subjects interact ($\sigma_{SA}^2 \neq 0$), then ω_A^2 is defined as

$$\omega_A^2 = \frac{\delta_A^2}{\sigma_e^2 + \sigma_{SA}^2 + \delta_A^2 + \sigma_S^2} \tag{8.17}$$

From the expected mean squares for the nonadditive model in Table 8.3, we can see that Equations 8.12 and 8.13 still provide an estimate of δ_A^2, and that MS_{SA} estimates $\sigma_e^2 + \sigma_{SA}^2$. However, in the nonadditive case, no direct estimate of σ_S^2 can be obtained, because such an estimate requires that we subtract σ_e^2 from MS_S and, as can be seen in the nonadditive EMS column of Table 8.3, no estimate of σ_e^2 is available. Therefore, neither $\hat{\omega}_A^2$ nor $\hat{\omega}_S^2$ can be calculated.[3] The best we can do is calculate boundaries between which these variance proportions must fall. Such boundaries require us to establish boundaries for σ_S^2, the quantity we cannot directly estimate. The obvious lower bound for σ_S^2 is zero. Note that when we assume $\sigma_S^2 = 0$, the denominator of Equation 8.17 is decreased; in other words, ω_A^2 achieves its upper bound when σ_S^2 is set to zero.

To find the upper bound on σ_S^2, and thus the lower bound on ω_A^2, note that $E(MS_S/a) = \sigma_e^2/a + \sigma_S^2$. Therefore, the largest value σ_S^2 can have is $E(MS_S/a)$, which will happen when the error variance, σ_e^2, is zero. Then the lower bound for ω_A^2 is obtained by substituting MS_S/a for the estimate of σ_S^2 in Equation 8.17. Now we have

$$\hat{\delta}_A^2 = \left[\frac{a-1}{a} \right] \left[\frac{MS_A - MS_{SA}}{n} \right]$$

$$\hat{\sigma}_e^2 + \hat{\sigma}_{SA}^2 = MS_{SA}$$

$$\text{lower bound of } \sigma_S^2 = 0$$

$$\text{upper bound of } \sigma_S^2 = \frac{MS_S}{a}$$

Substituting into Equation 8.17 and simplifying, we have the lower and upper bounds on ω_A^2:

$$\hat{\omega}_{A,L}^2 = \frac{(a-1)(MS_A - MS_{SA})}{an MS_{SA} + (a-1)(MS_A - MS_{SA}) + n MS_S} \tag{8.18a}$$

and

$$\hat{\omega}_{A,U}^2 = \frac{(a-1)(MS_A - MS_{SA})}{an MS_{SA} + (a-1)(MS_A - MS_{SA})} \tag{8.18b}$$

The bounds on ω_S^2 are

$$\hat{\omega}_{S,L}^2 = 0 \tag{8.19a}$$

---

[3] Nonadditivity does not always imply that values of $\hat{\omega}^2$ cannot be calculated. Whether they can be rests solely on the pattern of components of variance in the EMS. For example, if both S and A have random effects, estimating the variance proportions is possible even in the presence of an $S \times A$ interaction.

and

$$\omega_{S,U}^2 = \frac{n\text{MS}_S}{an\text{MS}_{SA} + (a-1)(\text{MS}_A - \text{MS}_{SA}) + n\text{MS}_S} \qquad (8.19b)$$

8.5.3 An Example of the Use of ω^2: Reliability and the Analysis of Variance

Suppose that we have a judges rate n individuals on some personality trait. Perhaps we need such ratings because we want to correlate the personality measure with some performance measure, or perhaps we want to pick individuals having various degrees of the trait for a subsequent experiment.

True differences among individuals may be obscured by random variability in our measuring "instrument," the set of judges. To the extent that the judges agree in their ratings, we have confidence that they are rating the same characteristics of the individuals. If there is high agreement, most of the variability in the judgments will be due to differences among the individuals being judged and not the judges themselves. If this is the case, we say that the judges are consistent or that the judgments are reliable. The problem of determining the reliability of a measuring instrument arises in many other situations. For example, the n objects to be measured may be individuals whose math abilities are of interest, or stimuli whose attractiveness we wish to assess. The instrument used to measure math ability may be a set of test items, whereas that for measuring the attractiveness of stimuli might be a panel of judges. One way of assessing reliability is to correlate the measurements produced by each pair of judges or items and examine the average of the $a(a-1)/2$ correlation coefficients. An approach derived from the analysis of variance that uses ω^2 as an index of reliability can be shown to provide equivalent information.

Suppose we wish to determine reliability when n objects or individuals are measured by a judges or test items. The more the measurements reflect the characteristics of the measured objects rather than error, the more reliable they are. Therefore, a reasonable definition of r_{11}, the reliability coefficient for a set of measurements, is

$$r_{11} = \hat{\omega}_{objects}^2 = \frac{\hat{\sigma}_{objects}^2}{\hat{\sigma}_Y^2} \qquad (8.20)$$

That is, r_{11} estimates the proportion of the variance in the measurements that is due to the objects.

In order to calculate r_{11}, we need to estimate the components of Equation 8.20. This, in turn, requires us to state a model for the set of measurements. A common assumption is that the measurements of a given object vary only because of measurement error, so the jth measurement of the ith item represents the sum of two components: the true score for that object, μ_i, and measurement error. That is,

$$Y_{ij} = \mu_i + \epsilon_{ij} \qquad (8.21)$$

$$= \mu + (\mu_i - \mu) + \epsilon_{ij}$$

$$= \mu + \eta_i + \epsilon_{ij}$$

We assume that the ϵ_{ij} are independently and normally distributed with variance σ_e^2. If the objects to be measured are thought of as randomly sampled from a large population, it is appropriate to treat objects as a random-effect factor; we would assume that the η_i are independently and normally distributed with variance $\sigma_{\text{objects}}^2$. Because the ϵ_{ij} and μ_i terms are independent, it follows that

$$\sigma_Y^2 = \sigma_{\text{objects}}^2 + \sigma_e^2$$

Therefore,

$$r_{11} = \hat{\omega}_{\text{objects}}^2 = \frac{\hat{\sigma}_{\text{objects}}^2}{\hat{\sigma}_{\text{objects}}^2 + \hat{\sigma}_e^2} \tag{8.22}$$

Table 8.8 presents a data set for 10 individuals (the objects to be measured) who have each been rated by 5 judges in a gymnastics event. A gymnast $\times$ judge ($G \times J$) analysis of variance is first carried out. There are no significant differences among judges, and therefore it is reasonable to assume that the model of Equation 8.21 describes the structure of the data. Therefore, the J and $G \times J$ terms have been pooled, giving rise to the revised ANOVA table in the next panel. Calculations of variance estimates and of r_{11} are illustrated. The reliability coefficient of .919 seems reasonably high and suggests good agreement among the judges.

This approach can be extended in several ways. If we wish to find the reliability not for Y_{ij} but for $\bar{Y}_{i.}$, the *average* of the measurements produced by a set of a judges, we start by taking the average of both sides of Equation 8.21. It follows that $\sigma_{\bar{Y}}^2 = \sigma_{\text{objects}}^2 + \sigma_e^2/a$. Therefore, the reliability coefficient for the average is

$$r_{aa} = \frac{\hat{\sigma}_{\text{objects}}^2}{\hat{\sigma}_{\text{objects}}^2 + \hat{\sigma}_e^2/a}$$

Also, in the example of Table 8.8, the preliminary analysis under the full additive model of Section 8.2.1 made it clear that Equation 8.21 was the correct model upon which to base our reliability estimate. However, in some instances, the preliminary analysis might reveal a significant effect of judges or items. For example, some items might be consistently harder than others for all the individuals being tested. In such a case, the appropriate model would be Equation 8.7 and the appropriate reliability coefficient would just be $\hat{\omega}_S^2$. The calculation of this quantity was described in Sections 8.5.1 and 8.5.2.

8.6 MISSING DATA

Suppose that a subject does not show up to be tested in condition A_j or it is found, after the fact, that the data in condition A_j were not recorded properly for one subject. If the data can reasonably be described by the additive model, it is possible to estimate the missing score using the $a - 1$ scores that the subject provided in the other treatment conditions and the $n - 1$ scores provided by the other subjects in condition A_j.

Assume the additive model and call the missing score X_{ij}. Then, because $Y_{ij} = \mu + \eta_1 + \alpha_j + \epsilon_j$, and $E(\epsilon_{ij}) = 0$,

TABLE 8.8 AN EXAMPLE OF A RELIABILITY (r_{11}) CALCULATION

Data and ANOVA under the complete model

Gymnasts	Judges				
	J_1	J_2	J_3	J_4	J_5
G_1	9.0	9.0	9.5	9.0	10.0
G_2	9.0	8.5	8.0	8.5	8.5
G_3	8.0	8.5	9.0	8.5	8.5
G_4	7.5	7.5	7.5	6.5	7.0
G_5	9.0	9.5	9.5	10.0	9.0
G_6	7.0	7.5	7.5	7.5	7.0
G_7	6.5	6.5	6.5	7.0	6.5
G_8	7.0	7.0	6.5	6.5	7.0
G_9	6.5	6.5	6.0	6.0	6.0
G_{10}	9.0	8.5	8.5	9.0	8.5

SV	df	MS	EMS	F
Gymnasts	9	6.769	$\sigma_e^2 + 5\sigma_G^2$	
Judges	4	.013	$\sigma_e^2 + 10\sigma_J^2$	.097
$G \times J$	36	.129	σ_e^2	

Reliability calculations

SV	df	MS	EMS	F
Gymnasts	9	6.769	$\sigma_e^2 + 5\sigma_G^2$	57.609
Error	40	.118	σ_e^2	

Note: $\text{MS}_{\text{error}} = \dfrac{(a-1)\text{MS}_A + (n-1)(a-1)\text{MS}_{SA}}{n(a-1)}$

$$= \frac{(4)(.013) + (36)(.129)}{40} = .1175$$

To calculate r_{11}, we estimate σ_G^2:

$$\hat{\sigma}_G^2 = \frac{\text{MS}_G - \text{MS}_e}{a} = \frac{6.769 - .118}{5} = 1.330$$

and

$$r_{11} = \frac{\hat{\sigma}_G^2}{\hat{\sigma}_e^2 + \hat{\sigma}_G^2} = \frac{1.330}{.118 + 1.330} = .919$$

$$E(X_{ij}) = \mu + \eta_i + \alpha_j \tag{8.23}$$

$$= \mu + (\mu_i - \mu) + (\mu_j - \mu)$$

$$= \mu_i + \mu_j - \mu$$

Our best estimate of μ is the grand total of all the scores divided by the number of scores, an. In this case, the grand total is $T_{..} + X_{ij}$, where $T_{..}$ is the sum of the $an - 1$ scores that were actually obtained. Similarly, our best estimates of μ_i and μ_j are given by $\hat{\mu}_i = (T_i + X_{ij})/a$, where T_i is the sum of the $a - 1$ scores actually provided by subject i, and $\hat{\mu}_j = (T_j + X_{ij})/n$, where T_j is the sum of the $n - 1$ scores provided by the other subjects in treatment condition A_j. From Equation 8.23, $\hat{X}_{ij}$, our best estimate of the missing score is given by

$$\hat{X}_{ij} = \hat{\mu}_i + \hat{\mu}_j - \hat{\mu} = \frac{T_i + \hat{X}_{ij}}{a} + \frac{T_j + \hat{X}_{ij}}{n} - \frac{T_{..} + \hat{X}_{ij}}{an}$$

which simplifies to

$$\hat{X}_{ij} = \frac{nT_i + aT_j - T_{..}}{(n - 1)(a - 1)} \tag{8.24}$$

Note that because the missing score is estimated from the remaining scores, it does not contribute a degree of freedom, so $df_{SA} = (n - 1)(a - 1) - 1$.

To illustrate the procedure, suppose that Y_{71} were missing from the data of Table 8.1. The total for the remaining three scores in that row is 17.8. The total for the remaining seven scores in column 1 is 20. The grand total of the 31 remaining scores is 138.3. Substituting into Equation 8.24, our estimate of the missing score is

$$\hat{X}_{71} = \frac{(8)(17.8) + (4)(20) - 138.3}{21} = 4.005$$

The procedure can be extended to situations in which we have more than one missing value—say both Y_{14} and Y_{53} are missing. Begin by "guessing" the value of one of the scores (e.g., assigning Y_{53} an arbitrary value such as its column mean). Treating this guess as though it were a real score, we may obtain $\hat{Y}_{14}^{(1)}$, a first approximation to Y_{14} by using Equation 8.24. Then, treating $\hat{Y}_{14}^{(1)}$ as though it were a real score, we can use Equation 8.24 to obtain $\hat{Y}_{53}^{(1)}$, an approximation to Y_{53}. The procedure is then repeated, using $\hat{Y}_{53}^{(1)}$ in Equation 8.24 to produce $\hat{Y}_{14}^{(2)}$, a second approximation to Y_{14}, then using $\hat{Y}_{14}^{(2)}$ to produce $\hat{Y}_{53}^{(2)}$, and so on. This can be continued until two successive cycles show as small a change in the estimates as desired. Note that when the data are analyzed, the df for the error term are reduced by 1 for each score that is estimated.

This method is called an *iterative procedure* (because of the iteration, or repetition, of steps). The procedure can be extended to situations with more than two missing scores. However, unless a computer program is written to implement it, calculations will quickly get out of hand.

This procedure for estimating missing data rests upon the assumption of the additive model, Equation 8.7. If subjects and treatments interact, the procedure may no longer be valid. Alternatives exist, but they also require rather strong assumptions. Our recommendation is to use the procedure we have described because it is fairly simple and provides a reasonable approximation for many data sets.

8.7 DEALING WITH NONADDITIVITY

We have noted several disadvantages when the population of scores is best described by a nonadditive model. The F ratios tend to be smaller if σ^2_{SA} terms contribute to both the numerator and denominator, exact estimates of the variance components and of ω^2 are sometimes impossible under nonadditivity, and procedures for estimating missing scores are more complicated for nonadditive data.

Two obvious questions, then, are how to detect nonadditivity and, if it is found, whether it is possible to transform a nonadditive data set into an additive one. One approach to detection is to calculate the predicted value of each of the an scores, assuming additivity:

$$\hat{Y}_{ij} = \hat{\mu} + \hat{\eta}_i + \alpha_j$$
$$= \bar{Y}_{..} + (\bar{Y}_{i.} - \bar{Y}_{..}) + (\bar{Y}_{.j} - \bar{Y}_{..})$$

Most computer packages are capable of saving (for further analysis), printing, and plotting both the predicted score and the residual, $Y_{ij} - \hat{Y}_{ij}$. Systematic relations between residual and predicted values—for example, a curvilinear relation—suggest an interaction between subjects and treatments. Such a relation suggests that the data might be transformed to additivity by some operation such as taking the log of all scores or raising all scores to some power. Or one subject might exhibit a very different pattern from the others, in which case a reanalysis omitting that individual might be of interest. Or one residual might be much larger than all the others, which would indicate the possibility of an erroneous measurement.

Such residual plots are useful, but in themselves do not test whether σ^2_{SA} is zero. Such a test is difficult because without within-cell variability no direct estimate of σ^2_e is available. Tukey (1949) proposed a test for nonadditivity that is based on partitioning SS_{SA} into two components, one that represents a specific type of nonadditivity and the other that serves as the error term for the test. However, the test is tedious to calculate, is not available in most computer packages, and does not detect all forms of nonadditivity.

Transformations that reduce the amount of nonadditivity in the data have also been proposed. The success of the transformation would be signaled by a lack of association between residuals and predicted scores or by nonsignificance in the Tukey test just cited. As we noted in Chapter 4 when discussing transformations as a remedy for heterogeneity of variance, transformations often will lead to scales that are less familiar and understandable than the original and should therefore be used with caution. Hoaglin, Mosteller, and Tukey (1983) present an excellent discussion of residuals and of transformations, as do Box, Hunter, and Hunter (1978); Box et al. also discuss Tukey's test for nonadditivity.

8.8 MULTIFACTOR REPEATED-MEASURES DESIGNS

Although we have so far considered only the $S \times A$ repeated-measures design, there is no reason why additional within-subjects factors cannot be included in the design. In multifactor designs, n randomly sampled subjects are each tested at every combination of levels of the other factors. Therefore, if there are two within-subjects factors A and B with a and b levels, respectively, each subject is tested ab times. If A and B both have fixed effects, the $S \times A \times B$ design is a simple extension of the $S \times A$ design. If A or B are assumed to have random effects (i.e., the levels selected to be included in the experiment have been randomly selected), significance tests and estimation of variance components are somewhat more complicated.

8.8.1 The $S \times A \times B$ Design, A and B Fixed

Consider an experiment in which three subjects are required to detect targets in a visual array. The targets are of three shapes (A), and all three are presented several times in a random order during an experimental session. An average response time is obtained for each subject for each shape (level of A). The experiment is repeated on three successive days (B) to determine if there are practice effects and if these depend upon the shape of the target. Data from this hypothetical experiment are presented in Table 8.9 along with an ANOVA table and computational formulas. Determining the sources of variance, degrees of freedom, and sums of squares is straightforward. These terms are the same as the analogous terms in the multifactor design with between-subjects factors (see Chapter 4), with the qualification that there is now only one score in each cell.

The F tests presented in Table 8.9 are based on the nonadditive model, assuming A and B both have fixed effects. Note that A is tested against SA, B against SB, and AB against SAB. In view of our earlier discussion of sphericity, it is reasonable to ask whether the df need to be adjusted by ϵ. To answer this, we have redone the analysis, using the BMDP2V computer program this time. The BMDP2V output, presented in Table 8.10, shows that the degrees of freedom correction is performed for each test of a null hypothesis in which the within-subjects factor has more than two levels (and therefore more than one set of difference scores). Although there is substantial nonsphericity, the general conclusions are unchanged; however, the effect of B (days) is somewhat less impressive than it originally appeared when we adjust for the nonsphericity.

The justification for the error terms employed in Tables 8.9 and 8.10 is found in the expected mean squares for the nonadditive model with A and B fixed. These EMS

TABLE 8.9   DATA AND ANALYSIS OF AN $S \times A \times B$ DESIGN

	B_1			B_2			B_3			
	A_1	A_2	A_3	A_1	A_2	A_3	A_1	A_2	A_3	
S_1	3.1	2.9	2.4	1.9	2.0	1.7	1.6	1.9	1.5	19.0
S_2	5.7	6.8	5.3	4.5	5.7	4.4	4.4	5.3	3.9	46.0
S_3	9.7	10.9	8.0	7.4	10.5	6.6	6.9	8.9	6.0	74.9
$T_{.jk}$	18.5	20.6	15.7	13.8	18.2	12.7	12.9	16.1	11.4	$T_{...} = 139.9$
$T_{.j.} =$	45.2, 54.9, 39.8						$T_{..k} = 54.8, 44.7, 40.4$			

SV	df	SS	MS	F
Total	$abn - 1 = 26$	$\sum_i \sum_j \sum_k (Y_{ijk} - \bar{Y}_{...})^2 = 207.681$		
S	$n - 1 = 2$	$ab\sum_i (\bar{Y}_{i..} - \bar{Y}_{...})^2 = 173.667$	86.834	
A	$a - 1 = 2$	$nb\sum_j (\bar{Y}_{.j.} - \bar{Y}_{...})^2 = 13.010$	6.505	$\dfrac{MS_A}{MS_{SA}} = 3.93$
SA	$(n-1)(a-1) = 4$	$b\sum_i \sum_j (\bar{Y}_{ij.} - \bar{Y}_{i..} - \bar{Y}_{.j.} + \bar{Y}_{...})^2 = 6.624$	1.656	
B	$b - 1 = 2$	$na\sum_k (\bar{Y}_{..k} - \bar{Y}_{...})^2 = 12.143$	6.071	$\dfrac{MS_B}{MS_{SB}} = 22.55$
SB	$(n-1)(b-1) = 4$	$a\sum_i \sum_k (\bar{Y}_{i.k} - \bar{Y}_{i..} - \bar{Y}_{..k} + \bar{Y}_{...})^2 = 1.078$	.269	
AB	$(a-1)(b-1) = 4$	$n\sum_j \sum_k (\bar{Y}_{.jk} - \bar{Y}_{.j.} - \bar{Y}_{..k} + \bar{Y}_{...})^2 = .508$	.127	$\dfrac{MS_{AB}}{MS_{SAB}} = 1.56$
SAB	$(n-1)(a-1)(b-1) = 8$	$SS_{tot} - SS_S - SS_A - \cdots - SS_{SB} - SS_{AB} = .652$	.081	

Computing formulas

$$C = \frac{T_{...}^2}{nab} = \frac{139.9^2}{27} = 724.889$$

$$SS_{tot} = \sum_i \sum_j \sum_k Y_{ijk}^2 - C = 932.57 - 724.889 = 207.681$$

$$SS_S = \frac{1}{ab}\sum_i T_{i..}^2 - C = \frac{1}{9}(19^2 + 46^2 + 74.9^2) - 724.889 = 174.667$$

$$SS_A = \frac{1}{nb}(\sum_j T_{.j.}^2) - C = \frac{1}{9}(45.2^2 + 54.9^2 + 39.8^2) - 724.889 = 13.010$$

$$SS_B = \frac{1}{na}(\sum_k T_{..k}^2) - C = \frac{1}{9}(54.8^2 + 44.7^2 + 40.4^2) - 724.889 = 12.143$$

$$SS_{AB} = \frac{1}{n}(\sum_j \sum_k T_{.jk}^2) - C - SS_A - SS_B = \frac{1}{3}(18.5^2 + \cdots + 11.4^2) - 724.889 - 13.010 - 12.143 = .508$$

$$SS_{SA} = \frac{1}{b}(\sum_i \sum_j T_{ij.}^2) - C - SS_S - SS_A = \frac{1}{3}(6.6^2 + \cdots + 20.6^2) - 724.889 - 173.667 - 13.010 = 6.624$$

$$SS_{SB} = \frac{1}{a}(\sum_i \sum_k T_{i.k}^2) - C - SS_S - SS_B = \frac{1}{3}(8.4^2 + \cdots + 21.8^2) - 724.889 - 173.667 - 12.143 = 1.078$$

$$SS_{SAB} = SS_{tot} - SS_A - SS_B - SS_S - SS_{AB} - SS_{SA} - SS_{SB} = .652$$

TABLE 8.10   BMDP2V OUTPUT FOR THE DATA OF TABLE 8.9

	SOURCE	SUM OF SQUARES	DEGREES OF FREEDOM	MEAN SQUARE	F	TAIL PROB.	GREENHOUSE-GEISSER PROB.	HUYNH FELDT PROB.
	MEAN	724.88926	1	724.88926	8.35	.1018		
1	ERROR	173.66741	2	86.83370				
	B	12.14296	2	6.07148	22.55	.0066	.0414	.0409
2	ERROR	1.07704	4	.26926				
	A	13.00963	2	6.50481	3.93	.1138	.1831	.1744
3	ERROR	6.62370	4	1.65593				
	AB	.50815	4	.12704	1.56	.2744	.3322	.3034
4	ERROR	.65185	8	.08148				

ERROR TERM   EPSILON FACTORS FOR DEGREES OF FREEDOM ADJUSTMENT

	GREENHOUSE-GEISSER	HUYNH-FELDT
2	.5011	.5044
3	.5150	.5620
4	.3200	.6390

are presented in Table 8.11, together with the rules for establishing them. Those rules apply to all variations of the design in which each subject is exposed to all combinations of levels of factors. Although the exact terms and coefficients will depend upon how many factors are pesent and whether they are fixed or random, the rules will take care of these design variations. In the next chapter, we will extend the rules to designs in which there are both between- and within-subjects variables.

8.8.2 The $S \times A \times B$ Design, A Fixed and B Random

There are situations in which random-effect factors other than subjects may be included in a design. Suppose that n subjects are each required to read b words under each of a treatments, where A may be an independent variable such as print size or type; reading time is measured. It is desirable to select the words so that they can be reasonably viewed as a random sample from a large population of words. Otherwise, we have no statistical grounds for generalizing our conclusions about the effects of the treatment A on reading time to words that were not included in the experiment.

There are many types of experiments in which we would like to generalize the results to stimuli other than those actually used. We may, for example, be concerned with ratings of pictures that vary along some dimension, the number of trials taken to solve problems that differ in difficulty, time to comprehend sentences with different structures, or galvanic skin responses to each of a series of words. What can reasonably be viewed as a random set of b stimuli is not a simple issue and will be discussed in more detail in Section 8.9. Until then, we assume that B is a random-effect factor and consider the implications of this assumption for data analysis.

TABLE 8.11 EXPECTED MEAN SQUARES FOR THE $S \times A \times B$ DESIGN, NONADDITIVE MODEL

SV	df	EMS A,B fixed	EMS A fixed, B random
S	$n - 1$	$\sigma_e^2 + ab\sigma_S^2$	$\sigma_e^2 + a\sigma_{SB}^2 + ab\sigma_S^2$
A	$a - 1$	$\sigma_e^2 + b\sigma_{SA}^2 + nb\theta_A^2$	$\sigma_e^2 + b\sigma_{SA}^2 + n\sigma_{AB}^2 + \sigma_{SAB}^2 + nb\theta_A^2$
SA	$(n - 1)(a - 1)$	$\sigma_e^2 + b\sigma_{SA}^2$	$\sigma_e^2 + \sigma_{SAB}^2 + b\sigma_{SA}^2$
B	$b - 1$	$\sigma_e^2 + a\sigma_{SB}^2 + na\theta_B^2$	$\sigma_e^2 + a\sigma_{SB}^2 + na\sigma_B^2$
SB	$(n - 1)(b - 1)$	$\sigma_e^2 + a\sigma_{SB}^2$	$\sigma_e^2 + a\sigma_{SB}^2$
AB	$(a - 1)(b - 1)$	$\sigma_e^2 + \sigma_{SAB}^2 + n\theta_{AB}^2$	$\sigma_e^2 + \sigma_{SAB}^2 + n\sigma_{AB}^2$
SAB	$(n - 1)(a - 1)(b - 1)$	$\sigma_e^2 + \sigma_{SAB}^2$	$\sigma_e^2 + \sigma_{SAB}^2$

Forming F Ratios

Assuming H_0 is true, the ratio of EMS for numerator and denominator must be 1.

Rules for EMS

1. σ_e^2 contributes to all lines.
2. The component corresponding to the SV under consideration (the null hypothesis component) should be present. For example, $E(\mathrm{MS}_A)$ includes a θ_A^2 component.
3. Any other variance component will be included if (1) its subscripts include the letter(s) in the present SV, and (2) if *all* of its other subscripts represent random factors. For example, under the model in which A and B are both fixed, σ_{SA}^2 contributes to $E(\mathrm{MS}_A)$ because SA includes A and the remaining subscript (S) is a random factor. When S and B are both random factors, $E(\mathrm{MS}_A)$ includes SA, AB, and SAB components because when A is ignored, the remaining letter(s) represent random factors.
4. The coefficient multiplying each variance component reflects the number of scores involved in the subscripted combinations. For example, there are n scores in each AB combination, so θ_{AB}^2 is multiplied by n; however, there are nb scores at each level of A, so θ_A^2 is multiplied by nb.

Comparing the two sets of EMS terms in Table 8.11, it is apparent that if B is random, additional components of variance contribute to some of the EMS terms. The reason for this is that our data represent a sample of the population of interaction effects at any level of A. The mean of this subset of interaction effects will not be zero and will tend to have a different value at each level of A, depending upon the particular interaction effects sampled. Therefore, σ_{AB}^2 will influence the value of MS_A. For further discussion of this point, review the presentation in Section 8.3.1.

Because of the additional variance components found in the EMS terms when B is random, there is no obvious error term against which to test A, the variable of primary interest. We require a mean square that has an expectation of $\sigma_e^2 + b\sigma_{SA}^2 + n\sigma_{AB}^2 + \sigma_{SAB}^2$. As can be seen in Table 8.11, there is no source of variance that has this expectation. The problem disappears if σ_{SA}^2 or σ_{AB}^2 is zero throughout Table 8.11; in either of those cases, an error term is readily found. This possible solution to the problem of testing the effects of A is discussed in Section 8.8.3. If this solution is not possible, an approach developed in Section 8.8.4 can always be used.

8.8.3 Consequences of Additivity for F Tests

Let us focus our attention on the EMS of Table 8.11 for the case in which B is assumed to have random effects. It should be immediately apparent that there is no problem in testing any main effect if there is complete additivity—that is, if it is assumed that there is no variance due to interactions in the population. In that case, three mean squares in the table are estimates of σ_e^2 plus a null hypothesis term, and four terms estimate only σ_e^2. With several terms that have precisely the same expectation, it makes sense to average them to achieve a single, more reliable estimate on more df and consequently greater power than any one of them alone would afford. We have already introduced the method of averaging called *pooling* (Section 5.8). In the present example, we obtain the sum of the four SS for interaction and divide by the total of the associated df.

A complete absence of interaction variability would be unlikely. A more promising possibility is that either σ_{AB}^2 or σ_{SA}^2 is zero. Suppose that A and B do not interact. Then we can delete σ_{AB}^2 from all EMS in Table 8.11, and now MS_{SA} provides an appropriate error term against which to test A.

In the developments thus far, we have been able to provide a test of A by assuming that one or more interaction components is zero. Unfortunately, wishing will not make it so, and there are dangers inherent in tests of A based on the false assumption that certain variables do not interact. For example, if we assume $\sigma_{AB}^2 = 0$, and if that assumption is incorrect, the ratio of EMS for the test of A against SA will be

$$\frac{\sigma_e^2 + \sigma_{SAB}^2 + b\sigma_{SA}^2 + n\sigma_{AB}^2 + bn\theta_A^2}{\sigma_e^2 + \sigma_{SAB}^2 + b\sigma_{SA}^2}$$

A significant F ratio may reflect the AB variance rather than a nonzero value of θ_A^2.

This raises the following issue. On what basis is it reasonable to conclude that one or more components of variance is zero when such a conclusion, if erroneous, may result in a biased F test of other null hypotheses? Statisticians do not completely agree on the answer to this question; we propose the following guidelines, however, based on studies by Bozivich, Bancroft, and Hartley (1956) and Srivastava and Bozivich (1961). We propose that before a variance component is deleted from the EMS, there should be prior grounds for believing that the interaction does not contribute to variability, *and* a preliminary F test should fail to be significant at the .25 level. Prior belief in additivity may derive from knowing the experimental situation or from analyzing results of preliminary tests in related experiments. The preliminary test is a test of the interaction null hypothesis. Before deleting σ_{AB}^2 from our set of expectations, for example, we should want some a priori reason for believing it to be zero and should want the ratio MS_{AB}/MS_{SAB} to be nonsignificant at the .25 level.

8.8.4 Quasi-F Ratios

There rarely will be prior grounds for simplifying the design model. The most reasonable expectation usually will be that the effects of treatment variables A will vary over subjects S and stimuli B. Thus, the issue remains: how do we test the hypothesis that $\theta_A^2 = 0$ when we have the EMS of Table 8.11 (B random)? The approach generally taken is to calculate a quasi F (F') ratio—a ratio of combinations of mean squares whose expectations are equal under the null hypothesis. One possibility is

$$F_1' = \frac{\text{MS}_A}{\text{MS}_{AB} + \text{MS}_{SA} - \text{MS}_{SAB}} \tag{8.25}$$

To understand the reasoning, note that the ratio of EMS is

$$\frac{\sigma_e^2 + \sigma_{SAB}^2 + b\sigma_{SA}^2 + n\sigma_{AB}^2 + bn\theta_A^2}{\sigma_e^2 + \sigma_{SAB}^2 + b\sigma_{SA}^2 + n\sigma_{AB}^2}$$

which equals 1 if H_0^2 ($\theta_A = 0$) is true.

Note that this is not a sufficient condition for the ratio to be a proper F statistic. Both numerator and denominator must be at least approximately distributed as χ^2/df. Satterthwaite (1946) has shown that under the usual assumptions of analysis of variance, a linear combination of mean squares has approximately this distribution; the appropriate df for the denominator are

$$\text{df} = \frac{(\text{MS}_{AB} + \text{MS}_{SA} - \text{MS}_{SAB})^2}{\text{MS}_{AB}^2/\text{df}_{AB} + \text{MS}_{SA}^2/\text{df}_{SA} + \text{MS}_{SAB}^2/\text{df}_{SAB}} \tag{8.26}$$

which should be rounded to the nearest integer. Other quasi F's can be computed. One that has been considered by several statisticians is

$$F_2' = \frac{\text{MS}_A + \text{MS}_{SAB}}{\text{MS}_{AB} + \text{MS}_{SA}} \tag{8.27}$$

As with F_1', the ratio of EMS is again 1 under the null hypothesis. The numerator df are

$$\text{df}_{\text{num}} = \frac{(\text{MS}_A + \text{MS}_{SAB})^2}{\text{MS}_A^2/\text{df}_A + \text{MS}_{SAB}^2/\text{df}_{SAB}} \tag{8.28}$$

In general, given a linear *combination of mean squares*, $CMS = \text{MS}_1 \pm \text{MS}_2 \pm \cdots \pm \text{MS}_K$, we get

$$\text{df}_{\text{CMS}} = \frac{(\text{MS}_1 \pm \text{MS}_2 \pm \cdots \pm \text{MS}_K)^2}{\text{MS}_1^2/\text{df}_1 + \text{MS}_2^2/\text{df}_2 + \cdots + \text{MS}_K^2/\text{df}_K} \tag{8.29}$$

The ratios formed in Equations 8.25 and 8.27 involve linear combinations of mean squares in which the weights by which the mean squares are multiplied are 1 and -1. In some circumstances, we will use other weights to combine mean squares or variances. A case in point is the Brown-Forsythe statistic, $F*$, which was presented in Chapter 4 as a way of testing treatment effects when variances are heterogeneous. To obtain a denominator mean square for $F*$, we combined group variances instead of mean squares, and the weights were a function of the group sizes. Nevertheless, $F*$ is a quasi-F statistic because its denominator is a linear combination of mean squares (CMS) of the general form

$$\text{CMS} = w_1 V_1 + w_2 V_2 + \cdots + w_k V_k + \cdots + w_K V_K \tag{8.30}$$

where the w's are any real numbers and the V's are either mean squares or variances of sets of numbers. The df associated with CMS are

$$\mathrm{df}_{\mathrm{CMS}} = \frac{(\mathrm{CMS})^2}{\Sigma_k(w_k^2 V_k^2/\mathrm{df}_k)} \tag{8.31}$$

The df_k are the degrees of freedom associated with the kth mean square or variance.

In many studies, observations are missing, and therefore it will be impossible to calculate MS_{SAB}. This means that neither F_1' nor F_2' can be calculated. For example, in the experiment suggested earlier in this section, time to read words (B) as a function of print size (A) was measured. If some subjects failed to correctly read words under some conditions, not all $S \times A \times B$ combinations are available. A conservative remedy in this situation is essentially to calculate F_1' or F_2' without $\mathrm{MS}_{S \times A \times B}$. We can see from Equations 8.25 and 8.27 that $\mathrm{MS}_A/(\mathrm{MS}_{AB} + \mathrm{MS}_{SA})$ must be less than F_1' or F_2', which is why we call this approach conservative. In actual practice, this minimum quasi-F (min F'; Clark 1973, Forster and Dickinson 1976) is computed as follows:

1. Find the average of the scores in each $S \times A$ combination; presumably there will be a few scores in each combination even if some values are missing. Compute $F_1 = \mathrm{MS}_A/\mathrm{MS}_{SA}$.
2. Find the average of the scores in each $A \times B$ combination. Compute $F_2 = \mathrm{MS}_A/\mathrm{MS}_{AB}$.
3. Compute

$$\mathrm{min}\ F' = \frac{F_1 F_2}{F_1 + F_2} \tag{8.32a}$$

and

$$\mathrm{df}_{\mathrm{error}}' = \frac{(F_1 + F_2)^2}{F_1^2/\mathrm{df}_{AB} + F_2^2/\mathrm{df}_{SA}} \tag{8.32b}$$

The calculations are equivalent to dropping MS_{SAB} from the usual quasi-F statistic and its error df.

Following are some recommendations for dealing with the $S \times A \times B$ design with B random. Although what follows specifies the design under consideration, the ideas generalize to any design in which no single mean square provides an appropriate error term for a test of a null hypothesis of interest.

1. Test MS_{SA} and MS_{AB} against MS_{SAB}. If both tests are nonsignificant at the .25 level, then σ_{SA}^2 and σ_{AB}^2 are assumed to be zero. Then MS_{AB}, MS_{SA}, and MS_{SAB} have the same expectation and may be pooled to provide a single error term against which to test A. If only one term is nonsignificant, the corresponding variance component is set to zero. For example, if MS_{SA} is not significant at the .25 level, we conclude that $\sigma_{SA}^2 = 0$ and MS_{AB} is an appropriate error term against which to test A.
2. If the preliminary tests in step 1 do not yield an error term against which to test A, construct a quasi-F ratio. The issue here is whether to use F_1' or F_2'. F_2' was consistently more powerful under conditions examined in simulation studies by Hudson and Krutchkoff (1968) and Davenport and Webster (1973); however,

the advantage appears to be slight except when df are small ($n = a = b = 3$) or the variance components σ_{SA}^2, σ_{AB}^2, and σ_{SAB}^2 are all very small. Another advantage of F_2' is that it takes on only positive values, whereas F_1' (when MS_{SAB} is large) can be negative. However, the variance of F_1' "is considerably smaller" than that of F_2' (Davenport and Webster 1973). Presumably this is because a ratio based on a chi-square variable and an approximately distributed chi-square variable is a better approximation to the F ratio than one based on two approximately distributed chi-square variables. Neither procedure is clearly preferable under all circumstances, but we lean toward F_2' because its power is somewhat greater under conditions in which power will tend to be low.

3. When observations are missing, steps 1 and 2 are impossible. The Forster and Dickinson study indicates that min F' yields power fairly close to that of F_2' and therefore should be used. We agree, but offer the caution that these authors ran their computer studies only for $a = 2$. The power loss may be greater with more treatment levels.

What about the effects of violations of the sphericity assumption upon the distribution of F'? Somewhat surprisingly, Maxwell and Bray (1986) have found evidence that nonsphericity does not inflate the Type 1 error rate for F'. Their article presents an interesting discussion of the reasons for this.

One last comment is in order. If B is a random-effect variable, b should be as large as possible. Interactions of A with both S and B contribute to the error variance against which treatment effects are to be evaluated. It is important to have sufficient df associated with both variables to ensure powerful tests of A. Furthermore, the limited evidence we have suggests that the distribution of F' more closely approximates that of F as a, b, and n increase.

8.8.5 Testing Contrasts in the $S \times A \times B$ Design

In Section 8.4.4, we noted that many statistical packages allow the user to test contrasts in the $S \times A$ design by typing a command followed by sets of contrast weights. We also discussed an alternate method in which a contrast could be tested by transforming each subject's scores into a single contrast score, and then calculating a test statistic based on the new variable. Both methods readily generalize to the $S \times A \times B$ design. Because the second approach may make the nature of the test clearer, we illustrate it with the data of Table 8.9. Recall that we had three subjects required to recognize each of three types of visual targets (A) on each of three days (B); A and B are fixed-effect variables. Suppose we hypothesized that response times to the second and third targets (A_2 and A_3) would not differ but that their average would be smaller than the average response time to the first target. This suggests the null hypotheses:

$$H_{01}: \mu_{2.} - \mu_{3.} = 0$$

and

$$H_{02}: \mu_{1.} - \frac{1}{2}(\mu_{2.} + \mu_{3.}) = 0$$

We can estimate the two population contrasts as

$$\hat{\psi}_1 = \bar{Y}_{.2.} - \bar{Y}_{.3.}$$

and

$$\hat{\psi}_2 = \bar{Y}_{.1.} - \frac{1}{2}(\bar{Y}_{.2.} + \bar{Y}_{.3.})$$

We might also wish to test whether the contrasts in question vary significantly over the levels of B. For example, we might believe that the difference between the average score for the first target and the combined average for the second and third targets will decrease over days. This implies that we wish to test the null hypothesis

$$H_{03}: \mu_{11} - \frac{1}{2}(\mu_{21} + \mu_{31}) = \mu_{12} - \frac{1}{2}(\mu_{22} + \mu_{32}) = \mu_{13} - \frac{1}{2}(\mu_{23} + \mu_{33})$$

Examples of such contrasts were presented and discussed in Chapter 6 for between-subjects designs. Numerator sums of squares are the same in the repeated-measures design as they were in Chapter 6. However, the denominator is somewhat different because we have repeated measures. Contrast scores for each subject will have the form

$$C_{1i} = \bar{Y}_{i2.} - \bar{Y}_{i3.}$$

for the first contrast and

$$C_{2i} = \bar{Y}_{i1.} - \frac{1}{2}(\bar{Y}_{i2.} + \bar{Y}_{i3.})$$

for the second. Each contrast can then be tested by using a single-sample t

$$t = \frac{\hat{\psi}}{\hat{\sigma}_{\hat{\psi}}} = \frac{\bar{C}}{\hat{\sigma}_C/\sqrt{n}}$$

If we wish to test the third null hypothesis, we can calculate contrast scores at each $S \times B$ combination. These will be of the form

$$C_{ik} = Y_{i1k} - \frac{1}{2}(Y_{i2k} + Y_{i3k})$$

These contrast scores have been calculated for the data set of Table 8.9 and are presented in Table 8.12. To illustrate, the value for subject 1 at B_1 is

$$C_{11} = 3.1 - \frac{1}{2}(2.9 + 2.4) = .45$$

Note that such contrast scores can easily be calculated by almost any statistical package using some kind of "compute" or "transformation" command. Once we have the scores, we can test H_{03} by running a simple repeated-measures ANOVA on the bn contrast scores.

The BMDP2V output for a repeated-measures ANOVA performed on these data is given in panel b of Table 8.12. The first two SV terms in the ANOVA table are mean (with $SS_{MEAN} = T^2/nb$ and df = 1) and its error term with $n - 1 = 2$ df. This F ratio tests H_{02}, which was stated above. Note that because the scores are actually contrasts, rather than the original values of Y, the mean reflects the average contrast score, $\bar{C}$,

TABLE 8.12 CONTRASTS AMONG MEANS IN THE $S \times A \times B$ DESIGN

(a) We wish to contrast the three target (A) means in Table 8.9. Specifically, we want to know if the mean at A_1 differs significantly from the combined average for A_2 and A_3, and if that contrast varies over days (B). The corresponding null hypotheses are

$$H_{02}: \mu_{1.} - \frac{1}{2}(\mu_{2.} + \mu_{3.})$$

and

$$H_{03}: \mu_{1k} - \frac{1}{2}(\mu_{2k} + \mu_{3k}) \text{ is the same for all } k \text{ (levels of } B)$$

To test these two hypotheses, we calculate the contrast scores,

$$Y_{i1k} - \frac{1}{2}(Y_{i2k} + Y_{i3k})$$

for each subject and level of B (combination of i and k). These scores are

	Contrast Scores			Row mean
	B_1	B_2	B_3	
S_1	.45	.05	−.10	.133
S_2	−.35	−.55	−.20	−.367
S_3	.25	−1.15	−.55	−.483
Column mean	.117	−.550	−.283	−.239

To test H_{02}, perform a one-sample t test on the row means:

$$t = \frac{\hat{\psi}}{\hat{\sigma}_{\hat{\psi}}} = \frac{-.239}{.327/\sqrt{3}} = -1.26$$

We cannot reject H_{02} at $\alpha = .05$.

To test H_{03}, test the B effect in the table against $S \times B$. Both tests are easily performed by any computer program capable of doing a repeated-measures ANOVA. The output from BMDP2V is the result of such an ANOVA performed on the nine contrast scores above. The first F (MEAN) is just the square of the t of -1.26 and tests H_{02}. The second F (B) tests H_{03}.

(b) BMDP2V output

	SOURCE	SUM OF SQUARES	DEGREES OF FREEDOM	MEAN SQUARE	F	TAIL PROB.	GREENHOUSE GEISSER PROB.	HUYNH FELDT PROB.
	MEAN	.51361	1	.51361	1.60	.3339		
1	ERROR	.64389	2	.32194				
	B	.67556	2	.33778	2.53	.1951	.2219	.1951
2	ERROR	.53444	4	.13361				

ERROR TERM	EPSILON FACTORS FOR DEGREES OF FREEDOM ADJUSTMENT

	GREENHOUSE-GEISSER	HUYNH-FELDT
2	.7448	1.0000

and the error term represents the variance of contrast scores over subjects. The F test of mean is not significant and therefore H_{02} cannot be rejected. Note that this F is the square of the t obtained using the contrast scores. The final two sources in the table, B, and its error term SB [df $= b - 1$ and $(n - 1)(b - 1)$] test whether the contrast varies across levels of B; that is, does the contrast between the third target and the average of the other two vary over days? As can be seen from the BMDP2V output, we cannot reject the null hypothesis that the population contrast has the same value on each day. Again, bear in mind that because the data are contrasts, not the original Y_{ijk}, B reflects variation in the contrast over days and SB reflects the interaction of subjects and B with respect to these contrast scores.

Of course, as we discussed in Chapter 6, whenever a family of comparisons is tested, the Type 1 error rate for the family must be controlled. We recommend that contrasts be planned and tested using the Bonferroni criterion for the reasons given in Section 8.4.4.

8.9 FIXED OR RANDOM EFFECTS?

It should be clear from the presentation to this point that designating variables fixed or random has important implications for both significance testing and parameter estimation, as well as for the degree to which we can generalize our results. Therefore let us look further into classifying effects.

Despite the importance of the decision of how effects should be classified, it is not always simple to make. At one extreme, we have variables that should be clearly viewed as fixed in effect. The levels of the variable have been arbitrarily selected for inclusion in the experiment, and because of the way they were selected, there is no basis for viewing them as a random sample from a population of levels. This class includes most treatment variables such as type of drug or amount of reward, and also such characteristics of individuals as clinical category.

At the other extreme, we have random sampling from some large well-defined population. This is rarely realized in practice, and it is therefore difficult to determine to what population of subjects it is appropriate to generalize the conclusions from our current experiment. Can we reasonably view our subjects as a random sample of adults? college students? college sophomores? college sophomores currently enrolled in introductory psychology? college sophomores currently enrolled in introductory psychology at the university in which the experiment was run? The answer largely depends on the research. In studies of a sensory process, like visual acuity, we might generalize to the population of adults having normal vision. In studies of human learning, we might define our population more narrowly, reserving judgment on whether our conclusions will hold for populations having a markedly different average level of ability from that characterizing those within the institution at which the current experiment was run. When in doubt, we prefer restricting generalizations to the more narrowly defined population.

Even though the population is rarely as well defined as we should like, it should be clear from the preceding comments that we do view subjects as a random-effect variable. Our justification is that subjects are not arbitrarily selected. Other individuals are provided an equal opportunity to participate and might well serve if replications of the experiment were run. Furthermore, those who are subjects are not necessarily more representative of the potential subject pool; there is some likelihood, although it is small, that extreme behaviors will be sampled. This is important because we assume such sampling variability in applying our statistical procedures.

Classifying stimuli such as words or pictures provides greater difficulty. For many experiments, we can argue, on much the same grounds that we presented in discussing subjects, that such stimuli are random samples from a population (perhaps ill-defined) of stimuli. That is, the stimuli are not arbitrarily selected and there are many other stimuli that have an equal opportunity for being chosen under the procedure used. In many other experiments, however, the choice of stimulus is so constrained that it is difficult to imagine a population from which this set of materials is one relatively small sample. In studies involving word pairs, for example, restrictions are often placed on the grammatical class, length in both syllables and letter, familiarity, and number of associates, of each word, and also on the rated similarity, associative strength, and other forms of possible relation between the two words. The experimenter may have great difficulty in meeting those restrictions. Under such conditions, it is not clear that stimuli should be treated as having random effects. Two rough guidelines may prove useful. First, under the constraints extant, could independent investigators produce other samples of stimuli? Second, if we assume that the answer to the first question is positive—there is a reasonably large population available—was there an equal likelihood that all members of that population could be included in the experiment? If this answer also proves positive, it seems reasonable to treat the stimuli as having random effects with all that this implies for our data analyses and the scope of our conclusions.

8.10 NONPARAMETRIC PROCEDURES FOR REPEATED-MEASURES DESIGNS

In Chapter 4 we presented a test based on ranks, the Kruskal-Wallis H test, which could be used to test the equality of population means in a between-subjects one-factor design. We demonstrated that this test was often more powerful than the F when the treatment populations had heavy-tailed or skewed distributions. We now will consider tests based on ranks for the $S \times A$ design. Section 8.10.1 presents several tests of the omnibus $H_0: \mu_1 = \cdots = \mu_j = \cdots = \mu_a$. Section 8.10.2 describes a test that is more powerful than these tests when $a = 2$, and can also be applied to testing any single df contrast. Section 8.10.3 considers the special case in which the an measures are all 0's or 1's.

8.10.1 Testing the Equality of the *a* Treatment Means

Several alternatives to the standard F test of A have been developed. We will describe two of these and discuss their merits relative to each other and to the F test.

Friedman's Chi Square, χ_F^2

Table 8.13 presents data for eight subjects on each of four trials. These data have been transformed by ranking the four scores for each subject, yielding a matrix of values of R_{ij}, $i = 1$ to 8 (n, in general) and $j = 1$ to 4 (a, in general). The test statistic may be written as

$$\chi_F^2 = \frac{SS_A}{a(a + 1)/12} \tag{8.33}$$

This statistic was first presented by Friedman (1937). Its denominator is the variance of a set of a consecutive integers, or untied ranks. Recall that the ratio of a sum of squares to the variance of its population is distributed as chi square (see Chapter 3) if the scores are independently and normally distributed; for correlated observations, sphericity is required. Therefore, the ratio in Equation 8.33 has an exact chi-square distribution on $a - 1$ df if sphericity holds for the population of R_{ij} and the sampling distribution of the mean of the ranks is normal. The central limit theorem assures us that for reasonably large n, the normality requirement will be met and, therefore, the use of the chi-square table (Appendix D.4) to evaluate χ_F^2 will be appropriate. A conservative strategy for using the chi-square table is to require $an > 30$. Odeh (1977) has tabled exact p values for χ_F^2 for values of a and n as large as 6; these may be used to evaluate significance in small samples.

TABLE 8.13 AN ANALYSIS OF A REPEATED-MEASURES DESIGN USING FRIEDMAN'S χ^2 TEST

	Original data				Ranked data			
	A_1	A_2	A_3	A_4	A_1	A_2	A_3	A_4
S_1	18	16	14	17	4	2	1	3
S_2	20	17	11	12	4	3	1	2
S_3	15	19	16	10	2	4	3	1
S_4	18	14	16	11	4	2	3	1
S_5	22	15	17	13	4	2	3	1
S_6	24	21	10	12	4	3	1	2
S_7	20	23	17	19	3	4	1	2
S_8	18	15	13	12	4	3	2	1
					$R_j = 29$	23	15	13 $R_{..} = 80$

where R_j = the sum of the ranks for treatment A_j.

To test H_0, calculate

$$SS_A = \frac{29^2 + 23^2 + 15^2 + 13^2}{8} - \frac{(29 + 23 + 15 + 13)^2}{32} = 20.5$$

and $\chi_F^2 = SS_A/[a(a + 1)/12] = 12.3$.

For $\alpha = .01$, the critical value of χ^2 on 3 df is 11.345; therefore, the null hypothesis of no effect of A can be rejected.

If ties are present in the data, the calculations are modified. Because major mainframe and microcomputer statistical packages can do the Friedman test, we see no point in burdening readers with a somewhat complicated formula. Those who are unable to constrain their curiosity, or who still rely entirely on a calculator, can find the general form of Equation 8.33 and worked-out examples in most standard nonparametric statistics textbooks (for example, Lehmann 1975, p. 265). The computer packages usually provide exact p values and Kendall's coefficient of concordance, an average of the correlations of ranks for pairs of treatments.

Rank-Transformation F Test (F_r)

In the last decade, there has been considerable interest in a rather simple alternative to ANOVA. The approach has been applied to several experimental designs; Iman, Hora, and Conover (1984), and Hora and Iman (1988) have considered its application to the repeated-measures design. There are two steps: (1) rank all an scores from smallest to largest, assigning average rank values in case of ties. Note that we do not rank each set of a scores separately as in Friedman's procedure. (2) Do the $S \times A$ ANOVA presented in this chapter on the rank values. This means that once the Y_{ij} have been converted to R_{ij}, the transformed values can be submitted to any program that analyzes data from an $S \times A$ design. Table 8.14 presents the R_{ij} transforms of the data of Table 8.13, together with the results of an ANOVA performed by BMDP2V.

Recommendations

Much of what we know about the performances of these tests relative to each other and to the standard F test is based on a concept called *asymptotic relative efficiency* (ARE). Consider two tests and assume that some specifiable power is achieved by test 1 when n_1 subjects are run in the study. Now assume that test 2 requires n_2 subjects to achieve the same power. As n_1 gets large, n_1/n_2 approaches a finite constant value independent of the specified value of power. The ratio n_1/n_2 is the ARE of the second test to the first. Thus if $n_1/n_2 = 2$, the first test requires twice as many subjects as test 2 to achieve the same power; the second test is said to be twice as efficient (asymptotically—that is, for large sample size). It has been known for some time that the ARE of χ_F^2 to F (the usual F based on the original scores) is $[a/(a + 1)](.956)$ when the scores for each subject are sampled from a normal population. This means that when $a = 2$, and assuming normality, the F requires about two-thirds (actually .637) as many scores as X_F^2 to achieve the same power. However, if a is large, the ARE is only slightly below 1.

As we might guess from the discussion of the F and H tests in Chapter 4, things look somewhat different when we sample from population distributions that are skewed or have heavy tails. Based on AREs derived by Kepner and Robinson (1988), we would recommend the rank-transform F statistic (F_r) when distributions are heavy-tailed and symmetric, a is less than 5, and the average correlation for pairs of treatments is about .2 or less. If the distribution appears skewed, F_r should be used if the average correlation is close to zero. In all other situations involving heavy-tailed or skewed distributions, χ_F^2 appears to be the more powerful test. As in Chapter 4, we emphasize using

TABLE 8.14 AN ANALYSIS OF THE DATA OF TABLE 8.13 USING THE RANK-TRANSFORMATION F TEST

	Original data				Ranked data			
	A_1	A_2	A_3	A_4	A_1	A_2	A_3	A_4
S_1	18	16	14	17	23.0	16.0	10.5	19.5
S_2	20	17	11	12	27.5	19.5	3.5	6.0
S_3	15	19	16	10	13.0	25.5	16.0	1.5
S_4	18	14	16	11	23.0	10.5	16.0	3.5
S_5	22	15	17	13	30.0	13.0	19.5	8.5
S_6	24	21	10	12	32.0	29.0	1.5	6.0
S_7	20	23	17	19	27.5	31.0	19.5	25.5
S_8	18	15	13	12	23.0	13.0	8.5	6.0
					$R_j = 199.0$	157.5	95.0	76.5 $R_{..} = 528$

ANOVA on the ranked scores (BMDP2V)

	SOURCE	SUM OF SQUARES	DEGREES OF FREEDOM	MEAN SQUARE	F	TAIL PROB.	GREENHOUSE GEISSER PROB.	HUYNH FELDT PROB.
	MEAN	8712.00000	1	8712.00000	119.23	.0000		
1	ERROR	511.50000	7	73.07143				
	A	1198.56250	3	399.52083	8.37	.0007	.0010	.0007
2	ERROR	1001.93750	21	47.71131				

ERROR TERM	EPSILON FACTORS FOR DEGREES OF FREEDOM ADJUSTMENT	
	GREENHOUSE-GEISSER	HUYNH-FELDT
2	.9409	1.0000

such devices as stem-and-leaf plots of your data, as well as theory and data from related experiments to help make judgments about the nature of the population distribution. For example, we find rating data to be short-tailed; in such cases, the standard F test is more powerful. On the other hand, the Friedman statistic may prove most powerful for response-time data.

Two points about these tests should be noted. First, the research on the relative power of these procedures has been done under the assumption of sphericity. We suspect that nonsphericity will cause an inflation of the Type 1 error rates associated with both χ_F^2 and F_r. The ϵ adjustment will provide an adequate remedy as in the case of the standard F test. Depending on which nonparametric test you wish to use, submit one of the two sets of ranks (Table 8.13 or 8.14) to a statistical package that calculates the ϵ adjustment for df (e.g., BMDP2V, SPSS^X MANOVA). To carry out the adjusted (for nonsphericity) χ_F^2 test, reject H_0 if $\epsilon\chi_F^2$ is greater than the critical χ^2 value on $(a-1)\epsilon$ df in Appendix Table D.4. For the rank-transformation test, just compare F_r with the critical F value on $(a-1)\epsilon$ and $(a-1)(n-1)\epsilon$ df. The second point to note is that neither χ_F^2 nor F_r is very powerful when $a = 2$. We turn now to a very useful procedure for testing difference scores or, more generally, contrasts.

8.10.2 The Wilcoxon Signed-Rank (WSR) Test

In Chapter 3, we noted that the one-sample t test can be used to test the null hypothesis of no difference between two treatments when the scores are correlated. Assuming that the population of difference scores is symmetrically and independently distributed, the WSR test can be used for the same purpose. Table 8.15 presents an example in which the difference between scores in two conditions $(Y_{i1} - Y_{i2})$ has been calculated for 20 subjects. If any of these difference scores had been zero, they would have been discarded prior to the analysis. Thus the presence of zero differences contributes to a reduction in the power of the WSR test but does not do so if the t test is used. The first step in the analysis of the remaining n pairs is to rank the difference scores, from smallest to largest, ignoring the sign of the difference. This has been done in Table 8.15. When several absolute values of differences are tied, each receives the median rank. For example, the data include two $+7$'s and a -7. If the signs are ignored, these scores are tied for 12th, 13th, and 14th place in rank order and are therefore all assigned the median value of 13. After ranks are assigned, a $+$ or $-$ is attached, depending upon the sign of the original difference score. Find the sum of the negative ranks, T_-. Then turn to Appendix Table D.11 and determine whether the absolute value of T_- ($|T_-|$) is less than the critical value (T_c) for $n = 20$ and $\alpha = .05$. T_c is 60 and $|T_-| = 34.5$, which is less than 60 and therefore results in rejection of H_0.

The null hypothesis will be rejected when the test statistic is less than T_c; however, the test statistic depends upon the alternative hypothesis in the following way:

H_1	Test statistic
$\mu > 0$	$|T_-|$
$\mu < 0$	T_+
$\mu \neq 0$	Smaller of T_+ and $|T_-|$

The logic of the WSR test is as follows. If the population of difference scores is symmetric about zero, then all patterns of signed ranks are equally likely. Because each rank has an equal probability of having a plus or minus sign, there are 2^n possible patterns. For example, if there were just three difference scores, the possible patterns would be

Rank = 1	2	3	T_+
$+$	$+$	$+$	6
$-$	$+$	$+$	5
$+$	$-$	$+$	4
$+$	$+$	$-$	3
$-$	$-$	$+$	3
$-$	$+$	$-$	2
$+$	$-$	$-$	1
$-$	$-$	$-$	0

TABLE 8.15   AN EXAMPLE OF THE WILCOXON SIGNED-RANK TEST

Difference	Signed rank
-7	-13
-4	-9
-2	-6.5
-1	-3
-1	-3
1	3
1	3
1	3
2	6.5
4	9
4	9
5	11
8	13
7	13
9	15
10	16
11	17
13	18
15	19
20	20

Calculate the sum of the negative ranks:

$$T- = -(13 + 9 + 6.5 + 3 + 3) = -34.5$$

The null and alternative hypotheses are

$$H_0: \mu_D = 0 \quad \text{and} \quad H_1: \mu_D > 0$$

The test is one-tailed and for $n = 20$ and $\alpha = .05$, $T_c = 60$. Because $|T-| \leqslant 60$, reject H_0.

Thus, the probability that $T_+ \leqslant 2$ would be 3/8.

If n is more than 50, we may take advantage of the central limit theorem by converting T to a z statistic and finding the p value in the table of the normal distribution, Appendix D.2. Because T is a sum, its sampling distribution should approach normality as n grows large. Therefore, we need only subtract $E(T)$ from T and divide by σ_T. The expected value of the sum of all the ranks is $n(n + 1)/2$; in the current example, this is $34.5 + 175.5 = (20)(21)/2$. If the null hypothesis is true, the expected value of T would be half of the sum of all the ranks, or $E(T) = n(n + 1)/4$. In general, for $n > 50$, we may test H_0 by

$$z = \frac{T - n(n + 1)/4}{\sqrt{n(n + 1)(2n + 1)/24}} \tag{8.34}$$

One reminder is in order with respect to the WSR test. Researchers often take the descriptive phrase "distribution-free" too literally. Although the test does not rest on the assumption that the population of difference scores is normally distributed, that distribution should be symmetric, or at least not markedly skewed. To see why this is so, consider the following set of scores and their assigned ranks:

Scores:	-4	-3	-2	$+2$	$+3$	$+4$
Ranks:	-5.5	-3.5	-1.5	$+1.5$	$+3.5$	$+5.5$

The median is zero, and $T_+ = |T_-| = n(n + 1)/4$. Suppose we now skew the distribution of scores as follows:

Scores:	-6	-5	-2	$+2$	$+3$	$+4$
Ranks:	-6	-5	-1.5	$+1.5$	$+3$	$+4$

The median is still zero but T_+ no longer equals $|T_-|$. Thus a true null hypothesis about a population median may be rejected if the distribution is skewed. In summary, the WSR test is useful when difference-score data suggest that the population distribution is at least approximately symmetric but is somewhat heavier-tailed than the normal.

Computer simulations attest to the potential usefulness of the WSR test. Blair and Higgins (1985) found that the t test had a slight power advantage over the WSR test when the sampled distribution is normal or light-tailed. When the sampled distribution was heavy-tailed and n was 10, the WSR test usually had greater power. For samples of size 25 or more from heavy-tailed distributions, the WSR test always had greater power, often by .2 or more. It is important to also note that, even when the population distributions are normal, the ARE of the WSR to the t test is .956 (Hodges and Lehmann 1956). Furthermore, for some distributions, the ARE falls to .864, but never lower. Thus, for some light-tailed distributions the t can yield the same power as the WSR test with about 86% as many subjects, provided the samples are large. On the other hand, for some long-tailed distributions, the ARE of the WSR to the t test can rise to infinity. In view of these findings, any doubts the researcher has usually should be resolved in favor of using the WSR test. There are two important qualifications about this advice, however. First, the Type 1 error rate of the t test will be less affected by departures from symmetry. Second, the power of the WSR procedure is weakened by the presence of tied scores. Nevertheless, the WSR test has a clear place in our arsenal of analytic procedures.

Finally, we note that although we have discussed the WSR test in terms of difference scores, it can be applied to any contrast among means. For example, given four levels of a variable A, we might wish to test whether the combined average of the A_1 and A_2 populations of scores differs from that for A_3 and A_4. In that case, we calculate $(1/2)(Y_{i1} + Y_{i2}) - (1/2)(Y_{i3} + Y_{i4})$ and proceed to rank and sign these contrast scores.

8.10.3 Cochran's Q Test

A common research situation is one in which each subject responds on several trials, or under several different conditions, and each response is classified in one of two ways. For example, suppose we record a success or failure for each subject on each of four mathematical problems that varied in their conceptual distance from a practice problem.

The question is whether probability of success depends upon the problem type. In general, $Y_{ij} = 1$ or 0, indicating a success or failure by subject i under condition j. If π_j is the probability of a success in the population of responses under A_j, then the null hypothesis is

$$H_0: \pi_1 = \pi_2 = \cdots = \pi_j \cdots = \pi_a$$

The Q statistic is simply defined as

$$Q = \frac{\text{SS}_A}{\text{MS}_{A/S}} \tag{8.35}$$

$\text{MS}_{A/S}$ is the average of n variances, where each variance is based on the a scores for a subject. If we denote the sum of scores for subject i as T_i, then

$$\text{MS}_{A/S} = \frac{a\Sigma T_{i.} - \Sigma T_{i.}^2}{an(a-1)} \tag{8.36}$$

We can also calculate $\text{MS}_{A/S}$ from the output of an ANOVA:

$$\text{MS}_{A/S} = \frac{\text{SS}_A + \text{SS}_{SA}}{n(a-1)} \tag{8.37}$$

Cochran (1950) proved that the ratio in Equation 8.35 is distributed as χ^2 when n is large and the population correlation for any pair of conditions is the same as for any other pair. Therefore, the null hypothesis is rejected when Q exceeds the critical value of χ^2 on $a - 1$ df.

As we noted in Chapter 3, the chi-square distribution rests on the assumption that the variable of interest is normally distributed. The central limit theorem tells us that this assumption is essentially true for the $T_{.j}$ when n is large. This raises the issue of how large an n is large. The answer depends on the values of the π_j; when π_j is closer to .5, n can be smaller because the distribution is more symmetric. Another factor is the number of subjects who exhibit no variability (that is, all 0's or 1's); such subjects contribute nothing to the value of Q and therefore do not contribute to the effective n. Based on a review of several simulation studies, Myers, DiCecco, White, and Borden (1982) recommend that the effective n (for $a = 3$) be at least 16. When n is small, empirical rejection rates of true null hypotheses are less than the nominal α, and power is quite low.

Table 8.16 presents data for 16 subjects on four math problems of different types. Although the analysis can be done easily by hand, and we have provided the necessary row and column totals, we also include an output obtained by submitting these data to BMDP2V. We do this to make three points. First, note in panel c how we have used the BMDP2V output, and Equation 8.37, to calculate Q; for large data sets, this can save considerable effort. Second, note that both the Q and F tests (the F test of P is presented in the BMDP2V output) yield a significant result at a p value between .02 and .05. In fact, Myers et al. (1982) report that the two tests have very similar Type 1 error rates for $n \geqslant 16$; for smaller n, the F test's Type 1 error rate is inflated. Finally, we will show how the ϵ adjustment reported in the BMDP2V output can also be used to adjust the p value associated with the Q test when there are violations of the assumption of homogeneous correlations for pairs of conditions.

TABLE 8.16 AN EXAMPLE OF COCHRAN'S Q TEST

(a) The data

	P_1	P_2	P_3	P_4	$T_{i.}$
			Problems		
S_1	1	0	1	0	2
S_2	1	0	1	0	2
S_3	1	0	0	0	1
S_4	0	0	0	1	1
S_5	1	0	0	0	1
S_6	1	0	1	0	2
S_7	1	1	0	1	3
S_8	1	0	1	0	2
S_9	0	0	1	1	2
S_{10}	1	1	1	0	3
S_{11}	1	0	1	0	2
S_{12}	0	1	0	1	2
S_{13}	0	1	1	0	2
S_{14}	1	0	1	0	2
S_{15}	1	1	0	0	2
S_{16}	1	0	0	0	1
$T_{.j} =$	12	5	9	4	30

(b) Output from BMDP2V

	SOURCE	SUM OF SQUARES	DEGREES OF FREEDOM	MEAN SQUARE	F	TAIL PROB.	GREENHOUSE GEISSER	HUYNH FELDT
1	MEAN	14.06250	1	14.06250	146.74	.0000		
	ERROR	1.43750	15	.09583				
2	P	2.56250	3	.85417	3.22	.0314	.0404	.0314
	ERROR	11.93750	45	.26528				

ERROR TERM	EPSILON FACTORS FOR DEGREES OF FREEDOM ADJUSTMENT

	GREENHOUSE-GEISSER	HUYNH-FELDT
2	.8444	1.0000

(c) Calculating Cochran's Q statistic

Using Equation 8.36 for $MS_{P/S}$ gives

$$SS_P = \frac{12^2 + 5^2 + 9^2 + 4^2}{16} - \frac{30^2}{64} = 2.5625 \qquad MS_{P/S} = \frac{(4)(30) - (62)}{(4)(16)(3)} = .3021$$

Alternatively, we can use Equation 8.37 for $MS_{P/S}$ and the output from BMDP2V:

$$SS_P = 2.5625 \qquad MS_{P/S} = \frac{SS_P + SS_{SP}}{n(a-1)} = \frac{2.5625 + 11.9375}{48} = .3021$$

TABLE 8.16   *(continued)*

Then

$$Q = \frac{SS_P}{MS_{P/S}} = \frac{2.5625}{.3021} = 8.4823$$

Evaluated against χ^2 on 3 df, the result is significant at the .01 level.

The ϵ values in the BMDP2V output are quite high. However, there will be situations in which the variation among the correlations for pairs of treatments will result in, sometimes severely, inflated Type 1 error rates (Myers et al. 1982). For such situations, a modified Q test appears to provide an honest Type 1 error rate. We calculate

$$Q^* = \epsilon Q$$

and evaluate it against the critical value of χ^2 with df equal to the integer nearest to $\epsilon(a - 1)$. For the example of Table 8.16, we have the ϵ value from the BMDP2V output; using the Greenhouse-Geisser adjustment, $Q^* = (.8444)(8.482) = 7.162$ and df $= (.8444)(3) = 2.533$, which we round to 3.0. In the present instance, as with the F test, this has little effect on the p value; with more severe nonsphericity, it is an important adjustment.

8.11 CONCLUDING REMARKS

It is easy to understand why the psychological literature abounds with examples of designs involving repeated measurements. Repeated-measures designs potentially have greater precision than designs that employ only between-subjects factors and are particularly useful when the supply of subjects is limited relative to the number of treatment combinations to be studied, or when the experimenter's goal is to collect data on some performance measure as a function of time.

However, experimenters must understand the potential problems associated with repeated-measures designs, the assumptions that are made, and the consequences of violating these assumptions. We have, for example, (1) indicated that tests on within-subjects factors having more than two levels will be positively biased when the sphericity assumption is violated and have recommended that a degree-of-freedom correction be used to counteract this bias; (2) shown that there will not always be obvious error terms in designs that contain several random-effect factors, but that hypotheses may often be tested using quasi-F tests; and (3) indicated that there are a number of negative consequences of nonadditivity and that sometimes it may be desirable to transform the data to a scale that is less nonadditive. These techniques provide a good starting point for coping with some of the problems that may result from using the repeated-measures design. Even more importantly, awareness of potential problems is necessary in order to decide whether to use the repeated-measures design and to evaluate the results of significance tests intelligently.

It should also be emphasized that when the independent variable is something other than time or trial number, it is important to randomize the order of presentation of treatments independently for each subject. Proper randomization will guard against confounding the effects of time and treatments (what inference can be drawn in the extreme case in which treatment 1 is always presented first, treatment 2 always second, and so on?) and also will minimize the possibility of severe heterogeneity of covariance and violations of the sphericity assumption. Scores for treatments close together in time should be more highly correlated than scores for treatments further apart. Randomizing the order of treatments for each subject allows each pair of treatments to be given an equal opportunity to appear any given length of time apart.

It is also helpful to provide sufficient time between presentations of treatments to minimize "carryover" effects. If participating in a particular experimental condition results in the subject becoming fatigued or highly aroused, these effects should be allowed to wear off before the next condition is presented. Even if the different orders of presentation balance so that treatments and trials are not confounded, carryover effects, if present, will result in increased variability among orders of presentation and thus reduce the efficiency of the design.

The "pure" repeated-measures designs represent only a subset of designs that employ within-subjects factors. We will go on to discuss designs that combine between-subjects and within-subjects factors, as well as designs in which the orders of treatments are systematically counterbalanced. A solid understanding of the contents of the present chapter should provide the necessary preparation for these extensions of the repeated-measures design.

EXERCISES

8.1 The following terms provide a useful review of some concepts in the chapter. Define, describe, or identify each of them:

additive model	quasi-F test
fixed-effect variable	min F'
random-effect variable	Friedman's chi-square (χ_F^2)
sphericity	rank-transformation F test
compound symmetry	asymptotic relative efficiency (ARE)
ϵ adjustment of degrees of freedom	Wilcoxon signed-rank test
Greenhouse-Geisser strategy	Cochran's Q test
reliability (r_{11})	

8.2 The following data set consists of three scores for each of four subjects:

	A_1	A_2	A_3
S_1	12	14	7
S_2	9	8	2
S_3	10	9	4
S_4	8	6	9

(a) Carry out the ANOVA.

(b) Assuming additivity, present the EMS.

(c) Use your answer to part (b) to estimate ω^2.

(d) If we had run a completely randomized, or CR, design with four different subjects at each level of A, the error variance would have been larger. We can estimate what $MS_{S/A}$ would have been from the ANOVA of part (a). The appropriate formula is

$$\text{estimate of } MS_{S/A} = \frac{n(a-1)}{an-1}MS_{SA} + \frac{n-1}{an-1}MS_S$$

Calculate $\hat{MS}_{S/A}$. (Note that this estimate is a weighted average of the S and SA mean squares.) Then estimate the power each design would have had to reject H_0 at the .05 level, assuming the effects in the data represent the true population effects.

8.3 Consider the following three data sets, all of which have the same column means:

	1				2				3		
	A_1	A_2	A_3		A_1	A_2	A_3		A_1	A_2	A_3
S_1	44	62	72	S_1	32	50	60	S_1	34	62	82
S_2	25	40	48	S_2	25	40	48	S_2	45	40	28
S_3	22	37	44	S_3	22	37	44	S_3	12	37	54
S_4	16	20	19	S_4	28	32	31	S_4	16	20	19

(a) Calculate MS_{SA} for each of the three data sets. You should find that the error mean squares are the same for sets (1) and (2), and these are smaller than that for set (3).

(b) Now estimate $MS_{S/A}$ as in Exercise 8.2. Is there a difference in the relative efficiencies of the repeated measures design ($MS_{S/A}/MS_{SA}$) for data sets (1) and (2)? Why do you think this is?

(c) Compare the relative efficiencies for data sets (1) and (3). Explain the results of this comparison.

8.4 Consider the following data set:

Subject	A_1	A_2	A_3
1	1.7	2.4	2.7
2	4.6	6.3	7.0
3	6.9	6.8	10.2
4	3.6	6.1	7.5
5	4.3	4.4	8.2
6	5.1	5.2	5.8

(a) (i) Calculate the variance-covariance matrix. (ii) Find the variances for $Y_{1i} - Y_{2i}$, $Y_{i1} - Y_{i3}$, and $Y_{i2} - Y_{i3}$ from this matrix. (iii) Now carry out the ANOVA and show that $MS_{SA} = (1/2)$ [average of the three variances calculated in part (ii)].

(b) Carry out the t test of the difference between the A_1 and A_2 means, (i) using MS_{SA} and (ii) using the appropriate variance from part (a). Find bounds on the p value for each procedure. Which analysis do you think gives the truer p value? Why?

(c) Calculate the t to test the null hypothesis that

$$\mu_1 - \frac{1}{2}(\mu_2 + \mu_3) = 0$$

Use the variance-covariance matrix to get the denominator for the t.

8.5 Consider the following data set:

	A_1	A_2	A_3	A_4
S_1	1.8	2.2	3.2	2.4
S_2	2.4	1.5	1.9	2.7
S_3	1.9	1.7	2.5	3.5
S_4	2.7	2.6	2.4	3.1
S_5	4.7	4.8	4.4	4.8
S_6	3.6	3.1	4.2	5.4
S_7	4.4	4.2	4.1	4.9
S_8	5.8	6.1	6.4	6.6

(a) Carry out the ANOVA on these data and find bounds on the p value, assuming sphericity.

(b) Find bounds on the p value assuming extreme nonsphericity.

(c) Following is the variance-covariance matrix:

$$\begin{array}{c|cccc} & A_1 & A_2 & A_3 & A_4 \\ \hline A_1 & 2.113 & & & \\ A_2 & 2.283 & 2.634 & & \\ A_3 & 1.887 & 2.211 & 2.111 & \\ A_4 & 1.955 & 2.082 & 1.948 & 2.176 \end{array}$$

Calculate the six possible variances of difference scores and verify that $MS_{S/A}$ equals one-half times the average of these values.

(d) Use the preceding variance-covariance matrix to calculate the Greenhouse-Geisser and Huynh-Feldt adjustments for df. What are the bounds on the p value for each of these adjusted F tests?

8.6 Huynh and Feldt (1970) present the following variance-covariance matrix. Does it satisfy compound symmetry? Does it satisfy sphericity? Calculate the Greenhouse-Geisser adjustment for degrees of freedom.

$$\begin{array}{c} \\ A_1 \\ A_2 \\ A_3 \end{array} \begin{array}{ccc} A_1 & A_2 & A_3 \\ \begin{bmatrix} 1.0 & .5 & 1.5 \\ & 3.0 & 2.5 \\ & & 5.0 \end{bmatrix} \end{array}$$

8.7 Assume that in Exercise 8.5, we had planned all six possible pairwise comparisons of means. Carry out the test of A_1 against A_4 and report whether it is significant. Let EF = .05.

8.8 From the variance-covariance matrix provided in Exercise 8.5, calculate the variance of the contrast:

$$\hat{\psi} = \frac{1}{2}(\bar{Y}_{.1} + \bar{Y}_{.2}) - \bar{Y}_{.4}$$

Then test for significance, assuming it is the only contrast performed and has been planned before seeing the data. Note that you can check your result by calculating a contrast score for each subject and directly calculating the variance of these contrast scores.

8.9 We wish to test whether the means in Exercise 8.3 represent a significant linear trend; this is the only contrast planned. Carry out the test.

8.10 An educational psychologist wishes to develop a measure of articulation that can then be used in examining the relation between reading comprehension and the ability to articulate words. She has 40 third graders read aloud each of 20 words and measures the time required for the response. A subjects × words ANOVA yields the following results:

SV	df	MS	F
Subjects (S)	39	208,305.017	244.158
Words (W)	19	739.141	.866
$S \times W$	741	853.157	

Calculate the reliability (r_{11}) of the measure. State in your own words just what this means.

8.11 Consider the following ANOVA table based on the administration of a battery of five measures of reading skill to 10 students:

SV	df	MS	F
S	9	6.769	52.473
Items	4	.846	6.558
$S \times I$	36	.129	

(a) Calculate the reliability of the items. Justify your choice of an approach.

(b) Another approach to calculating r_{11} would be to find the 10 correlations for the pairs of judges and average these. The variance-covariance matrix is

$$
\begin{array}{cccccc}
 & J_1 & J_2 & J_3 & J_4 & J_5 \\
J_1 & 1.169 & & & & \\
J_2 & 1.067 & 1.100 & & & \\
J_3 & 1.197 & 1.289 & 1.614 & & \\
J_4 & 1.281 & 1.317 & 1.558 & 1.781 & \\
J_5 & 1.272 & 1.256 & 1.522 & 1.522 & 1.622
\end{array}
$$

Each diagonal element is the variance of the 10 scores for that judge and the off-diagonal elements are covariances. Ordinarily, we would calculate the correlation for J_1 and J_2 as $\hat{\sigma}_{12}/\hat{\sigma}_1\hat{\sigma}_2$. However, we assume additivity in estimating reliability, and this in turn implies that $\sigma_1 = \sigma_2 = \cdots = \sigma_5 = \sigma$. Therefore, we can calculate the correlation of ratings for judges 1 and 2 as $\hat{\sigma}_{12}/\hat{\sigma}^2$, and similarly for the other nine pairs of judges. Calculate these interjudge correlations and then find their average. The result should equal that obtained in part (a) if you had chosen the proper analysis there.

8.12 Four subjects each were tested on three successive days. Unfortunately, the record for subject 1 on day 2 was accidentally deleted from the computer disk. The remaining scores are

	D_1	D_2	D_3
S_1	19	—	42
S_2	24	27	45
S_3	21	30	47
S_4	16	21	39

(a) Use Equation 8.24 to estimate the missing score.

(b) Pretend that both Y_{12} and Y_{43} were missing. Now go through an iterative estimation procedure as follows. First "guestimate" Y_{12}; the actual guess is arbitrary, but let's use $Y_{12}^{(1)} = 30.5$, the mean of the remaining scores for S_1. Now use Equation 8.24 to calculate our first estimate of Y; call this $Y_{43}^{(1)}$. You have now finished one estimation cycle. Next calculate $Y_{12}^{(2)}$, treating $Y_{43}^{(1)}$ as though it were an observed score. Do three iteration cycles; you should find that the estimates obtained at the end of the third cycle ($Y_{12}^{(3)}$ and $Y_{43}^{(3)}$) do not differ very much from those obtained at the end of the second.

8.13 Ten randomly sampled clerical workers (W) are observed on each of five randomly sampled occasions (O) with each of four word processing programs (P). The programs have been selected for comparison purposes to decide which ones to buy for the entire working force. A score is obtained for each worker with each processor on each occasion. An analysis of variance is then carried out. The results are

SV	df	MS	EMS
W	9	2580	
P	3	2610	
O	4	690	
WP	27	330	
WO	36	370	
PO	12	640	
WPO	108	320	

(a) Write out the EMS for the table, first specifying which factors have random effects and which have fixed effects.

(b) If you have the correct EMS, no single SV provides a proper error term against which to test P. Explain in terms of your EMS.

(c) One remedy for the problem noted in part (b) is to delete certain components of variance from the EMS on the basis of preliminary tests. By testing the two-factor interactions against the WPO, justify deleting some component(s) of variance from the EMS of part (a). Having done this, can you now test P? If so, carry out the test.

(d) Another remedy for the problem of testing P is to use the original EMS as the basis for constructing a quasi-F test. Perform this test.

8.14 In research on personality, there has been much discussion of the relative importance of traits and situations. The basic research design involves n subjects and t tasks representing a random sample of situations. Measures are obtained for each subject on each task on each of b randomly sampled occasions. We assume a completely additive model:

$$Y_{ijk} = \mu + \eta_i + \alpha_j + \beta_k + \epsilon_{ijk}$$

(a) Assuming this model and that η, α, and β represent subjects, tasks, and occasions, respectively, present the SV, df, and EMS.

(b) Present formulas in terms of the mean squares for estimating each of the variance components (σ_s^2, σ_{task}^2, $\sigma_{occasion}^2$).

(c) Suppose we have evidence from previous studies that subjects and tasks interact. Now our model is

$$Y_{ijk} = \mu + \eta_i + \alpha_j + \beta_k + (\eta\alpha)_{ij} + \epsilon_{ijk}$$

(d) Present a revised ANOVA table (SV, df, and EMS) and revised estimates of the variance components.

8.15 Each of five subjects is tested at four equally spaced points in time on a visual detection task. The numbers of errors for each test are

Subject	Time 1	2	3	4
1	9	6	7	5
2	11	8	6	6
3	6	8	7	5
4	13	10	10	9
5	12	8	9	6

(a) Time 1 provides a baseline. The experimenter wishes to test whether the mean at time 1 differs significantly from the combined mean for the other three times. State the null hypothesis and carry out the test.

(b) Test whether there is a significant linear trend.

(c) Test whether the means depart significantly from a straight line.

8.16 In the following data set, B represents pictures that are rated before (A_1) and after (A_2) reading a persuasive communication. Carry out the ANOVAs requested. You may do it by hand, by SYSTAT, or using any other program. The data are

	A_1			A_2		
	B_1	B_2	B_3	B_1	B_2	B_3
S_1	2	3	4	8	9	8
S_2	3	3	4	6	7	9
S_3	7	4	6	4	9	9
S_4	2	2	4	7	9	8
S_5	1	2	2	5	4	5

(a) Assume B is fixed and do an $S \times A \times B$ ANOVA.

(b) Find the mean at A_1 and at A_2 for each subject. Now do an ANOVA for this $S \times A$ design.

(c) In (a), how would the EMS change if B were treated as a random-effects variable (as seems more reasonable)? Do you see any particular problems?

8.17 (a) For the preceding data set, find the reliability (r_{11}) on occasion A_1. The model is

$$Y_{i1k} = \mu_{.1.} + \epsilon_{i1k}$$

(b) For each subject, obtain three average scores by averaging the two values at each level of B. The resulting EMS are (assuming chance variability among the pictures):

SV	EMS
Ss	$\sigma_{e'}^2 + \sigma_S^2$
Res	$\sigma_{e'}^2$

where $\sigma_{e'}^2 = \sigma_e^2/a$. Note that the usual EMS under this model are divided by $a\ (=2)$ because we are now dealing with variances of means based on a scores each. Find the reliability of the mean (r_{aa}). In terms of the EMS above, why does reliability increase when we average over many occasions?

8.18 We have the following data from an $S \times A \times B$ design:

	A_1			A_2			A_3		
	B_1	B_2	B_3	B_1	B_2	B_3	B_1	B_2	B_3
S_1	4	7	6	16	17	19	20	21	15
S_2	8	10	7	10	15	12	17	14	19
S_3	11	14	12	11	9	7	10	23	12

(a) Perform an ANOVA assuming A and B are both fixed effect variables.

(b) Retest the A effects under the assumption that B is a random effects variable. Justify the new F test by presenting EMS for numerator and denominator.

(c) Retest the A effects assuming A and B both have random effects. Again present the relevant EMS.

(d) We wish to determine whether the means at A_1 and A_2 differ significantly; that is we wish to test $H_0: \mu_{.1.} = \mu_{.2.}$. Carry out the test (Assume A and B have fixed effects in this and the remaining parts.)

(e) Now test $H_0: \mu_{..1} = \mu_{..2}$.

(f) Suppose B is a quantitative variable such as trial. We can now investigate questions of trend. Test whether there is a linear component of B [lin(B)].

(g) Test whether the linear component of B varies significantly as a function of the level of A [$A \times$ lin(B)].

8.19 Following are response times (in milliseconds) obtained under four different conditions for eight subjects:

Subjects	A_1	A_2	A_3	A_4
1	2036	2220	2211	2316
2	2034	2042	2094	2077
3	2198	2612	2272	2348
4	2593	2629	2652	2647
5	2347	2408	2416	2479
6	2308	2352	2463	2358
7	2454	2501	2475	2461
8	2462	2394	2491	2659

Carry out the ANOVA, Friedman's χ^2 test, and the rank-transform test on these scores. Find the bounds on the p value in each case.

8.20 (a) For the data of Exercise 8.19, use the Wilcoxon signed-rank procedure to test whether the A_1 and A_2 conditions are significantly different.

(b) For the data of Exercise 8.19, use the Wilcoxon signed-rank procedure to test whether there is a significant linear trend. (*Hint:* The logic parallels that for difference scores except that the contrast score for each subject is now a linear combination of four scores.)

8.21 Twenty people underwent a one-week program aimed to help them quit cigarette smoking. The researchers running the program checked on the progress of the individuals after three, six, and nine months. The results follow, with a 1 signifying that the individual has smoked at least once during the preceding three-month period and a 0 indicating that the individual has not smoked during that period.

Period	Subjects																			
	1	2	3	4	5	6	7	8	9	10	11	12	13	14	15	16	17	18	19	20
1	1	0	0	0	1	0	0	0	0	0	0	0	1	1	0	0	1	1	1	0
2	1	1	1	1	1	0	1	0	0	1	1	0	1	0	0	1	0	0	1	1
3	1	1	1	1	1	0	0	1	1	1	1	1	0	1	0	1	1	1	1	0

The investigators want to know if there has been a significant change in the percentage of smokers over the three periods in the follow-up study. Carry out an analysis to answer this question and state your conclusion.

Chapter 9

Mixed Designs: Combining Between-Subjects and Within-Subjects Factors

9.1 INTRODUCTION

In a study designed to investigate different methods of teaching probability, three groups of six subjects were taught by different methods. One group was taught by a standard instructional method (A_1), a second group was given additional problems (A_2), and a third group received additional problems from a computer that provided immediate feedback (A_3). All three groups were tested at the end of the instructional period, and then once every two weeks until four tests had been given. Although different problem sets were used on the four tests, they were equated for difficulty. This design permits us to compare the instructional methods (A), and also to see the time course (B) of performance following the end of instruction for each method. Table 9.1 presents the data. Notice that the design is a mixture of the between-subjects design of Chapter 4 and the within-subjects design of Chapter 8. If we average the four test scores for each subject, we have a between-subjects design and can conduct the ANOVA exactly as in Chapter 4. If, on the other hand, we retain the four test scores but ignore the instructional factor, we have an $S \times B$ design in which 24 subjects have scores at the four levels of B.

The design of Table 9.1 is an example of a mixed (sometimes called a "split-plot") design with one between-subjects factor, A (method of instruction in the example), and one within-subjects factor, B (time of test). In general, n subjects are tested at A_1, n other subjects are tested at A_2, and so on. All an subjects are tested at each of the b levels of the independent variable B. Therefore, $abn - 1$ df must be accounted for in the analysis of variance. Notice that two subscripts are necessary to identify each subject; for example, S_{62} refers to the sixth subject at the second level of A. In general, we refer to Y_{ijk}, where i indexes the subject ($i = 1, 2, \ldots, n$), j indexes the level of the between-subjects variable ($j = 1, 2, \ldots, a$), and k indexes the level of the within-subjects variable ($k = 1, 2, \ldots, b$). In the current example, $n = 6$, $a = 3$, and $b = 4$.

Designs of the sort exemplified by Table 9.1 are a compromise between a desire to employ within-subjects factors to reduce error variance (and thus increase power and

TABLE 9.1   DATA FOR A DESIGN WITH ONE BETWEEN-SUBJECTS (A) AND ONE WITHIN-SUBJECTS (B) VARIABLE

Method of instruction		B_1	B_2	B_3	B_4	$T_{ij.}$
				Time of test		
	S_{11}	82	48	41	53	224
	S_{21}	72	70	61	45	248
A_1	S_{31}	43	35	30	12	120
	S_{41}	77	41	51	31	200
	S_{51}	43	43	21	29	136
	S_{61}	67	39	30	40	176
	$T_{.1k}$	384	276	234	210	$T_{.1.} = 1104$
	S_{12}	71	53	50	62	236
	S_{22}	89	67	76	68	300
A_2	S_{32}	82	84	83	71	320
	S_{42}	56	56	55	45	212
	S_{52}	64	44	44	52	204
	S_{62}	76	74	64	74	288
	$T_{.2k}$	438	378	372	372	$T_{.2.} = 1560$
	S_{13}	84	80	75	77	316
	S_{23}	84	72	63	81	300
A_3	S_{33}	76	54	57	61	248
	S_{43}	84	66	61	77	288
	S_{53}	67	69	55	69	260
	S_{63}	61	67	55	61	244
	$T_{.3k}$	456	408	366	426	$T_{.3.} = 1656$
	$T_{..k}$	1278	1062	972	1008	$T_{...} = 4320$

the precision of estimation) and the reality that certain variables simply cannot be treated as within-subjects factors. Examples of variables that are inherently between-subjects factors are those whose levels are selected rather than manipulated (e.g., individual differences variables such as sex, age, or clinical diagnosis) and manipulated variables that entail carryover effect (e.g., training method).

9.2 ONE BETWEEN-SUBJECTS AND ONE WITHIN-SUBJECTS VARIABLE

We begin our discussion of mixed designs by analyzing the data of Table 9.1. First, we will examine how the total sum of squares is partitioned into sources of variance that provide the components of F tests. We do this by analogy to the analyses of the between-subjects design of Chapter 4 and the within-subjects designs of Chapter 8. Following this, we will present a more formal structural model to justify the partitioning and expected mean squares to justify the error terms used in the F tests.

9.2.1 Analyzing the Data

Mixed designs might at first glance seem more complicated than pure between-subjects or repeated-measures designs. However, we can obtain the appropriate model for any mixed design—and hence the entire ANOVA table for the design—using what we already know about the pure designs.

A simple way to find the appropriate sources of variance for any mixed design involves two steps: (1) first ignore the between-subjects factors and obtain the SV terms for the resulting repeated-measures design, then (2) partition the SV terms that contain the "subjects" variable, S, to reflect the effects of the between-subjects factors.

If we ignore the between-subjects variable, A, in Table 9.1, we have an $S \times B$ repeated-measures design with an (3×6) subjects and b (4) levels of the within-subjects factor, B. For this simplified design, we know from Chapter 8 that the appropriate sources of variance and degrees of freedom are those presented in panel (a) of Table 9.2. Note that the total sum of squares is completely accounted for by variability between subjects (SS_S) and within subjects (SS_{WSs}); that is,

$$SS_{tot} = SS_S + SS_{WSs}$$

In a similar manner, the total degrees of freedom can be divided into between- and within-subjects components:

$$abn - 1 = (an - 1) + an(b - 1)$$

Furthermore, the within-subject variability (SS_{WSs}) can be partitioned into a term due to B and one due to the interaction of S and B:

$$SS_{WSs} = SS_B + SS_{SB}$$

The df_{WSs} is similarly partitioned:

$$an(b - 1) = (b - 1) + (an - 1)(b - 1)$$

TABLE 9.2 SOURCES OF VARIANCE, DEGREES OF FREEDOM, AND CALCULATIONS FOR THE ONE BETWEEN-SUBJECTS AND ONE WITHIN-SUBJECTS FACTOR DESIGN

	(a) Appropriate SV and df terms if factor A is ignored	
SV	df	SS
Total	$abn - 1$	$\Sigma\Sigma\Sigma(Y_{ijk} - \bar{Y}_{...})^2$
Between subjects (S)	$an - 1$	$b\Sigma\Sigma(\bar{Y}_{ij.} - \bar{Y}_{...})^2$
Within subjects (WSs)	$an(b - 1)$	$\Sigma\Sigma\Sigma(Y_{ijk} - \bar{Y}_{ij.})^2$
B	$b - 1$	$an\Sigma(\bar{Y}_{..k} - \bar{Y}_{...})^2$
SB	$(an - 1)(b - 1)$	$SS_{tot} - SS_S - SS_B$

(b) Complete partitioning of the variability

SV	df	SS
Total	$abn - 1$	$\Sigma\Sigma\Sigma(Y_{ijk} - \bar{Y}_{...})^2$
Between subjects (S)	$an - 1$	$b\Sigma\Sigma(\bar{Y}_{ij.} - \bar{Y}_{...})^2$
A	$a - 1$	$bn\Sigma(\bar{Y}_{.j.} - \bar{Y}_{...})^2$
S/A	$a(n - 1)$	$b\Sigma\Sigma(\bar{Y}_{ij.} - \bar{Y}_{.j.})^2$
Within subjects (WSs)	$an(b - 1)$	$\Sigma\Sigma\Sigma(Y_{ijk} - \bar{Y}_{ij.})^2$
B	$b - 1$	$an\Sigma(\bar{Y}_{..k} - \bar{Y}_{...})^2$
SB	$(an - 1)(b - 1)$	$SS_{tot} - SS_S - SS_B$
AB	$(a - 1)(b - 1)$	$n\Sigma\Sigma(\bar{Y}_{.jk} - \bar{Y}_{...})^2 - SS_A - SS_B$
SB/A	$a(n - 1)(b - 1)$	$SS_{SB} - SS_{AB}$

(c) Calculations of sums of squares for the data of Table 9.1

$$C = \frac{T^2_{...}}{abn} = \frac{4320^2}{72} = 259{,}200$$

$$SS_{tot} = \Sigma\Sigma\Sigma Y^2_{ijk} - C = 82^2 + 48^2 + \cdots + 55^2 + 61^2 - C = 280{,}176 - C = 20{,}956$$

$$SS_S = \Sigma\Sigma\frac{T^2_{ij.}}{b} - C = \frac{224^2 + 248^2 + \cdots + 260^2 + 244^2}{4} - C = 273{,}648 - C = 14{,}448$$

$$SS_A = \Sigma\frac{T^2_{.j.}}{bn} - C = \frac{1104^2 + 1560^2 + 1.656^2}{24} - C = 266{,}448 - C = 7248$$

$$SS_{S/A} = SS_S - SS_A = 7200$$

$$SS_{WSs} = SS_{tot} - SS_S = 6508$$

$$SS_B = \Sigma\frac{T^2_{..k}}{an} - C = \frac{1278^2 + \cdots + 1008^2}{18} - C = 262{,}332 - C = 3132$$

$$SS_{AB} = \Sigma\Sigma\frac{T^2_{.jk}}{n} - C - SS_A - SS_B = \frac{384^2 + 276^2 + \cdots + 366^2 + 426^2}{6} - C - SS_A - SS_B$$

$$= 270{,}636 - C - SS_A - SS_B = 1056$$

$$SS_{SB/A} = SS_{WSs} - SS_B - SS_{AB} = 2320$$

The model and analysis suggested by this partitioning are obviously incomplete because we have so far ignored the A variable. Some of the between-subjects variability represented by the factor S occurs because scores come from different levels of A and some occurs because, even at a given level of A, subjects differ from one another. If we ignore factor B—say we average the b scores for each subject and analyze the resulting an subject averages—we have a one-factor between-subject design with n scores at each level of A. Just as we partitioned the total variability in the one-factor design of Chapter 4, we can partition the between-subjects variability here:

$$SS_S = SS_A + SS_{S/A}$$

The variability associated with the $S \times B$ term of panel a can be partitioned in an analogous fashion by crossing A and S/A with B; this yields

$$SS_{SB} = SS_{AB} + SS_{SB/A}$$

In similar fashion,

$$df_S = df_A + df_{S/A}$$

$$an - 1 = (a - 1) + a(n - 1)$$

and

$$df_{SB} = df_{AB} + df_{SB/A}$$

$$(an - 1)(b - 1) = (a - 1)(b - 1) + a(n - 1)(b - 1)$$

These partitionings of the S and SB terms complete the breakdown of SS_{tot} into its components. The result is presented in panel (b) of Table 9.2. The computing forms of the sums of squares are presented in panel (c).

Table 9.3 presents a summary of the analysis of the data in Table 9.1. The SV follow those in Table 9.2. Scanning them, you may wonder why there is no SA term present. The answer lies in the distinction between *crossing* and *nesting*. When data are obtained for all combinations of levels of one factor with all levels of another, we say that the two factors cross. For example, in the current design, A and B cross, as do S and B; however, A and S do not cross. Here, A is a between-subjects factor; that is, each subject provides data at a single level of A. We describe this situation by saying that subjects are "nested within levels of A," and we indicate nesting by using a "/" in our SV terms, as in S/A. It is not possible for two factors to interact with one another unless they cross. We do not consider the possibility of an interaction between S and A because the question of whether the difference between any two subjects is greater at A_1 than at A_2 is meaningless unless the subjects provide data both at A_1 and at A_2.

How do we interpret the nested terms? It is as though the sums of squares were obtained separately at each level of A and then pooled. Thus, $SS_{S/A}$ could be obtained by computing, for each level of A, the variability of the n subject means about the mean for that level of A, and then adding up the terms for the a levels of A. The $n - 1$ df obtained from each of the a levels of A combine to make up the $a(n - 1)$ df associated with S/A. Similarly, we could obtain $SS_{SB/A}$ by computing SS_{SB} separately at each level of A [accounting for $(n - 1)(b - 1)$ df at each level of A] and then summing the resulting a terms [thus accounting for the $a(n - 1)(b - 1)$ df associated with SB/A].

The sums of squares calculations are presented in Table 9.2. Although these are not too laborious to carry out on a calculator, with more data, and particularly with more variables, we strongly advise the use of a computer program. In addition to saving labor, packages such as SAS, SPSS^X, BMDP, and SYSTAT are capable of providing much additional information, such as summary statistics, plots of scores and residuals, tests of contrasts, and adjustments for nonsphericity.

TABLE 9.3 SUMMARY OF THE ANOVA OF THE DATA IN TABLE 9.1

SV	df	SS	MS	Error term	F	EMS
Total	71	20,596				
Between						
subjects	17	14,448				
A	2	7,248	3,624.000	S/A	7.55[a]	$\sigma_e^2 + b\sigma_{S/A}^2 + bn\theta_A^2$
S/A	15	7,200	480.000			$\sigma_e^2 + b\sigma_{S/A}^2$
Within						
subjects	54	6,508				
B	3	3,132	1,044.000	SB/A	20.25[a]	$\sigma_e^2 + \sigma_{SB/A}^2 + an\theta_B^2$
AB	6	1,056	176.000	SB/A	3.41[a]	$\sigma_e^2 + \sigma_{SB/A}^2 + n\theta_{AB}^2$
SB/A	45	2,320	51.556			$\sigma_e^2 + \sigma_{SB/A}^2$

[a] $p < .01$

The mean squares in Table 9.3 are, as always, ratios of sums of squares to degrees of freedom. The basis for the choice of error terms also follows our standard rule: the error term has the same expectation as the numerator mean square, assuming H_0 is true. The validity of the entries in the EMS column rests upon the validity of certain underlying assumptions, which we will specify next.

9.2.2 A Structural Model

As usual, the F tests (Table 9.3) are based on expected mean squares whose derivation requires a structural model. Table 9.4 defines the parameters that are components of that model. The following assumptions should be kept in mind when considering that table:

1. Y_{ijk}, the observed score for the ith subject at the jth level of A, is made up of a true-score component, μ_{ijk}, and a random-error component, ϵ_{ijk}. The ϵ_{ijk} are independently and normally distributed with mean zero and variance σ_e^2.
2. A and B are fixed-effect variables.
3. The n subjects in each of the a groups are randomly sampled from a treatment population consisting of an infinite number of subjects.

Assumptions 2 and 3 justify the definitions and constraints in Table 9.4. For example, we define μ_{ij} as the arithmetic mean of b values of μ_{ijk} because the average is taken over the levels of B, a fixed-effect variable. In contrast, we define μ_j as the expectation of the μ_{ij}. This is because S is a random-effects variable; therefore, to obtain the mean of the population of scores at A_j, we must average (i.e., take the expectation) over an infinite population of subjects. To consider one other example of the application of assumptions 2 and 3, note that we have the constraint that $\Sigma\beta_k = 0$, whereas $E(\eta_{i/j}) = 0$; in one case we sum fixed effects (the β_k), and in the other we take the expectation of random effects (the $\eta_{i/j}$). The reason for the differences in definitions and constraints as a function of the nature of the effects was discussed in Chapter 8.

TABLE 9.4 PARAMETERS OF THE MODEL FOR THE MIXED DESIGN OF TABLE 9.1

Population means[a]

$$\mu_{ijk} = E(Y_{ijk}), \quad \mu_{ij} = \sum_k \frac{\mu_{ijk}}{b}, \quad \mu_{jk} = E(\mu_{ijk}), \quad \mu_j = \sum_k \frac{\mu_{jk}}{b}, \quad \mu_k = \sum_j \frac{\mu_{jk}}{a},$$

$$\mu = \sum_j \frac{\mu_j}{a} = \sum_k \frac{\mu_k}{b}$$

Population effects and constraints

$$\epsilon_{ijk} = Y_{ijk} - \mu_{ijk}, \quad E(\epsilon_{ijk}) = 0, \quad \eta_{i/j} = \mu_{ij} - \mu_j, \quad E(\eta_{i/j}) = 0,$$

$$\alpha_j = \mu_j - \mu, \quad \sum_j \alpha_j = 0, \quad \beta_k = \mu_k - \mu, \quad \sum_k \beta_k = 0,$$

$$(\alpha\beta)_{jk} = \mu_{jk} - \mu_j - \mu_k + \mu, \quad \sum_j (\alpha\beta)_{jk} = \sum_k (\alpha\beta)_{jk} = 0,$$

$$(\eta\beta)_{ik/j} = \mu_{ijk} - \mu_{ij} - \mu_{jk} + \mu_j, \quad E[(\eta\beta)_{ik/j}] = \sum_k [(\eta\beta)_{ik/j}] = 0$$

Population variances

$$E(\epsilon_{ijk}^2) = \sigma_e^2, \quad E(\eta_{i/j}^2) = \sigma_{S/A}^2, \quad E[(\eta\beta)_{ik/j}^2] = \sigma_{SB/A}^2,$$

$$\sum_j \frac{\alpha_j^2}{a-1} = \theta_A^2, \quad \sum_k \frac{\beta_k^2}{b-1} = \theta_B^2, \quad \text{and} \quad \sum_j \sum_k \frac{(\alpha\beta)_{jk}^2}{(a-1)(b-1)} = \theta_{AB}^2$$

[a] E stands for the expectation over the population of subjects at A_j.

Given the definitions and constraints in Table 9.4, we can specify the structural model,

$$Y_{ijk} = \mu + \alpha_j + \eta_{i/j} + \beta_k + (\alpha\beta)_{jk} + (\eta\beta)_{ik/j} + \epsilon_{ijk} \tag{9.1}$$

Note that the effects, with the exception of the error component, ϵ_{ijk}, correspond to sources of variance in Table 9.3. In particular, the nested effects, $\eta_{i/j}$ and $(\eta\beta)_{ik/j}$, correspond to the sources S/A and SB/A, which were discussed in Section 9.2.1. There is no source corresponding directly to ϵ_{ijk} because we have only one score for each combination of S, A, and B. We saw a similar situation in the repeated-measures designs of Chapter 8.

Equation 9.1, together with the definitions of variances and constraints in Table 9.4, allows us to derive expected mean squares for the design. These were presented in Table 9.3. These expectations can be generated by a set of rules that are general enough to encompass more complex designs as well. We will present these rules in Section 9.3, but before doing so let us look again at the expected mean squares in Table 9.3.

The first thing to note is that we have two error terms; the EMS indicate that $MS_{S/A}$ is the appropriate error term with which to test whether the A effect is significant and that $MS_{SB/A}$ is the appropriate error term for B and AB. An important difference between the expectations of these two error terms is that $E(\mathrm{MS}_{S/A})$ involves the variance of the η_{ij} ($\sigma^2_{S/A}$), whereas $E(\mathrm{MS}_{SB/A})$ involves the variance of the $(\eta\beta)_{ik/j}$ ($\sigma^2_{SB/A}$); $\sigma^2_{S/A}$ will typically be greater than $\sigma^2_{SB/A}$. Therefore, within-subjects factors usually are tested against error terms that are smaller than the error terms used with between-subjects factors. In other words, tests of the within-subjects factors usually have greater power. Also note that Equation 9.1 presents a nonadditive model; the SB/A interaction term is present, and its variance contributes to within-subjects expectations. Unless other models are specified, most computer packages also will do the analysis based on this model.

The assumption of nonadditivity—that the effects of the within-subjects factor may not be the same for all subjects—is the most reasonable one to make unless there is strong evidence to the contrary. Therefore, when we consider tests of within-subjects factors, we must be concerned with violations of the sphericity assumption. In Chapter 8, we discussed how violations of sphericity may result in positively biased F tests. In mixed designs, the positive bias may occur for tests that involve the within-subjects factors. Conservative F tests parallel to those discussed in Chapter 8 can be carried out for the current design. The statistics MS_B and MS_{AB} are distributed on $(b-1)\epsilon$ and $(a-1)(b-1)\epsilon$ df, respectively, and $\mathrm{MS}_{SB/A}$, the error term for both A and AB, is distributed on $a(n-1)(b-1)\epsilon$ df. The values that ϵ can take vary between 1 (when there is no violation of the sphericity assumption) and $1/(b-1)$ (when there is maximum violation). The conservative test of the B effect would therefore use 1 and $a(n-1)$ df and that for the AB effect would use $a-1$ and $a(n-1)$ df.

In the ANOVA of Table 9.3, the B effect would still be significant at the .01 level even if the criterion F were based on 1 and 15 (3 × 5) df. However, the conservative test of interaction on 2 and 15 df is not significant at the .05 level. In this case, it would be worthwhile to compute the ϵ adjustment or to run the ANOVA with a program such as BMDP2V or SPSS^X MANOVA, which carries out the test based on the proper adjustment of degrees of freedom. If the equations in Table 8.6 are used to calculate ϵ, the overall $b \times b$ variance-covariance matrix must first be obtained separately at each level of A; these are then averaged. For example, S_{rc}, the element in the rth row and the cth column of the overall matrix, is obtained by finding the covariance between the scores at B_r and B_c separately at each level of A and then taking the average of these a covariances.

9.3 RULES FOR GENERATING EXPECTED MEAN SQUARES

Corresponding to α_j, $\eta_{i/j}$, β_k, $(\alpha\beta)_{jk}$, $(\eta\beta)_{ik/j}$, and ϵ_{ijk} in the structural model of Equation 9.1, there are six components of variance—θ^2_A, $\sigma^2_{S/A}$, θ^2_B, θ^2_{AB}, $\sigma^2_{SB/A}$, and σ^2_e—that might contribute to the EMS terms. We can use Equation 9.1 to determine which components actually contribute to any EMS. For example, if we are concerned with $E(\mathrm{MS}_A)$, we need only determine which components of variance contribute to the variance of the $\overline{Y}_{.j.}$'s. For the design discussed in Section 9.2, the structural equation is

$$Y_{ijk} = \mu + \alpha_j + \eta_{i/j} + \beta_k + (\alpha\beta)_{jk} + (\eta\beta)_{ik/j} + \epsilon_{ijk}$$

and

$$\bar{Y}_{.j.} = \frac{1}{nb}\sum_i\sum_k Y_{ijk}$$

As Table 9.4 states, if B is a fixed-effect variable,

$$\sum_k \beta_k = 0, \qquad \sum_k (\alpha\beta)_{jk} = 0, \qquad \text{and} \qquad \sum_i\sum_k (\eta\beta)_{ik/j} = 0$$

Therefore, it follows that

$$\bar{Y}_{.j.} = \mu + \alpha_j + \frac{1}{n}\sum_i \eta_{i/j} + \bar{\epsilon}_{.j.} \tag{9.2}$$

Because μ is a constant, it will not contribute to $E(MS_A)$. However, if there is a main effect of A, α_j will not be the same at all levels of A. In addition, for any given set of data, there is no reason to expect the average individual differences component or the average random-error component to be the same at each level of A. We therefore conclude that θ_A^2, $\sigma_{S/A}^2$, and σ_e^2 contribute to the variability in the $\bar{Y}_{.j.}$'s and, therefore, to $E(MS_A)$.

This kind of analysis can be done for the expectation of any mean square. However, a set of simple rules are available that allow us to decide quickly which components of variance contribute to any EMS and what the appropriate coefficients are for each component. These rules are listed in Table 9.5.

Consider how these rules might be used to find $E(MS_B)$. We first list σ_e^2 because it contributes to every EMS, and then add every additional component of variance that has subscripts containing B, multiplying each component by its appropriate coefficient. The result is

$$\sigma_e^2 + na\theta_B^2 + n\theta_{AB}^2 + \sigma_{SB/A}^2$$

From Rules 5 and 6, we drop the θ_{AB}^2 component, because once B is deleted from the subscripts A and B we are left with A, a subscript that denotes a fixed-effect variable. Note that we retain the $\sigma_{SB/A}^2$ component because when we delete B from the "essential" subscripts of that component we are left with S, a subscript that denotes a random effect variable. The final result is

$$E(MS_B) = \sigma_e^2 + na\theta_B^2 + \sigma_{SB/A}^2$$

The reader should apply the rules to the remaining sources of variance and compare the results with those in Table 9.3. We will see that the rules also apply to designs with more variables and with additional levels of nesting.

9.4 COMPARISONS AMONG MEANS IN MIXED DESIGNS

As we noted in previous chapters, many statistical packages can test contrasts among means if you just provide the weights involved in the contrasts. If you are not using such a package, the tests can still be done in a fairly straightforward way. Procedures

TABLE 9.5 RULES FOR OBTAINING EXPECTED MEAN SQUARES (EMS)

Rule 1. Decide for each independent variable (including subjects) whether it is fixed or random. Assign a letter to designate each variable. Assign another letter to be used as a coefficient that represents the number of levels of each variable. In the example of Table 9.3, the variables are designated A, B, and S; the coefficients are a, b, and n; A and B are fixed-effect variables and S is random.

Rule 2. List σ_e^2 as part of each EMS.

Rule 3. For each EMS, list the null hypothesis component—that is, the component corresponding directly to the SV under consideration. Thus we add $nb\theta_A^2$ to the EMS for the A line, $b\sigma_{S/A}^2$ to the EMS for the S/A line. Note that a component consists of three parts:

1. A coefficient representing the number of scores at each level of the effect (for example, nb scores at each level of A, or b scores for each subject).

2. A σ^2 or θ^2, depending on whether the effect is assumed to be random or fixed [σ^2 is the variance of the population of effects; for example, $\sigma_{S/A}^2 = E(\eta_{i/j}^2)$, $\theta_A^2 = \Sigma_j\alpha_j^2/(a-1)$].

3. As subscripts, those letters that designate the effect under consideration.

Rule 4. Now add to each EMS all components whose subscripts contain all the letters designating the SV in question. Since the subscript SB/A contains the letters S and A, for example, add $\sigma_{SB/A}^2$ to the EMS for the S/A line (this is later deleted according to Rule 6).

Rule 5. Next, examine the components for each SV. If a slash appears in the subscript, define only the letters to the left of the slash as "essential." If there are several slashes (as in the next chapter), only the letters preceding the leftmost slash are essential. If there is no slash in the subscript, all letters are considered essential.

Rule 6. Among the essential letters, ignore any that are necessary to designate the SV. If the source is A, in considering $n\theta_{AB}^2$, for example, ignore the A. If the source is S/A, in considering the $\sigma_{SB/A}^2$ component, S and B are essential subscripts and S is to be ignored. If any of the remaining essential letters designate fixed variables, delete the entire component from the EMS.

---

developed in Chapter 6 for between-subjects designs, and in Chapter 8 for within-subjects designs, can be extended to mixed designs.

In order to have an example in mind, let's take another look at Table 9.1. Recall that we had three groups of six subjects, and the groups had either a standard instructional approach (A_1), additional problems (A_2), or additional problems presented by a computer (A_3). Subjects were tested at four points in time (B_1 to B_4) beginning with a test at the end of the instructional period. In this section, we will use the example of Table 9.1 to illustrate tests of four kinds of contrasts. We indicate the first kind by $\psi_{p(A)}$, where ψ stands for a contrast and the subscript $p(A)$ specifies the pth contrast among the means of the between-subjects factor, A. For example, we might want to test the difference between the means of A_2 and A_3, the groups receiving additional problems. Or we might wish to pit the mean of our control, A_1, against the average of the combined A_2 and A_3 groups. In both these cases, we are contrasting the marginal means, those obtained by averaging the bn scores at a level of A.

We also might wish to test whether the contrast among the A_j differs as a function of the level of B. For example, the data of Table 9.1 suggest that the difference between the A_1 mean and the average of the A_2 and A_3 means increases as B, time since training, increases. This suggests a second kind of contrast, $\psi_{p(A) \times B}$, the variation in the pth contrast of the A means as a function of the level of B.

Still a third possible contrast is $\psi_{q(B)}$, which would involve a contrast among the marginal means at the different levels of B. For example, it appears from Table 9.1 that there is a trend for the average score to decrease as the delay since instruction increases. Using the weights for polynomial analysis that we discussed in Chapter 7, we can test whether this decline (the linear trend component) is a significant contributor to the overall variability due to B.

Finally, note that the decline seems more pronounced for the control group (A_1) than for either of the groups who received additional problems (A_2 and A_3). In terms of our discussion of trend analysis in Chapter 7, it appears that there may be a significant $\psi_{A \times \text{lin}(B)}$ source of variance.

In the remainder of this section, we will use the data of Table 9.1 to illustrate the calculations involved in testing whether these contrasts are significant. Bear in mind that the discussion in Chapter 6 of familywise error rates applies here as well. For example, if all pairwise comparisons of the $\overline{Y}_{.j.}$ are made, the Tukey or Bonferroni procedure should be used.

9.4.1 Comparisons among the Levels of the Between-Subjects Variable, A

Suppose we wish to compare the average score at A_1 with the average for the combined A_2 and A_3 conditions. The three marginal means receive the weights 1, $-1/2$, and $-1/2$, respectively. The null hypothesis is

$$H_0: \mu_{1.} - \frac{1}{2}(\mu_{2.} + \mu_{3.}) = 0$$

The general form of the null hypothesis is

$$H_0: \psi_{p(A)} = 0$$

where

$$\psi_{p(A)} = \sum_j w_j \mu_{j.}$$

w_j is a weight, and $\mu_{j.}$ is the mean of the b populations of scores at A_j. The sum of squares for this contrast is distributed on 1 df and therefore can be tested using either the t or its square, the F. Throughout this section we will use F ratios to test the hypotheses. From Chapter 6,

$$\text{SS}_{p(A)} = \frac{(\Sigma w_j \overline{Y}_{.j.})^2}{\Sigma w_j^2 / nb} = \frac{(\Sigma w_j T_{.j.})^2}{nb \Sigma w_j^2} \tag{9.3}$$

and

$$F_{1, a(n-1)} = \frac{\text{SS}_{p(A)}}{\text{MS}_{S/A}} \tag{9.4}$$

Panel (a) of Table 9.6 presents the cell totals from Table 9.1, together with contrasts at each level of B. For example, the value of 63 in the B_1 row is obtained from

$$\sum_j w_j T_{.j1} = (-1)(384) + \frac{1}{2}(438) + \frac{1}{2}(456) = 63$$

In panel (b) we apply Equations 9.3 and 9.4 to the data in panel (a). It is clear that the control group, A_1, differs significantly from the average of the two groups that received extra problems.

Next let's look at the interaction of $p(A)$ and B. For our example, we will test whether the difference between $\bar{Y}_{.1.}$ and the average of $\bar{Y}_{.2.}$ and $\bar{Y}_{.3.}$ varies significantly as a function of time (B); then the null hypothesis is

$$H_0: \mu_{11} - \frac{1}{2}(\mu_{21} + \mu_{31}) = \mu_{12} - \frac{1}{2}(\mu_{22} + \mu_{32}) = \mu_{13} - \frac{1}{2}(\mu_{23} + \mu_{33})$$

$$= \mu_{14} - \frac{1}{2}(\mu_{24} + \mu_{34})$$

In general, the null hypothesis is

$$H_0: \psi_{p(A)/B_1} = \psi_{p(A)/B_2} = \cdots = \psi_{p(A)/B_k} = \cdots = \psi_{p(A)/B_b}$$

where

$$\psi_{p(A)/B_k} = \sum_j w_j \mu_{jk}$$

The sum of squares for this comparison of b contrasts was also presented in Chapter 6 and is restated here:

$$SS_{p(A) \times B} = \frac{\sum_k (\sum_j w_j T_{.jk})^2}{n \sum_j w_j^2} - SS_{p(A)} \tag{9.5}$$

The F test is

$$F_{b-1, a(n-1)(b-1)} = \frac{SS_{p(A) \times B}/(b-1)}{MS_{SB/A}} \tag{9.6}$$

Note that the numerator sum of squares is distributed on $b-1$ df. The reason is that we are calculating the variability of b contrasts about their average. Calculations based on the data of Table 9.1 are presented in panel c of Table 9.6. The result is significant, probably because the difference between A_1 and the average of A_2 and A_3 is much smaller at B_1 than at the other levels of B. In terms of our example, the advantage of additional problems (A_2 and A_3) is greater as the time since instruction (B) increases.

9.4.2 Comparisons Among the Levels of the Within-Subjects Variable, B

Contrasts among the b means can be tested directly by using statistical packages. They also may be tested by transforming the data into contrast scores as we illustrated in Chapter 8. For example, suppose we want to test for linear trend over delay of test (B)

TABLE 9.6 COMPARING THE EFFECTS OF THE LEVELS OF A IN TABLE 9.1

(a) The totals for the AB combinations ($T_{.jk}$) and the weights for the contrast (w_j) are presented; $n = 6$.

	A_1	A_2	A_3	$\Sigma_j w_j T_{.jk}$
B_1	384	438	456	63
B_2	276	378	408	117
B_3	234	372	366	135
B_4	210	372	426	189
$T_{.j.}$	1104	1560	1656	
w_j	-1	1/2	1/2	

$$\sum_j w_j T_{.j.} = (-1)(1104) + \frac{1}{2}(1560) + \frac{1}{2}(1656)$$

$$= 63 + 117 + 135 + 189 = 504$$

(b) To test $H_0: \Sigma_j w_j \mu_j = 0$,

$$F_{1,a(n-1)} = \frac{SS_{p(A)}}{MS_{S/A}} = \frac{(\Sigma_j w_j T_{.j.})^2 / nb\Sigma_j w_j^2}{MS_{S/A}}$$

For the data of Table 9.1,

$$F_{1,15} = \frac{[(-1)(1104) + (1/2)(1560) + (1/2)(1656)]^2 / (6)(4)(3/2)}{480}$$

$$= \frac{254,016/36}{480} = \frac{7056}{480} = 14.7$$

(c) To test $H_0: \Sigma_j w_j \mu_{j1} = \Sigma_j w_j \mu_{j2} = \cdots = \Sigma_j w_j \mu_{jp}$, first calculate

$$SS_{p(A) \times B} = \frac{\Sigma_k (\Sigma_j w_j T_{.jk})^2}{n\Sigma_j w_j^2} - SS_{p(A)}$$

$$= \frac{63^2 + 117^2 + 135^2 + 189^2}{(6)(3/2)} - 7056$$

$$= 900$$

$$MS_{p(A) \times B} = \frac{SS_{p(A) \times B}}{b - 1} = \frac{900}{3} = 300$$

$$F_{b-1,a(n-1)(b-1)} = \frac{MS_{p(A) \times B}}{MS_{SB/A}} = \frac{300}{51.556} = 5.82$$

in the data set of Table 9.1. Then the weights are 3, 1, -1, -3; these are the orthogonal polynomial coefficients of Appendix Table D.7. Panel (a) of Table 9.7 contains 18 contrast scores (C_{ij}), one for each subject at each level of A. These were computed by combining the four scores for each subject by using the linear coefficients. For example, the four scores for the first subject in A_1 are 82, 48, 41, and 53. Then the contrast score is

$$(3)(82) + (1)(48) + (-1)(41) + (-3)(53) = 94$$

The test for a linear trend is a test of whether the average of all an (6×3 in our example) contrast scores differs significantly from zero. In general, the null hypothesis is

$$H_0: \psi_{q(B)} = 0$$

TABLE 9.7   COMPARING THE EFFECTS OF THE LEVELS OF B FOR THE DATA OF TABLE 9.1

(a) The weights w_k are 3, 1, -1, -3. Using these, we calculate a contrast score C_{ij} for each subject:

	C_{1j}	C_{2j}	C_{3j}	C_{4j}	C_{5j}	C_{6j}	$\Sigma_i C_{ij}$	$\hat{\sigma}^2_{C_j}$
A_1	94	90	98	128	64	90	564	420.8
A_2	30	54	34	34	36	16	204	148.8
A_3	26	18	42	26	8	12	132	148.8

$$\Sigma\Sigma C_{ij} = 900 \quad \hat{\sigma}^2_C = 239.4667$$

(b) The test of whether $\psi_{q(B)}$ differs from zero is

$$F_{1,a(n-1)} = \frac{(\Sigma_j \Sigma_i C_{ij})^2 / an}{\hat{\sigma}^2_C}$$

From the values in panel (a), we have

$$F_{1,15} = \frac{900^2 / 18}{239.667} = 187.761$$

(c) To test whether the contrast varies as a function of the level of A, calculate

$$SS_{q(B) \times A} = \frac{1}{n}\Sigma_j (\Sigma_i C_{ij})^2 - \frac{1}{an}(\Sigma_j \Sigma_i C_{ij})^2$$

$$= \frac{564^2 + 204^2 + 132^2}{6} - \frac{900^2}{18} = 17,856$$

$$MS_{q(B) \times A} = \frac{SS_{q(B) \times A}}{a - 1} = \frac{17,856}{2} = 8928$$

$$F_{a-1, a(n-1)} = \frac{MS_{q(B) \times A}}{\hat{\sigma}^2_C} = \frac{8928}{239.467} = 37.283$$

where

$$\psi_{q(B)} = \sum w_k \mu_{.k}$$

Another null hypothesis of interest is the interaction of A and $q(B)$. In our example, the significant AB interaction (see the ANOVA of Table 9.2) may be due to differences among the three groups in the rate at which performance drops off over the four test periods. The null hypothesis is that the three linear components do not differ; that is,

$$
\begin{aligned}
H_0 : (3)\mu_{11} &+ (1)\mu_{12} + (-1)\mu_{13} + (-3)\mu_{14} \\
&= (3)\mu_{21} + (1)\mu_{22} + (-1)\mu_{23} + (-3)\mu_{24} \\
&= (3)\mu_{31} + (1)\mu_{32} + (-1)\mu_{33} + (-3)\mu_{34}
\end{aligned}
$$

The general form of this null hypothesis is

$$H_0 : \psi_{q(B)/A_1} = \psi_{q(B)/A_2} = \cdots = \psi_{q(B)/A_j} = \cdots = \psi_{q(B)/A_a}$$

where

$$\psi_{q(B)/A_j} = \sum_k w_k \mu_{jk}$$

We can test both $q(B)$ and $A \times q(B)$ by performing a one-factor analysis of variance on the contrast scores. The $q(B)$ term is tested in panel (b) of Table 9.7 and the $A \times q(B)$ term in panel (c). The denominator for both F tests is the average within-group variance of contrast scores, which is distributed on $a(n-1)$ df as in any one-factor experiment. The numerator for testing whether there is a linear trend (in general, for testing the null hypothesis $\psi_{q(B)} = 0$) is an times the squared grand mean of the contrast scores, which is distributed on 1 df. The numerator for testing whether the linear trends are the same for all three groups is the usual mean square for a main effect except that the scores are the contrast scores obtained by summing the weighted Y_{ijk}. This term is distributed on $a - 1$ df. The result of the calculations in Table 9.7 is that both $q(B)$ and $A \times q(B)$ are very significant sources of variance. We conclude that, averaging over all three levels of A, performance in the population declines markedly since the end of the instruction period; however, the rate of decline varies as a function of A, presumably because it is greater at A_1 than at A_2 or A_3.

9.5 TESTING SIMPLE EFFECTS

In the example of Table 9.1, we might wish to test the effects of type of instruction (A) only immediately after instruction; that is, only at B_1. Or we might wish to test the effects of time (B) only for the control group (A_1). In general, we may wish to test the simple effects of A at B_k (A/B_k) or the simple effects of B at A_j (B/A_j). Let us consider the appropriate F test for each in turn.

To test the simple main effects of A at B_k, we recommend ignoring the data at all other levels of B. Then there are a groups of n subjects, each subject having exactly one score, the score at B_k. The numerator sum of squares is

$$SS_{A/B_k} = \frac{\Sigma_j T_{.jk}^2}{n} - \frac{T_{..k}^2}{an} \qquad (9.7)$$

and is distributed on $a - 1$ degrees of freedom. The denominator mean square ($MS_{S/A/B_k}$; subjects within A at B_k) is just the average of the variances within the a cells at B_k and therefore is distributed on $a(n - 1)$ degrees of freedom.

If the populations of scores corresponding to the ab cells have homogeneous variances, a more powerful test of the simple effects of A at B_k is possible. The numerator of the test statistic is the sum of squares of Equation 9.7, but its denominator is $MS_{W\text{cells}}$, the average of the ab cell variances. An easy way to calculate this error term is to pool the S/A and SB/A sources of variance in the original analysis of variance of Table 9.3. The reason for the potential power advantage when $MS_{W\text{cells}}$ is the error term is that $MS_{W\text{cells}}$ has more df than $MS_{S/A/B_k}$; that is, $ab(n - 1)$ df [the sum of $a(n - 1)$ and $a(n - 1)(b - 1)$] is greater than $a(n - 1)$ df.

Researchers should be cautious about using $MS_{W\text{cells}}$ as an error term for testing the simple effects of A. Suppose we wish to test the A effects at B_1. If the cell variances at the other levels of B are greater than those at B_1, we may actually lose power by averaging them into our error term. If they are smaller, we well may have an inflated Type 1 error rate. A more extensive discussion of the consequences of heterogeneity of variance for testing simple effects, and of why these consequences will be more severe than when main effects are tested, may be found in Chapter 5.

To test the effects of B at A_j, we view the design as a simple subjects $\times$ B design, in which there are n subjects with b scores for each. The error term for testing the simple effect of B is just the $S \times B/A_j$ (subjects by B at A_j) mean square. This will be on $(n - 1)(b - 1)$ df. As in testing simple effects of A, a potentially more powerful alternative exists; the simple effect of B may be tested against SB/A, the omnibus error term against which the main effects of B (and the AB interaction) are tested. However, we again urge caution in using this error term. Although SB/A is distributed on more df than the error term based only on the A_j data, Type 1 and 2 error rates may be severely distorted if the subjects $\times$ B interaction variability is at all heterogeneous across the levels of A.

9.6 DESIGNS IN WHICH THE WITHIN-SUBJECTS FACTOR IS PRETEST-POSTTEST

A common mixed design is one in which there is a between-subjects treatment (A), and two measures are obtained from each subject, a pretest score and a posttest score. Subjects are generally assigned randomly to the levels of A, and the pretest scores are obtained before the treatment is applied, so there are no systematic differences in the pretest scores across levels of A. The posttest scores reflect the effects of the treatment, if there are any. Although this design is often analyzed as a mixed design, other analyses are preferable. Because the treatment is applied to the posttest scores but not the pretest scores, the two types of scores are described by different structural models. Let the subscripts 1 and 2 refer to pretest and posttest scores, respectively. Then β_1 is the effect associated with the pretest scores, and the model for the pretest scores is

$$Y_{ij1} = \mu + \eta_{i/j} + \beta_1 + (\eta\beta)_{i1/j} + \epsilon_{ij1}$$

Note that A has no effect upon the pretest scores. Turning now to the posttest scores, A can have an effect so that the model for these scores is

$$Y_{ij2} = \mu + \alpha_j + \eta_{i/j} + \beta_2 + (\eta\beta)_{i2/j} + \epsilon_{ij2}$$

Assuming that we are interested in whether the variable A affects posttest performance, possible analyses for this design include:

1. *Analysis of covariance.* When its assumptions are met, the most powerful analysis is provided by the analysis of covariance that will be discussed in Chapter 13. Briefly, this analysis rests on the assumption that the posttest scores are linear functions of the pretest scores; it also is assumed that the slopes of these functions are the same at each level of A. The analysis takes advantage of this relationship, reducing error variance in the posttest scores by removing variability accounted for by the pretest scores.

2. *Analysis of gain scores.* Another possible analysis is based on gain scores. For each subject the pretest score is subtracted from the posttest score, and then a one-factor (A) ANOVA is performed on these gain scores. This approach assumes that the effect of each treatment is to add a constant to the pretest score. Because this model is less likely to be true than that assumed in the analysis of covariance, it will generally provide a less powerful test.

3. *Analysis of posttest scores only.* Because the effect of the treatment is only on the posttest scores, one could ignore the pretest scores and simply perform a one-factor ANOVA on the posttest scores. The resulting F would clearly test the effect of A upon posttest scores. However, although this approach does not violate any assumptions, it ignores data (the pretest scores) that could help reduce error variance, and therefore this approach will produce less powerful tests than those noted earlier.

4. *Using the mixed-design analysis.* We raised the issue of testing the effects of A using pretest and posttest scores in this section, because as several statisticians have pointed out (Huck and McLean 1975, Jennings 1988), such data are frequently analyzed as if generated by a mixed design with A as the between-subjects factor and trials as the within-subjects factor. If this is done, the F test for the A main effect will be very conservative because the pretest scores cannot be affected by the treatment. A better test of A is given by the F for the $A \times$ trials interaction. This test can be shown to be identical to that obtained by performing a one-factor ANOVA on the gain (posttest-pretest) scores. However, as indicated above, the analysis of covariance will generally provide a more powerful test. More detailed discussions of these issues are provided by Huck and McLean (1975), Jennings (1988), and Maxwell and Howard (1981).

We should emphasize that the discussion in this section has presupposed random assignment of subjects to levels of A. Without random assignment, there may be systematic differences in the pretest scores across levels of A. If so, interpretation becomes much more difficult, both for analysis of covariance and for analyses of gain scores (see, for example, Cronbach and Furby 1970, and Linn and Slinde 1977). Unless the investigator understands the issues involved (which requires understanding the material on regression covered in Chapters 12–16), he or she might be advised to analyze only the posttest data.

9.7 ADDITIONAL MIXED DESIGNS

The approach taken with the design discussed in the preceding section may be extended to any mixed design. In the present section, two additional designs are presented as illustrations.

9.7.1 Two Between-Subjects and One Within-Subjects Factor

Let us consider a variation of the experiment that has carried us through the chapter so far. This time, assume two instructional methods; this variable is again denoted by A. Further assume that the period of instruction is 6 weeks for half of the subjects taught by each method and 12 weeks for the remaining half of the subjects. Let B stand for this second between-subjects variable. Assume that we have six subjects in each of the four (2×2) groups. Finally, assume that we test each subject three times, at equally spaced intervals following the end of instruction. Let this within-subject variable, time, be denoted by C. Table 9.8 presents the data from this hypothetical experiment. In general, we have ab between-subject cells with n subjects in each and c scores for each subject. In Table 9.8, $a = b = 2$, $c = 3$, and $n = 6$. In referring to any score in such a design, the following notation will be used:

$$i = 1, 2, \ldots, n$$

$$j = 1, 2, \ldots, a$$

$$k = 1, 2, \ldots, b$$

$$m = 1, 2, \ldots, c$$

The approach we took in developing the ANOVA table for the one between- and one within-subjects design can be easily extended to the case in which there are two between-subjects factors. If we ignore the between-subjects factors and write down the SV and df for the resulting repeated-measures design, we obtain the result presented in panel (a) of Table 9.9. This preliminary analysis of the data accounts for all $abcn - 1$ df; there are $abn - 1$ df for between-subjects variability and $abn(c - 1)$ for within-subjects variability—that is, $c - 1$ df for each of the abn subjects. The analysis of panel (a) is, however, incomplete because it fails to take account of the between-subjects factors A and B. The between-subjects variability occurs partly because of the main effects of A and B, partly because of the joint effect of A and B, and partly because subjects tested at a given combination of the levels of A and B differ from one another. The partitioning of the between-subjects variability (SS_S) and degrees of freedom (df_S) can therefore be represented by

$$SS_S = SS_A + SS_B + SS_{AB} + SS_{S/AB}$$

and

$$abn - 1 = a - 1 + b - 1 + (a - 1)(b - 1) + ab(n - 1)$$

Crossing C with each of the preceding terms, the SC variability and degrees of freedom of panel (a) can be partitioned as

$$SS_{SC} = SS_{AC} + SS_{BC} + SS_{ABC} + SS_{SC/AB}$$

TABLE 9.8   DATA FOR A MIXED DESIGN WITH TWO BETWEEN-SUBJECTS FACTORS AND ONE WITHIN-SUBJECTS FACTOR

		C_1	C_2	C_3	$T_{i1k.}$			C_1	C_2	C_3	$T_{i2k.}$
	S_{111}	59	46	48	153		S_{121}	78	64	66	208
	S_{211}	72	53	50	175		S_{221}	72	73	64	209
A_1B_1	S_{311}	46	35	34	115	A_2B_1	S_{321}	56	36	32	124
	S_{411}	68	67	58	193		S_{421}	60	44	40	144
	S_{511}	57	40	44	141		S_{521}	51	43	47	141
	S_{611}	49	38	30	117		S_{621}	54	48	46	148
$T_{.11m} =$		351	279	264	$T_{.11.} = 894$	$T_{.21m} =$		371	308	295	$T_{.21.} = 974$
	S_{112}	56	49	38	143		S_{122}	77	74	71	222
	S_{212}	63	49	42	154		S_{222}	87	82	83	252
A_1B_2	S_{312}	74	70	68	212	A_2B_2	S_{322}	64	57	50	171
	S_{412}	81	74	85	240		S_{422}	65	60	57	182
	S_{512}	55	50	49	154		S_{522}	71	66	68	205
	S_{612}	61	44	40	145		S_{622}	89	84	82	255
$T_{.12m} =$		390	336	322	$T_{.12.} = 1048$	$T_{.22m} =$		453	423	411	$T_{.22.} = 1287$

Totals for the *AC* and *BC* cells are

	C_1	C_2	C_3	$T_{.j..}$			C_1	C_2	C_3	$T_{.j..}$
A_1	741	615	586	1942		B_1	722	587	559	1868
A_2	824	731	706	2261		B_2	843	759	733	2335
$T_{...m} =$	1565	1346	1292	$T_{....} = 4203$		$T_{...m} =$	1565	1346	1292	$T_{....} = 4203$

and

$$(abn-1)(c-1)=(a-1)(c-1)+(b-1)(c-1)+(a-1)(b-1)(c-1)+ab(n-1)(c-1)$$

The complete partitioning of the total variability is presented in panel (b) of Table 9.9, and SS calculations for the data of Table 9.8 are presented in panel (c). Table 9.10 presents a summary of the ANOVA, and the EMS. Note that there are two error terms in Table 9.10, one for the between-subjects terms (*A, B,* and *AB*) and another for the within-subjects terms (*C* and its interactions). We next present a structural model that underlies the expected mean squares, which in turn justify these error terms.

The appropriate structural equation, consistent with the analysis of Table 9.10 is an extension of the model of Equation 9.2 for the design with one between- and one within-subjects variable. Here we have added main and interaction effects of the second between-subjects variable:

$$Y_{ijkm} = \mu + \alpha_j + \beta_k + (\alpha\beta)_{jk} + \eta_{i/jk} + \gamma_m + (\alpha\gamma)_{jm} + (\beta\gamma)_{km} \qquad (9.8)$$

$$+ (\alpha\beta\gamma)_{jkm} + (\eta\gamma)_{im/jk} + \epsilon_{jkm}$$

TABLE 9.9 SOURCES OF VARIANCE, DEGREES OF FREEDOM, AND CALCULATIONS FOR THE DESIGN OF TABLE 9.8

(a) Appropriate SV and df terms if factors A and B are ignored

SV	df	SS
Total	$abcn - 1$	$\Sigma\Sigma\Sigma\Sigma(Y_{ijkm} - \bar{Y}_{....})^2$
Between-subjects (S)	$abn - 1$	$c\Sigma\Sigma\Sigma(\bar{Y}_{ijk.} - \bar{Y}_{....})^2$
Within-subjects (WSs)	$abn(c - 1)$	$\Sigma\Sigma\Sigma\Sigma(Y_{ijkm} - \bar{Y}_{ijk.})^2$
C	$c - 1$	$abn\Sigma(\bar{Y}_{...m} - \bar{Y}_{....})^2$
SC	$(abn - 1)(c - 1)$	$SS_{tot} - SS_S - SS_C$

(b) Complete partitioning of the variability

SV	df	SS
Total	$abcn - 1$	$\Sigma\Sigma\Sigma\Sigma(Y_{ijkm} - \bar{Y}_{....})^2$
Between subjects (S)	$abn - 1$	$c\Sigma\Sigma\Sigma(\bar{Y}_{ijk.} - \bar{Y}_{....})^2$
A	$a - 1$	$bcn\Sigma(\bar{Y}_{.j..} - \bar{Y}_{....})^2$
B	$b - 1$	$acn\Sigma(\bar{Y}_{..k.} - \bar{Y}_{....})^2$
AB	$(a - 1)(b - 1)$	$cn\Sigma\Sigma(\bar{Y}_{.jk.} - \bar{Y}_{....})^2 - SS_A - SS_B$
S/AB	$ab(n - 1)$	$c\Sigma\Sigma\Sigma(\bar{Y}_{ijk.} - \bar{Y}_{.jk.})^2$
Within Subjects (WSs)	$abn(c - 1)$	$\Sigma\Sigma\Sigma\Sigma(Y_{ijkm} - \bar{Y}_{ijk.})^2$
C	$c - 1$	$abn\Sigma(\bar{Y}_{...m} - \bar{Y}_{....})^2$
SC	$(abn - 1)(c - 1)$	$SS_{tot} - SS_S - SS_C$
AC	$(a - 1)(c - 1)$	$bn\Sigma\Sigma(\bar{Y}_{.j.m} - \bar{Y}_{....})^2 - SS_A - SS_C$
BC	$(b - 1)(c - 1)$	$an\Sigma\Sigma(\bar{Y}_{..km} - \bar{Y}_{....})^2 - SS_B - SS_C$
ABC	$(a - 1)(b - 1)(c - 1)$	$n\Sigma\Sigma\Sigma(\bar{Y}_{.jkm} - \bar{Y}_{....})^2 - SS_A - SS_B - SS_C$ $- SS_{AB} - SS_{AC} - SS_{BC}$
SC/AB	$ab(n - 1)(b - 1)$	$SS_{SC} - SS_{AC} - SS_{BC} - SS_{ABC}$

(c) Computations of sums of squares for the data of Table 9.8

$$C = \frac{T^2_{....}}{abcn} = \frac{4203^2}{72} = 245{,}350.125$$

$$SS_{tot} = \Sigma\Sigma\Sigma\Sigma Y^2_{ijkm} - C = 59^2 + 46^2 + \cdots + 82^2 - C = 261{,}459 - C = 16{,}108.875$$

$$SS_S = \Sigma\Sigma\frac{T^2_{ijk.}}{c} - C = \frac{153^2 + 175^2 + \cdots + 255^2}{3} - C$$

TABLE 9.9 *(continued)*

$$= 258,857.667 - C = 13,507.542$$

$$SS_A = \sum \frac{T_{.j..}^2}{bcn} - C = \frac{1942^2 + 2261^2}{36} - C$$

$$= 246,763.472 - C = 1413.347$$

$$SS_B = \sum \frac{T_{..k.}^2}{acn} - C = \frac{1868^2 + 2335^2}{36} - C$$

$$= 248,379.139 - C = 3029.014$$

$$SS_{AB} = \sum\sum \frac{T_{.jk.}^2}{cn} - C - SS_A - SS_B = \frac{894^2 + \cdots + 1287^2}{18} - C - SS_A - SS_B$$

$$= 250,143.6 - C - SS_A - SS_B = 351.125$$

$$SS_{S/AB} = SS_S - SS_A - SS_B - SS_{AB} = 8714.556$$

$$SS_{WSs} = SS_{tot} - SS_S = 2601.333$$

$$SS_C = \sum \frac{T_{...m}^2}{abn} - C = \frac{1565^2 + 1346^2 + 1293^2}{24} - C$$

$$= 247,091.875 - C = 1741.75$$

$$SS_{AC} = \sum\sum \frac{T_{.j.m}^2}{bn} - C - SS_A - SS_C = \frac{741^2 + \cdots + 706^2}{12} - C - SS_A - SS_C$$

$$= 248,539.583 - C - SS_A - SS_C = 34.361$$

$$SS_{BC} = \sum\sum \frac{T_{..km}^2}{an} - C - SS_B - SS_C = \frac{722^2 + \cdots + 733^2}{12} - C - SS_B - SS_C$$

$$= 250,196.083 - C - SS_B - SS_C = 75.194$$

$$SS_{ABC} = \sum\sum\sum \frac{T_{.jkm}^2}{n} - C - SS_A - SS_B - SS_C - SS_{AB} - SS_{AC} - SS_{BC}$$

$$= \frac{351^2 + \cdots + 411^2}{6} - C - SS_A - SS_B - SS_C - SS_{AB} - SS_{AC} - SS_{BC}$$

$$= 252,001.167 - C - SS_A - SS_B - SS_C - SS_{AB} - SS_{AC} - SS_{BC} = 6.25$$

$$SS_{SC/AB} = SS_{WSs} - C - SS_C - SS_{AC} - SS_{BC} - SS_{ABC} = 743.778$$

TABLE 9.10   SUMMARY OF THE ANOVA OF THE DATA IN TABLE 9.8

SV	df	SS	MS	Error term	F	EMS
Total	71	16,108.875				
Between subjects	23	13,507.542				
A	1	1413.347	1413.347	S/AB	3.24^a	$\sigma_e^2 + c\sigma_{S/AB}^2 + bcn\theta_A^2$
B	1	3029.014	3029.014	S/AB	6.95	$\sigma_e^2 + c\sigma_{S/AB}^2 + acn\theta_B^2$
AB	1	351.125	351.125	S/AB	.81	$\sigma_e^2 + c\sigma_{S/AB}^2 + cn\theta_{AB}^2$
S/AB	20	8714.056	435.703			$\sigma_e^2 + c\sigma_{S/AB}^2$
Within subjects	48	2601.333				
C	2	1741.750	870.875	SC/AB	46.84^b	$\sigma_e^2 + \sigma_{SC/AB}^2 + abn\theta_C^2$
AC	2	34.361	17.181	SC/AB	.92	$\sigma_e^2 + \sigma_{SC/AB}^2 + bn\theta_{AC}^2$
BC	2	75.194	37.597	SC/AB	2.02	$\sigma_e^2 + \sigma_{SC/AB}^2 + an\theta_{BC}^2$
ABC	2	6.250	3.125	SC/AB	.17	$\sigma_e^2 + \sigma_{SC/AB}^2 + n\theta_{ABC}^2$
SC/AB	40	743.778	18.594			$\sigma_e^2 + \sigma_{SC/AB}^2$

[a] $p < .05$
[b] $p < .01$

where α, β, and γ are the effects associated with levels of A, B, and C; $(\alpha\beta)$, $(\alpha\gamma)$, $(\beta\gamma)$, and $(\alpha\beta\gamma)$ are interaction effects defined in the usual manner; $\eta_{i/jk}$ is the unique contribution associated with the ith subject at the jth level of A and the kth level of B; $(\eta\gamma)_{im/jk}$ is the nested subject $\times$ C interaction effect; and ϵ_{ijkm} is the measurement error associated with sampling Y_{ijkm}. It is assumed that A, B, and C are fixed-effect variables; as usual, this implies that for each variable the sum of effects over the appropriate indices is 0. The ϵ_{ijkm}, $\eta_{i/jk}$, and $(\eta\gamma)_{im/jk}$ effects are assumed to be randomly sampled from infinite populations of such effects that are normally distributed with means equal to 0 and variances of σ_e^2, $\sigma_{S/AB}^2$, and $\sigma_{SC/AB}^2$, respectively. These assumptions, together with Equation 9.10, provide the basis for the EMS presented in Table 9.10.

9.7.2 One Between-Subjects and Two Within-Subjects Factors

Still another version of the mixed design involves more than one within-subjects variable. For example, we might have two instructional methods (A) with six subjects assigned to each. Each subject is tested at three different times (B) following the end of instruction. Two tests (C) are given at each time. One test directly measures knowledge of arithmetic operations, whereas the other requires the student to understand a problem presented in a story. In general, we have *an* (2 × 6 in the example) subjects with *bc* (3 × 2) scores for each. The data obtained using this design are presented in Table 9.11. The indices of notation are

$$i = 1, 2, \ldots, n$$
$$j = 1, 2, \ldots, a$$
$$k = 1, 2, \ldots, b$$
$$m = 1, 2, \ldots, c$$

Once again, we can begin our partitionings of SS_{tot} and df_{tot} by ignoring the variable *A* and considering an $S \times B \times C$ design in which 12 subjects are tested under all combinations of *B* and *C*. We have

$$SS_{tot} = SS_S + SS_B + SS_{SB} + SS_C + SS_{SC} + SS_{BC} + SS_{SBC}$$

Of course, the variability among the *an* subjects is in part due to *A*; therefore, as in the one-factor between-subjects design of Chapter 4,

$$SS_S = SS_A + SS_{S/A}$$

Crossing *S* with *B, C,* and *BC* in turn, we have

$$SS_{SB} = SS_{AB} + SS_{SB/A}$$
$$SS_{SC} = SS_{AC} + SS_{SC/A}$$

and

$$SS_{SBC} = SS_{ABC} + SS_{SBC/A}$$

The complete partitioning is presented in panel (a) of Table 9.12, and the calculations for the sums of squares are presented in panel (b). Table 9.13 presents BMDP2V's summary of the ANOVA of the data of Table 9.11. One difference from the previous analyses of this chapter is that there are four error terms. In descending order, these are S/A, SB/A, SC/A, and SBC/A. Table 9.14 presents expected mean squares that provide the justification for these error terms. Use the rules of Table 9.5 to verify these expectations, and then make sure that you agree that the BMDP2V error terms are the appropriate ones.

Aside from saving time and reducing the risk of computing error, the statistical package provides information, such as the ϵ adjustment, that would be difficult to calculate by hand. In addition, such packages usually have several options for generating additional statistics, plots, and significance tests. With respect to the ϵ adjustments, you may have noted that these are different for the tests of *B* and *BC*. If *C* had more than two levels, we would have had still another value of ϵ. There is a different matrix of variances and covariances associated with each of the three within-subjects sources.

The expected mean squares of Table 9.14 were derived from the structural model for the one between-subjects and two within-subjects factor design of Equation 9.9:

$$Y_{ijkm} = \mu + \alpha_j + \eta_{i/j} + \beta_k + \gamma_m + (\beta\gamma)_{km} + (\alpha\beta)_{jk} + (\alpha\gamma)_{jm} \qquad (9.9)$$
$$+ (\alpha\beta\gamma)_{jkm} + (\eta\beta)_{ik/j} + (\eta\gamma)_{im/j} + (\eta\beta\gamma)_{ikm/j} + \epsilon_{ijkm}$$

TABLE 9.11 DATA FOR A MIXED DESIGN WITH ONE BETWEEN-SUBJECTS FACTOR AND TWO WITHIN-SUBJECTS FACTORS

		C_1					C_2				
		B_1	B_2	B_3	$T_{ij.1}$	B_1	B_2	B_3	$T_{ij.2}$	$T_{ij..}$
	S_{11}	80	48	45	173	76	45	41	162	335
	S_{21}	46	37	34	117	42	34	33	109	226
A_1	S_{31}	51	49	36	136	45	38	30	113	249
	S_{41}	72	57	50	179	66	51	42	159	338
	S_{51}	68	40	33	141	58	38	30	126	267
	S_{61}	65	44	36	145	56	37	28	121	266
$T_{.1km} =$		382	275	234		343	243	204		

$T_{.1.1} = 891$ $T_{.1.2} = 790$ $T_{.1..} = 1681$

		B_1	B_2	B_3	$T_{ij.1}$	B_1	B_2	B_3	$T_{ij.2}$	$T_{ij..}$
	S_{12}	70	55	52	177	68	57	56	181	358
	S_{22}	88	69	66	223	91	74	70	235	458
A_2	S_{32}	58	60	54	172	50	41	38	129	301
	S_{42}	63	57	52	172	61	58	56	175	347
	S_{52}	78	81	75	234	79	78	74	231	465
	S_{62}	84	82	80	246	80	73	76	229	475
$T_{.2km} =$		441	404	379		429	381	370		

$T_{.2.1} = 1224$ $T_{.2.2} = 1180$ $T_{.2..} = 2404$

$T_{..km} =$	823	679	613		772	624	574	

$T_{...1} = 2115$ $T_{...2} = 1970$ $T_{....} = 4085$

Subject × B totals ($T_{ijk.}$)

		B_1	B_2	B_3	$T_{i1..}$			B_1	B_2	B_3	$T_{ij.2}$	$T_{i...}$
	S_{11}	156	93	86	335		S_{12}	138	112	108	358	693
	S_{21}	88	71	67	226		S_{22}	179	143	136	458	684
A_1	S_{31}	96	87	66	249	A_2	S_{32}	108	101	92	301	550
	S_{41}	138	108	92	338		S_{42}	124	115	108	347	685
	S_{51}	126	78	63	267		S_{52}	157	159	149	465	732
	S_{61}	121	81	64	266		S_{62}	164	155	156	475	741
$T_{.jk.} =$		725	518	438				870	785	749		

$T_{.1..} = 1681$ $T_{.2..} = 2404$ $T_{....} = 4085$

TABLE 9.12 SOURCES OF VARIANCE, DEGREES OF FREEDOM, AND CALCULATIONS FOR THE DESIGN OF TABLE 9.11

(a) Complete partitioning of the variability

SV	df	SS
Total	$abcn - 1$	$\Sigma\Sigma\Sigma\Sigma(Y_{ijkm} - \bar{Y}_{....})^2$
Between subjects (S)	$an - 1$	$bc\Sigma\Sigma(\bar{Y}_{ij..} - \bar{Y}_{....})^2$
A	$a - 1$	$bcn\Sigma(\bar{Y}_{.j..} - \bar{Y}_{....})^2$
S/A	$a(n - 1)$	$bc\Sigma\Sigma(\bar{Y}_{ij..} - \bar{Y}_{.j..})^2$
Within subjects (WSs)	$an(bc - 1)$	$\Sigma\Sigma\Sigma\Sigma(Y_{ijkm} - \bar{Y}_{ij..})^2$
B	$b - 1$	$acn\Sigma(\bar{Y}_{..k.} - \bar{Y}_{....})^2$
AB	$(a - 1)(b - 1)$	$cn\Sigma\Sigma(\bar{Y}_{.jk.} - \bar{Y}_{....})^2 - SS_A - SS_B$
SB/A	$a(n - 1)(b - 1)$	$c\Sigma\Sigma\Sigma(\bar{Y}_{ijk.} - \bar{Y}_{ij..} - \bar{Y}_{.jk.} + \bar{Y}_{.j..})^2$
C	$c - 1$	$abn\Sigma(\bar{Y}_{...m} - \bar{Y}_{....})^2$
AC	$(a - 1)(c - 1)$	$bn\Sigma\Sigma(\bar{Y}_{.j.m} - \bar{Y}_{....})^2 - SS_A - SS_C$
SC/A	$a(n - 1)(c - 1)$	$b\Sigma\Sigma\Sigma(\bar{Y}_{ij.m} - \bar{Y}_{ij..} - \bar{Y}_{.j.m} + \bar{Y}_{.j..})^2$
BC	$(b - 1)(c - 1)$	$an\Sigma\Sigma(\bar{Y}_{..km} - \bar{Y}_{....})^2 - SS_B - SS_C$
ABC	$(a - 1)(b - 1)(c - 1)$	$n\Sigma\Sigma\Sigma(\bar{Y}_{.jkm} - \bar{Y}_{....})^2 - SS_A - SS_B - SS_C - SS_{AB} - SS_{AC} - SS_{BC}$
SBC/A	$a(n - 1)(b - 1)(c - 1)$	$SS_{WSs} - SS_B - SS_{AB} - SS_C - SS_{AC} - SS_{BC} - SS_{ABC} - SS_{SB/A} - SS_{SC/A}$

(b) Computations of sums of squares for the data of Table 9.11

$$C = \frac{T^2_{....}}{abcn} = \frac{4085^2}{72} = 231{,}767.014$$

$$SS_{tot} = \Sigma\Sigma\Sigma\Sigma Y^2_{ijkm} - C = 80^2 + 48^2 + \cdots + 76^2 - C = 251{,}469 - C = 19{,}701.986$$

$$SS_S = \Sigma\Sigma\frac{T^2_{ij..}}{bc} - C = \frac{335^2 + 226^2 + \cdots + 475^2}{6} - C$$

$$= 245{,}396.5 - C = 13{,}629.486$$

$$SS_A = \Sigma\frac{T^2_{.j..}}{bcn} - C = \frac{1681^2 + 2404^2}{36} - C$$

$$= 239{,}027.139 - C = 7260.125$$

$$SS_{S/A} = SS_S - SS_A = 6369.361$$

$$SS_{WSs} = SS_{tot} - SS_S = 6072.5$$

$$SS_B = \sum \frac{T_{..k.}^2}{acn} - C = \frac{1595^2 + 1303^2 + 1187^2}{24} - C$$

$$= 235{,}450.125 - C = 3683.111$$

$$SS_{AB} = \sum\sum \frac{T_{.jk.}^2}{cn} - C - SS_A - SS_B = \frac{725^2 + \cdots + 749^2}{12} - C - SS_A - SS_B$$

$$= 243{,}326.584 - C - SS_A - SS_B = 616.333$$

$$SS_{SB/A} = \sum\sum\sum \frac{T_{ijk.}^2}{c} - C - SS_S - SS_B - SS_{AB}$$

$$= \frac{156^2 + 93^2 + \cdots + 156^2}{2} - C - SS_s - SS_B - SS_{AB}$$

$$= 2{,}501{,}707.5 - C - SS_S - SS_B - SS_{AB} = 1011.556$$

$$SS_C = \sum \frac{T_{...m}^2}{abn} - C = \frac{2115^2 + 1970^2}{36} - C$$

$$= 232{,}059.028 - C = 292.014$$

$$SS_{AC} = \sum\sum \frac{T_{.j.m}^2}{bn} - C - SS_A - SS_C = \frac{891^2 + \cdots + 1180^2}{18} - C - SS_A - SS_C$$

$$= 239{,}364.278 - C - SS_A - SS_C = 45.125$$

$$SS_{SC/A} = \frac{\sum\sum\sum T_{ij.m}^2}{b} - C - SS_S - SS_C - SS_{AC}$$

$$= \frac{173^2 + 162^2 + \cdots + 229^2}{3} - C - SS_S - SS_C - SS_{AC}$$

$$= 246{,}101.667 - C - SS_S - SS_C - SS_{AC} = 368.028$$

$$SS_{BC} = \sum\sum \frac{T_{..km}^2}{an} - C - SS_B - SS_C = \frac{823^2 + \cdots + 574^2}{12} - C - SS_B - SS_C$$

$$= 235{,}747.917 - C - SS_B - SS_C = 5.778$$

$$SS_{ABC} = \sum\sum\sum \frac{T_{.jkm}^2}{n} - C - SS_A - SS_B - SS_C - SS_{AB} - SS_{AC} - SS_{BC}$$

$$= \frac{382^2 + \cdots + 370^2}{6} - C - SS_A - SS_B - SS_C - SS_{AB} - SS_{AC} - SS_{BC}$$

$$= 243{,}676.5 - C - SS_A - SS_B - SS_C - SS_{AB} - SS_{AC} - SS_{BC} = 7.000$$

$$SS_{SBC/A} = SS_{W\,Ss} - SS_B - SS_{AB} - SS_{SB/A} - SS_C - SS_{AC} - SS_{SC/A} - SS_{BC} - SS_{ABC} = 83.556$$

TABLE 9.13 PART OF A BMDP2V ANOVA OF THE DATA OF TABLE 9.11

	SOURCE	SUM OF SQUARES	DEGREES OF FREEDOM	MEAN SQUARE	F	TAIL PROB.	GREENHOUSE GEISSER PROB.	HUYHN FELDT PROB.
	MEAN	231767.01389	1	231767.01389	363.88	.0000		
	A	7260.12500	1	7260.12500	11.40	.0070		
1	ERROR	6369.36111	10	636.93611				
	C	292.01389	1	292.01389	7.93	.0183		
	CA	45.12500	1	45.12500	1.23	.2941		
2	ERROR	368.02778	10	36.80278				
	B	3683.11111	2	1841.55556	36.41	.0000	.0000	.0000
	BA	616.33333	2	308.16667	6.09	.0086	.0270	.0210
3	ERROR	1011.55556	20	50.57778				
	CB	5.77778	2	2.88889	.69	.5124	.4604	.4844
	CBA	7.00000	2	3.50000	.84	.4473	.4089	.4268
4	ERROR	83.55556	20	4.17778				

ERROR TERM	EPSILON FACTORS FOR DEGREES OF FREEDOM ADJUSTMENT	
	GREENHOUSE-GEISSER	HUYNH-FELDT
3	.5756	.6676
4	.6627	.8014

TABLE 9.14 EXPECTED MEAN SQUARES FOR THE DESIGN OF TABLE 9.11

SV	df	EMS
A	$a - 1$	$\sigma_e^2 + bc\sigma_{S/A}^2 + bcn\theta_A^2$
S/A	$a(n - 1)$	$\sigma_e^2 + bc\sigma_{S/A}^2$
B	$b - 1$	$\sigma_e^2 + c\sigma_{SB/A}^2 + acn\theta_B^2$
AB	$(a - 1)(b - 1)$	$\sigma_e^2 + c\sigma_{SB/A}^2 + cn\theta_{AB}^2$
SB/A	$a(n - 1)(b - 1)$	$\sigma_e^2 + c\sigma_{SB/A}^2$
C	$c - 1$	$\sigma_e^2 + b\sigma_{SC/A}^2 + abn\theta_C^2$
AC	$(a - 1)(c - 1)$	$\sigma_e^2 + b\sigma_{SC/A}^2 + bn\theta_{AC}^2$
SC/A	$a(n - 1)(c - 1)$	$\sigma_e^2 + b\sigma_{SC/A}^2$
BC	$(b - 1)(c - 1)$	$\sigma_e^2 + \sigma_{SBC/A}^2 + an\theta_{BC}^2$
ABC	$(a - 1)(b - 1)(c - 1)$	$\sigma_e^2 + \sigma_{SBC/A}^2 + n\theta_{ABC}^2$
SB/A	$a(n - 1)(b - 1)(c - 1)$	$\sigma_e^2 + \sigma_{SBC/A}^2$

As usual, we need specific assumptions about the population parameters to be able to derive the EMS and to ensure that the ratios of mean squares testing the null hypotheses are distributed as F. It is assumed that A, B, and C are fixed-effect factors. It is further assumed that subjects is a random-effect variable and that $\eta_{i/j}$, $(\eta\beta)_{ik/j}$, $(\eta\gamma)_{im/j}$, $(\eta\beta\gamma)_{ikm/j}$, and ϵ_{ijkm} are randomly sampled from normally distributed populations with means equal to 0 and variances equal to σ_{SA}^2, $\sigma_{SB/A}^2$, $\sigma_{SC/A}^2$, $\sigma_{SBC/A}^2$ and σ_e^2, respectively.

9.8 CONCLUDING REMARKS

In mixed designs, the analyses for both between- and within-subjects factors are essentially the same as in the corresponding pure designs, with the same advantages, assumptions, and costs. Effects of within-subjects factors can potentially be tested with more precision because individual differences components are removed from the error terms. However, as in pure within-subjects designs, the possibility of subjects × treatments interactions is introduced for these factors as well as the possibility of violations of the sphericity assumption.

For within-subjects factors, if there are large subjects × treatments interactions, efficiency will be lowered, and if the sphericity assumption is violated, F tests may be positively biased unless the appropriate degree-of-freedom corrections are employed. Conservative F tests that adjust for this bias have been discussed in Chapter 8 and in the current chapter and are calculated automatically by several computer programs. It should be emphasized that there is no positive bias when testing the effects of between-subjects factors in a mixed design, nor when testing the effects of within-subjects factors that have only two levels.

EXERCISES

9.1 Consider the following data set:

		B_1	B_2	B_3
	S_{11}	23	16	12
A_1	S_{21}	27	1	14
	S_{31}	22	10	10
	S_{12}	25	16	19
A_2	S_{22}	17	33	22
	S_{32}	9	17	22

(a) Present the complete ANOVA table with all numerical results; assume A and B both have fixed-effects.

(b) In this chapter, we stated that $SS_{SB/A}$ was equivalent to the result of calculating the SB sum of squares separately for each level of A and then summing the a terms. To demonstrate this to yourself, ignore the A_2 data and calculate the sum of squares for SB at A_1 ($SS_{SB/A}$). Do the same thing at A_2 and check the sum of the two terms against $SS_{SB/A}$ calculated in part (a). Of course, $MS_{SB/A}$ is just the average of the two SB mean squares.

9.2 A_1 and A_2 in the preceding problem might have been two litters of three animals; in that case, we would expect A to have random effects. Assume B represents trials and is a fixed-effect variable.

(a) What are the expected mean squares for the various sources of variance?

(b) Recalculate any F ratios that are not the same as in Exercise 9.1.

9.3 Suppose A_1 and A_2 in Exercise 9.1 are subjects of two different ages and the levels of B correspond to three problems sampled randomly from some very large population of problems. Present the EMS under this model and recalculate F tests where necessary.

9.4 For the data of Table 9.1, (a) test the hypothesis that the population mean at B_2 is the same as the population mean at B_1; (b) test the hypothesis that the contrast mentioned in (a) is the same at all levels of A.

9.5 In a small-scale pilot study of the effects of diet upon the ability to withstand physical stress, nine volunteers were divided into three groups of three subjects and each given a different diet (D). They then underwent a battery of physical tests on each of four successive days (T for trials). Scores for each day were combined into a single score, with higher scores representing better performance. The results were

		T_1	T_2	T_3	T_4
	S_{11}	17	12	14	10
D_1	S_{21}	14	11	10	8
	S_{31}	18	13	12	10
	S_{12}	12	15	9	7
D_2	S_{22}	14	18	16	14
	S_{32}	10	12	12	9
	S_{13}	17	15	18	16
D_3	S_{23}	15	14	17	18
	S_{33}	15	15	16	14

(a) Carry out pairwise contrasts among the three diet means (each mean is an average over the four trials) using Tukey's procedure for controlling EF at .10.

(b) After inspecting the data, the researcher notes that D_2 seems to yield poorer performance. Test the difference between the D_2 mean and the average of the other two means, taking into consideration the fact that the test is post hoc (EF = .10).

(c) Another researcher argues that the interesting questions concern the trends over trials. (i) Test for a linear component of the trial function [lin(B)]. (ii) Test whether diet affects the linear trend. Present a brief interpretation of the results of both tests.

9.6 We have two groups (A_1, A_2) of four subjects each. Each subject is tested six times, with two levels of B and three levels of C crossing to form a 2×3 within subjects. The data are

Level of A	B_1			B_2		
	C_1	C_2	C_3	C_1	C_2	C_3
1	3	5	6	7	10	8
1	4	5	7	9	12	11
1	2	4	5	8	10	10
1	3	3	8	4	7	5
2	2	3	6	16	15	18
2	4	2	5	12	14	14
2	7	3	8	18	12	15
2	2	4	4	20	19	19

(a) Carry out an ANOVA on these data, assuming all main and interaction effects involving A, B, and C are fixed.

(b) Suppose C represents randomly chosen items. How will your analysis (F tests, EMS) differ from that in part (a)? Provide formulas (in terms of mean squares) for the F tests of the A, B, and AB effects, and the df for each test.

(c) Preliminary tests of interactions involving subjects or C may reveal an alternative to quasi-F tests. Explain why this is so and carry out this alternative procedure.

(d) Obtain a mean of the three scores at each level of B for each subject. You will have eight subjects with two "scores" each. Get SS, MS, and F ratios based on these mean "scores." Note that you have two levels of A and two levels of B, where B is a within-subjects variable; C no longer is part of the output. What is the relationship between the SS based on the mean of the three C scores and that obtained above? Does this analysis provide valid tests of A, B, and AB if C is fixed? if C is random? Explain briefly.

9.7 In a series of studies of cerebral hemispheric functioning, there were 72 subjects, half male and half female. Each sex (X) was further divided into three groups according to handedness (H): right-handed, not inverted; left, not inverted; left, inverted.

(a) On each of 20 trials, a dot appeared in each of the 20 possible positions of a 5×4 grid; the display was presented in the right visual field. On a second series of 20 trials, the display was presented in the left visual field. Location time and accuracy were recorded. The mean time for each visual field (V) was analyzed, providing two time scores per subject. Write out the SV, df, and EMS.

(b) In a second experiment, a different one of 20 words was presented in each location in the grid, once to the right and once to the left visual field. Thus, the subject saw 20 words twice. On each of these 40 trials, time to read the word aloud was recorded. Present the SV, df, and EMS for the ANOVA. What would be the F ratio, and its df, for the tests of X, H, and V, and their interactions?

9.8 In a study of the development of the concepts of conservation and weight, two standardized and quantifiable tasks (T, two levels) are presented to each child in the study. Mastering the first task requires that a child grasp the notion of conservation of quantity, whereas the second task depends on conservation of weight. The score is the number of trials required for the mastery of the task. Both age (A, two levels, 7 and 10 years) and gender (G, two levels) are included as major variables in the design. Finally, subjects are drawn from two school systems differing in the cultural (C, two levels) level of the community.

(a) Write down the sources of variance associated with the design of this study as well as the df's and EMS's (using numbers where possible).

(b) Will it be possible, after the study has been run, to obtain an unbiased estimate of ω_C^2 (the proportion of the total population variance attributable to the effects of C)? If so, *briefly* indicate how. If not, indicate why not, and specify conditions under which it would be possible.

9.9 An investigator is interested in the extent to which children are attentive to violent acts on television. She runs 120 subjects: 20 males and 20 females (sex, X) at each of three age levels (age, A). Each child views two scenes at each of three levels of violence (V, three levels), one involving animal cartoon characters and the other involving human characters (C, two levels). The dependent variable is a measure of attention during presentation of the scene.

(a) Write down the SV's, df's, and EMS's for this design.

(b) There are different designs for this type of study. Perhaps the children are available for only short periods of time, but the investigator has access to large numbers of subjects. How many subjects would be needed to have the same number of scores if each subject contributed only one score? What are the advantages and disadvantages of running the experiment as described, as opposed to using a between-subjects design?

Chapter **10**

Hierarchical Designs

10.1 INTRODUCTION

Now that we have examined the models and analyses that are appropriate when subjects are nested within levels of the variable *A* and are crossed with the levels of a second variable *B,* the pattern can be extended in two ways. One design has subjects nested within levels of a variable *G,* while the levels of *G* are in turn nested within the levels of *A.* Another involves nesting several within-subjects variables. We refer to these two kinds of design that involve nesting as *hierarchical,* referring to the hierarchy of variables that typifies them. The statistical models and computations are straightforward extensions of earlier developments.

The *S/G/A* designs have an important place in both psychological and educational research. To cite one example, consider a group dynamics experiment designed for studying the effects of stress on attitude change in the members of four-man conference teams. We might have several conference groups under high stress and an equal number under low stress; the resultant design might be characterized as *subjects within conference groups within levels of stress.* A similar example can be taken from educational psychology. Several first-grade classes are taught reading under one method, and an equal number are taught under a second method; all students are tested at the end of the term. The design might be characterized as *subjects within classes within methods.* Still another example might be taken from the animal laboratory. Different methods of rearing rats might be compared, with each method applied to several different litters. We might characterize this design as *subjects within litters within methods of rearing.* In all these examples, there are three potential sources of variability among the means computed for the levels of *A.* First, there may be a treatment effect in the sampled population. Second, differences in the compositions of the groups nested within the different levels of *A* may be a factor. Third, variability among treatment-level means will partly reflect that there are different subjects at each level of *A.*

The primary new aspect of the hierarchical design is the assumption that social, environmental, and genetic units are a source of variability separable from error variability. Even though the same experimental treatment is applied to two individuals in different social groups (or school classes or litters), their scores will differ, not only because they are different individuals but because they are subject to different social interactions (or genetic contributions). Once the possibility of such group effects are recognized, they must be incorporated within the structural model and, therefore, the ANOVA.

Nesting of within-subject variables is also a common occurrence in psychological research. Subjects are frequently tested with several sets of stimuli. Experimenters may measure response time to rare words and to common words (words within frequency level), for example, or may obtain ratings of pictures depicting several forms of social interaction (pictures within social interaction levels). If stimuli, say words or pictures, are viewed as having random effects, and this will usually be appropriate, there will be several potential causes of the variability among the means of the different conditions in which stimuli are nested. First, there may be reliable effects of the conditions, as when word frequency affects response time. Second, means at the different treatment levels may vary due to chance variability among the stimuli selected for the experiment. Third, as in repeated-measurement designs generally, variability due to interactions of subjects with within-subject variables may contribute to variability among the obtained treatment means. Recognizing stimulus variability necessitates finding an error term that includes such variability. In many instances, it will be necessary to calculate quasi-F ratios to incorporate both stimuli-within-treatment and subjects × treatments variability in the error term.

The importance of studying hierarchical designs goes beyond application of the design. The presentation of these designs and their analyses should further an understanding of how to establish structural models and then translate these models into analysis of variance tables. Every design that the researcher may encounter cannot possibly be considered here. The material of this chapter should, however, help establish certain fundamental but widely applied principles of data analysis.

10.2 GROUPS WITHIN TREATMENTS

10.2.1 Partitioning Variance

Table 10.1 presents a data set from an experiment of the sort described in the introduction. In this example we have three ($g = 3$) discussion groups of four ($n = 4$) individuals at each of two ($a = 2$) educational levels. Thus there are 24 subjects (agn). The dependent variable Y is a measure of attitude obtained from each individual after a 1-hour discussion session on some topic. An individual score might be represented as Y_{ijk}; in general,

$$i = 1, 2, \ldots, n \quad \text{(number of subjects in a group)}$$

$$j = 1, 2, \ldots, g \quad \text{(number of groups within a treatment level)}$$

$$k = 1, 2, \ldots, a \quad \text{(number of treatment levels)}$$

The deviation of any individual's score may be viewed as the sum of the deviation of that score from the group mean and the deviation of the group mean from the grand mean:

$$Y_{ijk} - \bar{Y}_{...} = (Y_{ijk} - \bar{Y}_{.jk}) + (\bar{Y}_{.jk} - \bar{Y}_{...})$$

A further breakdown follows from the observation that the group means (the $\bar{Y}_{.jk}$) vary because they represent different groups within a level of A and different levels of A:

TABLE 10.1 DATA FOR A GROUPS-WITHIN-TREATMENTS DESIGN

	A_1				A_2		
	G_{11}	G_{21}	G_{31}		G_{12}	G_{22}	G_{32}
	10	7	21		27	7	20
	23	4	14		24	21	29
	11	18	10		15	26	27
	17	11	19		19	17	22
$T_{.j1} =$	61	40	64	$T_{.j2} =$	85	71	98
		$T_{..1} = 165$				$T_{..2} = 254$	

$$T_{...} = 419$$

$$SS_A = \frac{(254 - 165)^2}{(2)(12)} = 330.042$$

$$Y_{ijk} - \bar{Y}_{...} = (Y_{ijk} - \bar{Y}_{.jk}) + (\bar{Y}_{.jk} - \bar{Y}_{..k}) + (\bar{Y}_{..k} - \bar{Y}_{...})$$

The ANOVA follows directly from this partitioning of deviations about the grand mean. Summing and squaring both sides of the preceding equation (and noting that cross-product terms equal zero), we have

$$\sum_i \sum_j \sum_k (Y_{ijk} - \bar{Y}_{...})^2 = \sum_i \sum_j \sum_k (Y_{ijk} - \bar{Y}_{.jk})^2 + n \sum_j \sum_k (\bar{Y}_{.jk} - \bar{Y}_{..k})^2 \qquad (10.1)$$

$$+ ng \sum_k (\bar{Y}_{..k} - \bar{Y}_{...})^2$$

That is,

$$SS_{tot} = SS_{S/G/A} + SS_{G/A} + SS_A$$

The notation $S/G/A$ is read as "subjects within groups within levels of A," or more briefly, "within groups."

Table 10.2 presents the ANOVA table for the data of Table 10.1. The SV are consistent with the partitioning just presented. We have included the lines for the total and between groups (Between G) as an aid in calculating their components. However, only the last three lines are essential to the inference process. The df follow directly from the SV. For the nested term G/A, we can use the fact that the group variability is a composite of A and G/A; therefore,

$$df_{Between\ G} = df_A + df_{G/A}$$

$$ag - 1 = a - 1 + df_{G/A}$$

Then

$$df_{G/A} = (ag - 1) - (a - 1) = a(g - 1)$$

Alternatively, we note that the variability of the group means at each level of A is based on $g - 1$ df; pooling over the a levels of A gives $a(g - 1)$. The SS raw-score formulas, presented below the ANOVA table, are algebraically equivalent to the defining formulas in Equation 10.1.

Justification of the EMS requires specification of the underlying structural model that relates each score to population parameters. We do that next, first pausing to note that, given these EMS, the F ratios calculated in Table 10.2 meet the usual requirement of equality of numerator and denominator expectations under H_0.

10.2.2 The Analysis of Variance Model

In establishing the relation of Y_{ijk} to population parameters, we begin as we did when partitioning deviations about the mean; ignore the variable A and view the design as

TABLE 10.2   ANOVA OF THE DATA OF TABLE 10.1

SV	df	SS	MS	F	EMS	Error term
Total	23	1131.96				
Between groups	5	506.71				
A	1	330.04	330.04	7.47	$\sigma_e^2 + n\sigma_{G/A}^2 + gn\theta_A^2$	G/A
G/A	4	176.67	44.17	1.27	$\sigma_e^2 + n\sigma_{G/A}^2$	$S/G/A$
$S/G/A$ (within-groups)	18	625.25	34.74		σ_e^2	

		Calculations	

SV	df	SS
Total	$agn - 1 = (2)(3)(4) - 1 = 23$	$\sum_i \sum_j \sum_k Y_{ijk}^2 - C = 8477 - 7315.04$ $= 1131.96$
Between groups	$ag - 1 = (2)(3) - 1 = 5$	$\dfrac{\sum_j \sum_k T_{.jk}^2}{n} - C = \dfrac{61^2 + 40^2 + \cdots + 98^2}{4} - 7315.04$ $= 506.71$
A	$a - 1 = 1$	$\dfrac{\sum_k T_{..k}^2}{gn} - C = \dfrac{165^2 + 278^2}{(3)(4)} - 7315.04$ $= 330.04$
G/A	$a(g-1) = 2(3-1) = 4$	$SS_{\text{Between } G} - SS_A = 506.71 - 330.04$ $= 176.67$
$S/G/A$	$ag(n-1) = (2)(3)(4-1) = 18$	$SS_{S/G/A} = SS_{\text{tot}} - SS_{\text{Between } G}$ $= 1131.96 - 506.71 = 625.25$

Note: $C = T_{...}^2 / agn = (419)^2 / (2)(3)(4) = 7315.04$.

having one factor, G, with ag levels and n subjects at each level. In accord with the one-factor model (Chapter 4), this view suggests

$$Y_{ijk} = \mu + \gamma_{jk} + \epsilon_{ijk} \tag{10.2}$$

where $\gamma_{jk} = \mu_{jk} - \mu$, the overall effect of the jth group at the kth level of A, and $\epsilon_{ijk} = Y_{ijk} - \mu_{jk}$, the residual error component. Equation 10.2 disregards the possibility of an effect due to A_k. Presumably, group means differ not only because the groups have different compositions, but also because some groups are at one level of A and others are at a different level. This line of reasoning suggests that part of the γ_{jk} effect is due to α_k, the effect due to the level of A in which the group exists. Accordingly, we subtract the contribution of α_k:

$$\gamma_{jk} - \alpha_k = (\mu_{jk} - \mu) - (\mu_k - \mu)$$

$$= \mu_{jk} - \mu_k$$

$$= \gamma_{j/k}$$

where $\gamma_{j/k}$ is the pure effect of the jkth group, uninflated by any contribution due to α_k. We can now substitute in Equation 10.2 for γ_{jk}, obtaining

$$Y_{ijk} = \mu + \alpha_k + \gamma_{j/k} + \epsilon_{ijk} \tag{10.3}$$

Each score is contributed to by a treatment effect, a group effect, and a residual component reflecting error of measurement and individual differences.

A common error in analyzing group designs is the failure to consider group effects in the model. In this case, the analysis proceeds as though the design were a completely randomized one-factor design with gn subjects in each of a treatment groups. This failure to separate the γ component from ϵ may result in an inflated F ratio as will be shown in the next section.

To complete the presentation of the underlying theory, and to arrive at the EMS, one must consider the nature of the effects. It is assumed that the levels of A have been arbitrarily chosen by the experimenter, and consequently, that α_k is a fixed variable. Then, $\Sigma_k \alpha_k = 0$ and the variance component is defined as $\theta_A^2 = \Sigma_k^a \alpha_k^2/(a - 1)$. The group effect $\gamma_{j/k}$ is viewed as a random variable, since the groups are clearly a random sample from the population of all possible groups of size n that could be composed. As usual, ϵ_{ijk} is also a random variable. The $\gamma_{j/k}$ and ϵ_{ijk} are assumed to be sampled from normally distributed populations with mean zero, and respective variances $\sigma_{G/A}^2$ and σ_e^2.

Given these assumptions, the EMS of Table 10.2 can be derived by the methods of Chapter 4. More simply, we can use the rules of thumb presented in Chapter 9. Considering the A line first, set down σ_e^2 and the null hypothesis term θ_A^2. The latter term is multiplied by ng, the number of scores at each level of A. Because the subscript G/A includes the letter A, and the essential letter G represents a random-effect variable, $\sigma_{G/A}^2$ is included in the expectation. Its coefficient is n, the number of scores in each group. The remaining two lines should pose no problems. Note, however, that σ_e^2 combines both variance due to individual difference and error of measurement. Only when the design involves repeated measures on subjects is there a need to distinguish between these two variance components.

10.2.3 Pooling Group and Individual Variability

The ANOVA of Table 10.2 failed to yield a significant A effect. However, the test against an error mean square on 4 df is likely to have a very high Type 2 error rate. Admittedly, most experiments are likely to involve more groups, and therefore more error df, than our example. Nevertheless, although exaggerated in our example, the problem of a relatively low value of $\mathrm{df}_{G/A}$ is a common one in hierarchical designs.

One possible solution to the problem of low power of the proper F test is to assume a different structural model. If the experimenter assumes that the group variable does not contribute to the total variability, the effect of G/A is deleted from the model and $\sigma^2_{G/A}$ is deleted from the EMS of Table 10.2. Then $\mathrm{MS}_{G/A}$ and $\mathrm{MS}_{S/G/A}$ both estimate σ^2_e. It is therefore reasonable to pool them to obtain a single estimate of error variance. In the example of Table 10.2, pooling leads to a new ANOVA table:

SV	df	SS	MS	F
A	1	330.04	330.04	9.05
S/A	22	801.92	36.45	

where S/A is the pool of G/A and $S/G/A$; the mean square was obtained by dividing the pooled SS by the pooled df. The new F ratio is larger than that in Table 10.2. Even if it had been slightly smaller, the result might be significant because the criterion F (with $\mathrm{df}_{\mathrm{error}} = 22$) now is 4.30 (instead of 7.71). Clearly, pooling can be useful. However, it should not be mindlessly done with all data sets obtained using the hierarchical design. Recall that in Chapter 5 pooling interaction and within-cell terms involved the risk of an inflated Type 2 error rate in the test of the main effects of interest against the pooled term. In the design of Table 10.1, there is a risk of an inflated Type 1 error rate. To see why, consider the pooled term

$$\mathrm{MS}_{S/A} = \frac{\mathrm{SS}_{G/A} + \mathrm{SS}_{S/G/A}}{\mathrm{df}_{G/A} + \mathrm{df}_{S/G/A}}$$

$$= \frac{(\mathrm{df}_{G/A})(\mathrm{MS}_{G/A}) + (\mathrm{df}_{S/G/A})(\mathrm{MS}_{S/G/A})}{\mathrm{df}_{G/A} + \mathrm{df}_{S/G/A}}$$

Taking the expectations of the two sides yields

$$E(\mathrm{MS}_{S/A}) = \left(\frac{\mathrm{df}_{G/A}}{\mathrm{df}_{G/A} + \mathrm{df}_{S/G/A}}\right) E(\mathrm{MS}_{G/A}) + \left(\frac{\mathrm{df}_{S/G/A}}{\mathrm{df}_{G/A} + \mathrm{df}_{S/G/A}}\right) E(\mathrm{MS}_{S/G/A})$$

Thus, $E(\mathrm{MS}_{S/A})$ is a weighted average of $E(\mathrm{MS}_{G/A})$ and $E(\mathrm{MS}_{S/G/A})$, where the weights are proportions of df associated with the two terms. If we assume that the two expectations to the right of the equals sign are unequal, their average must lie between them. Thus, $E(\mathrm{MS}_{S/A})$ will be less than $E(\mathrm{MS}_{G/A})$, the larger of the two terms being averaged. But this means that the pooled error term has a smaller expectation than the error term that is generally appropriate for testing the A main effect (see Table 10.2). If the two terms being pooled do not have identical expectations—that is, if $\sigma^2_{G/A} \neq 0$—the F test of A against the pooled error term will be positively biased; under H_0 ($\theta^2_A = 0$), the EMS ratio will be greater than 1. Furthermore, the ratio of mean squares will not have an F distribution over replications of the experiment, since under H_0, we get $E(MS_A) \neq E(MS_{S/A})$.

In Chapter 5 we recommend that pooling be done only when (1) there are prior grounds for believing that the pooled mean squares estimate the same population variance, and (2) the preliminary test is not significant at the .25 level. This rule of thumb is based on computer simulation studies by Bozivich, Bancroft, and Hartley (1956) and Srivastava and Bozivich (1961). These studies employed exactly the design and model of this section. Thus, the .25 rule for a preliminary test to decide whether to pool seems quite sensible here. Generally, groups will differ from each other and pooling will not be proper. Therefore, it is important to have as large a value of g as is practical in running these experiments so as to have a large value of $a(g - 1)$.

10.3 GROUPS VERSUS INDIVIDUALS

Many researchers in clinical, social, and educational psychology have been interested in assessing the effects of interpersonal interactions upon individual performance. For example, we might compare the effectiveness of group study with individual study. Assume that there are 15 students assigned to study a topic individually (individual condition, C_I) and another 15 students randomly assigned to five discussion groups of three students each (group condition, C_G). After the study session, all 30 students are tested individually on the subject matter studied. The data and group means are presented in panel a of Table 10.3. Although the subjects in C_I studied individually, we randomly partitioned their scores into groups of 3. This simplifies the notation, allowing us to denote each score by Y_{ijk}, $i = 1, 2, \ldots, n$, $j = 1, 2, \ldots, g$, and $k = 1, 2, \ldots, c$. In the example, $n = 3$, $g = 5$, and $c = 2$ (I or G).

The sum of squares for the individual-versus-group condition (SS_C) is calculated in panel a of Table 10.3, below the data. The denominator against which this is tested has usually been calculated as $MS_{S/C}$. In the data set of Table 10.3, this means calculating the variance of the 15 scores in each condition and averaging the two variances. The result, here distributed on 28 df, is 8.37. To the extent that the group factor in condition C_G contributes to the variance of the 15 scores in that condition, the resulting F test of C will be positively biased. Myers, DiCecco, and Lorch (1981) have shown that this inflation in Type 1 error rate can be quite marked and increases with the number of scores. Using expected mean squares, they justified two methods of analysis that yield unbiased F tests. We present these methods here. The calculations are in Table 10.3.

The Pseudogroup Procedure

We carry out the ANOVA as if the randomly constructed posthoc groups (the pseudogroups) in C_I were real groups and test C against G/C on 1 and $2(g - 1)$ df. The ratio of EMS is 1, assuming H_0, as we require of our F ratio. There is heterogeneity of variance here because the pseudogroup means vary less than the real group means do, but Myers, DiCecco, and Lorch (1981) reported simulation data demonstrating that Type 1

TABLE 10.3 DATA FROM A GROUPS-VERSUS-INDIVIDUALS DESIGN AND TWO ANALYSES

(a) The data

Individual condition (C_I)

$Y_{ijl} =$	9 9 11	15 16 12	12 8 15	16 15 16	14 11 13	
$\bar{Y}_{.jl} =$	9.67	14.33	11.67	15.67	12.67	$\bar{Y}_{..I} = 12.8$

Group condition (C_G)

$Y_{ijG} =$	11 16 15	17 18 19	11 13 15	17 18 19	10 13 13	
$\bar{Y}_{.jG} =$	14	18	13	18	12	$\bar{Y}_{..G} = 15$

$$SS_C = \frac{(15 - 12.8)^2}{2/15} = 36.3$$

(b) Pseudogroup analysis

$SS_{G/C} = n\sum(\bar{Y}_{.jl} - \bar{Y}_{..I})^2 + n\sum(\bar{Y}_{.jG} - \bar{Y}_{..G})^2$

$\quad = 3[(9.67 - 12.8)^2 + \cdots + (12.67 - 12.8)^2 + (14 - 15)^2 + \cdots + (12 - 15)^2] = 161$

$df_{G/C} = 2(g - 1) = 8$

Then the pseudogroup F ratio, distributed on 1 and 8 df, is

$$F_{PG} = \frac{36.3}{161/8} = 1.80$$

(c) Quasi-F test

The mean square based on the scores in condition C_I is

$$MS_{S/C_I} = \frac{\Sigma\Sigma(Y_{ijl} - \bar{Y}_{..I})^2}{gn - 1} = \frac{(9 - 12.8)^2 + \cdots + (13 - 12.8)^2}{14} = 7.6$$

The mean square based on the group means in condition C_G is

$$MS_{G/C_G} = n\sum \frac{(\bar{Y}_{.jG} - \bar{Y}_{..G})^2}{g - 1}$$

$$= \frac{3[(14 - 15)^2 + \cdots + (12 - 15)^2]}{4} = (3)(8.0) = 24$$

Let N_I be the number of scores in C_I, N_G be the number of scores in C_G, and $N = N_I + N_G$; $N_I = N_G = 15$ in the example. Then the error mean square is

$$MS_{QF} = \frac{N_I}{N}(MS_{S/C_I}) + \frac{N_G}{N}(MS_{G/C_G})$$

$$= \frac{15}{30}(7.6) + \frac{15}{30}(24.0) = 15.8$$

The quasi-F statistic is

$$F_{QF} = \frac{MS_C}{MS_{QF}} = \frac{36.3}{15.8} = 2.30$$

The df associated with the error term, MS_{QF}, are

$$df_{QF} = \frac{MS_{QF}^2}{\left(\dfrac{N_I}{N}\right)^2 \left[\dfrac{MS_{G/C_G}^2}{g-1}\right] + \left(\dfrac{N_G}{N}\right)^2 \left[\dfrac{MS_{S/C_I}^2}{N_I-1}\right]}$$

$$= \frac{(15.8)^2}{(1/2)^2[(24^2/4 + 7.6^2/14)]} = 6.74$$

which we have rounded to 7 df.

---

error rates are little affected by this when the number of pseudogroups are equal to the number of experimental groups and are of the same size. This is consistent with our earlier observation (Chapters 3 and 4) that heterogeneity of variance is not a problem except when the numbers of observations vary over conditions.

The Quasi-*F* Procedure

This approach uses the quasi-*F* statistic, which was first presented in Chapter 8.[1] The general formulas for the error MS (MS_{QF}) and *df* are presented in Table 10.3. In words, the steps are:

1. Find the variance of the N_I (15 in the example) scores in C_I; this is MS_{S/C_I}.
2. Find *n* times the variance of the N_G (5 in the example) group means in C_G; this is MS_{G/C_G}.
3. The error term for the quasi-*F* test is MS_{QF}, a weighted average of MS_{S/C_I} and MS_{G/C_G}; the weighting is defined in Table 10.3.
4. As illustrated in Table 10.3, df_{QF} are calculated, rounded to the nearest integer, and used as the error df.

The pseudogroup and quasi-*F* tests are similar in their power to reject false null hypotheses, and both tests have approximately correct Type 1 error rates (Myers et al. 1981). The pseudogroup approach appears simpler and should be used when the N_I observations can be partitioned into a number of pseudogroups equal to the number of real groups. The quasi-*F* approach is applicable regardless of the value of N_I. Also, Myers et al. (1981) have pointed out that the groups-versus-individuals design is a special case of designs in which group size is a variable; they have shown how the quasi-*F* statistic can be used to analyze data from the more general design.

---

[1] The statistic presented here is actually identical to the Brown-Forsythe statistic (*F**) presented in Chapter 4 as a way of testing effects in the presence of heterogeneity of variance.

10.4 EXTENSIONS OF THE GROUPS-WITHIN-TREATMENTS DESIGN

10.4.1 A Within-Group Variable

There are many possible extensions of the basic hierarchical design of Section 10.2. One possibility is to have subjects in the same group represent different levels of a variable. For example, suppose there are g conference groups at each of a levels of stress, giving a total of ag groups. Further suppose that the subjects within each conference group vary with respect to anxiety level as measured by the Taylor manifest anxiety scale. Then within each group there are b levels of B (anxiety level) with n subjects at each level. In such a design, there are bn subjects in a conference group, gbn subjects at a level of A (stress), and $abgn$ subjects in the entire study. The indices of notation for such a design are

$$i = 1, 2, \ldots, n \qquad k = 1, 2, \ldots, a$$
$$j = 1, 2, \ldots, g \qquad m = 1, 2, \ldots, b$$

Table 10.4 presents an artificial example of this design with $a = 3$, and $g = b = n = 2$.

TABLE 10.4   DATA MATRIX FOR A HIERARCHICAL DESIGN INCLUDING A WITHIN-GROUP VARIABLE

		Scores B_1		Scores B_2			Totals $T_{.jk1}$	$T_{.jk2}$	$T_{.jk.}$
A_1	G_{11}	4	6	8	14		10	22	32
	G_{21}	5	9	10	15		14	25	39
						$T_{..1m} =$	24	47	$T_{..1.} = 71$
A_2	G_{12}	3	10	12	15		13	27	40
	G_{22}	11	6	14	17		17	31	48
						$T_{..2m} =$	30	58	$T_{..2.} = 88$
A_3	G_{13}	20	23	19	26		43	45	88
	G_{23}	18	17	24	22		35	46	81
						$T_{..3m} =$	78	91	$T_{..3.} = 169$
						$T_{...m} =$	132	196	$T_{....} = 328$

Whether we analyze the data using a calculator or a statistical computer package, we need to be very clear about the structure of the design. It may help to realize that the layout of Table 10.4 is very similar to that for the mixed design of Chapter 9. Rather than partitioning SS_{tot} into between- and within-subjects components, we divide it into between- and within-groups components. Part of the variability among the ag group means may be due to variation among groups with respect to the level of A. This accounts for the partitioning of $SS_{Between\,G}$ (between-groups sum of squares) into A and G/A sources in Table 10.5. A potential source of the variability among the bn scores within each group is the factor B. This is also reflected in Table 10.5. The AB and GB/A terms fall out naturally as a result of crossing each between-group term with B. Finally, we have the variability of the n scores within each GB combination within each level of A ($S/GB/A$).

The df follow readily from the SV. All but two of these should be very familiar by now. The df for GB/A ($df_{GB/A}$) make sense if we understand that this SV represents the interaction of g groups and b levels of B, pooled over the a levels of A; that is, $df_{GB/A} = a \times (g - 1)(b - 1)$. Similarly, $df_{S/GB/A}$ represents the variability of n scores in each combination of G and B within each level of A, or $gba \times (n - 1)$.

Sums of squares can be computed by hand for a small data set such as this one. Because the student may feel more comfortable working through our data set on a calculator, we have presented computing, or raw-score, formulas beneath the ANOVA in Table 10.5. Generally, however, the amount of data and the complexity of the design, together with the possibility of doing supplementary analyses, will warrant a computer analysis. Several packages expedite the analysis of hierarchical designs. BMDP8V is one such package; the user must specify the variables in a "model" statement and also indicate which are fixed and which are random. In return, the program selects the correct error term, calculates the F ratios, and gives EMS (which are very useful for creating quasi-F ratios in some designs) and numerical estimates of the variance components (the σ^2 and θ^2 terms). The BMDP manual should be consulted for further details.

The F tests of Table 10.5 follow directly from the EMS presented there. They in turn are derived from a model consistent with the partitioning we have presented:

$$Y_{ijkm} = \mu + \alpha_k + \beta_m + \gamma_{m/k} + (\alpha\beta)_{km} + (\gamma\beta)_{jm/k} + \epsilon_{ijkm} \tag{10.4}$$

Note that there are no interactions involving subjects, nor is there any AG or ABG effect. This is because subjects cross with none of the three other variables, and G does not cross with A. Note, as a further help in establishing models for designs involving nesting, that the interaction of a nested effect with another variable will also be nested. For example, the interaction of G/A with B is the nested interaction GB/A.

It is assumed that A and B are fixed-effect variables, but that the groups are a random sample from a large population of such groups. Consequently,

$$\sum_k \alpha_k = \sum_m \beta_m = \sum_k \sum_m (\alpha\beta)_{km} = 0$$

TABLE 10.5 ANOVA AND CALCULATIONS OF SUMS OF SQUARES FOR THE DATA OF TABLE 10.4

SV	df	SS	MS	F	Error term	EMS
Between G	5	705.83				
A	2	685.58	342.79	50.78^a	G/A	$\sigma_e^2 + bn\sigma_{G/A}^2 + bgn\theta_A^2$
G/A	3	20.25	6.75			$\sigma_e^2 + bn\sigma_{G/A}^2$
Within G	18	313.50				
B	1	170.67	170.67	69.45^a	GB/A	$\sigma_e^2 + n\sigma_{GB/A}^2 + agn\theta_B^2$
AB	2	14.58	14.58	2.13	GB/A	$\sigma_e^2 + n\sigma_{GB/A}^2 + gn\theta_{AB}^2$
GB/A	3	10.25	3.42	.35	$S/GB/A$	$\sigma_e^2 + n\sigma_{GB/A}^2$
$S/GB/A$	12	118.00	9.83			σ_e^2
Total	23	1019.33				

$^a p < .05.$

Calculations

$$C = \frac{T_{....}^2}{abgn} = \frac{328^2}{24} = 4482.667$$

$$SS_{tot} = \sum_i^n \sum_j^a \sum_k^g \sum_m^b Y_{ijkm}^2 - C = 4^2 + 5^2 + \cdots + 22^2 - C$$

$$= 5502.000 - 4482.667 = 1019.333$$

$$SS_{Between\,G} = \sum_j^g \sum_k^a \frac{T_{.jk.}^2}{bn} - C = \frac{32^2 + 39^2 + \cdots + 81^2}{4} - C$$

$$= 5188.5 - 4482.667 = 705.833$$

$$SS_A = \sum_j^a \frac{T_{.j..}^2}{bgn} - C = \frac{71^2 + 88^2 + 169^2}{8} - C$$

$$= 5168.25 - 4482.667 = 685.583$$

$$SS_{G/A} = SS_{Between\,G} - SS_A = 20.250$$

$$SS_{Within\,G} = SS_{tot} - SS_{Between\,G} = 313.500$$

$$SS_B = \sum_m^b \frac{T_{...m}^2}{agn} - C = \frac{132^2 + 196^2}{24} - C$$

$$= 4653.333 - 4482.667 = 170.667$$

$$SS_{AB} = \sum_j^a \sum_m^b \frac{T_{.j.m}^2}{gn} - C - SS_A - SS_B$$

$$= \frac{24^2 + 47^2 + \cdots + 91^2}{4} - S_C - SS_A - SS_B$$

$$= 5353.500 - 4482.667 - 685.583 - 170.667 = 14.583$$

$$SS_{GB/A} = \sum_{j}^{g}\sum_{k}^{a}\sum_{m}^{b} \frac{T^2_{.jkm}}{n} - C - SS_{\text{Between }G} - SS_B - SS_{AB}$$

$$= \frac{10^2 + 22^2 + \cdots + 46^2}{2} - 4482.667 - 705.833 - 170.667 - 14.583$$

$$= 5384 - 4482.667 - 705.833 - 170.667 - 14.583 = 10.250$$

$$SS_{S/GB/A} = SS_{\text{Within }G} - SS_B - SS_{AB} - SS_{GB/A} = 118.000$$

The variance components for the fixed effects are

$$\theta^2_A = \frac{\sum_k \alpha^2_k}{a - 1} \qquad \theta^2_B = \frac{\sum_m \beta^2_m}{b - 1}$$

$$\theta^2_{AB} = \frac{\sum_k \sum_m (\alpha\beta)^2_{km}}{(a - 1)(b - 1)}$$

The terms $\gamma_{j/k}$, $(\gamma\beta)_{jm/k}$, and ϵ_{ijkm} comprise random samples from normally distributed populations with mean zero, and respective variances $\sigma^2_{G/A}$, $\sigma^2_{GB/A}$, and σ^2_e.

10.4.2 Repeated Measurements in Groups-Within-Treatments Designs

The design just presented can be extended by requiring several measures from each subject. For example, suppose we again have g groups under high stress A_1 and g groups under low stress A_2. Within each group there are n high-anxious subjects B_1 and n low-anxious subjects B_2. Each member of the group is tested on each of four trials; *trials* is the within-subject variable we shall label C in this section. We have g groups, generally, sampled randomly from a large population of such groups, at each of a levels of the independent variable A. Within each group are b arbitrarily chosen levels of the independent variable B; there are n different subjects at each of these levels. Thus, we have bn subjects in each of ag groups for a total of $abgn$ subjects, each providing one score at each of c levels of C. The analysis of variance is a direct extension of that presented in Table 10.5. The total variation of that table is now the between S variation; we add a within-S source on $abgn(c - 1)$ df. Note that $df_{\text{Between }S}$ and $df_{\text{Within }S}$ sum to $abcgn - 1$, the appropriate df_{total} for the design under consideration. To partition the within-S variability, first write C on $c - 1$ df. The remaining SV are generated by crossing C with each of the between-S sources of Table 10.5:

SV	df
C	$c - 1$
AC	$(a - 1)(c - 1)$
GC/A	$a(g - 1)(c - 1)$
$\cdots$	
$SC/GB/A$	$abg(n - 1)(c - 1)$

Note the general form of interactions involving nested variables. The interaction of G/A and C, for example, is represented by GC/A, not G/AC; the df also suggest that within each of a levels of A, we have a nested interaction on $(g - 1)(c - 1)$ df. The SS and EMS follow from the preceding developments.

10.5 NESTING WITHIN-SUBJECTS VARIABLES

10.5.1 The Design

In the designs of previous chapters, subjects were nested within levels of some independent variable A or within combinations of levels of several factors. In the designs of preceding sections of this chapter, subjects were nested within levels of a variable G, which in turn was nested within levels of A. We now turn to designs in which nesting is within subjects. In such experiments, the subject is faced with a randomly ordered set of stimuli that is nested within levels of some variable. For example, suppose that a researcher is interested in effects of problem difficulty on time to solution. Each subject is required to solve 12 problems, 4 of which are easy, 4 of intermediate difficulty, and 4 difficult. In this example, problems are nested within difficulty levels. The study is further complicated by having subjects nested within levels of experience; five subjects are novices and five are experts. The data from such a study are presented in Table 10.6.

The study just described is a specific example of a general design in which there are n subjects at each of a levels of A for a total of an subjects. Each subject is tested with b different stimuli at each of c levels of C; in the foregoing example, $b = 4$ (problems) and $c = 3$ (difficulty levels). This is very similar to the design of Section 9.7.2 except that B and C crossed there. Here, B is nested within levels of C (B/C). We assume that the effects of subjects (S/A) and stimuli (B/C) are random and that the

TABLE 10.6   DATA FOR A WITHIN-SUBJECTS DESIGN WITH LEVELS OF B NESTED WITHIN LEVELS OF C

		C_1				C_2				C_3			
		B_{11}	B_{21}	B_{31}	B_{41}	B_{12}	B_{22}	B_{32}	B_{42}	B_{13}	B_{23}	B_{33}	B_{43}
	S_{11}	4.4	5.2	5.8	4.5	5.5	5.7	4.2	4.8	5.4	4.2	5.6	5.6
	S_{21}	4.4	6.0	6.1	4.3	5.7	5.3	4.0	6.4	5.5	3.7	5.2	6.7
A_1	S_{31}	5.0	5.9	6.3	4.8	5.3	6.8	4.6	6.2	6.3	5.1	4.9	5.3
	S_{41}	4.4	5.0	5.5	3.4	5.2	5.2	3.5	4.9	5.8	4.6	5.4	4.5
	S_{51}	4.8	5.5	5.3	5.4	4.9	5.1	4.0	5.2	5.1	4.5	5.2	5.1
	S_{12}	5.0	7.8	9.8	6.0	8.4	8.7	5.9	8.4	8.9	5.1	7.9	9.5
	S_{22}	3.9	6.6	8.6	5.2	6.1	8.3	4.9	6.1	7.6	3.9	7.7	9.0
A_2	S_{32}	3.9	5.6	6.5	3.6	6.5	7.2	4.1	6.2	7.6	2.8	5.6	7.1
	S_{42}	6.1	7.1	8.3	5.7	6.9	9.1	5.2	6.5	9.5	6.4	7.5	8.9
	S_{52}	6.0	7.8	9.1	6.6	7.9	9.9	5.9	8.0	8.8	5.4	7.0	9.2

effects of A and B are fixed. In what follows, we shall try to develop an approach that is general enough to let us deal with design variations of any degree of complexity.

10.5.2 Partitioning the Total Variability

Our first concern is to determine SV and df for a general partitioning of the total variability. We do this in the following series of steps.

Step 1. Partition the total variability into two main sources, between-subjects and within-subject variability:

SV	df
Total	$abcn - 1$
Between S	$an - 1$
Within S	$an(bc - 1)$

Step 2. Further partition the between-subjects variability:

SV	df
Between S	$an - 1$
A	$a - 1$
S/A	$a(n - 1)$

Step 3. We make a first try at partitioning the within-subject variability by viewing the design as involving bc levels of stimuli:

SV	df
Within S	$an(bc - 1)$
Stimuli	$bc - 1$
$A \times$ stimuli	$(a - 1)(bc - 1)$
$S \times$ stimuli/A	$a(n - 1)(bc - 1)$

Note that once we have written *stimuli,* we merely cross it with the sources generated in step 2.

Step 4. We now must partition the variability due to stimuli and its interactions:

SV	df
Stimuli	$bc - 1$
C	$c - 1$
B/C	$c(b - 1)$

Crossing each of the above with A yields

SV	df
$A \times$ Stimuli	$(a-1)(bc-1)$
AC	$(a-1)(c-1)$
AB/C	$c(a-1)(b-1)$

Crossing C and B/C with S/A yields

SV	df
$S \times$ Stimuli/A	$a(n-1)(bc-1)$
SC/A	$a(n-1)(c-1)$
SB/AC	$ac(n-1)(b-1)$

Note that the interaction of S/A and B/C is SB/AC. This source corresponds to the *subject* × *stimuli SS* computed within each AC cell and then pooled over cells: note the correspondence between this verbal statement and the df.

In summary, we always first partition variability into between-subjects and within-subjects terms. We then partition the between-subjects variability, following the lines developed in Chapter 5 and in Sections 10.2–10.4. In partitioning within-subjects variability, we begin with our smallest experimental units, for example stimuli. The variability among those units is then further partitioned and interactions with between-subjects sources are then noted.

A useful check on the SV is to begin by listing all factors, both random and fixed; be sure to note any nesting. We have A, S/A, C, B/C. Now consider all cross products of the factors. We write "no" next to the cross product if no legal SV can be formed by crossing the two quantities; for example, if S is nested in A, it cannot also cross with it.

$A \times S/A$	No
$A \times C$	AC
$A \times B/C$	AB/C
$S/A \times C$	SC/A
$S/A \times B/C$	SB/AC
$C \times B/C$	No

The four factors we started with plus the four more cross-product terms generated above are the terms of interest in the ANOVA of this design. We should also consider crossing more than two factors. No legal terms would follow in this design; for example, $A \times C \times B/C$ is not legal because B cannot be nested in C and still cross with it. The ability to specify factors is important because this information is required by computer packages that deal directly with nested variables. For example, instructions to BMDP8V for the design under consideration would include the statement

```
MODEL IS 'A, S(A), C, B(C)'.
```

The parentheses are used in place of our slash to indicate nesting.

10.5.3 The Analysis of Variance

Table 10.7 presents a portion of the output from an ANOVA performed by BMDP8V. An important, and unusual, aspect of this program is that it specifies the EMS. To understand how this is done, look at the EMS for the A source. Each number in parentheses indicates a variance component and the number adjacent to it represents the coefficient by which the component is multiplied. Thus

$$E(MS_A) = 60(2) + 12(4) + 5(8) + (9)$$

is interpreted as

$$E(MS_A) = 60\theta_A^2 + 12\sigma_{S/A}^2 + 5\sigma_{AB/C}^2 + \sigma_{SB/AC}^2$$

The error variance, σ_e^2, is an implicit component of all EMS.

There is no F reported for A, C, or AC because no mean square in the table meets the criterion for a proper error term. For example, the error term for the A source should have the expectation $12(4) + 5(8) + (9)$; no single source in Table 10.7 has this expectation. We can construct quasi-F tests, however, by combining mean squares. Table 10.8 presents formulas and numerical results for two possible quasi-F tests of the A main effect. The tests of C and AC are left as an exercise for the reader.

The BMDP8V output in Table 10.7 also contains estimates of components of variance. These are readily obtained by

$$\text{estimate} = \frac{MS_{\text{numerator}} - MS_{\text{error}}}{\text{coefficient}}$$

TABLE 10.7   BMDP8V ANALYSIS OF DATA OF TABLE 10.6

SOURCE	ERROR TERM	SUM OF SQUARES	D.F.	MEAN SQUARE	F	PROB.	EXPECTED MEAN SQUARE
1 MEAN		4356.07500	1	4356.0750			120(1) + 12(4) + 10(5) + (9)
2 A		94.69633	1	94.6963			60(2) + 12(4) + 5(8) + (9)
3 C		4.11350	2	2.0567			40(3) + 10(5) + 4(7) + (9)
4 S(A)	SB(AC)	41.76033	8	5.2200	20.10	0.0000	12(4) + (9)
5 B(C)	SB(AC)	111.46950	9	12.3855	47.70	0.0000	10(5) + (9)
6 AC		2.80417	2	1.4021			20(6) + 4(7) + 5(8) + (9)
7 SC(A)	SB(AC)	5.00067	16	0.3125	1.20	0.2866	4(7) + (9)
8 AB(C)	SB(AC)	25.26550	9	2.8073	10.81	0.0000	5(8) + (9)
9 SB(AC)		18.69500	72	0.2597			(9)

ESTIMATES OF VARIANCE COMPONENTS

(1)	36.15608
(2)	1.44881
(3)	−.25954
(4)	.41337
(5)	1.21258
(6)	−.07290
(7)	.01322
(8)	.50953
(9)	.25965

TABLE 10.8 QUASI-F TESTS FOR THE A SOURCE OF VARIANCE IN TABLE 10.7

These tests follow from the entries in the expected-mean-square column of Table 10.7.

$$F_1' = \frac{MS_A}{MS_{S/A} + MS_{AB/C} - MS_{SB/AC}}$$

$$= \frac{94.6963}{5.22 + 2.8073 - .2597}$$

$$= \frac{94.6963}{7.7676} = 12.19$$

$$df_A = 1$$

$$df_{error} = \frac{MS_{error}^2}{MS_{S/A}^2/df_{S/A} + MS_{AB/C}^2/df_{AB/C} + MS_{SB/AC}^2/df_{SB/AC}}$$

$$= \frac{7.7676^2}{5.22^2/8 + 2.8073^2/9 + .2597^2/72}$$

$$= \frac{60.3356}{4.2826} = 14.09 \ (\text{or } 14)$$

$$F_2' = \frac{MS_A + MS_{SB/AC}}{MS_{S/A} + MS_{AB/C}}$$

$$= \frac{94.6963 + .2597}{5.22 + 2.8073} = 11.83$$

$$df_A = \frac{(MS_A + MS_{SB/AC})^2}{MS_A^2/df_A + MS_{SB/AC}^2/df_{SB/AC}} = \frac{(94.6963 + .2597)^2}{94.6963^2/1 + .2597^2/72}$$

$$= \frac{9016.6419}{8967.39} = 1.0055 \ (\text{or } 1)$$

$$df_{error} = \frac{(MS_{S/A} + MS_{AB/C})^2}{MS_{S/A}^2/df_{S/A} + MS_{AB/C}^2/df_{AB/C}} = \frac{(5.22 + 2.8073)^2}{5.22^2/8 + 2.8073^2/9}$$

$$= \frac{64.4375}{4.2817} = 15.05 \ (\text{or } 15)$$

Both quasi-F's are significant at the .01 level. Note that they give very similar results; F_2' is slightly smaller than F_1' but has an error term with one more df.

where "coefficient" refers to the coefficient of the component being estimated. For example, the result in the BMDP output is

$$\hat{\theta}_A^2 = \frac{MS_A - (MS_{S/A} + MS_{AB/C} - MS_{SB/AC})}{bcn}$$

$$= \frac{94.6963 - 7.7676}{60} = 1.4488$$

10.6 CONCLUDING REMARKS

Although there are countless variations of the hierarchical design, all yield to the same principles of analysis that have been earlier applied repeatedly (Chapters 8 and 9). The first step is to have a sound understanding of the layout of the design. Which variables are nested in which others? Which variables cross each other? Which are between-subjects and which are within-subject variables? The answers to these questions direct the partitioning of total variability and df. To generate EMS, we must determine which variables are to be viewed as random, which as fixed. This is not always an easy decision; the answer depends on how levels of the variable have been selected and on the range of generalization we intend. Once variables have been classified, applying the rules of thumb (Section 9.3) readily yields the EMS.

The nature of the data analysis is particularly sensitive to the choice of a structural model. If certain variables are assumed to have negligible effects, it will frequently be possible to have tests of greater power and to avoid quasi-F ratios. Although such consequences of simplifying the model are desirable, we have espoused a more conservative approach to model construction, and therefore to data analysis. It is preferable to assume a general model, incorporating all the effects we can conceive. Sometimes prior information and preliminary tests of certain terms will let us delete certain parameters from the model and the corresponding variances from the EMS but assumptions alone are not sufficient grounds for such a procedure. Wishing some variance component to be zero will not make it so, and the price of wrongly assuming that the component is zero is ordinarily a Type 1 error in testing treatment effects of interest.

The presence of other random-effect variables besides subjects, a characteristic of the designs of this chapter, raises additional considerations in planning the experiment. One important point is that merely running many subjects will not ensure sufficient power to test null hypotheses of interest. In the designs of Sections 10.2–10.4, the value of g—the number of social groups, classes, litters, and so on—is the critical determinant of error df and thus of power. Unless there are grounds for pooling to obtain an error term on more df, there is little the experimenter can do after the data are collected. Thus it is important to work out the actual analysis of sources and df before collecting data and to modify the design in whatever ways seem necessary to obtain powerful tests of effects of interest. In the extreme case in which $g = 1$, there is not only a loss of power but a confounding of groups and levels of A. If one class is taught by one method and another by a second, is a difference in class means due to the different methods or to differences in the personal interactions within the two classes? To determine the effect of the treatment, we need some measure of variability among classes taught by the same method. Experimenters often do not realize that the failure to replicate groups within levels is not particularly different from running one subject at each level of A in a simple completely randomized one-factor design.

Similar comments hold for the designs of Section 10.5. When stimuli are a random sample from some population, considering test power requires that there be an adequate number of stimuli, or as in the case of the design of Table 10.6, an adequate number of sets of stimuli.

EXERCISES

Note: The designs are getting more complicated (and probably more typical of the real world). Therefore, it's important to understand them before attempting to figure out the analysis. In each of the following problems, indicate which variables are between-subjects variables and which are within-subjects variables; which ones are nested, and what they are nested in; and which variables have fixed effects and which have random effects. Also, a few of these problems will require quasi-F (F') tests of the effects of interest. When this is the case, present the test and the associated degrees of freedom.

10.1 A group of personality researchers hypothesized that the self-image children have is very much related to their socioeconomic background, but that this is less the case for males than for females. To examine this question, they selected three school districts (D), each representing a different social stratum, for inclusion in the study. Five sixth-grade classes were randomly sampled from each school district, and 10 students of each sex from each class were then randomly selected and asked to fill out a self-evaluation form. The researchers performed the following analysis:

SV	df
Total	299
District (D)	2
Sex (X)	1
$D \times X$	2
S/DX	294

  (a) Present an alternative ANOVA table (SV, df, EMS).
  (b) What inferences might be changed by doing the analysis this way? Why?
  (c) Under what conditions would the first analysis, presented in part (a), be justified?

10.2 A therapist meets with 12 groups (G) for an hour each week. Each group consists of three males and three females (sex, X). Six of the groups are engaged in a type of directed therapy, the other six in nondirected therapy (T). Self-ratings are collected after a year of therapy and analyzed. Present the SV, df, EMS, and F tests.

10.3 An educational psychologist divides the 240 students in an introductory psychology course into 40 six-person discussion groups (G). There are four graduate student discussion leaders (L), each responsible for 10 groups. Half of the groups for each leader are taught by a didactic method in which the leader lectures and responds to questions from the six group members; the other five groups for each leader are taught by an interactive method in which the leader is strictly a resource person, monitoring the discussion and speaking only when the group can go no further in discussing a problem. Call this variable method (M). The dependent variable is the score on the midterm.

Present SV, df, EMS, and F tests. (Note: A critical decision is whether to treat L as a fixed- or random-effect variable. Think carefully about this point.)

10.4 Six high schools (Sc) are chosen at random for an experiment testing the effectiveness of an educational software package that is to be used at three of the schools, but not at the other three; call this variable P (package). The 120 students in the study come from 12 classes (C), two from each school. The measures are the scores on two tests (T), a midterm and final given to each subject.

(a) Present the SV, df, EMS, and F tests.

(b) Redo your answer, assuming the design is changed so that the software package was used in one of the two classes at each of the six schools.

10.5 A list of 50 stimulus items to be used in a memory task is constructed in the following way: First of all five large pools of items are selected that differ with respect to meaningfulness (M). Then 10 items (I) are randomly selected from each pool. The list is presented to 20 subjects, so that we have a total of 1000 scores.

(a) Present SV, df, EMS, and F tests for the appropriate ANOVA.

(b) Present the ANOVA assuming that the list is presented to each subject on each of three days (D).

10.6 In the following study, the researchers wished to determine whether the amount of time taken to recognize that a string of letters was a real word was influenced by whether or not it was a homophone of a common word (homophones are strings of letters that are pronounced the same—e.g., "bare" and "bear" sound alike, as do "break" and "brake"). They selected 25 homophones and 25 nonhomophones, mixed them in with nonword filler items, and presented them one at a time to 44 subjects and measured the word/nonword recognition times. The layout of the design is as follows:

H_1 (homophones)	H_2 (nonhomophones)
Word$_1$ W_2 $\cdots$ W_{25}	W_{26} $\cdots$ W_{50}

S_1
S_2
.
.
.
S_{44}

The study therefore involved three factors: (1) homophony, H, consisting of two fixed categories, homophones and nonhomophones, (2) words, W, nested within levels of homophony, consisting of random samples of all possible homophones and nonhomophones, and (3) subjects, S, again considered to be a random sample.

(a) Present the SV, df, EMS, and F tests appropriate for this design.

(b) It is possible in this kind of design that if some of the other effects were small it might be possible to pool some of the terms so that the major factor of interest (H) might be tested with more power. Indicate what tests you might want to make to pursue this.

10.7 In Exercise 9.7, we described a study of cerebral hemispheric functioning. In that study, the investigators ran 72 subjects who differed with respect to sex (X) and handedness $(H,$ three levels). Each subject had to read 20 words presented once in each visual field. These results may be affected by the repetition of the same word. Therefore, we change the design and present 20 words in the right visual field and a different 20 in the left. Again, reading time is the dependent variable. Now present SV, df, EMS, F tests.

10.8 Thirty-two subjects are randomly assigned to eight four-person groups (G). Two types of instructions are used to orient the groups in different ways (O). Four of the groups are given task-oriented instructions and the other four is given ego-oriented instructions. Each group is required to solve six problems (P), three under stress and three under no stress (St). The score for each group on each problem is the number of trials it takes the group to solve the problem (note that it is the group that gets scores, not subjects). Present SV, df, EMS, and F tests.

10.9 A researcher wishes to study the effects of viewing violence on television upon the behavior of children. She has a large sample of adult viewers rate the level of violence of a large number of episodes randomly selected from typical Saturday morning cartoon presentations. She then chooses 15 episodes; 5 of these are viewed as representative of a population of low-violence episodes; 5 more are medium, and 5 more are high in violence. Thus there are three levels of violence (V) with five different episodes (E) at each level. She decides to use 6-, 8-, and 10-year-old subjects in order to examine the interaction of age (A) and violence. Each of several dependent measures will be subjected to ANOVA.

One way to run the study would be to minimize the number of subjects needed. We might use 10 subjects at each age level. Each child would be tested on each of 15 days with a different episode viewed on each day; the order of episodes would be random.

(a) Present SV, df, EMS, and F tests, including quasi-F's and their df where needed.

(b) We want power and simple tests of A, V, and AV. Describe various preliminary tests and their possible consequences for testing the A source of variance.

10.10 There are many alternative designs for the study in Exercise 10.9. Perhaps the children are available only for a single session, but we have access to a large supply of subjects. Then each child sees only one of the 15 episodes. We have the same total number of observations as before if we use 150 children at each age level with one score each. Answer parts (a) and (b) of Exercise 10.9, assuming this design.

10.11 The members of each of 10 groups of three monkeys have been raised together. In five of the groups, the group members were separated at six months of age; in the other five groups, separation occurred at one year. At age 2, all monkeys were tested on four problems in the presence of their original cohorts and on four other problems with no other monkeys present. Thus we have two levels of age (A) of separation, eight different problems (P), and test environment (E). We also have 30 monkeys who constitute 10 different groups.

(a) Present SV, df, EMS, and F tests for this design.

(b) If you carried out the analysis correctly, you required some quasi-F tests. There are alternative models that, if valid, would eliminate the need for quasi-F tests. What assumptions would be involved? How would they be tested? What error terms could be used if these preliminary tests supported the alternative model?

10.12 The design in Exercise 10.11 involved 240 observations. We could do the same study using 240 monkeys reared in 80 three-monkey groups. Half of these groups would be separated at six months of age, and half at one year. Of the 40 groups at A_1, the members of five would be tested on problem 1, the members of another five groups would be tested on problem 2, and so on. The same would be done with the 40 groups at A_2. Problems 1–4 are those tested in the social environment, and problems 5–8 are those tested in the individual environment.

(a) Present SV, df, EMS, and F tests.

(b) What are the pros and cons of the two designs presented? There are both statistical and practical considerations.

Chapter **11**

Latin Squares and Related Designs

11.1 INTRODUCTION

An important consideration in designing an experiment is the efficiency of the design; all other things being equal, we want to design our research to minimize the effects of chance factors. When chance variability is small, we have greater statistical power; that is, we are more likely to detect effects of the independent variable. One way to increase design efficiency was described in Section 5.10. There we showed that assigning subjects to blocks on the basis of a measure related to the dependent variable could greatly reduce σ_e^2 and thus increase the likelihood of detecting treatment effects if these exist in the population. The repeated-measurements design of Chapter 8 is still more efficient. This design permits further reduction in variance due to individual differences; because every subject experiences all conditions, differences among subjects do not contribute to differences among the means for the different conditions. In this chapter we consider a class of experimental designs that have the potential for still greater reductions in error variance. We will begin by considering a research example that should provide a sense of the design, its potential benefits, and some potential problems.

Suppose we wish to compare the relative effects of five different drug dosages upon the ability of monkeys to learn a discrimination. Our measure will be the number of correct responses in a block of 20 trials. Monkeys are expensive subjects, so a between-subjects design will not be used. In fact, we will use only five subjects, testing each under a different dosage on a different day. One way to do this is to select a different random order of the five dosages for each subject. If we use a random-number table or a computer-generated sequence of random numbers, the resulting five sequences of dosages might be

Subject	Days				
	1	2	3	4	5
1	A_3	A_2	A_1	A_4	A_5
2	A_2	A_1	A_3	A_5	A_4
3	A_3	A_2	A_4	A_1	A_5
4	A_4	A_5	A_2	A_3	A_1
5	A_2	A_3	A_5	A_4	A_1

where A_j is the jth dosage. Averaging over the five subjects, some dosages are presented relatively earlier in the sequence than others; A_2, for example, never appears later than the third day, whereas only subject 4 experiences A_5 before the third day. This may not have any effect on performance; on the other hand, boredom may slow down, or practice may speed up, responses late in the series of days. Because all dosages have an equal chance of presentation on each of the five days, no dosage has a systematic advantage over many replications of the experiment; therefore, the statistical test of A is unbiased. Nevertheless, this chance variability due to the day of presentation does reduce the power of the test to detect effects of A. In a more efficient design, the variability due to days would be removed from the data much as the variability due to subjects was removed in the repeated-measurements design. This could be done by having each dosage appear equally often on each of the five days. An example of this would be

Subject	Days				
	1	2	3	4	5
1	A_2	A_5	A_1	A_3	A_4
2	A_1	A_4	A_5	A_2	A_3
3	A_3	A_1	A_2	A_4	A_5
4	A_5	A_3	A_4	A_1	A_2
5	A_4	A_2	A_3	A_5	A_1

This design is called a *Latin square;* it is characterized by the fact that each level of A appears exactly once in each row and column.

Because each subject and each level of A is represented exactly once in each column of the Latin-square design, variability due to columns (days in this example) can be extracted from the total sum of squares. The Latin-square design allows us to remove variability due to two sources of error, subjects and days in our example. In contrast, in a subjects $\times$ treatments design, only error variance due to subjects is removed; random variability due to days contributes to σ_e^2. Because the error variance is potentially smaller in the Latin square than in the subjects $\times$ treatments design, the Latin square is potentially a more efficient design.

We say "potentially a more efficient design" because there are also some potential problems that do not arise in the subjects $\times$ treatments design. The problems arise because the Latin square is what is often called an incomplete block design. This means that each subject (block) in our example is not tested under the complete set of 25 possible combinations of A and days. Because of this, we do not have enough total degrees of freedom to perform independent tests of the A, subjects, and days effects, and their interactions. For example, there are 24 $(25 - 1)$ df available for assessing the variability among the cell means. Twelve $[3 \times (5 - 1)]$ of these are associated with the main effects of subjects, days, and drugs. This leaves only 12 df to account for all the possible interactions, any one of which requires at least 16 $[(5 - 1) \times (5-1)]$ df. Our inability to calculate interaction sums of squares poses no problem if row, column, and treatment effects are *additive*—that is, if there are no interactions among these variables in the population. However, if such interactions are present in the population, they may affect our conclusions. We will consider this issue at several points in this chapter.

A necessary step in using the experimental designs of this chapter is the selection of the Latin square. We begin by demonstrating how this is done. Then we proceed to consider the data analysis. Finally, we will describe some modifications and extensions of the basic design.

11.2 SELECTING A LATIN SQUARE

The expected mean squares we will present in this chapter are based on the assumption that the Latin square used in the experiment has been randomly selected from the population of possible squares. In this section we consider procedures for selecting a square so that the assumption will be justified. This is straightforward for 2×2 squares; only two such squares are possible. There are 12 possible 3×3 squares. We could enumerate these and select a random number from 1 to 12 to choose the one for our experiment. However, it is simpler to obtain a square for an experiment by permuting the rows and columns of a *standard square*—that is, a square whose first row and first column is in the standard order: $< A_1, A_2, A_3 >$. The only possible 3×3 standard square is

$$\begin{bmatrix} A_1 & A_2 & A_3 \\ A_2 & A_3 & A_1 \\ A_3 & A_1 & A_2 \end{bmatrix}$$

To construct a random member of the possible set of 12 squares, begin by permuting all rows but the first. To do this, draw the numbers 2 and 3 in random order and reorder the rows accordingly; for example, if the sequence is $< 3,2 >$, the new square is

$$\begin{bmatrix} A_1 & A_2 & A_3 \\ A_3 & A_1 & A_2 \\ A_2 & A_3 & A_1 \end{bmatrix}$$

Now draw the numbers 1, 2, and 3 in random order. This time permute the columns. We might have $< 3, 1, 2 >$, yielding

$$\begin{bmatrix} A_3 & A_1 & A_2 \\ A_2 & A_3 & A_1 \\ A_1 & A_2 & A_3 \end{bmatrix}$$

Note that the row permutation stage involves two possible orders of rows; this combines with six possible orders of columns to generate the entire population of 12 possible squares.

There are four 4×4 standard squares:

$$\begin{bmatrix} A_1 & A_2 & A_3 & A_4 \\ A_2 & A_3 & A_4 & A_1 \\ A_3 & A_4 & A_1 & A_2 \\ A_4 & A_1 & A_2 & A_3 \end{bmatrix} \qquad \begin{bmatrix} A_1 & A_2 & A_3 & A_4 \\ A_2 & A_4 & A_1 & A_3 \\ A_3 & A_1 & A_4 & A_2 \\ A_4 & A_3 & A_2 & A_1 \end{bmatrix}$$

$$\begin{bmatrix} A_1 & A_2 & A_3 & A_4 \\ A_2 & A_1 & A_4 & A_3 \\ A_3 & A_4 & A_1 & A_2 \\ A_4 & A_3 & A_2 & A_1 \end{bmatrix} \qquad \begin{bmatrix} A_1 & A_2 & A_3 & A_4 \\ A_2 & A_1 & A_4 & A_3 \\ A_3 & A_4 & A_2 & A_1 \\ A_4 & A_3 & A_1 & A_2 \end{bmatrix}$$

Select a number at random from 1 to 4 and begin with the corresponding standard square. Then permute all rows except the first and all columns, as we did in the example of the 3×3 square. Note that this procedure generates 3!4! possible squares for each of the four standard squares. Therefore, we are selecting one square at random from the population of 576 ($4 \times 3! \times 4!$) squares.

The number of standard squares increases rapidly as a, the number of treatment levels, increases. Therefore, the procedure used for $a = 4$ is impractical for larger squares. A reasonable approach is to arbitrarily select a standard square, permute all rows, then all columns, and finally all letters. We did this to arrive at the 5×5 square presented in Section 11.1. We start with the standard square

$$\begin{bmatrix} A_1 & A_2 & A_3 & A_4 & A_5 \\ A_2 & A_4 & A_5 & A_3 & A_1 \\ A_3 & A_1 & A_2 & A_5 & A_4 \\ A_4 & A_5 & A_1 & A_2 & A_3 \\ A_5 & A_3 & A_4 & A_1 & A_2 \end{bmatrix}$$

A table of random numbers yields the values $< 2, 4, 3, 1, 5 >$; we permute the rows accordingly:

$$\begin{bmatrix} A_2 & A_4 & A_5 & A_3 & A_1 \\ A_4 & A_5 & A_1 & A_2 & A_3 \\ A_3 & A_1 & A_2 & A_5 & A_4 \\ A_1 & A_2 & A_3 & A_4 & A_5 \\ A_5 & A_3 & A_4 & A_1 & A_2 \end{bmatrix}$$

We turn again to the random-number table and this time get $< 4, 1, 2, 5, 3 >$, resulting in the following column permutation:

$$\begin{bmatrix} A_3 & A_2 & A_4 & A_1 & A_5 \\ A_2 & A_4 & A_5 & A_3 & A_1 \\ A_5 & A_3 & A_1 & A_4 & A_2 \\ A_4 & A_1 & A_2 & A_5 & A_3 \\ A_1 & A_5 & A_3 & A_2 & A_4 \end{bmatrix}$$

Draw one more set of random numbers; this time we have $< 4, 2, 5, 1, 3 >$. We will replace the A_1's in the preceding square by the A_4's, the A_2's will be in the same cells, the A_3's will be replaced by the A_5's, and so on. The square to be used in our experiment is

$$\begin{bmatrix} A_5 & A_2 & A_1 & A_4 & A_3 \\ A_2 & A_1 & A_3 & A_5 & A_4 \\ A_3 & A_5 & A_4 & A_1 & A_2 \\ A_1 & A_4 & A_2 & A_3 & A_5 \\ A_4 & A_3 & A_5 & A_2 & A_1 \end{bmatrix}$$

This procedure should be used whenever there are five or more treatment levels. Although not all squares have an equal opportunity to be sampled, the approach adequately approximates random sampling from the complete set of squares of size $a \times a$.

11.3 THE SINGLE LATIN SQUARE

We begin our discussion of the data analysis using the above layout and the example introduced in Section 11.1. Recall that each level of A is a drug dosage, each row is a subject, and the columns are days. This single Latin square is the simplest use of the Latin-square principle; extensions including running more than one subject through each sequence of treatments and using more than one $a \times a$ square. When we use only a single square with one subject in each sequence, the square should be at least 5×5. Anything smaller will have too few df associated with the error MS to provide adequate power to detect treatment effects.

11.3.1 Calculations

Table 11.1 presents the data collected using the 5×5 square obtained in Section 11.2. Table 11.2 presents the analysis of variance of these data, together with the expected mean squares that justify the error term for each F test. These expected mean squares, as well as the partitioning of variability into the sources presented, reflect a structural model that we will consider in the next section. The sums of squares for the total, subjects (S), drugs (A in general), and days (C in general) are calculated exactly as in the preceding chapters. For example,

$$SS_A = \sum_j \frac{T_{.j}^2}{a} - C$$

$$= \frac{84^2 + 80^2 + \cdots + 74^2}{5} - \frac{391^2}{25}$$

$$= 6128.20 - 6115.24 = 12.96$$

The residual sum of squares is exactly that, a residual of the total variability after removal of SS_S, SS_A, and SS_C. All F ratios are constructed by testing the numerator source of variance against the residual term. Note that the term of primary interest, A,

TABLE 11.1   EXAMPLES OF A SINGLE LATIN SQUARE (C = DAYS, A = DRUG)

	C_1	C_2	C_3	C_4	C_5	$T_{i..}$
S_1	$(A_1)_{17}$	$(A_2)_{18}$	$(A_4)_{18}$	$(A_3)_{19}$	$(A_5)_{20}$	92
S_2	$(A_3)_{14}$	$(A_1)_{16}$	$(A_5)_{16}$	$(A_2)_{18}$	$(A_4)_{17}$	81
S_3	$(A_4)_{13}$	$(A_3)_{15}$	$(A_1)_{18}$	$(A_5)_{16}$	$(A_2)_{18}$	80
S_4	$(A_2)_{14}$	$(A_5)_{14}$	$(A_3)_{16}$	$(A_4)_{17}$	$(A_1)_{19}$	80
S_5	$(A_5)_{8}$	$(A_4)_{10}$	$(A_2)_{12}$	$(A_1)_{14}$	$(A_3)_{14}$	58

$T_{..k} =$ 66   73   80   84   88

$T_{...} = 391$

The totals for the levels of A are

	A_1	A_2	A_3	A_4	A_5
$T_{.j.} =$	84	80	78	75	74

TABLE 11.2   THE ANOVA FOR THE DESIGN OF TABLE 11.1[a]

SV	df	SS	MS	F	EMS
S	$a - 1 = 4$	122.56	30.64	148.23^b	$\sigma_e^2 + a\sigma_S^2$
C	$a - 1 = 4$	61.76	15.44	74.67^b	$\sigma_e^2 + a\theta_C^2$
A	$a - 1 = 4$	12.96	3.24	15.67^b	$\sigma_e^2 + a\theta_A^2$
Residual	$(a - 1)(a - 2) = 12$	2.48	.2067		σ_e^2
Total	$a^2 - 1 = 24$	199.76			

[a] MS_{res} is the error term for all F tests.
[b] $p < .01$.

is significant at the .01 level. If the large variability due to days (SS_C) had not been extracted, the effects of A may not have been detected. The success of the design lies in the fact that performance improved over days and the Latin-square design enabled us to separate this variable from the error variance.

Most computer packages can analyze data obtained using the Latin-square design. Typically, each score is entered with its row, column, and treatment level coded. For the data set of Table 11.1, we would have a data file looking like this:

```
1   1   1   17
1   2   2   18
1   3   4   18
            .
            .
            .
5   5   3   14
```

SPSS^X MANOVA would then require the following MANOVA specification:

```
MANOVA Y BY S(1,5), C(1,5), A(1,5)/
    DESIGN=S,C,A/
```

The first line specifies the variables and their levels (1 through 5), and the second line specifies the model; note that only main effects are included. BMDP2V's "design" paragraph does much the same thing:

```
/DESIGN    GROUPING IS '3G,Y'.
           INCLUDED ARE 1,2,3.
```

The first line states that the first three values in each row of the data file are grouping codes, whereas the fourth is the score. The second line states that the model includes only the main effects of the three grouping variables (named in an earlier "variable" paragraph). SYSTAT requires a "category" and a "model" statement that parallel the two statements shown for SPSS^X and BMDP:

```
CATEGORY S=5, C=5, A=5
MODEL Y=CONSTANT+S+C+A
```

Of course, the manuals should be consulted for the full set of statements required to run the analysis.

11.3.2 The Additive Model

Let us now consider the model that justifies the analysis of Table 11.2. The advantages of the Latin square are clear and the interpretation of the analysis straightforward if the following structural model can reasonably be assumed:

$$Y_{ijk} = \mu + \eta_i + \alpha_j + \gamma_k + \epsilon_{ijk} \qquad (11.1)$$

where i indexes the subjects (rows of the square), j indexes the treatment levels, k indexes the columns, and there are a levels of each of the three variables.

As in Chapter 8, the label "additive" refers to the additivity of main effects; no interactions are assumed to be present in the model. We assume that A (treatments) and

C (columns) are fixed-effect variables so that $\Sigma\alpha_j = \Sigma\gamma_k = 0$. Also, the η_i and ϵ_{ijk} are normally distributed random effects with zero means and variances σ_S^2 and σ_e^2, respectively. The expected mean squares derived from this model are presented in Table 11.2.

If Equation 11.1 adequately describes the population of scores, the Latin-square design provides the basis for a very efficient test of effects because we have removed error variance due to both the row and column variables. It should be emphasized that this efficiency is realized only if the proper analysis is carried out. Frequently, researchers counterbalance treatment levels over positions in time, or with respect to some other variable, and then partition the total SS into only three components: SS_A, SS_S, and SS_{SA}. If C contributes more than chance variability, this analysis results in a negatively biased F test. The error term in this analysis is a pool of the C and residual mean squares, and therefore

$$E(MS_{SA}) = \frac{(a-1)E(MS_C) + (a-1)(a-2)E(MS_{res})}{(a-1) + (a-1)(a-2)}$$

$$= \sigma_e^2 + \frac{n\theta_C^2}{a-1}$$

A failure to detect an effect of A when tested against this error term might reflect an absence of A effects in the population or large enough C effects to obscure effects of A.

If the data are properly analyzed, the Latin square should be a more efficient design than the subjects $\times$ treatments design of Chapter 8. Let us consider this point next.

11.3.3 Relative Efficiency

We argued in the introduction to this chapter that the Latin-square design should have less error variance associated with it than the subjects $\times$ treatments design because the former permits us to remove variance due not only to individual differences (or, more generally, rows) but also due to columns. This intuitive argument can be supported in a more formal way. In Appendix 11.1 we use the EMS for the two designs to derive a relation between the two error terms. The result of that derivation is that

$$E(MS_{SA}) = E[MS_{res} + \frac{1}{a}(MS_C - MS_{res})] \tag{11.2}$$

From Equation 11.2, we can see that MS_{SA} (the error term in the subjects $\times$ treatments design) will be larger than MS_{res} (the Latin-square error term) provided that MS_C is larger than MS_{res}. The Latin-square error term has fewer df than the subjects $\times$ treatments error term, however. To account for this, Fisher (1952) proposed the following measure of relative efficiency of design 1 to design 2:

$$RE = \left[\frac{MS_{error\,2}}{MS_{error\,1}}\right] \left[\frac{df_1 + 1}{df_1 + 3}\right] \left[\frac{df_2 + 3}{df_2 + 1}\right] \tag{11.3}$$

Note the adjustment for error degrees of freedom. In the present case, assume that we have used the Latin-square design and want to decide whether it will be more efficient than the subjects $\times$ treatments design in future studies. Then we would estimate the magnitude of MS_{SA} from Equation 11.2 and use it in place of $MS_{error 2}$ in Equation 11.3. This gives us

$$RE = \left[\frac{MS_{res} + (1/a)(MS_C - MS_{res})}{MS_{res}} \right] \left[\frac{(a-1)(a-2)+1}{(a-1)(a-2)+3} \right] \left[\frac{(a-1)^2+3}{(a-1)^2+1} \right] \quad (11.4)$$

11.3.4 Missing Scores

As in Chapter 8, missing scores are easily estimated by assuming the additive model, Equation 11.1. If the missing value is labeled X_{ijk}, its expected value under the additive model is

$$E(X_{ijk}) = \mu + \eta_i + \alpha_j + \gamma_k$$

and therefore

$$\hat{X} = \left[\frac{T_{...} + \hat{X}}{a^2} \right] + \left[\frac{T_{i..} + \hat{X}}{a} - \frac{T_{...} + \hat{X}}{a^2} \right] + \left[\frac{T_{.j.} + \hat{X}}{a} - \frac{T_{...} + \hat{X}}{a^2} \right]$$

$$+ \left[\frac{T_{..k} + \hat{X}}{a} - \frac{T_{...} + \hat{X}}{a^2} \right]$$

Simplifying and solving give

$$\hat{X} = \frac{a(T_i + T_j + T_k) - 2T_{...}}{(a-1)(a-2)}$$

where $T_{...}$ = obtained grand total

$T_{i..}$ = obtained total for subject i

$T_{.j.}$ = obtained total for A_j

$T_{..k}$ = obtained total for C_k

If several scores are missing, the iterative procedure of Chapter 8 can be used.

11.3.5 Nonadditivity

In the immediately preceding sections, we assumed no interactions among S, C, and A. Because this model is unrealistic in many situations, it is important to examine the consequences of nonadditivity. We begin this examination by stating a completely nonadditive model:

$$Y_{ijk} = \mu + \eta_i + \alpha_j + \gamma_k + (\eta\alpha)_{ij} + (\eta\gamma)_{ik} + (\alpha\gamma)_{jk} + (\eta\alpha\gamma)_{ijk} + \epsilon_{ijk} \quad (11.5)$$

As in the additive case, η_i, α_j, and γ_k reflect S, C, and A effects, and we assume that subjects are randomly sampled but that the levels of A and C are fixed. Because A and C have fixed effects, it can be shown that

$$\sum_j \alpha_j = \sum_k \gamma_k = \sum_j (\alpha\gamma)_{jk} = \sum_k (\alpha\gamma)_{jk} = 0 \quad (11.6)$$

To understand the implications of Equations 11.5 and 11.6 for the expected mean squares, let us analyze the contributions of interaction effects in a specific design. Suppose we have

$$
\begin{array}{c}
 \quad C_1 \quad C_2 \quad C_3 \\
\begin{array}{c} S_1 \\ S_2 \\ S_3 \end{array}
\left[\begin{array}{ccc}
A_1 & A_3 & A_2 \\
A_3 & A_2 & A_1 \\
A_2 & A_1 & A_3
\end{array} \right]
\end{array}
$$

Then, according to the model of Equation 11.5, the mean at A_1 is

$$
\bar{Y}_{.1.} = \mu + \alpha_1 + \frac{\eta_1 + \eta_2 + \eta_3}{3} + \frac{\gamma_1 + \gamma_2 + \gamma_3}{3} + \frac{(\eta\alpha)_{11} + (\eta\alpha_{21}) + (\eta\alpha)_{31}}{3}
$$
$$
+ \frac{(\eta\gamma)_{11} + (\eta\gamma)_{23} + (\eta\gamma)_{32}}{3} + \frac{(\alpha\gamma)_{11} + (\alpha\gamma)_{13} + (\alpha\gamma)_{12}}{3}
$$
$$
+ \frac{(\eta\alpha\gamma)_{111} + (\eta\alpha\gamma)_{213} + (\eta\alpha\gamma)_{312}}{3} + \frac{\epsilon_{111} + \epsilon_{213} + \epsilon_{312}}{3}
$$

When we apply the constraints of Equation 11.6, the sums of the γ and $\alpha\gamma$ effects become zero; therefore, we can rewrite the preceding equation as

$$
\bar{Y}_{.1.} = \mu + \alpha_1 + \frac{\eta_1 + \eta_2 + \eta_3}{3} + \frac{(\eta\alpha)_{11} + (\eta\alpha)_{21} + (\eta\alpha)_{31}}{3}
$$
$$
+ \frac{(\eta\gamma)_{11} + (\eta\gamma)_{23} + (\eta\gamma)_{32}}{3} + \frac{(\eta\alpha\gamma)_{111} + (\eta\alpha\gamma)_{213} + (\eta\alpha\gamma)_{312}}{3} + \frac{\epsilon_{111} + \epsilon_{213} + \epsilon_{312}}{3}
$$

The grand mean (μ) and the average of the three sampled subject effects $[(\eta_1 + \eta_2 + \eta_3)/3]$ will appear in the expression for all three treatment level means; consequently, they make no contribution to the variance of the $\bar{Y}_{.j.}$ and play no role in $E(\mathrm{MS}_A)$. On the other hand, $\Sigma_i(\eta\alpha)_{i1}$ does not equal zero because we are summing over only a subset of the infinite population of $S \times A$ interaction effects at A_1. Furthermore, there will be different $S \times A$ interaction effects at A_2 and at A_3; we will have $(\eta\alpha)_{i2}$ and $(\eta\alpha)_{i3}$ at those levels. Consequently, some of the variability among the $\bar{Y}_{.j.}$ will be due to the difference in the interaction effects at the three levels of A, and $E(\mathrm{MS}_A)$ will include a σ_{SA}^2 component.

The $S \times C$ interaction effects also contribute to the variability among the means at the levels of A. This happens because we have different sets of $(\eta\gamma)_{ik}$ at each level of A, and the average of each set will not be zero. Therefore, σ_{SC}^2 contributes to $E(\mathrm{MS}_A)$, as does σ_{SCA}^2 for similar reasons. In general, whether the effects are fixed or random, σ_{SC}^2 will influence MS_A, σ_{SA}^2 will influence MS_C, θ_{AC}^2 will influence MS_S, and the three-way interaction will contribute to the mean squares for all three variables. Such confounding of effects is typical of incomplete block designs—that is, designs in which all combinations of two variables are not present at each level of a third variable.

Wilk and Kempthorne (1957) originally derived the expected mean squares for the Latin-square design. Table 11.3 presents their results for two cases, when C has fixed effects and when C has random effects. In both cases, S represents subjects and is assumed to have random effects, and A, the treatment variable, is assumed to have fixed effects. Note that the EMS depart from the rules of thumb of Chapter 9 in two respects: components of variance are present that the rules would not suggest but that follow from the preceding discussion of confounding, and the coefficients depart from those stated in the rules of thumb.

The table should clarify the consequences of nonadditivity. First, as we argued in Chapter 8, the addition of a component of variance to numerator and denominator tends to reduce the F ratio when the null hypothesis is false. In short, the increase in variance will tend to reduce power. The presence of some interactions in the population also can cause bias in the F test. When C is a fixed-effect variable, the interaction variance component, θ_{AC}^2, will contribute to MS_{res} but not to MS_A or MS_C. Thus, if $\theta_{AC}^2 > 0$, the test of A will be negatively biased. If this interaction variance is large relative to variance among the population means for the treatment levels, Type 2 error rates will be high.

TABLE 11.3 EXPECTED MEAN SQUARES FOR THE SINGLE LATIN-SQUARE, NONADDITIVE MODEL

	S Random, C and A Fixed
SV	EMS
S	$\sigma_e^2 + \theta_{AC}^2 + \left[\dfrac{a-1}{a}\right]\sigma_{SCA}^2 + a\sigma_S^2$
A	$\sigma_e^2 + \sigma_{SA}^2 + \sigma_{SC}^2 + \left[\dfrac{a-2}{a}\right]\sigma_{SCA}^2 + a\theta_A^2$
C	$\sigma_e^2 + \sigma_{SA}^2 + \sigma_{SC}^2 \left[\dfrac{a-2}{a}\right]\sigma_{SCA}^2 + a\theta_C^2$
Residual	$\sigma_e^2 + \sigma_{SA}^2 + \sigma_{SC}^2 + \theta_{AC}^2 + \left[\dfrac{a-2}{a}\right]\sigma_{SCA}^2$

	S and C Random, A Fixed
SV	EMS
S	$\sigma_e^2 + \sigma_{SC}^2 + \sigma_{AC}^2 + \left[\dfrac{a-1}{a}\right]\sigma_{SCA}^2 + a\sigma_S^2$
A	$\sigma_e^2 + \sigma_{SA}^2 + \sigma_{AC}^2 + \sigma_{SC}^2 + \left[\dfrac{a-1}{a}\right]\sigma_{SCA}^2 + a\theta_A^2$
C	$\sigma_e^2 + \sigma_{SA}^2 + \sigma_{SC}^2 + \left[\dfrac{a-1}{a}\right]\sigma_{SCA}^2 + a\sigma_C^2$
Residual	$\sigma_e^2 + \sigma_{SA}^2 + \sigma_{AC}^2 + \sigma_{SC}^2 + \left[\dfrac{a-1}{a}\right]\sigma_{SCA}^2$

Despite such potential bias, the Latin-square design will often provide a more powerful test than the subjects × treatments design. For this to be the case, θ_{AC}^2 should be small and θ_C^2 should be large. In short, the negative bias should not overwhelm any potential treatment effects, and blocking with respect to columns should have a large payoff in the form of a reduction of error variance. These conditions will often be met.

In some experiments, the effects of C will be random. For example, suppose a researcher wishes to examine the role of noise upon text comprehension. Four passages are written; these are C_1 through C_4 and may reasonably be viewed as a random sample from the potential population of passages of this length and comprehension difficulty. Each of four subjects reads each passage with a different noise background (level of A) and is then tested for comprehension. The design might look like

	C_1	C_2	C_3	C_4
Subject 1	A_3	A_2	A_4	A_1
Subject 2	A_2	A_1	A_3	A_4
Subject 3	A_1	A_4	A_2	A_3
Subject 4	A_4	A_3	A_1	A_2

The temporal order of presentation is randomized independently for each subject.

This use of the Latin square is quite common; it differs from the preceding development in that C is now viewed as having random effects. In that case, σ_{AC}^2 contributes to both the A and residual mean squares, and the test of A against the residual is unbiased regardless of which variance components are present in the population.

Let us sum up the consequences of nonadditivity in analyzing data from a design employing a single Latin square. When C has random effects, or when θ_{AC}^2 is zero, the F test of treatment effects is unbiased. Under other conditions, the bias will be negative because θ_{AC}^2 will contribute to the residual MS but not to that for A. Even if such bias is present, the Latin square usually will provide a more powerful test of the A SV than will other designs because row and column main effects will not contribute to the Latin square's error variance. However, estimates of θ_A^2 will be systematically too small in the presence of negative bias.

Provided there are sufficient data, nonadditivity may be detected by the methods cited in Chapter 8. As with the repeated-measures design, we can calculate the values of Y_{ijk} that are expected under the additive model:

$$\hat{Y}_{ijk} = \bar{Y}_{...} + (\bar{Y}_{i..} - \bar{Y}_{...}) + (\bar{Y}_{.j.} - \bar{Y}_{...}) + (\bar{Y}_{..k} - \bar{Y}_{...}) \tag{11.7}$$

If the data are additive, the residuals, $Y_{ijk} - \hat{Y}_{ijk}$, plotted as a function of $\hat{Y}_{ijk}$ will vary randomly about a line with slope of zero. One or two deviant points suggests that possibility of miscalculation of the data point or the need to replace the subject. A systematic pattern such as a curved function suggests nonadditivity that may be corrected by a power transformation (that is, Y^p). We again emphasize that transformations result in tests of null hypotheses on a scale other than the original data scale. If the original scale has no firm theoretical or practical basis, and the new scale is as readily interpreted, such transformations make sense, but not otherwise.

The residual plot can be supplemented by a test of nonadditivity (Tukey 1965). The test requires extensive calculations (which are not generally available in computer packages), is not sensitive to all patterns of nonadditivity, and requires a large enough square to ensure power. Our inclination is to rely primarily on plots of residuals to inform us about the presence and nature of nonadditivity in our data.

Finally, we should note that nonadditivity opens the possibility of nonsphericity. The ϵ adjustment of degrees of freedom, described in Section 8.4.3 is appropriate for the Latin square as well.

11.3.6 Investigating Main and Interaction Effects of Several Treatments

Even in a single Latin square, the interaction of two variables, A and B, can be analyzed provided that all possible combinations appear exactly once in each row and column of the square. Consider a study of visual perception in which targets are presented for a brief period of time. The targets are either large or small and are presented for 50 or 100 milliseconds. Each subject experiences a block of 20 trials under each of the four possible treatment combinations, and the dependent variable is the number of correctly identified targets in each block. In this case, the Latin-squared treatments form a 2×2 design (size $\times$ duration). An example of the design using one subject in each of four sequences of trial blocks, together with a set of scores, is presented in Table 11.4. The analysis of this data set appears in Table 11.5. The calculations are essentially the same as those for a single Latin square. The only difference is that we now calculate SS_A, SS_B, and SS_{AB} rather than a single treatment sum of squares.

TABLE 11.4 A LATIN-SQUARE DESIGN ALLOWING TESTS OF INTERACTIONS

	C_1	C_2	C_3	C_4	$T_{i..}$
S_1	$(A_1B_2)_{11}$	$(A_2B_1)_{16}$	$(A_1B_1)_{20}$	$(A_2B_2)_{15}$	62
S_2	$(A_2B_1)_6$	$(A_2B_2)_{13}$	$(A_1B_2)_{13}$	$(A_1B_1)_{18}$	50
S_3	$(A_2B_2)_2$	$(A_1B_1)_{13}$	$(A_2B_1)_9$	$(A_1B_2)_{10}$	34
S_4	$(A_1B_1)_5$	$(A_1B_2)_6$	$(A_2B_2)_4$	$(A_2B_1)_7$	22
$T_{..m} =$	24	48	46	50	$T_{....} = 168$

Treatment totals ($T_{.jk.}$)

	B_1	B_2	$T_{.j..}$
A_1	56	40	96
A_2	38	34	72
$T_{..k} =$	94	74	

TABLE 11.5   THE ANOVA FOR THE DESIGN OF TABLE 11.5[a]

SV	df	SS	MS	F	EMS
S	$ab - 1 = 3$	232	77.333	116.00	$\sigma_e^2 + ab\sigma_S^2$
C	$ab - 1 = 3$	110	36.667	55.00	$\sigma_e^2 + ab\theta_C^2$
A	$a - 1 = 1$	36	36.000	54.00	$\sigma_e^2 + ab^2\theta_A^2$
B	$b - 1 = 1$	25	25.000	37.50	$\sigma_e^2 + a^2b\theta_B^2$
AB	$(a-1)(b-1) = 1$	9	9.000	13.50	$\sigma_e^2 + ab\theta_{AB}^2$
Residual	$(ab-1)(ab-2) = 6$	4	.667		σ_e^2
Total	$(ab)^2 - 1 = 15$	416			

[a] MS_{res} is the error term for all F tests. All effects are significant at the .01 level.

In general, we have ab rows and ab columns; because $a = b = 2$ in our example, $ab = 4$. The model underlying the sources of variance and the expected mean squares in the ANOVA of Table 11.5 is

$$Y_{ijkm} = \mu + \eta_i + \alpha_j + \beta_k + (\alpha\beta)_{jk} + \gamma_m + \epsilon_{ijkm} \tag{11.8}$$

where i indexes subjects $(i = 1, 2, \ldots, ab)$, j indexes the levels of A (size; $j = 1, 2, \ldots, a$), k indexes the levels of B (duration; $k = 1, 2, \ldots, b$), and m indexes the columns (trial block; $m = 1, 2, \ldots, ab$). We assume that the α_j, β_k, γ_m, and $(\alpha\beta)_{jk}$ are fixed-effect variables, and therefore $\Sigma_j\alpha_j = \Sigma_k\beta_k = \Sigma_m\gamma_m = \Sigma_j(\alpha\beta)_{jk} = \Sigma_k(\alpha\beta)_{jk} = 0$. The η_i and ϵ_{ijkm} are random-effect variables with expected values of zero and variances σ_S^2 and σ_e^2, respectively.

11.4 USING SEVERAL SQUARES

In Section 11.3 we considered experiments in which a single Latin square was drawn from the population of such squares. This design has the limitation that the number of subjects equals the number of treatment levels. One way to increase the number of subjects, and therefore the power of the F test, is to use several squares. This design is particularly useful when there are fewer than five treatment levels. The several-squares design has a second important advantage. If there are interactions present in the population, using several different squares increases the likelihood that positive and negative interaction effects will cancel each other.

11.4.1 The Basic Design and Analysis

Table 11.6 presents an example of the design in which 12 subjects are tested on four tasks requiring different types of motor skills (A). Each task is performed on a different day (C). Three 4×4 Latin squares have been constructed following the procedure described in Section 11.2, and each subject has been randomly assigned to one

TABLE 11.6 AN EXAMPLE USING SEVERAL LATIN SQUARES

Square		C_1	C_2	C_3	C_4	$T_{i..m}$	$T_{...m}$
	S_{11}	$(A_1)_{28}$	$(A_3)_{47}$	$(A_4)_{54}$	$(A_2)_{47}$	176	
Q_1	S_{21}	$(A_3)_{42}$	$(A_4)_{55}$	$(A_2)_{44}$	$(A_1)_{43}$	184	792
	S_{31}	$(A_4)_{53}$	$(A_2)_{48}$	$(A_1)_{47}$	$(A_3)_{60}$	208	
	S_{41}	$(A_2)_{49}$	$(A_1)_{50}$	$(A_3)_{59}$	$(A_4)_{66}$	224	
	S_{12}	$(A_2)_{40}$	$(A_1)_{41}$	$(A_3)_{54}$	$(A_4)_{61}$	196	
Q_2	S_{22}	$(A_4)_{47}$	$(A_3)_{52}$	$(A_1)_{41}$	$(A_2)_{48}$	188	804
	S_{32}	$(A_3)_{45}$	$(A_2)_{46}$	$(A_4)_{63}$	$(A_1)_{46}$	200	
	S_{42}	$(A_1)_{40}$	$(A_4)_{65}$	$(A_2)_{54}$	$(A_3)_{61}$	220	
	S_{13}	$(A_2)_{41}$	$(A_1)_{46}$	$(A_4)_{61}$	$(A_3)_{56}$	204	
Q_3	S_{23}	$(A_1)_{31}$	$(A_2)_{48}$	$(A_3)_{51}$	$(A_4)_{62}$	192	780
	S_{33}	$(A_4)_{47}$	$(A_3)_{48}$	$(A_1)_{37}$	$(A_2)_{44}$	176	
	S_{43}	$(A_3)_{49}$	$(A_4)_{62}$	$(A_2)_{51}$	$(A_1)_{46}$	208	
$T_{..k.} =$		512	608	616	640		

$$T_{....} = 2376$$

The treatment (A) totals are

	A_1	A_2	A_3	A_4
$T_{.j..} =$	496	560	624	696

of the 12 sequences of tasks. In general, we have q $a \times a$ squares randomly sampled from the population of $a \times a$ squares; thus, $q = 3$ and $a = 4$ in this example.

Table 11.7 presents two different partitionings of the total sums of squares, with different F tests in each ANOVA. These two analyses correspond to two possible structural models of the data. Both models assume that the squares and the subjects are randomly sampled, A and C have fixed effects, and the effects of S, C, and A are additive. The model underlying the first ANOVA of Table 11.9 is

$$Y_{ijkm} = \mu + \eta_{i/m} + \alpha_j + \gamma_k + \pi_m + (\alpha\pi)_{jm} + (\gamma\pi)_{km} + \epsilon_{ijkm} \tag{11.9}$$

where i indexes subjects within a square ($i = 1, 2, \ldots, a$), j indexes the level of A (task; $j = 1, 2, \ldots, a$), k indexes columns (day; $k = 1, 2, \ldots, a$), and m indexes the square ($m = 1, 2, \ldots, q$). Note that the model allows for the possibility that treatment (tasks) and column (days) effects depend upon the particular square [$(\alpha\pi)_{jm}$ and $(\gamma\pi)_{km}$ effects]. If such interaction effects are present in the population, and if the number of sampled squares is small relative to the number of possible squares, σ^2_{AQ} and σ^2_{CQ} will contribute to the A and C variability, respectively. This can be seen in the

TABLE 11.7   TWO ANALYSES FOR THE DESIGN OF TABLE 11.6

ANOVA following Equation 11.9

SV	df	SS	MS	F	Error term	EMS[a]
Squares (Q)	$q - 1 = 2$	18.000	9.000	.123	S/Q	$\sigma_e^2 + a\sigma_{S/Q}^2 + a^2\sigma_Q^2$
S/Q	$q(a - 1) = 9$	658.000	73.111			$\sigma_e^2 + a\sigma_{S/Q}^2$
C	$a - 1 = 3$	793.333	264.444	95.192^b	CQ	$\sigma_e^2 + Ka\sigma_{CQ}^2 + qa\theta_C^2$
A	$a - 1 = 3$	1838.667	612.889	324.452^b	AQ	$\sigma_e^2 + Ka\sigma_{AQ}^2 + qa\theta_A^2$
CQ	$(a - 1)(q - 1) = 6$	16.667	2.778	.962	Res	$\sigma_e^2 + a\sigma_{CQ}^2$
AQ	$(a - 1)(q - 1) = 6$	11.333	1.889	.654	Res	$\sigma_e^2 + a\sigma_{AQ}^2$
Residual	$q(a - 1)(a - 2) = 18$	52.000	2.889			σ_e^2
Total	$qa^2 - 1 = 47$	3388.000				

ANOVA following Equation 11.10

SV	df	SS	MS	F	Error term	EMS
Squares (Q)	$q - 1 = 2$	18.000	9.000	.123	S/Q	$\sigma_e^2 + a\sigma_{S/Q}^2 + a^2\sigma_Q^2$
S/Q	$q(a - 1) = 9$	658.000	73.111			$\sigma_e^2 + a\sigma_{S/Q}^2$
C	$a - 1 = 3$	793.333	264.444	99.167^b	Pool	$\sigma_e^2 + qa\theta_C^2$
A	$a - 1 = 3$	1838.667	612.889	229.833^b	Pool	$\sigma_e^2 + qa\theta_A^2$
Pooled residual	$(qa - 2)(a - 1) = 30$	80.000	2.667			σ_e^2
Total	$qa^2 - 1 = 47$	3388.000				

[a] The coefficient $K = 1 - (q/q_{\text{pop}})$. It results because the number of possible squares (q_{pop}) is finite (Wilk and Kempthorne 1957). If q_{pop} is large, K will be close to 1.

[b] $p < .01$

expected mean squares of the first ANOVA in Table 11.7. Therefore, the appropriate error terms for A and C are the AQ and CQ mean squares.

We may test the AQ and CQ terms against the residual error term. If this test provides no evidence that A and C interact with Q, then MS_{AQ}, MS_{CQ}, and MS_{res} may be pooled to provide a single error term on $(qa - 2)(a - 1)$ df. In essence, we assume the revised model

$$Y_{ijkm} = \mu + \eta_{i/m} + \alpha_j + \gamma_k + \pi_m + \epsilon_{ijkm} \tag{11.10}$$

This model will often be appropriate. The first ANOVA of the data set of Table 11.6 reveals that the CQ and AQ terms were not significant when tested against the residual term. Consequently, we redid the analysis, testing C and A against the pooled residual term, $(MS_{AQ} + MS_{CQ} + MS_{res})/[(qa - 2)(a - 1)]$. The results of this second analysis are also presented in Table 10.7.

The calculations of the sums of squares of Table 11.7 are straightforward extensions of earlier developments in this and other chapters. The S/Q ("subjects within squares") term is calculated as $SS_{\text{Between } S} - SS_Q$, just as when we obtained the between-subjects error term of Chapter 9. The residual term is again just the difference between the total sum of squares and the sum of all other sums of squares. Most computer packages will perform the analysis, provided each score is properly coded and the appropriate model is specified. If there are many scores, the investigator may prefer not to have to code each score. An alternative is to create two data files in which each line contains a code for the square (1, 2, or 3 in our example) followed by the a scores for a single subject. In the first, C, file, the scores are ordered by level of C, whereas they are ordered by level of A in the second, or A, file. For the data of Table 11.6, the two files would look like

	File C		File A
1	28 47 54 47	1	28 47 47 54
1	42 55 44 43	1	43 44 42 55
	. . .		. . .
3	49 62 51 46	3	46 51 49 62

Each file is then submitted to any program capable of analyzing a data set from a design with one between-subjects variable (Q, the square) and one within-subjects variable (C or A, depending on the file). The C and CQ sums of squares are obtained from the analysis of file C, and the A and AQ sums of squares are obtained from the analysis of file A. Other sums of squares will be identical in both outputs.

11.4.2 Including Between-Subjects Variables

Suppose that we wanted to compare the effects of two levels of practice upon performance in the four motor tasks used in the example of the preceding section. Further suppose that we decide to run 16 subjects, 8 under each level of practice. One way to do this is illustrated in Table 11.8. We randomly divide our 16 subjects into four groups, then assign each group to a different randomly drawn 4 × 4 square. The eight subjects in the first two squares have 1 hour of practice, and those in the second two squares have 2 hours of practice. In general, we have a between-squares variable, B (hours of practice) with b levels, q squares (two in our example) at each level of B, and a subjects (four in our example) in each square with a (4) scores for each subject. Therefore, there is a total of bqa (2 × 2 × 4 = 16) subjects and bqa^2 (64) scores when this design is used.

To understand the model and the partitioning of sums of squares, we should note which variables are nested in which other variables and which variables cross with each other. Squares (Q) are nested within levels of B because the squares at B_1 differ from those at B_2. In turn, subjects are nested within squares. Both B and Q cross with A and

TABLE 11.8 A DESIGN WITH SEVERAL DIFFERENT SQUARES AND A BETWEEN-SQUARES FACTOR

			C_1	C_2	C_3	C_4	$T_{i.k.p}$
B_1	Q_{11}	S_{111}	$(A_4)_{37}$	$(A_1)_{39}$	$(A_3)_{37}$	$(A_2)_{31}$	144
		S_{211}	$(A_3)_{42}$	$(A_2)_{36}$	$(A_4)_{34}$	$(A_1)_{45}$	157
		S_{311}	$(A_1)_{33}$	$(A_3)_{43}$	$(A_2)_{42}$	$(A_4)_{34}$	152
		S_{411}	$(A_2)_{37}$	$(A_4)_{41}$	$(A_1)_{33}$	$(A_3)_{32}$	143
B_1	Q_{21}	S_{121}	$(A_4)_{38}$	$(A_3)_{43}$	$(A_2)_{38}$	$(A_1)_{32}$	151
		S_{221}	$(A_2)_{42}$	$(A_1)_{37}$	$(A_4)_{37}$	$(A_3)_{49}$	165
		S_{321}	$(A_1)_{33}$	$(A_4)_{47}$	$(A_3)_{45}$	$(A_2)_{35}$	160
		S_{421}	$(A_3)_{39}$	$(A_2)_{42}$	$(A_1)_{35}$	$(A_4)_{36}$	152
B_2	Q_{12}	S_{112}	$(A_1)_{30}$	$(A_4)_{55}$	$(A_3)_{46}$	$(A_2)_{36}$	167
		S_{212}	$(A_3)_{47}$	$(A_2)_{41}$	$(A_4)_{49}$	$(A_1)_{45}$	182
		S_{312}	$(A_4)_{45}$	$(A_1)_{42}$	$(A_2)_{47}$	$(A_3)_{43}$	177
		S_{412}	$(A_2)_{37}$	$(A_3)_{50}$	$(A_1)_{33}$	$(A_4)_{47}$	167
B_2	Q_{22}	S_{122}	$(A_2)_{36}$	$(A_1)_{41}$	$(A_3)_{49}$	$(A_4)_{49}$	175
		S_{222}	$(A_1)_{37}$	$(A_2)_{42}$	$(A_4)_{52}$	$(A_3)_{58}$	189
		S_{322}	$(A_4)_{44}$	$(A_3)_{59}$	$(A_1)_{41}$	$(A_2)_{33}$	185
		S_{411}	$(A_2)_{37}$	$(A_4)_{41}$	$(A_1)_{33}$	$(A_3)_{32}$	143

B	Q	Totals for BCQ Cells $(T_{..kmp})$				Total for ABQ Cells $(T_{.jk.p})$				$T_{..k.p}$
		C_1	C_2	C_3	C_4	A_1	A_2	A_3	A_4	
1	11	149	159	146	142	150	146	154	146	596
1	21	152	169	155	152	137	157	176	158	628
2	12	159	188	175	171	150	161	186	196	693
2	22	163	198	185	180	154	159	207	206	726
		C Totals $(T_{...m})$				A Totals $(T_{.j...})$				$T_{.....}$
		623	714	661	645	591	623	723	706	2643

with C but, as in all previous analyses of data from Latin-square designs, there are no terms involving both A and C.

One approach to partitioning the variance is demonstrated in Table 11.9. The partitioning assumes the structural model

$$Y_{ijkmp} = \mu + \eta_{i/p/k} + \alpha_j + \beta_k + \gamma_m + \pi_{p/k} + (\alpha\beta)_{jk} + (\beta\gamma)_{km} \qquad (11.11)$$
$$+ (\alpha\pi)_{jp/k} + (\gamma\pi)_{mp/k} + \epsilon_{ijkmp}$$

TABLE 11.9 THE ANALYSIS FOR THE DESIGN OF TABLE 11.8

SV	df	SS	MS	F	Error term	EMS
B	$b - 1 = 1$	594.141	25.492	.769	Q/B	$\sigma_e^2 + a\sigma_{S/Q/B}^2 + a^2\sigma_{Q/B}^2 + a\theta_B^2$
Q/B	$b(q - 1) = 2$	66.032	33.016	2.791	$S/Q/B$	$\sigma_e^2 + a\sigma_{S/Q/B}^2 + a^2\sigma_{Q/B}^2$
$S/Q/B$	$bq(a - 1) = 2$	141.938	11.828			$\sigma_e^2 + a\sigma_{S/Q/B}^2$
A	$a - 1 = 3$	763.297	254.432	12.846^a	AQ/B	$\sigma_e^2 + Kab\sigma_{AQ/B}^2 + abq\theta_A^2$
C	$a - 1 = 3$	281.797	92.932	75.925^a	CQ/B	$\sigma_e^2 + Kab\sigma_{CQ/B}^2 + abq\theta_C^2$
AB	$(a - 1)(b - 1) = 3$	290.297	96.766	4.885^b	AQ/B	$\sigma_e^2 + Kab\sigma_{AQ/B}^2 + aq\theta_{AB}^2$
BC	$(b - 1)(a - 1) = 3$	64.297	21.432	17.510^a	CQ/B	$\sigma_e^2 + Kab\sigma_{CQ/B}^2 + aq\theta_{BC}^2$
AQ/B	$b(a - 1)(q - 1) = 6$	118.844	19.807	.850	Res	$\sigma_e^2 + a\sigma_{AQ/B}^2$
CQ/B	$b(a - 1)(q - 1) = 6$	7.344	1.224	.053	Res	$\sigma_e^2 + a\sigma_{CQ/B}^2$
Residual	$bq(a - 1) = 24$	559.375	23.307			σ_e^2
Total	$bqa^2 - 1 = 63$	2887.362				

$^a p < .01$

$^b p < .05$

Calculations for the nested terms

$$C = \frac{T_{.....}^2}{bqa^2} = \frac{(2643)^2}{64} = 109,147.641$$

$$SS_{Q/B} = \frac{1}{a^2} \left[\sum_{k=1}^{b} \sum_{p=1}^{q} T_{..k.p}^2 \right] - C - SS_B$$

$$= \frac{596^2 + 628^2 + 693^2 + 726^2}{16} - C - SS_B$$

$$= 109,807.813 - 109,147.641 - 594.141 = 66.032$$

$$SS_{S/Q/B} = \frac{1}{a} \left[\sum_{i=1}^{a} \sum_{k=1}^{b} \sum_{p=1}^{q} T_{i.k.p}^2 \right] - \frac{1}{a^2} \left[\sum_{k=1}^{b} \sum_{p=1}^{q} T_{..k.p}^2 \right]$$

$$= \frac{144^2 + 157^2 + \cdots + 177^2}{4} - 109,807.813$$

$$= 109,949.75 = 109,807.813 = 141.938$$

$$SS_{AQ/B} = \frac{1}{a} \left[\sum_{j=1}^{a} \sum_{k=1}^{b} \sum_{p=1}^{1} T_{.jk.p}^2 \right] - \frac{1}{aq} \left[\sum_{j=1}^{a} \sum_{k=1}^{b} T_{.jk..}^2 \right] - SS_{Q/B}$$

$$= \frac{150^2 + 146^2 + \cdots + 206^2}{4} - \frac{(150 + 137)^2 + (146 + 157)^2 + \cdots + (196 + 206)^2}{8} - 66.032$$

$$= 110{,}980.25 - 110{,}795.375 - 66.032 = 118.844$$

$$SS_{CQ/B} = \frac{1}{a} \left[\sum_{k=1}^{b} \sum_{m=1}^{a} \sum_{p=1}^{q} T^2_{..kmp} \right] - \frac{1}{aq} \left[\sum_{k=1}^{b} \sum_{m=1}^{a} T^2_{..km.} \right] - SS_{Q/B}$$

$$= \frac{149^2 + 159^2 + \cdots + 180^2}{4} - \frac{(149 + 152)^2 + (159 + 169)^2 + \cdots + (171 + 180)^2}{8} - 66.032$$

$$= 110{,}161.25 - 110{,}087.875 - 66.032 = 7.344$$

where i indexes subjects within squares within levels of B ($i = 1, 2, \ldots, a$), j indexes the level of A (task; $j = 1, 2, \ldots, a$), k indexes the level of B (practice; $k = 1, 2, \ldots, b$), m indexes the level of C (day; $m = 1, 2, \ldots, a$), and p indexes the square within a level of B ($p = 1, 2, \ldots, q$). The subject ($\eta_{i/p/k}$) and square ($\pi_{p/k}$) effects are assumed to be random. They are also assumed to be normally distributed with mean zero and variances, $\sigma^2_{S/Q/B}$ and $\sigma^2_{Q/B}$, respectively. The effects of A, B, and C (α, β, and γ) are assumed to be fixed. The sources of variance and expected mean squares in Table 11.9 correspond to the effects stated in Equation 11.11.

Calculations of the sums of squares for the less familiar nested terms follow the ANOVA table. As in the designs of Chapters 8 and 9, we can view the total variability as consisting of two components, $SS_{\text{Between } S}$ (between-subjects sum of squares) and $SS_{\text{Within } S}$ (within-subjects sum of squares). As a check on the calculations in Table 11.9, compute

$$SS_{\text{Between } S} = \frac{1}{a} \left[\sum_{i=1}^{a} \sum_{k=1}^{b} \sum_{p=1}^{q} T^2_{i.k.p} \right] - C = \frac{144^2 + 157^2 + \cdots + 177^2}{4} - C$$

$$= 109{,}949.75 - 109{,}147.641 = 802.109$$

This should equal $SS_B + SS_{Q/B} + SS_{S/Q/B}$, which (within rounding error) it does. Similarly, the remaining terms in the ANOVA of Table 11.9 should add to $SS_{\text{Within } S}$, which equals $SS_{\text{tot}} - SS_{\text{Between } S}$. For larger data sets than that in our example, we again note that computer programs should be used. As in earlier examples, each score can be coded. The codes should reflect the level of B, the square, the subject, the level of C, and the level of A. Alternatively, two files can be set up; each row will contain all the scores for a subject, ordered either by level of A or C.

As in the case of the several-squares design of Section 11.4.1, an alternative partitioning of the sums of squares is possible. When $p > .25$ for tests of CQ/D and AQ/D against the residual error term, it is reasonable to pool these three terms to construct a single pooled error term on $(bqa - 2)(a - 1)$ df. We have not bothered to do this in our example because the tests of interest in Table 11.13 are clearly significant.

11.5 REPLICATING A SINGLE LATIN SQUARE

In many situations, it is easier to use a single square with n subjects tested in each treatment sequence (replicated-square design) than to use the designs of Section 11.4 in which several squares are used with one subject in each sequence. The several-squares design may require programming many sequences of stimuli for computer-controlled experiments or scoring data protocols based on many stimulus sequences. To reduce these complications, researchers often use a single Latin square, randomly assigning n subjects to each row. Because the error term used to test the effect of A may depend upon whether C has fixed or random effects, both cases will be considered.

11.5.1 Analysis When C Has Fixed Effects

Consider the experiment described in Section 11.4 in which performances on four motor tasks are compared. Instead of using three squares, we now use one square with three subjects tested on each of the four sequences of tasks. Table 11.10 presents the design and the data from 12 subjects. In general, we have an $a \times a$ Latin square with n subjects undergoing each sequence of treatments. In the example, $a = 4$, the variables A and C are tasks and days, and n—the number of subjects in each sequence—is 3. Because three subjects are in each row (sequence of tasks over days), we say that the square has been replicated three times.

Table 11.11 presents the ANOVA and expected mean squares. In many respects, the analysis parallels that of Chapter 8 for mixed designs involving between- and within-subjects factors. As in the mixed design, the total variability (SS_{tot}) is partitioned into between- and within-subjects components. The between-subjects sum of squares is then partitioned into variability due to groups (R, row) and subjects within groups (S/R). The major difference between the mixed design and the replicated Latin-square design is that the ANOVA table for the latter contains no RC, RA, or RCA terms. As we pointed out earlier in this chapter, those terms cannot be evaluated independently of the main effects. If such interactions are present, they are partly confounded with main effects, as in the single Latin square. They also contribute to the variability among cell means that remains after removing variability due to main effects. That residual variability is labeled $SS_{B\,cells\,res}$ (between-cells residual). Its status as residual variability after adjustment for main effects is reflected in its df:

$$df_{B\,cells\,res} = (a^2 - 1) - (3)(a - 1) = (a - 1)(a - 2)$$

where a^2 is the number of cell means in the Latin square, and $(3)(a - 1)$ reflects subtraction of SS_A, SS_C, and SS_R, each of which is on $a - 1$ df. The last line of the table, $W\ cells\ res$ (within-cells residual), is actually the interaction of subjects $\times A$, nested within levels of R. This residual sum of squares can also be viewed as subjects $\times C$ nested within rows because, within any row of the square, the subjects $\times A$ and subjects $\times C$ sums of squares are identical. The df, $a(n - 1)(a - 1)$, reflect the fact that the interaction in any row will be on $(n - 1)(a - 1)$ df and these are pooled over the a rows of the Latin square.

Although there is no AC interaction term in the usual sense, the tests of R and of B cells res may be interpreted as tests of the AC interaction term. Variation among the

TABLE 11.10 A REPLICATED-SQUARES DESIGN

Sequence	Subject	C_1	C_2	C_3	C_4	$T_{i.m}$
		A_2	A_4	A_3	A_1	
	S_{11}	37	50	42	36	165
R_1	S_{21}	41	49	51	40	181
	S_{31}	38	46	32	42	158
	$T_{.jk1} =$	116	145	125	118	$T_{...1} = 504$
		A_1	A_3	A_2	A_4	
	S_{12}	32	44	36	47	159
R_2	S_{22}	48	50	47	53	198
	S_{32}	42	44	45	57	188
	$T_{.jk2} =$	122	138	128	157	$T_{...2} = 545$
		A_4	A_2	A_1	A_3	
	S_{13}	45	40	42	48	175
R_3	S_{23}	53	57	54	60	224
	S_{33}	41	41	43	44	169
	$T_{.jk3} =$	139	138	139	152	$T_{...3} = 568$
		A_3	A_1	A_4	A_2	
	S_{14}	50	53	57	51	211
R_4	S_{24}	50	48	61	51	210
	S_{34}	46	39	48	47	180
	$T_{.jk4} =$	146	140	166	149	$T_{...4} = 601$
C Totals $= T_{.k.} =$		523	561	558	576	$T_{....} = 2,218$
A Totals $= T_{.j..} =$		$A_1\,519$	$A_2\,531$	$A_3\,561$	$A_4\,607$	

row means reflects different effects of AC combinations; in other words, if there are AC effects in the population, part of their contribution to the data will be reflected in MS_R. Furthermore, significant residual variability among the a^2 cell means (after adjusting for the contribution of the main effects) also reflects part of the variability of AC interaction effects. The $(AC)'$ label in the SV column of Table 11.11 (as well as the expected mean squares) indicates these partial interaction effects.

The model underlying this interpretation is

$$Y_{ijkm} = \mu + \eta_{i/m} + \alpha_j + \gamma_k + (\alpha\gamma)_{jk} + \epsilon_{ijkm} \qquad (11.12)$$

TABLE 11.11   THE ANALYSIS FOR THE DESIGN OF TABLE 11.10

SV	df	SS	MS	F	Error term	EMS
$R\,(AC')$	$a - 1 = 3$	353.383	117.944	1.130	S/R	$\sigma_e^2 + a\sigma_{S/R}^2 + na\theta_{AC}^2$
S/R	$a(n-1) = 8$	835.167	104.396			$\sigma_e^2 + a\sigma_{S/R}^2$
C	$a - 1 = 3$	117.167	39.056	4.762^a	W cells res	$\sigma_e^2 + na\theta_C^2$
A	$a - 1 = 3$	323.167	107.722	13.135^a	W cells res	$\sigma_e^2 + na\theta_A^2$
B cells res $(AC')^b$	$(a-1)(a-2) = 6$	44.833	7.472	.911	W cells res	$\sigma_e^2 + n\theta_{AC}^2$
W cells res	$a(n-1)(a-1) = 24$	196.833	8.201			σ_e^2
Total	$na^2 - 1 = 47$	1870.550				

$^a p < .01.$

^{b} The between-cells residual term ($SS_{B\,cells\,res}$) can be calculated as $SS_{RC} - SS_A$. That is, calculate the sum of squares for rows by columns as you would for any two-factor interaction, and subtract SS_A.

where i indexes the subject within the row ($i = 1, 2, \ldots, n$), j indexes the level of A ($j = 1, 2, \ldots, a$), k indexes the level of C ($k = 1, 2, \ldots, a$), and m indexes the row within the square ($m = 1, 2, \ldots, a$).

We assume that subjects are randomly sampled from their respective populations. This means that the $\eta_{i/m}$ are a random sample from a normally distributed population having mean zero and variance $\sigma_{S/R}^2$. All other variables are assumed to have fixed effects. Given this model, the expected mean squares of Table 11.11 follow. The F ratios are based on these expected mean squares; as usual, the error term is selected so that the ratio of numerator to denominator EMS in one under the null hypothesis.

The values of the F ratios indicate that the effects of A are significant, reflecting differences in task difficulty. There also is a significant improvement in scores over days (C). As we noted, the MS_R and $MS_{B\,cells\,res}$ may be inflated if AC interaction effects are present in the sampled population. These potential contributions of AC interaction effects are reflected in the expectations of the two mean squares in question. For the present data set, the F values for R and B cells res are quite small; therefore, we conclude that the effects of the task (A) do not vary over the days of the experiment.

11.5.2 Analysis When C Has Random Effects

Let us consider a second application of the replicated-squares design. Suppose the columns of the Latin-square design of Table 11.10 represent four different stories, instead of days. Further suppose that the levels of A are levels of noise intensity present during reading; performance on a comprehension test is the dependent variable. Twelve subjects experience four combinations of story and noise level in a Latin-square design.

Looking at the first row of the design in Table 11.10, we find that the first group of three subjects read the first story at the second level of noise intensity $(A_2 C_1)$, the second text at the fourth noise level $(A_4 C_2)$, the third story at the third level of noise $(A_3 C_3)$, and the fourth story at the first level of noise $(A_1 C_4)$.

In this experiment, it is reasonable to assume that the γ_k are a random sample from a normally distributed population having mean zero and variance σ_C^2; that is, the stories can be viewed as a random sample from a population of such stories. Furthermore, the $(\alpha\gamma)_{jk}$ are also randomly sampled from a normally distributed population of such interaction effects having mean zero and variance σ_{AC}^2. The EMS resulting from these assumptions are presented in Table 11.12. Note the difference between these and those presented in Table 11.11 for the example in which C was assumed to have fixed effects. In the present example, σ_{AC}^2 contributes to $E(\text{MS}_A)$. Therefore, the $\text{MS}_{B\,\text{cells res}}$ is the appropriate error term against which to test A. Unless the square is at least 5×5 (and bigger is better), the error df will be small and power to test the A effects may be low. We recommend that investigators using this design first test $\text{MS}_{B\,\text{cells res}}$ against $\text{MS}_{W\,\text{cells res}}$. If that result is not significant at the .25 level, we have strong evidence that $\sigma_{AC}^2 = 0$, and we can use $\text{MS}_{W\,\text{cells res}}$ (or the pool of the between- and within-cell terms) as the error term for the test of A and C. In our experience, we usually have found that we can use the within-cell residual to test main effects.

11.5.3 Using the Computer to Analyze the Data

As in previous designs, computer packages can be used to analyze the data of Table 11.10 by coding each individual score. The code must include the levels of all variables whose effects are to be calculated in the analysis. In the example of the replicated-squares design, the row, subject, column, and level of C must accompany each score. With large data sets this can be a nuisance. We have found it easier to create two data

TABLE 11.12   EXPECTED MEAN SQUARES FOR THE REPLICATED-SQUARES DESIGN WHEN C IS RANDOM

SV	df	EMS	Error term
$R(AC')$	$a - 1$	$\sigma_e^2 + a\sigma_{S/R}^2 + na\sigma_{AC}^2$	S/R
S/R	$a(n - 1)$	$\sigma_e^2 + a\sigma_{S/R}^2$	
C	$a - 1$	$\sigma_e^2 + na\theta_C^2$	W cells res
A	$a - 1$	$\sigma_e^2 + n\sigma_{AC}^2 + na\theta_A^2$	B cells res
B cells res (AC')	$(a - 1)(a - 2)$	$\sigma_e^2 + n\sigma_{AC}^2$	W cells res
W cells res (AC')	$a(n - 1)(a - 1)$	σ_e^2	

files, submit each to a package that carries out a repeated-measures ANOVA, and combine the results.

In the example of Table 11.10, both files would have 12 lines (one for each subject) of five numbers each. The first number would be a digit from 1 to 4, representing the row of the Latin square (level of the between-subjects variable R). The remaining four numbers would be the four scores. The difference in the two files is that in one the scores are ordered by level of C (call this file C) and in the other they are ordered by level of A (call this file A). For example, assume the Latin square of Table 11.14. Then the first line of data in the C file would be

$$1 \quad 37 \; 50 \; 42 \; 36$$

For the A file the numbers are rearranged so that the score obtained under treatment A_1 appears first, followed in order by those obtained under treatments A_2 through A_4:

$$1 \quad 36 \; 37 \; 42 \; 50$$

Each file is then analyzed as a mixed design with R as the between-subjects variable and C (in file C) or A (in file A) as the within-subjects variable. The between-cells residual is then calculated in one of two ways. Either subtract SS_A (obtained from the file A ANOVA) from SS_{RC} (file C ANOVA), or subtract SS_C (file C ANOVA) from SS_{RA} (file A ANOVA). From the two ANOVAs and this last adjustment of the RC or RA term, we have the ANOVA summary of Table 11.11.

We have presented this method of analysis because the design is frequently used and frequently analyzed improperly. Researchers typically either ignore R and C, analyzing the data as if they were from an $S \times A$ design, or else they ignore C, analyzing the data as if they were from a mixed design with one between-subject (R) and one within-subject variable (A). In either case, information about C and AC is lost and, perhaps more important, the F test often will be biased. The proper analysis involves somewhat more effort than the analyses usually carried out, but it is worth the effort.

11.5.4 Including Between-Subjects Variables

In Section 11.4.2 we considered an experiment in which each subject was tested on four motor tasks. There was also a between-subjects variable; half of the subjects were tested after 1 hour of practice on each task, whereas the other half received 2 hours of practice. Instead of using four different Latin squares, as we did in Section 11.4.2, we could run this experiment using a single 4×4 Latin square. We again have 16 subjects, 8 in each level of practice. The design is laid out in Table 11.13 with four subjects in each of the four sequences; these four subjects are divided among the two levels of practice. More generally, we have a sequences (four in our example), with bn subjects in each sequence; $b = 2$ and $n = 2$ in our example.

The sources of variance in Table 11.14 are the same as those in Table 11.12 with the addition of B (practice, in our example) and its interactions with C, A, and the between-cells residual. The underlying model is

$$Y_{ijkmp} = \mu + \eta_{i/kp} + \alpha_j + \beta_k + \gamma_m + (\alpha\beta)_{jk} + (\alpha\gamma)_{jm} + (\beta\gamma)_{km} \qquad (11.13)$$
$$+ (\alpha\beta\gamma)_{jkm} + \epsilon_{ijkmp}$$

TABLE 11.13 A REPLICATED-SQUARES DESIGN WITH A BETWEEN-SUBJECTS VARIABLE

	B_1						B_2				
	C_1	C_2	C_3	C_4			C_1	C_2	C_3	C_4	
R_1	A_4	A_2	A_1	A_3		R_1	A_4	A_2	A_1	A_3	
S_{111}	53	56	58	61	228	S_{121}	71	68	57	68	264
S_{211}	52	50	43	42	187	S_{221}	70	72	70	85	297
	105	106	101	103	415		141	140	127	153	561
R_2	A_3	A_1	A_4	A_2		R_2	A_3	A_1	A_4	A_2	
S_{112}	41	39	40	34	154	S_{122}	49	49	73	61	232
S_{212}	50	51	58	56	215	S_{222}	68	64	81	66	279
	91	90	98	90	369		117	113	154	127	511
R_3	A_2	A_4	A_3	A_1		R_3	A_2	A_4	A_3	A_1	
S_{113}	43	49	52	47	191	S_{123}	37	60	52	35	184
S_{213}	59	62	58	51	230	S_{223}	48	73	72	58	251
	102	111	110	98	421		85	133	124	93	435
R_4	A_1	A_3	A_2	A_4		R_4	A_1	A_3	A_2	A_4	
S_{114}	50	54	48	55	207	S_{124}	33	61	60	83	237
S_{214}	45	54	57	69	225	S_{224}	44	67	57	76	244
	95	108	105	124	432		77	128	117	159	481

	AB Totals						BC Totals				
	A_1	A_2	A_3	A_4			C_1	C_2	C_3	C_4	
B_1	384	403	412	438	1637	B_1	393	415	414	415	1637
B_2	410	469	522	587	1988	B_2	420	514	522	532	1988
	794	872	934	1025	3625		813	929	936	947	3625

where i indexes the subject within a row $\times$ B combination ($i = 1, 2, \ldots, n$), j indexes the level of A (task; $j = 1, 2, \ldots, a$), k indexes the level of B (practice; $k - 1, 2, \ldots, b$), m indexes the level of C (day; $m = 1, 2, \ldots, a$), and p indexes the row within the square ($p = 1, 2, \ldots, a$).

We assume that subjects are randomly sampled from their respective populations. This means that the $\eta_{i/m}$ are a random sample from a normally distributed population having mean zero and variance $\sigma_{S/R}^2$. All other variables are assumed to have fixed effects. As in the design of Section 11.4, AC interaction effects can be tested assuming Equation 11.13 is a valid representation of the structure of the data. As indicated in the SV column and expected mean squares of Table 11.14, both the row (R) and between-cells residual (B cells res) terms reflect AC interaction effects. The $B \times R$ and $B \times B$

TABLE 11.14 THE ANALYSIS FOR THE DESIGN OF TABLE 11.13

SV	df	SS	MS	F^a	EMS
B	$b - 1 = 1$	1925.016	1925.016	8.168	$\sigma_e^2 + a\sigma_{S/BR}^2 + a^2 n\theta_B^2$
$R\,(AC')$	$a - 1 = 3$	507.797	169.266	.718	$\sigma_e^2 + a\sigma_{S/BR}^2 + abn\theta_{AC}^2$
$BR\,(ABC')$	$(b-1)(a-1) = 3$	829.797	276.599	1.174	$\sigma_e^2 + a\sigma_{S/BR}^2 + an\theta_{ABC}^2$
S/BR	$ab(n-1) = 8$	1885.375	235.672		$\sigma_e^2 + a\sigma_{S/BR}^2$
C	$a - 1 = 3$	734.922	244.974	9.420^b	$\sigma_e^2 + abn\theta_C^2$
A	$a - 1 = 3$	1790.297	596.767	22.948^b	$\sigma_e^2 + abn\theta_A^2$
AB	$(a-1)(b-1) = 3$	533.296	177.765	6.836^b	$\sigma_e^2 + an\theta_{AB}^2$
BC	$(a-1)(b-1) = 3$	317.672	105.891	4.072	$\sigma_e^2 + an\theta_{BC}^2$
B cells res $(AC')^c$	$(a-1)(a-2) = 6$	147.969	24.662		$\sigma_e^2 + n\theta_{AC}^2$
B cells res $\times$ B $(ABC')^d$	$(a-1)(a-2)(b-1) = 6$	.470	.078		$\sigma_e^2 + n\theta_{AC}^2$
W cells res	$bn(a-1)^2 = 24$	624.125	26.005		σ_e^2
Total	$bna^2 - 1 = 63$	9296.736			

[a] Assuming the model of Equation 11.3, B, R, and BR are tested against S/BR, and all other terms are tested against the within-subjects residual.
[b] $p < .01$.
[c] $SS_{B\,\text{cells res}} = SS_{CR} - SS_A$.
[d] $SS_{B \times B\,\text{cells res}} = SS_{BCR} - SS_{AB}$.

cells res terms reflect *ABC* interaction effects under the model presented in Equation 11.13. This also is indicated in the SV and EMS columns of Table 11.14.

Main and interaction effects are calculated as in all preceding chapters. The only terms whose calculations might pose a problem are the between-cells residual and its interaction with *B*. As we state in the footnotes to Table 11.14, these can be viewed as differences between more standard terms. This observation again forms the basis for using computer packages to analyze the data. As with other designs we have discussed, create an A file and a C file. Each row of the file corresponds to a subject and contains values denoting the level of *B* and the row of the square, as well as the *a* scores for that subject. The scores are ordered either by level of *A* or by level of *C*. Any computer program capable of handling two between-subjects variables (*B*, *R*) and one within-subjects variable (*A* or *C* depending on the file) can then be used to do the two analyses. Following the footnotes in Table 11.14, we subtract SS_A (A file ANOVA) from SS_{CR} (C file ANOVA) and SS_{AB} from the SS_{BCR} to obtain $SS_{B\,\text{cells res}}$ and $SS_{B \times B\,\text{cells res}}$. The results will be the same if we calculate $SS_{AR} - SS_C$ and $SS_{ABR} - SS_{BC}$.

11.6 SEVERAL SQUARES OR REPLICATED SQUARES?

In Sections 11.4 and 11.5 we have considered two different approaches to experimental design, both based on the Latin-square principle. The replicated-squares design is very common, although the correct analysis of the data is considerably less so. The reason for the prominence of this design is that it is ordinarily more convenient to program and run experiments involving fewer sequences of treatments. Obtaining summary statistics may also be easier when there are fewer sequences. Nevertheless, our sense is that, whenever practical, the several-squares design should be preferred because interaction effects are likely to cancel each other when there are many different AC combinations present. The replicated-squares design runs the risk that some pattern of interaction unique to the one square selected will increase the difficulty of interpreting the results of F tests.

11.7 GRECO-LATIN SQUARES

The designs presented in this chapter are a subset of the possible variations of the Latin-square design. They are an even smaller subset of incomplete block designs. We have limited our presentation to those designs we view as most useful and most often used by researchers. However, we should note a few of the possible extensions. One which is occasionally referenced in the experimental literature is the Greco-Latin square. Suppose we have three levels of a variable A and three levels of a variable B. To have efficient tests of both, we wish to treat both A and B as within-subjects variables. But suppose it is impractical to test each subject under all nine combinations of A and B. If the levels of A are represented by the letters a, b, and c and those of B by α, β, and γ (hence "Greco-Latin"), the design might be

$$\begin{bmatrix} b\beta & a\alpha & c\gamma \\ c\alpha & b\gamma & a\beta \\ a\gamma & c\beta & b\alpha \end{bmatrix}$$

where a subject or group is tested with each sequence (row of the square). Note that the layouts of the Latin letters and the Greek letters each meet the requirements for a Latin square. Furthermore, each combination of levels of A and B appears exactly once in each row and column.

The above design has several problems. First, the introduction of another variable reduces the residual degrees of freedom; for a single square such as the preceding one, these degrees of freedom are $(a^2 - 1) - 4(a - 1) = (a - 1)(a - 3)$. This is zero when $a = 3$. To have any chance at rejecting a false null hypothesis, the Greco-Latin should involve at least six rows and columns. The second problem is perhaps more serious. The introduction of another factor in this way sets up the possibility of still more interactions that may be confounded with main effects of interest. This may cost us a clean test of such main effects. The third problem is that if AB interactions are present, we have no way of assessing them. It is rare that an investigator would want to forgo the opportunity to test for an interaction between two experimentally manipulated factors. In short, although there are circumstances in which this design has some appeal, we believe that investigators usually will be better off with some other design.

11.8 APPLICATIONS TO BETWEEN-SUBJECTS DESIGNS

Another extension of the designs considered so far is to experiments in which there are no repeated measures. Suppose we wish to compare the effects of four instructional methods (A). To increase the efficiency of our test of instructional effects, we decide to block subjects on the basis of previous experience with the subject matter (B) and on the basis of ability as measured by a pretest (C). If we have four instructional methods, and divide subjects into four levels with respect to both B and C, we require 64 cells with n subjects in each. We can reduce the experimental labor by using a Latin-square design with only 16 cells; B would be the rows, and C the columns.

Table 11.15 presents the design and cell totals; each cell total is based on nine subjects. Table 11.16 presents the analysis of variance. The SS_A, SS_B, and SS_C are calculated as usual, and $SS_{B \text{ cells res}} = SS_{\text{cells}} - SS_A - SS_B - SS_C$, where

$$SS_{\text{cells}} = \frac{T^2_{.jkm}}{9} - C$$

$$= \frac{12^2 + 24^2 + \cdots + 45^2)}{9} - \frac{516^2}{(16)(9)} = 35.5$$

To conserve space, we have not presented the individual scores. The within-cell mean square, $MS_{S/ABC}$, is just the average of the 16 cell variances, and is 9.5 in this example. The expected mean squares in Table 11.16 are based on the additive structural model,

TABLE 11.15   CELL TOTALS FOR A BETWEEN-SUBJECTS APPLICATION OF THE LATIN-SQUARE DESIGN

	C_1	C_2	C_3	C_4	$T_{..k.}$
B_1	$(A_3)_{12}$	$(A_1)_{24}$	$(A_2)_9$	$(A_4)_{48}$	93
B_2	$(A_1)_{36}$	$(A_3)_{18}$	$(A_4)_{57}$	$(A_2)_{21}$	132
B_3	$(A_2)_{27}$	$(A_4)_{69}$	$(A_1)_{33}$	$(A_3)_{15}$	144
B_4	$(A_4)_{51}$	$(A_2)_{24}$	$(A_3)_{27}$	$(A_1)_{45}$	147
$T_{...m} =$	126	135	126	129	$T_{....} = 516$

The totals for the levels of A are

	A_1	A_2	A_3	A_4
$T_{.j..} =$	138	81	72	225

TABLE 11.16 THE ANOVA FOR THE DESIGN OF TABLE 11.15

SV	df	SS	MS	F^a	EMS^b
B	$a - 1 = 3$	51.5	17.167	1.807	$\sigma_e^2 + a\sigma_B^2$
C	$a - 1 = 3$	1.5	.500	.053	$\sigma_e^2 + a\sigma_C^2$
A	$a - 1 = 3$	412.5	137.500	14.474^c	$\sigma_e^2 + a\sigma_A^2$
B cells res	$(a - 1)(a - 2) = 6$	35.5	6.083	.640	σ_e^2
S/ABC	$a^2(n - 1) = 128$	1216.0	9.500		σ_e^2
Total	$a^2n - 1 = 143$	1717.0			

[a] $MS_{S/ABC}$ is the error term for all F tests.
[b] EMS are based on Equation 11.14.
[c] $p < .01$.

$$Y_{ijkm} = \mu + \alpha_j + \beta_k + \gamma_m + \epsilon_{ijkm} \tag{11.14}$$

where α_j, β_k, and γ_m are assumed to be fixed-effect variables.

The design of Table 11.15 potentially reduces error variance due to two sources: subject matter knowledge and ability in our example. This can lead to substantial increases in power when the factor of primary interest is manipulated between subjects, as instructional methods must be in our example. However, there is a potential problem that can be most clearly seen in the expected mean squares presented in Table 11.17. As in the analysis of Table 11.16, we assume that all three factors are fixed; now, however, we also assume that interactions may exist in the sampled population. This generalization of the model of Equation 11.14 is

$$Y_{ijkm} = \mu + \alpha_j + \beta_k + \gamma_m + (\alpha\beta)_{jk} + (\alpha\gamma)_{jm} + (\beta\gamma)_{km} + (\alpha\beta\gamma)_{jkm} + \epsilon_{ijkm} \tag{11.15}$$

Looking at the expected mean squares in Table 11.17, we can see that if these interaction effects are present in the data, F tests using the within-cell mean square as an error term may be positively biased. Significant tests of A may reflect BC or ABC, rather than A, effects. Tests against the between-cells residual term will be negatively biased, and the extent of that bias will depend upon which interactions exist and how large their contribution to the data is. Such tests are also undesirable because the denominator will be based on few degrees of freedom unless many levels of all factors are used.

We recommend that this design be used in cases in which intuition and pilot data have already shown that unbiased F tests of the treatments of interest can be computed. This does not mean that complete nonadditivity is required. For example, if our blocking variables, B and C, do not interact with each other (that is, $\theta_{BC}^2 = 0$), we can see from the expected mean squares for A and S/ABC in Table 11.17 that the F test of A against S/ABC is unbiased even if other interaction components are present in the population. The F test of B cells res also can provide valuable information. If significant, we cannot know which interaction components are present, and the test of A against S/ABC is on shaky ground. However, if the F test of B cells res is clearly not significant, as in

TABLE 11.17   EXPECTED MEAN SQUARES FOR THE DESIGN OF TABLE 11.15,
NONADDITIVE MODEL

SV	EMS[a]
B	$\sigma_e^2 + n\theta_{AC}^2 + \left[\dfrac{n(a-2)}{a}\right]\theta_{ABC}^2 + an\theta_B^2$
C	$\sigma_e^2 + n\theta_{AB}^2 + \left[\dfrac{n(a-2)}{a}\right]\theta_{ABC}^2 + an\theta_C^2$
A	$\sigma_e^2 + n\theta_{BC}^2 + \left[\dfrac{n(a-2)}{a}\right]\theta_{ABC}^2 + an\theta_A^2$
B cells res	$\sigma_e^2 + n\theta_{AB}^2 + n\theta_{AC}^2 + n\theta_{BC}^2 + \left[\dfrac{n(a-3)}{a}\right]\theta_{ABC}^2$
S/ABC	σ_e^2

[a] EMS are based on Equation 11.15.

Table 11.16, it is reasonable to assume the additive model of Equation 11.14, and the test of A against S/ABC would appear to be unbiased.

11.9 CONCLUDING REMARKS

Several points about designs that use the Latin-square principle deserve emphasis. First, such designs are potentially very efficient because they permit blocking with respect to two variables, most often subjects and trials. One implication of this is that the design will be particularly useful when the pool of potential subjects is small, as in certain clinical settings, or when the subject's task is unpleasant or dangerous. Second, the potential benefits carry with them the risk that the F test may be biased. This is because variability due to interactions cannot be removed from the total pool of variability. This in turn is due to the fact that Latin squares are incomplete designs, and therefore there are insufficient degrees of freedom to permit orthogonal tests of both main effects and interactions of rows, columns, and treatments. Third, even if nonadditivity is present, its effects will often be outweighed by the reduction in error variance that results from blocking on the column variable. Fourth, experimenters should routinely plot residuals to obtain a sense of the degree, if any, of nonadditivity. In the replicated-squares design, $MS_{B\,cells\,res}/MS_{res}$ provides a ready test of nonadditivity. Given evidence of marked nonadditivity, the researcher might consider transforming the data to a scale for which the additive model is adequate. If such a transformation cannot be found, or if the researcher is reluctant to move to a new scale, at least the knowledge of nonadditivity can aid the experimenter in assessing the implications of the F ratios calculated and perhaps in qualifying the inferences to be drawn. Finally, we again wish to emphasize our sense that the Latin square, particularly the replicated-squares design, is often used by researchers, but the data rarely are properly analyzed.

If row and column variables are ignored, Type 1 and Type 2 error rates are likely to be distorted.

APPENDIX: The Relation Between Error Terms for the Repeated-Measures and Latin-Square Designs

Consider a repeated-measures (RM) design and a Latin-square (LS) design with a subjects and a levels of A. The expectations of the within-block sums of squares for the $S \times A$ and Latin-square designs will be the same. That is,

$$E[SS_{SA,RM} + SS_{A,RM}] = E[SS_{res} + SS_{A,LS} + SS_{C,LS}]$$

Because $E(SS) = (df)[E(MS)]$, we can rewrite this expression in terms of population variance components:

$$(a-1)^2\sigma_{SA}^2 + (a-1)(\sigma_{SA}^2 + a\theta_A^2) = (a-1)(a-2)\sigma_{LS}^2 + (a-1)(\sigma_{LS}^2 + a\theta_A^2)$$
$$+ (a-1)(\sigma_{LS}^2 + a\theta_C^2)$$

Subtracting $(a-1)a\theta_A^2$ from both sides and combining terms, we have

$$(a-1)(a)\sigma_{SA}^2 = (a-1)^2\sigma_{LS}^2 + (a-1)(\sigma_{LS}^2 + a\theta_C^2)$$

Dividing by $(a-1)(a)$ yields

$$\sigma_{SA}^2 = \frac{a-1}{a}\sigma_{LS}^2 + \frac{1}{a}(\sigma_{LS}^2 + a\theta_C^2)$$
$$= (\sigma_{LS}^2 - \frac{1}{a}\sigma_{LS}^2) + \frac{1}{a}(\sigma_{LS}^2 + a\theta_C^2)$$
$$= \sigma_{LS}^2 + \frac{1}{a}[(\sigma_{LS}^2 + a\theta_C^2) - \sigma_{LS}^2]$$

or

$$E(MS_{SA}) = E[MS_{res} + \frac{1}{a}(MS_C - MS_{res})]$$

which is Equation 11.2.

EXERCISES

11.1   Consider the following 4×4 Latin square:

	C_1	C_2	C_3	C_4
S_1	$(A_1)_{25}$	$(A_4)_{16}$	$(A_2)_{24}$	$(A_3)_{18}$
S_2	$(A_2)_{19}$	$(A_1)_{19}$	$(A_3)_{13}$	$(A_4)_{12}$
S_3	$(A_4)_{13}$	$(A_3)_{18}$	$(A_1)_{20}$	$(A_2)_{16}$
S_4	$(A_3)_{17}$	$(A_2)_{19}$	$(A_4)_{18}$	$(A_1)_{17}$

(a) Test the A and C sources of variance.

(b) Calculate the estimates of the 16 $S \times A$ interaction effects $[(\widehat{\eta\alpha})_{ij}]$. Then subtract each $(\widehat{\eta\alpha})_{ij}$ from Y_{ij}. Note that

$$Y_{ij} - (\widehat{\eta\alpha})_{ij} = Y_{ij} - (Y_{ij} - \bar{Y}_{i.} - \bar{Y}_{.j} + \bar{Y}_{..}) = \bar{Y}_{i.} + \bar{Y}_{.j} - \bar{Y}_{..}$$

Calculate SS_S, SS_A, and SS_C for these adjusted scores. Which, if any, change? What potential problem in the use of this design do these results reflect?

11.2 Suppose we had used a repeated-measures design in which the sequence of the A_j had been randomly assigned for each subject.

(a) Estimate MS_{SA} from the data set in Exercise 11.1.

(b) What is your estimate of the relative efficiency of the two designs?

(c) Suppose every subject had scores three points lower at C_1 and three points higher at C_2. Would the efficiency relative to the $S \times A$ design be less than, the same as, or greater than the value you calculated in part (b)? Why?

11.3 In many cases, researchers counterbalance the order of presentation of the treatment variable A, but then analyze the data as though they had an $S \times A$ design. Carry out this analysis for the data of Exercise 11.1. Do your conclusions change from those drawn using the proper Latin-square analysis? Explain why this is so. Under what conditions would you expect to have similar results (p values) for the two analyses?

11.4 In a study of decision making, two factors were manipulated. The task either resembled one seen during a practice session or did not (experience, E) and the amount of information available was either high or low (information level, I). Each subject was tested under all four combinations of E and I with the assignment of EI combinations to problems (P) counterbalanced through the use of a Latin square. Decision times were

	P_1	P_2	P_3	P_4
S_1	$(E_1 I_1)_{1.4}$	$(E_2 I_2)_{2.2}$	$(E_1 I_2)_{1.5}$	$(E_2 I_1)_{1.5}$
S_2	$(E_1 I_2)_{2.1}$	$(E_1 I_1)_{1.5}$	$(E_2 I_1)_{2.0}$	$(E_2 I_2)_{2.4}$
S_3	$(E_2 I_2)_{2.8}$	$(E_2 I_1)_{2.1}$	$(E_1 I_1)_{1.4}$	$(E_1 I_2)_{1.8}$
S_4	$(E_2 I_1)_{2.3}$	$(E_1 I_2)_{2.3}$	$(E_2 I_2)_{2.7}$	$(E_1 I_1)_{1.6}$

Carry out the analysis of variance.

11.5 Five subjects were asked to learn different lists of seven words on five different days. The lists differed with respect to the level of English language frequency (F) of the words. Errors made at each of the seven positions (P) in each list were recorded. Present SV and df, and indicate the error terms for testing F, P, and FP.

11.6 Each of 36 subjects was tested on four successive days in a signal detection task; the number of errors was recorded each day. A different payoff matrix was used on each day, and 36 sequences of payoffs were used, making nine 4

× 4 Latin squares. Present the SV and df and indicate the error term for both payoffs (P) and days (D).

11.7   A clinical investigator is concerned with how alcohol affects reaction time (RT) to perceptual inputs. He runs three groups of 20 subjects each on four tasks, each task on a different day. The groups differ in alcohol intake. The tasks (T) measure simple RT to a visual input, simple RT to an auditory input, choice RT to a visual input, and choice RT to an auditory input. The tasks are presented in 60 sequences making 15 Latin squares. The investigator is interested in the effects of both the task complexity (simple or choice RT) and the nature of the perceptual input (visual or auditory), the amount of alcohol intake, and the cumulative effects of alcohol over days. Present the SV, df, and error terms.

11.8   A different design could have been used in Exercise 11.6. We could have used one Latin square with nine subjects in each payoff sequence. Again present the SV, df, and error term(s) against which D and P will be tested.

11.9   In Exercise 11.7 we could have used a single Latin square, replicated 15 times. Redo the analysis, assuming this design.

11.10 A researcher wishes to investigate cognitive performance as a function of drug type (T) and dosage (D). Thirty-two subjects are randomly assigned to one of two drugs, and each subject is administered a different one of four dosages of the same drug on four different occasions (O). There are four sequences of dosages with eight subjects in each sequence. Note that half of the subjects in each sequence receive T_1 and half receive T_2.
(a) Give the SV, df, and error terms.
(b) Another way to run this study would be to create a single 8 × 8 Latin square with two subjects in each row (so there are still 128 total scores). The eight treatments would be all possible combinations of the drug (T) and dosage (D) levels. Write out the SV, df, and error terms for this case.
(c) What are the pros and cons of the two designs?

11.11 We are interested in gambling behavior under variations in initial stake (I), payoffs (P), and probability of winning (W). There are three levels of each of these variables. Eighty-one subjects are available for this experiment. There are many possible experimental designs we could use, ranging from a completely between-subjects design with three subjects in each of 27 cells to some of the Latin-square designs (or variations of them) described in this chapter. Suggest several alternative designs and discuss their relative merits.

Chapter 12

Bivariate Correlation and Regression

12.1 INTRODUCTION

Although the analysis of variance procedures we have so far discussed are widely used to determine whether independent and dependent variables are related, they have some important limitations. ANOVA treats independent variables that are inherently continuous as though they were categorical, thereby ignoring potentially useful information. In addition, the neat partitioning of variability into nonoverlapping components associated with main effects and interactions that is the hallmark of ANOVA is only possible when the independent variables are uncorrelated with one another.

We now begin a series of chapters that develop a more general framework for discussing both categorical and continuous independent variables as well as variables that are correlated. We start by considering *bivariate* (i.e., two-variable) correlation and regression and then extend our discussion to *multiple regression,* which deals with many variables. We eventually discuss analysis of variance as a special case of multiple regression.

In this chapter, we discuss how relations between pairs of variables can be measured and described. Variables may be related without being perfectly related. Tall fathers are more likely to have tall sons, but sons may be shorter or taller than their fathers. There is some relation between educational level and income, but many uneducated people are wealthy and many educated people are not. Given that variables may be more or less strongly related, it is desirable to develop measures of strength of relation.

Also, if two variables are related, it should be possible to use information about one of them to make predictions about the other. Knowing the father's height will usually help us in predicting the height of his son. Predictions of Y based on X are potentially more accurate when the two variables are strongly related. However, whatever the strength of relation, it will often be desirable to use information about X to make the most accurate predictions of Y that are possible. Equations that use information about one variable to make predictions about a second variable are referred to as *simple or bivariate regression equations.*

Table 12.1 contains two scores for each of 18 students who took an introductory statistics course: X is the score obtained on a math skills pretest administered at the beginning of the semester, and Y is the score obtained on the final examination. The 18 data points are plotted in Figure 12.1, using the SPSS^X PLOT procedure. Not surprisingly, the pretest and final scores tend to covary. We can measure the relation between pretest and final examination performance and use the pretest scores to predict final performance. We can find the equation for the straight line that produces the least amount of error in predicting the 18 final exam scores from the pretest scores. We can then use this *linear regression equation* to describe the relation between the two sets of scores and to predict final exam performance for other students who subsequently take the pretest. The SPSS^X PLOT procedure not only plots the data, it gives a measure of the strength of relation between the two variables and provides information about the regression equation for predicting Y from X.

In the first part of this chapter, we discuss the *Pearson product-moment correlation coefficient,* which is an index of the extent to which two variables have a *linear* relation. In the rest of the chapter, we concentrate on how to find the best regression equation for the data in a sample and how to make inferences about the equation that best fits the population from which the sample was selected.

TABLE 12.1   STATISTICS CLASS EXAMPLE DATA

Pretest score (X)	Final exam score (Y)
29	47
34	93
27	49
34	98
33	83
31	59
32	70
33	93
32	79
35	79
36	93
34	90
35	77
29	81
32	79
34	85
36	90
25	66
$\bar{X} = 32.28$	$\bar{Y} = 78.39$

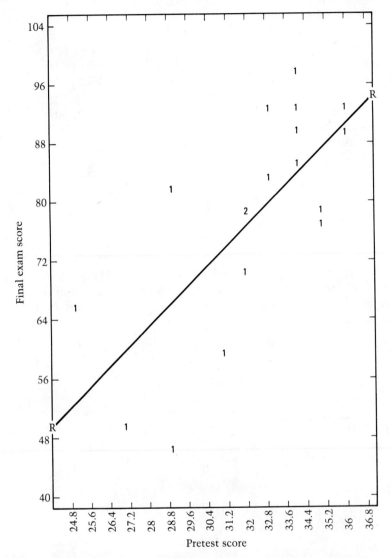

18 CASES PLOTTED. REGRESSION STATISTICS OF FINAL ON PRETEST:
CORRELATION .72509 R SQUARED .52576 S.E. OF EST 10.63832
2-TAILED SIG. .0007 INTERCEPT (S.E.) −36.08319 (27.29514) SLOPE (S.E.)
3.54647 (.84206)

Figure 12.1 Scatter diagram for the statistic class data. (Note that each 1
indicates the coordinates of the data point. The 2 indicates the location of two
data points that are too close together to be plotted separately.)

12.2 LINEAR FUNCTIONS AND SCATTER DIAGRAMS

Each panel in Figure 12.2 consists of a plot of (X, Y) data points called a scatter diagram. If all members of a set of data points fall exactly on a straight line, we say there is a perfect linear relation between X and Y. The perfect relation is said to be positive if the slope of the line is positive and negative if the slope is negative.

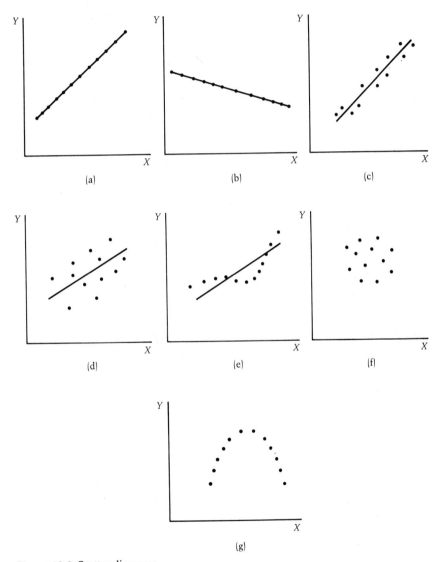

Figure 12.2 Scatter diagrams.

An equation of the form

$$Y = b_0 + b_1 X \qquad (12.1)$$

where b_0 and b_1 are constants, is referred to as a *linear equation* because all points (X, Y) that satisfy the equation fall on a straight line. The constant b_1 is called the slope of the line and indicates the rate of change of Y with X. From Equation 12.1, if X changes by an amount C, Y changes by an amount $b_1 \times C$, so that

$$\text{rate of change of } Y \text{ with } X = \frac{\text{corresponding change in } Y}{\text{change in } X}$$

$$= b_1$$

The constant b_0 is the *Y intercept*—that is, the value of Y when X is equal to zero.

If there is a perfect linear relation between X and Y, the equation of the line can be used to find the exact value of Y corresponding to any value of X. For example, if all the points fall on the line given by the equation $Y = 3X - 2$, then Y can be obtained by multiplying X by 3 and subtracting 2 from the result. Conversely, as $X = (Y + 2)/3$, X can be found if Y is known.

Panels a and b of Figure 12.2 depict examples of perfect positive and negative linear relations, respectively. In panel c, there is a systematic increase in Y as X increases. However, not all the data points fall on the straight line that seems to best capture this systematic increase, although they cluster closely around it. In this case, we say that there is a strong positive linear relation between X and Y but not a perfect one. In panel d there is less clustering around the line, indicating a weaker linear relation. In panel e it is possible to find a straight line that represents one component of the relation between X and Y. However, the points not only fail to cluster closely around the line, they seem to indicate that there is also a systematic nonlinear component. In panels f and g, there is no overall linear relation between X and Y; no straight line passing through the center of either "cloud" of data points is better than the line parallel to the X axis in characterizing the overall relation between X and Y. In g, however, X and Y are positively related for small values of X but negatively related for large values.

12.3 THE CORRELATION COEFFICIENT AS AN INDEX OF THE EXTENT TO WHICH THERE IS A LINEAR RELATION

We indicated in the previous section that the tighter the clustering of data points about the best-fitting straight line, the stronger the degree of linear relation. The notion of clustering around a straight line leads directly to a useful measure of linear relation. However, we will consider z scores instead of raw scores in developing this idea further. When raw scores are used, the appearance of a scatter diagram and the apparent degree of clustering around a straight line depends on the units in which X and Y are measured. This is not true for z scores.

In Chapter 3 we indicated that each member of a set of raw scores X_1, $X_2, \ldots, X_N$ can be converted to a z score by using $z_X = (X - \bar{X})/\hat{\sigma}_X$. The mean of a complete set of z scores is 0, and the standard deviation and variance are both 1, so that $\Sigma z^2/(N-1) = 1$. Also, any changes in units that involve multiplying all the scores by a constant (as when converting from inches to centimeters) or adding a constant to all the scores change the raw scores but not the sizes of their corresponding z scores.

If there is a strong positive relation between X and Y, Y tends to increase as X does. Therefore, large values of z_Y will be paired with large values of z_X, and small values of z_Y will be paired with small values of z_X. In fact, it can be shown (see Appendix 12.1) that if there is a perfect positive linear relation between X and Y, z_X is exactly equal to z_Y, and if there is a perfect negative linear relation, z_X is exactly equal to $-z_Y$. If there is no linear relation between X and Y, there is no overall tendency for large z_Y scores to be paired with large z_X scores.

The variance of $d = z_Y - z_X$, the difference between the z scores for X and Y, can be used as a measure of linear relation. This variance can be written as

$$\hat{\sigma}_d^2 = \Sigma \frac{(d - \bar{d})^2}{N - 1} \tag{12.2}$$

$$= \Sigma \frac{d^2}{N - 1} \quad \text{because } \bar{d} = \bar{z}_Y - \bar{z}_X = 0 - 0$$

$$= \Sigma \frac{(z_Y - z_X)^2}{N - 1}$$

If there is a perfect positive relation between X and Y, the variance of the d's must equal zero because each d is equal to zero. If there is a perfect negative linear relation, $d = z_Y - z_X = z_Y - (-z_Y) = 2z_Y$ so that

$$\hat{\sigma}_d^2 = \Sigma \frac{(2z_Y)^2}{N - 1} = 4\Sigma \frac{z_Y^2}{N - 1} = 4$$

Therefore, we can see that the variance of the d's is small for strong positive linear relations, is large for strong negative linear relations, and takes on intermediate values if there is little or no linear relation between X and Y. However, a closely related measure, r, the covariance of z_X and z_Y, is much more commonly used as a measure of linear relation because it has the desirable property of varying between $+1$ and -1. To compare $\hat{\sigma}_d^2$ and r, note that if we expand the square of $z_X - z_Y$ in Equation 12.2, we obtain

$$\hat{\sigma}_d^2 = \Sigma \frac{(z_X^2 + z_Y^2 - 2z_X z_Y)}{N - 1}$$

$$= 1 + 1 - 2\Sigma \frac{z_X z_Y}{N - 1}$$

$$= 2 \left[1 - \Sigma \frac{z_X z_Y}{N - 1} \right]$$

Defining r as

$$r = \Sigma \frac{z_X z_Y}{N - 1} \tag{12.3}$$

we have

$$\hat{\sigma}_d^2 = 2(1 - r)$$

so that

$$r = 1 - \frac{\hat{\sigma}_d^2}{2}$$

A comparison of the values of $\hat{\sigma}_d^2$ and r is presented here:

Nature of linear relation	$\hat{\sigma}_d^2$	r
Perfect positive, $z_Y = z_X$	0	1
Perfect negative, $z_Y = -z_X$	4	-1
None	2	0

The quantity $r = \Sigma z_X z_Y / (N - 1)$ is called the *Pearson product-moment correlation coefficient* or simply the correlation coefficient. The symbol r is used to denote the correlation coefficient in a sample, and the symbol ρ (the Greek letter rho) denotes the correlation in a population.

The correlation coefficient can always be obtained from Equation 12.3. However, other equivalent expressions are often encountered. If we substitute the expressions for z scores into Equation 12.3, r can be expressed as the covariance between X and Y divided by the standard deviations of X and Y.

$$r = \frac{1}{N - 1} \Sigma \left(\frac{X_i - \bar{X}}{\hat{\sigma}_X} \frac{Y_i - \bar{Y}}{\hat{\sigma}_Y} \right) \tag{12.4}$$

$$= \frac{\hat{\sigma}_{XY}}{\hat{\sigma}_X \hat{\sigma}_Y} \quad \text{where } \hat{\sigma}_{XY} \text{ is the covariance of } X \text{ and } Y$$

The covariance of X and Y (i.e., the *amount* of variance shared by X and Y) is not usually employed as a measure of relation because it varies with changes in the units of measurement. The correlation can be thought of as the "standardized" covariance; that is, the covariance of the z scores.

The expression for r most often encountered in elementary textbooks is the so-called computational formula,

$$r = \frac{\text{SP}_{XY}}{\sqrt{\text{SS}_X}\sqrt{\text{SS}_Y}} \tag{12.5}$$

$$= \frac{\Sigma X_i Y_i - (\Sigma X_i)(\Sigma Y_i)/N}{\sqrt{\Sigma X_i^2 - (\Sigma X_i)^2/N}\sqrt{\Sigma Y_i^2 - (\Sigma Y_i)^2/N}}$$

where $\text{SP}_{XY} = \Sigma_i (X_i - \bar{X})(Y_i - \bar{Y}) = \Sigma X_i Y_i - (\Sigma X_i)(\Sigma Y_i)/N$ is the sum of the cross products of the deviations about $\bar{X}$ and $\bar{Y}$, and the SS terms are the sums of squared deviations about the mean for X and Y. Equation 12.5 has the advantage of providing simpler calculations and less rounding error than other expressions for the correlation. This advantage is less important today than it was formerly, because most calculations are done on computers, and even inexpensive hand calculators may provide r at the press of a button. Any of the formulas for r should yield the value .7251, give or take rounding error, for the data in Table 12.1. For example, using Equation 12.5 we have

$$SS_X = \Sigma X^2 - \frac{(\Sigma X)^2}{N}$$

$$= 18{,}913 - \frac{581^2}{18} = 159.6111$$

$$SS_Y = \Sigma Y^2 - \frac{(\Sigma Y)^2}{N}$$

$$= 114{,}425 - \frac{1411^2}{18}$$

$$= 3818.2778$$

$$SP_{XY} = \Sigma XY - \frac{(\Sigma X)(\Sigma Y)}{N}$$

$$= 46{,}110 - \frac{(581)(1411)}{18}$$

$$= 566.0556$$

so that

$$r = \frac{566.0556}{\sqrt{158.6111}\sqrt{3818.2778}}$$

$$= .725$$

The large correlation coefficient indicates that the relation between pretest and final exam score for the 18 students who took the statistics class has a strong linear component; the higher the scores students had on the pretest, the better they tended to do on the final. The 18 students who took the class can be thought of as a sample from a hypothetical population consisting of students who might have taken the class and those who might do so in the future, and the sample coefficient provides an estimate of the correlation in this hypothetical population. Procedures for testing hypotheses about the population correlation coefficient and finding confidence intervals for it will be dealt with in Chapter 14.

It should be emphasized that finding a significant linear component does not preclude the possibility that there is also a systematic nonlinear component to the relation. If there is significant nonlinearity, the linear regression equation will not completely describe the relation between X and Y and the correlation coefficient will underestimate the extent to which Y systematically varies with X. Therefore, we should plot the scatter diagram to see whether there seems to be a strong departure from linearity and use the procedure discussed in Section 12.10 to test whether the nonlinearity is significant.

The scatter diagram in Figure 12.2 produced by the SPSS^X PLOT procedure does not indicate any strong nonlinearity in the relation between X and Y. However, we should be cautious about ruling out any nonlinearity in the population because we have only four pretest scores less than 30 in our sample, and the two scores of 29 have widely different final exam scores associated with them.

The PLOT procedure not only provides us with the scatter diagram, it tells us that the correlation is .72509 and that the slope and intercept of the linear regression equation are 3.55 and -36.08, respectively. It also provides some additional statistics that we will discuss later. In the next section, we discuss how the regression equation is obtained.

12.4 LINEAR PREDICTION AND THE LEAST-SQUARES CRITERION

12.4.1 The Optimal Linear Regression Equation

In order to find the linear equation

$$\hat{Y} = b_0 + b_1 X$$

that best predicts Y from X, we must find the values of b_0 and b_1 that result in the smallest amount of prediction error.

First of all, we have to decide what to use as our index of error. If, on a given trial, we predict that Y has a value of $\hat{Y}_i$ and it actually turns out to have a value of Y_i, our error is $Y_i - \hat{Y}_i$. The mean of these errors for a set of N predictions is $(1/N)\Sigma(Y_i - \hat{Y}_i)$. However, this mean isn't a good measure of error because positive and negative errors cancel one another. The index of error that is generally used is the mean of the *squared* errors, MSE $= (1/N)\Sigma(Y_i - \hat{Y}_i)^2$, because this measure will take on a value of zero only if prediction is perfect for the entire set of data points. Regression equations that have the property of minimizing the MSE are said to be optimal according to the *least-squares criterion*.

What we are seeking, then, are values of b_0 and b_1 that minimize the mean-squared error

$$\text{MSE} = \frac{1}{N} \Sigma(Y_i - \hat{Y}_i)^2$$

$$= \frac{1}{N} \Sigma(Y_i - b_0 - b_1 X_i)^2$$

Using differential calculus, we can show that these values are given by

$$b_1 = r\frac{\hat{\sigma}_Y}{\hat{\sigma}_X} \tag{12.6}$$

and

$$b_0 = \bar{Y} - b_1 \bar{X} \tag{12.7}$$

Note that b_1 can be expressed in any of a number of equivalent ways, including

$$r\frac{\sqrt{SS_Y}}{\sqrt{SS_X}}, \quad \frac{\hat{\sigma}_{XY}}{\hat{\sigma}_X^2}, \quad \frac{SP_{XY}}{SS_X}, \quad \text{and} \quad \frac{\Sigma X_i Y_i - (\Sigma X_i)(\Sigma Y_i)/N}{\Sigma X_i^2 - (\Sigma X_i)^2/N}$$

Applying Equations 12.6 and 12.7 to the statistics class data in Table 12.1 allows us to find the linear equation that best predicts final exam performance from pretest score. Substituting into one of the expressions for b_1, we have

$$b_1 = r\frac{\sqrt{SS_Y}}{\sqrt{SS_X}}$$

$$= (.725)\frac{\sqrt{3818.278}}{\sqrt{159.611}}$$

$$= 3.546$$

and from Equation 12.7, $b_0 = 78.39 - (3.546)(32.28) = -36.08$, so that our regression equation is

$$\hat{Y} = -36.08 + 3.55X$$

A difference of 1 point on the pretest translates into a predicted difference of about 3½ points on the final exam. Our prediction for the final exam score of a student who scored 30 on the pretest would be $-36.08 + (3.55)(30) = 70.42$, or 70, rounding to the nearest integer.

We would like to make several additional points in concluding this section. We first comment briefly on the calculus procedures that produced Equations 12.6 and 12.7, although knowing the details is not necessary for understanding any of the rest of the chapter. Finding the expressions for b_0 and b_1 that minimize the MSE involves finding the partial derivatives of MSE with respect to b_0 and b_1. Setting these derivatives equal to zero and simplifying produces a set of what are called *normal equations*:

$$b_0 N + b_1 \Sigma X_i - \Sigma Y_i = 0 \tag{12.8}$$

and

$$b_0 \Sigma X_i + b_1 \Sigma X_i^2 - \Sigma X_i Y_i = 0$$

Solving the normal equations simultaneously for b_0 and b_1 yields the expressions presented in Equations 12.6 and 12.7.

We would also like to point out that there are several additional ways of writing the regression equation that can be useful. Substituting Equation 12.7 into $\hat{Y} = b_0 + b_1 X$ yields

$$\hat{Y} = \bar{Y} + b_1(X_i - \bar{X}) \tag{12.9}$$

and substituting $b_1 = r\hat{\sigma}_Y/\hat{\sigma}_X$ into Equation 12.9 and dividing both sides by $\hat{\sigma}_Y$ yields $(\hat{Y} - \bar{Y})/\hat{\sigma}_Y = r(X - \bar{X})/\hat{\sigma}_X$ or

$$\hat{z}_Y = rz_X \tag{12.10}$$

the z-score form of the regression equation. Note that in z-score form, the regression line has slope r and passes through the origin.

12.4.2 Predicting X from Y

So far, we have discussed the regression equation for predicting Y from X that is optimal in the sense that it minimizes $(1/N)\Sigma(Y_i - \hat{Y}_i)^2$ (see panel a of Figure 12.3). Exactly the same reasoning can be used to find the regression equation for predicting X from Y. In this case, the index of error that is minimized is $(1/N)\Sigma(X_i - \hat{X}_i)^2$, the mean of the squares of the deviations indicated in panel b of Figure 12.3.

The expressions that have been developed for predicting Y from X can be transformed into expressions for predicting X from Y by simply interchanging X and Y. For example,

$$\hat{X}_i = r \frac{\hat{\sigma}_X}{\hat{\sigma}_Y}(Y_i - \bar{Y}) + \bar{X}$$

or

$$\hat{z}_X = rz_Y$$

Of course, whether it makes any sense to predict X from Y depends on the nature of the variables. It is unlikely that we would want to predict pretest scores from final exam scores because pretest scores are available first.

Figure 12.4 indicates how the regression lines that predict z_Y from z_X and z_X from z_Y differ from one another. Imagine that the elliptical "envelope" that has been drawn

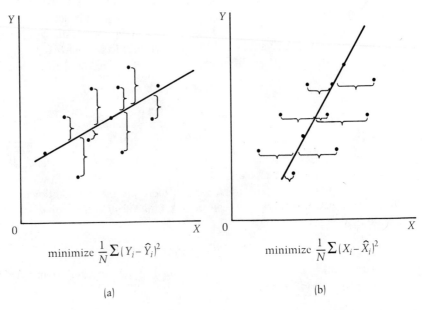

$$\text{minimize } \frac{1}{N}\Sigma(Y_i - \hat{Y}_i)^2 \qquad\qquad \text{minimize } \frac{1}{N}\Sigma(X_i - \hat{X}_i)^2$$

(a)   (b)

Figure 12.3 Graphical representation of (a) the regression of Y on X and (b) the regression of X on Y.

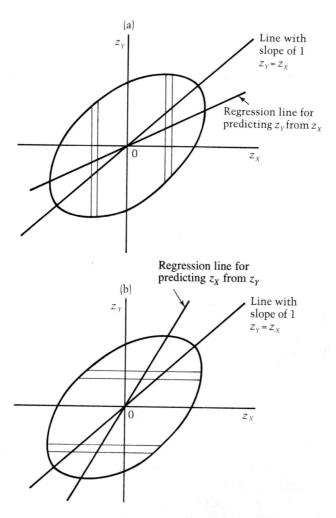

Figure 12.4 Regression lines for predicting (a) z_Y from z_X and
(b) z_X from z_Y when there is an imperfect relation between X and
Y.

in the figure to represent an imperfect linear relation contains a large number of data points. Imagine further that the ellipse is divided into a number of narrow vertical strips. Notice that even though the envelope is symmetrical about a straight line with a slope of 1 drawn through the origin (i.e., $z_Y = z_X$), the mean value of z_Y associated with any given value of z_X is closer to 0 than z_X is. The line that best fits the points representing the mean values of z_Y in the vertical strips will approximate $\hat{z}_Y = rz_X$, the regression equation for predicting z_Y from z_X.

On the other hand, if we divide the ellipse into a large number of narrow horizontal strips, the line that best fits the mean values of z_X in the strips will approximate the regression equation for predicting z_X from z_Y, $\hat{z}_X = rz_Y$. It should be apparent from Figure 12.4 that these two regression lines are not the same.

12.4.3 Regression to the Mean

In the previous section, we noted that when there is an imperfect linear relation between X and Y, the z_Y's associated with any given value of z_X will, on the average, be closer to zero than z_X is. This can also be seen from Equation 12.10, $\hat{z}_Y = rz_X$, which indicates that $\hat{z}_Y$ is closer to zero than z_X whenever $|r| < 1$. We refer to this phenomenon as *regression toward the mean*. The degree of regression toward the mean is larger for more extreme values of z_X. For $r = .5$, if $z_X = 2$, then $\hat{z}_Y = 1$, a difference of 1 standard deviation unit between z_X and $\hat{z}_Y$; however, if $z_X = .5$, then $\hat{z}_Y = .25$, a difference of only .25 unit. The only case in which there is no regression to the mean is when there is a perfect linear relation between X and Y. In this case, $|r|$ is equal to 1, and the z_Y's have exactly the same magnitudes as the corresponding z_X's.

The fact that optimal predictions for both X and Y regress toward the mean results in an apparent lack of symmetry that at first glance seems to be somewhat counterintuitive. Suppose, for example, that the correlation between the heights of fathers and their adult sons is .5 and that the mean height for males is 69 inches with a standard deviation of 2 inches. Using the regression equation, we find that the best prediction for the heights of sons whose fathers are 73 inches tall is 71 inches. However, it does not follow from this that the best prediction for the heights of fathers of 71-inch sons is 73 inches. Rather, from the regression equation (and as can be seen in Figure 12.4), the best prediction for the fathers' heights (70 inches) again regresses toward the mean.

Regression to the mean always occurs whenever there is an imperfect linear relation between two variables and any factors that prevent the correlation from being perfect will result in the regression phenomenon. Unfortunately, there is a good deal of confusion about regression toward the mean. In particular, investigators have often felt compelled to explain effects that are due to regression to the mean in terms of "interesting" variables—ignoring the fact that these effects could have been produced solely by random variability or measurement error or any other factor that results in a less-than-perfect correlation. Suppose, for example, that we administer similar aptitude tests to a group of children on two occasions separated by a few months of intense instruction. Suppose further, that we observe a tendency for children with extremely low scores on the first test to show improvement in standing on the second test and for children with extremely high scores on the first test to decline.

Although we might wish to attribute this "leveling effect" to the intervening instruction (i.e., conclude that the instruction is more effective for the students who scored lower on the first test than for those who scored higher), we really cannot. Instruction is only one of the possible factors that could result in a less-than-perfect correlation between the two sets of test scores and thereby produce regression effects. Even if instruction was equally effective for all children, we would expect regression toward the mean if the tests did not measure exactly the same thing or if there were random fluctuations in the children's alertness and their success in guessing. If our goal was to assess the effects of instruction, we would be much better off to include an appropriate control group and conduct an experiment.

Another example of regression toward the mean is the "sophomore slump." Individuals who perform particularly well in their first year of academics or athletics will probably perform relatively less well in their second year. People who fail to recognize this regression as a natural consequence of the lack of perfect consistency in performance often feel that it requires an explanation in terms of overconfidence and poor work habits. A good deal of work in the psychological literature indicates that people generally are unaware of or underestimate regression effects and seem to believe that behavior is more predictable than it actually is.

Regression toward the mean severely complicates the study of change. For example, consider the children we discussed several paragraphs back who took two aptitude tests separated by a period of instruction. It might, at first glance, seem desirable to subtract the score on the first test from the score on the second in the hope of obtaining a "pure" measure of improvement or change that is unrelated to initial performance. However, a little thought makes it obvious that the difference scores are *not* free of the influence of the scores on the first test. As we indicated at the beginning of this section, whenever there is less than a perfect correlation, regression effects will occur, and these effects will be larger for more extreme scores. In our example, children who had extremely high scores on the first test will tend to have lower scores on the second one; children with extremely low scores on the first test will tend to score relatively higher on the second. It follows that the difference scores will tend to be negatively correlated with the scores on the first test, and how other variables correlate with the difference scores will depend to some extent on how they correlate with initial performance.

The study of change constitutes a methodological minefield for the unwary in which, as Cohen and Cohen (1983) state, "intuitive 'doing what comes naturally' is almost certain to lead one astray." A thorough discussion of the issues involved in the study of change is beyond the scope of the current presentation. The interested reader is advised to consult sources such as Cohen and Cohen (1983), Cronbach and Furby (1970), and Nesselroade, Stigler, and Baltes (1980).

Finally, we note that the prediction equations that have been discussed in this chapter are referred to as "regression" equations in recognition of the regression effects that they exhibit.

12.4.4 Partitioning the Variability in Bivariate Regression

An important part of introducing analysis of variance in Chapter 4 was to show that the variability in the Y scores could be partitioned into meaningful components associated with "within" and "between" variability. We started by showing that the deviation of a score from the grand mean can be expressed as the sum of its deviation from the group mean and the deviation of the group mean from the grand mean:

$$Y_{ij} - \bar{Y}_{..} = (Y_{ij} - \bar{Y}_{.j}) + (\bar{Y}_{.j} - \bar{Y}_{..})$$

Squaring and simplifying, we showed that

$$\Sigma\Sigma(Y_{ij} - \bar{Y}_{..})^2 = \Sigma\Sigma(Y_{ij} - \bar{Y}_{.j})^2 + \Sigma\Sigma(\bar{Y}_{.j} - \bar{Y}_{..})^2$$
$$SS_Y \qquad = \qquad SS_{S/A} \qquad + \qquad SS_A$$

Partitioning the variability in Y can also be useful in understanding regression. We first note that the deviation of Y_i from the mean of all Y scores can be broken down into two components: the deviation of Y_i from its predicted value, $\hat{Y}_i$, and the deviation of the predicted value from the mean (see Figure 12.5); that is,

$$Y_i - \bar{Y} = (Y_i - \hat{Y}_i) + (\hat{Y}_i - \bar{Y})$$

Squaring both sides and simplifying (and noting that the cross-product term disappears when $\hat{Y}_i$ is obtained from the optimal regression equation), we have

$$\Sigma(Y_i - \bar{Y})^2 = \Sigma(Y_i - \hat{Y}_i)^2 + \Sigma(\hat{Y}_i - \bar{Y})^2 \qquad (12.11)$$

$$SS_Y = SS_{error} + SS_{reg}$$

Here, $\Sigma(Y_i - \hat{Y}_i)^2$ is *error variability*, variability not accounted for by the regression equation, and $\Sigma(\hat{Y}_i - \bar{Y})^2$ is the amount of variability that *is* accounted for by the regression; together, these two components make up SS_Y.

Some insight into these SS terms can be obtained by expressing them in terms of the slope or the correlation coefficient. Substituting Equation 12.9 into the expression for SS_{reg} yields

$$SS_{reg} = \Sigma[b_1(X_i - \bar{X})]^2 \qquad (12.12)$$

$$= b_1^2 \Sigma(X_i - \bar{X})^2$$

$$= b_1^2 SS_X$$

We know from Equation 12.6 that $b_1 = r\sqrt{SS_Y}/\sqrt{SS_X}$. Therefore, $b_1^2 SS_X = r^2 SS_Y$. Substituting into Equation 12.12, we have

$$SS_{reg} = r^2 SS_Y \qquad (12.13)$$

and, because $SS_{error} = SS_Y - SS_{reg}$, we have

$$SS_{error} = (1 - r^2)SS_Y \qquad (12.14)$$

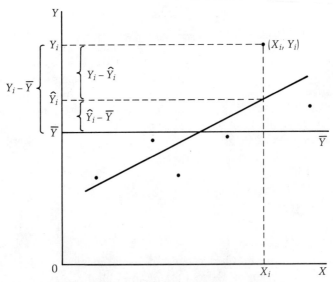

Figure 12.5 Partitioning $Y_i - \hat{Y}_i$ into components.

SS_{error} can be thought of as the variability that remains unaccounted for if the regression equation is used to predict each value of Y. If $r^2 = 1$, we see from Equation 12.14 that SS_{error} is equal to 0. If $r^2 = 0$, $SS_{error} = SS_Y$, the total variability in Y. $MSE = SS_{error}/N$ is referred to as the *sample variance of estimate* for predicting Y from X and is denoted by S_{YX}^2. We show in Appendix 12.2 that S_{YX}^2 is a biased estimate of σ_e^2, the population variance of estimate, but that an unbiased estimate is provided by dividing the error variance by its df. S_{YX}^2 is associated with $N - 2$ df; there are N data points, and 2 df are used up in estimating b_0 and b_1. Therefore, the unbiased estimate of σ_e^2 is given by $\hat{\sigma}_e^2 = MS_{error} = SS_{error}/(N - 2)$, so

$$\hat{\sigma}_e = \sqrt{MS_{error}} = \sqrt{\frac{SS_{error}}{N - 2}} \qquad (12.15)$$

$$= \sqrt{\frac{SS_Y(1 - r^2)}{N - 2}}$$

When we dealt with ANOVA, we saw how the ratio $F = MS_A/MS_{S/A}$ could be used to test whether there was an effect of the independent variable, A. Analogously, the ratio $F = MS_{reg}/MS_{error}$ can be used to test whether X provides a significant degree of predictability. In fact, the results of a regression analysis can be summarized in much the same way as an ANOVA.

SV	df	SS	MS	F
Regression	1	$\Sigma(\hat{Y}_i - \bar{Y})^2 = r^2 SS_Y = b_1^2 SS_X$	$\dfrac{SS_{reg}}{1}$	$\dfrac{MS_{reg}}{MS_{error}}$
Error	$N - 2$	$\Sigma(Y_i - \hat{Y}_i)^2 = (1 - r^2) SS_Y$	$\dfrac{SS_{error}}{N - 2}$	
Total	$N - 1$	$\Sigma(Y_i - \bar{Y})^2 = SS_Y$		

Table 12.2 provides output for the statistics class data obtained from the SPSS[X] REGRESSION procedure. Again, the output provides the correlation between the two variables and its square. The "standard error" of 10.638 is the value of $\hat{\sigma}_e$ given by Equation 12.15. The "analysis of variance" table tells us that SS_{reg} is 2007.50 and SS_{error} is 1810.78 (note that SPSS[X] uses the term *residual* rather than *error*). These are exactly the values that would be obtained by using Equations 12.13 and 12.14. The table also obtains the mean squares by dividing the SS terms by their df. There are 17 df associated with SS_Y because there are 18 scores and 1 df is used to estimate the mean of the Y scores. Of these 17 df, one is used to estimate the slope and is thus associated with SS_{reg} and the remaining 16 are associated with SS_{error}. The ratio $F = MS_{reg}/MS_{error}$ has a value of 17.738, indicating that the regression is highly significant.

Under the heading "variables in the equation," the output provides us with the best estimates of the slope (the coefficient of pretest) and the Y intercept (the constant in the equation), gives their estimated standard errors, and tests them for significance. It also

TABLE 12.2   SPSS^X OUTPUT FOR STATISTICS CLASS EXAMPLE

```
                    **** MULTIPLE REGRESSION ****

VARIABLE LIST NUMBER 1  LISTWISE DELETION OF MISSING DATA
EQUATION NUMBER 1  DEPENDENT VARIABLE.. FINAL
BEGINNING BLOCK NUMBER 1.  METHOD: ENTER  PRETEST
VARIABLE(S) ENTERED ON STEP NUMBER 1..  PRETEST

MULTIPLE R            .72509    ANALYSIS OF VARIANCE
R SQUARE              .52576                DF    SUM OF SQUARES   MEAN SQUARE
ADJUSTED R SQUARE     .49612    REGRESSION    1       2007.49741    2007.49741
STANDARD ERROR      10.63832    RESIDUAL     16       1810.78037     113.17377
                               F = 17.73819   SIGNIF F = .0007


          --------------VARIABLES IN THE EQUATION-----------------

VARIABLE           B        SE B      BETA        T     SIG T

PRETEST        3.54647     .84206    .72509     4.212    .0007
(CONSTANT)   -36.08319   27.29514             -1.322    .2048
```

Note: Because we deal with only two variables here, the MULTIPLE R referred to in the output is the same as the correlation coefficient r.

gives the regression coefficient for the standardized (z-score) version of the regression equation, which it calls "beta." From Equation 12.10, we know that in bivariate regression, beta takes on the value of the correlation coefficient.

12.4.5 The Coefficient of Determination, r^2

SS_{reg} is a measure of how much of the variability in Y is accounted for by the linear regression equation. Therefore,

$$\frac{SS_{reg}}{SS_Y} = \frac{\text{amount of variability in } Y \text{ accounted for by } X}{\text{total variability in } Y} \qquad (12.16)$$

$$= \text{proportion of the variability in } Y \text{ accounted for by } X$$

We also know from Equation 12.13 that $r^2 = SS_{reg}/SS_Y$. Therefore, r^2 can be interpreted as the proportion of the variability in Y accounted for by the regression on X.

The quantity r^2 is called the *coefficient of determination* and is considered by many investigators to be a better measure of linear relation than r. We can see that r^2 is a measure of how well the linear regression equation fits the data. However, we would argue that if the relation is linear, S_{YX} or $\hat{\sigma}_e$ provide more information than r or r^2 about the accuracy of the predictions based on X. The quantity $\hat{\sigma}_e$ estimates the standard deviation of the prediction errors in the population and is expressed in the same units as Y. On the other hand, r and r^2 are dimensionless quantities (i.e., are not expressed in terms of any units) that are measures of the variance accounted for relative to the total variance of Y. As we shall see in Chapter 14, measures of correlation such as r and r^2 have also been criticized because they are "sample-specific"; that is, their values are determined not only by the nature of the relation between X and Y but also by the variabilities of X and Y in the sample.

For the statistics class data, the regression equation does not produce terribly accurate predictions of final exam performance. The standard error of estimate is 10.64 points on the final, indicating that there is a good deal of variability about the regression line. If we can assume that the errors are normally distributed in the population, about one-third of the final exam scores will differ from their predicted values by at least 10.64 points. The coefficient of determination provides us with less useful information. The correlation between the pretest and the final exam score is .725, so the coefficient of determination is $(.725)^2 = .53$. All this tells us is that the variability accounted for by the linear regression equation, $SS_{reg} = 2007.50$, constitutes .53 of the total variability in Y.

Perhaps because of its definition as "the proportion of variance in Y explained or accounted for by X" (which somehow seems to promise more than it actually delivers), r^2 has frequently been misinterpreted and some of these misinterpretations have resulted in rather bizarre claims finding their way into the literature. For example, the claim has been made in a number of psychology textbooks that children achieve about 50% of their adult intelligence by age 4. The origin of this statement can be traced to a longitudinal study that found IQ scores at age 17 to have a correlation of about .7 with IQ at age 4. The resulting r^2 of about .5 (or 50%) provides an indication of how predictable adult IQ is from IQ at age 4 using a linear equation. However, it says absolutely nothing about the relative levels of intelligence at age 4 and age 17.

12.4.6 A Second Example

Panel a of Table 12.3 contains the data for our second example. In a visual "search" experiment, 20 subjects each look at a computer screen and are presented with an array of letters. They are asked to respond as quickly as they can whether or not a specified "target" letter is present in the array. Groups of five subjects are assigned to array sizes of two, four, six, or eight letters, and the times (in milliseconds) to respond correctly that the target letter is present are recorded. Panel b of Table 12.3 presents the associated regression statistics for the data, and Figure 12.6 displays the corresponding scatter diagram. The correlation between reaction time and array size is .873, and the linear regression equation that best predicts reaction time (Y) from array size (X) is

$$\hat{Y} = 381.90 + 23.92X$$

Although the correlation and regression equations are calculated in exactly the same way for the search experiment and statistics class examples, in one sense the two examples are fundamentally quite different. In the statistics class example, (X, Y) pairs were sampled, and therefore both X and Y are random variables. Students can be thought of as sampled from a hypothetical population, and the correlation between pretest and final exam score in the sample can be thought of as an estimate of the correlation in this hypothetical population. In contrast, in the search experiment example the values of array size are *selected* by the experimenter, not randomly sampled from a population; therefore, array size is a fixed-effect variable. The obtained correlation can

TABLE 12.3 SEARCH EXPERIMENT EXAMPLE

(a) Data

Array size X	Reaction time (ms) Y
2	418
2	428
2	410
2	445
2	471
4	475
4	455
4	418
4	524
4	516
6	537
6	500
6	480
6	511
6	529
8	550
8	617
8	590
8	608
8	548

(b) SYSTAT output for the search experiment data

```
DEP VAR:TIME   N:20    MULTIPLE R:.873    SQUARED MULTIPLE R:.763
ADJUSTED SQUARED MULTIPLE R:.749    STANDARD ERROR OF ESTIMATE:31.452
```

VARIABLE	COEFFICIENT	STD ERROR	STD COEF	TOLERANCE	T	P(2 TAIL)
CONSTANT	381.900	17.227	0.000	1.0000000	22.169	0.000
SIZE	23.920	3.145	0.873	1.0000000	7.605	0.000

ANALYSIS OF VARIANCE

SOURCE	SUM-OF-SQUARES	DF	MEAN-SQUARE	F-RATIO	P
REGRESSION	57216.640	1	57216.640	57.839	0.000
RESIDUAL	17806.360	18	989.242		

be considered to estimate the average correlation that would result if the experiment was repeated many times, but only if the same values of X were used each time.

The descriptive statistics of regression and correlation are identical for both situations. However, in dealing with inference, one must consider different models and make somewhat different assumptions when both X and Y are random variables than

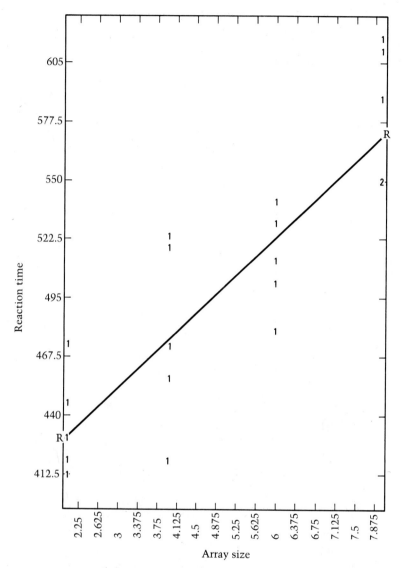

20 CASES PLOTTED. REGRESSION STATISTICS OF TIME ON SIZE:
CORRELATION .87330 R SQUARED .76265 S.E. OF EST 31.45222
2-TAILED SIG. .0000 INTERCEPT (S.E.) 381.90000(17.22709) SLOPE(S.E.)
23.92000(3.14522)

Figure 12.6 SPSS^X scatter diagram for the search experiment data. (Note that the 2 located at array size 8 and reaction time of about 550 milliseconds means that two data points were too close together to be plotted separately.)

when Y is random and X is fixed. Fortunately, under standard assumptions, the hypothesis tests and expressions for confidence intervals turn out to be the same in the two situations.

12.5 INFERENCE IN LINEAR REGRESSION

12.5.1 A Model for Linear Regression

We have so far concerned ourselves with the descriptive statistics of correlation and regression. However, in order to conduct hypothesis tests and form confidence intervals, we must state a formal model. Our strategy will be to begin with a regression model that assumes that X is a fixed-effect variable and is measured without error, and then discuss what happens when X is random and when X is measured with error.

Consider the model

$$Y_i = \beta_0 + \beta_1 X_i + \epsilon_i \tag{12.17}$$

where Y_i = value of the dependent variable on the ith trial

$\beta_0,\ \beta_1$ = the Y intercept and slope of the regression line

X_i = constant that is the value taken on by the predictor variable on the ith trial

ϵ_i = random-error component

We assume that the error component ϵ is independently and normally distributed with mean 0 and variance σ_e^2. Therefore,

$$E(\epsilon_i) = 0$$
$$\mathrm{Var}(\epsilon_i) = \sigma_e^2$$
$$\mathrm{Cov}(\epsilon_i, \epsilon_{i'}) = 0 \qquad \text{except when } i = i'$$

These assumptions imply that the conditional population mean of Y corresponding to any given value of X, $\mu_{Y.X}$, lies on the straight line $\mu_{Y.X} = \beta_0 + \beta_1 X$, and that the deviation of Y from its conditional population mean is due solely to random error. This is very important; if the true relation between X and Y is curvilinear, our measure of error will include more than chance variability, and the significance tests developed from this model may be biased.

In this model, X is a fixed-effect variable that is measured without error; if we replicated the study, exactly the same values of X would be used. The model is appropriate for the search experiment data in Table 12.3 because the values of X are chosen by the experimenter. It would not be appropriate for the statistics class data presented in Table 12.1 in which both X and Y are random variables.

In the next two sections, we discuss how to make statistical inferences about the slope and intercept of the regression equation and about the predictions made by it. In every case, we can test hypotheses about P, the parameter of interest, by finding

$$t = \frac{\hat{P} - P}{\hat{\sigma}_{\hat{P}}}$$

and obtain the $1 - \alpha$ confidence interval for P, using

$$\hat{P} \pm t_{\alpha/2} \hat{\sigma}_{\hat{P}}$$

where $\hat{P}$ is the estimate of P obtained from the sample and $\hat{\sigma}_{\hat{P}}$ [which, when there is subscripting, will sometimes be written as $\hat{\mathrm{SE}}(\hat{P})$] is the estimated standard error. For example, to test the hypothesis $H_0: \beta_1 = 0$, we form the statistic $t = b_1/\hat{SE}(b_1)$. We already know how to find b_1; all we require to test the hypothesis is to find $\hat{SE}(b_1)$.

Although the details of how to find the standard errors for b_1, b_0, and $\hat{Y}$ are relegated to Appendix 12.2, the basic idea is quite simple because b_1, b_0, and $\hat{Y}$ all can be expressed as linear combinations of the Y scores. In Chapter 2, we discussed how to find the variance of linear combinations, and the variances of the linear combinations that correspond to b_1, b_0, or $\hat{Y}$ can be found in much the same way. Once the appropriate variances are found, the standard errors can be obtained by taking their square roots.

In the remainder of this section, we will show that b_1, b_0, and $\hat{Y}$ can be expressed as linear combinations of the Y's. Also, we will develop an expression called the *leverage* that will prove to be useful later in the chapter. Starting with one of the forms of Equation 12.6, we first show that the expression for b_1 can be broken into two parts. One of the parts can be shown to be equal to zero, and the other is a linear combination of the Y's. That is,

$$b_1 = \frac{\sum_i (X_i - \bar{X})(Y_i - \bar{Y})}{\sum_i (X_i - \bar{X})^2} = \frac{\sum_i (X_i - \bar{X})(Y_i - \bar{Y})}{SS_X}$$

$$= \frac{\sum_i (X_i - \bar{X})Y_i}{SS_X} - \frac{\sum_i (X_i - \bar{X})\bar{Y}}{SS_X}$$

But $\sum_i (X_i - \bar{X})\bar{Y} = \bar{Y}\sum_i (X_i - \bar{X}) = 0$. Therefore,

$$b_1 = \frac{\sum_i (X_i - \bar{X})Y_i}{SS_X}$$

or

$$b_1 = \sum_i f_i Y_i \qquad \text{where } f_i = \frac{X_i - \bar{X}}{SS_X} \qquad (12.18)$$

Also, from Equations 12.7 and 12.18, we have

$$b_0 = \bar{Y} - b_1 \bar{X}$$

$$= \frac{1}{N}\sum_i Y_i - \bar{X}b_1$$

$$= \frac{1}{N}\sum_i Y_i - \bar{X}\sum_i f_i Y_i$$

$$= \sum_i \left[\frac{1}{N} - \bar{X}f_i \right] Y_i$$

so that

$$b_0 = \sum_i g_i Y_i \qquad \text{where } g_i = \frac{1}{N} - \frac{\bar{X}(X_i - \bar{X})}{SS_X} \qquad (12.19)$$

We have just shown that both b_0 and b_1 are linear combinations of the Y's. It follows that any predicted score given by $\hat{Y} = b_0 + b_1 X$ must also be a linear combination of the Y's. The linear combination can be obtained by substituting Equations 12.18 and 12.19 into the regression equation. That is,

$$\hat{Y} = b_0 + b_1 X$$
$$= \sum_i g_i Y_i + X \sum_i f_i Y_i$$
$$= \sum_i [g_i + X f_i] Y_i$$

Substituting the expressions for f_i and g_i, we have

$$\hat{Y} = \sum_i \left[\frac{1}{N} - \frac{\bar{X}(X_i - \bar{X})}{SS_X} + \frac{X(X_i - \bar{X})}{SS_X} \right] Y_i$$

so that

$$\hat{Y} = \sum_i h_i Y_i \qquad \text{where } h_i = \frac{1}{N} + \frac{(X - \bar{X})(X_i - \bar{X})}{SS_X} \qquad (12.20)$$

Suppose that (X_j, Y_j) is one of the data points that we used to obtain the regression equation. We can express the value of Y_j predicted by Equation 12.20 as

$$\hat{Y}_j = \sum_i h_{ij} Y_i \qquad \text{where } h_{ij} = \frac{1}{N} + \frac{(X_j - \bar{X})(X_i - \bar{X})}{SS_X}$$

$$= h_{jj} Y_j + \sum_{i \neq j} h_{ij} Y_i \qquad \text{where } h_{jj} = \frac{1}{N} + \frac{(X_j - \bar{X})^2}{SS_X}$$

The expression h_{jj} is called the "leverage" of case j and will be encountered frequently in the chapter. The leverage is a measure of how important the jth case [i.e., the data point (X_j, Y_j)] is in determining the prediction of Y_j made by the equation. Note that $h_{jj} Y_j$ can be interpreted as the component of $\hat{Y}_j$ that is directly due to Y_j.

All the expressions for the standard errors that we need to conduct hypothesis tests and form confidence intervals follow directly from the linear combinations we have developed in this section.

12.5.2 Inference about β_1 and β_0

We estimate the regression parameters β_1 and β_0, using the least squares estimates b_1 and b_0 obtained from Equations 12.6 and 12.7. It can be shown (see Appendix 12.2) that b_1 and b_0 are unbiased estimates of β_1 and β_0 and that the estimated standard errors of b_1 and b_0 are given by

$$\hat{SE}(b_1) = \frac{\hat{\sigma}_e}{\sqrt{SS_X}} \qquad \text{where } \hat{\sigma}_e = \sqrt{\frac{SS_{error}}{N - 2}} \qquad (12.21)$$

and

$$\hat{SE}(b_0) = \hat{\sigma}_e \sqrt{\frac{1}{N} + \frac{\bar{X}^2}{SS_X}} \qquad (12.22)$$

Under the assumptions of the model, $(b_1 - \beta_1)/\hat{SE}(b_1)$ and $(b_0 - \beta_0)/\hat{SE}(b_0)$ are distributed as t with $N - 2$ df, so we can find confidence intervals for, and test hypotheses about, β_1 and β_0. For example, we found for the search experiment data of Table 12.3 that

$$b_1 = 23.92 \quad \text{and} \quad b_0 = 381.90$$

The estimated standard errors of b_0 and b_1 can readily be calculated from Equations 12.21 and 12.22, although it should be noted that these values are directly available from the SYSTAT output in Table 12.3. To perform the calculations, we note that

$$\hat{\sigma}_e = \sqrt{\frac{SS_{\text{error}}}{N - 2}} = \sqrt{\frac{1,770,806.36}{18}} = 31.452$$

and

$$SS_X = \sum (X_i - \bar{X})^2$$
$$= 5[(2 - 5)^2 + (4 - 5)^2 + (6 - 5)^2 + (8 - 5)^2]$$
$$= 100.00$$

so that

$$\hat{SE}(b_1) = \frac{31.452}{\sqrt{100.00}} = 3.1452$$

Also,

$$\hat{SE}(b_0) = \hat{\sigma}_e \sqrt{\frac{1}{N} + \frac{\bar{X}^2}{SS_X}} = (31.452)\sqrt{\frac{1}{20} + \frac{25}{100}}$$
$$= 17.227$$

Therefore, the 95% confidence interval for β_1 is given by

$$b_1 \pm t_{.025}\hat{SE}(b_1) = 23.92 \pm (2.101)(3.145)$$
$$= 23.92 \pm 6.61$$

and the 95% confidence interval for β_0 is

$$381.90 \pm (2.101)(17.227) = 381.90 \pm 36.19$$

Because neither interval contains zero, we can reject the hypotheses $\beta_0 = 0$ and $\beta_1 = 0$. Alternatively, we could perform the hypothesis tests directly; for example, we would have for the slope

$$t = \frac{b_1}{\hat{SE}(b_1)} = \frac{23.92}{3.145} = 7.61$$

which indicates that the slope is highly significant.

We should point out that this t test is exactly equivalent to the F test for the significance of the regression (where $F = MS_{\text{reg}}/MS_{\text{error}}$) that we discussed in Section 12.4.4. From the SYSTAT output in Table 12.3, we see that the ratio of MS_{reg} to

MS_{error} (or of $MS_{REGRESSION}$ to $MS_{RESIDUAL}$ as it would be referred to by SYSTAT) has a value of 57.839, the square of the value (7.61) we obtained for the t.

We can show that the tests are equivalent by recalling from Section 12.4.4 that SS_{reg} can be expressed as $b_1^2 SS_X$ and has 1 df, so $MS_{reg} = b_1^2 SS_X$. Also, we know that $MS_{error} = \hat{\sigma}_e^2$. Therefore,

$$F = \frac{MS_{reg}}{MS_{error}} = \frac{b_1^2 SS_X}{\hat{\sigma}_e^2}$$

In the current section, we noted that the null hypothesis $H_0: \beta_1 = 0$ can be tested by using the statistic

$$t = \frac{b_1}{\hat{\sigma}_e / \sqrt{SS_X}}$$

Multiplying the numerator and denominator by $\sqrt{SS_X}$, we have

$$t = \frac{b_1 \sqrt{SS_X}}{\hat{\sigma}_e} = \sqrt{F}$$

We conclude this section by noting that although the expressions for the standard errors of b_0 and b_1 (Equations 12.21 and 12.22) are certainly not obvious, they have characteristics that make sense intuitively. For example,

1. We would expect that the greater the degree of scatter in the data points about the best regression line, the more uncertainty we should have about the precise location of the regression line. Therefore, the more scatter there is, the more variability there should be in our estimates of b_0 and b_1. Therefore, the standard errors of b_0 and b_1 should vary directly with a measure of this scatter such as $\hat{\sigma}_e$.

2. We would expect to get more stable estimates of the regression parameters if the sample contained both large and small values of X than if the sample contained only a narrow range of X values. Therefore, the standard errors should vary *inversely* with some measure of the variability in the X scores, such as $\hat{\sigma}_X^2$ or SS_X.

12.5.3 Inference about the Population Regression Line

According to the model stated in Equation 12.17, the expected value of Y for any X score, X_j, is given by $E(\beta_0 + \beta_1 X_j + \epsilon)$. Because X is a fixed-effect variable and $E(\epsilon) = 0$, this expectation is $\beta_0 + \beta_1 X_j$. We can think of this expected value as $\mu_{Y.X_j}$, the conditional population mean of the Y scores corresponding to the value of X_j. We show in Appendix 12.2 that $\hat{Y}_j = b_0 + b_1 X_j$ is an unbiased estimator of $\mu_{Y.X_j}$ and that the estimated standard error of $\hat{Y}_j$ is given by

$$\hat{SE}(\hat{Y}_j) = \hat{\sigma}_e \sqrt{h_{jj}} \tag{12.23}$$

where $h_{jj} = 1/N + (X_j - \bar{X})^2 / SS_X$ is the leverage of X_j (see Section 12.5.1).

For the search experiment data in Table 12.3, the best estimate of the population mean of the Y scores corresponding to $X = 4$ is given by $381.90 + (23.92)(4) = 477.58$. The estimated standard error of $\hat{Y}$ at $X = 4$ is given by

$$\hat{SE}(\hat{Y}) = 31.452\sqrt{\frac{1}{20} + \frac{(4 - 5)^2}{100.00}}$$

$$= 7.70$$

Therefore, the 95% confidence interval for $\mu_{Y.X}$ at $X = 4$ is given by

$$\hat{Y} \pm t_{.025}\hat{SE}(\hat{Y}) = 477.58 \pm (2.101)(7.70) = 477.58 \pm 16.19$$

Note that, as we can see from Equation 12.23, the standard error of $\hat{Y}_j$ depends on the value of X_j; it is smallest when $X_j = \bar{X}$ and becomes larger as X_j deviates from $\bar{X}$. For example, the value of $\hat{SE}(\hat{Y})$ when $X = 8$ is

$$(31.452)\sqrt{1/20 + (8 - 5)^2/100.00} = 11.77$$

This is considerably larger than the value of 7.70 obtained for $X = 4$.

12.5.4 Obtaining a Confidence Interval for Y_{new_j}, a New Value of Y at X_j

Suppose after having found the sample regression equation, we wish to predict the Y score for a new individual who has an X score of X_j. That is, we want to estimate a particular score, Y_{new_j}, selected from the population of Y scores that has $\mu_{Y.X_j}$ as its mean, where

$$Y_{new_j} = \mu_{Y.X_j} + \epsilon \tag{12.24}$$

$$= \beta_0 + \beta_1 X_j + \epsilon$$

and ϵ is the error component associated with Y_{new_j}.

Because $E(Y_{new_j}) = \mu_{Y.X_j}$, an unbiased estimate of Y_{new_j} is given by $\hat{Y}_j = \hat{\mu}_{Y.X_j} = b_0 + b_1 X_j$. From Equation 12.24, the variance of Y_{new_j} about the conditional mean $\mu_{Y.X_j}$ is $\text{Var}(\epsilon) = \sigma_e^2$. Therefore, if we wish to obtain a confidence interval for the single score Y_{new_j}, the standard error we should use is not the same one that would be used to obtain a confidence interval for $\mu_{Y.X_j}$, the conditional mean of all of the Y scores at X_j. The standard error we want for the interval around Y_{new_j} must contain an additional component to reflect the variance of the individual Y scores about the conditional mean. The appropriate standard error is given by

$$\hat{SE} = \hat{\sigma}_e\sqrt{1 + h_{jj}} \tag{12.25}$$

where h_{jj} is the leverage of X_j, $1/N + (X_j - \bar{X})^2/SS_X$.

For the search experiment data, the predicted reaction time for a new subject presented with an array size of 4 is $381.90 + (4)(23.92) = 477.58$, the same as the predicted conditional mean at $X = 4$. However, the estimated standard error is

$$(31.4522)\sqrt{1 + \frac{1}{20} + \frac{(4 - 5)^2}{100.00}} = 32.38$$

so that the 95% confidence interval for Y_{new} at $X = 4$ is $477.58 \pm (2.101)(32.38) = 477.58 \pm 68.03$, as opposed to the 95% confidence interval for the conditional population mean, $\mu_{Y.X}$, that we previously found to be 477.58 ± 16.19.

Confidence intervals for Y scores are sometimes misinterpreted. We cannot conclude that 95% of the population of scores at $X = 4$ lie within the interval 477.58 ± 68.03. The correct interpretation is the following: Assume that (1) we compute a new regression line for each of many samples of size N, using the same values of X for each sample, (2) for each sample we estimate the Y score for an additional individual at a value of X that is the same for all samples, and (3) following each prediction we observe the actual score of the additional individual. If the assumptions of the model are valid, the confidence intervals will contain the actual scores in 95% of the cases.

12.6 THE ONE-FACTOR DESIGN WITH TWO GROUPS: A SPECIAL CASE OF REGRESSION ANALYSIS

It is important to understand that ANOVA is simply a special case of regression analysis. Although we will discuss this issue at length once we have introduced multiple regression, we are already in a position to demonstrate that the ANOVA approach to testing the hypothesis $\mu_1 = \mu_2$ is equivalent to a regression test of the hypothesis that $\beta_1 = 0$.

We first consider the relation between the ANOVA model and the regression model. The ANOVA model for the one-factor between-subjects design is

$$Y_{ij} = \mu + \alpha_j + \epsilon_{ij}$$

with the usual assumptions that the ϵ_{ij} are normally and independently distributed with mean 0 and variance σ_e^2 within each population. To consider the two-group design as a problem in regression analysis, we assign the value X_1 to the subjects in one group and the value X_2 to the subjects in the second group. The only restriction on the values of X_1 and X_2 is that they must not be the same. The regression model may then be written

$$Y_{ij} = \beta_0 + \beta_1 X_{ij} + \epsilon_{ij}$$

where $X_{i1} = X_1$ and $X_{i2} = X_2$ for all i, or simply as

$$Y_{ij} = \beta_0 + \beta_1 X_j + \epsilon_{ij}$$

The distributional assumptions about ϵ_{ij} are the same as for the analysis of variance model. Because there are only two populations of Y values, the population means must lie on a straight line. The population mean of the first group is $\mu_1 = \beta_0 + \beta_1 X_1$, and the mean of the second group is $\mu_2 = \beta_0 + \beta_1 X_2$. Therefore,

$$\mu_1 - \mu_2 = \beta_1(X_1 - X_2)$$

Because $X_1 \neq X_2$, testing the hypothesis that $\beta_1 = 0$ is equivalent to testing the hypothesis $\mu_1 = \mu_2$.

We can also show that in this design, $SS_{reg} = SS_A$ and $SS_{error} = SS_{S/A}$. For simplicity, we assume there are N subjects, $n = N/2$ in each group. We first note that because the regression line passes through the means of both samples, our best prediction of Y for subjects in group 1 is $\bar{Y}._1$, and for subjects in group 2 it is $\bar{Y}._2$, so $\hat{Y}_{ij} = \bar{Y}._j$. Therefore,

$$SS_{error} = \sum_i \sum_j (Y_{ij} - \hat{Y}_{ij})^2$$

$$= \sum \sum (Y_{ij} - \bar{Y}._j)^2$$

$$= SS_{S/A}$$

and

$$SS_{reg} = \sum \sum (\hat{Y}_{ij} - \bar{Y})^2$$

$$= \sum \sum (\bar{Y}._j - \bar{Y}..)^2$$

$$= SS_A$$

The same null hypothesis is tested by using either

$$F = \frac{MS_{reg}}{MS_{error}} \quad \text{or} \quad F = \frac{MS_A}{MS_{S/A}} \quad \text{with 1 and } N - 2 \text{ df}$$

It is important to note that the equality between SS_{reg} and SS_A when Y is regressed on X will usually not hold when there are more than two groups of scores. For the general case in which there are a groups of scores, $SS_{reg} = SS_A$ only if, when Y is plotted against X, the a group means all fall on a straight line. In Chapter 16, we will show how multiple regression can be used to obtain SS_A in the general case.

12.7 REGRESSION ANALYSIS WHEN *X* IS A RANDOM VARIABLE

In the regression model we introduced in Section 12.5.1, X is fixed and measured without error. In other words, the values of X are known constants. However, in many cases, both X and Y vary from sample to sample. This will be the case if a sample of individuals is selected and values of X and Y are obtained for each individual.

If X is a random variable, it can be shown that the calculations for hypothesis tests and confidence intervals are the same as those for the fixed-X model under certain assumptions. We must assume that for each value of X, the sample value of Y is selected from a normal population with mean $\mu_{Y.X} = \beta_0 + \beta_1 X$ and constant variance σ_e^2, and the probability distribution of X does not involve the regression parameters β_0, β_1, and ϵ.

Of course, the sampling distributions that form the basis for hypothesis tests and confidence intervals are now thought of as resulting from repeated sampling of (X, Y) pairs in which the values of X as well as Y change from sample to sample. Also, the standard errors of b_0, b_1, and Y will vary, depending on the values of X that are selected, since they all involve SS_X.

12.8 REGRESSION WHEN X IS SUBJECT TO RANDOM ERROR

We have so far assumed that X is measured without error. Because this assumption is often not realistic, we consider what happens when the obtained value of X is made up of two components: X', its true value (i.e., the value it would have if it could be measured without error) and u, an error component, so that

$$X = X' + u$$

If we can assume that the measurement is unbiased and that the error component u is uncorrelated with X', we have $E(u) = 0$ and $\sigma_X^2 = \sigma_{X'}^2 + \sigma_u^2$.

Snedecor and Cochran (1967) point out that if ϵ, u, and X' are all normally and independently distributed, Y and X will follow a bivariate normal distribution (see Section 14.3) and the regression of Y on X will be linear with a slope of

$$\beta_1 = \frac{\beta_1'}{1 + \lambda}$$

where β_1' is the "true" slope (i.e., the slope of the equation that would be obtained by regressing Y on X') and $\lambda = \sigma_u^2/\sigma_{X'}^2$. Even if X' is not normal, the result holds for large samples and approximately for small samples when λ is small. This means that when X is measured with error, the obtained slope b_1 underestimates the magnitude of the true slope because it estimates β_1 rather than the true slope β_1'. If there is a great deal of measurement error, λ will be large and β_1 will be much closer to zero than the true slope.

Fortunately, in experimental research, measurement error is usually quite small. Consider a situation in which $\sigma_{X'} = 15$. Even if $\sigma_u = 5$ (this implies that about one-third of the measured values of X will be in error by at least 5 units), λ will be small. In this case, $\lambda = \sigma_u^2/\sigma_{X'}^2 = 25/225 = 1/9$, so $\beta_1 = \beta_1'/(1 + \lambda) = \beta_1'/(1 + 1/9) = .9\beta_1'$.

However, if measuring instruments with low reliabilities (i.e., large degrees of measurement error) are used, the magnitudes of the regression coefficients may be seriously underestimated. Also, if the amount of measurement error differs across conditions, this must be kept in mind when comparing slopes obtained in the different conditions.

The situation is more complicated if it cannot be assumed that X' is independent of u. For a more thorough discussion, see Draper and Smith (1981, pp. 122–125).

12.9 CHECKING FOR VIOLATIONS OF ASSUMPTIONS USING RESIDUALS

12.9.1 Introduction

Because our conclusions can be seriously in error if there are severe violations of the assumptions, in the next few sections we discuss how to check for such violations. One should not rely only on summary statistics such as the correlation coefficient or the slope of the regression line. It is important to plot the data and to use the diagnostics that are usually provided by statistical packages.

A dramatic illustration of the dangers of relying exclusively on summary measures has been provided by Anscombe (1973). He developed four hypothetical data sets that each provide *identical* output from a typical regression program. For each of the data sets, $N = 11$, $\bar{X} = 9.0$, $\bar{Y} = 7.5$, $b_1 = 0.50$, $b_0 = 3.0$, $SS_X = 110$, $SS_{reg} = 27.50$, $SS_{error} = 13.75$, $\hat{SE}(b_1) = 0.118$, and $r^2 = .667$ (see Table 12.4 and Figure 12.7). Looking only at the summary measures, it would seem reasonable to arrive at the same conclusions for all four data sets. However, set b suggests a curvilinear (quadratic) relation, whereas set c suggests that 10 of the 11 points are well fit by a linear equation but that one point is quite far off the line that seems to fit the remaining 10. Without knowing more about how the data were generated, we cannot say whether it is appropriate to delete the offending point; if we do, the slope changes from .50 to .35. Finally, the regression for set d depends very heavily on the eighth case. If this point was deleted, we could not even estimate the slope and the correlation would be undefined because the variance of X would be zero. We cannot have much confidence in measures of b_1 and r that depend so heavily on a single case.

Valuable information about whether the assumptions are valid may be obtained by studying residuals—that is, differences between the observed and predicted values of Y. The residuals, $e_i = Y_i - \hat{Y}_i = Y_i - (b_0 + b_1 X_i)$, provide information about the population error components, $\epsilon_i = Y_i - (\beta_0 + \beta_1 X_i)$.

The standard statistical software packages such as SAS, SPSS^X, BMDP, and SYS-TAT all provide residuals and allow them to be plotted in a variety of ways. If the distribution of residuals differs strongly from that assumed for the error components, the

TABLE 12.4   FOUR HYPOTHETICAL DATA SETS FROM ANSCOMBE (1973)

	Data set					
	a–c	a	b	c	d	d
	Variable					
Case number	X	Y	Y	Y	X	Y
1	10.0	8.04	9.14	7.46	8.0	6.58
2	8.0	6.95	8.14	6.77	8.0	5.76
3	13.0	7.58	8.74	12.74	8.0	7.71
4	9.0	8.81	8.77	7.11	8.0	8.84
5	11.0	8.33	9.26	7.81	8.0	8.47
6	14.0	9.96	8.10	8.84	8.0	7.04
7	6.0	7.24	6.13	6.08	8.0	5.25
8	4.0	4.26	3.10	5.39	19.0	12.50
9	12.0	10.84	9.13	8.15	8.0	5.56
10	7.0	4.82	7.26	6.42	8.0	7.91
11	5.0	5.68	4.74	5.73	8.0	6.89

Note: The values of X are the same for data sets *a*, *b*, and *c*.

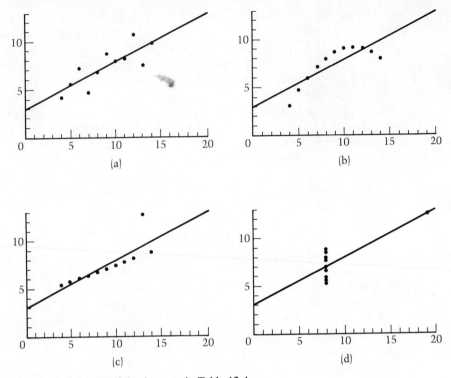

Figure 12.7 Graphs of the data sets in Table 12.4.

assumptions of the model may not be satisfied. Moreover, the nature of the difference can tell us which assumptions have been violated and suggest remedial measures.

Residuals cannot provide information about the assumption $E(\epsilon_i) = 0$ because when a least-squares regression equation is used, the residuals are constrained to sum to zero: $\Sigma e_i = \Sigma Y_i - N b_0 - b_1 \Sigma X_i = 0$ follows directly from one of the normal equations in Equation 12.8. The residuals can, however, provide useful information about whether there are violations of the assumptions of linearity, homogeneity of variance, normality, and independence of error.

12.9.2 Linearity and Homogeneity of Variance

If the assumptions of linearity and homogeneity of variance are both valid, when residuals are plotted against either X or $\hat{Y}$, data points should lie within a horizontal band as indicated in panel a of Figure 12.8.

Any other pattern suggests that the assumptions are not valid or that some kind of error has been made. For example, plots such as that in panel b indicate that the relation between Y and X is nonlinear and that the appropriate model should contain additional terms such as X^2. Plots like that in panel c, in which the residuals are more spread out for some values of X or Y than others, indicate that the variance of estimate is not constant. Either some variance-stabilizing transformation should be performed on

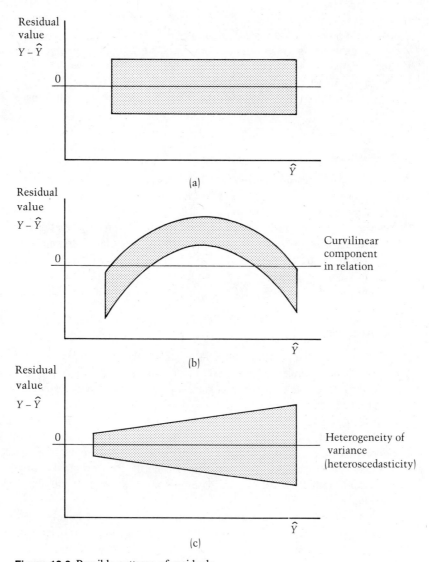

Figure 12.8 Possible patterns of residuals.

Y before the regression is performed, or else a procedure called *weighted least squares* (WLS) should be performed instead of the *ordinary least-squares* (OLS) procedure we have so far discussed. The WLS procedure is identical to OLS except that cases with lower error variance are weighted more heavily than cases that have high error variance on the rationale that cases with lower error variance contain more information. The standard statistical packages can all easily handle weighted least squares (for example, see the worked-out example in the SYSTAT manual).

We plot residuals against $\hat{Y}$ rather than against Y because it can be shown that e is not correlated with $\hat{Y}$ but has a correlation of $\sqrt{1 - r^2}$ with Y. It should also be noted that although plots of residuals against X and $\hat{Y}$ provide equivalent information for bivariate regression (because $\hat{Y}$ is simply a linear function of X), this will not be the case when there is more than one independent variable. Finally, we can get an idea about whether an additional variable, W, belongs in the model by plotting the residuals against W. If the residual varies systematically with W, then W should be included in the model; if it is not included, the error component, ϵ, will consist of more than chance variability.

12.9.3 Normality

Statistical packages are usually capable of constructing histograms and normal probability plots (see Chapter 2) of the residuals. A virtue of the normal probability plot is that if the residuals are normally distributed, the points fall on a straight line. It is easier to detect departures from a straight line than from a normal histogram.

Violations of the linearity and homogeneity of variance assumptions may cause the residuals to depart from normality, so generally the linearity and homogeneity of variance assumptions should be checked before looking for violations of the normality assumption.

12.9.4 Independence

We assume that the error components ϵ_i are independent of one another. The residuals cannot be strictly independent; there are N residuals and only $N - 2$ df (another way to think about this is that the residuals are all based on the same estimates of β_0 and β_1). Nonetheless, if N is reasonably large, this unavoidable dependency will be very small, so residuals can meaningfully be examined for evidence of lack of independence.

Data are usually collected and recorded sequentially. If the error components are independent, the residuals should not vary systematically over time. Systematic variation may reflect changes in subjects, measuring devices, or surroundings. When the residuals are plotted against time or case number, the result should again look like panel a of Figure 12.8.

It is possible for error components to exhibit different kinds of serial correlations: for example, the residual corresponding to case i may tend to be similar (or dissimilar) in size to those corresponding to trial $i - 1$ or $i - 2$. Several packages print values of the Durbin-Watson statistic that forms the basis for a test of serial correlation in adjacent residuals. The statistic

$$D = \frac{\Sigma_i(e_i - e_{i-1})^2}{\Sigma e_i^2} \tag{12.26}$$

will be small when sequentially adjacent residuals are positively correlated and large when they are negatively correlated. The value of D is approximately $2(1 - r_{i,i-1})$, where $r_{i,i-1}$ is the correlation between sequentially adjacent residuals. It is therefore evident that D can range from 0 to 4, with larger deviations from a value of 2 providing stronger evidence of serial correlation. A more detailed discussion of the test and appropriate tables for assessing significance may be found in Draper and Smith (1981). Under some conditions, weighted least squares can be used to perform the regression analysis when the data are serially correlated (see Draper and Smith, pp. 156–157).

12.10 AN *F* TEST FOR THE LACK OF FIT OF THE LINEAR MODEL

The assumption of linearity (i.e., that in the population, the conditional means of Y are a perfect linear function of X) is basic to the inferential procedures that we have discussed in this chapter. Suspected departures from linearity that are observed in scatter diagrams or plots of residuals may be tested for significance by employing a procedure based on partitioning SS_{error} into two components.

If the linear model is appropriate, SS_{error} reflects variability around the conditional means, and $MS_{error} = SS_{error}/(N - 2)$ estimates σ_e^2. If the linear model is not appropriate, SS_{error} can be partitioned into a "pure-error" component, which reflects variability about the conditional means, and a "nonlinearity" component, which reflects the extent to which the conditional means are not a perfect linear function of X.

Assume that X takes on the values $X_1, X_2, \ldots, X_j, \ldots, X_a$, and that there are n_j values $Y_{1j}, \ldots, Y_{ij}, \ldots, Y_{n_j j}$ at X_j. The predicted Y score at X_j is obtained from the linear equation $\hat{Y}_j = b_0 + b_1 X_j$. The identity

$$Y_{ij} - \hat{Y}_j = \underset{\text{pure error}}{(Y_{ij} - \bar{Y}._j)} + \underset{\text{nonlinearity}}{(\bar{Y}._j - \hat{Y}_j)}$$
$$\underset{\text{error}}{\phantom{Y_{ij} - \hat{Y}_j}}$$

suggests the following partitioning of error variance:

$$\underset{SS_{error}}{\Sigma\Sigma(Y_{ij} - \hat{Y}_j)^2} = \underset{SS_{pure\ error}}{\Sigma\Sigma(Y_{ij} - \bar{Y}._j)^2} + \underset{SS_{nonlinearity}}{\Sigma\Sigma(\bar{Y}._j - \hat{Y}_j)^2}$$

The pure-error SS term is associated with $N - a$ df ($n_j - 1$ df at each of the a levels of X, where $\Sigma_j n_j = N$), and its mean square, $\Sigma\Sigma(Y_{ij} - \bar{Y}._j)^2/(N - a)$, estimates the variance of the scores around the conditional means of Y, whether or not the linear model fits.

The nonlinearity SS term is obtained by subtracting $SS_{pure\ error}$ from SS_{error}. It has $a - 2$ df because there are a means, and 2 df are used up in estimating the slope and intercept of the linear regression equation; equivalently, $(N - 2) - (N - a) = a - 2$. The corresponding MS term estimates a quantity that is the sum of σ_e^2 and a component that reflects the departure from linearity.

Therefore, the linearity assumption may be tested by using

$$F = \frac{MS_{nonlinearity}}{MS_{pure\ error}} \qquad \text{with } a - 2 \text{ and } N - a \text{ df} \qquad (12.27)$$

As an example, we can test the linear model for the search experiment data presented in Table 12.3. We found earlier that $SS_{error} = 17,806.36$. The pure error SS for the first array size can be obtained by using

$$\Sigma(Y_{i1} - \bar{Y}._1)^2 = \Sigma Y_{i1}^2 - \frac{(\Sigma Y_{i1})^2}{n_1} = 945,874 - \frac{(2172)^2}{5} = 2357.2$$

The corresponding pure-error SS terms for the three other array sizes are 7697.2, 2081.2, and 4143.2, so the total pure-error SS is 16,278.8. The nonlinearity SS can be obtained by subtracting 16,278.8 from SS_{error}. The results of the analysis summarized in the accompanying table indicate that the search experiment data do not show any significant departure from linearity.

SV	df	SS	MS	F
Nonlinearity	$a - 2 = 2$	1,527.56	763.78	.75
Pure error	$N - a = 16$	16,278.80	1,017.42	
Total error	$N - 2 = 18$	17,806.36		

It is not necessary for X to be a fixed-effect variable in order to test for linearity in this fashion; only that some of the values of X have more than one value of Y associated with them. In general, SS_{error} is either provided directly by statistical packages or is readily obtainable from $\hat{\sigma}_e$. $SS_{pure\ error}$ is obtained by finding the sums of squared deviations about the mean for every value of X for which there is more than one value of Y and summing them. It is also possible to test for specific kinds of deviations from linearity (e.g., does the relation between X and Y have a significant quadratic component?). We shall consider such tests in Chapter 15.

12.11 LOCATING OUTLIERS AND INFLUENTIAL DATA POINTS

Because the results of a regression analysis can be markedly affected by a few extreme data points, it is important to identify data points that are unusually influential. If these points can be located, checks can be made to determine whether they resulted from different processes than the rest of the data or from recording or transcription errors. If so, they can be corrected or deleted from the data. If the influential points cannot be attributed to an error or failure of some sort, the appropriate way to deal with them depends on the specific research problem. Given the presence of influential points, it is often appropriate to collect more data and/or to report analyses both with the influential cases included and with them deleted in order to make clear to the reader the impact of these points. Also, in situations in which predictions are important, the effects of influential points may be partially circumvented by isolating regions where the influence is relatively unimportant.

It also may be of interest to locate points that have outlying X and/or Y values, even though a data point may have a large residual or a value of X that differs considerably from most other values of X without influencing the regression equation inordinately. These points can be examined for errors or to determine whether there is something "special" about them. Also, depending on the research problem, it may be reasonable to treat cases with extreme values of X differently or to confine discussion to cases that do not have outlying values of X.

Table 12.5 contains output obtained from the SYSTAT package when we did a bivariate regression analysis of the search experiment data in Table 12.3 and asked that the residuals and variables in the model be saved. Eight quantities are presented. The first two are the predicted value ($\hat{Y}$) and residual ($Y - \hat{Y}$) for each case. The last two are simply Y and X, and the sixth (SEPRED) measure is $\hat{SE}(\hat{Y})$, which, as we have previously noted, takes on different values for different values of X. The remaining measures, LEVERAGE (the h_{jj} that we have encountered on several occasions),

TABLE 12.5   SYSTAT RESIDUAL OUTPUT FOR THE SEARCH EXPERIMENT DATA

		ESTIMATE SEPRED	RESIDUAL RT	LEVERAGE SIZE	COOK	STUDENT
CASE	1	429.740	-11.740	0.140	0.013	-0.393
CASE	1	11.768	418.000	2.000		
CASE	2	429.740	-1.740	0.140	0.000	-0.058
CASE	2	11.768	428.000	2.000		
CASE	3	429.740	-19.740	0.140	0.037	-0.666
CASE	3	11.768	410.000	2.000		
CASE	4	429.740	15.260	0.140	0.022	0.512
CASE	4	11.768	445.000	2.000		
CASE	5	429.740	41.260	0.140	0.163	1.458
CASE	5	11.768	471.000	2.000		
CASE	6	477.580	-2.580	0.060	0.000	-0.082
CASE	6	7.704	475.000	4.000		
CASE	7	477.580	-22.580	0.060	0.017	-0.731
CASE	7	7.704	455.000	4.000		
CASE	8	477.580	-59.580	0.060	0.122	-2.139
CASE	8	7.704	418.000	4.000		
CASE	9	477.580	46.420	0.060	0.074	1.585
CASE	9	7.704	524.000	4.000		
CASE	10	477.580	38.420	0.060	0.051	1.282
CASE	10	7.704	516.000	4.000		
CASE	11	525.420	11.580	0.060	0.005	0.371
CASE	11	7.704	537.000	6.000		
CASE	12	525.420	-25.420	0.060	0.022	-0.826
CASE	12	7.704	500.000	6.000		
CASE	13	525.420	-45.420	0.060	0.071	-1.546
CASE	13	7.704	480.000	6.000		
CASE	14	525.420	-14.420	0.060	0.007	-0.462
CASE	14	7.704	511.000	6.000		
CASE	15	525.420	3.580	0.060	0.000	0.114
CASE	15	7.704	529.000	6.000		
CASE	16	573.260	-23.260	0.140	0.052	-0.789
CASE	16	11.768	550.000	8.000		
CASE	17	573.260	43.740	0.140	0.183	1.558
CASE	17	11.768	617.000	8.000		
CASE	18	573.260	16.740	0.140	0.027	0.563
CASE	18	11.768	590.000	8.000		
CASE	19	573.260	34.740	0.140	0.115	1.206
CASE	19	11.768	608.000	8.000		
CASE	20	573.260	-25.260	0.140	0.061	-0.860
CASE	20	11.768	548.000	8.000		

COOK, and STUDENT, are, respectively, measures of (a) the extent to which the case is an outlier with respect to the distribution of X values, (b) the influence exerted by the case on the regression equation, and (c) the extent to which the case has an outlying residual. In the remainder of this section, we attempt to provide an explanation of these measures and indicate why they might be useful.

12.11.1 Influential Points

A simple way of assessing the influence of the jth case is to compare the results of the analysis when the jth case is present with those when it is deleted. The *deleted prediction* for the jth case may be defined as

$$\hat{Y}_j^{(-j)} = b_0^{(-j)} + b_1^{(-j)} X_j$$

where $\hat{Y}_j^{(-j)}$ is the prediction of Y_j made by using the regression equation in which the Y intercept and slope, $b_0^{(-j)}$ and $b_1^{(-j)}$, are obtained from the $N - 1$ cases that remain when case j is deleted. The *deleted residual* for the jth case, $e_j^{(-j)}$, is defined as the difference between Y_j and its deleted prediction so that

$$e_j^{(-j)} = Y_j - \hat{Y}_j^{(-j)}$$

The difference between $e_j = Y_j - \hat{Y}_j$, the residual for the jth case obtained when the prediction of Y_j is based on all N data points, and $e_j^{(-j)}$, the residual that results when the prediction is based on the $N - 1$ points that remain when the jth case is deleted,

$$e_j - e_j^{(-j)} = \hat{Y}_j^{(-j)} - \hat{Y}_j$$

is a useful index of the influence of the jth case. A better index that considers the effect of deleting the jth case on all N residuals has been proposed by Cook (1977) and is known as Cook's distance:

$$CD_j = \frac{\Sigma_i(\hat{Y}_i^{(-j)} - \hat{Y}_i)^2}{(p + 1)\hat{\sigma}_e^2} \tag{12.28}$$

where $p = 1$ for bivariate regression and, in general, p is the number of predictor variables in the regression equation. Cook and Weisberg (1982) indicate that a CD value greater than 1 is to be considered large. Cook's distance is provided by each of the four major software packages we have been considering. As we can see from Table 12.5, none of the 20 cases have Cook's distances that exceed .183; therefore no case is unduly influential.

12.11.2 Outliers

Whether a given residual is an extreme outlier depends not only on its absolute size but also on the distribution of the other residuals. Therefore, if one is interested in locating extreme outliers, it makes sense to use some sort of standardized measure in which the raw residual is divided by something like the standard deviation. Finding that a residual has a z score of 4.50 informs us more directly that it is an outlier than finding that it has an absolute value of 34.58. Although there is nothing terribly complicated about this basic idea, different statistical packages provide a variety of measures termed *standardized or studentized residuals*. Unfortunately, the packages are not always consistent in the terms they use to refer to these measures.

To discuss the measures, we note (see Appendix 12.3) that the standard error for a given residual e_j is given by

$$\hat{SE}(e_j) = \hat{\sigma}_e\sqrt{1 - h_{jj}} \tag{12.29}$$

where h_{jj} is the leverage of X_j.

Dividing the raw residual by its standard error results in the *internally studentized* residual (Velleman and Welsch 1981),

$$\text{ISTUD}_j = \frac{e_j}{\hat{\sigma}_e \sqrt{1 - h_{jj}}} \tag{12.30}$$

which can be shown to have a variance of 1 and follows a distribution called the Beta. Both BMDP and SPSS[X] refer to this as the *studentized residual*. Because tables of the Beta distribution are not readily available, a similar measure that has a *t* distribution is more useful; this is the *externally studentized* residual

$$\text{ESTUD}_j = \frac{e_j}{\hat{\sigma}_e^{(-j)} \sqrt{1 - h_{jj}}} \tag{12.32}$$

The only difference from the previous measure is that when the residual of the *j*th case is considered, $\hat{\sigma}_e^{(-j)}$, the standard error of estimate for the $N - 1$ cases that remain when case *j* is deleted, is used instead of $\hat{\sigma}_e$ in order to make the numerator and denominator independent of one another.

Note that the "deleted" standard error of estimate,

$$\hat{\sigma}_e^{(-j)} = \sqrt{\frac{\Sigma_{i \neq j}[Y_i - \hat{Y}_i^{(-j)}]^2}{N - 3}}$$

has only $N - 3$ df (whereas $\hat{\sigma}_e$ has $N - 2$ df) because the regression equation used in determining the residual is based only on $N - 1$ data points and 2 df are used to estimate $\beta_0^{(-j)}$ and $\beta_1^{(-j)}$. The externally studentized residual is what SYSTAT calls STUDENT in Table 12.5 and can also be obtained from SPSS[X] by asking for the "studentized deleted residual."

When members of a set of *N* residuals are tested for significance, Type 1 error should be controlled for the complete family of tests (see Chapter 6). This can be accomplished conveniently by using the Bonferroni inequality—that is, by conducting each test at the α/N level of significance. It can readily be seen from Table 12.5 that none of the externally studentized residuals (STUDENT) approaches significance. The residual for case 8 would just be significant at $\alpha = .05$ if it was the only one considered. However, the *t* value does not approach the critical values of about ± 3.6 that would be appropriate for maintaining an overall α of .05 for a family of 20 tests.

We conclude this section by considering two measures of the extent to which X_i may be considered an outlier:

A direct measure of the extent to which X_j differs from the center of the distribution of X scores is the *z* score of X_j. BMDP, SAS, and SPSS[X] all provide a measure called the *Mahalanobis distance* that for bivariate regression is essentially the squared *z* score

$$D_j = \left(\frac{X_j - \bar{X}}{\hat{\sigma}_X} \right)^2 \tag{12.33}$$

$$= \frac{(N - 1)(X_j - \bar{X})^2}{\text{SS}_X}$$

Another measure of the extent to which X_j is an outlier is the leverage measure, h_{jj} = $1/N + (X_j - \bar{X})^2/SS_X$, that we introduced earlier. This measure has been discussed by Belsley, Kuh, and Welsch (1980) and is provided by the SYSTAT output in Table 12.5.[1] The leverage is closely related to the Mahalanobis distance and can be expressed in terms of it:

$$h_{jj} = \frac{1}{N} + \frac{D_j}{N-1}$$

It can be shown that $\Sigma_j h_{jj} = p + 1$, where p is the number of predictor variables; therefore, for bivariate regression, the h_{jj} sum to 2 and have a mean value of $2/N$. Hoaglin and Welsch (1978) suggest that values of h_{jj} greater than $2(p + 1)/N$ should be considered large. Belsley, Kuh, and Welsch (1980) caution that this cutoff will identify too many points when there are only a few predictor variables, but recommend it because it is easy to remember and use. From Table 12.5, we see that the leverage values have a mean of $2/20 = .1$, as expected. We also see that there are no leverage values that exceed the cutoff of .2, and, therefore, no X scores are considered to be outliers.

Finally, it is of interest to note that Cook's distance, introduced earlier as a measure of influence, can be expressed as

$$CD_j = \left[\frac{1}{p+1} \right] \left[\frac{h_{jj}}{1 - h_{jj}} \right] (ISTUD_j)^2$$

This expression shows that the influence of a case is a function both of its leverage and its residual.

More detailed discussions of these issues are contained in Belsley, Kuh, and Welsch (1980), Cook and Weisberg (1982), Stevens (1984), and Velleman and Welsch (1981).

12.11.3 An Example with an Outlier

To illustrate how sensitive our analyses are to outliers, we reanalyze the search experiment data originally presented in Table 12.3 after we change one data point—say we change the reaction time for case 18 from 590 to 990 milliseconds. The revised data set is presented in panel a of Table 12.6, and the output provided by SYSTAT is presented in panel b.

Comparing Tables 12.3 and 12.6, we see that the single discrepant point has a noticeable effect on the results of the analysis. The original correlation between reaction time and array size was .873; in the new data set it drops to .658. The values of b_0 and b_1 change from 381.90 and 23.92 to 341.90 and 35.92, and their standard errors increase by about a factor of 3. Notice that near the bottom of panel b, SYSTAT warns us that case 18 has an unusually large residual.

---

[1] It should be noted that when asked for the leverage, the SPSS^X REGRESSION procedure provides the "centered leverage," $h_{jj} - 1/N$.

In panel c, we obtain an influence plot for the new data set. An *influence plot* is a scatter diagram in which the symbols used to indicate the locations of the data points inform us directly about their influence on the correlation. Using a 0 to represent a data

TABLE 12.6 SEARCH EXPERIMENT DATA WITH CASE 18 CHANGED

(a) Data

Array size	Reaction time (ms)
2	418
2	428
2	410
2	445
2	471
4	475
4	455
4	418
4	524
4	516
6	537
6	500
6	480
6	511
6	529
8	550
8	617
8	990
8	608
8	548

(b) SYSTAT output

```
DEP VAR:TIME    N:20    MULTIPLE R:.658    SQUARED MULTIPLE R:.433
ADJUSTED SQUARED MULTIPLE R:.402    STANDARD ERROR OF ESTIMATE:96.838
```

VARIABLE	COEFFICIENT	STD ERROR	STD COEF	TOLERANCE	T	P(2 TAIL)
CONSTANT	341.900	53.041	0.000	1.0000000	6.446	0.000
SIZE	35.920	9.684	0.658	1.0000000	3.709	0.002

ANALYSIS OF VARIANCE

SOURCE	SUM-OF-SQUARES	DF	MEAN-SQUARE	F-RATIO	P
REGRESSION	129024.640	1	129024.640	13.759	0.002
RESIDUAL	168798.360	18	9377.687		

```
WARNING: CASE 18 IS AN OUTLIER (STUDENTIZED RESIDUAL = 12.131)

DURBIN-WATSON D STATISTIC    1.904
FIRST ORDER AUTOCORRELATION   .028

RESIDUALS HAVE BEEN SAVED
```

TABLE 12.6   (continued)

(c) Influence plot

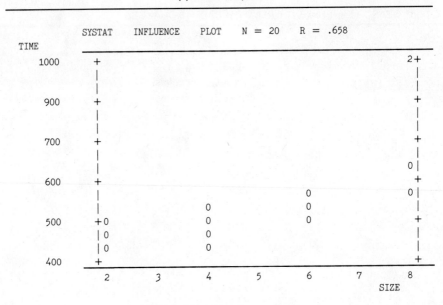

(d) Residual output

		ESTIMATE	RESIDUAL	LEVERAGE	COOK	STUDENT	SEPRED	TIME	SIZE
CASE	1	413.740	4.260	0.140	0.000	0.046	36.234	418.000	2.000
CASE	2	413.740	14.260	0.140	0.002	0.154	36.234	428.000	2.000
CASE	3	413.740	-3.740	0.140	0.000	-0.040	36.234	410.000	2.000
CASE	4	413.740	31.260	0.140	0.010	0.339	36.234	445.000	2.000
CASE	5	413.740	57.260	0.140	0.033	0.627	36.234	471.000	2.000
CASE	6	485.580	-10.580	0.060	0.000	-0.110	23.720	475.000	4.000
CASE	7	485.580	-30.580	0.060	0.003	-0.317	23.720	455.000	4.000
CASE	8	485.580	-67.580	0.060	0.017	-0.710	23.720	418.000	4.000
CASE	9	485.580	38.420	0.060	0.005	0.400	23.720	524.000	4.000
CASE	10	485.580	30.420	0.060	0.003	0.316	23.720	516.000	4.000
CASE	11	557.420	-20.420	0.060	0.002	-0.212	23.720	537.000	6.000
CASE	12	557.420	-57.420	0.060	0.012	-0.601	23.720	500.000	6.000
CASE	13	557.420	-77.420	0.060	0.022	-0.817	23.720	480.000	6.000
CASE	14	557.420	-46.420	0.060	0.008	-0.484	23.720	511.000	6.000
CASE	15	557.420	-28.420	0.060	0.003	-0.295	23.720	529.000	6.000
CASE	16	629.260	-79.260	0.140	0.063	-0.877	36.234	550.000	8.000
CASE	17	629.260	-12.260	0.140	0.002	-0.133	36.234	617.000	8.000
CASE	18	629.260	360.740	0.140	1.313	12.131	36.234	990.000	8.000
CASE	19	629.260	-21.260	0.140	0.005	-0.230	36.234	608.000	8.000
CASE	20	629.260	-81.260	0.140	0.067	-0.900	36.234	548.000	8.000

point tells us that deleting the data point would change the correlation by less than .1. The 2 in the upper right-hand corner marks the location of our discrepant point and tells us that deleting the point would change the correlation by at least .2 but less than .3. Notice that data points that are very close together will be represented by a single character. Because the Y scale is so condensed in the current example, SYSTAT only uses 13 characters to represent all 20 data points.

Finally, panel d presents the residuals and diagnostics for the regression. As we can readily see, case 18 is both an extreme outlier with respect to the distribution of residuals (ESTUD = 12.131) and an influential point (Cook's distance > 1).

What do we do about a situation like this? As mentioned earlier, we should first check to determine whether the data for case 18 was obtained and recorded appropriately. If an error is not found, there may be no simple answer; the best action to take will depend on the details of the research project and the purpose of the analysis. Although we might feel sorely tempted to delete the offending point, we cannot simply throw away data we don't like. In some circumstances it might be justifiable to trim or delete data if reasonable criteria can be established beforehand. Also, in any real study it would be desirable to collect a great deal more data than have been presented here so that the effects of a few outliers would be diluted. It may also be desirable to present the results both with and without the outliers so that the reader can form his or her own impression about the stability of the findings. Finally, one may want to use some of the "robust" correlation and regression procedures that are less sensitive to the presence of outliers (see, for example, Gideon and Hollister 1987, Rousseeuw and Leroy 1987, Wainer and Thissen 1976).

12.12 TESTING INDEPENDENT SLOPES FOR EQUALITY

Suppose two independent groups yield regression slopes b_{11} and b_{12}. To test the hypothesis H_0: $\beta_{11} = \beta_{12}$, we need only note that the test statistic

$$\frac{(b_{11} - b_{12}) - (\beta_{11} - \beta_{12})}{\hat{SE}(b_{11} - b_{12})}$$

follows a t distribution with $N_1 + N_2 - 4$ df. We know from Section 12.5.2 that $Var(b_1) = \sigma_e^2/SS_X$. Therefore,

$$Var(b_{11} - b_{12}) = Var(b_{11}) + Var(b_{12}) \qquad \text{because the groups are independent}$$

$$= \sigma_e^2 \left[\frac{1}{SS_{X_1}} + \frac{1}{SS_{X_2}} \right]$$

where SS_{X_1} and SS_{X_2} are the sums of squares of X in groups 1 and 2 so that

$$\hat{SE}(b_{11} - b_{12}) = \hat{\sigma}_e \sqrt{\frac{1}{SS_{X_1}} + \frac{1}{SS_{X_2}}}$$

where the best estimate of σ_e^2 is given by the weighted average of the estimates from group 1 and group 2:

$$\hat{\sigma}_e^2 = \frac{df_1\hat{\sigma}_{e_1}^2 + df_2\hat{\sigma}_{e_2}^2}{df_1 + df_2} = \frac{SS_{error_1} + SS_{error_2}}{N_1 + N_2 - 4}$$

Therefore, the test statistic

$$t = \frac{b_{11} - b_{12}}{\hat{\sigma}_e\sqrt{1/SS_{X_1} + 1/SS_{X_2}}} \qquad (12.34)$$

can be used to test for equality.

If we have more than two independent slopes, we can test the null hypothesis $\beta_{11} = \beta_{12} = \cdots = \beta_{1j} = \cdots = \beta_{1a}$ by using the test statistic

$$F = \frac{SS_1/(a-1)}{SS_2/(N-2a)} \qquad \text{with } a-1 \text{ and } N-2a \text{ df} \qquad (12.35)$$

SS_1 is the difference between the variability accounted for by the a regression lines when separate slopes are used and the variability accounted for by the regression lines if they are assumed to have a common slope estimated by

$$b_c = \frac{\Sigma_j\Sigma_i(X_{ij} - \bar{X}_{.j})Y_{ij}}{\Sigma_j\Sigma_i(X_{ij} - \bar{X}_{.j})^2} = \frac{\Sigma_j b_{1j}SS_{X_j}}{\Sigma_j SS_{X_j}} \qquad (12.36)$$

so that

$$SS_1 = \sum_j (b_{1j}^2 - b_c^2)SS_{X_j} \qquad (12.37)$$

There are $a-1$ df associated with SS_1; the common slope b_c is estimated from the a group slopes.

SS_1 is a measure of the variability of the slopes across groups. If the slopes are identical in the a groups, they will each equal b_c, so SS_1 is zero. Obviously, the slopes will vary to some extent because of sampling error. The question of whether they vary enough to reject the hypothesis that the population slopes are identical is addressed by the test statistic given in Equation 12.35.

SS_2 is the summed error variability in the a groups when slopes are determined separately for each of the groups. Summing the total variability in the Y scores for the a groups, we obtain $\Sigma_j SS_{Y_j}$. The variability accounted for by the a regression lines is $\Sigma_j b_{1j}^2 SS_{X_j}$, so

$$SS_2 = \sum_j SS_{error_j} = \sum_j (SS_{Y_j} - b_{1j}^2 SS_{X_j})$$

SS_2 is associated with $N-2a$ df; there are a total of N scores, but 2 df are required to estimate the slope and Y intercept for each of the a groups. $MS_2 = \Sigma_j SS_{error_j}/(N-2a)$ is the pooled estimate of σ_e^2 for the a groups.

Table 12.7 extends the search experiment data set. There are two conditions: one in which subjects search for specified letters in an array of letters, and one in which they search for letters in an array of digits. The two slopes are 23.92 and 6.95 for letter and digit arrays, respectively. As indicated in Table 12.7, the slopes are significantly different according to the calculations described in this section.

TABLE 12.7   REACTION TIME (RT) DATA FOR A SEARCH EXPERIMENT WITH TWO
CONDITIONS

Array size X	Array type	
	Letters	Digits
2	418	425
2	428	465
2	410	430
2	445	459
2	471	416
4	475	464
4	455	425
4	418	412
4	524	497
4	516	460
6	537	501
6	500	407
6	480	466
6	511	486
6	529	475
8	550	478
8	617	448
8	590	512
8	608	460
8	548	503

The summary statistics when we regress letter RT on X are $b_0 = 381.90$, $b_1 = 23.92$, $\hat{\sigma}_e = 31.452$, $SS_{reg} = 57{,}216.64$, and $SS_{error} = 17{,}806.36$. The corresponding statistics when we regress digit RT on X are $b_0 = 424.70$, $b_1 = 6.95$, $\hat{\sigma}_e = 28.492$, $SS_{reg} = 4830.25$, and $SS_{error} = 14{,}612.70$.

Testing equality of two slopes using the t statistic

Because the same X's are used for both regressions, SS_X is the same. $SS_X = 5[(2-5)^2 + (4-5)^2 + (6-5)^2 + (8-5)^2] = 5(9 + 1 + 1 + 9) = 100$. The t statistic of Equation 12.34 is

$$t = \frac{23.92 - 6.95}{\hat{\sigma}_e \sqrt{1/100 + 1/100}}$$

The best estimate of σ_e^2 is the weighted average of the $\hat{\sigma}_e^2$'s obtained in the two regressions; that is

$$\hat{\sigma}_e^2 = \frac{(18)(31.452^2) + (18)(28.492^2)}{20 + 20 - 4}$$

$$= 900.51$$

so the best estimate of σ_e is 30.01. Therefore,

$$t = \frac{23.92 - 6.95}{30.01\sqrt{.02}} = 4.00$$

TABLE 12.7 (continued)

with $20 + 20 - 4 = 36$ df. The t is large enough to reject the hypothesis that the two population slopes are equal at $p < .001$.

Testing equality of slopes using the F statistic of Equation 12.35

We begin by finding the common slope

$$b_c = \frac{\Sigma_j b_{1j} SS_{X_j}}{\Sigma_j SS_{X_j}}$$

$$= \frac{(23.92)(100) + (6.95)(100)}{100 + 100}$$

$$= 15.44$$

Notice that because the same X's are used for both regressions, the common slope is the simple average of the group slopes. We next find SS_1 from Equation 12.37:

$$SS_1 = \sum_j (b_{1j}^2 - b_c^2)SS_{X_j}$$

$$= (23.92^2 - 15.44^2)100 + (6.95^2 - 15.44^2)100$$

$$= 14,356.7$$

Because there are only two groups, $a - 1 = 1$ and $MS_1 = 14,356.7$. SS_2 is the sum of the SS_{error} terms for the individual groups. Therefore,

$$SS_2 = \Sigma SS_{error_j} = 17,806.36 + 14,612.70 = 32,419.06$$

$$MS_2 = \frac{SS_2}{N - 2a} = \frac{32,419.06}{40 - 4} = 900.5$$

Note that MS_2 is the weighted average of the estimates of σ_e^2 obtained from the different groups. Finally, we form the F ratio

$$\frac{MS_1}{MS_2} = \frac{14,356.7}{900.5} = 15.94$$

The obtained F is, within rounding error, the square of the t we obtained earlier. We reject the null hypothesis of equal slopes at $p < .001$.

12.13 CONFIDENCE INTERVALS AND HYPOTHESIS TESTS IN REPEATED-MEASURES DESIGNS

Up to this point, we have treated the search experiment as a between-subjects design: each subject was tested at only one array size. We could rerun the experiment as a repeated-measures design by testing each subject at all four array sizes. Let's say that we have 20 subjects search for particular target letters in arrays of letters and that we obtain 40 detection times from each subject, 10 at each of the four array sizes. The 20 subjects yield a total of 800 data points.

The correct linear regression equation could be obtained by using any of the following three procedures:

1. Regress reaction time on array size, using the combined data set (800 data points, 40 from each subject). Values of b_1 and b_0 could be obtained, as usual, by employing Equations 12.6 and 12.7.

2. Regress reaction time on array size, using the 80 data points obtained by pairing each array size with the mean detection time for each of the 20 subjects at that array size.

3. Regress mean reaction time on array size separately for each of the 20 subjects, basing each equation on four data points. Average the resulting 20 values of b_0 and b_1 to obtain values that best represent the entire group.

All three procedures yield exactly the same values for b_0 and b_1. This occurs because b_0 and b_1 are linear combinations of Y and will be true as long as every subject is tested at exactly the same levels of X.

However, procedure 3 has the advantage that the values of b_0 and b_1 obtained for each subject can be treated as scores in subsequent analyses in which we test hypotheses about β_0 and β_1, using the procedures that were developed in earlier chapters. In essence, when we test hypotheses about β_0 or β_1, we have one score for each subject. Suppose we use b_{1i} to represent the slope obtained for the ith subject. Then we could test the null hypothesis that β_1 is equal to some hypothesized value $\beta_{1\,hyp}$ using exactly the same t test that was introduced in Chapter 3.

$$t_{N-1} = \frac{\bar{b}_1 - \beta_{1\,hyp}}{\hat{\sigma}_{\bar{b}_1}} = \frac{\bar{b}_1 - \beta_{1\,hyp}}{\hat{\sigma}_{b_1}/\sqrt{N}}$$

where $\bar{b}_1$ is the average slope for the sample of N subjects and $\hat{\sigma}_{b_1}$ is the estimated population standard deviation of b_1 scores; that is,

$$\hat{\sigma}_{b_1} = \sqrt{\frac{\Sigma_i(b_{1i} - \bar{b}_1)^2}{N - 1}}$$

We could find the confidence interval for β_1 by using $\bar{b}_1 \pm t_{\alpha/2}\hat{\sigma}_{\bar{b}_1}$.

Suppose, for example, the values of b_1 for the 20 subjects are 19, 25, 27, 16, 14, 15, 18, 34, 19, 30, 25, 26, 19, 30, 27, 19, 24, 20, 23, and 25. The mean slope is 22.75, and the estimated population standard deviation $\hat{\sigma}_{b_1}$ is 5.45, so $\hat{\sigma}_{\bar{b}_1} = \hat{\sigma}_{b_1}/\sqrt{20} = 1.22$. Therefore, the 95% confidence interval about β_1 is given by 22.75 $\pm$ (2.093)(1.22) = 22.75 $\pm$ 2.55.

The individual values of b_0 or b_1 could also be used in matched- and independent-groups t tests and in repeated-measures and between-subjects ANOVAs if we wished to test for equality of slopes or intercepts across conditions. We might, for example, want to test whether slopes for arrays consisting of letters and digits are equal, as we did in the previous section for the between-subjects case.

12.14 CONCLUDING REMARKS

We have considered several measures that provide information about the strength of the linear relation between two variables. The intercept and slope of the regression equation, b_0 and b_1, characterize the linear component of the relation. The slope is particularly useful in developing explanatory models, being directly interpretable as a measure of how changes in the size of X translate into changes in the size of Y. The correlation coefficient, on the other hand, is an index of the extent to which the relation between X and Y is linear. It is perhaps best thought of as an index of the degree of "relative clustering" of the data points about the best-fitting straight line. Large values of r suggest that the degree of scatter of the data points about the best-fitting straight line is small compared to the degree of variability in X and Y; small values suggest that the data points do not cluster closely about any straight line.

Because the value of the correlation coefficient depends on the shape of the cloud of data points, and therefore on such factors as the variance of X, the slope of the regression line and the standard error of estimate, certain interpretive issues arise that will be discussed in Chapter 14. The chapter will also deal with inference about correlation, alternative measures of correlation, and the concepts of partial and part correlation.

In conclusion, we acknowledge that many new concepts and expressions have been presented. To assist the reader in integrating this new material, we present some of the more useful expressions we have developed in Table 12.8.

TABLE 12.8 SUMMARY OF SOME REGRESSION EXPRESSIONS

Expression	Point estimate	Expected value	Estimated standard error for finding confidence interval
Slope	$b_1 = \dfrac{\Sigma(X_i - \bar{X})(Y_i - \bar{Y})}{SS_X}$	β_1	$\hat{\sigma}_e \dfrac{1}{\sqrt{SS_X}}$
Y intercept	$b_0 = \bar{Y} - b_1\bar{X}$	β_0	$\hat{\sigma}_e \sqrt{\dfrac{1}{N} + \dfrac{\bar{X}^2}{SS_X}}$
Mean value of Y at X_j	$\hat{\mu}_{Y.X_j} = b_0 + b_1 X_j$	$\mu_{Y.X_j}$	$\hat{\sigma}_e \sqrt{h_{jj}}$, where $h_{jj} = \dfrac{1}{N} + \dfrac{(X_j - \bar{X})^2}{SS_X}$
New score at X_j	$\hat{Y}_{\text{new}_j} = \hat{\mu}_{Y.X_j}$ $= b_0 + b_1 X_j$	$\mu_{Y.X_j}$	$\hat{\sigma}_e \sqrt{1 + h_{jj}}$
Residual of jth case	$e_j = Y_j - \hat{Y}_j$	0	$\hat{\sigma}_e \sqrt{1 - h_{jj}}$

APPENDIX 12.1: Proof That $z_Y = \pm z_X$ When $Y = b_0 + b_1 X$

We want to show that if X and Y have a perfect linear relation, $z_Y = z_X$ when the relation is positive and $z_Y = -z_X$ when it is negative.

For any data point (X, Y) that falls on the line $Y = b_0 + b_1 X$, we have

$$Y = b_0 + b_1 X$$

Substituting into the usual expressions for the mean and standard deviation and simplifying, we have that $\bar{Y} = b_0 + b_1 \bar{X}$ and $\hat{\sigma}_Y = \pm b_1 \hat{\sigma}_X$, with $\hat{\sigma}_Y = +b_1 \hat{\sigma}_X$ when b_1 is positive and $\hat{\sigma}_Y = -b_1 \hat{\sigma}_X$ when b_1 is negative. Therefore, if there is a perfect linear relation between X and Y,

$$z_Y = \frac{Y - \bar{Y}}{\hat{\sigma}_Y}$$

$$= \frac{b_0 + b_1 X - (b_0 + b_1 \bar{X})}{\pm b_1 \hat{\sigma}_X}$$

$$= \frac{X - \bar{X}}{\pm \hat{\sigma}_X}$$

$$= \pm z_X$$

APPENDIX 12.2: Unbiased Estimators and Standard Errors

1. To show that b_1, b_0 and $\hat{Y}_j$ are unbiased estimates of β_1, β_2, and $\mu_{Y.X_j}$, respectively, and to find their standard errors,
2. To show that $E(\Sigma_i(Y_i - \hat{Y}_i)^2 = (N - 2)\sigma_e^2$ so that $MS_{error} = \Sigma_i(Y_i - \hat{Y}_i)^2/(N - 2)$ is an unbiased estimator of σ_e^2.

To show that b_1 is an unbiased estimator of β_1, we need to show that $E(b_1) = \beta_1$. Our model for Y_i (Equation 12.17) is

$$Y_i = \beta_0 + \beta_1 X_1 + \epsilon_i$$

$$= \mu_Y + \beta_1(X_i - \bar{X}) + \epsilon_i$$

From Equation 12.18, we have $b_1 = \Sigma_i f_i Y_i$, where $f_i = (X_i - \bar{X})/SS_X$. Therefore,

$$E(b_1) = E\left[\sum_i f_i[\mu_Y + \beta_1(X_i - \bar{X}) + \epsilon_i]\right]$$

$$= \mu_Y \sum_i f_i + E\left[\sum_i f_i \beta_1(X_i - \bar{X})\right] + \sum_i f_i E(\epsilon_i)$$

The first and third terms are equal to zero because $\Sigma_i f_i = \Sigma_i (X_i - \bar{X})/SS_X = 0$ and $E(\epsilon_j) = 0$. The second term is

$$E(b_1) = E\left[\sum_i \frac{X_i - \bar{X}}{SS_X} \beta_1 (X_i - \bar{X}) \right]$$

$$= E\left[\sum_i \frac{\beta_1 (X_i - \bar{X})^2}{SS_X} \right]$$

$$= E(\beta_1) \qquad \text{because } \sum_i (X_i - \bar{X})^2 = SS_X$$

$$= \beta_1$$

To show that b_0 is an unbiased estimate of β_0, we need to demonstrate that $E(b_0) = \beta_0$. We begin by noting that from the model $Y_i = \beta_0 + \beta_1 X_1 + \epsilon_i$. Therefore, we have

$$\bar{Y} = \frac{1}{N} \sum_i Y_i = \beta_0 + \beta_1 \bar{X} + \bar{\epsilon}$$

Therefore,

$$E(b_0) = E(\bar{Y} - b_1 \bar{X}) \quad \text{from Equation 12.7}$$

$$= E(\beta_0 + \beta_1 \bar{X} + \bar{\epsilon} - b_1 \bar{X})$$

$$= \beta_0 + \beta_1 \bar{X} + 0 - \beta_1 \bar{X}$$

$$= \beta_0$$

To show that $\hat{Y}$ is an unbiased estimator of $\mu_{Y.X}$, we must demonstrate that $E(\hat{Y}) = \mu_{Y.X}$. Because $\hat{Y} = b_0 + b_1 X$, we have

$$E(\hat{Y}) = E(b_0 + b_1 X)$$

$$= E(b_0) + X E(b_1)$$

$$= \beta_0 + \beta_1 X$$

$$= \mu_{Y.X}$$

To find the standard error of b_1, we first find the variance of b_1:

$$\text{Var}(b_1) = \text{Var}\left[\sum_i f_i Y_i \right]$$

Because the preceding is a linear combination of the Y_i (see Chapter 2) and assuming that $\text{Var}(Y_i) = \sigma_e^2$ for all i and $\text{Cov}(Y_i, Y_{i'}) = 0$ when $i \neq i'$, we have

$$\text{Var}(b_1) = \sigma_e^2 \sum_i f_i^2$$

But

$$\sum_i f_i^2 = \sum_i \left(\frac{X_i - \bar{X}}{SS_X} \right)^2$$

$$= \sum_i \frac{(X_i - \bar{X})^2}{SS_X^2}$$

$$= \frac{SS_X}{SS_X^2}$$

$$= \frac{1}{SS_X}$$

So

$$\text{Var}(b_1) = \frac{\sigma_e^2}{SS_X}$$

The standard error is the square root of the variance—that is, $\text{SE}(b_1) = \sigma_e/\sqrt{SS_X}$—and the estimated standard error is obtained by using the estimate of σ_e from the sample; thus,

$$\hat{\text{SE}}(b_1) = \frac{\hat{\sigma}_e}{\sqrt{SS_X}}$$

To find the standard error of b_0, we begin by finding $\text{Var}(b_0)$; that is,

$$\text{Var}(b_0) = \text{Var}\left[\sum_i g_i Y_i\right] \qquad \text{where } g_i = \frac{1}{N} - \frac{\bar{X}(X_i - \bar{X})}{SS_X}$$

Given our assumptions and noting that b_0 is a linear combination of the Y_i, we have

$$\text{Var}(b_0) = \sigma_e^2 \sum_i g_i^2$$

But

$$\sum_i g_i^2 = \sum_i \left[\frac{1}{N^2} + \frac{\bar{X}^2(X_i - \bar{X})^2}{SS_X^2} - \frac{2\bar{X}(X_i - \bar{X})}{NSS_X}\right]$$

$$= \frac{1}{N} + \frac{\bar{X}^2 \Sigma_i(X_i - \bar{X})^2}{SS_X^2} - \frac{2\bar{X}\Sigma_i(X_i - \bar{X})}{NSS_X}$$

$$= \frac{1}{N} + \frac{\bar{X}^2 SS_X}{SS_X^2} - \frac{2\bar{X}(0)}{NSS_X}$$

$$= \frac{1}{N} + \frac{\bar{X}^2}{SS_X}$$

Therefore, the variance of b_0 is given by

$$\text{Var}(b_0) = \sigma_e^2 \left[\frac{1}{N} + \frac{\bar{X}^2}{SS_X}\right]$$

and

$$\hat{\text{SE}}(b_0) = \hat{\sigma}_e \sqrt{\frac{1}{N} + \frac{\bar{X}^2}{SS_X}}$$

To find the standard error of $\hat{Y}_j$, we begin by recalling Equation 12.20,

$$\hat{Y}_j = \sum_i h_{ij} Y_i \quad \text{where } h_{ij} = \frac{1}{N} + \frac{(X_j - \bar{X})(X_i - \bar{X})}{SS_X}$$

Then

$$\text{Var}(\hat{Y}_j) = \text{Var}\left[\sum_i h_{ij} Y_i\right]$$

$$= \sigma_e^2 \sum_i h_{ij}^2$$

But it can be shown that $\sum_i h_{ij}^2 = h_{jj}$ (just expand and simplify to see that this is true). Therefore,

$$\text{Var}(\hat{Y}_j) = \sigma_e^2 h_{jj}$$

and so

$$\hat{SE}(\hat{Y}_j) = \hat{\sigma}_e \sqrt{h_{jj}}$$

To show that $E[\sum_i (Y_i - \hat{Y}_i)^2] = (N-2)\sigma_e^2$ so that $\text{MS}_{\text{error}} = \sum_i (Y_i - \hat{Y}_i)^2/(N-2)$ is an unbiased estimator of σ_e^2, we begin by recalling that, from Equation 12.9, $\hat{Y} = \bar{Y} + b_1(X - \bar{X})$ or $\hat{Y} = \bar{Y} + b_1 x$, where $x_i = X_i - \bar{X}$. Similarly, our model for Y can be expressed as $Y = \mu_Y + \beta_1 x + \epsilon$. We can then express ϵ as $Y - (\mu_Y + \beta_1 x)$ and write an identity that expresses this as the sum of three components:

$$Y - (\mu_Y + \beta_1 x) = [Y - (\bar{Y} + b_1 x)] + (\bar{Y} - \mu_Y) + (b_1 - \beta_1 x)$$

$$= (y - b_1 x) + (\bar{Y} - \mu_Y) + (b_1 - \beta_1)x \quad \text{where } y = Y - \bar{Y}$$

Squaring both sides and summing over the N values for the sample yields

$$\sum[Y - (\mu_Y + \beta_1 x)]^2 = \sum(y - b_1 x)^2 + N(\bar{Y} - \mu_Y)^2 + (b_1 - \beta_1)^2 \sum x^2 + 2(\bar{Y} - \mu_Y)\sum(y - b_1 x)$$

$$+ 2(b_1 - \beta_1)\sum x(y - b_1 x) + 2\sum(\bar{Y} - \mu_Y)(b_1 - \beta_1)\sum x$$

The three cross-product terms are all equal to zero. For the first cross-product term

$$\sum(y - b_1 x) = \sum y - b_1 \sum x = 0 - b_1(0) = 0$$

For the next

$$\sum x(y - b_1 x) = \sum xy - b_1 \sum x^2$$

But b_1 can be written as $\sum xy/\sum x^2$, so

$$\sum x(y - b_1 x) = \sum xy - \sum xy = 0$$

and for the last term, $\sum x = 0$. We therefore have

$$\sum[Y - (\mu_Y + \beta_1 x)]^2 = \sum(y - b_1 x)^2 + N(\bar{Y} - \mu_Y)^2 + (b_1 - \beta_1)^2 \sum x^2$$

The term on the left side is $\sum(Y - \mu_{Y.X})^2 = \sum \epsilon^2$. The first term on the right is $\sum(Y - \hat{Y})^2 = SS_{\text{error}}$. Solving for SS_{error}, we have

$$SS_{error} = \Sigma\epsilon^2 - N(\bar{Y} - \mu_Y)^2 - (b_1 - \beta_1)^2\Sigma x^2 \qquad (12.38)$$

Taking expectations of both sides, we have for the first term on the right,

$$E(\Sigma\epsilon^2) = N\sigma_e^2 \qquad \text{assuming the } \epsilon \text{ are independently distributed with variance } \sigma_e^2$$

Also, because the variance of a mean is the population variance divided by the sample size, the second term becomes

$$E[N(\bar{Y} - \mu_Y)^2] = N\left(\frac{\sigma_e^2}{N}\right) = \sigma_e^2$$

The expectation of the last term is

$$E[(b_1 - \beta_1)^2\Sigma x^2] = \Sigma x^2 E[(b_1 - \beta_1)^2]$$

But $Var(b_1) = E[(b_1 - \beta_1)^2]$, and we have shown earlier in the appendix that $Var(b_1) = \sigma_e^2/SS_X = \sigma_e^2/\Sigma x^2$. Therefore, substituting into the third term, we wind up with

$$E[(b_1 - \beta_1)^2\Sigma x^2] = \sigma_e^2$$

Putting these results together, we have, for Equation 12.38,

$$E(SS_{error}) = N\sigma_e^2 - \sigma_e^2 - \sigma_e^2$$

$$= (N - 2)\sigma_e^2$$

and so $SS_{error}/(N - 2) = MS_{error}$ is an unbiased estimator of σ_e^2.

APPENDIX 12.3: Proof That $\hat{SE}(e_j) = \hat{\sigma}_e\sqrt{1 - h_{jj}}$

To show that $\hat{SE}(e_j) = \hat{\sigma}_e\sqrt{1 - h_{jj}}$, we begin with

$$e_j = Y_j - \hat{Y}_j$$

$$= Y_j - \sum_i h_{ij}Y_i \qquad \text{from Equation 12.20}$$

So

$$Var(e_j) = Var(Y_j) + Var\left[\sum_i h_{ij}Y_i\right] - 2\,Cov\left[Y_j, \sum_i h_{ij}Y_i\right]$$

The first term is equal to σ_e^2. The second is equal to $h_{jj}\sigma_e^2$ from Appendix 12.2. The last term is equal to $-2h_{jj}\sigma_e^2$ because $Cov(Y_j, Y_j) = \sigma_e^2$ and $Cov(Y_j, Y_{j'}) = 0$ for $j \neq j'$. Therefore

$$Var(e_j) = \sigma_e^2(1 - h_{jj})$$

and

$$\hat{SE}(e_j) = \hat{\sigma}_e\sqrt{1 - h_{jj}}$$

EXERCISES

12.1 The following terms provide a useful review of many concepts in the chapter. Define, describe, or identify each of them:

linear equation	coefficient of determination, r^2
slope	conditional population mean, $\mu_{Y.X}$
Y intercept	$SS_{pure\ error}$
linear regression equation	$SS_{nonlinearity}$
bivariate regression	leverage, h_{jj}
correlation coefficient	standard error of a statistic
scatter diagram	residual
z score	outlier
covariance	influential point
least-squares criterion	Durbin-Watson test
mean-squared error	Cook's distance
regression of Y on X	deleted prediction
regression of X on Y	internally studentized residual
regression to the mean	externally studentized residual
SS_{reg}	Mahalanobis distance
SS_{error}	influence plot
standard error of estimate, $\hat{\sigma}_e$	

12.2 Given the following data:

X	Y
1	11
2	3
3	7
4	9
5	9
6	21

(a) Draw a scatter diagram for these data.

(b) What is the correlation between Y and X?

(c) Find the least-squares equation for the regression of Y on X.

(d) What is the proportion of the variance in Y accounted for by the regression on X? What is the amount of variance accounted for?

(e) What is the standard error of estimate for the regression?

(f) Find the equation for the regression of X on Y.

12.3 Given three variables, X, Y, and W, with $r_{XW} = .6$ and $r_{WY} = .3$, we wish to say something about r_{XY}. Consider the following reasoning:

$$\hat{z}_W = r_{WY}z_Y$$

so

$$z_Y = \frac{\hat{z}_W}{r_{WY}} \tag{1}$$

but

$$\hat{z}_W = r_{WX} z_X$$

substituting into Equation 1, we have

$$z_Y = \frac{r_{WX} z_X}{r_{WY}}$$

so

$$r_{XY} = \frac{\Sigma z_X z_Y}{N-1} = \frac{\Sigma z_X r_{WX} z_X}{(N-1) r_{WY}}$$

$$= \frac{(r_{WX}/r_{WY})\Sigma z_X^2}{N-1} = \frac{r_{WX}}{r_{WY}} = \frac{.6}{.3} = 2 \tag{2}$$

Obviously, r_{XY} cannot be 2, so the reasoning must be faulty. What step or steps were incorrect?

12.4 **(a)** We have three measures of neuroticism: X, Y, and W. Some important statistics are

	X	Y	W
Mean	28	32	35
$\hat{\sigma}^2$	20	10	15
$r_{XY} = .7$		$r_{XW} = .8$	

Discuss whether X is a better predictor of W or of Y. In what sense is X a better predictor of W than of Y? In what sense is X a better predictor of Y?

(b) There is a related measure, Q, for which $b_{XQ} = 2$ and $b_{YQ} = 1$. Does this mean that $r_{XQ} > r_{YQ}$?

(c) For a different set of data, $b_{YX} > 0$. Must b_{XY} also be greater than 0? Explain briefly.

12.5 Assume that the correlation between the adult heights of fathers and sons is .5 and, further, that the mean and standard deviation of the heights of adult males is 70.0 and 3.0 inches, respectively.

(a) Given the information that a father is 76 inches tall, what is the best linear prediction for the adult height of his son?

(b) Given that the adult height of a son is 73 inches, what is the best linear prediction for the height of his father?

(c) Given the phenomenon of "regression toward the mean," why wouldn't we expect all males to have about the same height in a few more generations?

12.6 **(a)** Calculate the value of b_1 for the following data:

$$X = 3, 3, 3, 4, 4, 4, 4, 5, 5, 5$$
$$Y = 0, 1, 1, 0, 2, 2, 2, 0, 1, 1$$

(b) We could present these data as a contingency table:

Y =	0	1	2
3	1	2	0
X = 4	1	0	3
5	1	2	0

Are X and Y independently distributed? Together with your answer to (a), what does this tell you about the relation between *linear independence* (i.e., zero regression or correlation) and *stochastic independence* [as defined by $p(Y|X) = p(Y)$ or $p(X$ and $Y) = p(X)p(Y)$]?

12.7 **(a)** After each of two practice landings, pilot trainees discuss their performance with their instructors. The instructors find that trainees who make poor landings the first time tend to make better landings the second time, whereas trainees who make good landings the first time tend to do worse the second time. The instructors conclude that the criticism that follows poor performance tends to make pilots do better and that the praise that follows good performance tends to make them do worse. Therefore, the instructors decide to be critical of all landings, good or bad. Is this a reasonable strategy?

(b) After the first examination in a course, students who scored in the bottom 25% of the distribution are given special tutoring. On the next examination, all of these students score above the average for the whole class. Can we conclude that the tutoring was effective or could the results simply be due to regression toward the mean?

(c) An educational psychologist wants to see if ability to spell has any effect on ability to read. To this end, he selects two groups of subjects, a group of poor spellers and a group of good spellers. However, he is worried that the poor spellers may not be as intelligent as the good spellers, so he creates a group of poor spellers and a group of good spellers that are equated on IQ. (To make this simple, let us assume that his procedure is to use only those subjects in both groups who scored 100 on an IQ test that he administered.) He now administers a reading test to both groups and finds that the good spellers do better on the average than the poor spellers. Does this mean that spelling ability affects reading ability?

12.8 After having determined a regression equation for predicting Y from X by using a sample of (X, Y) scores, we often would like to predict Y for some case that was not in the original sample.

(a) Discuss whether the precision of estimation is different if the new case has an X score close to the mean of the N X scores that went into determining the regression equation than if the new X score is outside the range of the original X score.

(b) Two new students sign up for the course that generated the data in Table 12.1. Given only the data in the table, what is your best prediction for the final exam scores for these two students? What is the 90% confidence interval for the mean final exam score for new students if no other information about them is available?

(c) Suppose that we are given the additional information that the two students obtained scores of 33 and 26, respectively, on the pretest.
   (i)   Now what are our best predictions for the final exam scores for the two students?
   (ii) What are the 90% confidence intervals for the conditional means of Y scores corresponding to $X = 33$ and $X = 26$?
   (iii) What are the 90% confidence intervals for the students' *actual* final exam scores?

12.9 Suppose we have a population described by the model

$$Y = \beta_0 + \beta_1 X + \epsilon$$

in which the ϵ's are normally and independently distributed with $E(\epsilon) = 0$ and $\text{Var}(\epsilon) = \sigma_e^2$ for all X. We have shown that the least-squares estimate of β_1 is given by

$$b_1 = \sum_i \frac{(X_i - \bar{X})(Y_i - \bar{Y})}{\text{SS}_X}$$

Now consider the estimate

$$b_1' = \frac{Y_{\max} - Y_{\min}}{X_{\max} - X_{\min}}$$

where $X_{\max}$ and $X_{\min}$ are the largest and smallest values of X in the sample and $Y_{\max}$ and $Y_{\min}$ are the corresponding values of Y (take averages if there are several Y scores corresponding to either $X_{\max}$ or $X_{\min}$).

(a) Is b_1' an unbiased estimator of β_1?
   For the data in Table 12.1, $b_1 = 3.55$ and $\widehat{\text{Var}}(b_1) = \hat{\sigma}_e^2/\text{SS}_X = \hat{\sigma}_e^2/159.6$.

(b) What is b_1'? Also, express $\text{Var}(b_1')$ in terms of σ_e^2. Which estimate, b_1 or b_1', would you prefer to use? Why?

12.10 In a search experiment, subjects are required to check for the presence of some target character in an array of characters. There are four different array sizes, $X = 2$, 4, 6, and 8. Ten subjects are assigned to each array size. The time to respond for each of the 40 subjects (Y) is recorded. The data for the four array sizes are

X_j:	2	4	6	8
$\bar{Y}_{.j}$:	480	520	540	540
$\hat{\sigma}_j^2$:	360	315	324	333

Two experimenters, Anne and Reg, have different views about the analysis. Anne uses the ANOVA design model

$$Y_{ij} = \mu + \alpha_j + \epsilon_{ij}$$

to test the hypothesis $H_0: \mu_1 = \mu_2 = \mu_3 = \mu_4$, and Reg assumes the linear regression model

$$Y_{ij} = \mu_Y + \beta_1(X_j - \bar{X}) + \epsilon_{ij}$$

and tests the hypothesis $H_0: \beta_1 = 0$.

(a) Are Anne and Reg testing equivalent hypotheses? Briefly, justify your answer. If your answer is "no," are the two null hypotheses related? That is, if Anne's is false, should Reg's be true? Or if Reg's is false, should Anne's be true?

(b) Use ANOVA to test $H_0: \mu_1 = \mu_2 = \mu_3 = \mu_4$; and use regression to test $H_0: \beta_1 = 0$.

(c) Must SS_A always be larger than SS_{reg}?

(d) You have been provided with enough information to determine whether there is a significant departure from linearity in the data. Find $SS_{nonlinearity}$ and $SS_{pure\ error}$ and perform the test, using $\alpha = .05$.

12.11 Groups of 40 males and 40 females each participate in the kind of search experiment described in the previous problem. For each group, $SS_X = 200$. For males, we obtain $b_1 = 30.0$ and $\hat{\sigma}_e = 15.5$; for females, $b_1 = 20.0$ and $\hat{\sigma}_e = 12.2$. Test whether the slopes for males and females differ significantly at $\alpha = .05$.

12.12 The search experiment is rerun as a repeated-measures study with eight males and eight females, each of whom are tested at all four array sizes. Slopes are determined separately for each of the 16 subjects. The slopes are

Males	Females
35	17
25	19
29	29
37	19
20	25
17	23
39	15
38	13

Test the hypothesis that the mean population slopes for males and females are identical (at $\alpha = .05$).

Chapter 13

Analysis of Covariance

13.1 INTRODUCTION

The analysis of covariance (ANCOVA) differs from the standard analysis of variance (ANOVA) because it not only uses information about the dependent variable (Y), it also uses information about an additional variable (X, called the covariate) that is correlated with the dependent variable. The goal of ANCOVA is to control for differences in the covariate, thereby achieving greater efficiency than the corresponding ANOVA and reducing the effects of chance differences between groups.

In earlier chapters, we considered several designs that are more efficient than the one-factor completely randomized design. In Chapter 8 we introduced repeated-measures designs in which variability due to individual differences could be partialed out if the subject was tested in all treatment conditions. However, there are many situations in which repeated-measures designs cannot be used because of carryover effects. For example, a repeated-measures design cannot be used to compare different methods for teaching arithmetic because there is no way that material learned using one of the methods can be removed so that it can be relearned using another method.

Another possibility would be to use the treatment × blocks design that was introduced in Chapter 5. In the treatment × blocks design, information about a concomitant measure, X, such as the score on a pretest, is used to add a blocking factor, B, yielding the model

$$Y_{ijk} = \mu + \alpha_j + \beta_k + (\alpha\beta)_{jk} + \epsilon_{ijk}$$

instead of the model for the one-factor completely randomized design

$$Y_{ij} = \mu + \alpha_j + \epsilon_{ij}$$

Greater efficiency is achieved by partialing the effects of B and the $A \times B$ interaction out of the error term at the cost of some degrees of freedom and of collecting and using the concomitant variable. The treatment × blocks design would be a good choice if the dependent and concomitant variables were moderately correlated and if Feldt's (1958) tables were used to determine the optimal number of blocks. However, if there is a fairly high correlation, an even greater increase in efficiency can be obtained by using ANCOVA (Section 13.8 presents a detailed discussion of the treatment × blocks design and ANCOVA).

The basic idea of ANCOVA is similar to that of treatment × blocks. However, instead of using information about X to form a categorical blocking factor and then partialing out the effect of the blocking factor and its interaction with A, we partial out in ANCOVA the variability accounted for by *regressing* Y on X. The model for a one-factor ANCOVA

$$Y_{ij} = \mu + \alpha_j + \beta(X_{ij} - \bar{X}_{..}) + \epsilon'_{ij} \tag{13.1}$$

adds a regression component to the one-factor ANOVA model

$$Y_{ij} = \mu + \alpha_j + \epsilon_{ij} \tag{13.2}$$

An increase in power may be achieved because, if SS_A and $SS_{S/A}$ are adjusted by removing the variability accounted for by the regression on X, the ratio of the resulting mean squares may be considerably larger than the ratio of the unadjusted mean squares. To see how this works, we will consider a numerical example in the next section.

Because ANOVA itself will be developed as a special case of multiple regression analysis in Chapter 16, it would be most efficient to introduce ANCOVA there. However, a reasonable treatment of ANCOVA can be presented by using the information about bivariate regression that was developed in Chapter 12, and we have decided to include a chapter on ANCOVA that does not depend on an understanding of multiple regression. In this chapter, we concentrate on the basic ideas behind ANCOVA and on its interpretation, especially on how correct interpretation depends on certain assumptions being met. Because we assume that ANCOVAs will be performed on computers, we will not concern ourselves much with calculations or computational formulas. Instead we will indicate in Section 13.9 how ANCOVAs can be performed by using several standard software packages.

13.2 AN EXAMPLE OF ANCOVA

Consider an experiment conducted to test the effectiveness of three software packages for teaching problem-solving skills to seventh graders. Thirty-six seventh graders are randomly selected and assigned to the software packages with the restriction that 12 children work with each of the packages. The dependent variable (Y) is the score obtained on a problem-solving achievement test administered after the children have worked with a package for six months. Panel a of Table 13.1 contains the achievement test scores (Y) along with the scores (X) obtained on a problem-solving pretest administered before the children were assigned to work with the software packages. Panel b contains the results of a one-factor ANOVA conducted on the Y scores. The ANOVA indicates that we cannot reject the null hypothesis that the software packages are equally effective in developing problem-solving skills.

The one-factor ANOVA is not very efficient because, as we can see from the within-group variability in achievement scores and from the variability in the pretest scores, children differ considerably from one another in problem-solving ability. Random assignment of children to treatment groups prevents systematic bias favoring any group. However, the random within- and between-group ability differences will obscure any effects of the treatments. The larger these differences are, the more difficult it will be to assess the effectiveness of the software packages.

Panel c contains the results of an ANCOVA using the pretest scores as a covariate. These results indicate that when the achievement scores are adjusted for the variability in the pretest scores, the null hypothesis can be rejected. In the remainder of the present section, we sketch out an overview of this ANCOVA. In the next section, we discuss the analysis in more detail.

In ANCOVA, what is partitioned is not the total variability in Y (which we refer to as SS_Y or $SS_{tot(Y)}$) but $SS_{tot(adj)}$, the total variability in Y that cannot be accounted for by regressing Y on X. If we regress Y on X using all N data points, we obtain a regression line with slope b_{tot}. Proceeding as in Chapter 12, we can compute $SS_{reg(tot)} = b_{tot}^2 SS_{tot(Y)}$, the sum of squares that represents the variability in Y accounted for by the regression on X. Subtracting $SS_{reg(tot)}$ from the original total variability, $SS_{tot(Y)}$, we have $SS_{tot(adj)}$, the adjusted total variability. Note that this is the sum of squared deviations of the N values of Y about the best-fitting straight line. Note also in panel c that $SS_{tot(adj)}$ is distributed on one less df than $SS_{tot(Y)}$. When we regress Y on X, we use up 2 df in estimating the slope and intercept of the regression line.

The reduction in the total variability that is achieved by the regression on X is partly in SS_A and partly in $SS_{S/A}$. We next find $SS_{S/A(adj)}$, the adjusted within-groups variability, using an approach similar to that involved in calculating $SS_{tot(adj)}$. $SS_{S/A(adj)}$ is the variability associated with the error component ϵ'_{ij} in the ANCOVA model given in Equation 13.1. It follows from the model that $SS_{S/A(adj)}$ is the within-group variabil-

TABLE 13.1   COMPARISON OF ANOVA AND ANCOVA

(a) Data and summary statistics for a one-factor between-subjects design

Y is the dependent variable and X is the covariate

	A_1		A_2		A_3	
	Y	X	Y	X	Y	X
	38	25	47	10	58	19
	61	35	73	28	74	37
	50	23	44	16	65	17
	44	11	85	30	91	40
	69	29	58	21	67	24
	72	36	64	18	45	25
	61	26	67	31	54	23
	41	10	69	34	65	31
	51	23	58	27	59	27
	57	29	81	35	57	14
	46	16	94	41	49	17
	62	27	43	15	74	41
mean	54.33	24.17	65.25	25.50	63.17	26.25
$\hat{\sigma}$	11.01	8.31	16.32	9.43	12.53	9.23

TABLE 13.1   (continued)

(b) Results of ANOVA on the data of panel a

SV	df	SS	MS	F	p
A	2	806.17	403.08	2.22	.124
S/A	33	5988.58	181.47		
Total	35	6794.75			

(c) Results of ANCOVA on the data of panel a

SV	df	SS	MS	F	p
A (adj)	2	539.50	269.75	4.07	.027
S/A (adj)	32	2118.84	66.21		
Total (adj)	34	2658.34			

$$SS_{tot(Y)} = SS_Y = 6794.50; \quad SS_{tot(X)} = SS_X = 2701.78; \quad b_{tot} = 1.237$$

$$SS_{reg(tot)} = b_{tot}^2 SS_{tot(X)} = 4134.18$$

$$SS_{tot(adj)} = SS_{tot(Y)} - SS_{reg(tot)} = 2658.34$$

$$SS_{Y_1} = 1332.66; \quad SS_{Y_2} = 2928.25; \quad SS_{Y_3} = 1727.67; \quad SS_{S/A(Y)} = \sum_j SS_{Y_j} = 5988.58;$$

$$SS_{X_1} = 759.66; \quad SS_{X_2} = 978.88; \quad SS_{X_3} = 936.25; \quad SS_{S/A(X)} = \sum_j SS_{X_j} = 2674.92;$$

$$b_1 = 1.036; \quad b_2 = 1.527; \quad b_3 = 0.999; \quad b_{S/A} = \frac{\sum_j b_j SS_{X_j}}{\sum_j SS_{X_j}} = 1.203$$

$$SS_{reg(S/A)} = \sum_j b_{S/A}^2 SS_{X_j} = b_{S/A}^2 SS_{S/A(X)} = 3869.74$$

$$SS_{S/A(adj)} = SS_{S/A(Y)} - SS_{reg(S/A)} = 2118.84$$

$$SS_{A(adj)} = SS_{tot(adj)} - SS_{S/A(adj)} = 539.50$$

$$F = \frac{SS_{A(adj)}/(a-1)}{SS_{S/A(adj)}/(N-a-1)} = \frac{539.50/2}{2118.84/32} = 4.07$$

ity left unaccounted for when regression lines that have different intercepts but a common slope are obtained for each of the treatment groups. It can be shown that the best estimate of this common slope, $b_{S/A}$ (called the pooled within-group slope), is a weighted average of the slopes obtained from separate regressions in each of the treatment groups. The slope for the jth group (b_j) is weighted by $SS_{X_j}/SS_{S/A(X)}$, where SS_{X_j}

is the numerator of the variance of X in group j and $SS_{S/A(X)} = \Sigma_j SS_{X_j}$. The equation for $b_{S/A}$ is

$$b_{S/A} = \frac{SS_{X_1}}{SS_{S/A(X)}}b_1 + \frac{SS_{X_2}}{SS_{S/A(X)}}b_2 + \frac{SS_{X_3}}{SS_{S/A(X)}}b_3$$

We can then obtain $SS_{reg(S/A)}$, the sum of squares for linear regression within groups, by summing the variabilities accounted for by the regressions with the common slope in each of the groups. This yields

$$SS_{reg(S/A)} = b^2_{S/A}SS_{X_1} + b^2_{S/A}SS_{X_2} + b^2_{S/A}SS_{X_3}$$

$$= b^2_{S/A}SS_{S/A(X)}$$

Then, just as we subtracted $SS_{reg(tot)}$ from $SS_{tot(Y)}$ to obtain $SS_{tot(adj)}$, we now subtract $SS_{reg(S/A)}$ from $SS_{S/A(Y)}$ to obtain an adjusted error sum of squares, $SS_{S/A(adj)}$. The difference between $SS_{tot(adj)}$ and $SS_{S/A(adj)}$ is $SS_{A(adj)}$. Finally, we test the overall null hypothesis $\mu_1 = \mu_2 = \mu_3$, using the statistic $F = MS_{A(adj)}/MS_{S/A(adj)}$.

Comparing the ANCOVA and ANOVA models (Equations 13.1 and 13.2), we see that

$$\epsilon_{ij} = \beta(X_{ij} - \bar{X}_{..}) + \epsilon'_{ij}$$

$$= \epsilon_{pre} + \epsilon_{res}$$

The error associated with the score for the ith subject in the jth condition in ANOVA is considered to be composed of two parts: a component predictable from our knowledge of X (ϵ_{pre}), and a residual component (ϵ_{res}); thus, $\sigma_e^2 = \sigma_{pre}^2 + \sigma_{res}^2$. Then the ratio of expected mean squares in ANOVA may be thought of as

$$\frac{\sigma_{pre}^2 + \sigma_{res}^2 + n\theta_A^2}{\sigma_{pre}^2 + \sigma_{res}^2} = 1 + \frac{n\theta_A^2}{\sigma_{pre}^2 + \sigma_{res}^2}$$

In ANCOVA, the ratio of expected mean squares may be viewed as

$$\frac{\sigma_{res}^2 + n\theta_A^2}{\sigma_{res}^2} = 1 + \frac{n\theta_A^2}{\sigma_{res}^2}$$

because σ_{pre}^2 is removed. The ratio of expected mean squares is larger in ANCOVA, and therefore the F test should be more powerful.

Graphing the regression of Y on X provides another way of looking at the potential advantage in efficiency of ANCOVA over ANOVA. Figure 13.1 schematically indicates how differences between two treatment groups may be easier to observe if the variability in Y predictable from X can be removed. The two ellipses in the central part of the figure represent clouds of data points for the two treatment groups. As can be seen from the two distributions plotted on the right vertical axis, the Y scores for the two groups overlap considerably. The distributions of the residuals for the two groups that are plotted along the left vertical axis indicate that when the effects of X are partialed out, the difference between the two treatment groups is more apparent.

The interpretation of ANCOVA as a more powerful test of the ANOVA null hypothesis $\mu_1 = \mu_2 = \mu_3$ rests on certain assumptions. We assume that there is a linear

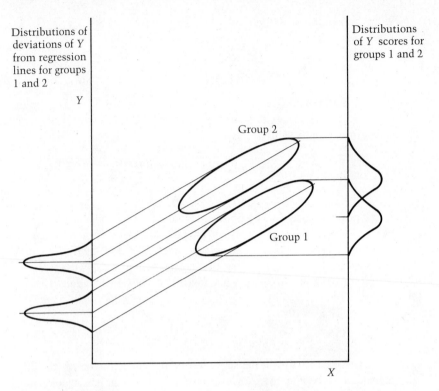

Figure 13.1 Distributions of Y scores (right side) and of deviations of Y scores from the regression lines (left side) in two groups.

relation between Y and X, that the slopes are identical in each of the treatment populations, and that the means of the X values are the same in all the treatment populations. Under other conditions—for example, if the three groups in the example of Table 13.1 differ systematically in their pretest (X) scores—the hypothesis tested may be quite different and interpretation may be very difficult. We will discuss these assumptions in Section 13.6.

13.3 HYPOTHESIS TESTS IN ANOVA AND ANCOVA AS COMPARISONS BETWEEN MODELS

We will find it particularly useful to consider tests of treatment effects in ANCOVA as comparisons between models. This idea will be introduced by first considering ANOVA because, although we have not emphasized this before, hypothesis tests in ANOVA can always be thought of as comparisons between two models.

For example, the A main effect may be tested by considering two models. One, the *full model* (F), contains a component corresponding to the A effect

$$Y_{ij} = \mu + \alpha_j + \epsilon_{ij} \tag{13.3}$$

The other, the *restricted model* (R), does not contain the A treatment effect

$$Y_{ij} = \mu + \epsilon_{ij} \tag{13.4}$$

The A main effect may be tested by determining whether the full model accounts for significantly more variability than the restricted model.

In the restricted model, ϵ_{ij}, the error associated with the ijth individual, is the only source of variability in Y. Therefore, the error variability associated with the restricted model, $SS_{error(R)}$, is equal to $SS_{tot} = \Sigma_i \Sigma_j (Y_{ij} - \bar{Y}_{..})^2$ with $df_{error(R)} = df_{tot} = N - 1 = an - 1$.

The full model is the standard model for one-factor ANOVA, which states that the scores in the data set come from populations with means $\mu + \alpha_1, \mu + \alpha_2, \ldots, \mu + \alpha_a$ and that both α_j, the effect of the treatment given to group j, and ϵ_{ij}, the error component associated with the ith subject in the jth group, contribute to the variability in Y. The model suggests that the variability in Y can be partitioned into two components: one associated with between-group differences (SS_A), and an error variability component that is the pooled variability around the a group means $(SS_{S/A})$. Therefore, $SS_{error(F)}$, the error variability associated with the full model is equal to $SS_{S/A} = \Sigma_j \Sigma_i (Y_{ij} - \bar{Y}_{.j})^2$ with $df_{error(R)} = df_{S/A} = N - a = a(n - 1)$.

The appropriate statistic for testing a full model against a restricted one is given by

$$F = \frac{[SS_{error(R)} - SS_{error(F)}]/[df_{error(R)} - df_{error(F)}]}{SS_{error(F)}/df_{error(F)}} \tag{13.5}$$

where $SS_{error(R)} - SS_{error(F)}$ represents the reduction in error variability achieved (or, equivalently, the additional variability accounted for) when the full model is used instead of the restricted model. Substituting for the error SS and df terms in Equation 13.5, we obtain

$$F = \frac{(SS_{total} - SS_{S/A})/[(an - 1) - a(n - 1)]}{SS_{S/A}/a(n - 1)}$$

$$= \frac{SS_A/(a - 1)}{SS_{S/A}/a(n - 1)}$$

$$= \frac{MS_A}{MS_{S/A}}$$

which is the usual statistic for testing for the treatment effect A.

For one-factor ANCOVA, we compare a full model (F) that contains the treatment component α_j,

$$Y_{ij} = \mu + \alpha_j + \beta(X_{ij} - \bar{X}_{..}) + \epsilon_{ij} \tag{13.6}$$

with a restricted model (R) that does not contain the treatment component:

$$Y_{ij} = \mu + \beta(X_{ij} - \bar{X}_{..}) + \epsilon_{ij} \tag{13.7}$$

We can again use the statistic given in Equation 13.5 to test whether including the treatment effect significantly increases the variability in Y accounted for by the model.

The restricted model given in Equation 13.7 states that with the exception of the random-error component ϵ_{ij}, Y can be totally accounted for by a regression on the covariate X (see panel a of Figure 13.2 in which this model is schematically represented for two groups). Therefore, for the restricted model, the error SS is all the variability in Y not accounted for by the regression on X. We refer to this as $SS_{tot(adj)}$, the total variability adjusted for the effects of the covariate.

In Section 12.4.1 we stated that when Y is regressed on X for a set of data points (X, Y), the regression coefficient (slope) is given by

$$b = \frac{SP}{SS_X} = \frac{\Sigma(X - \bar{X})(Y - \bar{Y})}{\Sigma(X - \bar{X})^2}$$

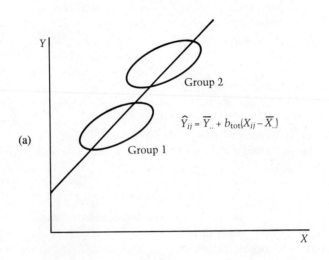

(a)

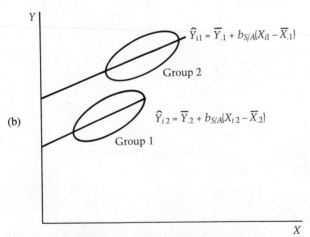

(b)

Figure 13.2 (a) Schematic representation of the regressions indicated by the model of Equation 13.7. All the data are used to obtain a single regression line.
(b) Schematic representation of the regressions indicated by Equation 13.6. Regression lines with equal slopes are obtained for each treatment group.

and the variability in Y accounted for by the regression on X can be expressed in terms of the correlation between X and Y, the regression coefficient, or the sum of cross products of X and Y, as follows:

$$SS_{reg} = r^2 SS_Y = b^2 SS_X = \frac{(SP)^2}{SS_X}$$

In the restricted model, the slope parameter β is estimated by b_{tot}, the sample regression coefficient that results when all of the Y's are regressed on the X's; that is,

$$b_{tot} = \frac{SP_{tot}}{SS_X} = \frac{\Sigma_i \Sigma_j (X_{ij} - \bar{X}_{..})(Y_{ij} - \bar{Y}_{..})}{\Sigma_i \Sigma_j (X_{ij} - \bar{X}_{..})^2}$$

and the variability accounted for by the regression is given by

$$SS_{reg(tot)} = b_{tot}^2 SS_X$$

Therefore, the error variability for the restricted model is given by

$$SS_{error(R)} = SS_{tot(adj)}$$
$$= SS_Y - SS_{reg(tot)}$$
$$= SS_Y - b_{tot}^2 SS_X$$

If we refer to SS_Y and SS_X as $SS_{tot(Y)}$ and $SS_{tot(X)}$, respectively, the equation becomes

$$SS_{tot(adj)} = SS_{tot(Y)} - b_{tot}^2 SS_{tot(X)} \tag{13.8}$$

For the data in Table 13.1,

$$SS_{tot(adj)} = 6794.75 - (1.237)^2 (2701.78)$$
$$= 2658.34$$

There are $N - 2 = 34$ error df associated with the regression of Y on X, so $df_{tot(adj)} = 36 - 2 = 34$. Comparing with the regular ANOVA on Y in which $SS_{tot} = 6794.75$ and $df_{tot} = 35$, we see that a substantial reduction in variability has been achieved at the cost of 1 df.

Next, we note that according to the full model of Equation 13.6, each of the groups in the design are fit by regression lines with a common slope (see panel b of Figure 13.2) but with different intercepts that reflect the treatment effects. The best estimate of the common slope can be shown to be $b_{S/A}$ (called the pooled within-group slope), the weighted average of the a group slopes:

$$b_{S/A} = \frac{\Sigma_j SP_j}{\Sigma_j SS_{X_j}} = \frac{\Sigma_j b_j SS_{X_j}}{\Sigma_j SS_{X_j}}$$

where b_j is the slope obtained by regressing Y on X for the jth group. The "within" slope, $b_{S/A}$, is obtained by weighting each b_j by SS_{X_j}, the sum of squares of X for that group or, equivalently, by adding up the sums of cross products in each group, $SP_j = \Sigma_i (X_{ij} - \bar{X}_{.j})(Y_{ij} - \bar{Y}_{.j})$, and dividing by the sum of the SS_{X_j}'s.

The error variability for the full model is the pooled within-group variability not accounted for by the regression, which we refer to as $SS_{S/A(adj)}$; $SS_{S/A(adj)}$ is the sum of the variabilities unaccounted for by the regression in each of the a groups; that is, $SS_{S/A(\text{adj})} = \Sigma_j SS_{\text{error}_j}$, where $SS_{\text{error}_j} = SS_{Y_j} - b_{S/A}^2 SS_{X_j}$. Therefore,

$$SS_{\text{error}(F)} = SS_{S/A(\text{adj})}$$
$$= \sum_j (SS_{Y_j} - b_{S/A}^2 SS_{X_j})$$
$$= \sum_j SS_{Y_j} - b_{S/A}^2 \sum_j SS_{X_j}$$

If we use the notation $SS_{S/A(Y)} = \Sigma_j SS_{Y_j}$ and $SS_{S/A(X)} = \Sigma_j SS_{X_j}$, the equation becomes

$$SS_{S/A(\text{adj})} = SS_{S/A(Y)} - b_{S/A}^2 SS_{S/A(X)} \qquad (13.9)$$

In our example, we see from Table 13.1 that

$$SS_{S/A/(\text{adj})} = 5988.58 - (1.203)^2 (2674.79)$$
$$= 2118.84$$

and note that $df_{S/A(\text{adj})} = N - a - 1 = 32$, compared to $df_{S/A} = N - a = 33$ in the standard ANOVA. In addition to the 3 df used up in obtaining estimates of the intercepts for the treatment conditions (corresponding to obtaining estimates of the three treatment population means in the standard ANOVA), 1 df is used up in estimating $b_{S/A}$, the common slope of the three regression equations. Compared with the regular ANOVA, we see that the adjustment reduces the within variability from 5988.58 to 2118.84.

By comparing Equations 13.6 and 13.7, we see that $SS_{\text{error}(R)} - SS_{\text{error}(F)}$ (or, equivalently, $SS_{\text{tot}(\text{adj})} - SS_{S/A(\text{adj})}$) reflects the extent to which the treatment component α_j is needed to account for the data. Accordingly, we refer to this difference as $SS_{A(\text{adj})}$, the treatment variability adjusted for the effects of the covariate. Here, $SS_{A(\text{adj})} = 2658.34 - 2118.84 = 539.50$. Substituting into Equation 13.5, we have

$$F = \frac{SS_{A(\text{adj})}/(a - 1)}{SS_{S/A(\text{adj})}/(N - a - 1)}$$
$$= \frac{590.50/2}{2118.84/32}$$
$$= 4.07$$

Because $F_{.05,2,30} = 3.32$, we see that according to the ANCOVA, the treatment effect is significant.

13.4 FINDING AND COMPARING ADJUSTED MEANS IN ANCOVA

13.4.1 Adjusting Group Means in Y for Differences in X

In the previous section, we showed how SS_A and $SS_{S/A}$ could be adjusted for differences in the covariate. Similarly, in certain situations, it is both possible and desirable to adjust the mean Y scores for covariate differences. In our example, children in groups 2 and 3 tend to score higher than those in group 1 on the dependent variable; however, they also tend to score higher on the covariate. It would be simpler to assess the effects of the treatments if the mean covariate value was the same for each group. We define the *adjusted Y mean* for group j, $\bar{Y}_{.j(adj)}$, as the score that would be predicted in group j if $X_{ij} = \bar{X}_{..}$, the grand mean of the covariate scores.

From Equation 12.9, the regression equation for the scores in group j is

$$\hat{Y}_{ij} = \bar{Y}_{.j} + b_1(X_{ij} - \bar{X}_{.j})$$

Substituting $b_1 = b_{S/A}$ and $X_{ij} = \bar{X}_{..}$, we have

$$\bar{Y}_{.j(adj)} = \bar{Y}_{.j} + b_{S/A}(\bar{X}_{..} - \bar{X}_{.j})$$

or

$$\bar{Y}_{.j(adj)} = \bar{Y}_{.j} - b_{S/A}(\bar{X}_{.j} - \bar{X}_{..}) \tag{13.10}$$

For our example, the adjusted group means are

$$\bar{Y}_{.1(adj)} = 54.33 - (1.203)(24.17 - 25.31) = 55.70$$

$$\bar{Y}_{.2(adj)} = 65.25 - (1.203)(25.50 - 25.31) = 65.02$$

$$\bar{Y}_{.3(adj)} = 63.07 - (1.203)(26.25 - 25.31) = 62.03$$

13.4.2 Testing Contrasts by Using Adjusted Means

If the assumptions of ANCOVA are met, more powerful tests of contrasts can be performed by using adjusted, instead of unadjusted, group means. These tests can be performed in much the same way as in Chapter 6. Suppose in our experiment we wished to consider a set of two a priori hypotheses, namely,

$$\psi_1 = \mu_1 - \frac{\mu_2 + \mu_3}{2} = 0 \quad \text{and} \quad \psi_2 = \mu_2 - \mu_3 = 0$$

These contrasts can be estimated by

$$\hat{\psi}_{1(adj)} = \bar{Y}_{.1(adj)} - \frac{\bar{Y}_{.2(adj)} + \bar{Y}_{.3(adj)}}{2}$$

and

$$\hat{\psi}_{2(adj)} = \bar{Y}_{.2(adj)} - \bar{Y}_{.3(adj)}$$

respectively. The hypothesis $H_0: \psi = 0$ can be tested by using the statistic

$$t = \frac{\hat{\psi}_{(adj)}}{\hat{\sigma}_{\hat{\psi}(adj)}} \tag{13.11}$$

where different denominators are recommended, depending on whether subjects have been assigned randomly to groups. If subjects are randomly assigned, the recommended denominator is

$$\hat{\sigma}_{\hat{\psi}(adj)} = \sqrt{MS_{S/A(adj)}\left[1 + \frac{MS_{A(X)}}{SS_{S/A(X)}}\right]\frac{\Sigma_j w_j^2}{n_j}} \tag{13.12}$$

with df $= N - a - 1$. In this expression, $MS_{S/A(adj)}$ is the error term for the ANCOVA, and $MS_{A(X)} = \sum_j \sum_i (\bar{X}_{.j} - \bar{X}_{..})^2/(a - 1)$ and $SS_{S/A(X)} = \sum_j SS_{X_j}$ are the MS_A and $SS_{S/A}$ terms that are obtained if an ANOVA is performed on the covariate.

However, if ANCOVA has been used with a nonrandomized design, the recommended denominator contains a correction factor for the covariate that depends on the specific contrast tested. For example, we might be interested in testing pairwise contrasts among several groups drawn from different clinical populations. In this case, we should use

$$\hat{\sigma}_{\hat{\psi}(adj)} = \sqrt{MS_{S/A(adj)} \left[1 + \frac{SS_{\hat{\psi}(X)}}{SS_{S/A(X)}} \right] \sum_j \frac{w_j^2}{n_j}} \qquad (13.13)$$

$$= \sqrt{MS_{S/A(adj)} \left[\sum_j \frac{w_j^2}{n_j} + \frac{(\sum_j w_j \bar{X}_{.j})^2}{SS_{S/A(X)}} \right]}$$

where $SS_{\hat{\psi}(X)} = (\sum_j w_j \bar{X}_{.j})^2/(\sum_j w_j^2/n_j)$ is the sum of squares that results if the contrast weights are applied to the covariate means.

For our example,

$$\hat{\psi}_{1(adj)} = 55.70 - \frac{65.02 + 62.03}{2} = -7.82$$

and

$$\hat{\psi}_{2(adj)} = 65.02 - 62.03 = 2.99$$

Using Equation 13.12, the denominator of the test statistic for contrasts in randomized designs, we have, for the first contrast,

$$\hat{\sigma}_{\hat{\psi}(adj)} = \sqrt{(66.21) \left[1 + \frac{13.36}{2674.92} \right] \left[\frac{1 + .25 + .25}{12} \right]}$$

$$= \sqrt{\frac{(66.54)(1 + .25 + .25)}{12}}$$

$$= 2.88$$

so that

$$t = \frac{-7.82}{2.88} = -2.71$$

and, for the second contrast,

$$t = \frac{2.99}{\sqrt{(66.54)(1 + 1)/12}} = .897$$

From Table D.8, the critical values of the Bonferroni t with EF $= .05$, $K = 2$, and df $= 32$ are approximately ± 2.36. Therefore, the first contrast is significantly different from zero, and the second is not.

Had we not employed a randomized design, we would use the expression given in Equation 13.13 as the denominator of the t. In this case, for the first contrast,

$$\sum_j \frac{w_j^2}{n_j} = \frac{1 + .25 + .25}{12} = .125$$

and

$$\frac{(\Sigma_j w_j \bar{X}_{.j})^2}{SS_{S/A(X)}} = \frac{[(1)(24.17) + (-.5)(25.50) + (-.5)(26.25)]^2}{2674.92}$$

$$= .001$$

so that

$$t = \frac{-7.82}{\sqrt{(66.21)(.125 + .001)}} = -2.71$$

Notice that, for the data in our example, both the randomized and nonrandomized t statistics have almost exactly the same value because the covariate means do not differ much from one another.

Post hoc contrasts can also be tested in much the same way as in Chapter 6. For the Scheffé test, the preceding t statistics can be referred to the criterion $\sqrt{(a-1)F_{EF,a-1,N-a-1}}$. For the Tukey post hoc test of pairwise differences, the same test statistics can be used with weights $+1$ and -1. If the covariate is a fixed-effect variable, the test statistic can be referred to $q_{EF,a,df}/\sqrt{2}$, where q is a critical value of the studentized range statistic (Table D.10) that we used with the Tukey test in Chapter 6. If the covariate is a random variable, as is usually the case, Bryant and Paulson (1976) have shown that q should be replaced by $Q_{EF,a,c,df}$, a value of the *generalized studentized range statistic* in which c is the number of covariates. Tables of the generalized studentized range statistic are available in Huitema (1980) and Kirk (1982).

13.5 TESTING HOMOGENEITY OF SLOPES

As we will discuss in Section 13.6, a necessary assumption for ANCOVA is that the slopes are equal in each of the treatment populations. In Section 12.12, we showed how to test for the homogeneity of regression slopes. In this section, we discuss the test as a comparison between a full model in which the slopes for the treatment groups can differ from one another,

$$Y_{ij} = \mu + \alpha_j + \beta_j(X_{ij} - \bar{X}_{..}) + \epsilon_{ij} \tag{13.14}$$

and a restricted model in which each group has the same slope,

$$Y_{ij} = \mu + \alpha_j + \beta(X_{ij} - \bar{X}_{..}) + \epsilon_{ij} \tag{13.15}$$

Note that in this case both the full and restricted models include the α_j component.

The restricted model for the homogeneity of slope test is the model for ANCOVA. Therefore, from Equation 13.9,

$$SS_{error(R)} = SS_{S/A(adj)}$$
$$= SS_{S/A(Y)} - b^2_{S/A}SS_{S/A(X)}$$

In Section 13.3, we showed that for the data in Table 13.1, $SS_{S/A(adj)}$ is equal to 2118.84 with 32 df.

The full model allows for individual regressions in each of the groups in which both the slopes and intercepts can vary. The error SS associated with the full model is the sum of the variabilities not accounted for by these individual group regressions, and we will refer to this as $SS_{S/A(indiv)}$. That is,

$$SS_{error(F)} = SS_{S/A(indiv)}$$
$$= \sum_j (SS_{Y_j} - b^2_j SS_{X_j})$$
$$= SS_{S/A(Y)} - \sum_j b^2_j SS_{X_j}$$

For our example (see Table 13.1),

$$\sum_j SS_{Y_j} = SS_{S/A(Y)} = 5988.58$$

and

$$\sum_j b^2_j SS_{X_j} = (1.036)^2(759.666) + (1.527)^2(978.879) + (0.999)^2(936.251)$$

$$= 4032.20$$

Therefore, $SS_{S/A(indiv)} = 5988.58 - 4032.20 = 1956.38$. There are $N - 2a = 30$ df associated with $SS_{S/A(indiv)}$; $2a = 6$ df are used up in estimating the slope and intercept for each of the three groups.

The difference between $SS_{S/A(adj)}$ and $SS_{S/A(indiv)}$ is the variability associated with the heterogeneous slopes, which we will refer to as SS_{het}. For our example, $SS_{het} = 2118.84 - 1956.38 = 162.46$. Substituting in Equation 13.5, we have

$$F = \frac{SS_{het}/(a - 1)}{SS_{S/A(indiv)}/(N - 2a)}$$

$$= \frac{162.46/2}{1956.38/30}$$

$$= 1.25$$

We cannot reject the hypothesis that the group slopes are identical because the obtained F is smaller than 3.32, the critical F for $\alpha = .05$.

13.6 ASSUMPTIONS AND INTERPRETATION IN ANCOVA

13.6.1 Introduction

When ANCOVA is used instead of ANOVA, increases in efficiency may be achieved at the cost of greater complexity and more assumptions. The standard assumptions for ANCOVA break down into two groups. As in ordinary ANOVA, some assumptions are necessary for the ratio of adjusted mean squares to be distributed as F. However, unless certain additional assumptions, such as homogeneity of regression and identity of the X distributions are met, the ANCOVA may test a different null hypothesis than ordinary ANOVA and the adjusted means may be biased estimates of the population means. In the sections that follow, we will discuss these assumptions and the consequences of violating them.

13.6.2 Normality and Homogeneity of Variance

In ANCOVA, it is assumed that the conditional distributions of Y at different values of X are normal and have equal variances. In general, the consequences of violating these assumptions are similar to those for ANOVA, with the exception that they depend to some extent on the distribution of the covariate (see Huitema 1980 for a more detailed discussion of these assumptions). ANCOVA is unlikely to be severely biased by violations of these normality and homogeneity of variance assumptions provided there are equal numbers of subjects in each group and the covariate itself is approximately normally distributed.

13.6.3 Linearity

As we have so far discussed, ANCOVA adjusts for differences in the covariate by removing the variability accounted for by a linear regression on X. If there is a systematic nonlinear component to the relation between X and Y, the use of linear regression will not remove all the variability in Y potentially accounted for by X. The effect of moderate nonlinearity is slight negative bias in the ANCOVA F test. Although it is rare in the behavioral sciences to observe strongly nonlinear relations, they can result in severely biased F tests if linear ANCOVA is used (Atiqullah 1964). However, if the nature of the nonlinearities can be specified, transformations of Y or polynomial ANCOVA (Section 13.11) may be used. It is recommended that the linearity assumption be checked as a preliminary step in using ANCOVA. Plotting the group scatter diagrams offers a quick check, and a significance test for nonlinearity is also available (see Section 12.10).

13.6.4 Assumption of Homogeneity of Regression Slopes

The idea of adjusting for differences in the covariate that underlies ANCOVA only makes sense if the same type of adjustment is appropriate for each treatment condition and each value of the covariate. Therefore, the assumption that the slopes are homogeneous should always be tested (see Section 13.5), and the results of the ANCOVA should be reported only if homogeneity cannot be rejected. In the one-factor ANCOVA design, we use the pooled within-group slope, $b_{S/A}$, to adjust the error SS's and the means for each group. If the slopes are heterogeneous, using an "average" adjustment is inappropriate for at least some of the groups. Furthermore, the interesting question is no longer what would happen if every subject had the same covariate value; rather, we should be concerned with what happens at different values of the covariate.

An analogy can be made between an adjusted A main effect in an ANCOVA in the presence of heterogeneous slopes and an ANOVA A main effect in the presence of an interaction between A and a second factor, B. If there is a large interaction, particularly if the curves cross, the F test of A may not adequately reflect the A effect at any level of B. Similarly, if the group regression coefficients vary, the effect of A varies with X and the ANCOVA F test may produce misleading results.

It probably helps to describe the situation by diagrams. In adjusting Y for the effects of the covariate X, we decide to interpret treatment effects in terms of differences in the group regression lines instead of differences in the group means (see Figure 13.1). Suppose that the lines in each panel of Figure 13.3 represent regression lines obtained separately for two treatment groups. Parallel lines, illustrated in panel a indicate that the treatment effects are the same for each value of the covariate. In this case, we can identify the treatment effects with the vertical distance between the lines. On the other hand, nonparallel lines, illustrated in panels b and c, indicate that the treatment effects are not the same for each value of the covariate. In panel b, the two lines intersect at $X = \bar{X}$, but differ considerably for high and low values of X. In panel c, there is a separation between the two lines at $X = \bar{X}$, but the separation is larger for large X and smaller for small X. For b and c, it is of more interest to ask for what values of X, if any, the separations are significant than to consider the separation at any single value of X. Johnson and Neyman (1936) have developed a procedure for establishing regions of significance on the covariate. A good presentation of the Johnson-Neyman technique and related procedures can be found in Huitema (1980).

13.6.5 Groups Formed by Random Assignment versus Nonequivalent Groups

In a discussion of ANCOVA, the distinction between randomized and nonequivalent group designs is extremely important. Under standard assumptions, the ANCOVA for a randomized design tests the same null hypotheses as the corresponding ANOVA. Also, both the adjusted and unadjusted group means are unbiased estimates of the treatment population means, although given a strong correlation between the dependent variable and the covariate, the adjusted means will be more efficient estimators. In contrast, when treatments are applied to preexisting groups, the treatment populations may differ

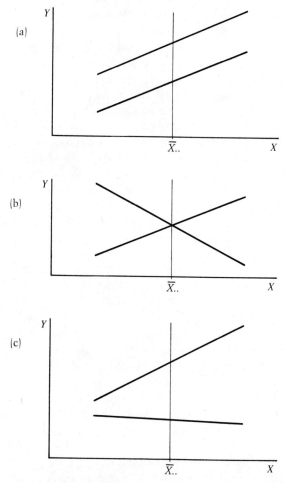

Figure 13.3 Regressions with homogeneous and heterogeneous slopes.

even if the treatments have equal or no effects. Although ANCOVA is often used in an attempt to control for preexisting differences, the resulting F tests and adjusted means may be strongly biased. A similar problem exists when the covariate is influenced by the treatment. We turn to this issue now.

13.6.6 Assumption of Independence of Treatment and Covariate

If ANCOVA is used with a design in which the covariate varies systematically with the treatment, interpretation will be extremely difficult. Using X as a covariate removes the component of Y predictable from X. If the treatment influences X or is otherwise predictable from X, performing an ANCOVA will not simply remove nuisance variability from Y; it will remove part of the effect of the treatment itself.

For example, suppose another experiment is conducted to evaluate three different teaching programs. The mean test score for program 1 is larger than the means for programs 2 and 3, and an ANOVA performed on the test scores is significant, indicating that the three programs are not all equally effective. However, the experimenter notes that students assigned to program 1 spend more time actually working with it than do students assigned to the other programs, and decides to perform an ANCOVA using "study time" as the covariate. The ANCOVA reveals no significant differences, and so the experimenter concludes that the three programs would be equally effective if study time was held constant. It is important to note that this interpretation is not necessarily correct. Statistically controlling for study time is not the same as experimentally controlling or manipulating it, and no causal statements are justified; we simply don't know from these data what would happen if study time was actually held constant. Program 1 may be more understandable and interesting to work with than the other programs, and these qualities may be the cause of both the superior test performance and the greater study time. Using study time as a covariate will tend to remove the effects of any variables correlated with study time, including the characteristics of the program that are actually responsible for the superior performance. It is therefore entirely possible that program 1 will produce superior performance even if study time is limited.

It is worth noting that "controlling" study time by throwing out data is also not appropriate here. Suppose that the mean daily study time is 40 minutes for the first group and 25 minutes for the other two groups. What if we analyze only the performance scores for students in the three groups who have comparable study times, say 30–35 minutes? This is a poor strategy because in selecting students who have comparable study times, we may be selecting students who widely differ on other important characteristics. There is no reason to think that students in the first group whose study times are below that group's average are comparable in ability and motivation to students who have above average study times in the other two groups. If one is interested in the effects of the program and of study time, there is simply no substitute for conducting a true experiment in which both variables are manipulated.

Even if the treatments don't influence the covariate, when they are applied to intact groups that differ from one another, ANCOVA presents the same kinds of difficulties that are associated with interpreting the results of observational studies. The covariate may be correlated with a host of other important variables, and performing an ANCOVA tends to remove the effects of all of these variables from Y. For example, suppose that three methods of teaching arithmetic are used in three different school districts. Method 1 produces the highest scores but happens to be used in the wealthiest neighborhood. Schools in this neighborhood tend to have better teachers and smaller class sizes; parents tend to be more educated, more concerned about education and more achievement-oriented. An ANOVA on the test scores is significant, but an ANCOVA using class size as the covariate is not. Although there is some sense in which the statement "adjusting for class size, there is no effect of teaching method" is true, it does not follow that if we equated class size in the three districts, the teaching methods would be equally effective. We do not know whether class size or one or more of the many variables correlated with it are responsible for the differences in performance.

The assumption of independence of the treatment and covariate can be tested by performing an ANOVA on the covariate. In nonequivalent-groups designs, a significant

ANOVA found for a covariate measured before the treatments have been administered indicates that the ANCOVA F's and adjusted means will almost certainly be biased. Unfortunately, a nonsignificant ANOVA cannot be taken as an indication that there will be no bias, although the bias is more likely to be small. In completely randomized designs, there will be no bias for covariates measured before treatments have been administered. However, ANOVAs should be performed on covariates measured during or following treatment.

13.6.7 Assumption That the Covariate Is Fixed and Measured Without Error

The assumption that the covariate is a fixed-effect variable can generally be violated without serious consequences, and, indeed, the most common applications of ANCOVA are with random covariates. For randomized designs, measurement error results in reduced power. However, for nonequivalent-groups designs, measurement error in the covariate can result in a large degree of bias and, therefore, severe difficulties in interpretation.

In a nonequivalent-groups design, if the mean covariate value varies across groups and if the covariate is measured with error, *the expected values of the adjusted $\bar{Y}$'s may differ even if the treatment has no effect.* Figure 13.4 illustrates the problem for two

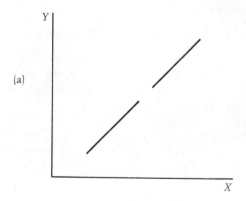

(a)

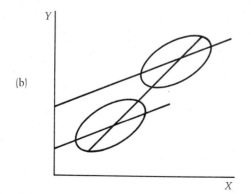

(b)

Figure 13.4 Effect of measurement error on regression slopes. Both samples are from populations in which the true scores of Y and X are perfectly correlated and no treatment is applied. X is measured without error in (a) and with error in (b).

groups in which the true scores of X and Y are perfectly correlated and there are no treatment effects. In panel a, X is measured without error, and both group equations have the same slope and intercept. Therefore, the adjusted means for these groups must be the same, and an ANCOVA would correctly reveal that there are no treatment effects. Panel b represents exactly the same situation, except that now X is measured with error. The effect of the measurement error will be to "spread out" the values of X and, as we discussed in Section 12.8, to reduce the slopes for both groups. As can be seen in panel b (and as we explain in detail in Appendix 13.2), the adjusted Y mean will now be larger for the group that has the larger $\bar{X}$, even though there is no treatment effect. Because groups do not differ systematically from one another in randomized designs, measurement error in X will result in reduced power but will not cause the adjusted means to be biased. However, in nonequivalent group designs, measurement error in X will introduce bias.

13.7 ANCOVA VERSUS THE USE OF GAIN SCORES IN PRETEST-POSTTEST DESIGNS

In Section 9.6, we pointed out that if pretest and posttest scores are on the same scale, an alternative to using the pretest score as the covariate in an ANCOVA is to perform an ANOVA on the gain scores, posttest − pretest. The use of gain scores is equivalent to using a regression adjustment in which β is constrained to have the value 1, rather than estimating β from the data, as is the case in ANCOVA. It can be shown that

$$\text{SS}_{S/A(\text{gain})} = \text{SS}_{S/A(\text{adj})} + (b_{S/A} - 1)^2 \text{SS}_{S/A(X)}$$

where $\text{SS}_{S/A(\text{gain})}$ is the within-group variability in the gain-score analysis and $\text{SS}_{S/A(\text{adj})}$ is the adjusted within-group variability for the ANCOVA. If $b_{S/A} = 1$, $\text{SS}_{S/A(\text{gain})} = \text{SS}_{S/A(\text{adj})}$ and $df_{S/A(\text{gain})} = df_{S/A(\text{adj})} + 1$ because no df have been lost in adjusting for the pretest. However, the gain score analysis is not recommended unless β is close to 1; the larger the departure from 1, the greater the power advantage for the ANCOVA. In fact, it can be shown that for randomized designs, the gain score analysis will be less precise[1] than ANOVA on the posttest scores alone for $\beta < .5$. For more details, see Huck and McLean (1975), Huitema (1980), and Reichardt (1979).

13.8 ANCOVA VERSUS TREATMENT BY BLOCKS

Both ANCOVA and the treatment × blocks design use information about an additional concomitant variable, X, to increase efficiency. It is therefore of interest to discuss the advantages and disadvantages of both designs. However, in doing so, we must consider

---

[1] Although power is certainly an important measure to be considered in comparing designs, another measure that has been used is the *imprecision* of the design. If we refer to the estimated magnitude of the treatment effect as d, imprecision is an index of the instability of d from sample to sample (see Appendix 13.1 for a more detailed definition). Imprecision can be a useful complement to power, especially when large changes in power cannot occur because of ceiling effects. For example, if design A has a power of .95 for a particular set of data, there is no way that design B can be demonstrated to be superior by having much more power; however, it can be considerably more precise.

how subjects are assigned to treatment conditions as well as what statistical analysis is subsequently used. We distinguish among four cases:

1. *Treatment × blocks with a priori blocking.* As described in Section 5.10, this design requires that subjects be divided into b blocks on the basis of their X scores, where the optimal value of b (Feldt 1958) depends on n and the correlation between X and the dependent variable. Within each block, subjects are randomly assigned to treatment conditions. Finally, a two-factor treatment × blocks ANOVA is performed.

2. *Treatment × blocks with post hoc blocking.* In this case, subjects are assigned to treatment conditions without regard to their X values. Blocks are formed after the Y scores have been collected, and the design is analyzed as a two-factor ANOVA.

3. *ANCOVA with random assignment of subjects to treatment conditions.* Subjects are assigned to treatment conditions without regard to their X values as in case 2; however, an ANCOVA is performed.

4. *ANCOVA with assignment of subjects to condition based on covariate value.* If the covariate values are available before the experiment is conducted, they can be used in assigning subjects to conditions even if an ANCOVA is to be used. For example, if the independent variable has a levels, the a subjects with the highest X scores can be randomly distributed across the treatment conditions, then the subjects with the next a highest scores, and so on. In the case of two treatment groups, several authors (Dalton and Overall 1977; Maxwell, Delaney, and Dill 1984) have recommended the *alternate ranks design* in which the subjects are ranked in X and then assigned to conditions A_1 and A_2 according to an $A_1 A_2 A_2 A_1 A_1 A_2 A_2 \ldots$ sequence. Whether the assignment of a scores to the levels of A is made randomly or based on the alternate ranks design, the means of X should vary less than in cases 2 and 3.

In a classic paper, Feldt (1958) compared cases 1 and 3. He found that ANCOVA was more precise than treatment × blocks with the optimal number of blocks when ρ, the population correlation between Y and X, was greater than .6; that blocking was more precise for ρ less than .4; and that there was little difference for $.4 < \rho < .6$. Because of these findings and of the complexity and stringent assumptions of ANCOVA, it has often been argued that blocking should be used instead of ANCOVA except when ρ is large.

However, in practice, it will not be possible for the experimenter to decide on the optimal number of blocks unless a good estimate of the correlation between X and Y is available before subjects are assigned to blocks and the Y scores are obtained. The ANCOVA designs do not require this information in advance. Also, when assignment to groups is considered as well as the method of analysis, there seem to be reasons to favor ANCOVA over blocking, even for moderate values of ρ. Maxwell et al. (1984) used simulation procedures to test the power and precision of nine methods, including the four listed here. Although in their simulations they used an independent variable with only two levels, it is likely that their findings hold more generally.

When information about X is not used in assigning subjects to treatment conditions, ANCOVA tends to have strong advantages over post hoc blocking. Maxwell et al. (1984) found case 3 to be more powerful and precise than case 2 even for values of ρ

as low as .28. Another disadvantage of post hoc blocking is that it will usually lead to unbalanced designs. If post hoc blocks are to be formed, they should be based on the X values of all members of the entire sample, without regard to group assignment. If this is done, it is unlikely that there will be equal numbers of scores in each of the ab cells. Experimenters should be very careful to avoid forming post hoc blocks within each group. As Myers (1979) has pointed out, this will lead to positively biased F tests.

Maxwell et al. also suggest that advantages in power and precision may be obtained by using the values of X to assign subjects to conditions. They found that when subjects were assigned to conditions using an alternate ranks procedure, both ANCOVA and treatment $\times$ blocks ANOVA were more powerful than when assignment was made without regard to X. Also, both when X was used in assigning subjects to treatments and when it was not, ANCOVA was found to be slightly more powerful than treatment $\times$ blocks ANOVA for $\rho = .28$, and it showed a larger advantage for $\rho = .50$.

In summary, if the concomitant measure is available prior to data collection, greater power may be achieved if it is used in assigning subjects to treatment conditions. In addition, if the relation between Y and X is linear and the other assumptions of ANCOVA are reasonably satisfied, greater power will generally result from using ANCOVA than from the treatment $\times$ blocks ANOVA.

13.9 USING SOFTWARE PACKAGES TO PERFORM ANCOVA

The standard statistical package routines that handle ANOVA will usually also perform ANCOVA. Usually all that is required is that we modify a statement in the command file that requests the corresponding ANOVA, or that we add a statement to the file.

For example, the SYSTAT instructions for performing an ANOVA on the data in Table 13.1 are

```
CATEGORY A = 3
MODEL Y = CONSTANT + A
ESTIMATE
```

Replacing the model statement[2] by

```
MODEL Y = CONSTANT + A + X
SAVE MEANS/ADJUSTED
```

will perform the ANCOVA and save the adjusted means into the file MEANS. The assumption that the group regression slopes are homogeneous can be tested by using the statements

```
CATEGORY A = 3
MODEL Y = CONSTANT + A + X + A*X
ESTIMATE
```

The homogeneity assumption is plausible if the $A \times X$ interaction is not significant.

---

[2]In version 4 of SYSTAT, it is possible to replace the first model statement by ANOVA Y and the second by the two statements, ANOVA Y and COVARIATE X.

Similar modifications will perform the desired ANCOVAs in the other packages we have considered. The SPSS^X MANOVA procedure will perform the ANCOVA, calculate the adjusted means, and test the homogeneity of regression assumption if, following the DATA LIST command, we have

```
MANOVA Y BY A(1,3) WITH X/
   PRINT = PMEANS/
   DESIGN/
ANALYSIS = Y/
   DESIGN = X, A, A*X/
```

In the SAS package, the commands

```
PROC GLM
   CLASS A;
   MODEL Y = A X /SOLUTION;
   LSMEANS A;
```

will produce both the one-factor ANOVA and ANCOVA as well as the adjusted means. To test the homogeneity of slopes, we would replace the model statement by

```
MODEL Y = A X A*X /SOLUTION;
```

Finally, in the BMDP series, several programs can perform ANCOVAs, although the details of the command statements vary somewhat from program to program. In the P2V program, for example, all that is required is that an additional line be added to the DESIGN paragraph. For the analysis in Table 13.1, this paragraph would become

```
/DESIGN  DEPENDENT IS Y.
         GROUPING IS A.
         COVARIATE IS X.
```

Although the ANCOVAs we have discussed so far in this chapter have included only a single factor, the packages can readily perform ANCOVAs for higher-order designs. For example, if an additional between-subjects factor C that had two levels was added to the design of Table 13.1, the SYSTAT statements would become

```
CATEGORY A = 3, C = 2
MODEL Y = CONSTANT + A + C + A*C + X
SAVE MEANS/ADJUSTED
ESTIMATE
```

The homogeneity of regression slope assumption could be tested by using the statements

```
CATEGORY A = 3, C = 2
MODEL Y = CONSTANT + A + C + A*C + X + A*X + C*X + A*C*X
ESTIMATE
```

The assumption is plausible if none of the interactions with X are significant.

13.10 ANCOVA IN HIGHER-ORDER DESIGNS

Sources of variance can be adjusted for the covariate in much the same way in higher-order designs as in the one-factor design we have so far considered. We indicated in the last section that any of the standard statistical software packages allow the use of

ANCOVA with higher-order designs. In Chapter 16 we will show that these analyses can also readily be performed by using multiple regression. In the remainder of this section, we briefly indicate how any source of variance in a higher-order design can be adjusted for the effect of the covariate by using bivariate regression.

An ANCOVA test for any main or interaction effect, E, can be thought of as a comparison between two models: a full model that contains E and a restricted model that does not. The test statistic is

$$F = \frac{SS_{E(adj)}/df_E}{SS_{error(adj)}/[df_{error} - 1]} \tag{13.16}$$

where

$$SS_{error(adj)} = SS_{error(Y)} - b^2_{error}SS_{error(X)} \tag{13.17}$$

$$= SS_{error(Y)} - \frac{SP^2_{error}}{SS_{error(X)}}$$

and

$$SS_{E(adj)} = SS_{(E+error)(adj)} - SS_{error(adj)} \tag{13.18}$$

where

$$SS_{(E+error)(adj)} = SS_{E(Y)} + SS_{error(Y)}$$
$$- b^2_{(E+error)}[SS_{E(X)} + SS_{error(X)}]$$
$$= SS_{E(Y)} + SS_{error(Y)}$$
$$- \frac{(SP_E + SP_{error})^2}{SS_{E(X)} + SS_{error(X)}}$$

Here, $SS_{(E+error)(adj)}$ is the pooled variability associated with E and the error component with which E would be tested in a standard ANOVA, adjusted for the effects of the covariate.

For example, suppose we had a between-subjects design with two factors, A and C. Tests of A, C, and $A \times C$ are comparisons between the full model

$$Y_{ijk} = \mu + \alpha_j + \gamma_k + (\alpha\gamma)_{jk} + \beta(X_{ijk} - \bar{X}_{...}) + \epsilon_{ijk}$$

and a restricted model from which the effect in question has been deleted. If we wish to test the A main effect, the restricted model is

$$Y_{ijk} = \mu + \gamma_k + (\alpha\gamma)_{jk} + \beta(X_{ijk} - \bar{X}_{...}) + \epsilon_{ijk}$$

The error SS for the full model is $SS_{S/AC(Y)}$ adjusted for the covariate. That is,

$$SS_{error(F)} = SS_{S/AC(adj)}$$
$$= SS_{S/AC(Y)} - b^2_{S/AC}SS_{S/AC(X)}$$

where $b_{S/AC}$ is the weighted average of the ac group slopes. As in the one-factor design, the weights are the group SS_X's, so

$$b_{S/AC} = \frac{\Sigma_j\Sigma_k SP_{jk}}{\Sigma_j\Sigma_k SS_{X_{jk}}} = \frac{\Sigma_j\Sigma_k b_{jk}SS_{X_{jk}}}{SS_{S/AC(X)}}$$

where

$$b_{jk} = \frac{SP_{jk}}{SS_{X_{jk}}} = \frac{\Sigma_i(X_{ijk} - \bar{X}_{.jk})(Y_{ijk} - \bar{Y}_{.jk})}{\Sigma_i(X_{ijk} - \bar{X}_{.jk})^2}$$

The restricted model does not contain the effect of A. Therefore, for this model, SS_{error} is the pooled variability of A and S/AC adjusted for the covariate

$$SS_{(A+S/AC)(adj)} = [SS_{A(Y)} + SS_{S/AC(Y)}] - b^2_{A+S/AC}[SS_{A(X)} + SS_{S/AC(X)}]$$

where $b_{A+S/AC}$ is the weighted average of $b_{S/AC}$ and b_A, the slope of the regression of $\bar{Y}_{.j.}$ on $\bar{X}_{.j.}$. That is,

$$b_{A+S/AC} = \frac{SP_A + SP_{S/AC}}{SS_{A(X)} + SS_{S/AC(X)}} = \frac{b_A SS_{A(X)} + b_{S/AC} SS_{S/AC(X)}}{SS_{A(X)} + SS_{S/AC(X)}}$$

where

$$b_A = \frac{SP_A}{SS_{A(X)}} = \frac{nb\Sigma_j(\bar{X}_{.j.} - \bar{X}_{...})(\bar{Y}_{.j.} - \bar{Y}_{...})}{nb\Sigma_j(\bar{X}_{.j.} - \bar{Y}_{...})^2}$$

Once $SS_{(A+S/AC)(adj)}$ has been found, $SS_{A(adj)}$ can be obtained from Equation 13.18 and the A effect tested by using Equation 13.16.

13.11 SOME EXTENSIONS OF ANCOVA

In this section we briefly introduce the ideas of adjustments for more than one covariate and adjustments based on polynomial regression. Because these procedures fit particularly well into a multiple regression framework, we defer more detailed consideration of them until Chapter 16. However, we should note that these extensions can easily be accommodated by the statistical packages we discussed in Section 13.9.

13.11.1 More than One Covariate

Suppose that our dependent variable is again problem-solving performance but that now we have available two possible covariates, scores on analytic reasoning (X) and verbal skills (W) tests. We have already discussed why we might wish to use either of the covariates in an ANCOVA and how to do so. However, we may choose to use both of them in the same analysis. This would involve testing a full model that contained a treatment component as well as the covariates

$$Y_{ij} = \mu + \alpha_j + \beta_{YX}(X_{ij} - \bar{X}_{..}) + \beta_{YW}(W_{ij} - \bar{W}_{..}) + \epsilon_{ij}$$

against a restricted model that contained the covariates but not the treatment component:

$$Y_{ij} = \mu + \beta_{YX}(X_{ij} - \bar{X}_{..}) + \beta_{YW}(W_{ij} - \bar{W}_{..}) + \epsilon_{ij}$$

If X and W are not redundant measures and are both correlated with Y, it is possible that using both covariates will result in a greater gain in efficiency than using either one by itself.

13.11.2 Polynomial ANCOVA

Suppose our dependent measure (Y) is performance, our possible covariate is a measure of anxiety (X), and that we know that the relation between X and Y contains both strong linear and quadratic components. It would be inappropriate to use ANCOVA based on the regression of Y on X alone because this will adjust only for the linear component of the relation. However, it is possible to use an ANCOVA model that contains both linear and quadratic components. In this case the full model would be

$$Y_{ij} = \mu + \alpha_j + \beta_1(X_{ij} - \bar{X}_{..}) + \beta_2(X_{ij}^2 - \bar{X}_{..}^2) + \epsilon_{ij}$$

and the restricted model would be

$$Y_{ij} = \mu + \beta_1(X_{ij} - \bar{X}_{..}) + \beta_2(X_{ij}^2 - \bar{X}_{..}^2) + \epsilon_{ij}$$

(note that $\bar{X}_{..}^2$ is the average of the X_{ij}^2's, not the square of $\bar{X}_{..}$). The linear and quadratic components should partial out more of the "nuisance" variability than the linear component alone and, therefore, provide a more powerful analysis.

13.12 CONCLUDING REMARKS

When used with an understanding of its assumptions and limitations, ANCOVA can be a useful tool. However, it is a "delicate instrument," and there is great potential for its abuse, especially now that it is so easy to perform ANCOVAs with the standard statistical packages. The greatest abuses occur when ANCOVA is used to try to equate groups that are basically different from one another. In such cases, Smith (1957) has argued that adjusted means might better be referred to as "fictitious means" and the use of ANCOVA can result in worse inferences than using no adjustment whatsoever. Although it seems trite to repeat it, one must always keep in mind that statistical control is not the same as experimental control and that correlation and prediction are not the same as causation.

APPENDIX 13.1: Definition of Design Imprecision

Imprecision is a measure based on the average variance of the differences between all pairs of treatment means. Cox (1957) defined *true imprecision* as

$$I_t = \frac{\text{average var}(\bar{Y}_j - \bar{Y}_{j'})}{\text{min var}(\bar{Y}_j - \bar{Y}_{j'})}$$

where the denominator represents the theoretical minimum for the variance of the difference between two treatment means. However, the preferred measure, *apparent imprecision* (I_a), where

$$I_a = \frac{\text{df}_{\text{error}} + 3}{\text{df}_{\text{error}} + 1} I_t$$

uses a df correction to allow a more meaningful comparison of designs that use the same number of subjects but have different error df's. For a more detailed discussion see Feldt (1958) and Myers (1979).

APPENDIX 13.2: Comment on Figure 13.4

We want to explain why the adjusted means must be different for the two groups in panel b of Figure 13.4. In panel a of Figure 13.4, all the data points in groups 1 and 2 fall on the line $Y = b_0 + b_1 X$. Therefore the adjusted means for both groups must be the same. The adjusted means are given by

$$\bar{Y}_{.1(adj)} = \bar{Y}_{.1} - b_1 (\bar{X}_{.1} - \bar{X}_{..})$$

and

$$\bar{Y}_{.2(adj)} = \bar{Y}_{.2} - b_1 (\bar{X}_{.2} - \bar{X}_{..})$$

Subtracting $\bar{Y}_{.1(adj)}$ from $\bar{Y}_{.2(adj)}$, we obtain

$$(\bar{Y}_{.2} - \bar{Y}_{.1}) = b_1 (\bar{X}_{.2} - \bar{X}_{.1}) = 0 \tag{13.19}$$

Now suppose that $\bar{X}_1$ and $\bar{X}_2$ remain the same but that random error is added to the X scores, reducing the slopes of the regression lines for both groups from b_1 to $b_1 - a$. The difference between the adjusted means is now

$$\bar{Y}_{.2(adj)} - \bar{Y}_{.1(adj)} = \bar{Y}_{.2} - (b_1 - a)(\bar{X}_{.2} - \bar{X}_{..}) - [\bar{Y}_{.1} - (b_1 - a)(\bar{X}_{.1} - \bar{X}_{..})]$$
$$= (\bar{Y}_{.2} - \bar{Y}_{.1}) - b_1 (\bar{X}_{.2} - \bar{X}_{.1}) + a(\bar{X}_{.2} - \bar{X}_{.1})$$

From Equation 13.19, we see that the first two terms on the right cancel out, so

$$\bar{Y}_{.2(adj)} - \bar{Y}_{.1(adj)} = a(\bar{X}_{.2} - \bar{X}_{.1})$$

Because $a \neq 0$ and $\bar{X}_2 \neq \bar{X}_1$, the two adjusted means must differ.

EXERCISES

13.1 The following terms provide a useful review of many concepts in the chapter. Define, describe, or identify each of them:

covariate
efficiency of a design
pooled within-group slope
$SS_{tot(adj)}$
$SS_{S/A(adj)}$
$SS_{A(adj)}$
full model
restricted model
homogeneity of regression slopes

$SS_{S/A(indiv)}$
SS_{het}
Johnson-Neyman procedure
nonequivalent groups
imprecision of a design
treatment × blocks design
a priori blocking
post hoc blocking
polynomial ANCOVA

13.2 Eighteen subjects are assigned randomly to three treatment conditions A_1, A_2, and A_3. After the treatment is applied, values of Y, the dependent variable, are obtained. However, before the treatment is applied, values of X, a variable closely related to Y, are recorded. The data are as follows:

A_1		A_2		A_3	
X	Y	X	Y	X	Y
12	26	11	32	6	23
10	22	12	31	13	35
7	20	6	20	15	44
14	34	18	41	15	41
12	28	10	29	7	28
11	26	11	31	9	30

(a) Perform an ANOVA on Y.
(b) Perform an ANOVA on X.
(c) Test for homogeneity of regression in the three groups.
(d) Perform an ANCOVA on Y, using X as the covariate.
(e) How do the hypotheses tested by the ANOVA and the ANCOVA differ?

13.3 For the data of the previous problem:
(a) What are the adjusted means for the three treatment groups?
(b) What is the interpretation of the adjusted means?
(c) For each of the contrasts

$$\psi_1 = \mu_2 - \mu_1 \quad \text{and} \quad \psi_2 = \mu_3 - \frac{\mu_1 + \mu_2}{2}$$

test $H_0: \psi = 0$ at EF = .05, controlling Type 1 error using the Bonferroni inequality.

13.4 If, for the data of Exercise 13.2, all the Y scores are regressed on all of the X scores, the regression equation obtained is

$$\hat{Y} = 9.974 + 1.816X$$

The residuals for this regression are as follows:

A_1	A_2	A_3
−5.771	2.045	2.128
−6.138	−0.771	1.412
−2.689	−0.872	6.780
−1.404	−1.670	3.780
−3.771	0.862	5.311
−3.955	1.045	3.678

Perform an ANOVA on these residuals. Is the ANOVA on the residuals equivalent to the ANCOVA of part (d) of Exercise 13.2? Why or why not?

13.5 The relation between Y and X for each treatment group can be described by the nonlinear function

$$Y_{ij} = K_j + 5X_{ij}^2$$

How would you perform an ANCOVA?

13.6 Perform an ANCOVA for the following design:

	A_1		A_2	
	X	Y	X	Y
B_1	23.8	7.9	28.5	25.1
	23.8	7.1	18.5	20.7
	22.6	7.7	20.3	20.3
	22.8	11.2	26.6	18.9
	22.0	6.4	21.2	25.4
	19.6	10.0	24.0	30.0
B_2	27.5	20.1	22.9	19.9
	28.1	17.7	25.2	28.2
	35.7	16.8	20.8	18.1
	27.7	30.5	13.5	13.5
	25.9	21.0	19.1	19.3
	27.9	29.3	32.2	35.1

13.7 Three subjects in each of three groups perform four problem-solving tasks. The time to solve for each task is recorded. IQ scores are also available for each subject. The data are as follows:

		Y data (time to solve)				X data (IQ)
		P_1	P_2	P_3	P_4	
A_1	S_{11}	34	46	48	64	108
	S_{21}	36	41	40	60	112
	S_{31}	28	37	35	52	124
A_2	S_{12}	46	60	63	84	116
	S_{22}	40	51	48	74	127
	S_{32}	55	72	73	96	103
A_3	S_{13}	45	70	74	88	106
	S_{23}	41	63	62	70	135
	S_{33}	49	71	70	85	112

(a) Is it possible to perform an ANCOVA in this case?

(b) If it is possible, perform the ANCOVA.

13.8 Discuss whether it is appropriate to perform ANCOVAs by using X as the covariate in each of the following cases:

(a) Measures of job satisfaction (Y) and performance evaluation by supervisors (X) are obtained for eight randomly sampled workers in each of the four departments of a company. The researchers desire to test whether job satisfaction is the same in each of the departments. The data are as follows:

D_1		D_2		D_3		D_4	
X	Y	X	Y	X	Y	X	Y
1.4	1.0	3.2	3.0	6.2	7.3	5.8	5.6
2.0	2.7	6.8	5.5	3.1	4.0	6.6	7.2
3.2	3.9	5.0	5.6	3.2	4.9	6.5	6.1
1.4	1.0	2.5	3.2	4.0	6.9	5.9	7.1
2.3	4.0	6.1	4.2	4.5	2.1	5.9	5.4
4.0	3.4	4.8	4.2	6.4	5.6	3.0	4.0
5.0	3.7	4.6	3.7	4.4	6.0	5.9	5.6
4.7	2.3	4.2	3.8	4.1	4.6	5.6	5.8

(b) Thirty children are each randomly assigned to one of three remedial math skills training programs. Before entering the programs, each child takes a standardized pretest (X). At the end of six months, a standardized achievement test (Y) is given to each of the children. The researchers wish to determine whether the training programs are all equally effective. The data are

A_1		A_2		A_3	
X	Y	X	Y	X	Y
29	61	39	79	41	78
37	73	34	66	36	66
26	54	35	76	29	56
32	63	39	84	33	61
31	62	35	73	42	70
37	76	27	75	35	65
33	72	35	66	32	59
39	80	29	85	42	80
33	73	34	62	39	65
36	72	26	79	36	64

Chapter **14**

More About Correlation

In Chapter 12, we introduced the Pearson product-moment correlation coefficient as a measure of linear relation between two variables. Although the correlation coefficient is frequently encountered, it often seems to be used and reported without a great deal of understanding. This can lead to serious interpretive problems because the correlation coefficient has some important limitations. Although it is a measure of the extent to which two variables are linearly related, it does not describe the linear component of the relation; different combinations of slope, variance of X and Y, and error variance can yield the same correlation. This means that the correlation coefficient is a *sample-specific* measure. That is, the correlation coefficient depends not only on the nature of the linear relation between X and Y, but also on the variability of X and Y in the sample. Nonetheless, it is important to understand the nature of the correlation coefficient, both because it can be useful in some situations and because it is so widely used. In the present chapter, we discuss the characteristics of the correlation coefficient in more detail and raise a number of issues that should be kept in mind when interpreting correlations. We also discuss significance tests and confidence intervals for correlation coefficients and introduce the concept of partial correlation, the correlation between two variables that have been adjusted for the effects of other variables. Finally, we introduce several special cases and alternative measures of correlation.

14.1 ISSUES IN THE INTERPRETATION OF THE CORRELATION COEFFICIENT

14.1.1 Correlation Does Not Imply Causation

It does not follow from establishing that two variables are correlated that changes in one of the variables are caused by changes in the other. Of course, there is nothing special about the correlation coefficient in this regard; no statistic implies causation. Inferring cause (see, for example, discussions by Einhorn and Hogarth 1986 and Cook and Campbell 1979) is essentially a logical process, not a statistical one. It is possible to have a correlation without causation (e.g., shoe size and verbal ability in children) and causation without a very high correlation (e.g., frequency of sexual intercourse and pregnancy).

Correlational analyses are frequently performed when the data have been obtained by sampling pairs of X and Y scores from a population. When the data are obtained in this way, it is particularly difficult to sort out causal influences. It is often straightforward to make causal inferences on the basis of a well-designed experiment because the effects of irrelevant variables will have been controlled by procedures such as randomization and matching. It is difficult if not impossible to sort out causal effects if the effects of irrelevant variables are not controlled, although procedures have been developed to help determine whether particular causal models are consistent with the obtained data.

A correlation between X and Y may occur because X exerts a causal influence on Y, because Y exerts a causal influence on X, or because other variables exert causal influences, directly or indirectly, on both X and Y. For example, upon finding a strong positive correlation between verbal ability and shoe size in a sample of elementary school children, we cannot conclude that inducing a change in one of these variables will cause a change in the other; rather the correlation would be expected to occur because both verbal ability and physical size increase with age in children. Of course, we can consider the possibility that there might be a correlation between verbal ability and shoe size even if we could somehow control for the effects of age. We will consider attempts to measure the relation between two variables while "partialing out" the effects of other variables in Section 14.5.

14.1.2 The Size of r Is Not Changed by Linear Transformations of X and Y

The size of the correlation coefficient is not changed by any linear transformation of X or Y of the form $X' = aX + b$, where a and b are constants, although the sign of the correlation can change if a is negative. This means that, for example, the correlation between height and weight will be the same whether height is measured in feet, inches, or centimeters and whether weight is measured in ounces or pounds.

The invariance of the size of r under linear transformations can perhaps most easily be seen by noting that r can be expressed as

$$r = \frac{\Sigma z_X z_Y}{N - 1}$$

and that the size of a z score is not affected by a linear transformation. Suppose, for example, that X is height measured in feet and X' is height measured in inches; therefore $X' = 12X$. Because $\bar{X}' = 12\bar{X}$ and $\hat{\sigma}_{X'} = 12\hat{\sigma}_X$, $z_{X'} = z_X$ and therefore $r_{X'Y} = r_{XY}$.

14.1.3 The Relation between r and b_1, $\hat{\sigma}_X$, $\hat{\sigma}_Y$, and $\hat{\sigma}_e$

The interpretation of r as a measure of relation is most straightforward when X and Y are equally variable. From Chapter 12, we know that

$$r = \frac{b_1 \hat{\sigma}_X}{\hat{\sigma}_Y} \tag{14.1}$$

so that if $\hat{\sigma}_X = \hat{\sigma}_Y$, $r = b_1$. In this case, r, the index of the extent to which there is a linear relation between X and Y, also directly informs us about a major feature of the relation; namely, how much, on the average, Y changes when there is a change of 1 unit in X.

In general, however, $\hat{\sigma}_X \neq \hat{\sigma}_Y$. When the standard deviations differ, r is no longer the slope of the regression equation. From Equation 14.1, we see that r depends not only on b_1, but also on $\hat{\sigma}_X$ and $\hat{\sigma}_Y$. As we shall see shortly, by packaging the information slightly differently we can show that r depends on b_1, $\hat{\sigma}_X$, and the variability about the regression line that we can represent by the sample standard error of estimate,

$$S_{YX} = \sqrt{\Sigma_i (Y_i - \hat{Y}_i)^2 / N} = \sqrt{SS_{\text{error}}/N}$$

or the estimate of the population standard error of estimate, $\hat{\sigma}_e = \sqrt{SS_{\text{error}}/(N - 2)}$.

We originally introduced r as an index of the degree to which a set of paired z scores cluster around a straight line. When the standard deviations of X and Y are not equal, it is apparent that, although r can be thought of as an index of the clustering of scores *relative to their standard deviations,* it does not reflect the "absolute" degree of clustering. The six scatter diagrams in Figure 14.1 are all based on the same set of 20 paired z scores, and each represents an r of .8. The diagrams differ in having different standard deviations for X and Y. The correlation coefficient remains the same despite the fact that the slope varies widely, as does the apparent degree of clustering. The fact that the same value of r can occur for different combinations of $\hat{\sigma}_X$, $\hat{\sigma}_Y$, b_1, and $\hat{\sigma}_e$ makes it important to plot the scatter diagram if one wishes to understand the relation between X and Y.

From equations that were developed in Chapter 12, we can express r^2 in ways that will facilitate our discussion of how different factors affect the correlation coefficient. We begin by recalling that

$$SS_Y = SS_{\text{reg}} + SS_{\text{error}}$$

and

$$SS_{\text{reg}} = r^2 SS_Y = b_1^2 SS_X$$

From

$$r^2 = \frac{SS_{\text{reg}}}{SS_Y} = \frac{SS_{\text{reg}}}{SS_{\text{reg}} + SS_{\text{error}}}$$

we have

$$r^2 = \frac{b_1^2 SS_X}{b_1^2 SS_X + SS_{\text{error}}} \tag{14.2}$$

From

$$r^2 = \frac{SS_{\text{reg}}}{SS_Y} = \frac{SS_Y - SS_{\text{error}}}{SS_Y}$$

we have

$$r^2 = 1 - \frac{SS_{\text{error}}}{SS_Y} \tag{14.3}$$

$$= 1 - \frac{\hat{\sigma}_e^2}{\hat{\sigma}_Y^2} \left[\frac{N - 2}{N - 1} \right]$$

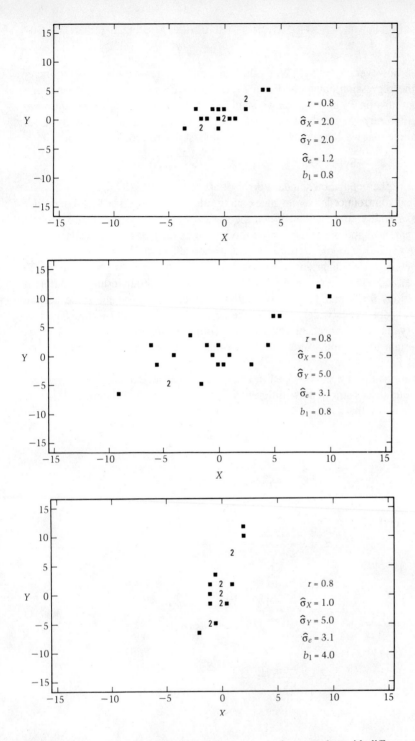

Figure 14.1 Scatter diagrams based on the same set of z scores but with different values for $\hat{\sigma}_X$ and $\hat{\sigma}_Y$.

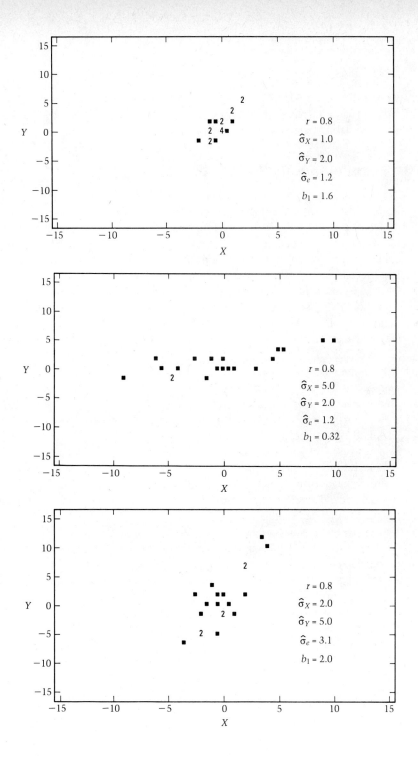

Figure 14.1 (*continued*)

From Equations 14.2 and 14.3, we can see that when all the data points fall on a straight line (i.e., $SS_{error} = 0$), $r^2 = 1$, so $r = \pm 1$. When $SS_{error} \neq 0$, all other things being equal, r increases as the slope and the standard deviation of X become larger and as the variability about the regression line becomes smaller.

14.1.4 The Sample-Specific Nature of *r*

The correlation coefficient is sometimes referred to as a *sample-specific* statistic because estimates of the population correlation, ρ, are more sensitive to the variability of X and Y in the sample than are estimates of other measures such as β_0 and β_1. Why this is the case can perhaps best be seen by observing what happens when only certain values of X are sampled. Consider a population of data points in which X and Y have an imperfect linear relation with a moderately high correlation, ρ, as indicated in Figure 14.2. Assume further that there is *homoscedasticity;* that is, σ_e is the same for all values of X. Suppose we select two samples, one from region B and the second from regions A and C. Although the two samples will have different values of $\hat{\sigma}_X$ and $\hat{\sigma}_Y$, they still estimate the intercept and slope, β_0 and β_1, that characterize the linear relation between X and Y.[1]

On the other hand, the r for a sample selected from region B will tend to *underestimate* ρ and the r for a sample selected from a region consisting of both A and C will tend to *overestimate* ρ. The standard deviation of Y will be smaller in region B (or for that matter, in any other single region) than it is in the entire population of data points. This is because there are relatively fewer large deviations of Y scores from their mean in a restricted region (see Figure 14.2). Therefore SS_Y will tend to be smaller for a sample selected from region B than for a sample of the same size selected from the entire population. In contrast, SS_{error} will not systematically differ by very much in the two samples. Therefore, the ratio SS_{error}/SS_Y will tend to be larger for the restricted sample than for the sample selected from the entire population, and from Equation 14.3 we can see that r would be expected to be smaller.

This type of bias is frequently referred to as the *restriction of range problem.* An oft-cited example of restriction of range is the low correlation between GRE scores and success in graduate school as measured by grades or faculty ratings (e.g., Dawes 1971), which has prompted calls (e.g., Marston 1971) for their abandonment as predictive measures. Even if the GRE score is an excellent measure of ability, this correlation would be expected to be quite low because only students with relatively high GRE scores get accepted into graduate programs.

Sampling from restricted regions of the distribution can also produce inflated estimates of ρ. Samples selected only from the *combined* regions A and C will tend to have smaller SS_{error}/SS_Y ratios than those selected from the entire population and, therefore, from Equation 14.3, larger values of r. If we consider only data points in A and C, there are relatively more Y scores that have large deviations from their mean than is the

---

[1] It should be noted that the accuracy of the estimates will depend on the characteristics of the sample; recall from Chapter 12 that the standard error of b_1 is $\hat{\sigma}_e/\sqrt{SS_X}$ and is therefore inversely proportional to the standard deviation of X.

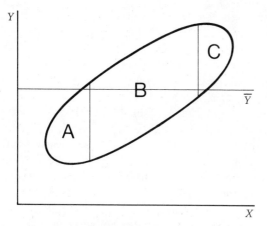

Figure 14.2 Scatter diagram with different regions marked to illustrate the effect of the variability of X and Y on the sample correlation coefficient.

case in the entire population. Therefore a sample selected from A and C will tend to have a larger SS_Y than a sample of the same size selected from the entire population. Because SS_{error} will not differ by much in the two cases, SS_{error}/SS_Y will tend to be smaller and r will tend to be larger than if the sample came from the whole population.

Investigators interested in the relation between two variables will sometimes drop the middle scores on one or both of them. For example, the gambling behavior of subjects scoring high on the MMPI psychopathic deviant scale may be compared with the gambling behavior of subjects who score low on the scale. Although this procedure is acceptable for determining whether there is a linear component to the relation (see the discussion of the extreme-groups design later in the chapter), the correlation between MMPI scale score and gambling behavior obtained from the extreme groups should definitely *not* be considered to be an estimate of the correlation in the whole population.

Suppose that the relation between X and Y is linear and that b_1 and $\hat{\sigma}_e$ are exactly the same for a restricted sample (i.e., a sample selected from only a particular range of X values) and an unrestricted sample. Then, starting with Equation 14.2, it can be shown that an estimate of the correlation coefficient in the unrestricted sample is given by

$$\hat{r} = \frac{r'(\hat{\sigma}_X/\hat{\sigma}_{X'})}{\sqrt{1 + r'^2[\hat{\sigma}_X^2/\hat{\sigma}_{X'}^2 - 1]}} \qquad (14.4)$$

where $\hat{\sigma}_X$ and $\hat{\sigma}_{X'}$ are the standard deviations of X for the unrestricted and the restricted sample, respectively, and r' is the correlation between X and Y in the restricted sample. For example, if $r' = .30$ and $\hat{\sigma}_{X'} = 10$ in the restricted sample, Equation 14.4 provides an estimate of $\hat{r} = .53$ for the correlation coefficient in an unrestricted sample that has $\hat{\sigma}_X = 20$.

So far, we have tried to indicate how r depends on the characteristics of the sample by noting what we would expect to happen for samples selected from particular restricted regions of the population. It is important to note that the same kind of bias will occur in samples randomly selected from the entire population. All other things being equal, samples that have larger values of $\hat{\sigma}_X$ because of sampling error will tend to produce higher estimates of ρ. One must be very cautious when comparing values of r that have been obtained from different groups or data sets. If $\hat{\sigma}_X$ varies, the correlations can differ without there being any fundamental difference in the nature of the underlying linear relation (i.e., in β_0, β_1, or σ_e).

Because the correlation coefficient is sample-specific and does not provide a description of the relation between X and Y, some writers (e.g., Achen 1982, Tukey 1969) have argued strongly against its use. If it can be established that there is a linear relation between X and Y, characteristics of the regression such as b_0, b_1, and $\hat{\sigma}_e$, the estimates of the intercept, slope, and standard error of estimate in the population, describe the nature of the relation more usefully than does r. We do not believe that the correlation coefficient should be abandoned; however, it should be used with caution and with an understanding of what it does and does not measure.

One possible reason for the popularity of the correlation coefficient is that in the social sciences we frequently work with variables that have arbitrary scales, so the standard deviation is as meaningful a unit as any other (what does a one-point difference on a seven-point anxiety scale really mean?). Finding the extent to which two variables have a linear relation may be about as good as we can do in this case. However, we can take little solace in the fact that use of dimensionless measures such as r or r^2 tends to obscure our lack of understanding of some of the variables we use. As Tukey (1969) puts it,

> Given two perfectly meaningless variables, one is reminded of their meaninglessness when a regression coefficient is given, since one wonders how to interpret its value. A correlation coefficient is less likely to bring up the unpleasant truth—we *think* we know what $r = -.7$ means.

14.2 OTHER FACTORS THAT AFFECT THE SIZE OF THE CORRELATION COEFFICIENT

14.2.1 Measurement Error

If X and Y are measured with error, the obtained correlation coefficient will tend to underestimate the "true" coefficient that would be obtained if X and Y could be measured without error. Suppose, instead of correlating the true values X' and Y', we correlate $X = X' + u$ and $Y = Y' + v$, where u and v are error components that are uncorrelated with X', Y', and each other. It can then be shown (see, for example, Nunnally 1978, chapter 6) that

$$r_{XY} = r_{X'Y'} \sqrt{r_{XX}} \sqrt{r_{YY}} \tag{14.5}$$

where $r_{X'Y'}$ is the correlation between the true values, r_{XX} is the reliability coefficient for X (i.e., r_{XX} estimates $\sigma_X^2 / \sigma_X^2 = 1 - \sigma_u^2 / \sigma_X^2$), and r_{YY} is the reliability coefficient for Y. From Equation 14.5, if the true correlation between X' and Y' is .6 and the reliability coefficients for both X and Y are .7, then the observed correlation would be expected to have a value of only .42. Given r_{XY} and the reliabilities of the measures of X and Y, it is possible to solve Equation 14.5 for $r_{X'Y'}$ and use it to estimate the correlation that would be obtained if the measures were perfectly reliable. If this is done, it should be kept in mind that the resulting value is an *estimated* correlation, not a correlation directly obtained from the observed scores. Estimated correlations obtained using Equation 14.5 should not be subjected to the usual tests of significance (see Section 14.3), nor should they be used in more complex procedures such as factor analysis.

14.2.2 The Shapes of the *X* and *Y* Distributions

The marginal distributions of X and Y place limits on the possible values of the correlation between X and Y. We indicated in Chapter 12 that the correlation is $+1$ if $z_X = z_Y$ for each data point and is -1 if $z_X = -z_Y$. If both X and Y have identical symmetrical distributions, for each value of z_X there will be z_Y's that take on the values z_X and $-z_X$. Therefore, it is conceivable that values of r from $+1$ to -1 may occur, depending on how the values of X and Y are paired. However, if X and Y have distributions that are asymmetrical or different, the full range of r cannot occur, no matter how the values of X and Y are paired.

In Figure 14.3, distribution A is positively skewed and distribution B is negatively skewed. If large scores in A are paired with small scores in B, and vice versa, it is possible to obtain a correlation of -1; however, it is not possible to obtain a correlation of $+1$. For one thing, there are no scores in B that have positive z scores as large as those in the tail of A; for another, if we attempted to pair large scores in A with large scores in B, it would soon become obvious that there are not enough large scores in A to match up with those in B. If we did the best we could (i.e., paired off scores with the same percentile ranks), the scatter diagram would show that we had a curvilinear relation with a correlation of less than $+1$. Similarly, if we had two variables whose marginal distributions both looked like either A or B, it would be possible to have a corre-

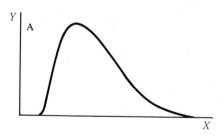

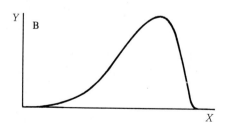

Figure 14.3 Two skewed distributions to illustrate how the distributions of X and Y limit the possible values of the correlation between X and Y.

lation of $+1$ but not one of -1. Because of these constraints, we should plot the distributions of X and Y as well as the scatter diagram when we try to understand the relation between X and Y.

14.2.3 Combining Data Across Groups

When the data from different groups are combined, the correlation in the resultant data set may not characterize the relation between X and Y in any of the groups. The problem can be particularly serious when the group means differ on X and/or Y. In panel a of Figure 14.4, it is apparent that the correlation will be higher if the groups are combined than in either of the separate groups, and in panel b it is apparent that the correlation will be lower if the groups are combined.

Suppose, for example, that we were interested in finding the correlation between height and weight. Because males tend to be both taller and heavier than females, we might have a situation like that depicted in panel a; the correlation is higher in a group

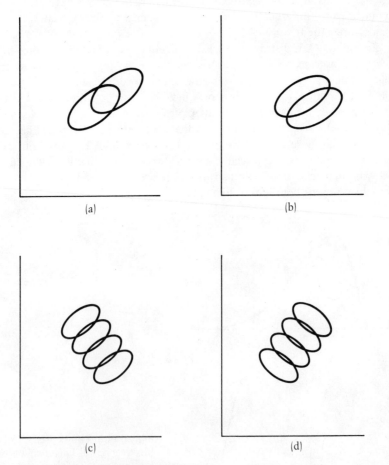

Figure 14.4 Scatter diagram illustrating how combining data from different groups can affect the correlation coefficient.

that contains both males and females than it is for either males or females considered separately. On the other hand, if we correlated scores on verbal and math skills tests and if the data conformed to the stereotypical view that females do better than males on the former but worse on the latter, we could have a situation like that depicted in panel b in which the correlation would be lower for the combined group than for either of the constituent groups.

In extreme cases, it is conceivable that X and Y could be positively correlated in each of a number of groups but negatively correlated when the groups were combined, as in panel c, or negatively correlated in each group and positively correlated when combined as in d.

The message here is simply that it is important for us to understand the effects of group differences when interpreting the correlation coefficient. Of course, this caveat can be extended with equal force to the interpretation of the regression coefficient. For example, the slope would be positive for each of the groups in panel c but negative for the combined data set.

14.3 INFERENCE ABOUT CORRELATION

14.3.1 A Model for Correlation

The model commonly assumed for inference about correlation asserts that the population of (X, Y) pairs has a *bivariate normal distribution*. Both X and Y are random variables and the density function that characterizes the distribution is

$$f(X, Y) = Ae^{-B/2}$$

where

$$A = \frac{1}{2\pi\sigma_X\sigma_Y\sqrt{1 - \rho^2}}$$

and

$$B = \frac{1}{1 - \rho^2}\left[\frac{(X - \mu_X)^2}{\sigma_X^2} + \frac{(Y - \mu_Y)^2}{\sigma_Y^2} - \frac{2\rho(X - \mu_X)(Y - \mu_Y)}{\sigma_X\sigma_Y}\right]$$

We can graphically represent the bivariate normal distribution in several ways. In panel a of Figure 14.5, the plane defined by the X and Y axes contains all possible pairings of X and Y. The bivariate normal density function can be thought of as a bell-shaped surface that rises above the X-Y plane. The intersections of this surface with planes perpendicular to the X-Y plane and parallel to either the X or Y axis all define normal distributions. Also, the intersections of the bivariate normal surface with planes parallel to, but above, the X-Y plane define a family of ellipses, as shown in panel b of Figure 14.5. Each point on a given ellipse will have the same probability density, and, as we can see by comparing panels a and b of Figure 14.5, the smaller ellipses in panel b correspond to larger values of probability density. The more eccentric the ellipses, the greater the correlation between X and Y. A set of concentric circles corresponds to a correlation of zero.

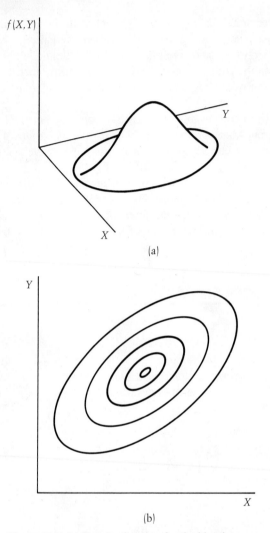

Figure 14.5 (a) Density function for the bivariate normal distribution. (b) Family of ellipses representing a joint distribution that has a bivariate normal distribution. All the points on an ellipse have the same probability density.

Some characteristics of a bivariate normal distribution in X and Y are as follows:

1. The marginal distributions of both X and Y are normal with variances σ_X^2 and σ_Y^2.
2. The conditional means of Y fall on the straight line with equation $\mu_{Y.X} = \mu_Y + \beta_{YX}(X - \mu_X)$, where $\beta_{YX} = \rho\sigma_Y/\sigma_X$. The conditional distributions of Y are normal with variance $\sigma_{Y.X}^2 = \sigma_Y^2(1 - \rho^2)$. The conditional means of X fall on the straight line given by the equation $\mu_{X.Y} = \mu_X + \beta_{XY}(Y - \mu_Y)$, and the conditional distributions of X are normal with variance $\sigma_{X.Y}^2 = \sigma_X^2(1 - \rho^2)$. Given bivariate normality, X and Y are independent if they have correlation of zero.

14.3.2 Testing Hypotheses and Finding Confidence Intervals about ρ

The hypothesis H_0: $\rho = 0$ may be tested with the test statistic

$$t = r\sqrt{\frac{N-2}{1-r^2}} \qquad (14.6)$$

with $N - 2$ df. It can be shown that this statistic is exactly equivalent to $b_1\sqrt{SS_X}/\hat{\sigma}_e$, the statistic we used to test the hypothesis H_0: $\beta_1 = 0$ in Chapter 12. It makes sense that if there is a linear relation between X and Y, both the correlation and regression coefficients should differ from 0.

At the beginning of Chapter 12, we discussed an example in which the correlation between the pretest and final exam score for 18 students who took an introductory statistics course was .725. For the data in Table 12.1,

$$t = .725\sqrt{\frac{16}{1-.725^2}} = 4.21$$

If the 18 students can be thought of as a random sample from a population of students who may have taken the course, we can see from Table D.3 that the hypothesis H_0: $\rho = 0$ can be rejected at $p < .001$.

Although this t test is quite robust with respect to the assumption of normality (see, for example, Edgell and Noon 1984), it cannot be used to test other hypotheses about ρ. This is because the shape and standard error of the sampling distribution of r depend on ρ. When ρ is substantially different from zero, the sampling distribution of r is skewed, even for large sample sizes.

However, Fisher (1921) showed that if we can assume bivariate normality, a logarithmic transformation of r (henceforth referred to as the Fisher Z transform)

$$Z_r = \frac{1}{2}\ln\left[\frac{1+r}{1-r}\right] \qquad (14.7)$$

is approximately normal with mean $(1/2)\ln[(1 + \rho)/(1 - \rho)]$ and standard error $1/\sqrt{N-3}$ for sample sizes as small as $N = 10$. This fact allows us to test hypotheses of the form H_0: $Z = Z_{\text{hyp}}$, using the normally distributed test statistic

$$z = \frac{Z_r - Z_{\text{hyp}}}{\sqrt{1/(N-3)}} = (Z_r - Z_{\text{hyp}})\sqrt{N-3} \qquad (14.8)$$

and indirectly allows us to test any hypothesis of the form H_0: $\rho = \rho_{\text{hyp}}$. The Fisher Z transform should not be confused with the z scores that appear often in the book. We will distinguish between the two by using uppercase Z's to refer to the Fisher transform. Values of Z corresponding to different values of r are presented in Table D.12.

Suppose we wish to test the null hypothesis that $\rho = .50$ at $\alpha = .05$ in a population of interest and that in our sample $N = 19$ and $r = .74$. From Table D.12, we see that the Z transform corresponding to a correlation of .74 (which we indicate by $Z_{.74}$) is equal to .950 and $Z_{.50} = .549$. Testing the hypothesis $\rho = .50$ is equivalent to testing the hypothesis $Z = .549$. Using the test statistic of Equation 14.8, we have $z =$

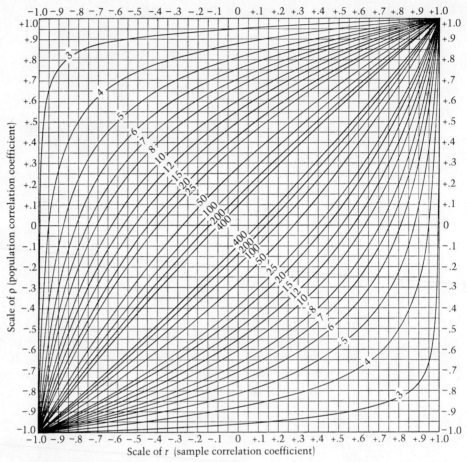

The numbers on the curves indicate sample size. The chart can also be used to determine upper and lower 2.5% significance points for r, given ρ.

Figure 14.6 Chart giving 95% confidence intervals for the population correlation coefficient ρ, given the sample coefficient r.

$(.950 - .549)\sqrt{16} = 1.60$. The obtained value of the test statistic does not exceed the upper critical value ($+1.96$), so we fail to reject the null hypothesis.

We can also use the Fisher Z transform to find confidence intervals for ρ by a two-step procedure. If we wanted to find the 95% confidence interval for ρ, the first step would be to find the corresponding confidence interval for Z,

$$Z_r \pm \frac{z_{.025}}{\sqrt{N-3}}$$

The second step would be to convert the upper and lower limits of the interval from the Z to the correlation scale.

In the previous example we had $r = .74$ and $N = 19$. The 95% confidence interval

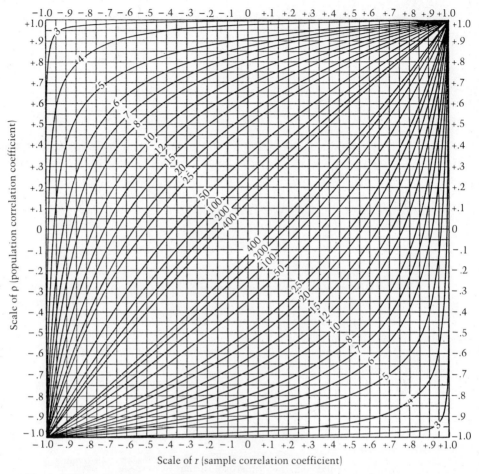

The numbers on the curves indicate sample size. The chart can also be used to determine upper and lower 0.5% significance points for r, given ρ.

Figure 14.6 (*continued*) Chart giving 99% confidence intervals.

for Z is given by $.950 \pm 1.96/4 = (.46, 1.44)$. From Table D.12, we see that a Fisher Z transform of .46 corresponds to a correlation of about .43 and that a Z of 1.44 corresponds to a correlation of about .89. Therefore, the 95% confidence interval about ρ is approximately $(.43, .89)$.

It is sobering to realize just how large confidence intervals about ρ are, even for moderately large sample sizes. For example, if $r = .50$ and $N = 103$, the 95% confidence interval about Z is $.549 \pm .196$, so the interval about ρ is given approximately by $(.34, .63)$. Approximate confidence intervals can readily be found by using the curves in Figure 14.6.

14.3.3 Testing whether Independent Correlations Are Significantly Different

It is possible to test whether the correlation of X and Y is the same in two different populations by using the test statistic

$$z = \frac{Z_{r_1} - Z_{r_2}}{\sqrt{1/(N_1 - 3) + 1/(N_2 - 3)}} \qquad (14.9)$$

We may, for example, want to know whether the correlation between number of years of education and income is the same for blacks as it is for whites or whether math and verbal aptitude scores are correlated equally for male and female ninth graders. If in random samples of 67 males and 84 females we have correlations of .50 and .28, respectively, the obtained value of the test statistic is

$$z = \frac{.549 - .288}{\sqrt{1/64 + 1/81}} = 1.56$$

so that the two correlations are not significantly different at $\alpha = .05$.

Of course, if we did find a significant difference, we would have to be very careful in interpreting what it meant. Even if the slope and variability around the regression line was the same for males and females, the males would have a higher correlation if they displayed greater variability in their aptitude scores. It would be a good idea to plot the scatter diagrams and to find the regression statistics for both groups instead of looking only at the correlations.

It is also possible to test for the homogeneity of k independent correlation coefficients. The hypothesis that the set of k ρ's are all equal employs the test statistic

$$\chi^2 = \sum_j (N_j - 3)Z_j^2 - \frac{[\Sigma_j(N_j - 3)Z_j]^2}{\Sigma_j(N_j - 3)} \qquad \text{with } k - 1 \text{ df} \qquad (14.10)$$

where Z_j is the Fisher Z transform of the jth correlation coefficient.

If the correlations between scores on math and verbal skills tests for random samples of males in three high schools are as follows:

School 1 $r = .45, \quad N = 50$

School 2 $r = .33, \quad N = 47$

School 3 $r = .29, \quad N = 50$

the obtained value of the test statistic is $\chi^2 = 20.43 - (51.91)^2/138 = .88$. Because $\chi^2_{.05,2} = 5.99$, we cannot reject the hypothesis that the correlations are identical in the three schools.

14.3.4 Testing Hypotheses about Correlation Matrices

When data on a number of variables have been collected from the same set of subjects, the correlations may be displayed in a *correlation matrix*. If the variables are $X, Y, U,$ and V, the matrix is

$$
\begin{array}{c}
\begin{array}{cccc} X & Y & U & V \end{array} \\
\begin{array}{c} X \\ Y \\ U \\ V \end{array}
\left[\begin{array}{cccc}
1 & r_{XY} & r_{XU} & r_{XV} \\
r_{YX} & 1 & r_{YU} & r_{YV} \\
r_{UX} & r_{UY} & 1 & r_{UV} \\
r_{VX} & r_{VY} & r_{VU} & 1
\end{array} \right]
\end{array}
$$

If there are k variables, the matrix consists of k^2 elements. The k elements on the major diagonal (the one that goes from upper left to lower right) are each equal to 1 because any variable is perfectly correlated with itself. The $k^2 - k = k(k - 1)$ off-diagonal elements can be divided into a set of $k(k - 1)/2$ correlations above the diagonal and a set of equal size below it. The matrix is *symmetric* because elements on opposite sides of the diagonal are identical; $r_{XY} = r_{YX}$ and, in general, $r_{ij} = r_{ji}$ for all i and j.

When k is large, the number of correlations can be very large; for example, if $k = 20$, $k(k - 1)/2 = 190$. If we tested each correlation for significance at $\alpha = .05$, the Type 1 error rate for the entire family of correlations would be very high. Although most software packages will dutifully print out the significance level for each correlation coefficient as though it was the only one tested, it is every bit as necessary to control Type 1 error here as when we perform multiple t tests on the differences between means. If we are interested in a limited number of tests based on a priori considerations, we can adjust the significance levels by using the Bonferroni inequality, just as we did in Chapter 6. We can use the Bonferroni procedure even if we wish to make a large number of significance tests, although it will be somewhat conservative because the tests are not independent.

Stieger (1980) has recommended a simple test of the hypothesis that all off-diagonal elements of the correlation matrix are equal to zero. If this hypothesis cannot be rejected, tests on the individual correlations in the matrix are not likely to be meaningful unless motivated by a priori considerations. The hypothesis can be tested by using the test statistic

$$
\chi^2 = (N - 3) \sum_{j > i} \sum Z_{ij}^2 \quad \text{with} \quad \frac{k(k - 1)}{2} \text{ df} \tag{14.11}
$$

where Z_{ij} is the Z transform of r_{ij} and the summation is over the $k(k - 1)/2$ squared transforms corresponding to the correlations above (or below) the diagonal.

For example, if our correlation matrix is

$$
\left[\begin{array}{cccc}
1.00 & .12 & .07 & -.36 \\
.12 & 1.00 & .29 & -.04 \\
.07 & .29 & 1.00 & .19 \\
-.36 & -.04 & .19 & 1.00
\end{array} \right]
$$

and is based on 40 subjects,

$$
\Sigma\Sigma Z_{ij}^2 = (.121)^2 + (.070)^2 + (-.377)^2 + (.300)^2 + (-.040)^2 + (.192)^2
$$
$$
= .290
$$

so that $\chi^2 = (37)(.290) = 10.73$. Because $\chi^2_{.05,6} = 12.592$, we cannot reject the hypothesis that all the off-diagonal correlations are 0. Notice that had we simply tested $r_{14} = -.36$ at $\alpha = .05$, we would reject the null hypothesis that $\rho_{14} = 0$ because the value of the test statistic, $z = -.377\sqrt{37} = -2.29$, is less than -1.96.

Also, it may be of interest to determine whether the correlation between X and Y is different from the correlation between X and W. For example, we want to know whether scores on an analytical reasoning test correlate differently with math and verbal ability scores in some population of interest. To test the hypothesis that $\rho_{XY} = \rho_{XW}$, we can use the statistic

$$t = (r_{XY} - r_{XW})\sqrt{\frac{(N-1)(1+r_{YW})}{2[(N-1)/(N-3)]|\mathbf{R}| + \bar{r}^2(1-r_{YW})^3}} \tag{14.12}$$

where the test statistic is distributed as t with $N - 3$ df if the null hypothesis is true, $\bar{r} = (r_{XY} + r_{XW})/2$, and $|\mathbf{R}| = 1 - r_{XY}^2 - r_{XW}^2 - r_{YW}^2 + 2r_{XY}r_{XW}r_{YW}$.

If, for example, we find in a random sample of 165 subjects that $r_{XY} = .38$, $r_{XW} = .20$, and $r_{YW} = .24$, then $|\mathbf{R}| = .794$ and

$$t = .18\sqrt{\frac{(164)(1.24)}{2(164/162)(.794) + .037}}$$

$$= 2.00$$

so that we can reject the null hypothesis that $\rho_{XY} = \rho_{XW}$ at $\alpha = .05$. Note that if we used the test for differences between independent correlations that we discussed in the last section, we would obtain a value of

$$z = \frac{.4001 - .2027}{\sqrt{1/162 + 1/162}}$$

$$= 1.78$$

and could not reject the null hypothesis.

Finally, it is possible to test hypotheses of the form $\rho_{XY} = \rho_{VW}$ or any hypothesis that states that members of some subset of the correlations in the matrix are equal to each other and/or some specified constant. We will not present these procedures here because they are computationally tedious and rarely used. Interested readers should consult Steiger (1980) for discussions of these procedures and for worked-out examples.

14.4 PARTIAL CORRELATIONS

14.4.1 The Partial Correlation Coefficient

We earlier indicated that two variables may be correlated because they are both influenced directly or indirectly by other variables. For example, verbal ability is correlated with shoe size in children because both verbal ability and shoe size increase with age. However, we might be interested in asking whether there is a relation between physical size and verbal ability even if the effects of age could somehow be controlled or "partialed out." Perhaps we believe that there is a relation between the rates of physical and mental maturation or that greater physical growth might be an indicator of better health or nutrition and therefore might be related to mental ability. How can we find a measure of the relation between size and verbal ability that is not contaminated by the effects of age?

One way of controlling for effects of a variable like age would be to hold it constant. We could select a sample of children who were all the same age and find the correlation between size and verbal ability in that sample. A second way would be to remove the components of the size and verbal ability variables that were predictable from age and to find the correlation between the parts of the variables that were left over (i.e., the residuals). The second approach is the one that underlies the idea of partial correlation.

If we use the notation $r_{XY} = \text{Corr}(X, Y)$ to stand for the correlation between two variables X and Y, then $r_{XY|W}$, the partial correlation between X and Y with the effects of W partialed out, is given by

$$r_{XY|W} = \text{Corr}(X|W, Y|W) \qquad (14.13)$$

where $X|W = X - \hat{X}$ is the residual that results when X is regressed on W, and $Y|W = Y - \hat{Y}$ is the residual that results when Y is regressed on W.

It is also possible to express $r_{XY|W}$ in terms of the simple correlations between X, Y, and W:

$$r_{XY|W} = \frac{r_{XY} - r_{XW}r_{YW}}{\sqrt{(1 - r_{XW}^2)(1 - r_{YW}^2)}} \qquad (14.14)$$

so that if the correlations of size and verbal ability with age were both .7 and the correlation between size and verbal ability was .5, $r_{\text{size,verbal}|\text{age}}$ would have a value of $[.5 - (.7)(.7)]/\sqrt{(1 - .49)(1 - .49)} = .02$.

As another example, consider the data in Table 14.1, which contains some motor vehicle statistics for the 48 contiguous states (i.e., excluding Alaska and Hawaii) obtained from the 1988 World Almanac. The three variables in the table are

pop   population of the state in thousands of people
deaths number of motor vehicle deaths in 1986
belt   has the value 1 if the state had a seat belt law as of July 1987 and the value 0 otherwise (a dichotomous variable)

The correlation between deaths and belt has a value of .309. This might seem somewhat disturbing at first glance because it indicates a tendency for there to be more motor vehicle deaths in states with seat belt laws. However, it turns out that $r_{\text{deaths,pop}} = .928$ and $r_{\text{belt,pop}} = .345$. There is not only an extremely strong tendency for larger states to have more deaths, the larger states are more likely to have enacted seat belt legislation. If we use Equation 14.14 to obtain the partial correlation, we find $r_{\text{deaths,belt}|\text{pop}} = -.032$, which indicates a small but negative relation between seat belt laws and motor vehicle deaths once the effects of population have been partialed out.

TABLE 14.1   MOTOR VEHICLE FATALITY DATA

State	Pop	Deaths	Belt
AL	3894	1080	0
AZ	2717	1002	0
AR	2286	601	0
CA	23668	5223	1
CO	2890	603	1
CN	3108	455	1
DE	594	138	0
FL	9747	2875	1
GA	5463	1540	0
ID	944	258	1
IL	11429	1616	1
IN	5490	1038	1
IA	2914	440	1
KS	2364	500	1
KY	3660	808	0
LA	4206	844	1
ME	1125	208	0
MD	4217	790	1
MA	5377	752	0
MI	9262	1624	1
MN	4076	572	1
MS	2521	766	0
MO	4917	1135	1
MT	787	222	1
NE	1570	290	0
NV	801	233	1
NH	921	172	0
NJ	7365	1040	1
NM	1303	498	1
NY	17558	2114	1
NC	5880	1640	1
ND	653	100	0
OH	10798	1673	1
OK	3025	710	1
OR	2633	619	1
PA	11865	1928	0
RI	947	114	0
SC	3121	1058	0
SD	690	134	0
TN	4591	1232	1
TX	14226	3568	1
UT	1461	311	1
VT	511	109	0
VA	5347	1118	1
WA	4132	714	1
WV	1950	437	0
WI	4706	757	0
WY	470	168	0

Equivalently, if we regress deaths and belt on pop, we find the two regression equations

$$\widehat{\text{deaths}} = 80.095 + (.187)\text{pop}$$

and

$$\widehat{\text{belt}} = .414 + (.0000362)\text{pop}$$

If we obtain the two sets of residuals and correlate them, we again obtain a partial correlation of $-.032$.

Partial correlations are often calculated on data obtained from observational studies, in an attempt to statistically control for the effects of irrelevant variables. Unfortunately, although it is easy to calculate a partial correlation, it is often difficult to understand what it means. It should be emphasized that when $r_{XY|W}$ is obtained, what is removed from X and Y are the components that are *predictable* from a linear regression on W. Therefore, one must be very cautious about making any kind of causal statement on the basis of a partial correlation. When the effects of W are partialed out, this not only removes any direct causal effects of W on X and Y but also some of the direct or indirect effects of any of the variables that are correlated with W. It should be noted that the issues involved are exactly the same ones that we encountered in the ANCOVA chapter when we considered the consequences of attempting to control for the effects of a covariate (also, see Pedhazur 1982 for a detailed discussion).

Suppose, for example, X measures parents' intelligence, Y measures their children's performance in school, and W is the number of books in the home. If it is found that $r_{XY|W}$ is considerably smaller than r_{XY}, this does not necessarily mean that the books themselves play any causally important role in children's performance or that parental intelligence is unimportant. Partialing books out of the correlation between parental intelligence and school performance removes more than the effects of the books; partialing removes any components of intelligence and school performance *predictable* from number of books. The number of books in the home is correlated with parental intelligence, as well as with other potentially important variables such as socioeconomic level and parental encouragement of achievement. When the number of books is partialed out of the relation between X and Y, some of the effects of these other variables are removed as well.

14.4.2 Partialing Out More than One Variable

Suppose we wish to partial out the effects of variables W and Q from the correlation between X and Y. The partial correlation $r_{XY|WQ}$ is given by $\text{Corr}(X|WQ, Y|WQ) = \text{Corr}(X - \hat{X}, Y - \hat{Y})$, where $\hat{X}$ is the optimal prediction of X based on a linear regression equation that contains both W and Q, and $\hat{Y}$ is the corresponding prediction of Y. The same logic holds no matter how many variables there are to be partialed out; the partial correlation can always be obtained by correlating the two sets of residuals that result when X and Y are regressed on these variables. We will discuss regression equations that include more than one predictor variable (multiple regression) in the next chapter.

There are expressions similar to Equation 14.14 for calculating these higher-order partial correlations. However, in practice the partial correlations will be most easily obtained by using statistical packages that either produce the partial correlations directly or provide the appropriate residuals that can then be correlated.

14.4.3 Significance Tests for Partial Correlation Coefficients

Significance tests and confidence intervals for partial correlation coefficients are completely analogous to those introduced in Section 14.3.

The null hypothesis H_0: $\rho_{XY|W} = 0$ can be tested by using

$$t = r_{XY|W} \sqrt{\frac{N-3}{1 - r_{XY|W}^2}}$$

with $N - 3$ df, and the more general null hypothesis $\rho_{XY|W} = \rho_{\text{hyp}}$ can be tested by using

$$z = (Z_r - Z_{\text{hyp}})\sqrt{N-4}$$

where Z_r and Z_{hyp} are the Fisher transforms of $r_{XY|W}$ and ρ_{hyp}.

In general, if r is the partial correlation with p variables partialed out, the corresponding expressions are

$$t = r\sqrt{\frac{N-2-p}{1-r^2}} \qquad \text{with } N - 2 - p \text{ df}$$

and

$$z = (Z_r - Z_{\text{hyp}})\sqrt{N-3-p}$$

14.4.4 The Semipartial (or Part) Correlation Coefficient

The semipartial correlation coefficient $r_{Y(X|W)}$ is the correlation between Y and $X|W$, where $X|W = X - \hat{X}$ is the residual when X is regressed on W. The coefficient can be obtained by regressing X on W and correlating the residuals with Y or by using Equation 14.15:

$$r_{Y(X|W)} = \frac{r_{XY} - r_{XW}r_{YW}}{\sqrt{1 - r_{XW}^2}} \tag{14.15}$$

Although we will not deal with the semipartial correlation any further here, it has a useful interpretation in terms of multiple regression and correlation that we will discuss in Chapter 15.

14.4.5 Constraints in Sets of Correlation Coefficients

Given three variables X, Y, and W, there are three correlation coefficients, r_{XY}, r_{XW}, and r_{YW}. The range of possible values that can be taken on by any one of these correlations is constrained by the values taken on by the other two.

As the most extreme example, if W has a perfect linear relation with both X and Y, X and Y must have a perfect linear relation. That is, if $|r_{XW}| = |r_{YW}| = 1$, then $|r_{XY}|$ must also equal 1. However, what can we say about the possible values of r_{XY} if r_{XW} and r_{YW} are both equal to some other value such as .7?

Because $r_{XY|W}$ is a correlation, it must take on a value between -1 and $+1$. If we solve Equation 14.14 for r_{XY}, we have

$$r_{XY} = r_{XW}r_{YW} + r_{XY|W}\sqrt{(1 - r_{XW}^2)(1 - r_{YW}^2)}$$

Therefore, the value of r_{XY} must lie between

$$r_{XW}r_{YW} - \sqrt{(1 - r_{XW}^2)(1 - r_{YW}^2)}$$

and

$$r_{XW}r_{YW} + \sqrt{(1 - r_{XW}^2)(1 - r_{YW}^2)}$$

Substituting into these expressions, we see that if $r_{XW} = r_{YW} = .7$, r_{XY} must lie between $-.02$ and 1.00. A strong negative correlation between X and Y is not possible if X and Y both have large positive correlations with W.

14.5 THE EXTREME-GROUPS DESIGN

Suppose we wish to evaluate the relation between X, a variable that is easy and inexpensive to measure, and Y, a variable that is expensive to measure so that relatively few observations are available. A strategy recommended by Alf and Abrahams (1975) involves first obtaining X scores from a sample of N subjects. A smaller number, n, of Y scores is then obtained from some subset of subjects who had very high or very low X scores (the proportions of high-scoring and low-scoring subjects need not be the same). After collecting the Y data, the full-range correlation (i.e., the correlation that would have resulted if all N of the Y scores had been available) can be estimated from the equation

$$\hat{r} = \frac{r'(\hat{\sigma}_X/\hat{\sigma}_{X'})}{\sqrt{1 + r'^2[(\hat{\sigma}_X/\hat{\sigma}_{X'})^2 - 1]}} \tag{14.4}$$

where $\hat{\sigma}_X$ and $\hat{\sigma}_{X'}$ are the standard deviations of X for the full sample of X scores and the combined high and low groups, respectively, and r' is the correlation between X and Y in the combined high and low groups. Having obtained an estimate of the full-range correlation, we can test the null hypothesis $H_0: \rho = 0$, employing the usual test statistic

$$t = \hat{r}\sqrt{\frac{N - 2}{1 - \hat{r}^2}}$$

Alf and Abrahams (1975) have shown that this strategy is uniformly more powerful than testing the hypothesis $\mu_{\text{high}} = \mu_{\text{low}}$, using a standard t test for means or randomly sampling n (X, Y) pairs and testing the resulting correlation for significance. However, it should be noted that the strategy depends on the assumption that there is no nonlinear component to the relation between X and Y and on the assumption of bivariate normality. If Y is nonlinearly related to X, the extreme-groups approach can be quite misleading. Clearly, if Y is a U-shaped function of X, the Y's corresponding to very high and low values of X may be similar to one another, but quite different from the Y's corresponding to medium values of X. Whenever there is reason to suspect a nonlinear relation, it is better to sample the available proportion randomly, whatever its size. In this way, information about the shape of the function relating X and Y can be obtained.

14.6 OTHER MEASURES OF CORRELATION

In this section, we introduce several classes of correlation measures other than the usual Pearson product-moment coefficient. We first discuss two measures of correlation used with ranked data. The first of these, the Spearman coefficient, is simply the usual Pearson r applied to ranks. The second, the Kendall tau coefficient, employs a different measure of agreement. We then discuss very briefly four measures used when one or both variables are dichotomous. The point-biserial and phi coefficients are simply the Pearson r with one and two dichotomous variables, respectively. The biserial and tetrachoric coefficients provide estimates of what the Pearson r would be, if instead of dichotomous variables we actually had normally distributed scores.

14.6.1 The Spearman Correlation Coefficient for Ranked Data (r_S)

Sometimes we wish to obtain correlations for data that occur in the form of ranks. We may, for example, have two judges rank a set of stimuli according to their difficulty and obtain the correlation between the two sets of rankings as a measure of reliability. Even if X and Y are continuous variables, we may wish to convert to ranks if X and Y have skewed distributions or if we desire measures of correlation that are more resistant to outliers than the usual Pearson r (see, for example, Gideon and Hollister 1987).

The special case of the Pearson r for ranked data is referred to as the Spearman correlation coefficient (r_S) or sometimes as the Spearman rho coefficient. Although the Spearman coefficient can always be obtained by applying any of the usual Pearson formulas to the ranked data, a fairly simple expression that takes advantage of some of the characteristics of ranks has been developed. If there are no ties, the ranks of N scores are the first N integers. Therefore, the mean rank for both X and Y is $(N + 1)/2$, and the sum of the squared ranks for both X and Y can be shown to be $N(N+1)(N+2)/6$. Substituting into the usual Pearson formula yields

$$r_S = 1 - \frac{6\Sigma_i D_i^2}{N(N^2 - 1)} \tag{14.16}$$

where D_i is the difference between the X and Y ranks for the ith case. If there are no ties, Equation 14.16 and the usual formulas for the Pearson r will take on the same value; for example, they will have a value of .89 for the data in Table 14.2. This will not be true when there are ties because tied ranks cause the variance of the N ranks to be smaller than the variance of the first N integers; therefore, Equation 14.16 is no

TABLE 14.2   CALCULATION OF r_S AND τ FOR A SET OF RANKED DATA

X	Rank of X	Y	Rank of Y	D	D_2
81	9	20	8	1	1
59	3	16	5	−2	4
37	1	12	2	−1	1
79	8	21	9	−1	1
63	5	19	7	−2	4
72	7	17	6	1	1
42	2	9	1	1	1
61	4	14	3	1	1
83	10	25	10	0	0
70	6	15	4	2	4

Calculation of r_S

$$r_S = 1 - \frac{6\Sigma_i D_i^2}{N(N^2 - 1)}$$

$$= 1 - \frac{(6)(18)}{(10)(99)}$$

$$= .89$$

Calculation of τ

Ordered ranks of X

Corresponding ranks of Y

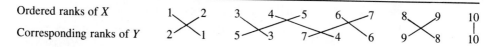

The number of inversions can most easily be obtained by drawing lines between the same ranks for X and Y. The number of times that pairs of lines cross one another is the number of inversions. For the current example there are six inversions. Therefore,

$$\tau = 1 - \frac{2(\text{number of inversions})}{N(N - 1)/2}$$

$$= 1 - \frac{(2)(6)}{(10)(9)/2}$$

$$= .73$$

longer appropriate. Although there are modifications of Equation 14.16 that adjust for ties, they are cumbersome and unnecessary, given that the appropriate value can always be obtained by finding the Pearson r for the ranked data and the fact that most statistical packages will do both the ranking and the calculations. In fact, many packages will calculate the Spearman rho and Kendall tau coefficients directly.

For $N > 10$, we can test the null hypothesis that the ranks of X and Y have a correlation of 0 in the population by using the test statistic

$$t = r_S \sqrt{\frac{N-2}{1-r_S^2}} \qquad (14.17)$$

with $N - 2$ df. The test statistic is the same one used to test the Pearson r for significance. Although this test is not appropriate for $N < 10$, tables of the critical values of r_S have been made available for small N and have been reproduced in a number of books (e.g., Siegel and Castellan 1988).

14.6.2 The Kendall Tau Coefficient (τ)

Kendall (1963) has developed a somewhat different approach to the problem of assessing agreement between two sets of ranks. Rather than using a measure of discrepancy that depends on the magnitude of the differences in the ranks of X and Y (i.e., the ΣD_i^2 quantity that appears in the formula for the Spearman coefficient), Kendall's approach depends on the number of agreements and disagreements in ranking.

Assume that we have N objects $O_1, O_2, \ldots, O_N$ that receive two sets of rankings, X and Y. If the X and Y rankings are exactly the same, there will be no disagreement and the Kendall τ will have a value of 1. We say that an *inversion* occurs if, for any two objects O_i and O_j, O_i is ranked higher than O_j in one set of rankings but lower than O_j in the other set. For N objects there are $N(N-1)/2$ possible pairings of objects; therefore, if there are no ties, there are a maximum of $N(N-1)/2$ possible inversions. The Kendall τ coefficient is defined as

$$\tau = 1 - \frac{2(\text{number of inversions})}{\text{maximum possible number of inversions}} \qquad (14.18)$$

$$= 1 - \frac{2(\text{number of inversions})}{\text{total number of pairs of objects}}$$

$$= \frac{\text{number of agreements in order} - \text{number of disagreements in order}}{\text{total number of pairs of objects}}$$

Various procedures exist for obtaining the number of rank inversions, but the simplest is the graphic method illustrated in Table 14.2. The graphic method is appropriate only if there are no tied ranks, and more general procedures are outlined in Siegel and Castellan (1988) and Hays (1988).

For an N of about 10 or more, the significance of τ can be tested by using the test statistic

$$z = \frac{\tau}{\sigma_\tau} \qquad (14.19)$$

where $\sigma_\tau = \sqrt{2(2N+5)/9N(N-1)}$ and z has a unit normal distribution under the null hypothesis. We can reject the hypothesis that τ is equal to zero in the population at $\alpha = .05$ for the data in Table 14.2 because the obtained value of the test statistic, $.73/\sqrt{2(25)/(90)(9)} = 2.94$, is greater than 1.96. Tables that can be used when $N < 10$ can be found in Siegel and Castellan (1988).

Although r_S and τ are based on different conceptions of discrepancy (ΣD_i^2 for the former and number of inversions for the latter) and generally produce different values for the same set of data, they have similar power and Type 1 error rates.

14.6.3 The Point-Biserial and Phi Correlation Coefficients

We are frequently concerned with dichotomies such as male/female, pass/fail, and experimental/control. It is therefore of interest to note that we can express each of these dichotomies in terms of a variable that can then be correlated with other variables. For example, we can correlate a dichotomous variable with a continuous variable (i.e., passing or failing a particular item with the overall test score) or correlate two dichotomous variables (e.g., male/female with pass/fail). We can find the correlation by assigning any two numbers to the categories that make up the dichotomy. Usually, the numbers 0 and 1 are used, but the size of correlation would be the same if the numbers were 31 and 57.

When the Pearson r formula is applied to a data set in which one variable is continuous and the other variable takes on the values 0 and 1 to represent male/female, pass/fail, or any other dichotomy, the result is called the point-biserial correlation coefficient. There are specialized formulas that take advantage of the fact that one of the variables is dichotomous, but they will always yield the same answer and therefore will not be presented here. The point-biserial coefficient can be tested for significance by using the statistic presented in Equation 14.6.

When the Pearson r formula is applied to two variables that are each coded 0 and 1 to represent dichotomies, the result is called the phi coefficient (ϕ). As in the case of the point-biserial coefficient, specialized formulas for ϕ exist, but always give the same result as applying the Pearson r to the dichotomous data. The ϕ is closely related to the χ^2 test for independence, and it can be shown that

$$\chi^2 = N\phi^2 \tag{14.20}$$

The χ^2 statistic with 1 df can be used to test the hypothesis that X and Y are independent in the population, whereas the ϕ coefficient (or ϕ^2, which can be interpreted as the proportion of variance accounted for) can be used as a measure of the strength of the relation between X and Y.

As a final comment, we note that Equation 14.20 provides us with an opportunity to emphasize that with large enough samples, even small effects may be statistically significant. Looking up the critical value in Table D.4, we find that we can reject the hypothesis that X and Y are independent at $\alpha = .05$ as long as $\chi^2 = N\phi^2 > 3.84$ or $\phi^2 > 3.84/N$. It follows that for $N = 1000$, we would have a statistically significant relation even if ϕ^2 was only about .00384. In this case the "significant" relation would account for only about one-third of 1% of the variance.

14.6.4 The Biserial and Tetrachoric Correlation Coefficients

The biserial and point-biserial correlation coefficients of X and Y are analogous in that for both measures, one variable is dichotomous and the other is continuous. The important difference between them is that whereas the point-biserial coefficient is simply the special case of the Pearson r that is obtained when one variable is dichotomous, the biserial correlation coefficient is an *estimate* of what the correlation would be if, instead of having a dichotomous variable, one actually had scores on an underlying normally distributed variable. Although we will not discuss the biserial coefficient in any detail, it should be emphasized that it acts quite differently than the point-biserial coefficient. Unlike the latter measure, the biserial coefficient is not very sensitive to the proportion of 0's and 1's in the dichotomous variable; moreover, the biserial coefficient can be shown to be always at least 25% larger than the point-biserial and under some circumstances can take on values greater than 1.

The tetrachoric correlation coefficient is also an estimate. Like the phi coefficient, the tetrachoric coefficient takes as data two dichotomous variables. However, rather than measuring the correlation between the dichotomous variables, the tetrachoric coefficient estimates the r that would result if scores on two normally distributed underlying variables were available.

The idea of representing a dichotomy in terms of an underlying normal variable may make sense in some cases but not in others. For the dichotomy pass/fail, one can imagine passing a test if one has more than a certain amount of a normally distributed ability and failing the test if one has less than that amount of ability. On the other hand, the idea of a normally distributed underlying variable does not make sense for a variable like gender. Even if there is an underlying continuous variable, the assumption of normality may be suspect. For example, the dichotomous data may have been collected for a sample of college students. Even if an underlying ability dimension is normally distributed in the general population, the lowest part of the distribution will not be represented in the population of college students.

The biserial and tetrachoric coefficients depend very strongly on the assumption of normality and should be used only with caution. Because these measures are rarely used, we do not deal with them further here other than to refer the interested reader to sources, such as Lindeman, Merenda, and Gold (1980), that provide more detailed discussions.

14.7 CONCLUDING REMARKS

We conclude this chapter by again mentioning several concerns about correlational measures that should be kept in mind. Although we outlined procedures for testing differences in correlations across groups, the sample-specific nature of the correlation coefficient makes such differences difficult to interpret. Also, we must be concerned with the meaningfulness of our variables. Interpreting a regression slope that is expressed in the units of X and Y forces us to consider what one-unit changes in X and Y actually mean. We may be less likely to consider our variables carefully if we use a dimensionless correlational measure. If we are interested in understanding the relation between two variables, it is a good idea to plot the scatter diagram and to consider the regression statistics in addition to the correlation coefficient.

EXERCISES

14.1 The following terms provide a useful review of many concepts in the chapter. Define, describe, or identify each of them:

sample-specific measure	partial correlation coefficient
linear transformation	semipartial correlation coefficient
homoscedasticity	Kendall tau coefficient
restriction of range problem	dichotomous data
reliability coefficient for X	point-biserial correlation coefficient
bivariate normal distribution	phi coefficient
Fisher Z transform	biserial correlation coefficient
correlation matrix	tetrachoric correlation coefficient

14.2 Criticize or comment on the following procedures and conclusions in the light of what you know about correlation and regression.

(a) A school committee member presents evidence that she claims demonstrates the failure of special enriched preschool programs for children from economically deprived homes. She states that a study of 5000 economically deprived children showed that, despite three years of enriched programs, the IQs of these children correlated very highly (.85) with their IQs *before* they entered the program.

(b) A clinical psychologist reads a description of a study in which a correlation of $-.8$ was obtained between a measure of anxiety and a measure of emotional stability. He decides to attempt to verify the result. He administers the same measures of anxiety and emotional stability to a random sample of patients in a VA hospital. The obtained correlation of $-.2$ between measures is not significant. He concludes that he has no evidence of any relationship (at least any linear relationship) between anxiety and emotional stability.

(c) A sociologist is interested in looking at the extent to which recent graduates of a training program for the hard-core unemployed are perceived as being useful members of society once they have secured jobs. The indices he uses are

(i) the individuals' self-image scores as measured by a test (X).

(ii) the sociologist's own estimate of how useful they are (Y).

(iii) estimates solicited from "authority figures" such as social workers and policemen (W).

He finds that these measures are uncorrelated; that is, $r_{XW} = r_{YW} = r_{XY} = 0$. He further finds that the discrepancies between the self-image scores and his estimates $(z_X - z_Y)$ correlates .5 with the discrepancies between authority figures and his estimates $(z_W - z_Y)$, and he writes an article about how, in our society, "self-distortion" is positively correlated with "authority distortion."

(d) It is reported in the press that getting a degree from a four-year college is worth several hundred thousand dollars in lifetime earnings.

14.3   Given the following results from two items on a test taken by 100 students:

	Item 1		
	Pass	Fail	
Pass			40
Fail			60
	70	30	

Item 2

(a) What are the maximum positive and negative values of ϕ?

(b) Under what conditions would performance on items 1 and 2 be uncorrelated (i.e., what would the cell values be)?

14.4   (a) What percentage of the variance of Y can be accounted for by X if the correlation between z_X and $z_X - z_Y$ is .8?

(b) Show that if $\hat{\sigma}_X = \hat{\sigma}_Y$, then the z scores of $X + Y$ and $z_X + z_Y$ are equal.

14.5   Sometimes we will find that a correlation has been computed between some variable X and another variable T, which is the sum of a number of variables, including X (e.g., $T = X + Y$). Under these circumstances, we can expect a positive correlation between X and T even if X is not related to Y because X is part of T. Show that, in general,

$$r_{X,T-X} = \frac{r_{XT}\hat{\sigma}_T - \hat{\sigma}_X}{\sqrt{\hat{\sigma}_X^2 + \hat{\sigma}_T^2 - 2r_{XT}\hat{\sigma}_X\hat{\sigma}_T}}$$

where $r_{X,T-X}$ is the correlation between X and the part of T not containing X.

14.6   Note that the previous question has implications for the interpretation of correlations involving change or difference scores. Suppose that X refers to pretest scores and T to posttest scores. Therefore, $T - X$ refers to change scores. If we assume that $\hat{\sigma}_{pre} = \hat{\sigma}_{post}$, the equation in Exercise 14.5 reduces to

$$r_{X,T-X} = r_{pre,change} = \frac{r_{pre,post} - 1}{\sqrt{2(1 - r_{pre,post})}}$$

We would not expect a perfect correlation between pre- and posttest scores for a lot of reasons, including random error. Suppose $r_{pre,post} = .70$. What do we expect for $r_{pre,change}$?

14.7   A researcher discovers that by administering only every other item in a standard IQ test, she obtains scores that correlate .80 with the scores obtained by administering the whole test. She states that, given this result, it is only reasonable to conclude that the half of the test she gave must be correlated highly with the other half. If we assume that the variance of the half of the test that was administered is equal to the variance of the other half, how highly do the two halves of the test correlate?

14.8   We know that the relation between X and Y is linear and that b_1 and $\hat{\sigma}_e$ are the same for an "unrestricted" sample randomly selected from the whole population and a restricted sample chosen such that $\hat{\sigma}_X = 20$. If $r_{XY} = .25$ in the restricted sample, what is our best estimate for the correlation in the unrestricted sample for which $\hat{\sigma}_X$ is known to be 60?

14.9 The correlation between two test scores X and Y is known to be .40. The reliabilities of X and Y are known to be .75 and .60. What is our best estimate of what the correlation coefficient would be if the reliabilities were both 1.00?

14.10 Given the linear regression equation

$$\hat{Y}_i = \bar{Y} + b_1 (X_i - \bar{X}) \qquad \text{where } b_1 = \frac{r_{XY}\hat{\sigma}_Y}{\hat{\sigma}_X}$$

If $r_{XY} = .6$, find the correlations between
(a) Y and $\hat{Y}$
(b) Y and $Y - \hat{Y}$
(c) $\hat{Y}$ and $Y - \hat{Y}$

14.11 It is found in an introductory calculus course with 40 students that scores on the final examination correlate $-.35$ with the number of hours studied.
(a) Is the correlation significantly different from 0 at $\alpha = .05$?
(b) Can we conclude that studying too much interferes with test taking?
(c) Each year, a random sample of 200 freshmen admitted to the Elite Institute of Technology must take a standardized skills test when they first enroll. Two years ago, the correlation between the test and first year GPA was .22. Last year, after the test had been revised, the correlation rose to .35. If the two entering classes can be considered random samples of EIT freshmen, do the two correlations differ significantly at $\alpha = .05$?

14.12 Given the following correlation matrix, $\mathbf{R}$, in which each correlation is based on 20 data points:

$$\mathbf{R} = \begin{bmatrix} 1.00 & .20 & .30 & .40 \\ .20 & 1.00 & .50 & .60 \\ .30 & .50 & 1.00 & .70 \\ .40 & .60 & .70 & 1.00 \end{bmatrix}$$

If r_{ij} refers to the correlation in the ith row and jth column of the matrix,
(a) test the hypothesis that all of the off-diagonal population correlations are 0; use $\alpha = .05$.
(b) find $r_{23|4}$ and test $H_0: \rho_{23|4} = 0$.
(c) find $r_{23|41}$ and test $H_0: \rho_{23|41} = .20$.
(d) test $H_0: \rho_{12} = \rho_{14}$.

14.13 Find the Spearman rank correlation coefficient for the data in Table 12.1. Note that there are ties, so you cannot use the expression given in Equation 14.16.

14.14 If W is positively correlated with X but uncorrelated with Y, what can we say about the relation between $r_{XY|W}$ and r_{XY}?

14.15 One hundred people suffering from a serious disease receive a new treatment while another 100 people do not receive the treatment. Suppose that outcomes are as expressed in the following 2×2 table:

| | Survive | | |
	Yes	No	
Treatment — Yes	60	40	100
Treatment — No	40	60	100
	100	100	

(a) What is the correlation between survival and treatment?

(b) Discuss whether the size of the relation between survival and treatment is large. Consider both the proportion of variance accounted for and the difference in survival rates for the treatment and the no-treatment groups.

14.16 Suppose we have three dichotomous variables: size (*S*), with levels big and small; verbal ability (*V*), with levels high and low; and Age (*A*), with levels young and old. Further suppose that we have 100 young and 100 old people and that for young and old people, considered separately, size and verbal ability are uncorrelated. However, there are more big people in the old group and more high verbal people in the old group because age is correlated with both size and verbal ability. We have, for example:

(1)   for the older group

	S — Big	S — Small	
V — High	81	9	90
V — Low	9	1	10
	90	10	

so that $\phi_{SV} = 0$

(2)   for the younger group

	S — Big	S — Small	
V — High	1	9	10
V — Low	9	81	90
	10	90	

so that $\phi_{SV} = 0$

From tables 1 and 2, we have

	A — Old	A — Young	
V — High	90	10	100
V — Low	10	90	100
	100	100	

so $\phi_{AV} = .8$

and

	A		
	Old	Young	
High	90	10	100
Low	10	90	100
	100	100	

S labels the rows (High, Low).

so $\phi_{AS} = .8$

If you were to collapse tables 1 and 2 into a single table without taking account of age, you would wind up with

	S		
	Big	Small	
High	82	18	100
Low	18	82	100
	100	100	

V labels the rows (High, Low).

so that $\phi_{SV} = .64$

(a) Explain how it can be that the correlations between S and V are 0 for the young and old groups taken separately but nonzero when the two groups are combined.

(b) Find the partial correlation between S and V with A partialed out.

(c) By writing down 2 × 2 tables (gender, with levels female and male, versus admissions outcome, with levels accept and reject) for two hypothetical departments in a university, illustrate how it might be possible for there to be a higher proportion of acceptances for females than males in each department but a higher proportion of acceptances for males than females in the university taken as a whole.

14.17 What constraints are placed on the possible values of r_{YW} if we know that $r_{XY} = .9$ and $r_{XW} = .6$?

14.18 One hundred patients are seen independently by two clinicians, after which each clinician recommends whether each patient should be hospitalized. The clinicians' recommendations are summarized in the following 2 × 2 table:

		Clinician A's recommendations		
		Hospitalize	Don't hospitalize	
Clinician B's recommendations	Hospitalize	1	9	10
	Don't hospitalize	9	81	90
		10	90	

(a) For what percentage of the patients do the recommendations of the two clinicians agree?

(b) Does this high percentage of agreement give you confidence that the clinicians are accurately evaluating the patients? Discuss why or why not.

Chapter 15

Multiple Regression

15.1 INTRODUCTION

In Chapter 12, we considered linear regression with a single predictor variable. Our concern was with estimating the parameters of the linear model

$$Y_i = \beta_0 + \beta_1 X_i + \epsilon_i$$

Estimates of β_0 and β_1 were obtained using the least-squares criterion; that is, b_0 and b_1 were obtained such that the prediction equation

$$\hat{Y}_i = b_0 + b_1 X_i$$

minimized MSE $= (1/N)\Sigma_i(Y_i - \hat{Y}_i)^2$, the measure of prediction error.

In Chapter 15, we extend this discussion to multiple linear regression. We consider models in which the criterion variable, Y, is expressed as a linear function of a number of predictor variables, $X_1, X_2, \dots, X_p$. Although the additional predictor variables result in some increase in complexity, the basic concepts underlying bivariate and multiple regression are the same.

If there are p predictor variables, the relevant data can be expressed as a $N \times (p + 1)$ array (see Table 15.1). As in bivariate regression, we can obtain the least-squares estimate of the linear model

$$Y_i = \beta_0 + \beta_1 X_{i1} + \beta_2 X_{i2} + \cdots + \beta_p X_{ip} + \epsilon_i$$

TABLE 15.1 DATA FOR A MULTIPLE REGRESSION WITH CRITERION VARIABLE Y AND PREDICTOR VARIABLES $X_1, X_2, \dots, X_p$

			Variable	
Case	Y	X_1	$X_2 \cdots$	X_p
1	Y_1	X_{11}	$X_{12} \cdots$	X_{1p}
2	Y_2	X_{21}	$X_{22} \cdots$	X_{2p}
3	Y_3	X_{31}	$X_{32} \cdots$	X_{3p}
.				
.				
.				
N	Y_N	X_{N1}	$X_{N2} \cdots$	X_{Np}

by finding $b_0, b_1, b_2, \ldots, b_p$ such that the linear regression equation

$$\hat{Y}_i = b_0 + b_1 X_{i1} + b_2 X_{i2} + \cdots + b_p X_{ip}$$

minimizes MSE $= (1/N)\Sigma_i(Y_i - \hat{Y}_i)^2$ for the N data points in our sample.

15.2 A REGRESSION EXAMPLE WITH TWO PREDICTOR VARIABLES

Table 15.2 provides some data for 25 students who took an undergraduate statistics course. The table presents the final exam score (Y), the score on a math skills pretest that was administered on the first day of class (X_1), and the quantitative SAT score (X_2) for each student, as well as some summary statistics and the correlations between each pair of variables. Because the pretest and final exam scores are highly correlated ($r = .686$), we can use the pretest score to predict final exam performance. If we regress the final exam scores on the pretest scores, using the procedures for bivariate regression discussed in Chapter 12, the resulting regression equation is

$$\hat{Y} = -6.58 + 2.44 X_1$$

where the slope of 2.44 indicates that for each increase of one point in the pretest score, our best prediction of the final exam score increases by 2.44 points.

However, we might also consider using the SAT score information to pedict performance on the final exam. Even though the SAT and pretest scores are quite highly correlated ($r = .533$), the combination of pretest and SAT scores might provide some predictability over and above that of the pretest scores alone. Table 15.3 contains the SYSTAT output for the regression of final exam scores on both pretest and SAT scores.

The *coefficient* column of the output tells us that the best regression equation that includes both pretest and SAT scores as predictors is

$$\hat{Y} = -57.42 + 1.72 X_1 + 0.12 X_2$$

This indicates that if X_2 is held constant, a one-unit change in X_1 results in a change of 1.72 units in $\hat{Y}$. Similarly, if X_1 is held constant, a one-unit change in X_2 results in a change of 0.12 unit in $\hat{Y}$.

The output also contains other useful information about the regression. We will use the remainder of this section to introduce each piece of information in the output. More detailed discussions will be provided in the following sections.

The *multiple R* or multiple correlation coefficient is .758. This means that the final exam scores predicted by the regression equation, $\hat{Y}$, have a correlation of .758 with the actual scores, Y, obtained by the students in the class. The squared multiple correlation coefficient is $.758^2 = .574$. Therefore, we can conclude that 57.4% of the variance in Y is accounted for by the regression on X_1 and X_2.

The *adjusted squared multiple R* is .536. As we will see in Section 15.4.4, the sample multiple correlation coefficient is a positively biased estimator of the population coefficient because the regression equation obtained from the sample fits the sample better than it fits the population. The adjusted R results from an attempt to remove the positive bias.

TABLE 15.2   DATA AND SUMMARY STATISTICS FOR 25 STUDENTS

		FINAL	PRETEST	SAT
CASE	1	55.	23.	610.
CASE	2	97.	38.	680.
CASE	3	75.	28.	710.
CASE	4	83.	36.	630.
CASE	5	79.	28.	600.
CASE	6	80.	30.	690.
CASE	7	42.	25.	510.
CASE	8	45.	27.	590.
CASE	9	65.	21.	520.
CASE	10	68.	29.	510.
CASE	11	78.	28.	690.
CASE	12	74.	34.	640.
CASE	13	43.	27.	550.
CASE	14	67.	33.	610.
CASE	15	65.	29.	580.
CASE	16	72.	36.	600.
CASE	17	99.	37.	620.
CASE	18	53.	22.	570.
CASE	19	89.	34.	640.
CASE	20	62.	32.	660.
CASE	21	50.	33.	610.
CASE	22	64.	30.	610.
CASE	23	25.	21.	530.
CASE	24	50.	23.	540.
CASE	25	52.	32.	570.

PEARSON CORRELATION MATRIX

	FINAL	PRETEST	SAT
FINAL	1.000		
PRETEST	0.686	1.000	
SAT	0.638	0.533	1.000

NUMBER OF OBSERVATIONS: 25

	FINAL	PRETEST	SAT
N OF CASES	25	25	25
MINIMUM	25.000	21.000	510.000
MAXIMUM	99.000	38.000	710.000
MEAN	65.280	29.440	602.800
STANDARD DEV	18.024	5.067	57.338

The *standard error of estimate* is 12.282. This provides a measure of how well the regression equation predicts the final exam scores. The equation provides a prediction, $\hat{Y}$, for each combination of X_1 and X_2. The standard error of estimate is obtained by summing the squared deviations of the actual final exam scores from the predicted

TABLE 15.3   SYSTAT OUTPUT FOR THE REGRESSION OF FINAL SCORE ON PRETEST AND SAT SCORE FOR THE DATA IN TABLE 15.2

DEP VAR:	FINAL	N:	25	MULTIPLE R:	.758	SQUARED MULTIPLE R:	.574

ADJUSTED SQUARED MULTIPLE R: .536   STANDARD ERROR OF ESTIMATE: 12.282

VARIABLE	COEFFICIENT	STD ERROR	STD COEF	TOLERANCE	T	P(2 TAIL)
CONSTANT	-57.423	26.478	0.000	.	-2.169	0.041
PRETEST	1.719	0.585	0.483	0.7154841	2.938	0.008
SAT	0.120	0.052	0.381	0.7154841	2.314	0.030

ANALYSIS OF VARIANCE

SOURCE	SUM-OF-SQUARES	DF	MEAN-SQUARE	F-RATIO	P
REGRESSION	4478.446	2	2239.22	314.845	0.000
RESIDUAL	3318.594	22	150.845		

scores, yielding $SS_{residual}$ (or SS_{error}) = $\Sigma(Y - \hat{Y})^2$, dividing this quantity by its df, and then taking the square root. If the underlying model for the population is $Y = \beta_0 + \beta_1 X_1 + \beta_2 X_2 + \epsilon$, then the standard error of estimate provides an estimate of the standard deviation of ϵ.

The *analysis of variance* table at the bottom of the output indicates that the total SS of 7797.040 associated with final exam scores can be partitioned into two components, one the SS accounted for by the regression and the other the SS left unaccounted for, with

$$SS_{reg} = \Sigma(\hat{Y} - \bar{Y})^2$$
$$= R^2 SS_Y$$
$$= 4478.446$$

and

$$SS_{residual} = \Sigma(Y - \hat{Y})^2$$
$$= (1 - R^2)SS_Y$$
$$= 3318.594$$

The F formed by taking the ratio of the regression and residual mean squares is highly significant. The null hypothesis rejected by this test states that the regression coefficients β_1 and β_2 are both zero or, equivalently, that the population multiple correlation coefficient is zero.

Just above the ANOVA table, we have the information about the constant (b_0), the coefficients of X_1 and X_2 in the regression equation (b_1 and b_2), and their standard errors. The STD COEF (standardized coefficients) column contains the values of the regression coefficients that would result if the regression was performed by using z scores. The use of standardized coefficients facilitates comparisons of the effects of different predictors by putting them on a common scale. However, as we point out in Section 15.7.4, standardized coefficients are sample-specific and should not be used to generalize across situations.

The *tolerance* of each predictor variable is a measure of how *nonredundant* that variable is with the other predictor variables in the equation; with only two predictor variables, the tolerance is 1 minus the square of the correlation between the two predictors. A tolerance of zero indicates that the variable in question is totally redundant with the other predictor variables (i.e., it can be expressed as a linear combination of them). Low tolerance values for predictor variables result in dramatically inflated standard errors for the regression coefficients (see Section 15.7.2).

Finally, for each coefficient, a t statistic is formed by dividing b by its standard error. This tests the hypothesis that the corresponding $\beta = 0$. The t of 2.938 and two-tailed significance level of .008 for pretest score indicates that when SAT score is held constant, the rate of change of Y with pretest score is significant. The t of 2.314 and p of .030 for SAT score indicate that there is a significant rate of change of Y with SAT score when pretest score is held constant, indicating a significant contribution of SAT score to the predictability of final performance over and above that provided by the pretest score. The significant t for b_0 indicates that we can reject the null hypothesis that $\beta_0 = 0$ in the population.

15.3 THE NATURE OF THE REGRESSION COEFFICIENTS

When a regression equation does a good job predicting, it is tempting to use it not only to predict but also as an explanatory model, or at least to think of the regression coefficients as measures of the importance of the corresponding X's in influencing Y. Although we discuss the interpretation of regression coefficients in more detail later (Section 15.7), it is important to point out some cautions about yielding to the temptation without a good deal of thought. One thing we must keep in mind is that regression equations deal with predictability, not causal influence. The number of books that parents own may be a perfectly good predictor of children's performance in elementary school even if the children don't read the books; that is, the books themselves may not directly influence school performance, but they may be correlated with factors that do.

Also, the size of a regression coefficient generally depends both on what other variables are contained in the equation and how well the equation matches the population model. In the current example, the coefficient of pretest score was 2.44 when pretest score was the only predictor in the regression equation but only 1.72 when SAT score was also included in the equation. To see why this might be the case, consider the following situation.

Suppose we have a population described by the model

$$Y = \beta_0 + \beta_1 X_1 + \epsilon \tag{15.1}$$

If we ignore the random error component, Y is a linear function of X_1 with slope β_1, where β_1 is the rate of change of Y with X_1; a one-unit change in X_1 corresponds to a change of β_1 in Y. If, however, the population model is

$$Y = \beta_0 + \beta_1 X_1 + \beta_2 X_2 + \epsilon \tag{15.2}$$

then if we ignore the random error component, β_1 is the rate of change of Y with X_1 *given that X_2 is held constant* (i.e., if X_1 is changed by one unit and X_2 is not changed, then Y will change by β_1 units) and β_2 is the rate of change of Y with X_2 given that X_1

is held constant. Note that β_1 in Equation 15.2 does not have the same interpretation as β_1 in Equation 15.1. In order to emphasize this distinction, some authors would write the coefficient of X_1 in Equation 15.2 as $\beta_{Y1.2}$, to indicate that X_2 was also in the equation, and would refer to it as a "partial regression coefficient."

Although the interpretation of the β's in the preceding linear models is quite simple, the interpretation of sample regression coefficients is less straightforward because we usually do not know what underlying model truly describes the population. If we select a sample from the population described by Equation 15.1 and regress Y on X_1, we can show that the regression coefficient, b_1, is an unbiased estimator of the population parameter, β_1, of Equation 15.1. Here, b_1 not only represents the rate of change of the predicted score $\hat{Y}$ with X_1 in the regression equation, it also estimates the rate of change of the actual score, Y, with X_1 in the population.

However, if we select a sample from the population described by Equation 15.2 and regress Y only on X_1, the sample regression coefficient will be a biased estimator of β_1 in Equation 15.2. Although in this case b_1 still represents the rate of change of $\hat{Y}$ with X_1, it is not necessarily a good estimator of how Y changes with X_1 in the population *if X_2 is held constant* because, as we show in Section 15.7.1, if X_2 is left out of the regression equation b_1 will generally reflect the effects of both X_1 and X_2.

In terms of our class example, suppose for the moment that (1) the pretest measured mechanical algebra skills, (2) the SAT measured abstract mathematical thinking skills, (3) people with better mechanical skills also tended to have better abstract thinking skills, and (4) that performance on the final depended on both types of skills. If we regressed final score only on pretest score, we would be mistaken if we interpreted the regression coefficient solely as a measure of the importance of mechanical skills in determining the grade on the final. The change in the predicted final exam score associated with a one-unit difference in the pretest score would reflect both the difference in mechanical skills and the associated difference in abstract skills. However, if we regressed final exam score on both pretest and SAT score the coefficient of pretest would no longer reflect the importance of abstract thinking skills. In this case, the coefficient of pretest score would represent the rate of change of the predicted score on the final with pretest score, *with SAT score held constant.*

15.4 THE MULTIPLE CORRELATION COEFFICIENT AND THE PARTITIONING OF VARIABILITY IN MULTIPLE REGRESSION

15.4.1 The Multiple Correlation Coefficient

In Chapter 12, we defined the correlation coefficient r as a measure of the linear relation between Y and X, and we introduced the coefficient of determination r^2 as the proportion of the variability in one of the variables accounted for by the regression on the other. Both of these concepts can be extended to deal with the relation between a criterion variable Y and a collection of predictors $X_1, X_2, \ldots, X_p$.

We define the multiple correlation coefficient, $R_{Y.12\ldots p}$, as $r_{Y\hat{Y}}$, the correlation between Y and $\hat{Y}$, where

$$\hat{Y} = b_0 + b_1 X_1 + b_2 X_2 + \cdots + b_p X_p$$

is the prediction of Y obtained from the multiple regression equation that contains the p predictors. If Y is perfectly predicted by the multiple regression equation, then $R = 1$. If the multiple regression equation predicts no better than the equation $\hat{Y} = \bar{Y}$, then $R = 0$. When there is a single predictor variable X, the multiple correlation coefficient reduces to $R_{Y.X} = |r_{YX}|$, the absolute value of the bivariate correlation coefficient.

The proportion of the variability in Y accounted for by the regression on p predictor variables is $r_{Y\hat{Y}}^2 = R_{Y.12 \ldots p}^2$. Therefore, we can write

$$R_{Y.12 \ldots p}^2 = \frac{SS_{reg}}{SS_Y}$$

where $SS_{reg} = \Sigma_i (\hat{Y}_i - \bar{Y})^2$ is the amount of variability in Y accounted for by the regression.

15.4.2 Partitioning SS$_Y$ into SS$_{reg}$ and SS$_{error}$

As was the case with bivariate regression, the variability of Y can be partitioned into a component accounted for by the regression, SS_{reg}, and a component not accounted for by the regression, SS_{error} or $SS_{residual}$,

$$\Sigma(Y_i - \bar{Y})^2 = \Sigma(\hat{Y}_i - \bar{Y})^2 + \Sigma(Y_i - \hat{Y}_i)^2$$

$$SS_Y \quad = \quad SS_{reg} \quad + \quad SS_{error}$$

where $SS_{reg} = R^2 SS_Y$ and $SS_{error} = (1 - R^2)SS_Y$. It is convenient to express the partitioning of variability in terms of an ANOVA table of the form

SV	df	SS	MS
Regression	p	$\sum_i (\hat{Y}_i - \bar{Y})^2$ or $R_{Y.12 \ldots p}^2 SS_Y$	$\dfrac{R_{Y.12 \ldots p}^2 SS_Y}{p}$
Error (or residual)	$N - 1 - p$	$\sum_i (Y_i - \hat{Y}_i)^2$ or $(1 - R_{Y.12 \ldots p}^2)SS_Y$	$\dfrac{(1 - R_{Y.12 \ldots p}^2)SS_Y}{N - 1 - p}$
Total	$N - 1$	$SS_Y = \Sigma(Y_i - \bar{Y})^2$	

SS_Y is associated with $N - 1$ df because 1 df is used to estimate the population mean. Of these $N - 1$ df, p are associated with the regression SS because coefficients for each of the p predictors must be estimated. The remaining $N - 1 - p$ df are associated with the error SS.

Under standard assumptions that will be discussed in Section 15.5, if the p regression coefficients $\beta_1, \beta_2, \ldots, \beta_p$ are all 0 in the population, the ratio

$$\frac{MS_{reg}}{MS_{error}} = \frac{R^2 SS_Y/p}{(1 - R^2)SS_Y/(N - 1 - p)} \tag{15.3}$$

$$= \frac{R^2/p}{(1 - R^2)/(N - 1 - p)}$$

will be distributed as F with p and $N - 1 - p$ df. Therefore, the ratio of mean squares can be used to test the null hypothesis that $\beta_1 = \beta_2 = \cdots \beta_p = 0$. In the class example, when we regressed final exam score on pretest and SAT score (so that $p = 2$), we found that $R = .758$. To the nearest integer, SS for final exam score is 7797. Therefore, SS_{reg} and SS_{error} should be approximately $(.758^2)(7797) = 4480$ and $(1 - .758^2)(7797) = 3317$, respectively. These values are all the same, within rounding error, as those given in the SYSTAT output in Table 15.3. Hence, from Equation 15.3,

$$\frac{MS_{reg}}{MS_{error}} = \frac{4480/2}{3317/(25 - 1 - 2)}$$

$$= 14.85$$

From Table D.5, $F_{.05,2,22}$ is approximately equal to 3.44; therefore we can reject the hypothesis that the population regression coefficients for both pretest and SAT score are zero.

MS_{error} is the square of the standard error of estimate provided in the SYSTAT output. If all important systematic sources of variability are included in the regression equation so that the residual variability is due only to random error, MS_{error} estimates the random error variance, σ_e^2. If important sources of variability are omitted from the equation, MS_{error} will reflect these sources as well as random error, resulting in a negatively biased F test.

15.4.3 Partitioning SS$_{reg}$

If the p predictor variables in a multiple regression are mutually uncorrelated, SS_{reg} can be partitioned into nonoverlapping components associated with each of the predictors (see Figure 15.1). That is

$$SS_{reg} = SS_{Y.1} + SS_{Y.2} + \cdots + SS_{Y.p}$$

where $SS_{Y.j} = r_{Yj}^2 SS_Y$ is the amount of variability in Y accounted for by the predictor X_j and r_{Yj} is the correlation between X_j and Y. Because $SS_{reg} = R_{Y.12\ldots p}^2 SS_Y$, it follows that for uncorrelated predictor variables,

$$R_{Y.12\ldots p}^2 = r_{Y1}^2 + r_{Y2}^2 + \cdots + r_{Yp}^2 \tag{15.4}$$

$$= \sum_j r_{Yj}^2$$

However, predictor variables are usually correlated with one another, and, if this is the case, their variabilities will overlap. The proportion of variability in Y accounted for by a set of correlated predictor variables is not the sum of the proportions associated with the individual predictors but must be adjusted for overlapping variability. In the current example, there is a considerable correlation (.533) between pretest and SAT scores.

Figure 15.1 is an attempt to represent the situation graphically. The circles represent the variabilities of Y and several predictors. Overlapping circles represent shared variability, indicating that the corresponding variables covary (i.e., are correlated). In panel a the variabilities of the predictors overlap Y but not one another, suggesting that the proportions of the variability of Y accounted for are additive. That is, the total variability in Y accounted for collectively by the predictors is the sum of the variabilities accounted for by the individual predictors. In panel b the predictor circles overlap, indicating that the total variability in Y accounted for is not the sum of the variabilities accounted for by the individual predictors; rather, the sum of the variabilities must be adjusted to take account of the correlations between predictors.

A general expression for R^2 that takes the correlations between predictors into account is given by

$$R^2_{Y.12\cdots p} = \frac{\Sigma_j r_{Yj} b_j \hat{\sigma}_j}{\hat{\sigma}_Y} \tag{15.5}$$

(a)

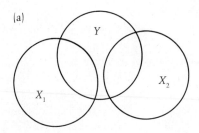

(b)

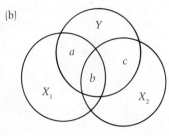

$R^2_{Y.1} = a + b$
$R^2_{Y.2} = b + c$ $r^2_{Y(2|1)} = c$
$R^2_{Y.12} = a + b + c$ $r^2_{Y(1|2)} = a$

Figure 15.1 Representation of variability in the criterion variable accounted for by uncorrelated and by correlated predictor variables. (a) Uncorrelated predictors: Variabilities accounted for by X_1 and X_2 do not overlap so that $R^2_{Y.12} = r^2_{Y1} + r^2_{Y2}$. (b) Correlated predictors: Variabilities accounted for by X_1 and X_2 overlap so that $R^2_{Y.12}$ is not the sum of the r^2_{Yj}. R^2 can be found from $R^2_{Y.12} = \Sigma_j b_j r_{Yj} \hat{\sigma}_j / \hat{\sigma}_Y$.

where b_j is the regression coefficient of X_j in the multiple regression equation and $\hat{\sigma}_j$ and $\hat{\sigma}_Y$ are the standard deviations of X_j and Y, respectively. For example, in the class example, final exam score (Y) was regressed on pretest score (X_1) and SAT score (X_2). We obtained

	Final	Pretest	SAT
b_j		1.719	.120
r_{Yj}		.686	.638
$\hat{\sigma}$	18.024	5.067	57.338

so that from Equation 15.5,

$$R^2 = \frac{(.686)(1.719)(5.067) + (.638)(.120)(57.338)}{18.024}$$

$$= .575$$

This value of R^2 is the same, within rounding error, as the value calculated by SYSTAT and displayed in Table 15.2.

The increase in R^2 when X_2 is added to a regression equation that already contains X_1 is $r^2_{Y(2|1)}$, the square of the semipartial correlation coefficient introduced in Chapter 14. In terms of panel b of Figure 15.1, we may think of $r^2_{Y(2|1)}$ as the proportion of the Y circle that overlaps with X_2 and does not overlap with X_1. The semipartial coefficient $r_{Y(2|1)}$ is the correlation of Y with $X_2 | X_1$, where the latter term represents what remains of X_2 after the effects of X_1 have been "partialed out"; that is, the residual that results when X_2 is regressed on X_1. Information about semipartials is directly provided by a number of statistical packages, although it is not contained in the SYSTAT output.

Using Equation 14.15, we find $r_{final(SAT|pretest)} = .322$. We know from Table 15.1 that $r_{final,pretest} = .686$, so

$$R^2_{final.pretest,SAT} = .686^2 + .322^2 = .574$$

In general, the squared semipartial coefficient $r^2_{Y(p+1|12\,\cdots\,p)}$ is the increase in R^2 that follows from adding X_{p+1} to a regression equation that already contains p predictors. That is,

$$R^2_{Y.12\,\cdots\,p+1} = R^2_{Y.12\,\cdots\,p} + r^2_{Y(p+1|12\,\cdots\,p)} \tag{15.6}$$

so that

$$r^2_{Y(p+1|12\,\cdots\,p)} = R^2_{Y.12\,\cdots\,p+1} - R^2_{Y.12\,\cdots\,p} \tag{15.6$'$}$$

where $r_{Y(p+1|12\,\cdots\,p)}$ is the correlation between Y and $X_{p+1}|X_1, X_2, \ldots, X_p$, and the latter term represents the residuals of the regression of X_{p+1} on $X_1, X_2, \ldots, X_p$.

Table 15.4 presents the amount of final exam score variability accounted by pretest and SAT score individually (3670.76 and 3176.03) and together (4478.45). The table also presents the increments in variability accounted for when SAT score is added to a

TABLE 15.4   VARIABILITY IN THE FINAL EXAM SCORE ACCOUNTED FOR BY DIFFERENT SV's

SV	df	SS
Pretest, SAT	2	$SS_{Y.12} = R^2_{Y.12}SS_Y = (.758)^2(7797.04)$ $= 4478.45$
Pretest	1	$SS_{Y.1} = r^2_{Y1}SS_Y = (.686)^2(7797.04)$ $= 3670.76$
SAT|pretest	1	$SS_{Y.2\|1} = (R^2_{Y.12} - r^2_{Y1})SS_Y$ $= (.758^2 - .686^2)(7797.04)$ $= 807.69$
SAT	1	$SS_{Y.2} = r^2_{Y2}SS_Y = (.638^2)(7797.04) = 3176.03$
Pretest|SAT	1	$SS_{Y.1\|2} = (R^2_{Y.12} - r^2_{Y2})SS_Y$ $= (.758^2 - .638^2)(7797.04)$ $= 1302.42$
Error	$N - 1 - 2 = 22$	$SS_{error} = (1 - R^2_{Y.12})SS_Y$ $= SS_Y - SS_{Y.12} = 7797.04 - 4478.446$ $= 3318.01$

regression equation that already contains pretest score (807.69) and when pretest score is added to an equation that contains SAT score (1302.42). The importance of these partitionings is that they allow us to ask more than whether the two predictors together account for a significant amount of variance. We will soon discuss "partial F tests" that will allow us to test whether the addition of one or more variables significantly increases the variability accounted for by a regression equation.

15.4.4 Cross-Validation and the Adjusted (or Shrunken) Multiple Correlation Coefficient

When a multiple regression equation is developed from a sample of data, the multiple correlation coefficient R or its square is often used as an index of how well the equation fits the population from which the sample was obtained. However, using R or R^2 as a measure of fit can be misleading. Not only is R a sample-specific measure because it is particularly sensitive to the variances of Y and the predictor variables in the sample (see Section 14.1.4), the sample multiple correlation coefficient is a positively biased estimator of the population coefficient. The sample multiple correlation coefficient, R_{samp}, tends to be larger than the population coefficient, R_{pop}, because the regression equation obtained from the sample fits the sample better than it fits the population.

Bias occurs because the coefficients obtained in a regression analysis are chosen in such a way as to maximize R for the sample. With enough predictors, the regression equation has to fit the sample no matter how the predictors and the criterion are related in the population. Just as any two data points can be fit by a straight line, any $p + 1$ data points can be fit perfectly by the p-dimensional surface that corresponds to a regression equation with p predictor variables. With p predictors, unless there are more than $p + 1$ data points, there is no error variability and R_{samp} must equal 1 even if $R_{pop} = 0$. With more data points, R_{samp} need not be 1 but will tend to be larger than R_{pop} as long as the N/p ratio (number of cases divided by number of predictor variables) is small because R "capitalizes on chance"; that is, it takes advantage of chance fluctuations in scores that allow for predictability in the sample but not in the population.

The bias in R can be reduced by working with larger samples. How large should samples be? Although the recommended sample size depends to some extent on the nature of the research problem and the purpose of the analysis, the N/p ratio should be large—perhaps 30 or more—if the size of R is to be taken very seriously.

One way of obtaining a more realistic estimate of the population R is to employ a procedure called *cross-validation*, which avoids capitalizing on chance by using two samples. One of the samples (the screening sample) is used to develop a regression equation. The regression equation developed in the screening sample is then used to predict Y scores for each case in a second (calibration) sample. The cross-validated R is the correlation between these predicted Y scores and the actual Y scores in the calibration sample. Because the regression weights are obtained from one sample and the value of R is obtained from a second sample, the cross-validated R does not systematically capitalize on sampling variability.

Another adjustment for the positive bias is based on a somewhat different formulation of R provided by Wherry (1931). A standard definition of the population correlation ρ is given by

$$\rho_{XY}^2 = 1 - \frac{\sigma_e^2}{\sigma_Y^2}$$

If we replace the population variances by their unbiased estimates, we have

$$R_{adj}^2 = 1 - \frac{SS_{error}/(N - 1 - p)}{SS_Y/(N - 1)}$$

Because $R^2 = SS_{reg}/SS_Y = 1 - SS_{error}/SS_Y$, we have

$$R_{adj}^2 = 1 - (1 - R^2)\left[\frac{N - 1}{N - 1 - p}\right] \tag{15.7}$$

The *adjusted* (or "shrunken") squared multiple correlation coefficient, R_{adj}^2, is provided by most of the standard statistical packages. For the current example, Equation 15.7 yields $R_{adj}^2 = 1 - (1 - .758^2)(24/22) = .536$, the value provided by the SYSTAT output in Table 15.3. Stevens (1986) offers a discussion of this statistic as well as a number of related measures. Unless there is a large number of cases per predictor variable, both the cross-validated R and the adjusted R will be smaller than the usual R and will provide a more realistic indication of how well the population is fit by the sample regression equation.

15.5 INFERENCE IN MULTIPLE REGRESSION

15.5.1 Models and Assumptions

As was the case for bivariate regression, we need to state a model and make certain assumptions about the data in order to make statistical inferences. Also, we again distinguish between situations in which the predictors are fixed-effect variables and situations in which the predictors are random variables. Fixed predictors generally occur in experimental studies in which the independent variables are manipulated; here Y is a random variable, but the values of the X's are selected by the researcher and are therefore considered to be fixed over replications of the experiment. Predictors are considered to be random variables when they, as well as Y, are randomly sampled. Fortunately, although somewhat different assumptions are made for fixed and random predictor variables, the procedures for testing hypotheses and forming confidence intervals are the same in both cases when the assumptions are met.

Whether X is fixed or random, we assume that the model is

$$Y = \beta_0 + \beta_1 X_1 + \beta_2 X_2 + \cdots + \beta_p X_p + \epsilon$$

$$= \mu_{Y.X_1 X_2 \cdots X_p} + \epsilon$$

where $\mu_{Y.X_1 X_2 \cdots X_p}$ is the mean of the population of Y scores corresponding to a particular set of values for the p predictor variables.

For the fixed-X situation, we assume that

1. None of the predictor variables is completely redundant; that is, no predictor variable X_p can be perfectly predicted by a linear equation that contains the other $p - 1$ predictors. If this condition is not satisfied, the set of normal equations that must be solved in order to obtain the sample regression coefficients will not have a unique solution (see Section 15.10).
2. The error components associated with each of the Y scores are normally and independently distributed with mean zero and variance σ_e^2.
3. The values of the predictor variables are fixed and measured without error. This means that X will be exactly the same for each replication of the experiment and can be treated as a constant.

For the observational situation, we assume 1 and 2 and further assume that the distributions of the predictor variables are independent of ϵ.

In the remainder of Section 15.5, we discuss and illustrate some types of statistical inferences that can be made in multiple regression. We will not derive expressions for standard errors in this section; rather, we state the results and concentrate on the logic and interpretation of the statistical tests. Also, we make only a few references to matrix expressions and, where possible, express results in terms of quantities that we have already encountered. However, the results can be obtained fairly easily by matrix algebra. For those readers who are willing to deal with some matrix algebra, we present a matrix approach to multiple regression in Section 15.10 and derive a number of basic results for inference with fixed-effect predictors. Derivations for random predictors are available in sources such as Graybill (1961).

15.5.2 Testing the Hypothesis $\beta_1 = \beta_2 = \cdots = \beta_p = 0$

As we indicated in Section 15.4.2, if the p regression coefficients $\beta_1, \beta_2, \ldots, \beta_p$ are all 0 in the population, the ratio

$$\frac{\text{MS}_{\text{reg}}}{\text{MS}_{\text{error}}} = \frac{R^2 \text{SS}_Y / p}{(1 - R^2) \text{SS}_Y / (N - 1 - p)}$$

$$= \frac{R^2 / p}{(1 - R^2) / (N - 1 - p)}$$

will be distributed as F with p and $N - 1 - p$ df under standard assumptions. Therefore, $\text{MS}_{\text{reg}} / \text{MS}_{\text{error}}$ can serve as the statistic to test the null hypothesis $\beta_1 = \beta_2 = \cdots = \beta_p = 0$. This test can be thought of as asking whether the model

$$Y = \beta_0 + \beta_1 X_1 + \beta_2 X_2 + \cdots + \beta_p X_p + \epsilon$$

accounts for Y in the population better than the restricted model

$$Y = \beta_0 + \epsilon$$

If the restricted model is appropriate, $\beta_1 = \beta_2 = \cdots = \beta_p = 0$, so the best predictor for Y is $\beta_0 = \mu_Y$, and $R_{\text{pop}} = 0$.

In the class example, when we regress final exam score on pretest and SAT score (so that $p = 2$), we find

$$\frac{\text{MS}_{\text{reg}}}{\text{MS}_{\text{error}}} = \frac{4478.446 / 2}{3318.594 / (25 - 1 - 2)}$$

$$= 14.845$$

and so we can reject the hypothesis that the population regression coefficients are zero for both pretest and SAT score. The test assumes that $(1 - R^2) \text{SS}_Y / (N - 1 - p)$ is an estimate of σ_ϵ^2, the variance of the random-error component. If important variables are left out of the regression equation, MS_{error} will reflect their effects as well as random error and the test will be negatively biased.

15.5.3 Testing the Hypothesis $\beta_j = 0$

Under standard assumptions, the ratio

$$\frac{b - \beta}{\hat{\text{SE}}(b)}$$

will be distributed as t with $N - 1 - p$ df. Therefore, if we can estimate the standard errors, we can test the hypothesis that the population intercept β_0 or any of the population regression coefficients β_j are equal to any constant. In practice, the null hypothesis $\beta_j = 0$ is usually tested. Rejection of this hypothesis implies that X_j makes a significant contribution to the predictability of Y when added to the other variables in the equation. In multiple regression, the estimated standard error is often written in terms of a matrix expression (see Section 15.10); however, we can express the estimated standard error of any regression coefficient b_j as

$$\hat{SE}(b_j) = \frac{\hat{\sigma}_e}{\sqrt{SS_j}} \sqrt{\frac{1}{1 - R_{j.}^2}} \qquad (15.8)$$

where

$$\hat{\sigma}_e = \sqrt{\frac{(1 - R_{Y.12\cdots p}^2)SS_Y}{N - 1 - p}}$$

and

$$SS_j = \sum_i (X_{ij} - \bar{X}_{.j})^2$$

and $R_{j.}$ is the multiple correlation coefficient that is obtained when X_j is regressed on the other $p - 1$ predictor variables in the equation. The quantity in the denominator of the second square-root expression, $1 - R_{j.}^2$, is referred to as the *tolerance* of X_j. If X_j has a tolerance of 0, it can be perfectly expressed as a linear combination of the other predictors in the regression equation. If any predictor variable has a tolerance of 0, we cannot obtain least-squares estimates of the regression coefficients (see Section 15.10); the set of equations that must be solved to find the b_j's does not have a unique solution. If the tolerance of X_j approaches 0, we can see from Equation 15.8 that the estimated SE will become very large and t will be small. This makes sense, because if X_j was redundant with the other predictors we would not expect it to provide a significant increase to the predictability of Y. The tolerance is usually provided as part of the output when a regression is performed. Many packages will allow you to set a minimum tolerance below which a predictor will not be added to the regression equation.

Also, once we have obtained the appropriate $\hat{SE}$'s, we can obtain the $100(1 - \alpha)\%$ confidence interval for each parameter by using

$$b \pm t_{\alpha/2}\hat{SE}(b)$$

In the class example, the 95% confidence intervals for β_0, β_1, and β_2 are -57.42 ± 54.92, 1.72 ± 1.21, and 0.12 ± 0.11, respectively.

15.5.4 Partial F Tests: Procedures for Testing a Subset of the β_j

We can use partial F tests to determine whether adding one or more predictors to a regression equation that already contains p predictors significantly increases the predictability of Y. If we consider just one additional predictor, X_{p+1}, a test of the model

$$Y = \beta_0 + \beta_1 X_1 + \beta_2 X_2 + \cdots + \beta_p X_p + \beta_{p+1} X_{p+1} + \epsilon$$

against the restricted model

$$Y = \beta_0 + \beta_1 X_1 + \beta_2 X_2 + \cdots + \beta_p X_p + \epsilon$$

is equivalent to testing the hypothesis $H_0: \beta_{p+1} = 0$.

We can represent the variability in Y accounted for by regression on the variables of the restricted model as

$$SS_{Y.12\cdots p} = R^2_{Y.12\cdots p}SS_Y$$

The variability accounted for by the larger model can be expressed as

$$SS_{Y.12\cdots p+1} = R^2_{Y.12\cdots p+1}SS_Y$$

Therefore the increment in variability accounted associated with the predictor X_{p+1} is given by

$$SS_{Y.p+1|12\cdots p} = (R^2_{Y.12\cdots p+1} - R^2_{Y.12\cdots p})SS_Y$$

$$= r^2_{Y(p+1|12\cdots p)}SS_Y$$

and is associated with a single df because only one additional regression coefficient must be estimated in the larger model.

Table 15.5 presents these results in the form of an ANOVA table. The hypothesis $H_0: \beta_{p+1} = 0$ can be tested with the ratio

$$F = \frac{MS_{Y.p+1|12\cdots p}}{MS_{error}}$$

The error term is the mean square associated with the variability not accounted for by the larger model; that is,

$$MS_{error} = \frac{(1 - R^2_{Y.12\cdots p+1})SS_Y}{N - p - 2}$$

The numerator of the F is associated with 1 df, the denominator with $N - p - 2$ df. Therefore, the F can be expressed as

$$F = \frac{(R^2_{Y.12\cdots p+1} - R^2_{Y.12\cdots p})SS_Y}{(1 - R^2_{Y.12\cdots p+1})SS_Y/(N - p - 2)} \tag{15.9}$$

When a partial F test is used to test whether a single population regression coefficient is zero, the results produced are exactly equivalent to those of the t test discussed in the preceding section. However, partial F tests can also be used to test hypotheses that state that some subsets of the β's are equal to zero, and these hypotheses cannot be tested with the t.

If, for example, we start with a model containing p predictor variables,

$$Y = \beta_0 + \beta_1 X_1 + \beta_2 X_2 + \cdots + \beta_p X_p + \epsilon$$

and consider adding k more predictor variables so that the model is

$$Y = \beta_0 + \beta_1 X_1 + \cdots + \beta_p X_p + \beta_{p+1} X_{p+1} + \cdots + \beta_{p+k} X_{p+k} + \epsilon$$

We can test the hypothesis $H_0: \beta_{p+1} = \beta_{p+2} = \cdots = \beta_{p+k} = 0$ by using the statistic

$$F = \frac{(R^2_{Y.12\cdots p+k} - R^2_{Y.12\cdots p})SS_Y/k}{(1 - R^2_{Y.12\cdots p+k})SS_Y/(N - 1 - p - k)} \tag{15.10}$$

TABLE 15.5 ANOVA TABLE FOR TESTING THE EFFECT OF ADDING k PREDICTOR VARIABLES TO A MODEL THAT ALREADY CONTAINS p PREDICTORS

SV	df	SS	MS
Larger model ($p+k$ predictors)	$p+k$	$R^2_{Y.12\cdots p+k}SS_Y$	$\dfrac{R^2_{Y.12\cdots p+k}SS_Y}{p+k}$
Smaller model	p	$R^2_{Y.12\cdots p}SS_Y$	$\dfrac{R^2_{Y.12\cdots p}SS_Y}{p}$
Increment	k	SS_{inc}	$\dfrac{SS_{inc}}{k}$
		where $SS_{inc} = (R^2_{Y.12\cdots p+k} - R^2_{Y.12\cdots p})SS_Y$	
Error (residual)	$N-1-(p+k)$	$(1-R^2_{Y.12\cdots p+k})SS_Y$	$\dfrac{SS_{error}}{N-1-p-k}$

The appropriate ANOVA table is given in Table 15.5. This general approach can be used to assess the effect of adding any set of predictors to the equation and tests the hypothesis that the regression coefficients for these added predictors are all equal to 0 in the population.

15.5.5 Inferences About the Prediction of Y

In bivariate regression, the expected value of Y corresponding to a value X_j of X is given by

$$\mu_{Y.X_j} = \beta_0 + \beta_1 X_j$$

which can be estimated by

$$\hat{Y}_j = \hat{\mu}_{Y.X_j} = b_0 + b_1 X_j$$

We showed in Chapter 12 that the estimated standard error associated with the prediction of Y at X_j is given by

$$\hat{SE}(\hat{Y}_j) = \hat{\sigma}_e \sqrt{h_{jj}} \tag{15.11}$$

where $\hat{\sigma}_e$ is the standard error of estimate and h_{jj}, the leverage of X_j, is given by $h_{jj} = 1/N + (X_j - \bar{X})^2/SS_X$.

In multiple regression, we can find the estimated standard error for the prediction of Y associated with any combination of scores on the p predictor variables in the regression equation. If the combination of predictor values is one that occurred for any of the cases in our sample, say case j, the estimated standard error for the prediction is again given by Equation 15.11, where h_{jj}, the leverage of the values of the predictor variables in case j, can be expressed as the jth diagonal element in the so-called "hat matrix" $\mathbf{H} = \mathbf{X}(\mathbf{X'X})^{-1}\mathbf{X'}$ (see Section 15.10). Although the expression for h_{jj} is now more complicated, it can be thought of and used in the same way as in bivariate regression. Table 15.6 provides the residual output for the multiple regression we performed on the data of Table 15.2, and the third column provides the leverage values for each of the 25 cases. If, for example, we wish to find the 95% confidence interval for a prediction based on the predictor values for case 4 ($X_1 = 36$ and $X_2 = 630$), we note that the value predicted by the regression equation is 79.809, $t_{.025,22} = 2.074$, and the estimated standard error is $\hat{SE}(\hat{Y}_j) = \hat{\sigma}_e\sqrt{h_{jj}} = (12.282)\sqrt{.113} = 4.12$. Therefore, the 95% confidence interval is given by

$$\hat{Y} \pm t_{.025}\hat{SE}(\hat{Y}) = 79.81 \pm 8.54$$

The interpretation of the interval is that in 95% of samples selected from the population, the interval formed in this way for $X_1 = 36$ and $X_2 = 630$ will contain $\mu_{Y|36,630}$, the population conditional mean of Y for those predictor values.

If we wish to find the standard error for the prediction of Y based on a combination of values for the p predictors, $X_{j1}, X_{j2}, \ldots, X_{jp}$, that did not occur in our sample, the estimated standard error for $\hat{Y}$ is not made available by the computer output and must be calculated. The estimated standard error of the predicted Y can be written as the matrix expression

$$\hat{SE}(\hat{Y}) = \hat{\sigma}_e\sqrt{\mathbf{x}_j'(\mathbf{X}'\mathbf{X})^{-1}\mathbf{x}_j} \qquad (15.12)$$

TABLE 15.6   SYSTAT RESIDUAL OUTPUT FOR THE REGRESSION OF FINAL EXAM SCORE ON PRETEST AND SAT SCORE FOR THE DATA IN TABLE 15.2

	ESTIMATE	RESIDUAL	LEVERAGE	COOK	STUDENT	SEPRED
CASE 1	55.072	-0.072	0.145	0.000	-0.006	4.675
CASE 2	89.227	7.773	0.170	0.033	0.686	5.071
CASE 3	75.627	-0.627	0.281	0.000	-0.059	6.514
CASE 4	79.809	3.191	0.113	0.003	0.270	4.121
CASE 5	62.470	16.530	0.044	0.029	1.407	2.576
CASE 6	76.673	3.327	0.165	0.006	0.290	4.988
CASE 7	46.549	-4.549	0.149	0.009	-0.394	4.743
CASE 8	59.555	-14.555	0.050	0.026	-1.230	2.739
CASE 9	40.869	24.131	0.174	0.327	2.379	5.117
CASE 10	53.424	14.576	0.184	0.130	1.337	5.272
CASE 11	73.235	4.765	0.206	0.016	0.427	5.578
CASE 12	77.567	-3.567	0.075	0.002	-0.296	3.373
CASE 13	54.771	-11.771	0.075	0.027	-0.996	3.371
CASE 14	72.260	-5.260	0.064	0.004	-0.434	3.112
CASE 15	61.797	3.203	0.048	0.001	0.262	2.677
CASE 16	76.220	-4.220	0.142	0.008	-0.364	4.623
CASE 17	80.332	18.668	0.147	0.156	1.717	4.710
CASE 18	48.569	4.431	0.132	0.008	0.380	4.470
CASE 19	77.567	11.433	0.075	0.025	0.967	3.373
CASE 20	76.522	-14.522	0.082	0.045	-1.249	3.506
CASE 21	72.260	-22.260	0.064	0.080	-1.997	3.112
CASE 22	67.104	-3.104	0.041	0.001	-0.252	2.480
CASE 23	42.066	-17.066	0.164	0.151	-1.569	4.975
CASE 24	46.699	3.301	0.117	0.004	0.280	4.209
CASE 25	65.757	-13.757	0.092	0.047	-1.186	3.723

where the vector $\mathbf{x}_j' = [1 \quad X_{j1} \quad X_{j2} \quad \cdots \quad X_{jp}]$. Some packages provide the matrix $(\mathbf{X}'\mathbf{X})^{-1}$ and therefore allow $\hat{SE}(\hat{Y})$ to be readily calculated by matrix multiplication, as indicated in Appendix C.

If we wish to find the confidence interval for a particular Y score, rather than for the conditional mean of Y at $X_1, X_2, \ldots, X_p$ in the population, the standard error must contain an additional component to reflect the fact that for any combination of predictor values, the Y scores are distributed around their conditional mean with variance σ_e^2. The appropriate estimated standard error is therefore given by

$$\hat{SE}(\hat{Y}) = \hat{\sigma}_e \sqrt{1 + \mathbf{x}_j'(\mathbf{X}'\mathbf{X})^{-1}\mathbf{x}_j} \tag{15.13}$$

15.6 SELECTING THE BEST REGRESSION EQUATION FOR PREDICTION

Sometimes we want to predict some criterion of interest by developing a regression equation that contains a subset of the potentially useful predictor variables that are available. In predicting, we are normally concerned both with the accuracy of the predictions and with the costs involved in making them. If our only concern was accuracy, we would be inclined to use as many valid predictors as possible in the regression equation; on the other hand, concerns about costs would motivate us to use fewer predictors. Because in many types of research most of the predictor variables are correlated with one another, including all of them in a regression equation would not only be expensive and cumbersome but would also introduce a good deal of redundancy. A number of automated procedures that allow a compromise between these concerns have been developed to produce the best possible predictions with regression equations that contain relatively few predictors. These procedures include forward selection, backward elimination, and stepwise regression and are available in many statistical packages. With these procedures, it is often possible to select a subset of the potential predictors that accounts for nearly as large a proportion of the variability in Y as does the entire pool of predictors. Before describing them, we should emphasize that these automated procedures have been developed solely to produce the best prediction equations according to certain criteria. These equations need not be best or even very good in any explanatory or theoretical sense. Running an automated regression routine may be useful for predicting, but it is a very poor way to develop theory.

15.6.1 Forward Selection

In the forward selection procedure, the regression equation is built up one variable at a time. On the first step, the predictor that has the highest correlation (positive or negative) is selected. If it fails to meet the criterion for inclusion, the procedure ends with no predictors in the equation, and the final equation is $\hat{Y}_i = \bar{Y}$. If the first predictor meets the criterion, on the next step a second predictor is selected and tested to determine whether it should be entered into the equation. The predictor selected is the one that would result in the greatest increment in R^2 if added to the equation. If the second predictor does not meet the criterion for inclusion, the procedure terminates with only a single predictor in the equation. If it does meet the criterion, on the third step, a third predictor is selected and tested, and so on. At each step, a partial F test (see Section 15.5.4) is performed on the selected variable, and the criterion for inclusion is stated in terms of the critical value or the significance level of the F. In forward selection and stepwise regression (which will be considered later), a liberal criterion for entering variables into the equation is often employed (Pedhazur 1982 recommends using an F of 2). This will generally allow the investigation of more variables than would normally be used, so that a number of possible equations can be considered.

It should be noted that for procedures like forward selection, the usual significance levels obtained from the F distribution are not appropriate. This is because at each step a number of possible predictors are examined and only one—the one that produces the greatest increment in R^2 or, equivalently, the one that has the largest partial F—is tested. If only a single predictor is to be chosen from a pool of m possible predictors, the situation is analogous to choosing the largest member of a family of m contrasts and testing it for significance. As in the case of contrasts, if a single predictor is to be chosen, it is appropriate to use the Bonferroni procedure to control Type 1 error; that is, use $\alpha^* = \alpha/m$, where α is the overall probability of a Type 1 error.

If a subset of k predictors is to be chosen, where $1 < k < m$, the distribution of R^2 is unknown. Wilkinson (1979) has discussed this problem and has provided tables of the upper 95th and 99th percentage points of the sample R^2 distribution in forward selection based on simulations (other tables and discussions of this problem can be found in Hocking 1983, Rencher and Pun 1980, and Wilkinson and Dallal 1982). These tables are much more conservative than the usual F tables. For example, with $N = 35$ and $\alpha = .05$, if all four members of a set of predictor variables are to be included in the regression equation, it is appropriate to use the standard F test to test R^2 for significance. When this is done, it is found that R^2_{samp} has to exceed .26 in order to reject the hypothesis that $R^2_{\text{pop}} = 0$. If the four predictors are to be selected from a larger set of 20 predictors by a forward selection procedure, according to Wilkinson's tables R^2 must exceed .51 in order to reject the null hypothesis. Many researchers do not seem to be aware of this problem; for a sample of 66 published papers that reported significant forward selection analyses according to the usual F tests, Wilkinson found that 19 were not significant when his tables were used.

15.6.2 Backward Elimination

Whereas forward selection begins with no predictors in the equation and adds them to the equation one by one, backward elimination begins with all the predictors in the equation and removes them one by one until the final equation is obtained. At each step, the predictor in the equation that produces the smallest increment in R^2 is tested to determine whether it should be removed from the equation. Again the criterion for removal is generally stated in terms of the significance level of a partial F test. If the selected variable is removed, another predictor is selected and tested on the next step.

The procedure terminates when a predictor that has been selected for testing is not removed from the equation; it and all the other predictors remaining in the equation are included in the final regression equation.

15.6.3 Stepwise Regression

Stepwise regression, the most popular procedure used to obtain the best prediction equation, is a combination of the forward selection and backward elimination procedures. The procedure is essentially the same as forward selection with the exception that after each new predictor has been added to the regression equation, all the predictors already in the equation are reexamined to determine whether they should be removed. A partial F test is performed on the predictor already in the equation that produces the smallest increment in R^2. If the predictor no longer satisfies the criteria for inclusion, it is removed from the equation. Statistical packages allow the user to set the significance levels (or critical F values) for entering or removing a variable. The F for entering variables into the equation should be set at least as high as the F for removing them. Otherwise, variables may be cycled in and out of the equation.

It is not difficult to see why it is sometimes desirable to remove a predictor that had been entered early in the analysis. For example, suppose that X_7 is highly predictable from X_4 and X_9 but is more highly correlated with Y than either of them. Even though X_7 may enter the equation early because of its high correlation with Y, it will become superfluous after X_4 and X_9 are entered. That is, even if X_7 contributes significantly to the predictability of Y by itself, it may not make a significant contribution over and above the predictability provided by the other two variables.

Again, it is important to emphasize that when predictor variables entered into the equation are selected from a larger pool, the significance levels printed out by stepwise programs are not "real" p values. Because many practitioners seem to be unaware of this fact, stepwise regression outputs are frequently misinterpreted. Wilkinson refuses to provide p values in the SYSTAT stepwise regression output because, as he states in his manual, stepwise regression programs are probably the most notorious source of "pseudo p-values" in the field of automated data analysis. As with forward selection, we recommend that Wilkinson's (1979) tables be used to test R^2 for significance.

Finally, we again emphasize that the sole motivation for the automated procedures described in this section is to develop useful prediction equations that include subsets of the available predictors. There is no reason to think that the equations they produce are "best" or even reasonable in any theoretical sense. Variables that are useful predictors need not be important components of a good theory or causal explanation of the situation. The automated procedures may include theoretically uninteresting variables in the regression equations they produce, and they may not include the important variables. Consider, for example, a stepwise regression with several predictors that are highly correlated with the criterion and with each other. The correlation between the criterion and the predictor included on the first step may be only marginally greater than the correlation between the criterion and the other predictors. Yet, including the first predictor may prevent any of the others from being entered into the equation on subsequent steps. Even though the other predictors add significantly to the predictability of Y in the absence of the first variable, they may not do so when the first variable is in the equation.

15.7 EXPLANATION VERSUS PREDICTION IN REGRESSION

In the previous section, we stated rather emphatically that equations that are good for prediction are not necessarily very useful for explanation or for advancing theory. However, as researchers, we would like to use regression analysis not only to predict but also to help understand and explain the phenomena under investigation. Understanding might involve learning about how the variables influence one another and how they cause changes in the criterion. Although it is tempting to interpret b_j as an index of the importance of X_j, things are not so simple; it is difficult to make conclusions about importance and causality in nonexperimental research. The remainder of this section will briefly outline some of the considerations that must be kept in mind when interpreting regression analyses.

In discussing causal interpretations, it is important to distinguish between experimental and nonexperimental research. In well-designed experiments with manipulated independent variables, it is possible to make causal statements because the systematic effects of irrelevant variables are controlled by using procedures such as matching and randomization. For example, in the teaching experiment discussed in Chapter 1, if students are randomly assigned to teaching methods and the quality of teachers is somehow controlled, it should be possible to interpret significant differences in students' performance as having been caused by the teaching methods. On the other hand, if the variables are merely observed in a correlational study, it will be difficult, if not impossible, to disentangle the effects of teaching methods from effects that result from differences in students, schools, and teachers.

Procedures beyond the scope of this book, such as path analysis and structural modeling (e.g., Bentler 1980, Joreskog and Sorbom 1986, Kenny 1979; also see a very interesting collection of papers in the summer 1987 edition of the *Journal of Educational Statistics*), have been developed in the hope of extracting causal information from correlational data. However, these procedures do not generate causal models. At best, they may be able to reject models that have been proposed when, under certain conditions, these models can be shown to be inconsistent with the data. Without well-developed theory, it is not possible to use regression coefficients derived from nonexperimental data as measures of causal influence.

15.7.1 Specification Errors

Under conditions we discussed earlier, if parameters of the population model

$$Y = \beta_0 + \beta_1 X_1 + \beta_2 X_2 + \cdots + \beta_p X_p + \epsilon$$

are estimated by the coefficients of the sample regression equation

$$\hat{Y} = b_0 + b_1 X_1 + b_2 X_2 + \cdots + b_p X_p$$

the b_j's can be shown to be unbiased estimators of the β_j's. However, as we indicated in Section 15.3, if the b_j's are obtained from a regression equation that does not include the same variables as the correct population model, they will generally be biased estimators of the β_j's in the population model. If we misspecify the population model by

(1) omitting relevant variables or (2) including irrelevant variables, the sample regression coefficients will be less useful estimators of the population coefficients.

Omitting Relevant Variables

Suppose that the true model in the population is

$$Y = \beta_{012} + \beta_{Y1.2}X_1 + \beta_{Y2.1}X_2 + \epsilon$$

Here we use the notation $\beta_{Y1.2}$ to emphasize that the regression coefficient of X_1 comes from a model that includes both X_1 and X_2. If we attempt to estimate $\beta_{Y1.2}$ and $\beta_{Y2.1}$ from a sample by using the regression equation

$$\hat{Y} = b_{012} + b_{Y1.2}X_1 + b_{Y2.1}X_2$$

it can be shown that $E(b_{Y1.2}) = \beta_{Y1.2}$ and $E(b_{Y2.1}) = \beta_{Y2.1}$, so the sample regression coefficients are unbiased estimators of the model parameters.

If, however, we regress Y only on X_1 and obtain the sample regression equation

$$\hat{Y} = b_{01} + b_{Y1}X_1$$

b_{Y1} will generally be a biased estimator of $\beta_{Y1.2}$. To show this, we first express b_{Y1} as

$$b_{Y1} = \frac{\Sigma(X_1 - \bar{X}_1)(Y - \bar{Y})}{SS_1} \qquad \text{where } SS_1 = \Sigma(X_1 - \bar{X}_1)^2$$

so that

$$E(b_{Y1}) = E\left[\frac{\Sigma(X_1 - \bar{X}_1)(Y - \bar{Y})}{SS_1}\right]$$

Because X is fixed, we can rewrite the preceding equation as

$$E(b_{Y1}) = \frac{\Sigma(X_1 - \bar{X}_1)E(Y - \bar{Y})}{SS_1}$$

Substituting the population model expressions for Y and $\bar{Y}$ and simplifying, we have

$$E(b_{Y1}) = \frac{\beta_{Y1.2}\Sigma(X_1 - \bar{X}_1)^2 + E[\beta_{Y2.1}\Sigma(X_1 - \bar{X}_1)(X_2 - \bar{X}_2)]}{SS_1}$$

$$= \beta_{Y1.2} + \beta_{Y2.1}E\left[r_{12}\sqrt{\frac{SS_2}{SS_1}}\right]$$

$$= \beta_{Y1.2} + \beta_{Y2.1}E(b_{21})$$

where r_{12} is the correlation between X_1 and X_2, SS_1 and SS_2 are the sums of squares of X_1 and X_2, and b_{21} is the regression coefficient obtained by regressing X_2 (the omitted variable) on X_1. Therefore, the expected value of b_{Y1} is not $\beta_{Y1.2}$, but also depends on $\beta_{Y2.1}$ as well as the sums of squares of X_1 and X_2 and the correlation between X_1 and X_2. Note that the biasing term disappears if $r_{12} = 0$.

Although things get more complicated when there are more predictors, omitting variables that are in the population model will result in biased parameter estimates unless all the omitted variables are uncorrelated with those that are included in the regression.

Including Irrelevant Variables

Because of the negative consequences of omitting relevant variables, researchers are sometimes inclined to include additional variables in their regression equations, just to make sure that nothing important has been left out. This may result in the addition of irrelevant variables to the equation—that is, variables that are not included in the population model. Because adding irrelevant variables does not bias parameter estimates, the consequences of including irrelevant variables are not as serious as those of omitting relevant ones. Nonetheless, including irrelevant variables will use up degrees of freedom and will tend to inflate the standard errors of the relevant variables that are in the equation, making parameter estimates less precise and significance tests less powerful.

15.7.2 Multicollinearity

Multicollinearity occurs when predictor variables included in a regression equation are themselves highly predictable from the other variables in the equation. Although high correlations among predictors do not generally result in much difficulty if the only goal is prediction, they can present difficulties both for estimating and for interpreting regression coefficients. If several highly correlated predictors are included in the regression equation, their combined effect will be split among them; the nature of the split will depend on the details of the data and may vary widely from sample to sample.

Including highly correlated predictor variables in the regression equation can result in dramatically inflated standard errors for the regression coefficients. As we pointed out earlier, one way of writing the expression for the standard error of b_j is

$$\hat{SE}(b_j) = \frac{\hat{\sigma}_e}{\sqrt{SS_j}} \sqrt{\frac{1}{1 - R_{j.}^2}} \tag{15.8}$$

where $R_{j.}^2$ is the squared multiple correlation coefficient for the regression of the predictor X_j on the other predictors in the equation. If $R_{j.}^2$ approaches 1 (which may well happen if X_j is highly correlated with some of the other predictors, but could happen even if none of the pairwise correlations are large), $1 - R_{j.}^2$, the *tolerance* of X_j will approach 0. It can be seen from Equation 15.8 that the standard error of b_j increases as the tolerance decreases, and for low tolerances the value of the sample regression coefficient may be meaningless.

A number of remedies have been suggested to reduce problems associated with multicollinearity. One recommendation is to delete some of the predictors, thus reducing $R_{j.}$. Unfortunately, this might result in specification errors that themselves can have serious consequences. Another recommendation is to combine clusters of highly related predictor variables into new variables that represent common underlying processes. Deciding which variables to combine is best done on the basis of theoretical considerations. Other procedures that are beyond the scope of the present chapter, such as principal components analysis and factor analysis, can provide suggestions about possible underlying processes; however, be aware that such methods are frequently abused. Finally, a procedure called *ridge regression* (see, for example, Draper and Smith 1981, Price 1977, Rozeboom 1979) is sometimes used to deal with multicollinearity. This procedure takes advantage of the fact that under certain conditions it is possible to obtain biased estimates with small standard errors that are more useful than unbiased estimates with large standard errors.

15.7.3 Interpretation of the Regression Coefficients as the Direct Effects of the X_j

It is tempting to think that if we were to change X_j by one unit, Y would change by b_j units. However, even if we have included the correct variables in the regression equation and there is a causal relationship between X_j and the criterion, the regression coefficient b_j does not represent the total effect of X_j on Y. Rather, the regression coefficient reflects the *direct effect* of X_j on Y—that is, the rate of change of Y with X_j, *holding all of the other variables in the equation constant.* In many situations it does not make a great deal of sense to consider the consequences of changes in X_j when closely related variables are held constant (for example, if we were using a regression equation to try to account for fatalities in car accidents, it is not clear what it would mean to change miles traveled while holding fuel consumption constant). Changes in X_j will generally influence some other relevant variables (whether or not they are included in the model), and the resultant changes in the criterion will reflect the changes in the other variables as well as those in X_j. When we inquire about the consequences of changing the value of X_j, we must be concerned not only with the direct effect of X_j on Y but also with the *indirect effects*—the effects on Y that occur because of changes in the other variables.

Given a valid causal model, path analysis can be used to calculate the total effect (the sum of the direct and indirect effects) of changing a variable. However, the conclusions may be misleading if important variables are omitted from the model or if the model is otherwise invalid. Unless a complete model exists and is fully understood, if we want to draw conclusions about the effects of changing one of the variables, we have to manipulate the variable without confounding it with the other variables and observe the results.

15.7.4 Standardized versus Unstandardized Regression Coefficients

It is important to remember that the sizes of the unstandardized regression coefficients that we have been discussing depend on the units in which the variables are measured. If, for example, X_j is height, a one-unit change will be 12 times as large (and therefore b_{Yj} will be 12 times as large) when the unit of measurement is feet rather than inches.

If the regression is performed with standardized (i.e., z) scores, the standardized regression coefficient of X_j (sometimes called the beta weight of X_j) is the number of standard deviations Y changes with a one-standard-deviation change in X_j, when all other variables are held constant. Although standardized regression coefficients offer the advantage of common (standard deviation) units, they have the problem of being sample-specific in much the same way as correlation coefficients and therefore should not be used to generalize across situations. The magnitudes of standardized coefficients depend not only on the variances and covariances of the variables included in the model; they also depend on the variances of variables that are not included in the model but which contribute to the error term and thereby to the variance of the criterion variable. The unstandardized coefficients, b_{Yj}, are preferable because they are fairly stable even when variances and covariances vary across settings. For some very illuminating simulations that indicate how standardized regression coefficients can vary across samples while unstandardized coefficients remain quite stable, see Chapter 4 of Hanushek and Jackson (1977).

15.8 MULTIPLE REGRESSION AND TREND ANALYSIS

Linear multiple regression is perhaps more general than one might think. Although what is produced is an equation that is a linear combination of the predictor variables, $X_1, X_2, \ldots, X_p$, the predictor variables themselves can be related nonlinearly. Consider a set of N data points (X_i, Y_i). We know that if we regress Y on X, we get an equation

$$\hat{Y} = b_0 + b_1 X$$

which is the best linear equation for predicting Y from X. However, if we regress Y on X and X^2, we get an equation of the form

$$\hat{Y} = b_0 + b_1 X + b_2 X^2$$

which represents the best quadratic equation (or polynomial equation of degree 2) for predicting Y from X. For any set of data points, we can test whether the slope of the best-fitting line differs significantly from 0 and whether there are significant quadratic, cubic, and other nonlinear components.

The trend analyses we discussed in Chapter 7 can be readily performed by using multiple regression. In Table 15.7, we again present the data from Table 7.1. From the ANOVA on these data reported in Table 7.2, we have $SS_A = 35.242$ and $MS_{S/A} = 2.257$. If we let X_j be the levels of the independent variable, we can obtain SS_{linear} by regressing the group means on the levels of X_j and then finding

$$SS_{\text{linear}} = R^2_{\bar{Y} \cdot X} SS_A$$

The linear trend can be tested for significance by using

$$F(1, \text{df}_{S/A}) = \frac{SS_{\text{linear}}}{MS_{S/A}}$$

For these data, the regression of $\bar{Y}_{\cdot j}$ on X_j yields $R^2_{\bar{Y} \cdot X} = .423$; therefore, $SS_{\text{linear}} = (.423)(35.242) = 14.900$ and $F(1, 45) = 14.900/2.257 = 3.904$, which is significant at $p < .01$.

The quadratic trend can be tested by first regressing $\bar{Y}_{.j}$ on X and X^2. The resulting multiple R is indicated by $R_{\bar{Y}.X,X^2}$, so that the increment in the variability accounted for when the quadratic component is considered in addition to the linear component is $SS_{quadratic} = (R^2_{\bar{Y}.X,X^2} - R^2_{\bar{Y}.X})SS_A$. We can test whether the quadratic trend is significant using the statistic

$$F(1, df_{S/A}) = \frac{SS_{quadratic}}{MS_{S/A}}$$

The cubic trend can be tested by regressing $\bar{Y}_{.j}$ on X, X^2, and X^3 and then using the test statistic

$$F(1, df_{S/A}) = \frac{SS_{cubic}}{MS_{S/A}}$$

where $SS_{cubic} = (R^2_{\bar{Y}.X,X^2,X^3} - R^2_{\bar{Y}.X,X^2})SS_A$, and so on. The SS terms for all the trend

TABLE 15.7 DATA SET AND TREND ANALYSIS FOR A ONE-FACTOR DESIGN (DATA OF TABLE 7.1)

Stimulus height (X)	7	9	11	13	15
Observed mean GSR ($\bar{Y}_{.j}$)	1.910	3.560	4.440	3.530	3.830
Variance estimate ($\hat{\sigma}_j^2$)	2.228	2.563	1.964	2.881	1.659

$$SS_A = n\sum(\bar{Y}_{.j} - \bar{Y}_{..})^2 = 35.242; \quad MS_{S/A} = \frac{1}{a}\sum_j \hat{\sigma}_j^2 = 2.257;$$

$$R^2_{\bar{Y}.X} = .423; \quad R^2_{\bar{Y}.X,X^2} = .841; \quad R^2_{\bar{Y}.X,X^2,X^3} = .940; \quad R^2_{\bar{Y}.X,X^2,X^3,X^4} = 1.000$$

Therefore,

$$SS_{lin} = (.423)(35.242) = 14.900$$

$$SS_{quad} = (.841 - .423)(35.242) = 14.727$$

$$SS_{cub} = (.940 - .841)(35.242) = 3.534$$

$$SS_{quart} = (1.000 - .940)(35.242) = 2.085$$

The ANOVA table is

SV	df	SS	MS	F
A	4	35.242	8.811	3.904[a]
lin(A)	1	14.900	14.900	6.602[b]
quad(A)	1	14.723	14.723	6.523[b]
cub(A)	1	3.534	3.534	1.566
quart(A)	1	2.085	2.085	.924
S/A	45	101.565	2.257	

[a] $p < .01$.

[b] $p < .05$.

components and the significance tests are reported in Table 15.7. These yield exactly the same results as the trend analysis performed with orthogonal polynomials in Chapter 7.

Although the two procedures yield the same results for the present data set, there is an advantage in working within the regression framework. The orthogonal polynomials we used in Chapter 7 assume equal numbers of subjects in each group and equal spacing between levels of the independent variable; if either of these equalities are violated, new orthogonal polynomials must be calculated. Unequal spacing will not cause any difficulties if we perform trend analysis by using hierarchical multiple regression; that is, if we test the increments in variability accounted for when higher-order components are added to regression equations that already contain the lower-order components. If we have unequal spacing among levels but equal n, we can proceed in exactly the same way as we did before.

We can perform trend analysis with unequal n by regressing the Y scores instead of the group means (that is, the Y's, not the $\bar{Y}_{.j}$'s) on the powers of X. If we do this, we can find the SS for each trend component by multiplying the increment in R^2 by SS_Y; for example,

$$SS_{\text{linear}} = R^2_{Y.X}SS_Y$$

$$SS_{\text{quadratic}} = (R^2_{Y.X,X^2} - R^2_{Y.X})SS_Y$$

$$SS_{\text{cubic}} = (R^2_{Y.X,X^2,X^3} - R^2_{Y.X,X^2})SS_Y$$

and so on. Table 15.8 presents an example of a trend analysis for a four-group between-subjects design with both unequal n and unequal spacing between the levels of the independent variable.

TABLE 15.8 TREND ANALYSIS FOR A DATA SET WITH UNEQUAL n AND UNEQUAL SPACING BETWEEN LEVELS OF THE INDEPENDENT VARIABLE

	X				Y	X	X^2	X^3
1	3	4	6					
3	9	10	18		3	1	1	1
7	14	14	30		7	1	1	1
10	19	21	27		10	1	1	1
4		24			4	1	1	1
		11			9	3	9	27
					14	3	9	27
					19	3	9	27
					10	4	16	64
					14	4	16	64
					21	4	16	64
					24	4	16	64
					11	4	16	64
					18	6	36	216
					30	6	36	216
					27	6	36	216

TABLE 15.8   (continued)

SV	df	SS	MS	F
X	3	630.93	210.31	7.42[a]
lin(X)	1	623.38	623.38	21.98[a]
quad(X)	1	1.74	1.74	<1
cub(X)	1	5.66	5.66	<1
S/A	11	312.00	28.364	

[a] $p < .01$.

$$R^2_{Y.X} = .661$$

$$R^2_{Y.X,X^2} = .663$$

$$R^2_{Y.X,X^2,X^3} = .669$$

$$SS_Y = 942.933$$

$$SS_{lin} = (.661)(942.933)$$
$$= 623.376$$

$$SS_{quad} = (.663 - .661)(942.933)$$
$$= 1.737$$

$$SS_{cub} = (.669 - .663)(942.933)$$
$$= 5.658$$

15.9 MULTIPLE REGRESSION IN REPEATED-MEASURES DESIGNS

When we analyze repeated-measures designs using multiple regression, hypothesis tests can be performed and confidence intervals can be found in the same ways that we indicated in Section 12.13 for bivariate regression.

Table 15.9 contains the data for a hypothetical reading experiment described by Lorch and Myers (1990). Each of 10 subjects reads a paragraph consisting of seven sentences, and reading times are recorded in milliseconds for each sentence. There are three predictor variables for each sentence: the serial position of the sentence in the text (SP), the number of words in the sentences (WORDS), and the number of new arguments in the sentence (NEW).

We are interested in whether each of the predictors makes a significant contribution to the prediction of reading time over and above that provided by the other two predictors (i.e., if the rate of change of reading time with the value of the predictor is significant when the values of the other two predictors are held constant).

There are several correct ways of testing hypotheses and forming confidence intervals for the regression coefficients of the predictors. One of them is to regress reading time on the three predictors separately for each subject and then to perform subsequent analyses on the regression coefficients. Table 15.10 contains the regression coefficients for each of the 10 subjects along with the mean, standard error, and t (i.e., the ratio of the mean to the standard error) for each coefficient. Because there are 10 subjects, there are 9 df associated with each t. Because $t_{crit.05,9} = \pm 2.262$, we can reject the hypotheses that $\beta_{SP} = 0$ and $\beta_{WORDS} = 0$ at $\alpha = .05$. We cannot reject the hypothesis that $\beta_{NEW} = 0$.

We should point out that, in the literature, regression analyses of data from repeated-measures designs have frequently been conducted inappropriately. Many researchers would incorrectly analyze the data presented in Table 15.9 by first averaging over subjects and then regressing the mean reading times for each sentence on the predictor variables. If this is done, the resulting significance tests will be positively biased. Even if the expected value of a regression coefficient is zero, the F may be large if the effect of the predictor varies across subjects. For a more detailed discussion, see Lorch and Myers (1990).

TABLE 15.9   VALUES OF THE PREDICTOR VARIABLES FOR THE SEVEN SENTENCES AND THE READING TIMES IN MILLISECONDS FOR EACH OF THE 10 SUBJECTS

SNT	SP	WORDS	NEW	S1	S2	S3	S4	S5	S6	S7	S8	S9	S10
1	1	13	1	3429	2795	4161	3071	3625	3161	3232	7161	1536	4063
2	2	16	3	6482	5411	4491	5063	9295	5643	8357	4313	2946	6652
3	3	9	2	1714	2339	3018	2464	6045	2455	4920	3366	1375	2179
4	4	9	2	3679	3714	2666	2732	4205	6241	3723	6330	1152	3661
5	5	10	3	4000	2902	2991	2670	3884	3223	3143	6143	2759	3330
6	6	18	4	6973	8018	6625	7571	8795	13188	11170	6071	7964	7866
7	7	6	1	2634	1750	2268	2884	3491	3688	2054	1696	1455	3705

TABLE 15.10   REGRESSION COEFFICIENTS FOR THE REGRESSION OF READING TIME ON SP, WORDS, AND NEW FOR EACH OF THE 10 SUBJECTS

Subject	SP	WORDS	NEW
1	.23124	.39103	.22161
2	.30533	.43415	.34637
3	.20637	.40360	− .25294
4	.48300	.50203	− .27683
5	− .06210	.28778	.92680
6	1.10982	.80850	− .23336
7	.25448	.57498	.79643
8	− .33147	.11341	.33124
9	.66786	.50078	.16320
10	.46921	.56964	− .50621
Mean	.33337	.45859	.15163
$\hat{SE}$	.12417	.05855	.14982
t	2.6849	7.8329	1.0121

15.10 DEVELOPING MULTIPLE REGRESSION WITH MATRIX NOTATION

15.10.1 Using Matrices to Represent Linear Equations in Regression

In this section, we discuss regression using matrix notation. Matrix notation is particularly useful for representing systems of linear equations, and matrix algebra allows us to perform operations on systems of equations conveniently. An introduction to matrix notation and matrix algebra is contained in Appendix C.

When we deal with regression, we deal with systems of linear equations. In developing bivariate regression in Chapter 12, we wrote linear equations for predicted scores as well as for b_0 and b_1. In the current chapter, we expressed the linear equation that best predicts Y from the predictors $X_1, X_2, \ldots, X_p$ in the sample as

$$\hat{Y}_i = b_0 + b_1 X_{i1} + b_2 X_{i2} + \cdots + b_p X_{ip}$$

This equation can be thought of as actually representing a system of N equations, one for each of the N data points in the sample. That is,

$$\hat{Y}_1 = b_0 + b_1 X_{11} + b_2 X_{12} + \cdots + b_p X_{1p}$$
$$\hat{Y}_2 = b_0 + b_1 X_{21} + b_2 X_{22} + \cdots + b_p X_{2p}$$
$$\hat{Y}_3 = b_0 + b_1 X_{31} + b_2 X_{32} + \cdots + b_p X_{3p}$$

$$\cdot$$
$$\cdot$$
$$\cdot$$

$$\hat{Y}_N = b_0 + b_1 X_{N1} + b_2 X_{N2} + \cdots + b_p X_{Np}$$

We can write this system of equations as the matrix equation

$$\hat{\mathbf{Y}} = \mathbf{X}\hat{\boldsymbol{\beta}} \tag{15.14}$$

where

$$\hat{\mathbf{Y}} = \begin{bmatrix} \hat{Y}_1 \\ \hat{Y}_2 \\ \cdot \\ \cdot \\ \cdot \\ \hat{Y}_N \end{bmatrix}$$

is an $N \times 1$ vector of the predicted Y scores (note that an $r \times c$ matrix is a matrix with r rows and c columns, and a matrix with one row or one column is referred to as a vector),

$$\mathbf{X} = \begin{bmatrix} 1 & X_{11} & X_{12} & \cdot & \cdot & \cdot & X_{1p} \\ 1 & X_{21} & X_{22} & \cdot & \cdot & \cdot & X_{2p} \\ \cdot & & \cdot & & & & \cdot \\ \cdot & & & & & & \cdot \\ \cdot & & & & & & \cdot \\ 1 & X_{N1} & X_{N2} & \cdot & \cdot & \cdot & X_{Np} \end{bmatrix}$$

is an $N \times (p + 1)$ matrix in which the first column consists entirely of 1's and each of the remaining columns consists of the N scores in the sample for one of the p predictor variables (e.g., X_{34} refers to the value of predictor X_4 for case 3), and

$$\hat{\beta} = \begin{bmatrix} b_0 \\ b_1 \\ b_2 \\ \cdot \\ \cdot \\ \cdot \\ b_p \end{bmatrix}$$

is a $(p + 1) \times 1$ vector consisting of b_0 and the regression coefficients $b_1, b_2, \ldots, b_p$ of the p predictor variables.

The coefficients $b_0, b_1, \ldots, b_p$ take on values that minimize the measure of error, $\text{MSE} = (1/N)\Sigma_i(Y_i - \hat{Y}_i)^2$. We can express MSE in matrix notation as

$$\text{MSE} = \frac{1}{N}(\mathbf{Y} - \hat{\mathbf{Y}})'(\mathbf{Y} - \hat{\mathbf{Y}})$$

$$= \frac{1}{N}(\mathbf{Y} - \mathbf{X}\hat{\beta})'(\mathbf{Y} - \mathbf{X}\hat{\beta})$$

where

$$\mathbf{Y} = \begin{bmatrix} Y_1 \\ Y_2 \\ \cdot \\ \cdot \\ \cdot \\ Y_N \end{bmatrix}$$

is an $N \times 1$ vector containing the N Y scores in the sample and $\hat{\beta}$, $\mathbf{X}$, and $\hat{\mathbf{Y}}$ are defined as before.

The vector $\hat{\beta}$ that minimizes the MSE can be found by using calculus. Taking the derivative of MSE with respect to $\hat{\beta}$, setting the result equal to $\mathbf{0}$, and simplifying, we have

$$\mathbf{X}'\mathbf{X}\hat{\beta} = \mathbf{X}'\mathbf{Y} \tag{15.15}$$

Equation 15.15 corresponds to the system of normal equations for the regression. We can solve the matrix equation for $\hat{\beta}$ by premultiplying both sides of Equation 15.15 by $(\mathbf{X}'\mathbf{X})^{-1}$, the inverse of $\mathbf{X}'\mathbf{X}$. When we do so, we obtain

$$(\mathbf{X}'\mathbf{X})^{-1}(\mathbf{X}'\mathbf{X})\hat{\beta} = (\mathbf{X}'\mathbf{X})^{-1}\mathbf{X}'\mathbf{Y}$$

but $(\mathbf{X}'\mathbf{X})^{-1}(\mathbf{X}'\mathbf{X}) = \mathbf{I}$, the identity matrix. Therefore,

$$\mathbf{I}\hat{\beta} = (\mathbf{X}'\mathbf{X})^{-1}\mathbf{X}'\mathbf{Y}$$

so

$$\hat{\beta} = (\mathbf{X}'\mathbf{X})^{-1}\mathbf{X}'\mathbf{Y} \tag{15.16}$$

Equation 15.16 represents a system of equations in which b_0, b_1, ... , b_p are expressed as linear combinations of the Y's.

In the case of bivariate regression, we have a single predictor variable X that takes on the values X_1, X_2, ... , X_N. Therefore,

$$\mathbf{X} = \begin{bmatrix} 1 & X_1 \\ 1 & X_2 \\ \cdot & \cdot \\ \cdot & \cdot \\ \cdot & \cdot \\ 1 & X_N \end{bmatrix} \quad \hat{\beta} = \begin{bmatrix} b_0 \\ b_1 \end{bmatrix} \quad \text{and} \quad \mathbf{Y} = \begin{bmatrix} Y_1 \\ Y_2 \\ \cdot \\ \cdot \\ \cdot \\ Y_N \end{bmatrix}$$

so that

$$\mathbf{X}'\mathbf{X} = \begin{bmatrix} N & \Sigma X \\ \Sigma X & \Sigma X^2 \end{bmatrix} \quad \text{and} \quad \mathbf{X}'\mathbf{Y} = \begin{bmatrix} \Sigma Y \\ \Sigma XY \end{bmatrix}$$

For the bivariate case, Equation 15.15 becomes

$$\begin{bmatrix} b_0 N & + & b_1 \Sigma X \\ b_0 \Sigma X & + & b_1 \Sigma X^2 \end{bmatrix} = \begin{bmatrix} \Sigma Y \\ \Sigma XY \end{bmatrix}$$

and corresponds exactly to Equation 12.8, the set of normal equations that was presented when we first introduced bivariate regression. Solving the normal equations yields the usual least-squares estimates of β_0 and β_1.

For bivariate regression, we have two normal equations (that is, two equations in the two unknowns b_0 and b_1); with p predictor variables, we have $p + 1$ normal equations. A virtue of matrix notation is that we can represent the set of normal equations by a single matrix equation that looks the same even though the dimensions of the matrices change with the number of cases (N) and predictor variables (p).

Even though the SS terms that occur in the ANOVA tables for multiple regression are scalar (i.e., not matrix) quantities, they can be obtained conveniently by using matrix expressions as follows:

$$SS_Y = \sum_i (Y_i - \bar{Y})^2 \tag{15.17}$$

$$= \mathbf{Y}'\mathbf{Y} - \frac{1}{N}\mathbf{Y}'\mathbf{E}\mathbf{Y}$$

$$SS_{reg} = \sum_i (\hat{Y}_i - \bar{Y})^2 \tag{15.18}$$

$$= Y'HY - \frac{1}{N}Y'EY$$

and

$$SS_{error} = \sum_i (Y_i - \hat{Y}_i)^2 \tag{15.19}$$

$$= Y'Y - Y'HY$$

where $\mathbf{E}$ is an $N \times N$ matrix that consists entirely of 1's and $\mathbf{H} = \mathbf{X(X'X)^{-1}X'}$. Equations 15.17–15.19 follow directly from the definitions of the SS terms. For example, we can write SS_Y as

$$SS_Y = \Sigma Y_i^2 - \frac{(\Sigma Y_i)^2}{N}$$

Substituting $\mathbf{Y'Y}$ for ΣY_i^2 and $\mathbf{Y'EY}$ for $(\Sigma Y_i)^2$, we obtain Equation 15.17. Also,

$$SS_{error} = (\mathbf{Y} - \mathbf{X}\hat{\beta})'(\mathbf{Y} - \mathbf{X}\hat{\beta})$$

$$= \mathbf{Y'Y} - \mathbf{Y'X}\hat{\beta} - \hat{\beta}'\mathbf{X'Y} + \hat{\beta}'\mathbf{X'X}\hat{\beta}$$

Because it can be shown that $\mathbf{Y'X}\hat{\beta} = \hat{\beta}'\mathbf{X'Y} = \hat{\beta}'\mathbf{X'X}\hat{\beta}$, we have

$$SS_{error} = \mathbf{Y'Y} - \mathbf{Y'X}\hat{\beta}$$

Substituting $\hat{\beta} = \mathbf{(X'X)^{-1}X'Y}$ (Equation 15.16), we have Equation 15.19:

$$SS_{error} = \mathbf{Y'Y} - \mathbf{Y'X(X'X)^{-1}X'Y}$$

$$= \mathbf{Y'Y} - \mathbf{Y'HY}$$

Equation 15.18 can be obtained by noting that $SS_Y = SS_{reg} + SS_{error}$.

The matrix $\mathbf{H}$ is convenient to work with because it is symmetric (i.e., $\mathbf{H'} = \mathbf{H}$) and idempotent (i.e., $\mathbf{HH} = \mathbf{H}$). It is known as the "hat" matrix and plays a prominent role in treatments of multiple regression. We can express any of the predicted Y's for our sample as a linear combination of the actual Y scores by using the elements of $\mathbf{H}$ as weights. If we start with the prediction equation $\hat{\mathbf{Y}} = \mathbf{X}\hat{\beta}$ and substitute $\hat{\beta} = \mathbf{(X'X)^{-1}X'Y}$, we obtain $\hat{\mathbf{Y}} = \mathbf{X(X'X)^{-1}X'Y}$ or

$$\hat{\mathbf{Y}} = \mathbf{HY} \tag{15.20}$$

that is,

$$
\begin{bmatrix} \hat{Y}_1 \\ \hat{Y}_2 \\ \cdot \\ \cdot \\ \cdot \\ \hat{Y}_N \end{bmatrix} = \begin{bmatrix} h_{11} & h_{12} & \cdots & h_{1N} \\ h_{21} & h_{22} & \cdots & h_{2N} \\ \cdot & & & \cdot \\ \cdot & & & \cdot \\ \cdot & & & \cdot \\ h_{N1} & h_{N2} & \cdots & h_{NN} \end{bmatrix} \begin{bmatrix} Y_1 \\ Y_2 \\ \cdot \\ \cdot \\ \cdot \\ Y_N \end{bmatrix}
$$

Therefore, for any predicted score, $\hat{Y}_j$,

$$\hat{Y}_j = \sum_i h_{ji} Y_i$$

$$= \sum_i h_{ij} Y_i \qquad \text{(because, since } \mathbf{H}' = \mathbf{H}, \ h_{ji} = h_{ij})$$

Note that this equation is exactly the same as Equation 12.20. For bivariate regression, we showed in Chapter 12 that the weights h_{ij} were given by

$$h_{ij} = \frac{1}{N} + \frac{(X_j - \bar{X})(X_i - \bar{X})}{SS_X}$$

Although the expressions for h_{ij} become more complicated as the number of predictor variables increases, they can be simply thought of as extensions of the expressions obtained in bivariate regression and can be obtained as elements of the matrix $\mathbf{H} = \mathbf{X}(\mathbf{X}'\mathbf{X})^{-1}\mathbf{X}'$. Also, in Chapter 12 we attached particular significance to h_{jj}, the "leverage" of the jth case. We noted that the leverage was one of the regression diagnostics printed out by statistical software packages and that it provided valuable information about the extent to which X_j was an outlier and the extent to which Y_j was determined by the jth case. The h_{jj} are simply the diagonal elements of $\mathbf{H}$.

15.10.2 Inference in Multiple Regression Expressed in Terms of Matrix Notation

The assumptions for the fixed-X case originally stated in Section 15.5 can be restated in matrix terms as follows. Given the model

$$\mathbf{Y} = \mathbf{X}\boldsymbol{\beta} + \boldsymbol{\epsilon}$$

we assume that

1. None of the predictor variables is completely redundant; we say that the predictor variables are linearly independent or that the matrix $\mathbf{X}$ is of full rank. If this condition is not satisfied, $(\mathbf{X}'\mathbf{X})^{-1}$ does not exist, so Equation 15.15 cannot be solved for $\hat{\beta}$.
2. $E(\epsilon) = \mathbf{0}$, where $\mathbf{0}$ is a column vector of 0's and $\text{Var}(\epsilon) = \sigma_e^2 \mathbf{I}$.
3. The values of the predictor variables are fixed and measured without error. This means that $\mathbf{X}$ is exactly the same for each replication of the experiment and can be treated like a constant.

We first show that

$$E(\hat{\beta}) = \beta$$

which indicates that the least-squares estimates of the elements of β are unbiased. Substituting $\hat{\beta} = (\mathbf{X}'\mathbf{X})^{-1}\mathbf{X}'\mathbf{Y}$, we have

$$E(\hat{\beta}) = E[(\mathbf{X}'\mathbf{X})^{-1}\mathbf{X}'\mathbf{Y}]$$

$$= (\mathbf{X}'\mathbf{X})^{-1}\mathbf{X}'E(\mathbf{Y}) \qquad \text{because } \mathbf{X} \text{ is fixed}$$

$$= (\mathbf{X}'\mathbf{X})^{-1}\mathbf{X}'E(\mathbf{X}\beta + \epsilon) \qquad \text{substituting for } \mathbf{Y}$$

$$= (\mathbf{X'X})^{-1}\mathbf{X'X}\boldsymbol{\beta} + (\mathbf{X'X})^{-1}\mathbf{X'}E(\boldsymbol{\epsilon})$$

$$= \boldsymbol{\beta} \qquad \text{because } (\mathbf{X'X})^{-1}\mathbf{X'X} = \mathbf{I} \text{ and we assume } E(\boldsymbol{\epsilon}) = \mathbf{0}$$

The standard errors necessary to test hypotheses and obtain confidence intervals can be found by noting that

1. For any constant matrix $\mathbf{C}$ that can multiply a random matrix $\mathbf{Y}$, it can be shown that

$$\text{Var}(\mathbf{CY}) = \mathbf{C}\text{Var}(\mathbf{Y})\mathbf{C'} \qquad (15.21)$$

This is the matrix extension of the result that $\text{Var}(CY) = C^2\text{Var}(Y)$ when C is a scalar (i.e., not a matrix) constant and Y is a scalar variable.

2. The variance-covariance matrix for $\hat{\boldsymbol{\beta}}$ is given by

$$\text{Var}(\hat{\boldsymbol{\beta}}) = E[(\hat{\boldsymbol{\beta}} - \boldsymbol{\beta})(\hat{\boldsymbol{\beta}} - \boldsymbol{\beta})'] \qquad (15.22)$$

$$= \begin{bmatrix} \text{Var}(b_0) & \text{Cov}(b_0, b_1) & \cdots & \text{Cov}(b_0, b_p) \\ \text{Cov}(b_1, b_0) & \text{Var}(b_1) & \cdots & \text{Cov}(b_1, b_p) \\ \cdot & \cdot & & \cdot \\ \cdot & \cdot & & \cdot \\ \cdot & \cdot & & \cdot \\ \text{Cov}(b_p, b_0) & \text{Cov}(b_p, b_1) & \cdots & \text{Var}(b_p) \end{bmatrix}$$

$$= (\mathbf{X'X})^{-1}\sigma_e^2$$

This last result is proved in the chapter appendix.

The variance of any estimated regression coefficient b_j can be expressed as a diagonal element of the matrix $(\mathbf{X'X})^{-1}$ multiplied by σ_e^2. Therefore, the estimated standard error of any of the b's can be expressed as

$$\hat{\text{SE}}(b_{j-1}) = \hat{\sigma}_e\sqrt{a_{jj}} \qquad (15.23)$$

where a_{jj} is the jth diagonal element of the matrix $(\mathbf{X'X})^{-1}$ and $\hat{\sigma}_e^2 = (1 - R_{Y.12\ldots p}^2)\text{SS}_Y/(N - 1 - p)$.

Also, we can obtain the standard error for the prediction of Y associated with any of the combinations of scores that occurred in our sample. From Equation 15.20, we have $\hat{\mathbf{Y}} = \mathbf{HY}$ where $\mathbf{H} = \mathbf{X}(\mathbf{X'X})^{-1}\mathbf{X'}$. Therefore,

$$\begin{aligned} \text{Var}(\hat{\mathbf{Y}}) &= \text{Var}(\mathbf{HY}) \\ &= \mathbf{H}\,\text{Var}(\mathbf{Y})\mathbf{H'} \qquad \text{from Equation 15.21} \\ &= \sigma_e^2\mathbf{HH'} \qquad \text{because we assume that } \text{Var}(\mathbf{Y}) = \sigma_e^2\mathbf{I} \\ &= \sigma_e^2\mathbf{H} \qquad \text{because } \mathbf{HH'} = \mathbf{H} \end{aligned}$$

Therefore,

$$\text{Var}(\hat{Y}_j) = \sigma_e^2 h_{jj}$$

where h_{jj} is the jth diagonal element of $\mathbf{H}$, and

$$\hat{\mathrm{SE}}(\hat{Y}_j) = \hat{\sigma}_e \sqrt{h_{jj}} \qquad (15.24)$$

To find the SE for the prediction of Y from a combination of predictor values X_{j1}, $X_{j2}, \ldots, X_{jp}$ that did not occur in our sample, we note that the prediction of Y can be obtained from

$$\hat{\mathbf{Y}} = \mathbf{x}_j' \hat{\boldsymbol{\beta}}$$

where $\mathbf{x}_j' = [1 \quad X_{j1} \quad X_{j2} \quad \cdots \quad X_{jp}]$. The variance of the prediction is

$$
\begin{aligned}
\mathrm{Var}(\hat{\mathbf{Y}}) &= \mathrm{Var}(\mathbf{x}_j' \hat{\boldsymbol{\beta}}) \\
&= \mathbf{x}_j' \mathrm{Var}(\hat{\boldsymbol{\beta}}) \mathbf{x}_j \qquad \text{from Equation 15.21}
\end{aligned}
$$

But from Equation 15.22, we have $\mathrm{Var}(\hat{\boldsymbol{\beta}}) = \sigma_e^2 (\mathbf{X}'\mathbf{X})^{-1}$, so

$$\mathrm{Var}(\hat{\mathbf{Y}}) = \sigma_e^2 \mathbf{x}_j' (\mathbf{X}'\mathbf{X})^{-1} \mathbf{x}_j$$

and the estimated standard error of the prediction is

$$\hat{\mathrm{SE}}(\hat{Y}) = \hat{\sigma}_e \sqrt{\mathbf{x}_j' (\mathbf{X}'\mathbf{X})^{-1} \mathbf{x}_j} \qquad (15.25)$$

15.11 OUTLIERS AND INFLUENTIAL POINTS IN MULTIPLE REGRESSION

In discussing bivariate regression, we introduced measures for identifying cases that had outlying values of X or had inordinate influence in determining the value of the regression coefficient for X. The corresponding measures for multiple regression are just generalizations of these measures. For example, in Chapter 12, the Mahalanobis distance for the jth case was defined as

$$D_j = \frac{(X_j - \bar{X})^2}{\hat{\sigma}_X^2}$$

which can be thought of as the squared z score of X_j and therefore as a standardized measure of the distance of X_j from the mean of the X's. The corresponding multivariate measure can be written in a similar form

$$D_j = (\mathbf{x}_j - \bar{\mathbf{x}})' \mathbf{S}^{-1} (\mathbf{x}_j - \bar{\mathbf{x}})$$

where $\mathbf{x}_j$ is the vector of predictor values for the jth case, $\bar{\mathbf{x}}$ is the vector of predictor means, and $\mathbf{S}^{-1}$ is the inverse of the estimated variance-covariance matrix for the predictors. The term D_j can be thought of as a measure of distance in a p-dimensional space from the point defined by the vector of scores for the jth case to the centroid (the vector of predictor means).

As in the case of bivariate regression, the leverage of the jth case can be expressed in terms of D_j:

$$h_{jj} = \frac{1}{N} + (N-1)D_j$$

$$= \frac{1}{N} + (N-1)(\mathbf{x}_j - \bar{\mathbf{x}})'\mathbf{S}^{-1}(\mathbf{x}_j - \bar{\mathbf{x}})$$

In bivariate regression, outliers in X can usually be identified by looking at the distribution of X or at the scatter diagram of X and Y. However, one must depend more heavily on measures such as D_j or h_{jj} to identify outliers in multiple regression because multivariate outliers can occur in subtle ways. For example, if there are 10 predictors, the jth case may be an outlier in the 10-dimensional predictor space because moderate deviations from the mean occur on 5 or 6 of the predictors; that is, the data point can be an outlier even though none of the deviations for individual predictors are large enough to be noticed.

As we mentioned earlier, it can be shown that $\Sigma_j h_{jj} = p + 1$, where p is the number of predictor variables; therefore, the mean value of h_{jj} is $(p + 1)/N$. Hoaglin and Welsch (1978) suggest that values of h_{jj} greater than $2(p + 1)/N$ should be considered to be large. Belsley, Kuh, and Welsch (1980) indicate that this cutoff will identify a few too many data points if p is small, but they recommend its use because it is easy to remember and use. It must be emphasized, however, that outlying cases are not necessarily influential (i.e., they need not exert inordinate influence on the values of the regression coefficients). In Chapter 12, we introduced Cook's distance, CD_j, as a measure of the change that would result in the regression coefficients if the jth case was omitted. CD_j can be written as

$$CD_j = \frac{\Sigma_i(\hat{Y}_i^{(-j)} - \hat{Y}_i)^2}{(p+1)\sigma_e^2}$$

or, in terms of the estimated regression coefficients, as

$$CD_j = \frac{(\hat{\beta} - \hat{\beta}^{(-j)})'(\mathbf{X}'\mathbf{X})(\hat{\beta} - \hat{\beta}^{(-j)})}{(p+1)\hat{\sigma}_e^2}$$

where $\hat{Y}_i^{(-j)}$ is the prediction of Y_i made using regression coefficients obtained with the jth case deleted and $\hat{\beta}^{(-j)}$ is the vector of these coefficients. Cook and Weisberg (1982) indicate that a Cook's distance of 1 should be considered to be large. Perhaps the most useful way to write Cook's distance is as

$$CD_j = \left(\frac{h_{jj}}{1 - h_{jj}}\right)\left(\frac{\text{ISTUD}_j^2}{(p+1)\hat{\sigma}_e^2}\right)$$

because it most clearly indicates that the influence of the jth case on the regression coefficients depends on the lack of fit of the model for the jth case as indicated by the internally studentized residual, ISTUD_j (see Section 12.11.2), and the extent to which the predictor values of the jth case make it an outlier. The expression $h_{jj}/(1 - h_{jj})$ can be shown to be the distance from the vector $\mathbf{x}_j$ to the vector of means based on all the data except the jth case (see Weisberg 1980, p. 109). A large Cook's distance can occur because of a large residual, a large leverage, or both.

Unfortunately, although regression diagnostics that consider the effect of deleting one point at a time work quite well when there is a single influential outlier, it is much more difficult to diagnose outliers when there are several of them. For a useful discussion of recent developments in the detection of multiple outliers and of robust regression, see Rousseeuw and Leroy (1987).

15.12 CONCLUDING REMARKS

In the analysis of observational data, naturally continuous variables are frequently transformed into categorical variables. For example, subjects may be assigned to one of a small number of categories on the basis of their scores on some aptitude test or personality scale. In the most extreme case, only two categories, "high" and "low" as determined by a median split, are used. Similarly, stimuli such as words may be categorized on the basis of characteristics such as length, rated pleasantness, or familiarity.

When subjects or stimuli are categorized on the basis of more than one attribute, correlations among the attributes will lead to unequal cell frequencies. Consider, for example, a study designed to examine performance in some laboratory task (Y) as a function of two personality measures, depression (X_1) and helplessness (X_2). Assume that after a median split on each of the two variables, 100 subjects are categorized so as to produce the following cell frequencies:

		X_2	
		High	Low
X_1	High	35	15
	Low	15	35

We should not be surprised to find unequal cell frequencies; it seems reasonable to think that depressed people would be more likely to have high helplessness scores than normals. Nonetheless, the researcher might suppress the correlation between helplessness and depression in the data by, for example, selecting 15 subjects from each cell to participate in the laboratory task, and then analyze the resulting data with a standard two-factor analysis of variance.

The apparent simplicity offered by this approach is more illusory than real. For one thing, thrusting orthogonality on the data (i.e., making X_1 and X_2 uncorrelated by selecting equal numbers of subjects from each cell) does not solve the interpretive problems associated with having correlated factors; it merely avoids them. Using equal cell frequencies in the analysis amounts to making inferences about the effects of X_1 and X_2 in a population in which they are uncorrelated—even though in the population of interest they *are* correlated. Perhaps the best way to think about the issue is to note that (as we will show in the next chapter) ANOVA with equal cell frequencies is exactly equivalent to multiple regression with uncorrelated categorical variables. In experimental research, manipulated independent variables are usually categorical, and in multifactor designs it makes sense to assign subjects to conditions in such a way as to keep the independent variables uncorrelated. However, in nonexperimental research, if the goal is to assess the effects of the variables of interest in the population, the sample correlations provide valid information about the population and should be retained in the analysis.

One possible approach is to use an unequal-n (sometimes called unbalanced or nonorthogonal) ANOVA using the data of all 100 subjects. We will see in the next chapter that this is exactly equivalent to using a multiple regression with correlated categorical variables. Interpretation is always difficult with correlated variables, but at least thinking of the analysis in terms of multiple regression—in which correlations are the rule rather than the exception—should keep us aware of the difficulties associated with correlated variables and the constraints imposed on our inferences.

Given that artificially forcing nonexperimental data into a factorial ANOVA design doesn't really avoid any interpretive difficulties, the most sensible approach is to give up thinking in terms of ANOVA and to perform a multiple regression on the original (noncategorized) scores. Dealing with the original scores may provide us with more power because artificially forcing scores into categories throws away potentially useful information. Also, using the original scores provides the opportunity to investigate the functional relationships between the X scores and the dependent variable Y.

In summary, forcing the data from nonexperimental research into simple equal-n ANOVAs may result more from wishful thinking and ignorance of the alternatives than from a thoughtful consideration of the issues involved. Thinking of such analyses in terms of multiple regression does not necessarily resolve any of the difficulties associated with nonexperimental data, but at least it clarifies the nature of the difficulties and gives us a fighting chance of dealing with them and arriving at reasonable conclusions about the population.

APPENDIX: Proof That $\mathrm{Var}(\hat{\beta}) = \sigma_e^2(X'X)^{-1}$

To prove $\mathrm{Var}(\hat{\beta}) = \sigma_e^2(X'X)^{-1}$, we first note that, because $\hat{\beta} = (X'X)^{-1}X'Y$, we have

$$\mathrm{Var}(\hat{\beta}) = \mathrm{Var}[(X'X)^{-1}X'Y]$$

$$= (X'X)^{-1}X' \, \mathrm{Var}(Y)[(X'X)^{-1}X']' \qquad \text{from Equation 15.21}$$

For any two matrices A and B that can be multiplied, $(AB)' = B'A'$, so $[(X'X)^{-1}X']' = X''[(X'X)^{-1}]'$. But $X'' = X$ and $[(X'X)^{-1}]' = (X'X)^{-1}$ because $(X'X)^{-1}$ is a symmetric matrix. Therefore,

$$\mathrm{Var}(\hat{\beta}) = (X'X)^{-1}X'\mathrm{Var}(Y)X(X'X)^{-1}$$

$$= (X'X)^{-1}X'X(X'X)^{-1}\mathrm{Var}(Y)$$

Because $(X'X)^{-1}X'X = I$ and we assume $\mathrm{Var}(Y) = \sigma_e^2 I$, we have

$$\mathrm{Var}(\hat{\beta}) = \sigma_e^2(X'X)^{-1}X'X(X'X)^{-1}$$

$$= \sigma_e^2(X'X)^{-1}$$

EXERCISES

15.1 The following terms provide a useful review of many concepts in the chapter. Define, describe, or identify each of them:

multiple regression	stepwise regression
multiple correlation coefficient	specification error
adjusted multiple R	multicollinearity
standard error of estimate	direct and indirect effects of X_j
standardized regression coefficient	trend analysis
unstandardized regression coefficient	matrix
tolerance	transpose of A
cross-validation	inverse of A
screening sample	hat matrix
calibration sample	outlier
partial F test	influential point
conditional mean of Y at X	leverage
forward selection	Cook's distance
backward elimination	

15.2 Consider the table containing some motor vehicle data for the 48 contiguous states (i.e., excluding Alaska and Hawaii) obtained from the 1988 World Almanac. The variables in the table are defined as follows:

POP	population of the state in thousands of people
TAX	fuel tax in cents/gallon
BELT	has the value 1 if seat belt law in effect as of July, 1987; otherwise 0
DEATHS	1986 motor vehicle deaths
DTHRATE	1986 motor vehicle deaths/100,000 people
PRLIC	proportion of the population with driver's licenses
PCFUEL	per capita fuel consumption (thousands of gallons/person in the state)

(a) Using any of the standard statistical packages, find the summary statistics and the correlation matrix for the variables mentioned.

(b) It seems reasonable to hypothesize that for a given population, the number of motor vehicle deaths should increase both with the number of people who drive (indexed by PRLIC) and the average amount driven (indexed by PCFUEL). To consider this, (i) regress DTHRATE on PRLIC, (ii) regress DTHRATE on PCFUEL, (iii) regress DTHRATE on both PRLIC and PCFUEL.

(c) Explain why the coefficients of PRLIC and PCFUEL in (i) and (ii) are different than in (iii).

(d) Plot DTHRATE against PRLIC and PCFUEL.

(e) In doing the regressions in part (b), some cases seem to have undue influence. Identify these cases for each of the regressions and speculate why the cases may have undue influence.

TABLE DATA FOR EXERCISE 15.2

STATE	POP	TAX	BELT	DEATHS	DTHRATE	PRLIC	PCFUEL
AL	3894	13.0	0	1080	27.735	0.638	0.615
AZ	2717	16.0	0	1002	36.879	0.881	0.702
AR	2286	13.5	0	601	26.290	0.773	0.653
CA	23668	9.0	1	5223	22.068	0.753	0.572
CO	2890	18.0	1	603	20.865	0.798	0.635
CN	3108	17.0	1	455	14.640	0.751	0.515
DE	594	13.0	0	138	23.232	0.764	0.608
FL	9747	9.7	1	2875	29.496	0.822	0.654
GA	5463	7.5	0	1540	28.190	0.724	0.740
ID	944	14.5	1	258	27.331	0.748	0.585
IL	11429	13.0	1	1616	14.139	0.615	0.485
IN	5490	14.0	1	1038	18.907	0.659	0.576
IA	2914	16.0	1	440	15.100	0.644	0.553
KS	2364	11.0	1	500	21.151	0.699	0.627
KY	3660	15.0	0	808	22.077	0.621	0.600
LA	4206	16.0	1	844	20.067	0.663	0.573
ME	1125	14.0	0	208	18.489	0.726	0.632
MD	4217	13.5	1	790	18.734	0.700	0.558
MA	5377	11.0	0	752	13.985	0.711	0.502
MI	9262	15.0	1	1624	17.534	0.681	0.511
MN	4076	17.0	1	572	14.033	0.614	0.552
MS	2521	9.0	0	766	30.385	0.727	0.610
MO	4917	10.0	1	1135	23.083	0.696	0.637
MT	787	17.0	1	222	28.208	0.747	0.695
NE	1570	16.7	0	290	18.471	0.694	0.621
NV	801	13.0	1	233	29.089	0.891	0.769
NH	921	14.0	0	172	18.675	0.819	0.591
NJ	7365	8.0	1	1040	14.121	0.802	0.551
NM	1303	11.0	1	498	38.219	0.771	0.738
NY	17558	8.0	1	2114	12.040	0.568	0.408
NC	5880	15.5	1	1640	27.891	0.714	0.646
ND	653	13.0	0	100	15.314	0.686	0.730
OH	10798	12.0	1	1673	15.494	0.679	0.533
OK	3025	10.0	1	710	23.471	0.753	0.708
OR	2633	11.0	1	619	23.509	0.737	0.589
PA	11865	12.0	0	1928	16.249	0.643	0.425
RI	947	15.0	0	114	12.038	0.661	0.451
SC	3121	13.0	0	1058	33.899	0.694	0.648
SD	690	13.0	0	134	19.420	0.703	0.730
TN	4591	17.0	1	1232	26.835	0.669	0.677
TX	14226	10.0	1	3568	25.081	0.756	0.760
UT	1461	14.0	1	311	21.287	0.677	0.628
VT	511	13.0	0	109	21.331	0.771	0.587
VA	5347	11.0	1	1118	20.909	0.722	0.623
WA	4132	18.0	1	714	17.280	0.729	0.544
WV	1950	10.5	0	437	22.410	0.671	0.479
WI	4706	17.5	0	757	16.086	0.693	0.514
WY	470	8.0	0	168	35.745	0.664	0.968

(f) Suppose we refer to the residuals when DTHRATE is regressed on PCFUEL as RDF and the residuals when PRLIC is regressed on PCFUEL as RLF. If we now regress RDF on RLF, we find that the regression coefficient for RLF is exactly the same as the regression coefficient of PRLIC when DTHRATE is regressed on both PRLIC and PCFUEL. Explain why this is the case.

(g) Can we conclude anything from the data set about whether seat belt laws reduce traffic fatalities?

15.3 In a visual "search" experiment, a subject is presented with a display containing an array of letters and makes a response when he or she detects the presence of a specific "target letter" that was specified beforehand. Arrays can differ in the number of letters they contain and (because of differences in brightness and contrast or the presence of visual "noise") how difficult it is to identify the letters. We simulated the results of such an experiment in which number of letters and identification difficulty were varied orthogonally, using the model

$$\text{TIME} = 400 + 30 \times \text{NUMBER} + 2 \times \text{DIFF} + \epsilon$$

where NUMBER stands for the number of letters in the array (2, 4, 6, or 8), DIFF stands for identification difficulty (10 or 20 units), and ϵ is a number selected randomly from a normal population with mean 0 and standard deviation 40, to generate the 24 cases in the following table:

TIME	NUMBER	DIFF
493	2	10
504	2	10
483	2	10
508	2	20
573	2	20
515	2	20
533	4	10
490	4	10
614	4	10
623	4	20
585	4	20
542	4	20
559	6	10
576	6	10
618	6	10
598	6	20
686	6	20
656	6	20
705	8	10
602	8	10
570	8	10
672	8	20
629	8	20
709	8	20

(a) Find the summary statistics and correlation matrix for these data.

(b) Regress TIME on NUMBER and DIFF. Are the effects of NUMBER and DIFF significant at $\alpha = .05$? Are these significance tests equivalent to the tests of the NUMBER and DIFF main effects in a standard ANOVA? Perform an ANOVA on TIME, using the factors NUMBER and DIFF and compare the results with those that follow from the regression.

(c) What are the estimates of the parameters of the model that are obtained from the regression? How do these compare with the actual parameter values ($\beta_0 = 400$, $\beta_1 = 30$, and $\beta_2 = 2$) that were used to generate the data? What are the 95% and 99% confidence intervals for β_0, β_1, and β_2?

It should be emphasized that in the real world one does not know what the parameters of the model are, or even the form of the model. One uses data to try to infer something about the underlying model.

15.4 In an experiment designed to determine the effects of drug dosage on performance, the following data are obtained:

Dosage in milligrams			
10	20	30	40
6.8	10.4	10.7	8.9
2.8	6.4	14.4	12.5
5.2	13.1	15.9	12.7
4.8	8.7	10.6	7.4
	12.4		8.5
	7.2		

(a) Do an ANOVA to test the effect of dosage on performance.

(b) Because dosage is a quantitative variable, it is decided to perform a trend analysis. Find the SS's for linear, quadratic, and cubic trends and determine which, if any, of the trend components are significant.

(c) Find the best-fitting polynomial regression equation of degree 2—that is, of the form

$$\hat{Y} = b_0 + b_1 X + b_2 X^2$$

that expresses performance as a function of dosage.

15.5 Calculate the adjusted R^2 given the following information:

(a) $R^2_{Y.1234} = .50$, $N = 10$

(b) $R^2_{Y.1234} = .50$, $N = 40$

(c) $R^2_{Y.1234} = .50$, $N = 200$

(d) $R^2_{Y.12} = .30$, $N = 40$

15.6 For the following 17 cases:

X_1	X_2	Y
2	5	42, 65
2	6	55
3	5	68, 55, 65
3	6	79, 59, 74, 67
3	7	97, 75, 80, 78
4	6	83, 72
4	7	92

(a) Fit the model $Y = \beta_0 + \beta_1 X_1 + \beta_2 X_2 + \epsilon$.

(b) Apply the method introduced in Section 12.10 to test for lack of fit. That is, test for nonlinearity, using "pure error" as the error term.

(c) From the data, does it seem that both X_1 and X_2 should be included in the regression model?

15.7 Given the following data set:

Y	X_1	X_2
4	2	-2
1	-1	-1
5	-2	0
7	-1	1
12	2	2

(a) Verify that X_1 and X_2 are uncorrelated.

(b) Verify that in this case (X_1 and X_2 uncorrelated) $R_{Y.12}^2 = r_{Y1}^2 + r_{Y2}^2$.

(c) For this data set, what is the relation between (i) the regression coefficient for X_1 when Y is regressed on X_1 alone and (ii) the regression coefficient for X_1 when Y is regressed on both X_1 and X_2? Is this true in general? What is the relation between the $\hat{SE}(b_1)$'s in (i) and (ii)?

15.8 Consider the data in Table 15.2. A new student enrolling in the course has a pretest score of 33 and a quantitative SAT score of 700.

(a) What is the best prediction for this student's score on the final examination?

(b) When we regress final exam score on pretest score and SAT, we find from the computer output that the standard error of estimate is 12.282 and that

$$(\mathbf{X'X})^{-1} = \begin{bmatrix} 4.6475795 & -0.0023286 & -0.0075299 \\ -0.0023286 & 0.0022683 & -0.0001069 \\ -0.0075299 & -0.0001069 & 0.0000177 \end{bmatrix}$$

Use this information to find the 95% confidence interval for the student's score on the final exam.

(c) What is the 95% confidence interval for the conditional mean of Y scores corresponding to pretest = 33 and SAT = 700?

15.9 Use the information provided in the previous problem to find the covariance matrix for b_0, b_1, and b_2. What is the corresponding correlation matrix for the coefficients?

15.10 Consider 15 cases selected randomly from the same population:

Y	X_1	X_2	X_3
.15	.07	.27	.66
.09	1.08	.51	1.61
−3.36	−.40	−.78	1.38
−1.34	−1.25	−2.92	−2.38
−.69	−1.31	−.69	.20
−1.15	2.16	−.31	−.60
−.51	2.26	−.83	−.16
−1.47	3.69	−.82	−.02
−.38	−1.44	.34	2.60
.78	−1.80	.23	2.56
−3.60	−.76	−.60	2.29
−2.40	−2.69	2.33	4.14
1.46	−1.33	.51	.20
−.71	−.45	.10	−.76
1.42	.09	.72	.07

(a) Using only the first four cases, regress Y on X_1, X_2, and X_3. What is the value of $R_{Y.123}$? Comment on why $R_{Y.123}$ must take on the value that it does here.

(b) Use the regression equation obtained in (a) to predict the values of Y for each of the 15 cases.

(c) Find the correlations between the scores predicted in (b) and the actual Y values (i) for the first four cases and (ii) for the remaining 11 cases. Comment on the difference between these correlations.

15.11 In the following data set, Y is a measure of verbal achievement for students in elementary school, X_1 and X_2 are measures of school and teacher quality, and X_3 and X_4 are measures of student and parent background:

Y	X_1	X_2	X_3	X_4		Y	X_1	X_2	X_3	X_4
7.4	7.6	5.32	17.2	2.9		4.6	4.2	4.70	−2.9	1.2
5.3	5.8	4.88	−1.7	2.0		7.0	5.0	4.72	10.9	2.2
7.3	5.9	5.14	22.3	6.9		6.9	4.4	4.90	14.8	1.4
8.1	5.8	5.15	24.2	6.5		6.6	5.3	5.16	9.0	3.2
7.4	6.1	5.08	16.3	3.0		4.5	5.4	5.04	−6.1	1.2
6.7	4.1	4.32	16.2	4.5		7.9	6.3	5.01	20.6	6.8
8.3	5.0	4.98	22.7	7.7		6.4	7.1	5.00	12.7	4.2
6.7	4.9	5.00	9.8	2.5		6.3	5.0	4.96	−1.1	1.7
8.2	6.2	5.32	19.9	6.5		8.6	5.4	5.11	25.1	8.6
7.4	4.8	5.60	10.0	1.0		8.2	4.7	5.10	22.8	7.7

(a) Do the school and teacher measures contribute to the predictability of Y? Do the student and parent measures? Consider regression equations containing different combinations of predictors in arriving at your answer.

(b) Perform a stepwise regression, using one of the standard packages. Do you think that the regression equation identified by the stepwise regression is a reasonable explanatory model of the situation?

15.12 (a) A researcher is interested in relating measures of mother-child attachment to measures of externalization and criticism obtained from a series of interviews. A regression of the attachment measure on both externalization and criticism yields significant t tests for both the predictor variables. What can be concluded?

(b) The researcher then decides to determine whether the joint effect of externalization and criticism is an important predictor of attachment. She creates a new variable by multiplying the externalization and criticism measures for each case. She then regresses attachment on the externalization and criticism measures as well as the new variable. The regression now shows that none of the t tests for the three predictors are significant. Is this an appropriate way to assess whether the joint effect of externalization and criticism is an important predictor of attachment? Why or why not? What is the most likely reason that the t tests for the coefficients of externalization and criticism are not significant in the second regression even though they were significant in the first regression described in part (a)? Considering the results of the two regressions together, what can be concluded?

Chapter **16**

Regression with Categorical Variables

16.1 INTRODUCTION

In this final chapter, we consider categorical variables with levels that differ qualitatively from one another. Examples of such variables are gender, with levels female and male; diagnosed mental illness, with levels schizophrenia, depression, and anxiety disorder; and treatment condition, with levels defined by the different conditions. Our development of regression to this point has focused on quantitative predictor and criterion variables. In Chapter 15, we discussed trend analysis with quantitative categorical variables from the point of view of multiple regression. However, qualitative categorical variables can also be incorporated into regression analyses, providing us with a general and powerful framework within which many of the analyses we have previously considered, including ANOVA and ANCOVA, can usefully be considered as special cases. This framework can both increase our understanding of how different kinds of analyses are related to one another and allow us to deal with data from designs that cannot be handled easily by the standard ANOVA approach.

In ANOVA, all factors are treated as though they were qualitative categorical variables. ANOVA is an elegant system for partitioning the variability in factorial designs. However, the system breaks down when confronted with data from multifactor designs with disproportionate cell frequencies (see Section 5.9). The usual sums of squares are no longer orthogonal, as we can see in a design with factors A and C, when SS_A, SS_C, and SS_{AC} no longer add to the between-cell sum of squares. Multiple regression is a system for evaluating the contributions of variables that are not orthogonal to one another. Therefore, the issues involved in dealing with nonorthogonality are easier to understand within the multiple regression framework, and different kinds of multiple regression analyses are the appropriate generalizations of ANOVA for nonorthogonal designs.

We will deal with nonorthogonal factorial designs in Section 16.3 and then go on to reconsider, within the multiple regression framework, several other types of analyses that we discussed previously. However, we first consider how categorical variables may be coded so that they can be included in regression analyses and then introduce a procedure that allows us to test a number of contrasts simultaneously. This latter procedure will prove to be valuable in Section 16.3 when we consider exactly what hypotheses on the cell means are tested by regression analyses of nonorthogonal designs.

16.2 ONE-FACTOR DESIGNS

16.2.1 Coding Categorical Variables

Any categorical variable can be coded by defining one or more *dummy variables* (also called *indicator variables*) that take on numerical values. These numbers are not measures of the categories; rather, they are best thought of as labels that together specify category membership.

A categorical variable A that has only two levels can be coded by a single dummy variable X that takes on any two numerical values, one to label instances of each of the levels. Furthermore, we can show that the overall test of the regression of Y on X is exactly equivalent to the ANOVA F test for the same data. For example, suppose we define a dummy variable X that takes on the values 1 and 2 for scores at A_1 and A_2, respectively. If we regress Y on X, the regression line must pass through the points $(1, \bar{Y}_{.1})$ and $(2, \bar{Y}_{.2})$ (see Figure 16.1). The regression line must pass through these two points because it is defined as the line that minimizes the mean of the squared deviations of the Y's and the group means minimize the mean of the squared deviations in each of the groups. Because the regression accounts for all the variability in the group means, SS_{reg} must equal the between-group variability, SS_A; also, SS_{error}, the variability unaccounted for by the regression, must equal $SS_{S/A}$. Therefore, MS_{reg}/MS_{error} is identical to $MS_A/MS_{S/A}$ (see Section 12.6 for a more formal development of this point).

However, if A has three levels, regression on a single dummy variable will not, in general, account for all of SS_A. Panel a of Table 16.1 presents scores at levels A_1, A_2, and A_3. If we code A with a single dummy variable, X_1, that takes on the values 1, 2, and 3, as in panel b, the points that represent the group means in the space defined by X and Y will be $(1, \bar{Y}_{.1})$, $(2, \bar{Y}_{.2})$, and $(3, \bar{Y}_{.3})$; see panel a of Figure 16.2. In general, it will not be possible to fit these three points perfectly with a straight line; therefore, if Y is regressed on X_1, SS_{reg} will be less than SS_A. However, we can account for all of SS_A by defining an additional dummy variable, X_2, that is linearly independent[1] of X_1. If we represent the three group means in the three-dimensional space defined by Y, X_1, and X_2, it is apparent that they can be perfectly fit by a regression plane; see panel b of Figure 16.2. Therefore, when Y is regressed on both dummy variables, all the between-group variability must be accounted for by the regression, so $SS_{reg} = SS_A$, and the F test for the overall regression of Y on X_1 and X_2 must have the same value as the ANOVA F. The more levels A has, the more dummy variables it will take to code it. In general, if A has a levels, it will take $a - 1$ linearly independent dummy variables to account for all of SS_A.

---

[1] As we shall see, the actual values of the $a - 1$ dummy variables can be chosen in a variety of ways. The only real restriction on a set of dummy variables is that no one of them may be obtained as a linear combination of the others. If this was to happen, the dummy variable would be completely redundant and would not contribute anything to the specification of the categories. Sets of variables that obey this restriction are termed *linearly independent*.

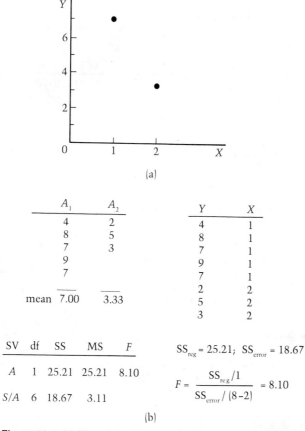

Figure 16.1 (a) Plot of the means of A against X for the data in panel b; the points representing the means can be fit perfectly by a straight line. (b) Data for graph and results of ANOVA and multiple regression.

16.2.2 Effect Coding

In discussing ANOVA as a special case of multiple regression, we will find it useful to consider one type of dummy variable coding that is called *effect* coding because, as we show below, it produces regression coefficients that estimate the ANOVA effects α_1, $\alpha_2, \ldots, \alpha_{a-1}$ (two other types, *dummy* and *contrast* coding are presented in the chapter appendix). Effect coding represents group membership with dummy variables that contain 1's, 0's, and -1's as illustrated by X_{E1} and X_{E2} in panel b of Table 16.1. Each dummy variable is defined as follows:

$$X_{Ej} = \begin{array}{ll} 1 & \text{for scores at level } A_j \\ -1 & \text{for scores at some arbitrary level of } A, \text{ here } A_3 \\ 0 & \text{otherwise} \end{array}$$

Because for the data of Table 16.1 there are three levels of A, we require two dummy variables to account for all the between-group variability. Scores at A_1 receive

TABLE 16.1   REGRESSION ON DUMMY VARIABLES FOR A ONE-FACTOR DESIGN

(a) Data and results of standard ANOVA for a one-factor design

	A_1	A_2	A_3
	4	2	4
	8	5	5
	7	3	3
	9		6
	7		
$\bar{Y}_{.j}$	7.00	3.33	4.50
$\hat{\sigma}_j$	1.87	1.53	1.29

SV	df	SS	MS	F
A	2	28.583	14.292	5.435
S/A	9	23.667	2.630	

(b) Dummy variable coding and some statistics obtained from the regressions of Y on X_1 and X_2 and on X_{E1} and X_{E2}

	Y	X_1	X_2	X_{E1}	X_{E2}
A_1	4	1	0	1	0
	8	1	0	1	0
	7	1	0	1	0
	9	1	0	1	0
	7	1	0	1	0
A_2	2	2	3	0	1
	5	2	3	0	1
	3	2	3	0	1
A_3	4	3	1	-1	-1
	5	3	1	-1	-1
	3	3	1	-1	-1
	6	3	1	-1	-1

Statistics for the regression of Y either on X_1 and X_2 or on X_{E1} and X_{E2}: $R^2 = .740$; $SS_{reg} = 28.583$; $SS_{error} = 23.667$; $F = MS_{reg}/MS_{error} = 5.435$.

a 1 on X_{E1} and a 0 on X_{E2}; scores at A_2 receive a 0 on X_{E1} and a 1 on X_{E2}; and scores at A_3 receive a -1 on both X_{E1} and X_{E2}. As we see in panel b of Table 16.1, the regression of Y on X_{E1} and X_{E2} produces a value of SS_{reg} that is equal to the SS_A obtained from the standard ANOVA and the overall F test for the regression is equal to the ANOVA F.

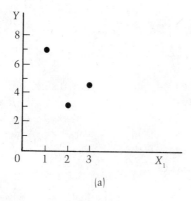

(a)

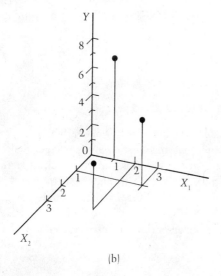

(b)

Figure 16.2 (a) Plot of the means of A against X_1 for the data of Table 16.1; in general, the points representing the three means cannot be fit perfectly by a straight line. (b) Plot of the means of A against X_1 and X_2; the points representing the three means can always be fit perfectly by a plane.

It is quite easy to find the regression coefficients. Because the regression accounts for all the between-group variability, the scores predicted by the regression equation for each of the groups are the group means. Therefore, for the scores at A_1, we have $\hat{Y} = \bar{Y}_{.1}$, $X_{E1} = 1$, and $X_{E2} = 0$. Substituting into the regression equation for Y on X_{E1} and X_{E2},

$$\hat{Y} = b_{E0} + b_{E1}X_{E1} + b_{E2}X_{E2}$$

we obtain

$$\bar{Y}_{.1} = b_{E0} + b_{E1}(1) + b_{E2}(0) = b_{E0} + b_{E1}$$

Similarly, for scores at the other two levels of A, we have

$$\bar{Y}_{.2} = b_{E0} + b_{E1}(0) + b_{E2}(1) = b_{E0} + b_{E2}$$

and

$$\bar{Y}_{.3} = b_{E0} + b_{E1}(-1) + b_{E2}(-1) = b_{E0} - b_{E1} - b_{E2}$$

Adding up the three equations, we obtain $3b_{E0} = \bar{Y}_{.1} + \bar{Y}_{.2} + \bar{Y}_{.3}$, so

$$b_{E0} = \frac{\bar{Y}_{.1} + \bar{Y}_{.2} + \bar{Y}_{.3}}{3} = \bar{Y}_U$$

Here, $\bar{Y}_U$ is the *unweighted* average of the group means. If there are equal numbers of scores in each of the groups, $\bar{Y}_U$ will equal $\bar{Y}_{..}$, the grand mean of all of the scores. However, for unequal-n designs, $\bar{Y}_{..}$ is the weighted average of the group means.

We can write the coefficients of the dummy variables as deviations from the unweighted average of the group means:

$$b_{E1} = \bar{Y}_{.1} - \bar{Y}_U \quad \text{and} \quad b_{E2} = \bar{Y}_{.2} - \bar{Y}_U$$

Note that for equal-n designs, the estimated main-effect components of the ANOVA, $\hat{\alpha}_j$, correspond exactly to the regression coefficients. That is,

$$b_{E1} = \hat{\mu}_1 - \hat{\mu} = \hat{\alpha}_1$$

$$b_{E2} = \hat{\mu}_2 - \hat{\mu} = \hat{\alpha}_2$$

Furthermore, because of the requirement that $\Sigma_j \alpha_j = 0$,

$$\hat{\alpha}_3 = -b_{E1} - b_{E2}$$

Although it is important to understand how categorical variables can be represented in regression, we should point out that many software packages will produce the required dummy variables for us, once we have appropriately specified category membership. In SYSTAT, for example, if we specify the levels of A as 1, 2, and 3 and give the instruction CATEGORY A $=$ 3 in the MGLH module, exactly the same dummy variables that we have discussed, X_{E1} and X_{E2}, will be produced. Also, including A in any subsequent MODEL statement will cause the dummy variables to be included in the regression.

16.2.3 Testing Hypotheses about a Set of Contrasts

We next introduce a procedure for simultaneously testing any set of k linearly independent contrasts on a set of a group means, where k can be any number less than or equal to $a - 1$.

In Chapter 6, we indicated that for any contrast, ψ,

$$SS_{\hat{\psi}} = \frac{(\Sigma_j w_j \bar{Y}_{.j})^2}{\Sigma_j w_j^2 / n_j} \tag{16.1}$$

and that the null hypothesis $H_0: \psi = \Sigma_j w_j \mu_j = 0$ can be tested by using the statistic $F = SS_{\hat{\psi}} / MS_{\text{error}}$.

We can test several contrasts simultaneously, using matrices and vectors analogous to the w_j and $\bar{Y}_{.j}$ of Equation 16.1. We begin by noting that the SS for any source of variance can be obtained from calculations employing a matrix of weights and a vector of group means. For example, reconsider the data of Table 16.1. Recall that ordinary ANOVA revealed that SS_A was 28.583. As shown in panel b of the table, a regression analysis provides the same result. Now, in an approach that parallels Equation 16.1, let us set up two contrasts among the means of A; any two contrasts, orthogonal or not, are permissible provided they are linearly independent. One possibility is

$$\psi_1 = (1)\mu_1 + (-1)\mu_2 + (0)\mu_3$$

and

$$\psi_2 = (2)\mu_1 + (-1)\mu_2 + (-1)\mu_3$$

We can form a 2×3 (in general, $k \times a$) matrix $\mathbf{W}$ with rows made up of the contrast weights

$$\mathbf{W} = \begin{bmatrix} 1 & -1 & 0 \\ 2 & -1 & -1 \end{bmatrix}$$

we also have a vector of group means

$$\bar{\mathbf{Y}} = \begin{bmatrix} 7.000 \\ 3.333 \\ 4.500 \end{bmatrix}$$

Now, we can define a matrix equation that is analogous to Equation 16.1:

$$SS_{\hat{\psi}} = (\mathbf{W}\bar{\mathbf{Y}})'[\mathbf{W}\mathbf{N}^{-1}\mathbf{W}']^{-1}(\mathbf{W}\bar{\mathbf{Y}}) \tag{16.2}$$

In this equation, $\mathbf{N}$ is an $a \times a$ matrix that indicates the number of scores in each of the a groups on its major diagonal; that is,

$$\mathbf{N} = \begin{bmatrix} n_1 & 0 & 0 & \cdots & 0 \\ 0 & n_2 & 0 & \cdots & 0 \\ 0 & 0 & n_3 & \cdots & 0 \\ \cdot & \cdot & \cdot & \cdots & 0 \\ \cdot & \cdot & \cdot & \cdots & 0 \\ \cdot & \cdot & \cdot & \cdots & 0 \\ 0 & 0 & 0 & \cdots & n_a \end{bmatrix}$$

so

$$\mathbf{N}^{-1} = \begin{bmatrix} 1/n_1 & 0 & 0 & \cdots & 0 \\ 0 & 1/n_2 & 0 & \cdots & 0 \\ 0 & 0 & 1/n_3 & \cdots & 0 \\ \cdot & \cdot & \cdot & \cdots & 0 \\ \cdot & \cdot & \cdot & \cdots & 0 \\ \cdot & \cdot & \cdot & \cdots & 0 \\ 0 & 0 & 0 & \cdots & 1/n_a \end{bmatrix}$$

For the data in Table 16.1, $n_1 = 5$, $n_2 = 3$, and $n_3 = 4$. Therefore,

$$\mathbf{N}^{-1} = \begin{bmatrix} 1/5 & 0 & 0 \\ 0 & 1/3 & 0 \\ 0 & 0 & 1/4 \end{bmatrix}$$

Substituting into Equation 16.2, we obtain

$$SS_{\hat{\psi}} = [3.667 \quad 6.167] \begin{bmatrix} .533 & .733 \\ .733 & 1.383 \end{bmatrix}^{-1} \begin{bmatrix} 3.667 \\ 6.167 \end{bmatrix}$$

$$= 28.58.$$

This calculation tells us that the two contrasts whose weights make up the matrix $\mathbf{W}$ together account for SS_A. In general, the null hypothesis

$$H_0: \mathbf{W}\boldsymbol{\mu} = 0$$

can be tested by the statistic $F = (SS_{\hat{\psi}}/k)/MS_{error}$. For the two contrasts in the current example, this statistic produces a result equal to the ANOVA F.

Why bother with this procedure? After all, we have just calculated the same value of SS_A by using standard ANOVA and by using a multiple regression in which A was represented by dummy variables; moreover, we stated in Chapter 6 that $a - 1$ linearly independent contrasts must account for all the between-group variability in an a-group design. The reason we introduce the simultaneous contrast procedure here is that it can be used very generally to determine the SS's that are associated with tests of particular hypotheses. Because of this property, we will find the procedure to be very useful when we consider nonorthogonal factorial designs in the next section. In nonorthogonal designs, unequal cell frequencies cause correlations between the different factors, resulting in overlapping SS's. We have to decide whether to adjust for this overlap, and, if so, what kind of adjustment to use. Equation 16.2 produces the SS's associated with particular sets of contrasts on the cell means. Therefore, we can use Equation 16.2 to determine which hypotheses are tested by different methods of adjusting for the overlap in variability among main effects and interactions.

16.3 REGRESSION ANALYSIS AND FACTORIAL DESIGNS

In Section 16.2, we saw how any categorical variable can be coded by a set of dummy variables and that a multiple regression analysis that uses these dummy variables as predictors provides all of the information, and more, that can be obtained from a one-factor ANOVA. In Section 16.3 we extend this discussion to multifactor designs, first considering orthogonal designs and then the issues that arise in analyzing data from nonorthogonal or unbalanced (unequal-n) designs.

16.3.1 Orthogonal Designs

A regression analysis of a factorial design can be performed if both the factors and their interactions are coded by sets of dummy variables. Panel a of Table 16.2 contains data from a 3×3 design with factors A and C, and panel c contains the results of an ANOVA on the data. Panel b contains the sets of effect dummy variables that code the design. Each set of dummy variables has as many members as the corresponding sources of variance have df's. Variables A and C are coded as though each was the only factor in the design, and the set of four dummy variables that code the $A \times C$ interaction is obtained by multiplying each dummy variable in the A set by each one in the C set. Together, the eight dummy variables code membership in the nine cells of the design.

With effect coding, the dummy variables within any one of the A, C, and AC sets are correlated. However, if the cell frequencies are all equal, the *sets* of dummy variables are not correlated—that is, the dummy variables in any set are uncorrelated with all the dummy variables in each of the other sets—and the SS's associated with the different sets do not overlap. Let's use the notation $R_{Y.A}$, $R_{Y.AC}$, and $R_{Y.A,AC}$ to represent the multiple correlation coefficients that result when Y is regressed on the sets of dummy variables that code A, AC, and both A and AC, respectively. Then, because the sets of dummy variables corresponding to A, C, and AC are uncorrelated, we have

$$R_{Y.A,C,AC}^2 = R_{Y.A}^2 + R_{Y.C}^2 + R_{Y.AC}^2$$

Multiplying each of the squared correlations by SS_Y, we have

$$SS_{\text{Between cell}} = SS_A + SS_C + SS_{AC}$$

Tests of the A and C main effects and the AC interaction are provided by

$$F = \frac{R_{Y.A}^2/(a-1)}{(1 - R_{Y.A,C,AC}^2)/(N - ac)}$$

$$F = \frac{R_{Y.C}^2/(c-1)}{(1 - R_{Y.A,C,AC}^2)/(N - ac)}$$

and

$$F = \frac{R_{Y.AC}^2/(a-1)(c-1)}{(1 - R_{Y.A,C,AC}^2)/(N - ac)}$$

As can be seen in Table 16.2, these test statistics have exactly the same values as the ANOVA F's for A, C, and $A \times C$.

Because it takes several dummy variables to define each of A, C, and $A \times C$, the regression coefficients of the individual dummy variables are not likely to be of great interest. However, if desired, expressions for the coefficients that result from a simultaneous regression on all eight dummy variables can be obtained by substituting into the equation

$$\hat{Y} = b_0 + b_1 X_1 + b_2 X_2 + b_3 X_3 + b_4 X_4 + b_5 X_5 + b_6 X_6 + b_7 X_7 + b_8 X_8$$

TABLE 16.2 EFFECT CODING FOR AN ORTHOGONAL 3 × 3 DESIGN

(a) Data

	C_1	C_2	C_3
A_1	53	88	56
	51	63	42
A_2	55	48	79
	78	42	50
A_3	79	80	69
	99	92	94

(b) Dummy variables formed by using effect coding

Effect:		A		C		AC			
	Y	X_1	X_2	X_3	X_4	X_5	X_6	X_7	X_8
$A_1 C_1$	53	1	0	1	0	1	0	0	0
	51	1	0	1	0	1	0	0	0
$A_1 C_2$	88	1	0	0	1	0	0	1	0
	63	1	0	0	1	0	0	1	0
$A_1 C_3$	56	1	0	-1	-1	-1	0	-1	0
	42	1	0	-1	-1	-1	0	-1	0
$A_2 C_1$	55	0	1	1	0	0	1	0	0
	78	0	1	1	0	0	1	0	0
$A_2 C_2$	48	0	1	0	1	0	0	0	1
	42	0	1	0	1	0	0	0	1
$A_2 C_3$	79	0	1	-1	-1	0	-1	0	-1
	50	0	1	-1	-1	0	-1	0	-1
$A_3 C_1$	79	-1	-1	1	0	-1	-1	0	0
	99	-1	-1	1	0	-1	-1	0	0
$A_3 C_2$	80	-1	-1	0	1	0	0	-1	-1
	92	-1	-1	0	1	0	0	-1	-1
$A_3 C_3$	69	-1	-1	-1	-1	1	1	1	1
	94	-1	-1	-1	-1	1	1	1	1

(c) ANOVA table for the data in panel a obtained by using either standard analysis of variance or multiple regression

SV	df	SS	MS	F
A	2	2862.333	1431.167	7.577
C	2	64.333	32.167	0.170
AC	4	1399.333	349.833	1.852
Error	9	1700.000	188.889	

This leads to nine equations, one for each cell, that can be solved for the b's. Using effect coding in an orthogonal design leads to regression coefficients that are equal to the best estimates of the main effect and interaction components in the corresponding ANOVA. Here, for example, $b_1 = \hat{\alpha}_1$, $b_2 = \hat{\alpha}_2$, $b_3 = \hat{\gamma}_1$, and $b_5 = (\widehat{\alpha\gamma})_{11}$.

16.3.2 Nonorthogonal Designs

In Section 5.9, we discussed some of the difficulties that stem from the fact that in nonorthogonal factorial designs, the between-cell variability cannot be neatly partitioned into nonoverlapping components associated with the main effects and interactions. Panel a of Table 16.3 contains the data for a nonorthogonal design with factors A and C, and panel b contains the effect coding for the design. Because of the unequal cell frequencies, the sets of dummy variables that code the A, C, and AC effects are no longer uncorrelated. Therefore,

$$R^2_{Y.A,C,AC} \neq R^2_{Y.A} + R^2_{Y.C} + R^2_{Y.AC}$$

and the variabilities associated with A, C, and AC overlap, as represented by Figure 16.3.

Multiple regression analysis allows a variety of possible adjustments for this overlap. For example, in considering the A effect:

1. We may decide to use only the variability that is uniquely associated with A— that is, the variability in A that does not overlap with the other effects in the design. This variability is represented by area t in the upper circle of Figure 16.3 and can be obtained from

$$SS_{A|C,AC} = (R^2_{Y.A,C,AC} - R^2_{Y.C,AC})SS_Y$$

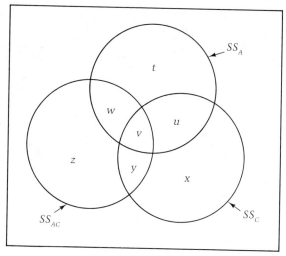

Figure 16.3 Partitioning of variability in a nonorthogonal two-factor design.

TABLE 16.3   EFFECT CODING FOR A NONORTHOGONAL 3 × 3 DESIGN

(a) Data

	C_1	C_2	C_3
A_1	53	88	56
	51	63	42
		50	
		71	
A_2	55	48	79
	78	42	50
	39		62
A_3	79	80	69
	99	92	94
			80
			77

(b) Dummy variables formed by using effect coding

Effect:	Y	A		C		AC			
		X_1	X_2	X_3	X_4	X_5	X_6	X_7	X_8
$A_1 C_1$	53	1	0	1	0	1	0	0	0
	51	1	0	1	0	1	0	0	0
$A_1 C_2$	88	1	0	0	1	0	0	1	0
	63	1	0	0	1	0	0	1	0
	50	1	0	0	1	0	0	1	0
	71	1	0	0	1	0	0	1	0
$A_1 C_3$	56	1	0	-1	-1	-1	0	-1	0
	42	1	0	-1	-1	-1	0	-1	0
$A_2 C_1$	55	0	1	1	0	0	1	0	0
	78	0	1	1	0	0	1	0	0
	39	0	1	1	0	0	1	0	0
$A_2 C_2$	48	0	1	0	1	0	0	0	1
	42	0	1	0	1	0	0	0	1
$A_2 C_3$	79	0	1	-1	-1	0	-1	0	-1
	50	0	1	-1	-1	0	-1	0	-1
	62	0	1	-1	-1	0	-1	0	-1
$A_3 C_1$	79	-1	-1	1	0	-1	-1	0	0
	99	-1	-1	1	0	-1	-1	0	0
$A_3 C_2$	80	-1	-1	0	1	0	0	-1	-1
	92	-1	-1	0	1	0	0	-1	-1
$A_3 C_3$	69	-1	-1	-1	-1	1	1	1	1
	94	-1	-1	-1	-1	1	1	1	1
	80	-1	-1	-1	-1	1	1	1	1
	77	-1	-1	-1	-1	1	1	1	1

2. We may decide to adjust the A effect only for the other main effect C, yielding the variability represented by areas t and w in the upper circle of Figure 16.3:

$$SS_{A|C} = (R_{Y.A,C}^2 - R_{Y.C}^2)SS_Y$$

3. We may decide not to adjust for the contributions of the other effects. This yields

$$SS_A = R_{Y.A}^2 SS_Y$$

Although each of these three expressions has the same value if the cell frequencies are all equal, they may differ considerably in nonorthogonal designs. Also, note that the covariation represented by the overlapping areas in Figure 16.3 can be positive or negative. Therefore, the adjusted variabilities may be smaller or larger than their unadjusted values. For the data in Table 16.3, $SS_{A|C,AC} = 4139.42$, $SS_{A|C} = 3609.95$, and $SS_A = 3581.08$.

On what basis are we to decide which, if any, of these SS's to use? As we stated in Section 5.9, our view is that the proper analysis depends upon the weights that should be given to the levels of the factors in the study. If the variables are manipulated, or if the data in the cells can be viewed as samples from naturally occurring, equal-sized treatment populations, it makes sense to give each cell in the design the same weight, even though chance variations in actual cell frequency may occur. That is, we may plan to have equal cell frequencies, but fail to obtain them because of chance occurrences such as equipment failures or subjects failing to show up. In this case, the overall A null hypothesis that is of interest states that the *unweighted* averages of the c population means at each level of A are equal. That is,

$$\mu_{1.} = \mu_{2.} = \cdots = \mu_{a.}$$

where

$$\mu_{j.} = \frac{1}{c}\sum_k \mu_{jk}$$

However, if the cell populations vary systematically in size, we may wish to test hypotheses in which the cell means are weighted according to population size. If we have reliable information about the relative sizes of the cell populations, we can use it to weight the cell means (this can be done conveniently in programs such as BMDP4V by using the WEIGHT command). If we do not have such information, we can use the cell frequencies as weights. In this case, the overall A null hypothesis of interest states that the weighted means of A are equal. That is,

$$\mu_{1*} = \mu_{2*} = \mu_{3*}$$

where

$$\mu_{j*} = \frac{1}{n_{j.}}\sum_k n_{jk}\mu_{jk}$$

The procedure introduced in Section 16.2.2 can be used to test any set of contrasts on the cell population means, μ_{jk}. With this procedure, we can show that the unweighted-means null hypothesis is tested if $SS_{A|C,AC}$ is used and the weighted-means hypothesis is tested with SS_A. We begin by noting that the overall A null hypothesis, $\mu_{1.} = \mu_{2.} = \mu_{3.}$, is true only if the two contrasts

$$\psi_1 = \mu_{1.} - \mu_{3.} \quad \text{and} \quad \psi_2 = \mu_{2.} - \mu_{3.}$$

are both 0. Therefore, a simultaneous test of both contrasts corresponds to a test of the overall hypothesis. The weights on the cell means for the two contrasts are

	μ_{11}	μ_{12}	μ_{13}	μ_{21}	μ_{22}	μ_{23}	μ_{31}	μ_{32}	μ_{33}
ψ_1	⅓	⅓	⅓	0	0	0	$-⅓$	$-⅓$	$-⅓$
ψ_2	0	0	0	⅓	⅓	⅓	$-⅓$	$-⅓$	$-⅓$

We can form a 2×9 matrix $\mathbf{W}$ from these contrast weights. We can also obtain $\mathbf{N}^{-1}$ and $\overline{\mathbf{Y}}$ from the data of Table 16.3:

$$\mathbf{N}^{-1} = \begin{bmatrix} ½ \\ & ¼ \\ & & ½ & & & & 0 \\ & & & ⅓ \\ & & & & ½ \\ & & & & & ⅓ \\ & & 0 & & & & ½ \\ & & & & & & & ½ \\ & & & & & & & & ¼ \end{bmatrix}$$

and

$$\overline{\mathbf{Y}} = \begin{bmatrix} 52.00 \\ 68.00 \\ 49.00 \\ 57.33 \\ 45.00 \\ 63.67 \\ 89.00 \\ 86.00 \\ 80.00 \end{bmatrix}$$

Substituting $\mathbf{W}$, $\overline{\mathbf{Y}}$, and $\mathbf{N}^{-1}$ into Equation 16.2, we find that $SS_{\hat{\psi}} = 4139.42$. This value is exactly what is obtained by using $SS_{A|C,AC} = (R^2_{Y.A,C,AC} - R^2_{Y.C,AC})SS_Y$, suggesting that if we adjust SS_A for the contributions of both C and AC, we test the overall unweighted means hypothesis.

On the other hand, the hypothesis that the *weighted* means of A are equal will be true only if

$$\mu_{1*} - \mu_{3*} = 0 \quad \text{and} \quad \mu_{2*} - \mu_{3*} = 0$$

Forming a 2×9 matrix $\mathbf{W}$ of the contrast weights

μ_{11}	μ_{12}	μ_{13}	μ_{21}	μ_{22}	μ_{23}	μ_{31}	μ_{32}	μ_{33}
$n_{11}/n_{1.}$	$n_{12}/n_{1.}$	$n_{13}/n_{1.}$	0	0	0	$-n_{31}/n_{3.}$	$-n_{32}/n_{3.}$	$-n_{33}/n_{3.}$
0	0	0	$n_{21}/n_{2.}$	$n_{22}/n_{2.}$	$n_{23}/n_{2.}$	$-n_{31}/n_{3.}$	$-n_{32}/n_{3.}$	$-n_{33}/n_{3.}$

and substituting $\mathbf{W}$ and the same $\overline{\mathbf{Y}}$ and $\mathbf{N}^{-1}$ as before into Equation 16.2, we obtain $SS_{\hat{\psi}} = 3581.08$, the same value obtained for $SS_A = R^2_{Y.A}SS_Y$. This suggests that not adjusting SS_A results in a test of the overall weighted means hypothesis. These kinds of results can be shown to hold generally.

The different kinds of adjustments form the basis of three regression analysis procedures that might be considered as generalizations of ANOVA for nonorthogonal factorial designs. Table 16.4 describes the three methods, indicates the hypotheses that are tested by each of them, and provides some information about software packages and

TABLE 16.4   THREE METHODS FOR ANALYZING FACTORIAL DESIGNS

Method 1: Adjusting for all main and interaction effects		
SV	df	SS
A	$a - 1$	$(R^2_{Y.A,C,AC} - R^2_{Y.C,AC})SS_Y$
C	$c - 1$	$(R^2_{Y.A,C,AC} - R^2_{Y.A,AC})SS_Y$
AC	$(a - 1)(c - 1)$	$(R^2_{Y.A,C,AC} - R^2_{Y.A,C})SS_Y$
Residual	$N - ac$	$(1 - R^2_{Y.A,C,AC})SS_Y$

Hypotheses tested

$$A: \quad \mu_{1.} = \mu_{2.} = \cdots = \mu_{a.}$$

where $\mu_{j.} = \Sigma_k \mu_{jk}/c$, the simple average of the c cell means in the jth row,

$$C: \quad \mu_{.1} = \mu_{.2} = \cdots = \mu_{.c}$$

$$AC: \quad \mu_{jk} - \mu_{j'k} - \mu_{jk'} + \mu_{j'k'} = 0 \quad \text{for all } j, j', k, k'$$

Usage: This method tests hypotheses about unweighted column and row means and will usually be the method of choice when unequal cell frequencies occur by chance.

Some names for the method: Overall and Spiegel's (1969) method 1, Yates's (1934) weighted squares of means, SPSS's classic regression approach.

Other information: default for BMDP2V and BMDP4V, default for SYSTAT MGLH, SAS Type III SS in GLM,[a] obtained by using SSTYPE = UNIQUE in SPSS^X MANOVA, default for SPSS^X MANOVA starting with release 2.1.

[a] For designs in which there are no empty cells, Type III and Type IV SS provide the same results.

TABLE 16.4 (continued)

Method 2: Adjusting for the effects of the same and lower order

SV	df	SS
A	$a - 1$	$(R^2_{Y.A,C} - R^2_{Y.C})SS_Y$
C	$c - 1$	$(R^2_{Y.A,C} - R^2_{Y.A})SS_Y$
AC	$(a - 1)(c - 1)$	$(R^2_{Y.A,C,AC} - R^2_{Y.A,C})SS_Y$
Residual	$N - ac$	$(1 - R^2_{Y.A,C,AC})SS_Y$

Hypotheses tested

$$A: \sum_k \left[n_{jk} - \frac{n_{jk}^2}{n_{.k}} \right] \mu_{jk} - \sum_{j \neq j'} \sum_k \left[\frac{n_{jk}n_{j'k}}{n_{.k}} \right] \mu_{j'k} = 0$$

$$\text{for } j = 1, 2, \ldots, a - 1$$

$$C: \sum_j \left[n_{jk} - \frac{n_{jk}^2}{n_{j.}} \right] \mu_{jk} - \sum_{k \neq k'} \sum_j \left[\frac{n_{jk}n_{jk'}}{n_{j.}} \right] \mu_{jk'} = 0$$

$$\text{for } k = 1, 2, \ldots, c - 1$$

$$AC: \quad \mu_{jk} - \mu_{j'k} - \mu_{jk'} + \mu_{j'k'} = 0 \quad \text{for all } j, j', k, k'$$

Usage: If there is *no interaction*, method 2 tests method 1's hypotheses with somewhat more power than method 1 itself. However, if there is a possibility of an interaction, the method should not be used (see hypotheses tested). If the cell frequencies are proportional, this method reduces to method 3.

Some names for the method: Overall and Spiegel's method 2, Yates's fitting constants method, SPSS's classic experimental design approach.

Other information: SAS Type II SS in GLM.

Method 3: Main effects not adjusted

SV	df	SS
A	$a - 1$	$R^2_{Y.A}SS_Y$
C	$c - 1$	$R^2_{Y.C}SS_Y$
AC	$(a - 1)(c - 1)$	$(R^2_{Y.A,C,AC} - R^2_{Y.A,C})SS_Y$
Residual	$N - ac$	$(1 - R^2_{Y.A,C,AC})SS_Y$

Hypotheses tested

$$A: \quad \mu_{1*} = \mu_{2*} = \cdots = \mu_{a*}$$

where $\mu_{j*} = \sum_k n_{jk}\mu_{jk}/n_{j.}$, the weighted average of the cell means at A_j,

$$C: \quad \mu_{*1} = \mu_{*2} = \cdots = \mu_{*b}$$

$$AC: \quad \mu_{jk} - \mu_{j'k} - \mu_{jk'} + \mu_{j'k'} = 0 \qquad \text{for all } j, j', k, k'$$

Usage: The main-effect hypotheses tested are about weighted row and column means. These tests may be desirable if cell frequencies are proportional to population sizes.

Some names for the method: Yates's method for proportional cell sizes.

Other information: Can be obtained in BMDP4V by using the statement BETWEEN ARE SIZES. in the WEIGHT paragraph.

procedures that use them. Table 16.5 contains the results obtained by using the three methods with the data of Table 16.3. In all three methods, interactions are adjusted for all other effects in the design, resulting in the usual interaction null hypothesis

$$H_0: \mu_{jk} - \mu_{j'k} - \mu_{jk'} + \mu_{j'k'} = 0 \qquad \text{for all } j, j', k, k'$$

Methods 1 and 3 correspond to adjusting each term for every other term and not adjusting main-effect SS's, respectively. In method 2, adjustments are made only for terms of the same or lower order, so that main effects are adjusted for other main effects but not for interactions. As indicated before, method 1 can be shown to test unweighted main effect null hypotheses and is recommended when unequal cell frequencies occur by chance. Method 3 tests weighted main-effect hypotheses and can be useful when cell frequencies are proportional to corresponding population sizes.

TABLE 16.5 RESULTS OBTAINED USING THE THREE METHODS WITH THE DATA OF TABLE 16.3

SV	df	SS	MS	F
		Method 1		
A	2	$SS_{A|C,AC} = 4139.423$	2069.712	11.639
C	2	$SS_{C|A,AC} = 21.074$	10.537	0.059
AC	4	$SS_{AC|A,C} = 1103.074$	275.769	1.551
Error	15	$(1 - R^2_{Y.A,C,AC})SS_Y = 2667.833$	177.822	
		Method 2		
A	2	$SS_{A|C} = 3609.950$	1804.975	10.150
C	2	$SS_{C|A} = 60.468$	30.234	0.170
AC	4	$SS_{AC|A,C} = 1103.074$	275.769	1.551
Error	15	$(1 - R^2_{Y.A,C,AC})SS_Y = 2667.833$	177.822	
		Method 3		
A	2	$SS_A = 3581.083$	1790.542	10.089
C	2	$SS_C = 31.601$	15.800	0.089
AC	4	$SS_{AC|A,C} = 1103.074$	275.769	1.551
Error	15	$(1 - R^2_{Y.A,C,AC})SS_Y = 2667.833$	177.822	

The method 2 approach that is favored by some statisticians (e.g., Cramer and Appelbaum 1980) corresponds to a hierarchical series of model tests that starts with higher-order effects. On the rationale that main effects are not very meaningful in the presence of an interaction, this approach first tests the interaction by comparing the model

$$Y_{ijk} = \mu_{..} + \alpha_j + \gamma_k + (\alpha\gamma)_{jk} + \epsilon_{ijk}$$

against

$$Y_{ijk} = \mu_{..} + \alpha_j + \gamma_k + \epsilon_{ijk}$$

If there is no interaction, tests of the main effects are then conducted by comparing

$$Y_{ijk} = \mu_{..} + \alpha_j + \gamma_k + \epsilon_{ijk}$$

against

$$Y_{ijk} = \mu_{..} + \gamma_k + \epsilon_{ijk}$$

and

$$Y_{ijk} = \mu_{..} + \alpha_j + \gamma_k + \epsilon_{ijk}$$

against

$$Y_{ijk} = \mu_{..} + \alpha_j + \epsilon_{ijk}$$

The method 2 tests correspond exactly to these model comparisons. The advantage of this approach is that if there is no interaction, tests of main effects are somewhat more powerful than the comparisons of

$$Y_{ijk} = \mu_{..} + \alpha_j + \gamma_k + (\alpha\gamma)_{jk} + \epsilon_{ijk}$$

against

$$Y_{ijk} = \mu_{..} + \gamma_k + (\alpha\gamma)_{jk} + \epsilon_{ijk}$$

and

$$Y_{ijk} = \mu_{..} + \alpha_j + \gamma_k + (\alpha\gamma)_{jk} + \epsilon_{ijk}$$

against

$$Y_{ijk} = \mu_{..} + \alpha_j + (\alpha\gamma)_{jk} + \epsilon_{ijk}$$

that correspond to the method 1 main-effect tests.

We do not recommend the method 2 approach because, if an interaction does exist, the hypotheses tested are data-dependent; that is, they depend on the cell frequencies, and do so in ways that are of little, if any, interest. For example, the null hypothesis for the effect of A can be shown to be

$$\sum_k \left[n_{jk} - \frac{n_{jk}^2}{n_{.k}} \right] \mu_{jk} - \sum_{j' \neq j} \sum_k \left[\frac{n_{jk} n_{j'k}}{n_{.k}} \right] \mu_{j'k} = 0$$

$$\text{for } j = 1, 2, \ldots, a - 1$$

(see Carlson and Timm 1974). Even interactions that do not approach significance can result in biased tests of the main effects (see, for example, Overall, Lee, and Hornick 1981).

Finally, it is possible that a logical or theoretical analysis of the research problem might dictate the order in which the sets of dummy variables are entered into the regression equation and, therefore, the nature of the adjustments. Suppose, for example, that A and C indicate levels of child and parental educational achievement. It is reasonable to assume that parental education may influence a child's educational achievement but not the reverse. In this case, it may be desirable to consider the unadjusted C effect but to adjust A for the effects of C.

In summary, adding the ability to use categorical variables in multiple regression analyses enables us to adjust SS's in ways that result in tests of the hypotheses of factorial ANOVA in nonorthogonal designs. We can use the method 1 approach to test hypotheses about unweighted cell means, and the method 3 approach to test hypotheses about weighted means. Also, if we have logically or theoretically determined orderings of factors, we can perform sequential adjustments.

16.4 USING CATEGORICAL AND CONTINUOUS VARIABLES IN THE SAME ANALYSIS

16.4.1 Testing Homogeneity of Regression Slopes by Using Multiple Regression

In Chapters 12 and 13, we discussed tests for equality of slopes when Y was regressed on X at different levels of some categorical variable A. In the context of the present chapter, we can see that hypotheses about homogeneity of regression concern the interaction between X and A and therefore can be tested by multiple regression. The advantage of using the multiple regression approach to test homogeneity of regression is that it can easily be extended to designs that contain more than one categorical variable.

Table 16.6 contains the data from the hypothetical search experiment discussed in Section 12.2.2. The dependent variable Y is the time in milliseconds it takes to respond to the presence of a target stimulus, and X is the number of characters in the stimulus array. There are two conditions: one in which the array is made up of letters, and a second in which it is made up of digits. The dummy variable X_1 codes array type, taking on the value 1 for scores obtained with letter arrays and -1 for scores obtained with the digit arrays. Finally, the variable X_2, obtained by multiplying corresponding values of X and X_1, codes the joint effect of array size and type. The increment in the Y variability accounted for when X_2 is added to a regression equation that contains X and X_1, $(R^2_{Y.X,X_1,X_2} - R^2_{Y.X,X_1})SS_Y$, reflects the interaction between array size and type. Therefore, the partial F test for X_2 (see Section 15.5.4),

$$F = \frac{(R^2_{Y.X,X_1,X_2} - R^2_{Y.X,X_1})SS_Y}{(1 - R^2_{Y.X,X_1,X_2})SS_Y/(N-4)}$$

or the t test for the regression coefficient of X_2 provides a test of the hypothesis that the slope is the same for letter and digit arrays.

For the data in Table 16.6, $R^2_{Y.X,X_1,X_2}SS_Y = 79,728.915$ and $R^2_{Y.X,X_1}SS_Y = 65,329.105$, so substituting into the expression for the partial F gives a value of 15.99 with 1 and 36 df. Therefore, we can reject the null hypothesis that the rate of change of search time with array size is equal for letters and digits. The partial F value is equivalent to the t value of 4.00 that was obtained for the test presented in Table 12.7.

The procedure can be extended to categorical variables with more than two levels. For example, the data in Table 13.1, which was used to introduce ANCOVA, contains values of Y and X at each of the three levels of the categorical variable A. We can use effect coding to represent A by defining dummy variables X_1 and X_2 such that

$$X_1 = \begin{matrix} 1 \\ 0 \\ -1 \end{matrix} \quad \begin{matrix} \text{for scores at } A_1 \\ \text{for scores at } A_2 \\ \text{for scores at } A_3 \end{matrix}$$

and

$$X_2 = \begin{matrix} 0 \\ 1 \\ -1 \end{matrix} \quad \begin{matrix} \text{for scores at } A_1 \\ \text{for scores at } A_2 \\ \text{for scores at } A_3 \end{matrix}$$

We can represent the joint effects of X and A by defining variables X_3 and X_4 such that, for each Y score, $X_3 = (X)(X_1)$ and $X_4 = (X)(X_2)$. The values of X_1, X_2, X_3, and X_4 are presented for each data point (X, Y) in Table 16.7. The hypothesis of no interaction between X and A (or, equivalently, that the population regression slopes are identical at A_1, A_2, and A_3) can be tested by using the partial F test for X_3 and X_4. Let the notation $R_{Y.X,A,AX}$ stand for $R_{Y.X,X_1,X_2,X_3,X_4}$, the multiple correlation coefficient for Y regressed on X, the dummy variables that code A, and the dummy variables that code the joint effects of A and X. Similarly, let $R_{Y.X,A}$ stand for $R_{Y.X,X_1,X_2}$. Then the partial F statistic that tests the interaction between A and X can be written as

$$F = \frac{(R^2_{Y.X,A,AX} - R^2_{Y.X,A})SS_Y/2}{(1 - R^2_{Y.X,A,AX})SS_Y/(N - 6)} \tag{16.3}$$

In Equation 16.3, the numerator has $a - 1$ df because we are considering the increment in variability accounted for by the two dummy variables that code the joint effects of A and X. The denominator has $N - p - 1 = 30$ df because there are 36 scores and 5 predictor variables in the full regression (or, equivalently, we note that A, X, and AX collectively account for 5 df). For the data of Table 13.1, we have $F = [162.46/2]/[1956.38/30] = 1.25$. Therefore, we cannot reject the hypothesis of homogeneity of regression at the three levels of A.

Again, we should point out that, although it is instructive to consider the analyses in terms of the regression on the quantitative and dummy predictor variables, we may not need to produce the dummy variable coding ourselves. Instead, we can specify that the appropriate regression analysis be done by using some sort of MODEL or DESIGN statement in a statistical software package (see Section 13.9). For example, if we have coded the levels of A as 1, 2, and 3, the statement CATEGORY A = 3 in the SYSTAT MGLH module will produce the appropriate dummy variables to code A, and the statement MODEL Y = CONSTANT + A + X + A*X will produce the appropriate

TABLE 16.6 USING MULTIPLE REGRESSION TO TEST THE EQUALITY OF SLOPES FOR THE DATA OF TABLE 12.7

	Time to respond Y	Array size X	Array type X_1	Type × size X_2
	418	2	1	2
	428	2	1	2
	410	2	1	2
	445	2	1	2
	471	2	1	2
	475	4	1	4
	455	4	1	4
	418	4	1	4
	524	4	1	4
	516	4	1	4
Letter	537	6	1	6
	500	6	1	6
arrays	480	6	1	6
	511	6	1	6
	529	6	1	6
	550	8	1	8
	617	8	1	8
	590	8	1	8
	608	8	1	8
	548	8	1	8
	425	2	−1	−2
	465	2	−1	−2
	430	2	−1	−2
	459	2	−1	−2
	416	2	−1	−2
	464	4	−1	−4
	425	4	−1	−4
	412	4	−1	−4
Digit	497	4	−1	−4
	460	4	−1	−4
arrays	501	6	−1	−6
	407	6	−1	−6
	466	6	−1	−6
	486	6	−1	−6
	475	6	−1	−6
	478	8	−1	−8
	448	8	−1	−8
	512	8	−1	−8
	460	8	−1	−8
	503	8	−1	−8

$$R^2_{Y.X,X_1,X_2}SS_Y = 79{,}728.915 \qquad R^2_{Y.X.X_1}SS_Y = 65{,}329.870$$

$$F = \frac{(R^2_{Y.X,X_1,X_2} - R^2_{Y.X,X_1})SS_Y}{(1 - R^2_{Y.X,X_1,X_2})SS_Y/(N-4)} = \frac{14{,}399.045}{32{,}419.060} = 15.99$$

TABLE 16.7   DATA OF TABLE 13.1 CODED FOR *A*
AND THE JOINT EFFECTS OF *X* AND *A*

Y	X	A		AX	
		X_1	X_2	X_3	X_4
38	25	1	0	25	0
61	35	1	0	35	0
50	23	1	0	23	0
44	11	1	0	11	0
69	29	1	0	29	0
72	36	1	0	36	0
61	26	1	0	26	0
41	10	1	0	10	0
51	23	1	0	23	0
57	29	1	0	29	0
46	16	1	0	16	0
62	27	1	0	27	0
47	10	0	1	0	10
73	28	0	1	0	28
44	16	0	1	0	16
85	30	0	1	0	30
58	21	0	1	0	21
64	18	0	1	0	18
67	31	0	1	0	31
69	34	0	1	0	34
58	27	0	1	0	27
81	35	0	1	0	35
94	41	0	1	0	41
43	15	0	1	0	15
58	19	−1	−1	−19	−19
74	37	−1	−1	−37	−37
65	17	−1	−1	−17	−17
91	40	−1	−1	−40	−40
67	24	−1	−1	−24	−24
45	25	−1	−1	−25	−25
54	23	−1	−1	−23	−23
65	31	−1	−1	−31	−31
59	27	−1	−1	−27	−27
57	14	−1	−1	−14	−14
49	17	−1	−1	−17	−17
74	41	−1	−1	−41	−41

analysis. The output is an ANOVA table in which A, X, and A*X (i.e., *A* × *X*) are included as sources of variance. Note also that the statement MODEL Y = CONSTANT + A + X will perform an ANCOVA (see the next section).

Finally, the procedure can be extended to factorial designs. If we have *Y* and *X* scores for each cell of a 2 × 4 design with factors *A* and *C*, we can test homogeneity of regression in the eight cells of the design, using the partial *F* test

$$F = \frac{(R^2_{Y.X,A,C,AC,AX,CX,ACX} - R^2_{Y.X,A,C,AC})SS_Y/7}{(1 - R^2_{Y.X,A,C,AC,AX,CX,ACX})SS_Y/(N - 16)} \tag{16.4}$$

where $R_{Y.X,A,C,AC,AX,CX,ACX}$ is the multiple correlation coefficient for the regression of Y on 15 predictors: X, the seven dummy variables that code A, C, and the joint effects of A and C, and the seven variables that code the joint effects of X and A, X and C, and X and AC.

16.4.2 ANCOVA as a Special Case of Multiple Regression

In Chapter 13, we portrayed ANCOVA as a kind of hybrid that incorporated elements of ANOVA to deal with the categorical factors and bivariate regression to adjust for the covariate. Using a multiple regression framework that can deal with both categorical variables and covariates allows a more integrated approach, which we believe is easier to understand.[2]

One-Factor ANCOVA

Performing an ANCOVA on a design that has a single factor A can now be seen as determining whether A has effects over and above those of the covariate X. First of all, we determine whether there is an $A \times X$ interaction by using Equation 16.3 (see Chapter 13 for a discussion of why the logic of ANCOVA requires homogeneity of regression). If there is no significant interaction, we can go ahead with the ANCOVA by performing a partial F test for the effects of A. We find the variability in Y accounted for by A and the covariate, $R^2_{Y.X,A}SS_Y$, and the variability accounted for by the covariate alone, $R^2_{Y.X}SS_Y$; we then use the statistic

$$F = \frac{(R^2_{Y.X,A} - R^2_{Y.X})SS_Y/df_A}{(1 - R^2_{Y.X,A})SS_Y/(N - 2 - df_A)} \tag{16.5}$$

This partial F test is exactly equivalent to the ANCOVA test for A that was presented in Chapter 13. The adjustments for the covariate that produced $SS_{A(adj)}$ and $SS_{S/A(adj)}$ in Chapter 13 simply involve partialing out the effects of X; thus,

$$SS_{A(adj)} = (R^2_{Y.X,A} - R^2_{Y.X})SS_Y$$

and

$$SS_{S/A(adj)} = (1 - R^2_{Y.X,A})SS_Y$$

and $b_{S/A}$, the pooled within-group slope of Chapter 13, is equal to b_X, the coefficient of the covariate in the regression equation

$$\hat{Y} = b_0 + b_X X + b_1 X_1 + \cdots + b_{a-1} X_{a-1} \tag{16.6}$$

---

[2] Note that we have presented the instructions necessary to perform an ANCOVA with a number of statistical packages in Section 13.9.

The adjusted means are the values predicted for each group by Equation 16.6 when $X = \bar{X}_{..}$.

If there is a significant $A \times X$ interaction, we do not proceed with the ANCOVA. Instead, the relation between X and Y can be characterized by obtaining the regression equations for each of the groups, and the Johnson-Neyman procedure (see, for example, Huitema 1980) can be used to identify regions of the covariate for which the groups differ significantly in Y.

Factorial ANCOVA

The multiple regression approach to ANCOVA can be readily extended to factorial designs. For example, if we consider the two-factor design described in the previous section, we can test for homogeneity of regression by using Equation 16.4. If homogeneity is not rejected, the ANCOVA tests for A, C, and AC are provided by

$$F = \frac{(R^2_{Y.X,A,C,AC} - R^2_{Y.X,C,AC})SS_Y/df_A}{(1 - R^2_{Y.X,A,C,AC})SS_Y/(N - 2 - df_A - df_C - df_{AC})}$$

$$F = \frac{(R^2_{Y.X,A,C,AC} - R^2_{Y.X,A,AC})SS_Y/df_C}{(1 - R^2_{Y.X,A,C,AC})SS_Y/(N - 2 - df_A - df_C - df_{AC})}$$

and

$$F = \frac{(R^2_{Y.X,A,C,AC} - R^2_{Y.X,A,C})SS_Y/df_{AC}}{(1 - R^2_{Y.X,A,C,AC})SS_Y/(N - 2 - df_A - df_C - df_{AC})}$$

Using More than One Covariate

A researcher may wish to adjust for several sources of unwanted variability by using several covariates. For example, suppose we have a one-factor design and information about two covariates X and W that are each linearly related to Y. Performing an ANCOVA that uses both covariates tests whether A has significant effects over and above both X and W. The appropriate test statistic is the partial F

$$\frac{(R^2_{Y.X,W,A} - R^2_{Y.X,W})SS_Y/df_A}{(1 - R^2_{Y.X,W,A})SS_Y/(N - 3 - df_A)} \tag{16.7}$$

Note that the denominator of Equation 16.7 has one less df than that of Equation 16.5 because of the additional covariate. The adjusted means are the scores predicted by the regression equation for each group if $X = \bar{X}_{..}$ and $W = \bar{W}_{..}$. Homogeneity of regression can be tested by using the partial F for the interactions between A and the covariates:

$$F = \frac{(R^2_{Y.X,W,A,AX,AW} - R^2_{Y.X,W,A})SS_Y/2(a - 1)}{(1 - R^2_{Y.X,W,A,AX,AW})SS_Y/(N - 3 - 3df_A)}$$

Nonlinear ANCOVA

The relations between the dependent variable and potential covariates are not always linear. For example, according to the Yerkes-Dodson law, we would expect a quadratic relation between measures of performance and motivation. If we use standard ANCOVA procedures when there is substantial nonlinearity, the ANCOVA F tests may have little power and the adjusted means may be biased estimates of the treatment means. Therefore, as indicated in Section 13.6.3, it is a good idea to check for severe violations of nonlinearity by plotting scatter diagrams for each group. Also, significance tests for nonlinearity are available (see Section 12.10).

If the relation between Y and X is nonlinear but monotonic (i.e., Y increases or decreases with X but not in a linear fashion), it may be worth checking to see if there is a simple transformation of X such as $\log X$ or X raised to some power (see, for example, Chapter 5 of Draper and Smith 1981), for which the relation between Y and the transformed X is approximately linear. If such a transformation can be found, the transformed value of X can be used as the covariate in a standard ANCOVA.

If the relation between Y and X is not monotonic, a simple transformation will not achieve linearity. However, in this case it may be worthwhile to use polynomial ANCOVA in which the ANCOVA model contains linear and higher-order polynomial components (see Section 13.10). For quadratic ANCOVA, it is assumed that the relation between Y and X is of the form

$$Y = b_0 + b_1 X + b_2 X^2$$

for cubic ANCOVA, the polynomial function contains an X^3 term, and so forth. Quadratic ANCOVA is conducted by including both X and X^2 as covariates so that for the one-factor design the ANCOVA test for A becomes

$$F = \frac{(R^2_{Y.X,X^2,A} - R^2_{Y.X,X^2})SS_Y/df_A}{(1 - R^2_{Y.X,X^2,A})SS_Y/(N - 3 - df_A)} \tag{16.8}$$

Higher-order polynomial ANCOVA can be performed by adding X^3, X^4, and so on, as covariates. However, it is important to keep in mind that although more complex models will fit better, using more covariates results in fewer error df's. Therefore, one should be careful not to use more complex models or more covariates than are necessary.

Finally, it should be noted that the powers of X (X, X^2, X^3, etc.) are highly correlated, and using them in the same multiple regression will result in multicollinearity (see Section 15.7.2), which may result in computational difficulties for some software packages. These problems can generally be avoided by including deviation scores. For example, $x = (X - \bar{X})$ and $x^2 = (X - \bar{X})^2$ may be used instead of X and X^2 in the regression.

16.5 CODING DESIGNS WITH WITHIN-SUBJECTS FACTORS

In a repeated-measures design, each subject is tested at every level of at least one independent variable, and subjects are considered to define levels of a factor, S, in the design. If there are n subjects, we can code S with $n - 1$ dummy variables in the same way as any other categorical variable.

Table 16.8 contains data for an $S \times A$ design originally presented in Table 8.1. We can code the eight levels of S with seven dummy variables (labeled in Table 16.11 as $S1$–$S7$), the four levels of A with three dummy variables, and the SA interaction with 21 dummy variables ($SA11$–$SA73$) formed by multiplying every dummy variable in the S set by every dummy variable in the A set. The SS's can be found from the relations

$$SS_S = R_{Y.S}^2 SS_Y$$

$$SS_A = R_{Y.A}^2 SS_Y$$

$$SS_{SA} = R_{Y.SA}^2 SS_Y$$

TABLE 16.8 DATA AND DUMMY VARIABLE CODING FOR AN $S \times A$ DESIGN

	Y	S1	S2	S3	S4	S5	S6	S7	A1	A2	A3	SA11	SA21	$\cdots$	SA73
A_1	1.4	1	0	0	0	0	0	0	1	0	0	1	0	$\cdots$	0
	2.0	0	1	0	0	0	0	0	1	0	0	0	1	$\cdots$	0
	1.4	0	0	1	0	0	0	0	1	0	0	0	0	$\cdots$	0
	2.3	0	0	0	1	0	0	0	1	0	0	0	0	$\cdots$	0
	4.7	0	0	0	0	1	0	0	1	0	0	0	0	$\cdots$	0
	3.2	0	0	0	0	0	1	0	1	0	0	0	0	$\cdots$	0
	4.0	0	0	0	0	0	0	1	1	0	0	0	0	$\cdots$	0
	5.0	−1	−1	−1	−1	−1	−1	−1	1	0	0	−1	−1	$\cdots$	0
A_2	3.2	1	0	0	0	0	0	0	0	1	0	0	0	$\cdots$	0
	2.5	0	1	0	0	0	0	0	0	1	0	0	0	$\cdots$	0
	4.2	0	0	1	0	0	0	0	0	1	0	0	0	$\cdots$	0
	4.6	0	0	0	1	0	0	0	0	1	0	0	0	$\cdots$	0
	4.8	0	0	0	0	1	0	0	0	1	0	0	0	$\cdots$	0
	5.0	0	0	0	0	0	1	0	0	1	0	0	0	$\cdots$	0
	6.8	0	0	0	0	0	0	1	0	1	0	0	0	$\cdots$	0
	6.1	−1	−1	−1	−1	−1	−1	−1	0	1	0	0	0	$\cdots$	0
A_3	3.2	1	0	0	0	0	0	0	0	0	1	0	0	$\cdots$	0
	3.1	0	1	0	0	0	0	0	0	0	1	0	0	$\cdots$	0
	4.1	0	0	1	0	0	0	0	0	0	1	0	0	$\cdots$	0
	4.0	0	0	0	1	0	0	0	0	0	1	0	0	$\cdots$	0
	4.4	0	0	0	0	1	0	0	0	0	1	0	0	$\cdots$	0
	6.2	0	0	0	0	0	1	0	0	0	1	0	0	$\cdots$	1
	4.5	0	0	0	0	0	0	1	0	0	1	0	0	$\cdots$	−1
	6.4	−1	−1	−1	−1	−1	−1	−1	0	0	1	0	0	$\cdots$	−1
A_4	3.0	1	0	0	0	0	0	0	−1	−1	−1	−1	0	$\cdots$	0
	5.8	0	1	0	0	0	0	0	−1	−1	−1	0	−1	$\cdots$	0
	5.6	0	0	1	0	0	0	0	−1	−1	−1	0	0	$\cdots$	0
	5.9	0	0	0	1	0	0	0	−1	−1	−1	0	0	$\cdots$	0
	5.9	0	0	0	0	1	0	0	−1	−1	−1	0	0	$\cdots$	0
	5.9	0	0	0	0	0	1	0	−1	−1	−1	0	0	$\cdots$	0
	6.5	0	0	0	0	0	0	1	−1	−1	−1	0	0	$\cdots$	−1
	6.6	−1	−1	−1	−1	−1	−1	−1	−1	−1	−1	1	1	$\cdots$	1

TABLE 16.9   DUMMY VARIABLE CODING FOR A MIXED DESIGN

(a) Data and ANOVA table

		C_1	C_2	C_3
	S_{11}	7	1	7
A_1	S_{21}	9	2	10
	S_{31}	7	3	8
	S_{12}	12	7	8
A_2	S_{22}	16	14	9
	S_{32}	19	11	12

SV	df	SS	MS	F
A	1	162.00	162.00	13.50
S/A	4	48.00	12.00	
C	2	85.33	42.67	17.66
AC	2	49.33	24.67	10.21
SC/A	8	19.33	2.42	

(b) Dummy variable coding for the design

	Y	A A1	S/A S/A11	S/A21	S/A12	S/A22	C C1	C2	AC AC11	AC12
$A_1 S_1$	7	1	1	0	0	0	1	0	1	0
	1	1	1	0	0	0	0	1	0	1
	7	1	1	0	0	0	−1	−1	−1	−1
$A_1 S_2$	9	1	0	1	0	0	1	0	1	0
	2	1	0	1	0	0	0	1	0	1
	10	1	0	1	0	0	−1	−1	−1	−1
$A_1 S_3$	7	1	−1	−1	0	0	1	0	1	0
	3	1	−1	−1	0	0	0	1	0	1
	8	1	−1	−1	0	0	−1	−1	−1	−1
$A_2 S_1$	12	−1	0	0	1	0	1	0	−1	0
	7	−1	0	0	1	0	0	1	0	−1
	8	−1	0	0	1	0	−1	−1	1	1
$A_2 S_2$	16	−1	0	0	0	1	1	0	−1	0
	14	−1	0	0	0	1	0	1	0	−1
	9	−1	0	0	0	1	−1	−1	1	1
$A_2 S_3$	19	−1	0	0	−1	−1	1	0	−1	0
	11	−1	0	0	−1	−1	0	1	0	−1
	12	−1	0	0	−1	−1	−1	−1	1	1

From Chapter 8, we know that for an $S \times A$ design the appropriate test for the A main effect is given by $F = MS_A/MS_{SA}$.

The coding procedure can be directly extended to designs in which there are several within-subjects variables, although the number of dummy variables required increases rather dramatically.[3] If we had an $S \times A \times B$ design with eight subjects, four levels of A and two of B, coding all the main effects and interactions would require $abn - 1 = 63$ dummy variables, as many dummy variables as df's for each source of variance. However, if a multiple regression program was used to analyze such a design, one would really only have to code the S, A, and B effects. Most software packages have some sort of COMPUTE or TRANSFORM instruction that will create the variables needed to code the interaction effects.

Finally, the coding procedures can be extended to mixed designs that contain both within-subjects and between-subjects factors. Panel a of Table 16.9 contains a set of hypothetical data and the ANOVA table for a design that has one between-subjects variable (A) and one within-subjects variable (C), and panel b presents dummy variables that code the design. The A, C, and AC SV's can be coded as in a factorial design, and S/A can be directly represented by coding subjects separately at each level of A as indicated in Table 16.9. It is not really necessary to code SC/A because $SS_{SC/A}$ can be obtained as a residual

$$SS_{SC/A} = SS_Y(1 - R_{Y.A,S/A,C,AC}^2)$$

However, SC/A could be coded by the eight dummy variables that would result from multiplying the values of variables in the C and S/A sets.

16.6 CONCLUDING REMARKS

The two goals we had in this chapter were to discuss how categorical variables can be coded so that they can be incorporated into multiple regression analysis and to reconsider within the multiple regression framework a number of the analyses we discussed earlier. We did not include this second goal in order to encourage our readers to perform ANOVAs and ANCOVAs by coding categorical variables in terms of dummy variables and then using multiple regression, although they could do so if the standard ANOVA and ANCOVA programs were not available. Rather, we believe that considering ANOVA and ANCOVA from the multiple regression perspective allows us to gain a deeper understanding of these analyses and their relations to one another.

Also, the generality and flexibility of the multiple regression framework offers some clear advantages. As we mentioned in Section 16.3, the standard ANOVA

---

[3] A procedure called sum or criterion coding can be used to find the sum of squares for any factor by using only a single coding variable, and it has been recommended for designs containing within-subjects factors (Edwards 1979, Pedhazur 1977). Although the idea of avoiding large numbers of dummy variables is appealing, this procedure has problems of its own: sums must be calculated and inserted into the coding variable, the df's displayed on the computer output are not the appropriate ones, and one must be very careful about the order in which coding variables are added to the regression equation. Therefore, we do not present criterion coding here.

approach breaks down for disproportionate-n designs. Thinking in terms of multiple regression, a system in which nonorthogonality is the rule, not the exception, facilitates consideration of the kinds of adjustments that might be made. Even though many of the standard ANOVA and MANOVA programs provide appropriate analyses of nonorthogonal designs, they are really multiple regression programs. We hope that this chapter provides some understanding of how these programs might work and what options they allow. Finally, the ability to include categorical and continuous variables in the same analysis not only allows a discussion of ANCOVA that we believe is quite easy to understand, but also makes it clear that it is not necessary to transform inherently continuous variables into categorical ones (by, for example, using median splits) in order to analyze the data.

APPENDIX: Some Other Forms of Coding

In this chapter, we have primarily used effect coding. However, there are other coding systems that might be used with categorical variables. Any of these will provide the appropriate SS's; however, some of these systems have the advantage of producing regression coefficients that may be of particular interest in certain situations.

Dummy Coding

In dummy coding, each of the $a - 1$ dummy variables used to code an a-level categorical variable consists only of 1's and 0's. In panel a of Table 16.10, X_{D1} and X_{D2} provide the dummy coding for a three-level factor A. The two variables are defined as follows:

$$X_{D1} = \begin{matrix} 1 & \text{for scores in } A_1 \\ 0 & \text{otherwise} \end{matrix}$$

$$X_{D2} = \begin{matrix} 1 & \text{for scores in } A_2 \\ 0 & \text{otherwise} \end{matrix}$$

Together, these two dummy variables completely specify membership *in all three groups*. For each of the first five Y scores, the $X_{D1} = 1$ indicates that the score belongs to A_1, and the $X_{D2} = 0$ confirms that it does not belong to A_2. Similarly, the second set of three scores is specified to belong to A_2. For the final four Y scores, we have $X_{D1} = X_{D2} = 0$. This indicates that these scores do not belong to A_1 or A_2; therefore, they must belong to A_3.

Regressing Y on X_{D1} and X_{D2} results in a regression plane in the three-dimensional space defined by X_{D1}, X_{D2}, and Y that must pass through the points $(1, 0, \bar{Y}_{.1})$, $(0, 1, \bar{Y}_{.2})$, and $(0, 0, \bar{Y}_{.3})$. For each of these points, the first two coordinates specify the group, and the third coordinate is the group mean—the value of Y that minimizes the mean-squared deviation for the group. Because the Y score predicted by the regression for each group must be the group mean, SS_{reg} accounts for all the variability in the group means and, as we can see from comparing panel b of Table 16.10 with Table 16.1, SS_{reg} is equal to SS_A, the between-group variability obtained from the ANOVA. Similarly, the variability not accounted for by the regression, SS_{error}, is the within-group variability, $SS_{S/A}$. It follows that the F test for the overall significance of the regression is exactly equivalent to the standard ANOVA F test.

TABLE 16.10   DUMMY AND CONTRAST CODING

(a) Dummy and contrast coding of the data in Table 16.1

	Y	Y_{D1}	X_{D2}	X_{C1}	X_{C2}
A_1	4	1	0	2	0
	8	1	0	2	0
	7	1	0	2	0
	9	1	0	2	0
	7	1	0	2	0
A_2	2	0	1	-1	1
	5	0	1	-1	1
	3	0	1	-1	1
A_3	4	0	0	-1	-1
	5	0	0	-1	-1
	3	0	0	-1	-1
	6	0	0	-1	-1

(b) Some statistics from the regressions of Y on the different sets of dummy variables (effect, dummy, and contrast)

$$R^2 = .547; \quad SS_{reg} = 28.583; \quad SS_{error} = 23.667; \quad F = \frac{MS_{reg}}{MS_{error}} = 5.435$$

(c) Interpretation and t tests for the b's in the regressions of Y on the dummy and contrast coding variables for the data of Table 16.1

Type of coding	Quantity	Interpretation	Value	SE	t
Dummy	Intercept	$\bar{Y}_{.3}$	4.500	0.811	5.550
	b_{D1}	$\bar{Y}_{.1} - \bar{Y}_{.3}$	2.500	1.088	0.591
	b_{D2}	$\bar{Y}_{.2} - \bar{Y}_{.3}$	-1.167	1.239	-0.942
Contrast	Intercept	$\bar{Y}_U$	4.944	0.478	10.335
	b_{C1}	$\dfrac{2\bar{Y}_{.1} - \bar{Y}_{.2} - \bar{Y}_{.3}}{6}$	1.028	0.318	3.233
	b_{C2}	$\dfrac{\bar{Y}_{.2} - \bar{Y}_{.3}}{2}$	-0.583	0.619	-0.942

In addition, tests of the regression coefficients may also provide useful information about the data. We can find expressions for the coefficients of the dummy variables by substituting into the regression equation $\hat{Y} = b_{D0} + b_{D1}X_{D1} + b_{D2}X_{D2}$ and solving for the b's. We shall find that when group membership is completely specified by the predictors, each regression coefficient is a linear combination of the group means. For scores at A_1, $\hat{Y} = \bar{Y}_{.1}$, $X_{D1} = 1$, and $X_{D2} = 0$. Substituting into the regression equation, we obtain

$$\bar{Y}_{.1} = b_{D0} + b_{D1}(1) + b_{D2}(0) = b_{D0} + b_{D1}$$

Similarly, for scores at A_2 and A_3, we obtain

$$\bar{Y}_{.2} = b_{D0} + b_{D1}(0) + b_{D2}(1) = b_{D0} + b_{D2}$$

and

$$\bar{Y}_{.3} = b_{D0}$$

Solving for the b's, we have

$$b_{D0} = \bar{Y}_{.3}$$
$$b_{D1} = \bar{Y}_{.1} - \bar{Y}_{.3}$$
$$b_{D2} = \bar{Y}_{.2} - \bar{Y}_{.3}$$

Note that the Y intercept, the value of $\hat{Y}$ for $X_{D1} = X_{D2} = 0$, is the mean of the scores at A_3 and that the coefficients of X_{D1} and X_{D2} are the differences between the other group means and the mean at A_3. This kind of coding would be particularly useful if A_3 was a control condition. Not only would the regression analysis test the A main effect, the regression coefficients would correspond to pairwise comparisons between the control group and the other three groups.

It is important to note that the interpretation we have given to the regression coefficients holds only if the *entire set* of dummy variables is used in the regression. The dummy variables are correlated because when one of the X's equals 1, the others must equal 0. Therefore, the coefficient of each dummy variable depends on what other variables are included in the equation.

These findings for dummy coding generalize to any number of groups and hold for both equal- and unequal-n designs. When dummy coding is used to represent an a-level categorical variable, membership in $a - 1$ of the levels is indicated by a 1 in one of the dummy variables and 0's in all the others. Membership in the remaining level is indicated by 0's in all the dummy variables. The Y intercept of the regression equation is the mean of the "control" group (the one specified by 0's in all the dummy variables) and the regression coefficients are the pairwise comparisons of the treatment levels and the control. Finally, it is important to note that if we wish to test the entire set of $a - 1$ pairwise comparisons with the control group, we must limit the amount of Type 1 error. As we discussed in Chapter 6, we can do this by referring the obtained coefficient t's to the Dunnett table (Table D.9).

Contrast Coding

If regressing Y on a complete set of dummy variables produces regression coefficients that are linear combinations of the group means, we should be able to choose the dummy variables in a way that gives us any linear combinations we want. Consider what happens if we use the weights of orthogonal contrasts as the numbers in the dummy variables. Recall from Chapter 6 that for equal-n designs, any set of numbers that sum to zero may be used as contrast weights and that two contrasts, ψ_1 and ψ_2, with weights

$$\psi_1: \quad w_{11}, w_{21}, w_{31}, \dots, w_{a1}$$

$$\psi_2: \quad w_{12}, w_{22}, w_{32}, \dots, w_{a2}$$

are orthogonal if they are uncorrelated. Because each set of weights sums to 0, the numerator of the correlation coefficient reduces to $\sum_j w_{j1} w_{j2}$, and the sets of weights will be uncorrelated if

$$\sum_j w_{j1} w_{j2} = 0$$

For the data in Table 16.1, the sets of numbers

$$
\begin{array}{ccc}
2 & -1 & -1 \\
0 & 1 & -1
\end{array}
$$

form the weights of a possible sets of contrasts. These contrasts can be estimated by

$$\hat{\psi}_1 = 2\bar{Y}_{.1} - \bar{Y}_{.2} - \bar{Y}_{.3} \quad \text{and} \quad \hat{\psi}_2 = \bar{Y}_{.2} - \bar{Y}_{.3} \tag{16.9}$$

In an equal-n design, these two contrasts would be orthogonal, and dummy variables based on the contrast weights would be uncorrelated. In this case, the variability accounted for by regressing Y on the two dummy variables could be partitioned into two components, one associated with each of the contrasts, and a test of each of the regression coefficients would be exactly equivalent to a test of the corresponding contrast.

In the present case, the cell frequencies are not equal, and, therefore, the contrasts are not orthogonal and the two dummy variables, X_{C1} and X_{C2}, are correlated. Nonetheless, as can be seen in Table 16.10, the regression of Y on the dummy variables accounts for all the between-group variability in A and tests of the regression coefficients correspond to tests of the contrasts.

One way of seeing this latter point is to substitute the values of the dummy variables into the regression equation and solve for the b's. Substituting into

$$\hat{Y} = b_{C0} + b_{C1}X_{C1} + b_{C2}X_{C2}$$

we have

$$\bar{Y}_{.1} = b_{C0} + b_{C1}(2) + b_{C2}(0) = b_{C0} + 2b_{C1}$$

$$\bar{Y}_{.2} = b_{C0} + b_{C1}(-1) + b_{C2}(1) = b_{C0} - b_{C1} + b_{C2}$$

$$\bar{Y}_{.3} = b_{C0} + b_{C1}(-1) + b_{C2}(-1) = b_{C0} - b_{C1} - b_{C2}$$

Solving, we obtain

$$b_{C0} = \frac{\bar{Y}_{.1} + \bar{Y}_{.2} + \bar{Y}_{.3}}{3} = \bar{Y}_U \tag{16.10}$$

$$b_{C1} = \frac{(2\bar{Y}_{.1} - \bar{Y}_{.2} - \bar{Y}_{.3})}{6}$$

$$b_{C2} = \frac{(\bar{Y}_{.2} - \bar{Y}_{.3})}{2}$$

Comparing Equations 16.9 and 16.10, we see that the regression weights are proportional to the estimated values of the contrasts. Therefore, in each case, the same null hypothesis, $H_0: \psi = 0$, is tested by $t = \hat{\psi}/\hat{SE}(\hat{\psi})$ and the test of the corresponding regression coefficient, $t = b/\hat{SE}(b)$.

EXERCISES

16.1   The following terms provide a useful review of many concepts in the chapter. Define, describe, or identity each of them:

categorical variable   approaches to dealing with
quantitative variable   nonorthogonal designs:

qualitative variable	method 1
nonorthogonal or unbalanced design	method 2
dummy variable	method 3
effect coding	dummy coding
linear independence	contrast coding

16.2 Attitude scores on a scale from 0 to 100 are solicited from groups of 10 females and 10 males. The data are as follows:

	Gender	
	Females	Males
	53	79
	51	50
	88	79
	63	99
	56	80
	42	92
	55	69
	78	94
	48	70
	92	72
Mean	62.60	78.40
Standard deviation	17.36	14.38

(a) Perform an independent-groups t test to determine whether there is a significant effect of gender on the attitude scores.

(b) How many dummy variables are needed to code the gender variable?

(c) Indicate how gender could be coded by using (i) effect $(1, -1)$ coding and (ii) dummy $(1, 0)$ coding.

(d) Regress attitude score on the appropriate dummy variables for (i) and (ii). How do the tests of whether b_1 is significantly different from 0 relate to the t test indicated in (a)?

(e) What is the interpretation of the regression coefficient for each of the two types of coding?

16.3 **(a)** For the data presented in Exercise 16.2, how would the results of the regression analysis change if $(16, 7)$ coding was used instead of $(1, 0)$ coding?

(b) How would the interpretation of the regression coefficients in Exercise 16.2 change if there were unequal numbers of males and females in the two groups?

16.4 Given the following data from an experiment with four treatment conditions:

A_1	A_2	A_3	A_4
53	42	48	79
51	55	42	99
88	78	79	80
63		50	92
56			69
			94
			70

(a) How many linearly independent dummy variables are needed to code the design?

(b) Code the design using effect coding.

(c) Code the design using dummy coding.

(d) Code the design using contrast coding.

(e) Regress the dependent variable on the dummy variables for each of (b), (c), and (d).

(f) What are the interpretations of the regression coefficients in each of the three regressions?

16.5 How would the interpretations of the regression weights in the previous problem change if there were equal numbers of scores in the four groups?

16.6 Given the following data from a 2 × 2 design with unequal n:

	A_1	A_2
C_1	53, 63, 51, 56, 88	48, 79, 42, 50
C_2	42, 55, 78	79, 92 99, 69, 70, 80, 94

(a) Code the design using effect coding.

(b) Are the dummy variables obtained in (a) correlated with one another?

(c) Regress the dependent variable on the three dummy variables obtained in (a).

(d) What are the interpretations of each of the regression coefficients?

(e) What hypotheses about the population means are tested by the t tests for the regression coefficients in (c).

(f) Regress the dependent variable on only the dummy variable that codes the A effect. Now what is the interpretation of the regression coefficient? What hypothesis about the population means is tested by the t test for the regression coefficient?

(g) Regress the dependent variable on only the dummy variable that codes the C effect. Now what is the interpretation of the regression coefficient? What hypothesis about the population means is tested by the t test for the regression coefficient?

16.7 Can the regressions performed in Exercise 16.6 be identified with any of the methods for analyzing factorial designs summarized in Table 16.4? Discuss.

16.8 Given the following data from a nonorthogonal 2×3 design:

	B_1	B_2	B_3
A_1	17, 33, 26, 27, 21	11, 18, 14	9, 12, 10
A_2	17, 8, 22, 12, 9	9, 7, 4, 3	13, 10, 6, 6, 8

(a) Code the design using effect coding.

(b) Suppose the unequal n's in the different cells have occurred by chance and we therefore wish to test hypotheses about unweighted means. Perform the appropriate regression analyses and test the A, B, and $A \times B$ tests. Compare your results with that provided by a standard ANOVA program.

(c) Suppose we have reason to believe that the populations corresponding to the six cells differ in size and that the cell n's are proportional to the population sizes. Perform the appropriate regression analyses and test the A, B, and $A \times B$ effects.

16.9 Table 16.7 contains the data originally introduced in Chapter 13 to introduce ANCOVA. The table indicates how the X, A, and $A \times X$ effects may be coded. Use multiple regression to perform an ANCOVA for the design. Verify that the results are the same as those obtained in Chapter 13.

16.10 Use multiple regression to test the design in Table 16.7 for homogeneity of regression slope.

Appendix **A**

Notation and Summation Operations

We must have a common language to talk about the derivations and computational formulas that relate to psychological experimentation. Such a language exists in the notational system presented here. If you try to master it now, your efforts will be amply repaid. You will find first a few simple rules, which are then applied to some elementary statistical quantities.

A.1 A SINGLE GROUP OF SCORES

A.1.1 Some Basic Rules

In a group of scores like Y_1, Y_2, Y_3, Y_4, ... , Y_n, the subscript has no purpose except to distinguish among the individual scores. The quantity n is the total number of scores in the group. Suppose that $n = 5$ and we want to show that all five scores are to be added together. We could write

$$Y_1 + Y_2 + Y_3 + Y_4 + Y_5$$

or, more briefly,

$$Y_1 + Y_2 + \cdots + Y_5$$

Still more briefly, we write

$$\sum_{i=1}^{5} Y_i$$

This expression is read "sum the values of Y for all i from 1 to 5." In general, $i = 1$, $2, \ldots, n$ (that is, i takes on the values of 1 to n), and the summation of a group of n scores is indicated by

$$\sum_{i=1}^{n} Y_i$$

The quantity i is the *index,* and 1 and n are the *limits* of summation.[1] When the context of the presentation permits no confusion, the index and limits are often dropped. Thus we may often indicate by ΣY that a group of scores are to be summed.

Three rules for summation follow.

■ **RULE 1.** The sum of a constant times a variable equals the constant times the sum of the variable; or

$$\Sigma CY = C\Sigma Y$$

The term C is a constant in the sense that its value does not change as a function of i; the value of Y depends on i, and Y is therefore a variable relative to i. The rule is easily proved.

$$\Sigma CY = CY_1 + CY_2 + CY_3 + \cdots + CY_n$$
$$= C(Y_1 + Y_2 + Y_3 + \cdots + Y_n)$$
$$= C\Sigma Y$$

■ **RULE 2.** The sum of a constant equals n times the constant, where n equals the number of quantities summed; or

$$\Sigma C = C + C + \cdots + C = nC$$

■ **RULE 3.** The summation sign operates like a multiplier on quantities within parentheses.

■ **EXAMPLE 1.**

$$\sum_{i}^{n}(X_i - Y_i) = \sum_{i}^{n}X_i - \sum_{i}^{n}Y_i$$

Proof.

$$\Sigma(X - Y) = (X_1 - Y_1) + (X_2 - Y_2) + \cdots + (X_n - Y_n)$$
$$= (X_1 + X_2 + \cdots + X_n) - (Y_1 + Y_2 + \cdots + Y_n)$$
$$= \Sigma X - \Sigma Y$$

■ **EXAMPLE 2.**

$$\Sigma(X - Y)^2 = \Sigma X^2 + \Sigma Y^2 - 2\Sigma XY$$

---

[1] To conserve space, when we wish to indicate an index of summation in a line of text or a fraction, we will often write it as a subscript. The expression $\Sigma_i Y_i$ should be considered equivalent to

$$\sum_{i} Y_i.$$

Proof.

$$\Sigma(X - Y)^2 = (X_1 - Y_1)^2 + \cdots + (X_n - Y_n)^2$$
$$= (X_1^2 + Y_1^2 - 2X_1 Y_1) + (X_2^2 + Y_2^2 - 2X_2 Y_2) + \cdots$$
$$+ (X_n^2 + Y_n^2 - 2X_n Y_n)$$
$$= (X_1^2 + X_2^2 + \cdots + X_n^2) + (Y_1^2 + Y_2^2 + \cdots + Y_n^2)$$
$$- 2(X_1 Y_1 + X_2 Y_2 + \cdots + X_n Y_n)$$
$$= \Sigma X^2 + \Sigma Y^2 - 2\Sigma XY$$

A.1.2 Applying the Summation Rules

We can apply the rules of summation to prove the properties of means and variances stated in Chapter 2 (Table 2.1). Throughout this section it should be clear that we are summing over i from 1 to n even though the index and limits are not explicitly presented in each expression.

Properties of the Mean

1. $\Sigma(Y - \bar{Y}) = 0$; the sum of all deviations of scores about their mean is zero. Applying Rule 3, we get

$$\Sigma(Y - \bar{Y}) = \Sigma Y - \Sigma \bar{Y}$$

However, $\bar{Y}$ is a constant; its value is the same regardless of the value of the index of summation. Therefore, applying Rule 2, we rewrite the last equation as

$$\Sigma(Y - \bar{Y}) = \Sigma Y - n\bar{Y}$$

Because $\bar{Y} = \Sigma Y / n$, we can rewrite this as

$$\Sigma(Y - \bar{Y}) = \Sigma Y - n\bar{Y} = \Sigma Y - n\left(\frac{\Sigma Y}{n}\right) = \Sigma Y - \Sigma Y = 0$$

2. $\Sigma(Y + k)/n = \bar{Y} + k$; if a constant is added to all scores, the mean is increased by that constant. Applying Rule 3 gives

$$\frac{\Sigma(Y + k)}{n} = \frac{\Sigma Y + \Sigma k}{n} = \frac{\Sigma Y}{n} + \frac{\Sigma k}{n}$$

Applying Rule 2 and noting that $\Sigma Y / n = \bar{Y}$, we have

$$\frac{\Sigma(Y + k)}{n} = \bar{Y} + \frac{nk}{n} = \bar{Y} + k$$

3. $\Sigma kY/n = k\bar{Y}$; if all scores are multiplied by a constant, the mean is multiplied by that constant. Applying Rule 1, we have

$$\frac{\Sigma kY}{n} = \frac{k\Sigma Y}{n} = k\bar{Y}$$

4. $\Sigma(Y - \bar{Y})^2$ is a minimum. Assume that there is some value $\bar{Y} + d$ such that the sum of squared deviations of all scores about it is smaller than the sum about any other value. This sum of squared distances is $\Sigma[Y - (\bar{Y} + d)]^2$. Expanding in accord with Rule 3, we have

$$\Sigma[Y - (\bar{Y} + d)]^2 = \Sigma[(Y - \bar{Y}) - d]^2 = \Sigma(Y - \bar{Y})^2 + \Sigma d^2 - 2\Sigma d(Y - \bar{Y})$$

Applying Rule 1, we rewrite the rightmost term as

$$2\Sigma d(Y - \bar{Y}) = 2d\Sigma(Y - \bar{Y}) = (2d)(0)$$

because $\Sigma(Y - \bar{Y}) = 0$. Applying Rule 2, we have

$$\Sigma d^2 = nd^2$$

Therefore,

$$\Sigma[Y - (\bar{Y} + d)]^2 = \Sigma(Y - \bar{Y})^2 + nd^2$$

which is as small as possible when $d = 0$; that is, when we sum the squared deviations of scores about their mean.

Properties of the Variance

1. Adding a constant to all scores leaves the variance unchanged. If a constant k is added to all scores the new variance is

$$\hat{\sigma}^2_{Y+k} = \frac{\Sigma[(Y + k) - (\bar{Y} + k)]^2}{n - 1} = \frac{\Sigma(Y - \bar{Y})^2}{n - 1} = \hat{\sigma}^2_Y$$

2. Multiplying all scores by a constant k is equivalent to multiplying the variance by k^2 and the standard deviation by k. We have

$$\hat{\sigma}^2_{kY} = \frac{\Sigma(kY - k\bar{Y})^2}{n - 1} = \frac{\Sigma k^2(Y - \bar{Y})^2}{n - 1}$$

By Rule 1 this becomes

$$\hat{\sigma}^2_{kY} = \frac{k^2\Sigma(Y - \bar{Y})^2}{n - 1} = k^2\hat{\sigma}^2_Y$$

z Scores

The properties proven allow us to show that the mean of a set of z scores is zero and its variance is 1. Recall that

$$z = \frac{Y - \bar{Y}}{\hat{\sigma}_Y}$$

To obtain the average of a set of n z scores, we sum them and divide by n, keeping in mind that $\Sigma(Y - \bar{Y}) = 0$. Then

$$\frac{\Sigma z}{n} = \frac{\Sigma(Y - \bar{Y})}{n\hat{\sigma}_Y} = \frac{(0)}{n\hat{\sigma}_Y} = 0$$

To prove that the variance (and therefore the standard deviation) of the z scores is 1, expand the formula for z as

$$z = \left[\frac{1}{\hat{\sigma}_Y}\right] Y - \left[\frac{1}{\hat{\sigma}_Y}\right] \bar{Y}$$

Note that $1/\hat{\sigma}_Y$ is a constant with respect to the index of summation i. Because adding (or subtracting) a constant from a variable does not change its variance (see the first property of the variance), the variance of z is the same as the variance of $(1/\hat{\sigma}_Y)Y$. But, from the second property of a variance, we know that the variance of a constant $(1/\hat{\sigma}_Y)$ times a variable (Y) is the squared constant times the variance of the variable. That is,

$$\hat{\sigma}_z^2 = \left[\frac{1}{\hat{\sigma}_Y}\right]^2 \hat{\sigma}_Y^2 = 1$$

A.1.3 Raw-Score Formulas

The summation rules can be applied to obtain raw-score formulas for quantities such as the variance and covariance. These raw-score or *computational* formulas contain sums of scores, squared scores, and cross products rather than sums of squared differences and cross products of difference scores. This allows them to minimize rounding error and makes them convenient to use with simple hand calculators that do not have variance and correlation keys.

The numerator of the expression for the variance of Y is $SS_Y = \Sigma(Y_i - \bar{Y})^2$. To get the raw-score formula for SS_Y, expand the quantity within the summation sign. Thus

$$\Sigma(Y - \bar{Y})^2 = \Sigma(Y^2 + \bar{Y}^2 - 2Y\bar{Y})$$

Applying Rule 3, we have

$$\Sigma(Y - \bar{Y})^2 = \Sigma Y^2 + \Sigma \bar{Y}^2 - \Sigma 2Y\bar{Y}$$

Noting that $\bar{Y}^2$ is a constant and applying Rule 2, we have

$$\Sigma(Y - \bar{Y})^2 = \Sigma Y^2 + n\bar{Y}^2 - \Sigma 2Y\bar{Y}$$

The quantity $2\bar{Y}$ is a constant and, by Rule 1, can be placed before the summation sign. Thus,

$$\Sigma(Y - \bar{Y})^2 = \Sigma Y^2 + n\bar{Y}^2 - 2\bar{Y}\Sigma Y$$

Now replace $\bar{Y}$ by $\Sigma Y/n$ to get

$$\Sigma(Y - \bar{Y})^2 = \Sigma Y^2 + \frac{n(\Sigma Y)^2}{n^2} - 2\left[\frac{\Sigma Y}{n}\right]\Sigma Y$$

Simplifying, we have

$$\Sigma(Y - \bar{Y})^2 = \Sigma Y^2 - \frac{(\Sigma Y)^2}{n} \tag{A.1}$$

Dividing the right-hand side of Equation A.1 by $n - 1$ gives the raw-score formula for $\hat{\sigma}_Y^2$.

We can find the raw-score formula for the covariance of X and Y,

$$\hat{\sigma}_{XY} = \frac{\Sigma(X - \bar{X})(Y - \bar{Y})}{n - 1}$$

by noting that Equation A.1 could be rewritten as

$$\Sigma(Y - \bar{Y})^2 = \Sigma(Y - \bar{Y})(Y - \bar{Y}) = \Sigma YY - \frac{(\Sigma Y)(\Sigma Y)}{n}$$

By analogy, the numerator of $\hat{\sigma}_{XY}$ has the raw-score formula

$$\Sigma(X - \bar{X})(Y - \bar{Y}) = \Sigma XY - \frac{(\Sigma X)(\Sigma Y)}{n}$$

Dividing by $n - 1$ yields the raw-score formula for $\hat{\sigma}_{XY}$.

A.2 SEVERAL GROUPS OF SCORES

The simplest possible experimental design involves several groups of scores. Thus one might have a groups of n subjects each, which differ in the amount of reward they receive for their performance on some learning task. In setting the data down on paper, there would be a column for each level of amount of reward—that is, for each experimental group. The scores for a group could be written in order within the appropriate column. In referring to a score, we should designate it by its position in the column (or experimental group) and by the position of the column. Table A.1 illustrates this procedure. Note that the first subscript refers to the position in the group (row), the second to the position of the group (column). Thus Y_{22} is the second score in group 2, and in general, Y_{ij} is the ith score in the jth group.

TABLE A.1   A TWO-DIMENSIONAL MATRIX

			Groups		
	Y_{11}	Y_{12}	$\cdots$	Y_{1j}	$\cdots$	Y_{1a}
	Y_{21}	Y_{22}	$\cdots$	Y_{2j}	$\cdots$	Y_{2a}
	.	.		.		.
	.	.		.		.
Subjects	Y_{i1}	Y_{i2}	$\cdots$	Y_{ij}	$\cdots$	Y_{ia}
	.	.		.		.
	Y_{n1}	Y_{n2}	$\cdots$	Y_{nj}	$\cdots$	Y_{na}

Suppose we want to refer to the mean of a single column. The term used previously, $\bar{Y}$, is obviously inadequate since it does not designate the row or column that we want. Even $\bar{Y}_1$ is not clear, since it might as easily refer to the mean of the first row as to the mean of the first column.[2] The appropriate designation is $\bar{Y}_{.1} = (1/n)\Sigma_i^n Y_{i1}$; the dot represents the summation over i, the index that ordinarily appears in that position. Similarly, the mean of row i would be designated by $\bar{Y}_{i.} = (1/a)\Sigma_j^a Y_{ij}$; summation is over the index j. The mean of all an scores would be designated by $\bar{Y}_{..} = (1/an)\Sigma\Sigma Y_{ij}$, or merely $\bar{Y}$.

Some examples using the double summation $(\Sigma_i \Sigma_j)$ may be helpful. Suppose we have

$$\sum_{j=1}^{a} \sum_{i=1}^{n} Y_{ij}^2$$

This is an instruction to set i and j initially at 1; the resulting score Y_{11} is then squared. Holding j at 1, we step i from 1 to n, squaring each score thus obtained and adding it to those previously squared. When n scores have been squared and summed, we reset the index i at 1 and step j to 2; the squaring and summing is then carried out for all Y_{i2}. The process continues until all an scores have been squared and summed. The process just described can be represented by

$$(Y_{11}^2 + Y_{21}^2 + \cdots + Y_{na}^2)$$

If we have

$$\sum_{j=1}^{a} \left[\sum_{i=1}^{n} Y_{ij} \right]^2$$

the notation indicates that a sum of n scores is to be squared. We again set j at 1, and after adding together all the Y_{i1}, square the total. The index j is then stepped to 2 and i is reset at 1; we get another sum of n scores, which is squared and added to the previous squared sum. We again continue until all an scores have been accounted for. The process can be represented by

$$(Y_{11} + Y_{21} + \cdots + Y_{n1})^2 + \cdots + (Y_{1a} + Y_{2a} + \cdots + Y_{na})^2$$

A third possibility is

$$\left[\sum_{j=1}^{a} \sum_{i=1}^{n} Y_{ij} \right]^2$$

which indicates that the squaring operation is carried out once on the total of an scores; we then have

$$[(Y_{11} + Y_{21} + \cdots + Y_{n1}) + \cdots + (Y_{1a} + Y_{2a} + \cdots + Y_{na})]^2$$

---

[2] In the design we used for an example, the mean of the first row would not be a quantity of interest, since we stipulated that the order within each column was arbitrary. There are designs, however, giving rise to tables like Table A.1 for which it is as interesting to obtain row means as it is to obtain column means.

Note that the indices within the parentheses show how many scores are to be summed prior to squaring, and the indices outside the parentheses show how many squared totals are to be summed. When no parentheses appear, as in $\Sigma\Sigma Y^2$, we treat the notation as if it were $\Sigma\Sigma(Y^2)$. When no indices appear outside the parentheses, it is understood that we are dealing with a single squared term, as in $(\Sigma\Sigma Y)^2$. When several indices appear together, whether inside or outside the parentheses, the product of their upper limits tells us the number of terms involved. Thus, $(\Sigma_{j=1}^{a}\Sigma_{i=1}^{n}Y)^2$ indicates that *an* scores are summed before the squaring.

Our three illustrations of the double summation can be further clarified if we use some numbers. Let us use the three groups of four scores each shown in Table A.2. Now,

$$\sum_{j}\sum_{i}Y_{ij}^2 = 30 + 70 + 93 = 193$$

and

$$\sum_{j}\left[\sum_{i}Y_{ij}\right]^2 = (10)^2 + (14)^2 + (19)^2 = 657$$

and

$$\left[\sum_{j}\sum_{i}Y_{ij}\right]^2 = (10 + 14 + 19)^2 = 1849$$

As another example of how to use double summation, we might derive a raw score formula for the average group variance, often referred to as the *within-group mean square*. This is the sum of the group variances divided by *a*, the number of groups, or

$$\frac{1}{a}\left[\frac{\Sigma_{i=1}^{n}(Y_{i1} - \bar{Y}_{.1})^2}{n-1} + \cdots + \frac{\Sigma_{i=1}^{n}(Y_{ia} - \bar{Y}_{.a})^2}{n-1}\right]$$

More briefly, this average is indicated by

$$\frac{1}{a(n-1)}\sum_{j}^{a}\sum_{i}^{n}(Y_{ij} - \bar{Y}_{.j})^2$$

Now, expanding the numerator (or "sums of squares") of this quantity, we get

$$\sum_{j=1}^{a}\sum_{i=1}^{n}(Y_{ij} - \bar{Y}_{.j})^2 = \sum_{j=1}^{a}\sum_{i=1}^{n}(Y_{ij}^2 + \bar{Y}_{.j}^2 - 2Y_{ij}\bar{Y}_{.j})$$

TABLE A.2   SOME SAMPLE DATA

	Group 1	Group 2	Group 3
	4	1	6
	1	7	4
	3	2	5
	2	4	4
$\Sigma_i Y_{ij} =$	10	14	19
$\Sigma_i Y_{ij}^2 =$	30	70	93

We "multiply through" by Σ_i, noting that $\bar{Y}_{.j}$ varies only with j; it is constant when i is the index of summation. Terms are also rearranged so that sums are premultiplied by constants:

$$\sum_j \sum_i (Y_{ij} - \bar{Y}_{.j})^2 = \sum_j \left[\sum_i Y_{ij}^2 + n\bar{Y}_{.j}^2 - 2\bar{Y}_{.j}\sum_i Y_{ij} \right]$$

Note that $\Sigma_i \bar{Y}_{.j} = n\bar{Y}_{.j}$. Although $\bar{Y}_{.j}$ is a variable relative to the index j, it is a constant relative to i, the index over which we are currently summing; therefore Rule 2 applies.

Substituting raw-score formulas for the group means gives

$$\sum_j \sum_i (Y_{ij} - \bar{Y}_{.j})^2 = \sum_j \left[\sum_i Y_{ij}^2 + n\frac{(\Sigma_i Y_{ij})^2}{n^2} - 2\left(\frac{\Sigma_i Y_{ij}}{n}\right)\sum_i Y_{ij} \right]$$

Simplifying gives

$$\sum_j \sum_i (Y_{ij} - \bar{Y}_{.j})^2 = \sum_j \left[\sum_i Y_{ij}^2 - \frac{(\Sigma_i Y_{ij})^2}{n} \right]$$

which can also be written

$$\sum_j \sum_i Y_{ij}^2 - \frac{\Sigma_j(\Sigma_i Y_{ij})^2}{n}$$

To simplify notation, we shall generally use T (for "total") to replace ΣY. The sum of scores, for example, for group j is

$$T_{.j} = \sum_i Y_{ij}$$

and the raw-score expression just derived can be rewritten as

$$\sum_j \sum_i Y_{ij}^2 - \frac{\Sigma_j T_{.j}^2}{n}$$

Appendix **B**

Expected Values and Their Applications

The view of a population parameter as the expected value of a statistic is inherent in most inferential procedures. Furthermore, many important results are derived by taking expectations of statistics. The following discussion provides an introduction to these matters. We begin by defining an expected value, and we then present some rules for working with expectations. We then apply these rules to derive some results that were presented earlier in this book.

B.1 DEFINITIONS AND BASIC RULES

We repeat the earlier definitions of expected values (see Chapter 3) for convenience in dealing with the other material in this appendix. The expected value of a random variable, Y, may be viewed as a weighted average of all possible values Y can take. The weights are probabilities, $p(y)$, when Y is discretely distributed and densities, $f(y)$, when Y is continuously distributed. In the discrete case,

$$E(Y) = \sum_y yp(y)$$

and in the continuous case,

$$E(Y) = \int_y yf(y)\,dy$$

$E(Y)$ is read as "the expected value of Y" or "the expectation of Y." The y under the summation and integral signs is meant to remind us that the sum or integral is over all possible values of Y.

The symbol E is often referred to as an *expectation operator,* meaning that it is an instruction to sum or integrate the variable indicated. The expectation operator follows a set of rules similar to those presented in Appendix A for the summation operator. The most important of these rules are presented next.

■ **RULE 1.** *The expectation of a constant times a variable equals the constant times the sum of the variable:*

$$E(CY) = CE(Y)$$

This may be seen by writing

$$E(CY) = \Sigma(Cy)p(y) = C\Sigma yp(y) = CE(Y)$$

■ **RULE 2.**   *The expectation of a constant is the constant:*

$$E(C) = C$$

If several events have the same numerical value C, the average value will equal C.

■ **RULE 3.**   *E acts like a multiplier.* For example,

$$E(X + Y) = E(X) + E(Y)$$

To prove this, begin with the definition of $E(X + Y)$:

$$E(X + Y) = \sum_x \sum_y (x + y)p(x, y)$$

where the expression on the right indicates that each possible value of $X + Y$ is multiplied by its joint probability, and these products are then summed. Distributing this expression, we obtain

$$E(X + Y) = \sum_x \sum_y xp(x, y) + \sum_x \sum_y yp(x, y)$$

$$= \sum_x x \left[\sum_y p(x, y) \right] + \sum_y y \left[\sum_x p(x, y) \right]$$

$$= \sum_x xp(x) + \sum_y yp(y) = E(X) + E(Y)$$

A special case of this expression occurs when one variable is replaced by a constant; then

$$E(Y + C) = E(Y) + E(C) = E(Y) + C$$

This equation provides an immediate basis for asserting that

$$E(Y - \mu) = 0$$

because

$$E(Y - \mu) = E(Y) - \mu = \mu - \mu = 0$$

Another application of Rule 3 is

$$E(X + Y)^2 = E(X)^2 + E(Y)^2 + 2E(XY)$$

This leads to a proof of the statement in Chapter 2 that the variance of Y, $E(Y - \mu)^2$, equals $E(Y^2) - \mu^2$:

$$\begin{aligned}
E(Y - \mu)^2 &= E(Y^2) + E(\mu)^2 - 2E(Y\mu) \\
&= E(Y^2) + \mu^2 - 2\mu E(Y), && \text{because } \mu \text{ is a constant} \\
&= E(Y)^2 + \mu^2 - 2\mu^2, && \text{because } \mu \text{ and } E(Y) \text{ are the same entity} \\
&= E(Y^2) - \mu^2
\end{aligned}$$

■ **RULE 4.**   *IF X and Y are independently distributed, then E(XY) = E(X)E(Y).* To prove this, we again begin with the definition of an expectation:

$$E(XY) = \sum_x \sum_y xyp(x, y)$$

$$= \sum_x \sum_y xyp(x)p(y)$$

because the joint probability $p(x, y) = p(x)p(y)$ if X and Y are independently distributed. Rearranging terms gives

$$E(XY) = [\Sigma xp(x)][\Sigma yp(y)] = E(X)E(Y)$$

A useful implication of this is that $E(X - \bar{X})(Y - \bar{Y}) = 0$ if X and Y are independent. This follows because $E(X - \bar{X})(Y - \bar{Y})$ then must equal $[E(X - \bar{X})][E(Y - \bar{Y})] = 0 \times 0$. Therefore, if X and Y are independent, their covariance (and consequently ρ) must equal zero.

B.2 APPLICATIONS TO ESTIMATION

We can now show that $\bar{Y}$ is an unbiased estimate of μ; that is, $E(\bar{Y}) = E(Y)$, or μ. We have

$$E(\bar{Y}) = E\left(\frac{\Sigma Y}{n}\right) = \frac{1}{n}E(\Sigma Y) \qquad \text{by Rule 1}$$

$$= \frac{1}{n}\Sigma E(Y)$$

$$= \frac{1}{n}(n)E(Y) = E(Y)$$

We next show that $\hat{\sigma}^2$ is an unbiased estimator of σ^2; that is, $E(\hat{\sigma}^2) = \sigma^2$. Begin by considering the sum of squares, the numerator of $\hat{\sigma}^2$:

$$E[\Sigma(Y - \bar{Y})^2] = E\Sigma[(Y - \mu) - (\bar{Y} - \mu)]^2$$

$$= E[\Sigma(Y - \mu)^2 + \Sigma(\bar{Y} - \mu)^2 - 2(\bar{Y} - \mu)\Sigma(Y - \mu)]$$

$$= E[\Sigma(Y - \mu)^2 + n(\bar{Y} - \mu)^2 - 2n(\bar{Y} - \mu)^2]$$

$$= E[\Sigma(Y - \mu)^2 - n(\bar{Y} - \mu)^2]$$

$$= \Sigma E(Y - \mu)^2 - nE(\bar{Y} - \mu)^2 \qquad \text{by Rule 3}$$

The average squared deviation of a quantity from its average is a variance; that is, $E(Y - \mu)^2 = \sigma^2$ and $E(\bar{Y} - \mu)^2 = \sigma^2/n$. Therefore,

$$E[\Sigma(Y - \bar{Y})^2] = n\sigma^2 - \frac{n\sigma^2}{n}$$

$$= (n - 1)\sigma^2$$

Therefore,

$$E\left(\frac{\Sigma(Y - \bar{Y})^2}{n - 1}\right) = E[\hat{\sigma}^2] = \sigma^2$$

B.3 THE MEAN AND VARIANCE OF THE BINOMIAL DISTRIBUTION

Consider a series of n Bernoulli trials and let $X = 1$ or 0, depending upon whether the trial outcome was a success or failure; $p(X = 1) = p$ and $p(X = 0) = q$. The total number of successes in the n trials is $Y = \Sigma X$. We want to derive expressions for $E(Y)$ and $\text{Var}(Y)$, the mean and variance of the binomial distribution. We have

$$E(Y) = E(\Sigma X) = \Sigma E(X)$$

$$= \Sigma x p(x) \qquad \text{by definition of an expected value}$$

$$= \Sigma[(1)(p) + (0)(q)] = \Sigma p = np$$

We derive the expression for the variance of the binomial distribution in a similar manner:

$$\text{Var}(Y) = \text{Var}(\Sigma X)$$

The variance of a sum of independent variables is the sum of their variances (see Section 2.5.2); therefore,

$$\text{Var}(Y) = \text{Var}(\Sigma X) = \Sigma \text{Var}(X)$$

The variance of X is $E[X - E(X)]^2 = E(X^2) - [E(X)]^2$; see the development under Rule 3, immediately preceding Rule 4. We showed above that $E(X) = p$, and

$$E(X^2) = (1^2)(p) + (0^2)(q) \qquad \text{by definition of an expected value}$$

$$= p$$

Therefore, $\text{Var}(X) = E(X^2) - [E(X)]^2 = p - p^2 = p(1 - p) = pq$. Finally, we have

$$\text{Var}(Y) = \Sigma \text{Var}(X) = \Sigma pq = npq$$

Appendix **C**

Matrix Algebra

C.1 INTRODUCTION

Linear combinations of variables and systems of linear equations appear frequently in this book. For example, in regression, the prediction of the criterion variable, $\hat{Y}$, is expressed in terms of a linear combination of the predictor variables, the regression coefficients are obtained by solving a system of $p + 1$ normal equations, and the coefficients themselves can be expressed as linear combinations of the Y's. The use of matrix notation and algebra to represent and manipulate linear combinations and systems of linear equations is appealing because the matrix expressions remain the same whether we have few or many variables, even though the "sizes" of the matrices change to represent the numbers of variables used.

In order to make this book accessible to readers with a variety of backgrounds, we have not used matrix algebra extensively. However, we have given some indication of the power of matrix algebra, especially in the last few sections of Chapter 15. The introduction to matrix algebra presented in this appendix is brief, but should be adequate to allow the reader to follow those sections of Chapter 15 that use matrices.

C.2 DEFINITIONS

A *scalar* is an ordinary number or symbol that represents a number, such as 124, -3, c, or x.

A *vector* is a one-dimensional array of elements, such as

$$
\mathbf{u} = \begin{bmatrix} 6 \\ 12 \\ 22 \\ 3 \\ 47 \end{bmatrix} \quad \text{or} \quad \mathbf{v}' = [33 \quad 19 \quad 44 \quad 71 \quad 24 \quad 7]
$$

Here, the vertical array, $\mathbf{u}$, is referred to as a *column* vector, and the horizontal array $\mathbf{v}'$ is a *row* vector. Often, the *prime* ($'$) is used to denote a row vector.

A *matrix* is a two-dimensional array of elements. For example, the data set obtained by testing each of five subjects under three treatment conditions can be represented by the matrix $\mathbf{Y}$, where

$$Y = \begin{bmatrix} 5 & 2 & 7 \\ 6 & 1 & 3 \\ 2 & 1 & 4 \\ 8 & 9 & 6 \\ 6 & 4 & 3 \end{bmatrix}$$

The number of rows and columns a matrix has is referred to as its *order*. The Y matrix we just considered is of order 5×3. The row and column vectors discussed can be regarded as matrices with one row and one column, respectively. Vector u is a 5×1 matrix, and v' is a 1×6 matrix. Consider the $m \times n$ matrix

$$A = \begin{bmatrix} a_{11} & a_{12} & \cdots & a_{1n} \\ a_{21} & a_{22} & \cdots & a_{2n} \\ a_{31} & & & \\ \cdot & & a_{rc} & \cdot \\ \cdot & & & \cdot \\ \cdot & & & \cdot \\ a_{m1} & a_{m2} & \cdots & a_{mn} \end{bmatrix}$$

Any element of A can be specified by indicating its location in the array. For example, a_{rc} refers to the element in the rth row and the cth column of A. The brackets around the collection of elements that make up A, u, and v' signify that they are matrix quantities and are subject to the operations of matrix algebra. We use boldface type to indicate that A represents a matrix rather than a scalar quantity. Frequently, but not always, vectors are denoted by lowercase, and matrices by uppercase, boldface symbols.

We are sometimes interested in the *transpose* of matrix A, which is denoted by the symbol A'. A' is obtained by interchanging the rows and columns of A. That is, the first row of A becomes the first column of A', the second row of A becomes the second column of A', and so on. For example, the transpose of the matrix Y is

$$Y' = \begin{bmatrix} 5 & 6 & 2 & 8 & 6 \\ 2 & 1 & 1 & 9 & 4 \\ 7 & 3 & 4 & 6 & 3 \end{bmatrix}$$

There are a number of adjectives that describe important characteristics of matrices. Any matrix with equal numbers of rows and columns is referred to as a *square* matrix. Any matrix A such that $A' = A$ is referred to as a *symmetric* matrix. In a symmetric matrix such as

$$\begin{bmatrix} 7 & 2 & 8 \\ 2 & 9 & 3 \\ 8 & 3 & 4 \end{bmatrix}$$

$a_{ij} = a_{ji}$; that is, the element in the ith row and jth column is equal to the element in the jth row and the ith column. If this is the case, the elements are symmetrical about the *major diagonal* of the matrix, the diagonal that extends from the upper left to the lower right element. A number of very important matrices, such as the sample correlation and variance-covariance matrices for k variables

$$\mathbf{R} = \begin{bmatrix} 1 & r_{12} & r_{13} & \cdots & r_{1k} \\ r_{21} & 1 & r_{23} & \cdots & r_{2k} \\ . & & & . \\ . & & & . \\ . & & & . \\ r_{k1} & r_{k2} & \cdots & & 1 \end{bmatrix} \quad \text{and} \quad \mathbf{S} = \begin{bmatrix} \hat{\sigma}_1^2 & \hat{\sigma}_{12} & \hat{\sigma}_{13} & \cdots & \hat{\sigma}_{1k} \\ \hat{\sigma}_{21} & \hat{\sigma}_2^2 & \hat{\sigma}_{23} & \cdots & \hat{\sigma}_{2k} \\ . & . & & & . \\ . & . & & & . \\ . & . & & & . \\ \hat{\sigma}_{k1} & \hat{\sigma}_{k2} & \hat{\sigma}_{k3} & \cdots & \hat{\sigma}_k^2 \end{bmatrix}$$

respectively, are symmetric. Matrices that have nonzero elements along their major diagonals but only zeros as off-diagonal elements are referred to as *diagonal* matrices.

A square matrix that is of special interest is the *identity matrix*, which has 1's along the major diagonal and 0's elsewhere; for example,

$$\mathbf{I} = \begin{bmatrix} 1 & 0 & 0 \\ 0 & 1 & 0 \\ 0 & 0 & 1 \end{bmatrix}$$

We shall always use the symbol $\mathbf{I}$ to denote matrices of this kind. As we shall see in the next section, multiplication by $\mathbf{I}$ leaves a matrix unchanged.

C.3 MATRIX ALGEBRA

Just as ordinary algebra involves the operations of addition, subtraction, multiplication, and division of scalar quantities, matrix algebra involves the corresponding matrix operations. We discuss these matrix operations after defining the equality of two matrices.

Equality Two matrices $\mathbf{A}$ and $\mathbf{B}$ are equal if $a_{ij} = b_{ij}$ for all i and j. That is, for two matrices to be equal they must be of the same order and have identical elements.

Addition Two matrices $\mathbf{A}$ and $\mathbf{B}$ may be added if they are both of the same order. The elements of the matrix $\mathbf{C}$ that represents their sum are given by $c_{ij} = a_{ij} + b_{ij}$. Therefore, if we have the matrices

$$\mathbf{A} = \begin{bmatrix} 6 & 9 \\ 7 & 1 \\ 8 & 11 \end{bmatrix} \quad \mathbf{B} = \begin{bmatrix} 6 & 14 \\ 9 & 7 \\ 4 & 12 \end{bmatrix} \quad \text{and} \quad \mathbf{D} = \begin{bmatrix} 9 & 11 & 7 \\ 1 & 8 & 12 \\ 4 & 4 & 9 \end{bmatrix}$$

then

$$C = A + B = \begin{bmatrix} (6+6) & (9+14) \\ (7+9) & (1+7) \\ (8+4) & (11+12) \end{bmatrix} = \begin{bmatrix} 12 & 23 \\ 16 & 8 \\ 12 & 23 \end{bmatrix}$$

Note that neither **A** nor **B** can be added to **D** because they are of different orders.

Subtraction One matrix may be subtracted from another of the same order by subtracting the corresponding elements of one matrix from the other. The elements of the matrix **F** that is obtained by subtracting **B** from **A** are given by $f_{ij} = a_{ij} - b_{ij}$:

$$F = \begin{bmatrix} (6-6) & (9-14) \\ (7-9) & (1-7) \\ (8-4) & (11-12) \end{bmatrix} = \begin{bmatrix} 0 & -5 \\ -2 & -6 \\ 4 & -1 \end{bmatrix}$$

Multiplication There are two kinds of multiplication in matrix algebra. *Scalar multiplication* occurs when a matrix is multiplied by a scalar. The result is the matrix obtained by multiplying each of the elements of the original matrix by the scalar. Therefore, if $L = 7D$, the elements of **L** are given by $l_{ij} = 7d_{ij}$:

$$L = 7D = \begin{bmatrix} (7)(9) & (7)(11) & (7)(7) \\ (7)(1) & (7)(8) & (7)(12) \\ (7)(4) & (7)(4) & (7)(9) \end{bmatrix} = \begin{bmatrix} 63 & 77 & 49 \\ 7 & 56 & 84 \\ 28 & 28 & 63 \end{bmatrix}$$

Note that it is possible to multiply a matrix by a scalar even though a scalar may not be added to or subtracted from a matrix.

In *matrix multiplication,* one matrix is multiplied by another to produce (depending on the orders of the two matrices) either a scalar or another matrix. Matrix multiplication is more complicated than scalar multiplication and involves restrictions on the rows and columns of the matrices that are multiplied. The basic operation involved is that of multiplying a row vector by a column vector to form what is called a *scalar product.* Given a $1 \times n$ vector $\mathbf{r}'$ and an $n \times 1$ column vector $\mathbf{c}$,

$$\mathbf{r}' = [r_1 \quad r_2 \quad r_3 \quad \cdots \quad r_n] \quad \text{and} \quad \mathbf{c} = \begin{bmatrix} c_1 \\ c_2 \\ \cdot \\ \cdot \\ \cdot \\ c_n \end{bmatrix}$$

the scalar product $\mathbf{r}' \cdot \mathbf{c}$ is defined as

$$\mathbf{r}' \cdot \mathbf{c} = \sum_{k=1}^{n} r_k c_k = r_1 c_1 + r_2 c_2 + \cdots + r_n c_n \tag{C.1}$$

That is, the scalar product is the scalar that results from multiplying the elements in the corresponding positions of $\mathbf{r}'$ and $\mathbf{c}$ and summing them. If

$$\mathbf{r}' = [2 \quad 0 \quad 9 \quad 1 \quad 0] \qquad \text{and} \qquad \mathbf{c} = \begin{bmatrix} 6 \\ 2 \\ 1 \\ 4 \\ 3 \end{bmatrix}$$

then the scalar product is $(2)(6) + (0)(2) + (9)(1) + (1)(4) + (0)(3) = 25$. Note that, according to this definition, we can only find $\mathbf{r}' \cdot \mathbf{c}$ if $\mathbf{r}'$ has as many columns as $\mathbf{c}$ has rows. If this condition is not met, the scalar product is undefined.

For matrices $\mathbf{A}$ and $\mathbf{D}$ that each have several rows and columns, the product $\mathbf{P} = \mathbf{D} \times \mathbf{A}$ (also written $\mathbf{DA}$) is defined as the matrix that has as its ijth element, p_{ij}, the scalar product of the ith row of D and the jth column of $\mathbf{A}$:

$$p_{ij} = \mathbf{d}_i' \cdot \mathbf{a}_j = \sum_{k=1}^{n} d_{ik} a_{kj} \tag{C.2}$$

Note that, according to this definition, in order for the product matrix $\mathbf{P}$ to exist, the number of columns in $\mathbf{D}$ must equal the number of rows in $\mathbf{A}$. If this is not the case, the product is undefined. If $\mathbf{P}$ does exist, it has the same number of rows as $\mathbf{D}$ and the same number of columns as $\mathbf{A}$. If we again consider the matrices $\mathbf{D}$, $\mathbf{A}$, and $\mathbf{B}$, where

$$\mathbf{D} = \begin{bmatrix} 9 & 11 & 7 \\ 1 & 8 & 12 \\ 4 & 4 & 9 \end{bmatrix} \qquad \mathbf{A} = \begin{bmatrix} 6 & 9 \\ 7 & 1 \\ 8 & 11 \end{bmatrix} \qquad \text{and} \qquad \mathbf{B} = \begin{bmatrix} 6 & 14 \\ 9 & 7 \\ 4 & 12 \end{bmatrix}$$

the elements of the product matrix $\mathbf{P} = \mathbf{D} \times \mathbf{A}$ are the scalar products of the rows of $\mathbf{D}$ and the columns of $\mathbf{A}$, indicated by Equation C.2. For example, p_{11}, the element in the first row and first column of $\mathbf{P}$ is given by

$$p_{11} = \mathbf{d}_1' \cdot \mathbf{a}_1$$
$$= d_{11} a_{11} + d_{12} a_{21} + d_{13} a_{31}$$
$$= (9)(6) + (11)(7) + (7)(8)$$
$$= 54 + 77 + 56$$
$$= 187$$

Similarly, p_{21} is given by $\mathbf{d}_2' \cdot \mathbf{a}_1$:

$$p_{21} = d_{21} a_{11} + d_{22} a_{21} + d_{23} a_{31}$$
$$= (1)(6) + (8)(7) + (12)(8)$$
$$= 6 + 56 + 96$$
$$= 158$$

The entire product matrix $\mathbf{P} = \mathbf{D} \times \mathbf{A}$ is given by

$$\mathbf{P} = \begin{bmatrix} (9)(6) + (11)(7) + (7)(8) & (9)(9) + (11)(1) + (7)(11) \\ (9)(6) + (8)(7) + (12)(8) & (1)(9) + (8)(1) + (12)(11) \\ (4)(6) + (4)(7) + (9)(8) & (4)(9) + (4)(1) + (9)(11) \end{bmatrix} = \begin{bmatrix} 187 & 169 \\ 158 & 149 \\ 124 & 139 \end{bmatrix}$$

You should verify that the product $\mathbf{D} \times \mathbf{B}$ is given by

$$\begin{bmatrix} \mathbf{d}_1' \cdot \mathbf{b}_1 & \mathbf{d}_1' \cdot \mathbf{b}_2 \\ \mathbf{d}_2' \cdot \mathbf{b}_1 & \mathbf{d}_2' \cdot \mathbf{b}_2 \\ \mathbf{d}_3' \cdot \mathbf{b}_1 & \mathbf{d}_3' \cdot \mathbf{b}_2 \end{bmatrix} = \begin{bmatrix} 181 & 287 \\ 126 & 214 \\ 96 & 192 \end{bmatrix}$$

Note that not all matrices may be multiplied by one another. For example, it is not possible to find the product $\mathbf{AB}$ because the number of columns in $\mathbf{A}$ does not equal the number of columns in $\mathbf{B}$. Also, matrix multiplication is not *commutative;* that is, it is not generally true that $\mathbf{DB} = \mathbf{BD}$. In fact, for the given matrices $\mathbf{D}$ and $\mathbf{B}$ we can find $\mathbf{DB}$, but the product $\mathbf{BD}$ is not defined because the number of columns in $\mathbf{B}$ does not equal the number of rows in $\mathbf{D}$. Therefore, when a matrix $\mathbf{X}$ is multiplied by another matrix $\mathbf{Y}$, we must distinguish between the products $\mathbf{XY}$ and $\mathbf{YX}$. In the former case, we could say that $\mathbf{Y}$ was *premultiplied* by $\mathbf{X}$ or that $\mathbf{X}$ was *postmultiplied* by $\mathbf{Y}$; in the latter, that $\mathbf{Y}$ was postmultiplied by $\mathbf{X}$ or $\mathbf{X}$ was premultiplied by $\mathbf{Y}$.

Although matrix multiplication is not commutative, it is *associative.* That is, $\mathbf{XYZ} = (\mathbf{XY})\mathbf{Z} = \mathbf{X}(\mathbf{YZ})$. If the product $\mathbf{XYZ}$ exists, it can be obtained by first finding the product $\mathbf{XY}$ and postmultiplying it by $\mathbf{Z}$ or by finding the product $\mathbf{YZ}$ and premultiplying it by $\mathbf{X}$.

Finally, as indicated in the previous section, multiplication of any matrix by an appropriate identity matrix leaves the original matrix unchanged. That is, $\mathbf{AI} = \mathbf{IA} = \mathbf{A}$. For the matrix $\mathbf{A}$ defined earlier, you should verify that

$$\overset{\mathbf{A}}{\begin{bmatrix} 6 & 14 \\ 9 & 7 \\ 4 & 12 \end{bmatrix}} \overset{\mathbf{I}}{\begin{bmatrix} 1 & 0 \\ 0 & 1 \end{bmatrix}} = \overset{\mathbf{I}}{\begin{bmatrix} 1 & 0 & 0 \\ 0 & 1 & 0 \\ 0 & 0 & 1 \end{bmatrix}} \overset{\mathbf{A}}{\begin{bmatrix} 6 & 14 \\ 9 & 7 \\ 4 & 12 \end{bmatrix}} = \overset{\mathbf{A}}{\begin{bmatrix} 6 & 14 \\ 9 & 7 \\ 4 & 12 \end{bmatrix}}$$

Note that because $\mathbf{A}$ is not a square matrix, the identity matrix that premultiplies $\mathbf{A}$ differs in size from the identity matrix that postmultiplies it.

Although the definition of matrix multiplication that we have considered in this section may at first glance seem to be arbitrary, it is very useful because it allows us to express linear combinations of variables and systems of equations in very simple ways. Suppose, for example, we consider the matrix $\mathbf{D}$ and two column vectors,

$$\mathbf{x} = \begin{bmatrix} x \\ y \\ z \end{bmatrix} \quad \text{and} \quad \mathbf{k} = \begin{bmatrix} 1 \\ 4 \\ 7 \end{bmatrix}$$

Then $\mathbf{Dx}$ is the 3×1 matrix

$$\begin{bmatrix} 9x & + & 11y & + & 7z \\ x & + & 8y & + & 12z \\ 4x & + & 4y & + & 9z \end{bmatrix}$$

and by the definition of matrix equality presented in the previous section, the matrix equation $\mathbf{Dx} = \mathbf{k}$ indicates that

$$9x + 11y + 7z = 1$$
$$x + 8y + 12z = 4$$
$$4x + 4y + 9z = 7$$

so that the matrix equation represents a set of three simultaneous scalar equations.

As another example, consider the development of the regression model in Chapter 12. If we define the matrices

$$\mathbf{Y} = \begin{bmatrix} Y_1 \\ Y_2 \\ \cdot \\ \cdot \\ Y_i \\ \cdot \\ \cdot \\ Y_N \end{bmatrix} \quad \mathbf{X} = \begin{bmatrix} 1 & X_1 \\ 1 & X_2 \\ \cdot & \cdot \\ \cdot & \cdot \\ \cdot & \cdot \\ 1 & X_i \\ \cdot & \cdot \\ \cdot & \cdot \\ \cdot & \cdot \\ 1 & X_N \end{bmatrix} \quad \boldsymbol{\beta} = \begin{bmatrix} \beta_0 \\ \beta_1 \end{bmatrix} \quad \hat{\boldsymbol{\beta}} = \begin{bmatrix} b_0 \\ b_1 \end{bmatrix} \quad \boldsymbol{\epsilon} = \begin{bmatrix} \epsilon_1 \\ \epsilon_2 \\ \cdot \\ \cdot \\ \epsilon_i \\ \cdot \\ \cdot \\ \epsilon_N \end{bmatrix}$$

then, by the rules of matrix multiplication and addition,

$$\mathbf{X}\boldsymbol{\beta} = \begin{bmatrix} (1)\beta_0 + \beta_1 X_1 \\ (1)\beta_0 + \beta_1 X_2 \\ \cdot \\ \cdot \\ \cdot \\ (1)\beta_0 + \beta_1 X_i \\ \cdot \\ \cdot \\ \cdot \\ (1)\beta_0 + \beta_1 X_N \end{bmatrix} \quad \text{and} \quad \mathbf{X}\boldsymbol{\beta} + \boldsymbol{\epsilon} = \begin{bmatrix} \beta_0 + \beta_1 X_1 + \epsilon_1 \\ \beta_0 + \beta_1 X_2 + \epsilon_2 \\ \cdot \\ \cdot \\ \cdot \\ \beta_0 + \beta_1 X_i + \epsilon_i \\ \cdot \\ \cdot \\ \cdot \\ \beta_0 + \beta_1 X_N + \epsilon_N \end{bmatrix}$$

Therefore, the single matrix equation $\mathbf{Y} = \mathbf{X}\boldsymbol{\beta} + \boldsymbol{\epsilon}$ represents the set of N equations $Y_i = \beta_0 + \beta_1 X_i + \epsilon_i$ formed as i takes on the values $1, 2, \ldots, N$. If we were dealing with a multiple regression model in which we had p predictor variables X_1, $X_2, \ldots, X_p$, the system of equations

$$Y_i = \beta_0 + \beta_1 X_{i1} + \beta_2 X_{i2} + \cdots + \beta_p X_{ip} + \epsilon_i$$

could still be represented by the single matrix equation $\mathbf{Y} = \mathbf{X}\beta + \epsilon$, except that β would be a $(p + 1) \times 1$ column vector with elements $\beta_0, \beta_1, \beta_2, \ldots, \beta_p$ and $\mathbf{X}$ would be an $N \times (p + 1)$ matrix in which there was a column of 1's as well as separate columns for each of the p predictor variables.

Also, as we indicate in Section 15.10, matrix algebra provides a useful approach to estimating parameters in regression analysis. In Section 15.10, we indicated that the matrix equation

$$\mathbf{X'X}\hat{\beta} = \mathbf{X'Y} \tag{C.3}$$

can be used to represent the system of normal equations for bivariate regression that was first presented in Equation 12.8. Equation C.3 can also be used to represent the system of $p + 1$ normal equations obtained in multiple regression with p predictor variables if the sizes of $\hat{\beta}$ and $\mathbf{X}$ are increased appropriately to take account of the number of predictor variables.

Finally, note that we could solve Equation C.3 for $\hat{\beta}$ if we could somehow divide both sides of the equation by $\mathbf{X'X}$ and therefore isolate $\hat{\beta}$ on the left-hand side of the equation. We now consider what corresponds to division in matrix algebra.

C.4 THE INVERSE OF A MATRIX

Dividing a scalar quantity B by another scalar quantity A is equivalent to multiplying B by $1/A$ or A^{-1}, the reciprocal or *inverse* of A. The product of A and its inverse, $(A^{-1})(A) = (A)(A^{-1})$, equals 1. Analogously, a square matrix $\mathbf{A}$ is said to have an inverse if we can find a matrix $\mathbf{A}^{-1}$ such that

$$\mathbf{AA}^{-1} = \mathbf{A}^{-1}\mathbf{A} = \mathbf{I}$$

We showed earlier that the matrix equation of the form

$$\mathbf{Ax} = \mathbf{k}$$

represented a system of scalar equations. If $\mathbf{A}$ has an inverse, the matrix equation can be solved by premultiplying both sides of the equation by $\mathbf{A}^{-1}$. This yields

$$\mathbf{A}^{-1}\mathbf{Ax} = \mathbf{A}^{-1}\mathbf{k}$$

$$\mathbf{Ix} = \mathbf{A}^{-1}\mathbf{k} \qquad \text{because } \mathbf{A}^{-1}\mathbf{A} = \mathbf{I}$$

and

$$\mathbf{x} = \mathbf{A}^{-1}\mathbf{k} \tag{C.4}$$

because multiplying by $\mathbf{I}$ leaves a matrix unchanged. If, for example,

$$\mathbf{A} = \begin{bmatrix} 2 & 1 \\ 1 & 2 \end{bmatrix} \qquad \mathbf{x} = \begin{bmatrix} x \\ y \end{bmatrix} \qquad \text{and} \qquad \mathbf{k} = \begin{bmatrix} 9 \\ 6 \end{bmatrix}$$

then the equation $\mathbf{Ax} = \mathbf{k}$ represents the system of scalar equations

$$2x + \ y = 9$$

$$x + 2y = 6$$

If the equation is solved by using the equation $\mathbf{x} = \mathbf{A}^{-1}\mathbf{k}$, the elements of $\mathbf{x}$, $x = 4$ and $y = 1$, represent the solution to the system of scalar equations. Similarly, consider the normal equations for multiple regression, which we can represent by the matrix equation

$$(\mathbf{X}'\mathbf{X})\hat{\beta} = \mathbf{X}'\mathbf{Y}$$

If the matrix $\mathbf{X}'\mathbf{X}$ has an inverse, the solution is provided by

$$\hat{\beta} = (\mathbf{X}'\mathbf{X})^{-1}\mathbf{X}'\mathbf{Y}$$

It should be emphasized that not all matrices have inverses. Matrices that have inverses are referred to as *nonsingular* and those that do not are referred to as *singular*. Using our definition of inverse, only square matrices may have inverses, and not all square matrices will have them.

One way of understanding why a square matrix $\mathbf{A}$ may not have an inverse is to consider a set of simultaneous equations that have the elements of $\mathbf{A}$ as their coefficients. For example, consider a matrix $\mathbf{A}$ in which the second and third rows are proportional to one another,

$$\mathbf{A} = \begin{bmatrix} 1 & 9 & -2 \\ 2 & 3 & 7 \\ 4 & 6 & 14 \end{bmatrix}$$

and the corresponding system of equations

$$x + 9y - \ 2z = k_1$$

$$2x + 3y + \ 7z = k_2$$

$$4x + 6y + 14z = k_3$$

Because the coefficients in the last two equations are proportional, the system of equations does not have a unique solution no matter what values the k's take on. If $k_3 = 2k_2$, the last two equations are completely redundant, and we have three unknowns and only two independent equations; if $k_3 \neq 2k_2$, the last two equations provide inconsistent information. In either case, we cannot solve the equations for x, y, and z. We can infer that $\mathbf{A}^{-1}$ does not exist, because, if it did, we would be able to find the solution to the system of simultaneous equations by using Equation C.4.

It can be shown that matrix $\mathbf{A}$ will have an inverse only if the rows and columns of $\mathbf{A}$ are *linearly independent*. The rows of a matrix are linearly independent if no row can be expressed as a linear combination of the other rows, and the columns are linearly independent if no column can be expressed as a linear combination of the other columns. In the matrix $\mathbf{A}$ we just considered, the second and third rows are *linearly dependent;* one row is twice another. In Section 15.7.2, we indicate that if any predictor variable in a multiple regression can be expressed as a linear combination of the other predictors, the normal equations cannot be solved to find the regression coefficients. In this case, the matrices **X** and **X'X** will be singular.

There are a variety of procedures for finding the inverses of matrices that are sometimes presented in textbooks. We will not provide any of them here because, other than perhaps for illustrative purposes with a 2 × 2 or 3 × 3 matrix, inverses will virtually never be calculated by hand or with a pocket calculator. Even algorithms for finding inverses that might be incorporated into a computer program should be used with great care because the extensive calculations involved can lead to serious computational inaccuracies, especially in the presence of a degree of multicollinearity. Consequently, many statistical software packages do not actually compute matrix inverses when, for example, obtaining regression coefficients. Instead they use alternative algorithms that produce the desired coefficients with maximal accuracy (for this reason, one should avoid nonstandard statistical programs that have not been checked for computational accuracy). Somewhat more detail about inverses than is presented here is available in sources such as Draper and Smith (1981), Myers (1979), and Pedhazur (1982).

EXERCISES

C.1 Let

$$\mathbf{X} = \begin{bmatrix} X_{11} & X_{12} \\ X_{21} & X_{22} \\ X_{31} & X_{32} \end{bmatrix} \quad \text{and} \quad \mathbf{Y} = \begin{bmatrix} Y_{11} & Y_{12} & Y_{13} & Y_{14} \\ Y_{21} & Y_{22} & Y_{23} & Y_{24} \end{bmatrix}$$

Verify that $(\mathbf{XY})' = \mathbf{Y'X'}$.

C.2 Given

$$\mathbf{c}' = [3 \quad 0 \quad 7 \quad 0] \qquad \mathbf{b} = \begin{bmatrix} 4 \\ 0 \\ 4 \\ 1 \end{bmatrix} \qquad \mathbf{u} = \begin{bmatrix} 3 \\ 5 \\ 1 \end{bmatrix}$$

$$\mathbf{D} = \begin{bmatrix} 1 & 4 & 7 \\ 2 & 5 & 8 \\ 3 & 6 & 9 \end{bmatrix} \qquad \mathbf{M} = \begin{bmatrix} 2 & 1 \\ 5 & 4 \\ 8 & 1 \end{bmatrix} \qquad \mathbf{L} = \begin{bmatrix} 4 & 3 \\ 1 & 2 \\ 7 & 5 \end{bmatrix}$$

Find each of the following or indicate that the operation is not possible:

(a) $\mathbf{M} - \mathbf{L}$	(b) $\mathbf{b} + \mathbf{D}$	(c) $\mathbf{DM}$	(d) $\mathbf{MD}$
(e) $\mathbf{M}'$	(f) $\mathbf{M}^{-1}$	(g) $\mathbf{c}'\mathbf{D}$	(h) $\mathbf{M}'\mathbf{D}$
(i) $\mathbf{b}'\mathbf{c}$	(j) $\mathbf{u}'\mathbf{Du}$	(k) $\mathbf{u}'\mathbf{D}'\mathbf{Du}$	
(l) $(\mathbf{Du})'\mathbf{Du}$			

C.3 Using the matrices defined in Exercise C.2, show that $(\mathbf{L'D})\mathbf{M} = \mathbf{L'}(\mathbf{DM})$.

C.4 Two scores, X and Y, are obtained from each of N subjects. Let $\mathbf{x}$ and $\mathbf{y}$ be $N \times 1$ column vectors whose elements are the deviations of the scores from their means. Express the variances of X and Y and the covariance between X and Y in terms of $\mathbf{x}$ and $\mathbf{y}$.

Appendix **D**

Statistical Tables

68	11	19	83	59	37	13	10	55	87	06	04	15	72	88	06	74	58	89	20	77	84	93	11	39
39	16	58	70	03	75	44	89	24	33	20	71	90	61	89	69	79	42	70	17	09	41	06	18	07
85	82	21	17	57	94	00	36	53	58	18	91	70	41	87	03	38	46	09	54	66	49	46	71	66
71	21	12	06	20	17	91	36	76	38	13	67	08	38	42	41	81	41	87	65	31	53	08	42	85
68	32	56	88	77	39	13	96	08	96	93	54	27	43	59	95	41	71	26	26	37	39	87	14	75
43	60	05	27	17	45	71	18	47	50	91	51	50	56	67	48	33	36	88	44	92	00	47	96	25
64	98	17	99	65	20	80	59	52	33	76	19	39	66	67	37	60	43	04	11	04	14	45	73	29
49	83	54	86	62	74	59	38	38	08	60	76	05	57	10	53	06	00	23	57	84	37	22	97	01
13	46	18	39	67	97	24	49	48	38	52	24	15	06	20	95	85	91	33	41	90	38	91	12	11
34	59	74	17	94	07	00	48	44	67	66	02	50	33	15	54	26	43	76	24	47	62	80	84	28
44	42	38	32	14	50	10	11	50	62	81	26	30	22	70	13	73	17	74	33	98	08	25	53	24
06	15	64	26	19	75	30	26	86	91	41	65	96	02	18	26	51	01	14	34	48	50	63	80	29
71	84	69	47	01	30	73	84	59	55	53	69	63	65	04	25	45	74	25	84	72	24	79	34	76
25	53	39	41	19	87	99	51	37	62	92	64	66	76	24	00	39	39	07	94	26	98	97	21	96
11	51	19	10	22	60	80	34	50	71	21	55	52	54	23	66	42	42	06	82	33	79	72	38	37
18	52	92	51	96	42	02	26	61	51	85	34	08	63	49	37	42	93	16	94	63	03	70	75	31
89	53	38	14	42	77	59	45	86	93	00	51	82	35	74	15	39	88	75	25	51	57	44	29	33
34	05	21	93	97	50	47	27	58	15	30	69	72	87	54	40	44	52	74	80	69	87	38	52	27
26	67	87	51	57	39	07	65	29	81	99	87	60	70	66	30	97	36	33	97	51	28	58	68	63
21	27	53	18	86	57	84	40	49	45	96	81	62	74	79	64	50	24	60	03	89	55	46	73	60
72	05	18	24	75	83	86	37	35	22	26	66	80	54	08	53	39	24	36	51	73	63	99	05	98
72	96	30	30	96	53	69	30	69	91	53	16	61	17	31	24	29	19	48	50	69	97	01	43	29
89	40	12	56	87	63	96	19	52	81	14	44	09	13	65	79	79	69	68	48	18	26	15	40	62
61	38	98	35	29	83	84	20	14	67	76	20	76	33	07	59	99	92	18	00	35	86	18	40	74
50	89	83	20	15	66	99	80	43	02	91	48	99	22	10	27	85	92	31	33	02	05	32	33	04
03	16	15	59	91	28	81	81	73	42	28	13	11	59	97	67	17	23	78	37	69	77	95	44	32
92	99	57	20	90	21	10	94	29	45	11	19	87	43	35	15	68	20	31	12	45	13	99	55	33
37	20	87	70	68	13	82	48	91	58	31	68	64	15	76	25	74	46	97	19	54	72	80	79	91
52	76	65	26	69	13	19	41	17	90	80	94	56	44	48	69	75	81	67	12	76	77	04	28	74
61	22	35	82	02	11	25	35	92	53	29	56	91	84	71	39	40	56	64	14	70	41	25	18	08
81	47	51	02	74	97	77	34	48	59	25	67	58	99	61	01	12	88	72	82	53	77	59	38	57
76	76	61	21	40	84	75	37	57	42	73	23	59	13	86	64	41	05	40	77	60	12	16	39	90
64	27	04	93	20	94	22	84	92	34	78	66	80	37	50	81	30	19	34	35	19	08	99	79	66
73	66	22	30	59	75	96	89	30	13	89	05	43	15	51	32	65	54	30	97	57	07	83	83	27
54	14	08	36	19	42	17	99	86	85	61	22	55	32	55	46	91	38	32	98	22	45	23	38	55
95	35	69	09	22	83	24	61	25	78	74	29	98	93	23	23	72	90	59	25	95	03	70	09	97
05	41	28	24	56	42	14	27	71	07	14	93	89	70	11	95	12	58	64	47	34	38	18	53	19
63	67	82	87	47	43	56	04	45	40	37	84	83	04	64	59	47	57	77	52	75	31	01	55	16
90	29	71	40	74	77	46	23	97	44	37	27	04	26	93	11	39	85	30	74	51	25	22	37	43
49	17	00	81	31	99	61	74	76	00	40	05	01	57	76	49	19	33	07	20	63	80	56	47	48

TABLE D.2 THE STANDARDIZED NORMAL DISTRIBUTION

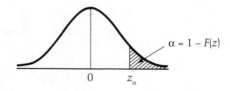

$\alpha = 1 - F(z)$

z	α	z	α	z	α	z	α	z	α
.00	.5000	.33	.3707	.66	.2546	.99	.1611	1.32	.0934
.01	.4960	.34	.3669	.67	.2514	1.00	.1587	1.33	.0918
.02	.4920	.35	.3632	.68	.2483	1.01	.1562	1.34	.0901
.03	.4880	.36	.3594	.69	.2451	1.02	.1539	1.35	.0885
.04	.4840	.37	.3557	.70	.2420	1.03	.1515	1.36	.0869
.05	.4801	.38	.3520	.71	.2389	1.04	.1492	1.37	.0853
.06	.4761	.39	.3483	.72	.2358	1.05	.1469	1.38	.0838
.07	.4721	.40	.3446	.73	.2327	1.06	.1446	1.39	.0823
.08	.4681	.41	.3409	.74	.2296	1.07	.1423	1.40	.0808
.09	.4641	.42	.3372	.75	.2266	1.08	.1401	1.41	.0793
.10	.4602	.43	.3336	.76	.2236	1.09	.1379	1.42	.0778
.11	.4562	.44	.3300	.77	.2206	1.10	.1357	1.43	.0764
.12	.4522	.45	.3264	.78	.2177	1.11	.1335	1.44	.0749
.13	.4483	.46	.3228	.79	.2148	1.12	.1314	1.45	.0735
.14	.4443	.47	.3192	.80	.2119	1.13	.1292	1.46	.0721
.15	.4404	.48	.3156	.81	.2090	1.14	.1271	1.47	.0708
.16	.4364	.49	.3121	.82	.2061	1.15	.1251	1.48	.0694
.17	.4325	.50	.3085	.83	.2033	1.16	.1230	1.49	.0681
.18	.4286	.51	.3050	.84	.2005	1.17	.1210	1.50	.0668
.19	.4247	.52	.3015	.85	.1977	1.18	.1190	1.51	.0655
.20	.4207	.53	.2981	.86	.1949	1.19	.1170	1.52	.0643
.21	.4168	.54	.2946	.87	.1922	1.20	.1151	1.53	.0630
.22	.4129	.55	.2912	.88	.1894	1.21	.1131	1.54	.0618
.23	.4090	.56	.2877	.89	.1867	1.22	.1112	1.55	.0606
.24	.4052	.57	.2843	.90	.1841	1.23	.1093	1.56	.0594
.25	.4013	.58	.2810	.91	.1814	1.24	.1075	1.57	.0582
.26	.3974	.59	.2776	.92	.1788	1.25	.1056	1.58	.0571
.27	.3936	.60	.2743	.93	.1762	1.26	.1038	1.59	.0559
.28	.3897	.61	.2709	.94	.1736	1.27	.1020	1.60	.0548
.29	.3859	.62	.2676	.95	.1711	1.28	.1003	1.61	.0537
.30	.3821	.63	.2643	.96	.1685	1.29	.0985	1.62	.0526
.31	.3783	.64	.2611	.97	.1660	1.30	.0968	1.63	.0516
.32	.3745	.65	.2578	.98	.1635	1.31	.0951	1.64	.0505

(*continued*)

z	α	z	α	z	α	z	α	z	α
1.65	.0495	1.98	.0239	2.31	.0104	2.64	.0041	2.97	.0015
1.66	.0485	1.99	.0233	2.32	.0102	2.65	.0040	2.98	.0014
1.67	.0475	2.00	.0228	2.33	.0099	2.66	.0039	2.99	.0014
1.68	.0465	2.01	.0222	2.34	.0096	2.67	.0038	3.00	.0013
1.69	.0455	2.02	.0217	2.35	.0094	2.68	.0037	3.01	.0013
1.70	.0446	2.03	.0212	2.36	.0091	2.69	.0036	3.02	.0013
1.71	.0436	2.04	.0207	2.37	.0089	2.70	.0035	3.03	.0012
1.72	.0427	2.05	.0202	2.38	.0087	2.71	.0034	3.04	.0012
1.73	.0418	2.06	.0197	2.39	.0084	2.72	.0033	3.05	.0011
1.74	.0409	2.07	.0192	2.40	.0082	2.73	.0032	3.06	.0011
1.75	.0401	2.08	.0188	2.41	.0080	2.74	.0031	3.07	.0011
1.76	.0392	2.09	.0183	2.42	.0078	2.75	.0030	3.08	.0010
1.77	.0384	2.10	.0179	2.43	.0075	2.76	.0029	3.09	.0010
1.78	.0375	2.11	.0174	2.44	.0073	2.77	.0028	3.10	.0010
1.79	.0367	2.12	.0170	2.45	.0071	2.78	.0027	3.11	.0009
1.80	.0359	2.13	.0166	2.46	.0069	2.79	.0026	3.12	.0009
1.81	.0351	2.14	.0162	2.47	.0068	2.80	.0026	3.13	.0009
1.82	.0344	2.15	.0158	2.48	.0066	2.81	.0025	3.14	.0008
1.83	.0336	2.16	.0154	2.49	.0064	2.82	.0024	3.15	.0008
1.84	.0329	2.17	.0150	2.50	.0062	2.83	.0023	3.16	.0008
1.85	.0322	2.18	.0146	2.51	.0060	2.84	.0023	3.17	.0008
1.86	.0314	2.19	.0143	2.52	.0059	2.85	.0022	3.18	.0007
1.87	.0307	2.20	.0139	2.53	.0057	2.86	.0021	3.19	.0007
1.88	.0301	2.21	.0136	2.54	.0055	2.87	.0021	3.20	.0007
1.89	.0294	2.22	.0132	2.55	.0054	2.88	.0020	3.21	.0007
1.90	.0287	2.23	.0129	2.56	.0052	2.89	.0019	3.22	.0006
1.91	.0281	2.24	.0125	2.57	.0051	2.90	.0019	3.23	.0006
1.92	.0274	2.25	.0122	2.58	.0049	2.91	.0018	3.24	.0006
1.93	.0268	2.26	.0119	2.59	.0048	2.92	.0018	3.25	.0006
1.94	.0262	2.27	.0116	2.60	.0047	2.93	.0017		
1.95	.0256	2.28	.0113	2.61	.0045	2.94	.0016		
1.96	.0250	2.29	.0110	2.62	.0044	2.95	.0016		
1.97	.0244	2.30	.0107	2.63	.0043	2.96	.0015		

Source: Adapted from Table 1 in Pearson, E. S. and Hartley, H. O. (1958). *Biometrika Tables for Statisticians*, Vol. 1, 2nd ed. Cambridge University Press: Cambridge, with the kind permission of the trustees of *Biometrika*.

TABLE D.3 PERCENTAGE POINTS OF THE *t* DISTRIBUTION

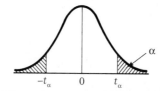

	Level of Significance for a One-Tailed Test									
	0.4	0.25	0.1	0.05	0.025	0.01	0.005	0.0025	0.001	0.0005
	Level of Significance for a Two-Tailed Test									
df	0.8	0.5	0.2	0.1	0.05	0.02	0.01	0.005	0.002	0.001
1	0.325	1.000	3.078	6.314	12.706	31.821	63.657	127.32	318.31	636.62
2	.289	0.816	1.886	2.920	4.303	6.965	9.925	14.089	22.326	31.598
3	.277	.765	1.638	2.353	3.182	4.541	5.841	7.453	10.213	12.924
4	.271	.741	1.533	2.132	2.776	3.747	4.604	5.598	7.173	8.610
5	0.267	0.727	1.476	2.015	2.571	3.365	4.032	4.773	5.893	6.869
6	.265	.718	1.440	1.943	2.447	3.143	3.707	4.317	5.208	5.959
7	.263	.711	1.415	1.895	2.365	2.998	3.499	4.029	4.785	5.408
8	.262	.706	1.397	1.860	2.306	2.896	3.355	3.833	4.501	5.041
9	.261	.703	1.383	1.833	2.262	2.821	3.250	3.690	4.297	4.781
10	0.260	0.700	1.372	1.812	2.228	2.764	3.169	3.581	4.144	4.587
11	.260	.697	1.363	1.796	2.201	2.718	3.106	3.497	4.025	4.437
12	.259	.695	1.356	1.782	2.179	2.681	3.055	3.428	3.930	4.318
13	.259	.694	1.350	1.771	2.160	2.650	3.012	3.372	3.852	4.221
14	.258	.692	1.345	1.761	2.145	2.624	2.977	3.326	3.787	4.140
15	0.258	0.691	1.341	1.753	2.131	2.602	2.947	3.286	3.733	4.073
16	.258	.690	1.337	1.746	2.120	2.583	2.921	3.252	3.686	4.015
17	.257	.689	1.333	1.740	2.110	2.567	2.898	3.222	3.646	3.965
18	.257	.688	1.330	1.734	2.101	2.552	2.878	3.197	3.610	3.922
19	.257	.688	1.328	1.729	2.093	2.539	2.861	3.174	3.579	3.883
20	0.257	0.687	1.325	1.725	2.086	2.528	2.845	3.153	3.552	3.850
21	.257	.686	1.323	1.721	2.080	2.518	2.831	3.135	3.527	3.819
22	.256	.686	1.321	1.717	2.074	2.508	2.819	3.119	3.505	3.792
23	.256	.685	1.319	1.714	2.069	2.500	2.807	3.104	3.485	3.767
24	.256	.685	1.318	1.711	2.064	2.492	2.797	3.091	3.467	3.745
25	0.256	0.684	1.316	1.708	2.060	2.485	2.787	3.078	3.450	3.725
26	.256	.684	1.315	1.706	2.056	2.479	2.779	3.067	3.435	3.707
27	.256	.684	1.314	1.703	2.052	2.473	2.771	3.057	3.421	3.690
28	.256	.683	1.313	1.701	2.048	2.467	2.763	3.047	3.408	3.674
29	.256	.683	1.311	1.699	2.045	2.462	2.756	3.038	3.396	3.659
30	0.256	0.683	1.310	1.697	2.042	2.457	2.750	3.030	3.385	3.646
40	.255	.681	1.303	1.684	2.021	2.423	2.704	2.971	3.307	3.551
60	.254	.679	1.296	1.671	2.000	2.390	2.660	2.915	3.232	3.460
120	.254	.677	1.289	1.658	1.980	2.358	2.617	2.860	3.160	3.373
∞	.253	.674	1.282	1.645	1.960	2.326	2.576	2.807	3.090	3.291

Source: Adapted from Table 12 in Pearson, E. S. and Hartley, H. O. (1958). *Biometrika Tables for Statisticians*, Vol. 1, 2nd ed. Cambridge University Press: Cambridge, with the kind permission of the trustees of *Biometrika*.

TABLE D.4 PERCENTAGE POINTS OF THE CHI-SQUARE DISTRIBUTION

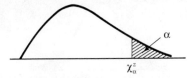

df \ α	0.995	0.990	0.975	0.950	0.900	0.750	0.500
1	392704.10^{-10}	157088.10^{-9}	982069.10^{-9}	393214.10^{-8}	0.0157908	0.1015308	0.454937
2	0.0100251	0.0201007	0.0506356	0.102587	0.210720	0.575364	1.38629
3	0.0717212	0.114832	0.215795	0.351846	0.584375	1.212534	2.36597
4	0.206990	0.297110	0.484419	0.710721	1.063623	1.92255	3.35670
5	0.411740	0.554300	0.831211	1.145476	1.61031	2.67460	4.35146
6	0.675727	0.872085	1.237347	1.63539	2.20413	3.45460	5.34812
7	0.989265	1.239043	1.68987	2.16735	2.83311	4.25485	6.34581
8	1.344419	1.646482	2.17973	2.73264	3.48954	5.07064	7.34412
9	1.734926	2.087912	2.70039	3.32511	4.16816	5.89883	8.34283
10	2.15585	2.55821	3.24697	3.94030	4.86518	6.73720	9.34182
11	2.60321	3.05347	3.81575	4.57481	5.57779	7.58412	10.3410
12	3.07382	3.57056	4.40379	5.22603	6.30380	8.43842	11.3403
13	3.56503	4.10691	5.00874	5.89186	7.04150	9.29906	12.3398
14	4.07468	4.66043	5.62872	6.57063	7.78953	10.1653	13.3393
15	4.60094	5.22935	6.26214	7.26094	8.54675	11.0365	14.3389
16	5.14224	5.81221	6.90766	7.96164	9.31223	11.9122	15.3385
17	5.69724	6.40776	7.56418	8.67176	10.0852	12.7919	16.3381
18	6.26481	7.01491	8.23075	9.39046	10.8649	13.6753	17.3379
19	6.84398	7.63273	8.90655	10.1170	11.6509	14.5620	18.3376
20	7.43386	8.26040	9.59083	10.8508	12.4426	15.4518	19.3374
21	8.03366	8.89720	10.28293	11.5913	13.2396	16.3444	20.3372
22	8.64272	9.54249	10.9823	12.3380	14.0415	17.2396	21.3370
23	9.26042	10.19567	11.6885	13.0905	14.8479	18.1373	22.3369
24	9.88623	10.8564	12.4011	13.8484	15.6587	19.0372	23.3367
25	10.5197	11.5240	13.1197	14.6114	16.4734	19.9393	24.3366
26	11.1603	12.1981	13.8439	15.3791	17.2919	20.8434	25.3364
27	11.8076	12.8786	14.5733	16.1513	18.1138	21.7494	26.3363
28	12.4613	13.5648	15.3079	16.9279	18.9392	22.6572	27.3363
29	13.1211	14.2565	16.0471	17.7083	19.7677	23.5666	28.3362
30	13.7867	14.9535	16.7908	18.4926	20.5992	24.4776	29.3360
40	20.7065	22.1643	24.4331	26.5093	29.0505	33.6603	39.3354
50	27.9907	29.7067	32.3574	34.7642	37.6886	42.9421	49.3349
60	35.5346	37.4848	40.4817	43.1879	46.4589	52.2938	59.3347
70	43.2752	45.4418	48.7576	51.7393	55.3290	61.6983	69.3344
80	51.1720	53.5400	57.1532	60.3915	64.2778	71.1445	79.3343
90	59.1963	61.7541	65.6466	69.1260	73.2912	80.6247	89.3342
100	67.3276	70.0648	74.2219	77.9295	82.3581	90.1332	99.3341
z	−2.5758	−2.3263	−1.9600	−1.6449	−1.2816	−0.6745	0.0000

df \ α	0.250	0.100	0.050	0.025	0.010	0.005	0.001
1	1.32330	2.70554	3.84146	5.02389	6.63490	7.87944	10.828
2	2.77259	4.60517	5.99147	7.37776	9.21034	10.5966	13.816
3	4.10835	6.25139	7.81473	9.34840	11.3449	12.8381	16.266
4	5.38527	7.77944	9.48773	11.1433	13.2767	14.8602	18.467
5	6.62568	9.23635	11.0705	12.8325	15.0863	16.7496	20.515
6	7.84080	10.6446	12.5916	14.4494	16.8119	18.5476	22.458
7	9.03715	12.0170	14.0671	16.0128	18.4753	20.2777	24.322
8	10.2188	13.3616	15.5073	17.5346	20.0902	21.9550	26.125
9	11.3887	14.6837	16.9190	19.0228	21.6660	23.5893	27.877
10	12.5489	15.9871	18.3070	20.4831	23.2093	25.1882	29.588
11	13.7007	17.2750	19.6751	21.9200	24.7250	26.7569	31.264
12	14.8454	18.5494	21.0261	23.3367	26.2170	28.2995	32.909
13	15.9839	19.8119	22.3621	24.7356	27.6883	29.8194	34.528
14	17.1170	21.0642	23.6848	26.1190	29.1413	31.3193	36.123
15	18.2451	22.3072	24.9958	27.4884	30.5779	32.8013	37.697
16	19.3688	23.5418	26.2962	28.8454	31.9999	34.2672	39.252
17	20.4887	24.7690	27.5871	30.1910	33.4087	35.7185	40.790
18	21.6049	25.9894	28.8693	31.5264	34.8053	37.1564	42.312
19	22.7178	27.2036	30.1435	32.8523	36.1908	38.5822	43.820
20	23.8277	28.4120	31.4104	34.1696	37.5662	39.9968	45.315
21	24.9348	29.6151	32.6705	35.4789	38.9321	41.4010	46.797
22	26.0393	30.8133	33.9244	36.7807	40.2894	42.7956	48.268
23	27.1413	32.0069	35.1725	38.0757	41.6384	44.1813	49.728
24	28.2412	33.1963	36.4151	39.3641	42.9798	45.5585	51.179
25	29.3389	34.3816	37.6525	40.6465	44.3141	46.9278	52.620
26	30.4345	35.5631	38.8852	41.9232	45.6417	48.2899	54.052
27	31.5284	36.7412	40.1133	43.1944	46.9630	49.6449	55.476
28	32.6205	37.9159	41.3372	44.4607	48.2782	50.9933	56.892
29	33.7109	39.0875	42.5569	45.7222	49.5879	52.3356	58.302
30	34.7998	40.2560	43.7729	46.9792	50.8922	53.6720	59.703
40	45.6160	51.8050	55.7585	59.3417	63.6907	66.7659	73.402
50	56.3336	63.1671	67.5048	71.4202	76.1539	79.4900	86.661
60	66.9814	74.3970	79.0819	83.2976	88.3794	91.9517	99.607
70	77.5766	85.5271	90.5312	95.0231	100.425	104.215	112.317
80	88.1303	96.5782	101.879	106.629	112.329	116.321	124.839
90	98.6499	107.565	113.145	118.136	124.116	128.299	137.208
100	109.141	118.498	124.342	129.561	135.807	140.169	149.449
z	+0.6745	+1.2816	+1.6449	+1.9600	+2.3263	+2.5758	+3.0902

For df > 100 take

$$\chi^2 = df \left[1 - \frac{2}{9\,df} + z\sqrt{\frac{2}{9\,df}} \right]^3 \quad \text{or} \quad \chi^2 = \frac{1}{2} \left(z + \sqrt{2\,df - 1} \right)^2$$

according to the degree of accuracy required. z is the standardized normal deviate corresponding to α and is shown in the bottom line of the table.

Source: Adapted from Table 8 in Pearson, E. S. and Hartley, H. O. (1958). *Biometrika Tables for Statisticians*, Vol. 1, 2nd ed. Cambridge University Press: Cambridge, with the kind permission of the trustees of *Biometrika*.

TABLE D.5 UPPER PERCENTAGE POINTS OF THE F DISTRIBUTION

df_2	α	df_1 1	2	3	4	5	6	8	12	24	∞
1	.001	405284	500000	540379	562500	576405	585937	598144	610667	623497	636619
	.005	16211	20000	21615	22500	23056	23437	23925	24426	24940	25465
	.01	4052	4999	5403	5625	5764	5859	5981	6106	6234	6366
	.025	647.79	799.50	864.16	899.58	921.85	937.11	956.66	976.71	997.25	1018.30
	.05	161.45	199.50	215.71	224.58	230.16	233.99	238.88	243.91	249.05	254.32
	.10	39.86	49.50	53.59	55.83	57.24	58.20	59.44	60.70	62.00	63.33
	.25	5.83	7.50	8.20	8.58	8.82	8.98	9.19	9.41	9.63	9.85
2	.001	998.5	999.0	999.2	999.2	999.3	999.3	999.4	999.4	999.5	999.5
	.005	198.50	199.00	199.17	199.25	199.30	199.33	199.37	199.42	199.46	199.51
	.01	98.49	99.00	99.17	99.25	99.30	99.33	99.36	99.42	99.46	99.50
	.025	38.51	39.00	39.17	39.25	39.30	39.33	39.37	39.42	39.46	39.50
	.05	18.51	19.00	19.16	19.25	19.30	19.33	19.37	19.41	19.45	19.50
	.10	8.53	9.00	9.16	9.24	9.29	9.33	9.37	9.41	9.45	9.49
	.25	2.56	3.00	3.15	3.23	3.28	3.31	3.35	3.39	3.44	3.48
3	.001	167.5	148.5	141.1	137.1	134.6	132.8	130.6	128.3	125.9	123.5
	.005	55.55	49.80	47.47	46.20	45.39	44.84	44.13	43.39	42.62	41.83
	.01	34.12	30.81	29.46	28.71	28.24	27.91	27.49	27.05	26.60	26.12
	.025	17.44	16.04	15.44	15.10	14.89	14.74	14.54	14.34	14.12	13.90
	.05	10.13	9.55	9.28	9.12	9.01	8.94	8.84	8.74	8.64	8.53
	.10	5.54	5.46	5.39	5.34	5.31	5.28	5.25	5.22	5.18	5.13
	.25	2.02	2.28	2.36	2.39	2.41	2.42	2.44	2.45	2.46	2.47
4	.001	74.14	61.25	56.18	53.44	51.71	50.53	49.00	47.41	45.77	44.05
	.005	31.33	26.28	24.26	23.16	22.46	21.98	21.35	20.71	20.03	19.33
	.01	21.20	18.00	16.69	15.98	15.52	15.21	14.80	14.37	13.93	13.46
	.025	12.22	10.65	9.98	9.60	9.36	9.20	8.98	8.75	8.51	8.26
	.05	7.71	6.94	6.59	6.39	6.26	6.16	6.04	5.91	5.77	5.63
	.10	4.54	4.32	4.19	4.11	4.05	4.01	3.95	3.90	3.83	3.76
	.25	1.81	2.00	2.05	2.06	2.07	2.08	2.08	2.08	2.08	2.08
5	.001	47.04	36.61	33.20	31.09	29.75	28.84	27.64	26.42	25.14	23.78
	.005	22.79	18.31	16.53	15.56	14.94	14.51	13.96	13.38	12.78	12.14
	.01	16.26	13.27	12.06	11.39	10.97	10.67	10.29	9.89	9.47	9.02
	.025	10.01	8.43	7.76	7.39	7.15	6.98	6.76	6.52	6.28	6.02
	.05	6.61	5.79	5.41	5.19	5.05	4.95	4.82	4.68	4.53	4.36
	.10	4.06	3.78	3.62	3.52	3.45	3.40	3.34	3.27	3.19	3.10
	.25	1.70	1.85	1.89	1.89	1.89	1.89	1.89	1.89	1.88	1.87

6	.001	35.51	27.00	23.70	21.90	20.81	20.03	19.03	17.99	16.89	15.75
	.005	18.64	14.54	12.92	12.03	11.46	11.07	10.57	10.03	9.47	8.88
	.01	13.74	10.92	9.78	9.15	8.75	8.47	8.10	7.72	7.31	6.88
	.025	8.81	7.26	6.60	6.23	5.99	5.82	5.60	5.37	5.12	4.85
	.05	5.99	5.14	4.76	4.53	4.39	4.28	4.15	4.00	3.84	3.67
	.10	3.78	3.46	3.29	3.18	3.11	3.05	2.98	2.90	2.82	2.72
	.25	1.62	1.76	1.78	1.79	1.79	1.78	1.78	1.77	1.75	1.74
7	.001	29.22	21.69	18.77	17.19	16.21	15.52	14.63	13.71	12.73	11.69
	.005	16.24	12.40	10.88	10.05	9.52	9.16	8.68	8.18	7.65	7.08
	.01	12.25	9.55	8.45	7.85	7.46	7.19	6.84	6.47	6.07	5.65
	.025	8.07	6.54	5.89	5.52	5.29	5.12	4.90	4.67	4.42	4.14
	.05	5.59	4.74	4.35	4.12	3.97	3.87	3.73	3.57	3.41	3.23
	.10	3.59	3.26	3.07	2.96	2.88	2.83	2.75	2.67	2.58	2.47
	.25	1.57	1.70	1.72	1.72	1.71	1.71	1.70	1.68	1.67	1.65
8	.001	25.42	18.49	15.83	14.39	13.49	12.86	12.04	11.19	10.30	9.34
	.005	14.69	11.04	9.60	8.81	8.30	7.95	7.50	7.01	6.50	5.95
	.01	11.26	8.65	7.59	7.01	6.63	6.37	6.03	5.67	5.28	4.86
	.025	7.57	6.06	5.42	5.05	4.82	4.65	4.43	4.20	3.95	3.67
	.05	5.32	4.46	4.07	3.84	3.69	3.58	3.44	3.28	3.12	2.93
	.10	3.46	3.11	2.92	2.81	2.73	2.67	2.59	2.50	2.40	2.29
	.25	1.54	1.66	1.67	1.66	1.66	1.65	1.64	1.62	1.60	1.58
9	.001	22.86	16.39	13.90	12.56	11.71	11.13	10.37	9.57	8.72	7.81
	.005	13.61	10.11	8.72	7.96	7.47	7.13	6.69	6.23	5.73	5.19
	.01	10.56	8.02	6.99	6.42	6.06	5.80	5.47	5.11	4.73	4.31
	.025	7.21	5.71	5.08	4.72	4.48	4.32	4.10	3.87	3.61	3.33
	.05	5.12	4.26	3.86	3.63	3.48	3.37	3.23	3.07	2.90	2.71
	.10	3.36	3.01	2.81	2.69	2.61	2.55	2.47	2.38	2.28	2.16
	.25	1.51	1.62	1.63	1.63	1.62	1.61	1.60	1.58	1.56	1.53
10	.001	21.04	14.91	12.55	11.28	10.48	9.92	9.20	8.45	7.64	6.76
	.005	12.83	9.43	8.08	7.34	6.87	6.54	6.12	5.66	5.17	4.64
	.01	10.04	7.56	6.55	5.99	5.64	5.39	5.06	4.71	4.33	3.91
	.025	6.94	5.46	4.83	4.47	4.24	4.07	3.85	3.62	3.37	3.08
	.05	4.96	4.10	3.71	3.48	3.33	3.22	3.07	2.91	2.74	2.54
	.10	3.28	2.92	2.73	2.61	2.52	2.46	2.38	2.28	2.18	2.06
	.25	1.49	1.60	1.60	1.60	1.59	1.58	1.56	1.54	1.52	1.48

df₂	α	1	2	3	4	5	6	8	12	24	∞
11	.001	19.69	13.81	11.56	10.35	9.58	9.05	8.35	7.63	6.85	6.00
	.005	12.23	8.91	7.60	6.88	6.42	6.10	5.68	5.24	4.76	4.23
	.01	9.65	7.20	6.22	5.67	5.32	5.07	4.74	4.40	4.02	3.60
	.025	6.72	5.26	4.63	4.28	4.04	3.88	3.66	3.43	3.17	2.88
	.05	4.84	3.98	3.59	3.36	3.20	3.09	2.95	2.79	2.61	2.40
	.10	3.23	2.86	2.66	2.54	2.45	2.39	2.30	2.21	2.10	1.97
	.25	1.46	1.58	1.58	1.58	1.56	1.55	1.54	1.51	1.49	1.45
12	.001	18.64	12.97	10.80	9.63	8.89	8.38	7.71	7.00	6.25	5.42
	.005	11.75	8.51	7.23	6.52	6.07	5.76	5.35	4.91	4.43	3.90
	.01	9.33	6.93	5.95	5.41	5.06	4.82	4.50	4.16	3.78	3.36
	.025	6.55	5.10	4.47	4.12	3.89	3.73	3.51	3.28	3.02	2.72
	.05	4.75	3.88	3.49	3.26	3.11	3.00	2.85	2.69	2.50	2.30
	.10	3.18	2.81	2.61	2.48	2.39	2.33	2.24	2.15	2.04	1.90
	.25	1.46	1.56	1.56	1.55	1.54	1.53	1.51	1.49	1.46	1.42
13	.001	17.81	12.31	10.21	9.07	8.35	7.86	7.21	6.52	5.78	4.97
	.005	11.37	8.19	6.93	6.23	5.79	5.48	5.08	4.64	4.17	3.65
	.01	9.07	6.70	5.74	5.20	4.86	4.62	4.30	3.96	3.59	3.16
	.025	6.41	4.97	4.35	4.00	3.77	3.60	3.39	3.15	2.89	2.60
	.05	4.67	3.80	3.41	3.18	3.02	2.92	2.77	2.60	2.42	2.21
	.10	3.14	2.76	2.56	2.43	2.35	2.28	2.20	2.10	1.98	1.85
	.25	1.45	1.55	1.55	1.53	1.52	1.51	1.49	1.47	1.44	1.40
14	.001	17.14	11.78	9.73	8.62	7.92	7.43	6.80	6.13	5.41	4.60
	.005	11.06	7.92	6.68	6.00	5.56	5.26	4.86	4.43	3.96	3.44
	.01	8.86	6.51	5.56	5.03	4.69	4.46	4.14	3.80	3.43	3.00
	.025	6.30	4.86	4.24	3.89	3.66	3.50	3.29	3.05	2.79	2.49
	.05	4.60	3.74	3.34	3.11	2.96	2.85	2.70	2.53	2.35	2.13
	.10	3.10	2.73	2.52	2.39	2.31	2.24	2.15	2.05	1.94	1.80
	.25	1.44	1.53	1.53	1.52	1.51	1.50	1.48	1.45	1.42	1.38
15	.001	16.59	11.34	9.34	8.25	7.57	7.09	6.47	5.81	5.10	4.31
	.005	10.80	7.70	6.48	5.80	5.37	5.07	4.67	4.25	3.79	3.26
	.01	8.68	6.36	5.42	4.89	4.56	4.32	4.00	3.67	3.29	2.87
	.025	6.20	4.77	4.15	3.80	3.58	3.41	3.20	2.96	2.70	2.40
	.05	4.54	3.68	3.29	3.06	2.90	2.79	2.64	2.48	2.29	2.07
	.10	3.07	2.70	2.49	2.36	2.27	2.21	2.12	2.02	1.90	1.76
	.25	1.43	1.52	1.52	1.51	1.49	1.48	1.46	1.44	1.41	1.36

df	α										
16	.001	16.12	10.97	9.00	7.94	7.27	6.81	6.19	5.55	4.85	4.06
	.005	10.58	7.51	6.30	5.64	5.21	4.91	4.52	4.10	3.64	3.11
	.01	8.53	6.23	5.29	4.77	4.44	4.20	3.89	3.55	3.18	2.75
	.025	6.12	4.69	4.08	3.73	3.50	3.34	3.12	2.89	2.63	2.32
	.05	4.49	3.63	3.24	3.01	2.85	2.74	2.59	2.42	2.24	2.01
	.10	3.05	2.67	2.46	2.33	2.24	2.18	2.09	1.99	1.87	1.72
	.25	1.42	1.51	1.51	1.50	1.48	1.47	1.45	1.43	1.39	1.34
17	.001	15.72	10.66	8.73	7.68	7.02	6.56	5.96	5.32	4.63	3.85
	.005	10.38	7.35	6.16	5.50	5.07	4.78	4.39	3.97	3.51	2.98
	.01	8.40	6.11	5.18	4.67	4.34	4.10	3.79	3.45	3.08	2.65
	.025	6.04	4.62	4.01	3.66	3.44	3.28	3.06	2.82	2.56	2.25
	.05	4.45	3.59	3.20	2.96	2.81	2.70	2.55	2.38	2.19	1.96
	.10	3.03	2.64	2.44	2.31	2.22	2.15	2.06	1.96	1.84	1.69
	.25	1.42	1.51	1.51	1.49	1.47	1.46	1.44	1.41	1.38	1.33
18	.001	15.38	10.39	8.49	7.46	6.81	6.35	5.76	5.13	4.45	3.67
	.005	10.22	7.21	6.03	5.37	4.96	4.66	4.28	3.86	3.40	2.87
	.01	8.28	6.01	5.09	4.58	4.25	4.01	3.71	3.37	3.00	2.57
	.025	5.98	4.56	3.95	3.61	3.38	3.22	3.01	2.77	2.50	2.19
	.05	4.41	3.55	3.16	2.93	2.77	2.66	2.51	2.34	2.15	1.92
	.10	3.01	2.62	2.42	2.29	2.20	2.13	2.04	1.93	1.81	1.66
	.25	1.41	1.50	1.49	1.48	1.46	1.45	1.43	1.40	1.37	1.32
19	.001	15.08	10.16	8.28	7.26	6.61	6.18	5.59	4.97	4.29	3.52
	.005	10.07	7.09	5.92	5.27	4.85	4.56	4.18	3.76	3.31	2.78
	.01	8.18	5.93	5.01	4.50	4.17	3.94	3.63	3.30	2.92	2.49
	.025	5.92	4.51	3.90	3.56	3.33	3.17	2.96	2.72	2.45	2.13
	.05	4.38	3.52	3.13	2.90	2.74	2.63	2.48	2.31	2.11	1.88
	.10	2.99	2.61	2.40	2.27	2.18	2.11	2.02	1.91	1.79	1.63
	.25	1.41	1.50	1.49	1.48	1.46	1.44	1.42	1.40	1.36	1.31
20	.001	14.82	9.95	8.10	7.10	6.46	6.02	5.44	4.82	4.15	3.38
	.005	9.94	6.99	5.82	5.17	4.76	4.47	4.09	3.68	3.22	2.69
	.01	8.10	5.85	4.94	4.43	4.10	3.87	3.56	3.23	2.86	2.42
	.025	5.87	4.46	3.86	3.51	3.29	3.13	2.91	2.68	2.41	2.09
	.05	4.35	3.49	3.10	2.87	2.71	2.60	2.45	2.28	2.08	1.84
	.10	2.97	2.59	2.38	2.25	2.16	2.09	2.00	1.89	1.77	1.61
	.25	1.40	1.49	1.48	1.47	1.45	1.44	1.42	1.39	1.35	1.29

df_2	α	1	2	3	4	5	6	8	12	24	∞
21	.001	14.59	9.77	7.94	6.95	6.32	5.88	5.31	4.70	4.03	3.26
	.005	9.83	6.89	5.73	5.09	4.68	4.39	4.01	3.60	3.15	2.61
	.01	8.02	5.78	4.87	4.37	4.04	3.81	3.51	3.17	2.80	2.36
	.025	5.83	4.42	3.82	3.48	3.25	3.09	2.87	2.64	2.37	2.04
	.05	4.32	3.47	3.07	2.84	2.68	2.57	2.42	2.25	2.05	1.81
	.10	2.96	2.57	2.36	2.23	2.14	2.08	1.98	1.88	1.75	1.59
	.25	1.40	1.49	1.48	1.46	1.44	1.43	1.41	1.38	1.34	1.29
22	.001	14.38	9.61	7.80	6.81	6.19	5.76	5.19	4.58	3.92	3.15
	.005	9.73	6.81	5.65	5.02	4.61	4.32	3.94	3.54	3.08	2.55
	.01	7.94	5.72	4.82	4.31	3.99	3.76	3.45	3.12	2.75	2.31
	.025	5.79	4.38	3.78	3.44	3.22	3.05	2.84	2.60	2.33	2.00
	.05	4.30	3.44	3.05	2.82	2.66	2.55	2.40	2.23	2.03	1.78
	.10	2.95	2.56	2.35	2.22	2.13	2.06	1.97	1.86	1.73	1.57
	.25	1.40	1.48	1.47	1.46	1.44	1.42	1.40	1.37	1.33	1.28
23	.001	14.19	9.47	7.67	6.69	6.08	5.65	5.09	4.48	3.82	3.05
	.005	9.63	6.73	5.58	4.95	4.54	4.26	3.88	3.47	3.02	2.48
	.01	7.88	5.66	4.76	4.26	3.94	3.71	3.41	3.07	2.70	2.26
	.025	5.75	4.35	3.75	3.41	3.18	3.02	2.81	2.57	2.30	1.97
	.05	4.28	3.42	3.03	2.80	2.64	2.53	2.38	2.20	2.00	1.76
	.10	2.94	2.55	2.34	2.21	2.11	2.05	1.95	1.84	1.72	1.55
	.25	1.39	1.47	1.47	1.45	1.43	1.41	1.40	1.37	1.33	1.27
24	.001	14.03	9.34	7.55	6.59	5.98	5.55	4.99	4.39	3.74	2.97
	.005	9.55	6.66	5.52	4.89	4.49	4.20	3.83	3.42	2.97	2.43
	.01	7.82	5.61	4.72	4.22	3.90	3.67	3.36	3.03	2.66	2.21
	.025	5.72	4.32	3.72	3.38	3.15	2.99	2.78	2.54	2.27	1.94
	.05	4.26	3.40	3.01	2.78	2.62	2.51	2.36	2.18	1.98	1.73
	.10	2.93	2.54	2.33	2.19	2.10	2.04	1.94	1.83	1.70	1.53
	.25	1.39	1.47	1.46	1.44	1.43	1.41	1.39	1.36	1.32	1.26
25	.001	13.88	9.22	7.45	6.49	5.88	5.46	4.91	4.31	3.66	2.89
	.005	9.48	6.60	5.46	4.84	4.43	4.15	3.78	3.37	2.92	2.38
	.01	7.77	5.57	4.68	4.18	3.86	3.63	3.32	2.99	2.62	2.17
	.025	5.69	4.29	3.69	3.35	3.13	2.97	2.75	2.51	2.24	1.91
	.05	4.24	3.38	2.99	2.76	2.60	2.49	2.34	2.16	1.96	1.71
	.10	2.92	2.53	2.32	2.18	2.09	2.02	1.93	1.82	1.69	1.52
	.25	1.39	1.47	1.46	1.44	1.42	1.41	1.39	1.36	1.32	1.25

df	p										
26	.001	13.74	9.12	7.36	6.41	5.80	5.38	4.83	4.24	3.59	2.82
	.005	9.41	6.54	5.41	4.79	4.38	4.10	3.73	3.33	2.87	2.33
	.01	7.72	5.53	4.64	4.14	3.82	3.59	3.29	2.96	2.58	2.13
	.025	5.66	4.27	3.67	3.33	3.10	2.94	2.73	2.49	2.22	1.88
	.05	4.22	3.37	2.98	2.74	2.59	2.47	2.32	2.15	1.95	1.69
	.10	2.91	2.52	2.31	2.17	2.08	2.01	1.92	1.81	1.68	1.50
	.25	1.38	1.46	1.45	1.44	1.42	1.41	1.38	1.35	1.31	1.25
27	.001	13.61	9.02	7.27	6.33	5.73	5.31	4.76	4.17	3.52	2.75
	.005	9.34	6.49	5.36	4.74	4.34	4.06	3.69	3.28	2.83	2.29
	.01	7.68	5.49	4.60	4.11	3.78	3.56	3.26	2.93	2.55	2.10
	.025	5.63	4.24	3.65	3.31	3.08	2.92	2.71	2.47	2.19	1.85
	.05	4.21	3.35	2.96	2.73	2.57	2.46	2.30	2.13	1.93	1.67
	.10	2.90	2.51	2.30	2.17	2.07	2.00	1.91	1.80	1.67	1.49
	.25	1.38	1.46	1.45	1.43	1.42	1.40	1.38	1.35	1.31	1.24
28	.001	13.50	8.93	7.19	6.25	5.66	5.24	4.69	4.11	3.46	2.70
	.005	9.28	6.44	5.32	4.70	4.30	4.02	3.65	3.25	2.79	2.25
	.01	7.64	5.45	4.57	4.07	3.75	3.53	3.23	2.90	2.52	2.06
	.025	5.61	4.22	3.63	3.29	3.06	2.90	2.69	2.45	2.17	1.83
	.05	4.20	3.34	2.95	2.71	2.56	2.44	2.29	2.12	1.91	1.65
	.10	2.89	2.50	2.29	2.16	2.06	2.00	1.90	1.79	1.66	1.48
	.25	1.38	1.46	1.45	1.43	1.41	1.40	1.38	1.34	1.30	1.24
29	.001	13.39	8.85	7.12	6.19	5.59	5.18	4.64	4.05	3.41	2.64
	.005	9.23	6.40	5.28	4.66	4.26	3.98	3.61	3.21	2.76	2.21
	.01	7.60	5.42	4.54	4.04	3.73	3.50	3.20	2.87	2.49	2.03
	.025	5.59	4.20	3.61	3.27	3.04	2.88	2.67	2.43	2.15	1.81
	.05	4.18	3.33	2.93	2.70	2.54	2.43	2.28	2.10	1.90	1.64
	.10	2.89	2.50	2.28	2.15	2.06	1.99	1.89	1.78	1.65	1.47
	.25	1.38	1.45	1.45	1.43	1.41	1.40	1.37	1.34	1.30	1.23
30	.001	13.29	8.77	7.05	6.12	5.53	5.12	4.58	4.00	3.36	2.59
	.005	9.18	6.35	5.24	4.62	4.23	3.95	3.58	3.18	2.73	2.18
	.01	7.56	5.39	4.51	4.02	3.70	3.47	3.17	2.84	2.47	2.01
	.025	5.57	4.18	3.59	3.25	3.03	2.87	2.65	2.41	2.14	1.79
	.05	4.17	3.32	2.92	2.69	2.53	2.42	2.27	2.09	1.89	1.62
	.10	2.88	2.49	2.28	2.14	2.05	1.98	1.88	1.77	1.64	1.46
	.25	1.38	1.45	1.44	1.42	1.41	1.39	1.37	1.34	1.29	1.23

TABLE D.5 (continued)

df_2	α	1	2	3	4	5	6	8	12	24	∞
40	.001	12.61	8.25	6.60	5.70	5.13	4.73	4.21	3.64	3.01	2.23
	.005	8.83	6.07	4.98	4.37	3.99	3.71	3.35	2.95	2.50	1.93
	.01	7.31	5.18	4.31	3.83	3.51	3.29	2.99	2.66	2.29	1.80
	.025	5.42	4.05	3.46	3.13	2.90	2.74	2.53	2.29	2.01	1.64
	.05	4.08	3.23	2.84	2.61	2.45	2.34	2.18	2.00	1.79	1.51
	.10	2.84	2.44	2.23	2.09	2.00	1.93	1.83	1.71	1.57	1.38
	.25	1.36	1.44	1.42	1.41	1.39	1.37	1.35	1.31	1.27	1.19
60	.001	11.97	7.76	6.17	5.31	4.76	4.37	3.87	3.31	2.69	1.90
	.005	8.49	5.80	4.73	4.14	3.76	3.49	3.13	2.74	2.29	1.69
	.01	7.08	4.98	4.13	3.65	3.34	3.12	2.82	2.50	2.12	1.60
	.025	5.29	3.93	3.34	3.01	2.79	2.63	2.41	2.17	1.88	1.48
	.05	4.00	3.15	2.76	2.52	2.37	2.25	2.10	1.92	1.70	1.39
	.10	2.79	2.39	2.18	2.04	1.95	1.87	1.77	1.66	1.51	1.29
	.25	1.35	1.42	1.41	1.39	1.37	1.35	1.32	1.29	1.24	1.15
120	.001	11.38	7.31	5.79	4.95	4.42	4.04	3.55	3.02	2.40	1.56
	.005	8.18	5.54	4.50	3.92	3.55	3.28	2.93	2.54	2.09	1.43
	.01	6.85	4.79	3.95	3.48	3.17	2.96	2.66	2.34	1.95	1.38
	.025	5.15	3.80	3.23	2.89	2.67	2.52	2.30	2.05	1.76	1.31
	.05	3.92	3.07	2.68	2.45	2.29	2.17	2.02	1.83	1.61	1.25
	.10	2.75	2.35	2.13	1.99	1.90	1.82	1.72	1.60	1.45	1.19
	.25	1.34	1.40	1.39	1.37	1.35	1.33	1.30	1.26	1.21	1.10
∞	.001	10.83	6.91	5.42	4.62	4.10	3.74	3.27	2.74	2.13	1.00
	.005	7.88	5.30	4.28	3.72	3.35	3.09	2.74	2.36	1.90	1.00
	.01	6.64	4.60	3.78	3.32	3.02	2.80	2.51	2.18	1.79	1.00
	.025	5.02	3.69	3.12	2.79	2.57	2.41	2.19	1.94	1.64	1.00
	.05	3.84	2.99	2.60	2.37	2.21	2.09	1.94	1.75	1.52	1.00
	.10	2.71	2.30	2.08	1.94	1.85	1.77	1.67	1.55	1.38	1.00
	.25	1.32	1.39	1.37	1.35	1.33	1.31	1.28	1.24	1.18	1.00

Source: Adapted from Table 18 in Pearson, E. S. and Hartley, H. O. (1958). *Biometrika Tables for Statisticians*, Vol. 1, 2nd ed. Cambridge University Press: Cambridge, with the kind permission of the trustees of *Biometrika*.

TABLE D.6 CHARTS OF THE POWER FUNCTION FOR ANALYSIS OF VARIANCE TESTS

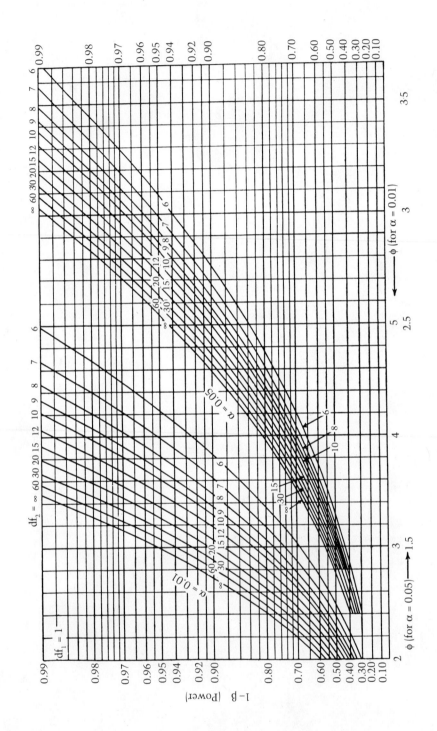

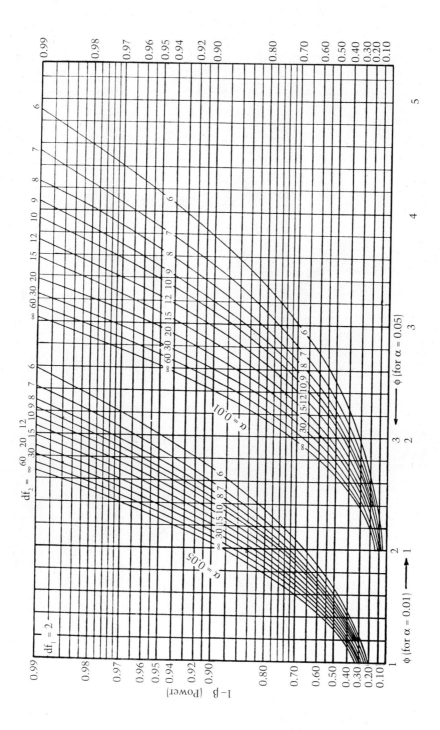

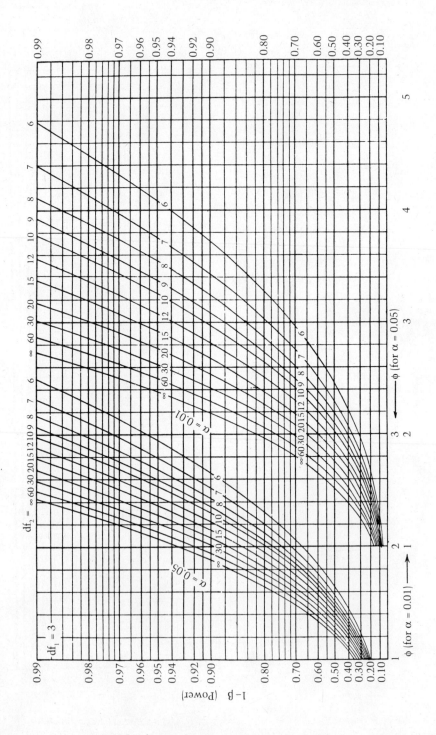

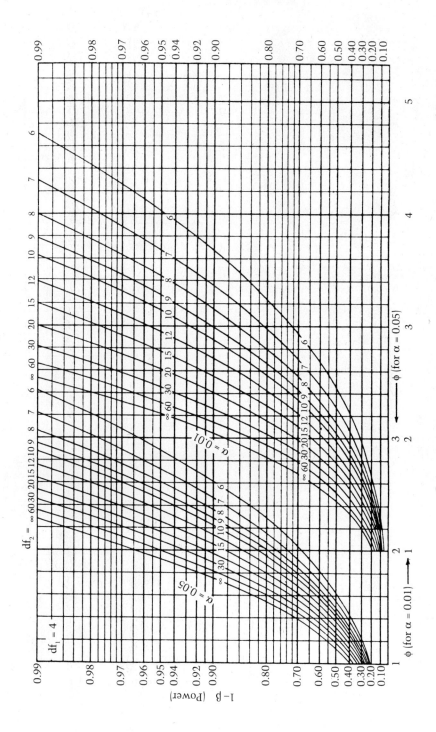

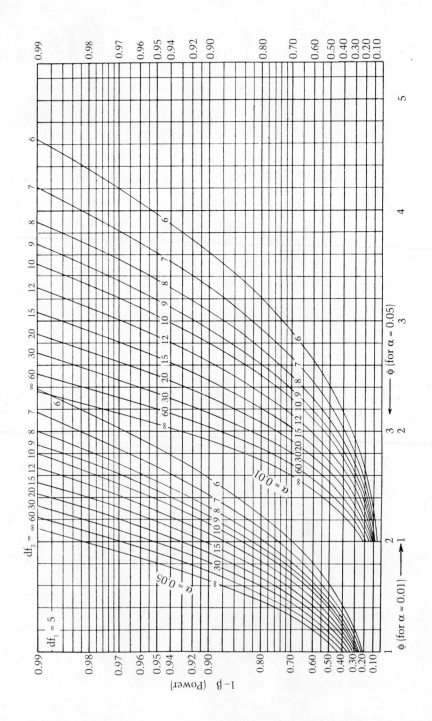

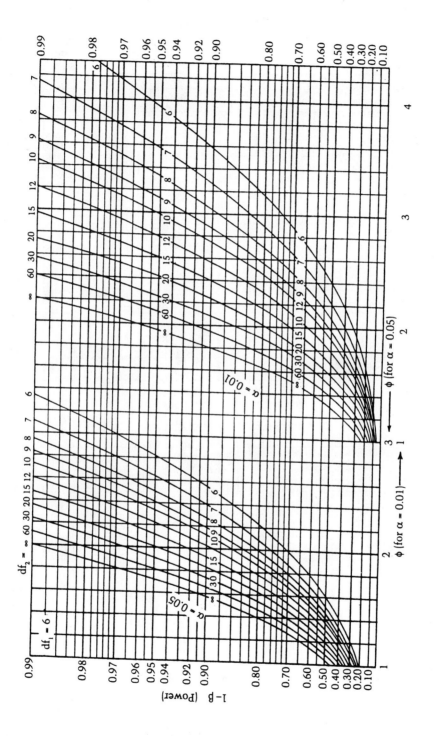

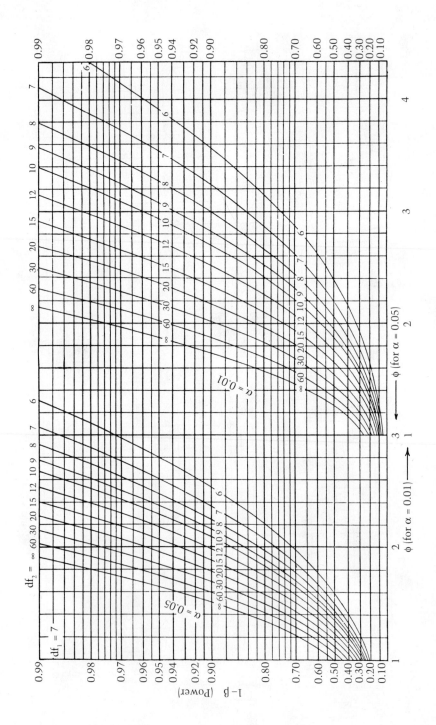

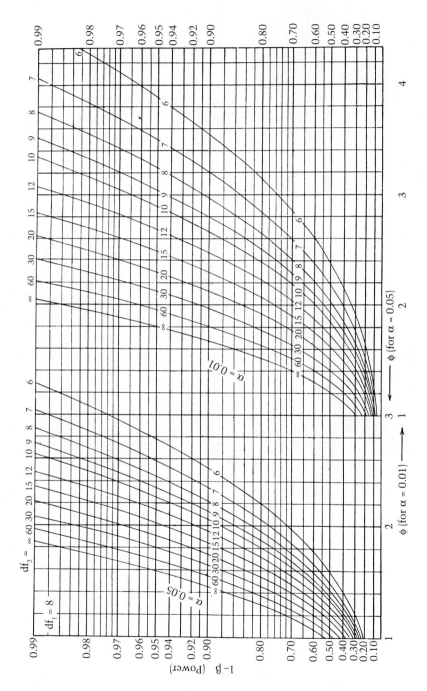

Source: Reproduced from Pearson, E. S. and Hartley, H. O. (1951), "Charts of the power function for analysis of variance tests derived from the non-central-*F* distribution," *Biometrika, 38*, 112–130.

TABLE D.7 COEFFICIENTS OF ORTHOGONAL POLYNOMIALS

k	Polynomial	X = 1	2	3	4	5	6	7	8	9	10	$\Sigma\xi'^2$	λ
3	Linear	−1	0	1								2	1
	Quadratic	1	−2	1								6	3
4	Linear	−3	−1	1	3							20	2
	Quadratic	1	−1	−1	1							4	1
	Cubic	−1	3	−3	1							20	10/3
5	Linear	−2	−1	0	1	2						10	1
	Quadratic	2	−1	−2	−1	2						14	1
	Cubic	−1	2	0	−2	1						10	5/6
	Quartic	1	−4	6	−4	1						70	35/12
6	Linear	−5	−3	−1	1	3	5					70	2
	Quadratic	5	−1	−4	−4	−1	5					84	3/2
	Cubic	−5	7	4	−4	−7	5					180	5/3
	Quartic	1	−3	2	2	−3	1					28	7/12
7	Linear	−3	−2	−1	0	1	2	3				28	1
	Quadratic	5	0	−3	−4	−3	0	5				84	1
	Cubic	−1	1	1	0	−1	−1	1				6	1/6
	Quartic	3	−7	1	6	1	−7	3				154	7/12
8	Linear	−7	−5	−3	−1	1	3	5	7			168	2
	Quadratic	7	1	−3	−5	−5	−3	1	7			168	1
	Cubic	−7	5	7	3	−3	−7	−5	7			264	2/3
	Quartic	7	−13	−3	9	9	−3	−13	7			616	7/12
	Quintic	−7	23	−17	−15	15	17	−23	7			2184	7/10
9	Linear	−4	−3	−2	−1	0	1	2	3	4		60	1
	Quadratic	28	7	−8	−17	−20	−17	−8	7	28		2772	3
	Cubic	−14	7	13	9	0	−9	−13	−7	14		990	5/6
	Quartic	14	−21	−11	9	18	9	−11	−21	14		2002	7/12
	Quintic	−4	11	−4	−9	0	9	4	−11	4		468	3/20
10	Linear	−9	−7	−5	−3	−1	1	3	5	7	9	330	2
	Quadratic	6	2	−1	−3	−4	−4	−3	−1	2	6	132	1/2
	Cubic	−42	14	35	31	12	−12	−31	−35	−14	42	8580	5/3
	Quartic	18	−22	−17	3	18	18	3	−17	−22	18	2860	5/12
	Quintic	−6	14	−1	−11	−6	6	11	1	−14	6	780	1/10

Source: Adapted from Table 47 in Pearson, E. S. and Hartley, H. O. (1958). *Biometrika Tables for Statisticians*, Vol. 1, 2nd ed. Cambridge University Press: Cambridge, with the kind permission of the trustees of *Biometrika*.

TABLE D.8   CRITICAL VALUES OF THE BONFERRONI t STATISTIC (note that the tabled values are two-tailed)

df	EF	2	3	4	5	6	7	8	9	10	15
	.01	7.453	8.575	9.465	10.215	10.869	11.453	11.984	12.471	12.924	14.819
3	.05	4.177	4.857	5.392	5.841	6.232	6.580	6.895	7.185	7.453	8.575
	.10	3.182	3.740	4.177	4.541	4.857	5.138	5.392	5.625	5.841	6.741
	.01	5.598	6.254	6.758	7.173	7.529	7.841	8.122	8.376	8.610	9.568
4	.05	3.495	3.961	4.315	4.604	4.851	5.068	5.261	5.437	5.598	6.254
	.10	2.776	3.186	3.495	3.747	3.961	4.148	4.315	4.466	4.604	5.167
	.01	4.773	5.247	5.604	5.893	6.138	6.352	6.541	6.713	6.869	7.499
5	.05	3.163	3.534	3.810	4.032	4.219	4.382	4.526	4.655	4.773	5.247
	.10	2.571	2.912	3.163	3.365	3.534	3.681	3.810	3.926	4.032	4.456
	.01	4.317	4.698	4.981	5.208	5.398	5.563	5.709	5.840	5.959	6.434
6	.05	2.969	3.287	3.521	3.707	3.863	3.997	4.115	4.221	4.317	4.698
	.10	2.447	2.749	2.969	3.143	3.287	3.412	3.521	3.619	3.707	4.058
	.01	4.029	4.355	4.595	4.785	4.944	5.082	5.202	5.310	5.408	5.795
7	.05	2.841	3.128	3.335	3.499	3.636	3.753	3.855	3.947	4.029	4.355
	.10	2.365	2.642	2.841	2.998	3.128	3.238	3.335	3.422	3.499	3.806
	.01	3.833	4.122	4.334	4.501	4.640	4.759	4.864	4.957	5.041	5.374
8	.05	2.752	3.016	3.206	3.355	3.479	3.584	3.677	3.759	3.833	4.122
	.10	2.306	2.566	2.752	2.896	3.016	3.117	3.206	3.285	3.355	3.632
	.01	3.690	3.954	4.146	4.297	4.422	4.529	4.622	4.706	4.781	5.076
9	.05	2.685	2.933	3.111	3.250	3.364	3.462	3.547	3.622	3.690	3.954
	.10	2.262	2.510	2.685	2.821	2.933	3.028	3.111	3.184	3.250	3.505
	.01	3.581	3.827	4.005	4.144	4.259	4.357	4.442	4.518	4.587	4.855
10	.05	2.634	2.870	3.038	3.169	3.277	3.368	3.448	3.518	3.581	3.827
	.10	2.228	2.466	2.634	2.764·	2.870	2.960	3.038	3.107	3.169	3.409
	.01	3.497	3.728	3.895	4.025	4.132	4.223	4.303	4.373	4.437	4.685
11	.05	2.593	2.820	2.981	3.106	3.208	3.295	3.370	3.437	3.497	3.728
	.10	2.201	2.431	2.593	2.718	2.820	2.906	2.981	3.047	3.106	3.334
	.01	3.428	3.649	3.807	3.930	4.031	4.117	4.192	4.258	4.318	4.550
12	.05	2.560	2.779	2.934	3.055	3.153	3.236	3.308	3.371	3.428	3.649
	.10	2.179	2.403	2.560	2.681	2.779	2.863	2.934	2.998	3.055	3.273
	.01	3.372	3.584	3.735	3.852	3.948	4.030	4.101	4.164	4.221	4.440
13	.05	2.533	2.746	2.896	3.012	3.107	3.187	3.256	3.318	3.372	3.584
	.10	2.160	2.380	2.533	2.650	2.746	2.827	2.896	2.957	3.012	3.223
	.01	3.326	3.530	3.675	3.787	3.880	3.958	4.026	4.086	4.140	4.349
14	.05	2.510	2.718	2.864	2.977	3.069	3.146	3.214	3.273	3.326	3.530
	.10	2.145	2.360	2.510	2.624	2.718	2.796	2.864	2.924	2.977	3.181
	.01	3.286	3.484	3.624	3.733	3.822	3.897	3.963	4.021	4.073	4.273
15	.05	2.490	2.694	2.837	2.947	3.036	3.112	3.177	3.235	3.286	3.484
	.10	2.131	2.343	2.490	2.602	2.694	2.770	2.837	2.895	2.947	3.146

df	EF	Number of contrasts (K)									
		2	3	4	5	6	7	8	9	10	15
	.01	3.252	3.444	3.581	3.686	3.773	3.846	3.909	3.965	4.015	4.208
16	.05	2.473	2.673	2.813	2.921	3.008	3.082	3.146	3.202	3.252	3.444
	.10	2.120	2.328	2.473	2.583	2.673	2.748	2.813	2.870	2.921	3.115
	.01	3.222	3.410	3.543	3.646	3.730	3.801	3.862	3.917	3.965	4.152
17	.05	2.458	2.655	2.793	2.898	2.984	3.056	3.119	3.173	3.222	3.410
	.10	2.110	2.316	2.458	2.567	2.655	2.729	2.793	2.848	2.898	3.089
	.01	3.197	3.380	3.510	3.610	3.692	3.762	3.822	3.874	3.922	4.104
18	.05	2.445	2.639	2.775	2.878	2.963	3.034	3.095	3.149	3.197	3.380
	.10	2.101	2.304	2.445	2.552	2.639	2.712	2.775	2.829	2.878	3.065
	.01	3.174	3.354	3.481	3.579	3.660	3.727	3.786	3.837	3.883	4.061
19	.05	2.433	2.625	2.759	2.861	2.944	3.014	3.074	3.127	3.174	3.354
	.10	2.093	2.294	2.433	2.539	2.625	2.697	2.759	2.813	2.861	3.045
	.01	3.153	3.331	3.455	3.552	3.630	3.697	3.754	3.804	3.850	4.023
20	.05	2.423	2.613	2.744	2.845	2.927	2.996	3.055	3.107	3.153	3.331
	.10	2.086	2.285	2.423	2.528	2.613	2.683	2.744	2.798	2.845	3.026
	.01	3.078	3.244	3.361	3.450	3.523	3.584	3.637	3.684	3.725	3.884
25	.05	2.385	2.566	2.692	2.787	2.865	2.930	2.986	3.035	3.078	3.244
	.10	2.060	2.252	2.385	2.485	2.566	2.634	2.692	2.742	2.787	2.959
	.01	3.030	3.189	3.300	3.385	3.454	3.513	3.563	3.607	3.646	3.796
30	.05	2.360	2.536	2.657	2.750	2.825	2.887	2.941	2.988	3.030	3.189
	.10	2.042	2.231	2.360	2.457	2.536	2.601	2.657	2.706	2.750	2.915
	.01	2.996	3.150	3.258	3.340	3.407	3.463	3.511	3.553	3.591	3.735
35	.05	2.342	2.515	2.633	2.724	2.797	2.857	2.910	2.955	2.996	3.150
	.10	2.030	2.215	2.342	2.438	2.515	2.579	2.633	2.681	2.724	2.885
	.01	2.971	3.122	3.227	3.307	3.372	3.426	3.473	3.514	3.551	3.691
40	.05	2.329	2.499	2.616	2.704	2.776	2.836	2.887	2.931	2.971	3.122
	.10	2.021	2.204	2.329	2.423	2.499	2.562	2.616	2.663	2.704	2.862
	.01	2.915	3.057	3.156	3.232	3.293	3.344	3.388	3.426	3.460	3.590
60	.05	2.299	2.463	2.575	2.660	2.729	2.785	2.834	2.877	2.915	3.057
	.10	2.000	2.178	2.299	2.390	2.463	2.524	2.575	2.620	2.660	2.811
	.01	2.860	2.995	3.088	3.160	3.217	3.265	3.306	3.342	3.373	3.494
120	.05	2.270	2.428	2.536	2.617	2.683	2.737	2.783	2.824	2.860	2.995
	.10	1.980	2.153	2.270	2.358	2.428	2.486	2.536	2.579	2.617	2.761

TABLE D.9 DISTRIBUTION OF DUNNETT'S d STATISTIC FOR COMPARING TREATMENT MEANS WITH A CONTROL (note that the tabled values are two-tailed)

df for MS_{error}	EF	\multicolumn{9}{c}{Number of means (including control)}								
		2	3	4	5	6	7	8	9	10
6	.10	1.94	2.34	2.56	2.71	2.83	2.92	3.00	3.07	3.12
	.05	2.45	2.86	3.18	3.41	3.60	3.75	3.88	4.00	4.11
	.02	3.14	3.61	3.88	4.07	4.21	4.33	4.43	4.51	4.59
	.01	3.71	4.22	4.60	4.88	5.11	5.30	5.47	5.61	5.74
7	.10	1.89	2.27	2.48	2.62	2.73	2.82	2.89	2.95	3.01
	.05	2.36	2.75	3.04	3.24	3.41	3.54	3.66	3.76	3.86
	.02	3.00	3.42	3.66	3.83	3.96	4.07	4.15	4.23	4.30
	.01	3.50	3.95	4.28	4.52	4.17	4.87	5.01	5.13	5.24
8	.10	1.86	2.22	2.42	2.55	2.66	2.74	2.81	2.87	2.92
	.05	2.31	2.67	2.94	3.13	3.28	3.40	3.51	3.60	3.68
	.02	2.90	3.29	3.51	3.67	3.79	3.88	3.96	4.03	4.09
	.01	3.36	3.77	4.06	4.27	4.44	4.58	4.70	4.81	4.90
9	.10	1.83	2.18	2.37	2.50	2.60	2.68	2.75	2.81	2.86
	.05	2.26	2.61	2.86	3.04	3.18	3.29	3.39	3.48	3.55
	.02	2.82	3.19	3.40	3.55	3.66	3.75	3.82	3.89	3.94
	.01	3.25	3.63	3.90	4.09	4.24	4.37	4.48	4.57	4.65
10	.10	1.81	2.15	2.34	2.47	2.56	2.64	2.70	2.76	2.81
	.05	2.23	2.57	2.81	2.97	3.11	3.21	3.31	3.39	3.46
	.02	2.76	3.11	3.31	3.45	3.56	3.64	3.71	3.78	3.83
	.01	3.17	3.53	3.78	3.95	4.10	4.21	4.31	4.40	4.47
11	.10	1.80	2.13	2.31	2.44	2.53	2.60	2.67	2.72	2.77
	.05	2.20	2.53	2.76	2.92	3.05	3.15	3.24	3.31	3.38
	.02	2.72	3.06	3.25	3.38	3.48	3.56	3.63	3.69	3.74
	.01	3.11	3.45	3.68	3.85	3.98	4.09	4.18	4.26	4.33
12	.10	1.78	2.11	2.29	2.41	2.50	2.58	2.64	2.69	2.74
	.05	2.18	2.50	2.72	2.88	3.00	3.10	3.18	3.25	3.32
	.02	2.68	3.01	3.19	3.32	3.42	3.50	3.56	3.62	3.67
	.01	3.05	3.39	3.61	3.76	3.89	3.99	4.08	4.15	4.22
13	.10	1.77	2.09	2.27	2.39	2.48	2.55	2.61	2.66	2.71
	.05	2.16	2.48	2.69	2.84	2.96	3.06	3.14	3.21	3.27
	.02	2.65	2.97	3.15	3.27	3.37	3.44	3.51	3.56	3.61
	.01	3.01	3.33	3.54	3.69	3.81	3.91	3.99	4.06	4.13
14	.10	1.76	2.08	2.25	2.37	2.46	2.53	2.59	2.64	2.69
	.05	2.14	2.46	2.67	2.81	2.93	3.02	3.10	3.17	3.23
	.02	2.62	2.94	3.11	3.23	3.32	3.40	3.46	3.51	3.56
	.01	2.98	3.29	3.49	3.64	3.75	3.84	3.92	3.99	4.05

df for MS$_{error}$	EF	Number of means (including control)								
		2	3	4	5	6	7	8	9	10
16	.10	1.75	2.06	2.23	2.34	2.43	2.50	2.56	2.61	2.65
	.05	2.12	2.42	2.63	2.77	2.88	2.96	3.04	3.10	3.16
	.02	2.58	2.88	3.05	3.17	3.26	3.33	3.39	3.44	3.48
	.01	2.92	3.22	3.41	3.55	3.65	3.74	3.82	3.88	3.93
18	.10	1.73	2.04	2.21	2.32	2.41	2.48	2.53	2.58	2.62
	.05	2.10	2.40	2.59	2.73	2.84	2.92	2.99	3.05	3.11
	.02	2.55	2.84	3.01	3.12	3.21	3.27	3.33	3.38	3.42
	.01	2.88	3.17	3.35	3.48	3.58	3.67	3.74	3.80	3.85
20	.10	1.72	2.03	2.19	2.30	2.39	2.46	2.51	2.56	2.60
	.05	2.09	2.38	2.57	2.70	2.81	2.89	2.96	3.02	3.07
	.02	2.53	2.81	2.97	3.08	3.17	3.23	3.29	3.34	3.38
	.01	2.85	3.13	3.31	3.43	3.53	3.61	3.67	3.73	3.78
24	.10	1.71	2.01	2.17	2.28	2.36	2.43	2.48	2.53	2.57
	.05	2.06	2.35	2.53	2.66	2.76	2.84	2.91	2.96	3.01
	.02	2.49	2.77	2.92	3.03	3.11	3.17	3.22	3.27	3.31
	.01	2.80	3.07	3.24	3.36	3.45	3.52	3.58	3.64	3.69
30	.10	1.70	1.99	2.15	2.25	2.33	2.40	2.45	2.50	2.54
	.05	2.04	2.32	2.50	2.62	2.72	2.79	2.86	2.91	2.96
	.02	2.46	2.72	2.87	2.97	3.05	3.11	3.16	3.21	3.24
	.01	2.75	3.01	3.17	3.28	3.37	3.44	3.50	3.55	3.59
40	.10	1.68	1.97	2.13	2.23	2.31	2.37	2.42	2.47	2.51
	.05	2.02	2.29	2.47	2.58	2.67	2.75	2.81	2.86	2.90
	.02	2.42	2.68	2.82	2.92	2.99	3.05	3.10	3.14	3.18
	.01	2.70	2.95	3.10	3.21	3.29	3.36	3.41	3.46	3.50
60	.10	1.67	1.95	2.10	2.21	2.28	2.35	2.39	2.44	2.48
	.05	2.00	2.27	2.43	2.55	2.63	2.70	2.76	2.81	2.85
	.02	2.39	2.64	2.78	2.87	2.94	3.00	3.04	3.08	3.12
	.01	2.66	2.90	3.04	3.14	3.22	3.28	3.33	3.38	3.42
120	.10	1.66	1.93	2.08	2.18	2.26	2.32	2.37	2.41	2.45
	.05	1.98	2.24	2.40	2.51	2.59	2.66	2.71	2.76	2.80
	.02	2.36	2.60	2.73	2.82	2.89	2.94	2.99	3.03	3.06
	.01	2.62	2.84	2.98	3.08	3.15	3.21	3.25	3.30	3.33
∞	.10	1.64	1.92	2.06	2.16	2.23	2.29	2.34	2.38	2.42
	.05	1.96	2.21	2.37	2.47	2.55	2.62	2.67	2.71	2.75
	.02	2.33	2.56	2.68	2.77	2.84	2.89	2.93	2.97	3.00
	.01	2.58	2.79	2.92	3.01	3.08	3.14	3.18	3.22	3.25

Source: Adapted from tables in Dunnett, C. W. (1955). A multiple comparison procedure for comparing several treatments with a control. *Journal of the American Statistical Association, 50*, 1096–1121, and from Dunnett, C. W. (1964). New tables for multiple comparisons with a control. *Biometrics, 20*, 482–491, with permission of the author and the editors.

TABLE D.10 CRITICAL VALUES OF THE STUDENTIZED RANGE DISTRIBUTION

Error df	EF	Number of ordered means						
		2	3	4	5	6	7	8
2	.01	14.04	19.02	22.29	24.72	26.63	28.20	29.53
	.05	6.09	8.33	9.80	10.88	11.74	12.44	13.03
	.10	4.13	5.73	6.77	7.54	8.14	8.63	9.05
3	.01	8.26	10.62	12.17	13.33	14.24	15.00	15.64
	.05	4.50	5.91	6.83	7.50	8.04	8.48	8.85
	.10	3.33	4.47	5.20	5.74	6.16	6.51	6.81
4	.01	6.51	8.12	9.17	9.96	10.58	11.10	11.55
	.05	3.93	5.04	5.76	6.29	6.71	7.05	7.35
	.10	3.02	3.98	4.59	5.04	5.39	5.68	5.93
5	.01	5.70	6.98	7.80	8.42	8.91	9.32	9.67
	.05	3.64	4.60	5.22	5.67	6.03	6.33	6.58
	.10	2.85	3.72	4.26	4.66	4.98	5.24	5.46
6	.01	5.24	6.33	7.03	7.56	7.97	8.32	8.61
	.05	3.46	4.34	4.90	5.31	5.63	5.90	6.12
	.10	2.75	3.56	4.07	4.44	4.73	4.97	5.17
7	.01	4.95	5.92	6.54	7.01	7.37	7.68	7.94
	.05	3.34	4.17	4.68	5.06	5.36	5.61	5.82
	.10	2.68	3.45	3.93	4.28	4.56	4.78	4.97
8	.01	4.75	5.64	6.20	6.63	6.96	7.24	7.47
	.05	3.26	4.04	4.53	4.89	5.17	5.40	5.60
	.10	2.63	3.37	3.83	4.17	4.43	4.65	4.83
9	.01	4.60	5.43	5.96	6.35	6.66	6.92	7.13
	.05	3.20	3.95	4.42	4.76	5.02	5.24	5.43
	.10	2.59	3.32	3.76	4.08	4.34	4.55	4.72
10	.01	4.48	5.27	5.77	6.14	6.43	6.67	6.88
	.05	3.15	3.88	4.33	4.65	4.91	5.12	5.31
	.10	2.56	3.27	3.70	4.02	4.26	4.47	4.64
11	.01	4.39	5.15	5.62	5.97	6.25	6.48	6.67
	.05	3.11	3.82	4.26	4.57	4.82	5.03	5.20
	.10	2.54	3.23	3.66	3.97	4.21	4.40	4.57
12	.01	4.32	5.05	5.50	5.84	6.10	6.32	6.51
	.05	3.08	3.77	4.20	4.51	4.75	4.95	5.12
	.10	2.52	3.20	3.62	3.92	4.16	4.35	4.51
13	.01	4.26	4.96	5.40	5.73	5.98	6.19	6.37
	.05	3.06	3.74	4.15	4.45	4.69	4.89	5.05
	.10	2.51	3.18	3.59	3.89	4.12	4.31	4.46
14	.01	4.21	4.90	5.32	5.63	5.88	6.09	6.26
	.05	3.03	3.70	4.11	4.41	4.64	4.83	4.99
	.10	2.49	3.16	3.56	3.85	4.06	4.27	4.42
15	.01	4.17	4.84	5.25	5.56	5.80	5.99	6.16
	.05	3.01	3.67	4.08	4.37	4.60	4.78	4.94
	.10	2.48	3.14	3.54	3.83	4.05	4.24	4.39
16	.01	4.13	4.79	5.19	5.49	5.72	5.92	6.08
	.05	3.00	3.65	4.05	4.33	4.56	4.74	4.90
	.10	2.47	3.12	3.52	3.80	4.03	4.21	4.36

TABLE D.10 (continued)

Error df	EF	2	3	4	5	6	7	8
		Number of ordered means						
17	.01	4.10	4.74	5.14	5.43	5.66	5.85	6.01
	.05	2.98	3.63	4.02	4.30	4.52	4.71	4.86
	.10	2.46	3.11	3.50	3.78	4.00	4.18	4.33
18	.01	4.07	4.70	5.09	5.38	5.60	5.79	5.94
	.05	2.97	3.61	4.00	4.28	4.50	4.67	4.82
	.10	2.45	3.10	3.49	3.77	3.98	4.16	4.31
19	.01	4.05	4.67	5.05	5.33	5.55	5.74	5.89
	.05	2.96	3.59	3.98	4.25	4.47	4.65	4.79
	.10	2.45	3.09	3.47	3.75	3.97	4.14	4.29
20	.01	4.02	4.64	5.02	5.29	5.51	5.69	5.84
	.05	2.95	3.58	3.96	4.23	4.45	4.62	4.77
	.10	2.44	3.08	3.46	3.74	3.95	4.12	4.27
24	.01	3.96	4.55	4.91	5.17	5.37	5.54	5.69
	.05	2.92	3.53	3.90	4.17	4.37	4.54	4.68
	.10	2.42	3.05	3.42	3.69	3.90	4.07	4.21
30	.01	3.89	4.46	4.80	5.05	5.24	5.40	5.54
	.05	2.89	3.49	3.85	4.10	4.30	4.46	4.60
	.10	2.40	3.02	3.39	3.65	3.85	4.02	4.16
40	.01	3.83	4.37	4.70	4.93	5.11	5.27	5.39
	.05	2.86	3.44	3.79	4.04	4.23	4.39	4.52
	.10	2.38	2.99	3.35	3.61	3.80	3.96	4.10
60	.01	3.76	4.28	4.60	4.82	4.99	5.13	5.25
	.05	2.83	3.40	3.74	3.98	4.16	4.31	4.44
	.10	2.36	2.96	3.31	3.56	3.76	3.91	4.04
120	.01	3.70	4.20	4.50	4.71	4.87	5.01	5.12
	.05	2.80	3.36	3.69	3.92	4.10	4.24	4.36
	.10	2.34	2.93	3.28	3.52	3.71	3.86	3.99
∞	.01	3.64	4.12	4.40	4.60	4.76	4.88	4.99
	.05	2.77	3.31	3.63	3.86	4.03	4.17	4.29
	.10	2.33	2.90	3.24	3.48	3.66	3.81	3.93

Error df	EF	9	10	11	12	13	14	15
		Number of ordered means						
2	.01	30.68	31.69	32.59	33.40	34.13	34.81	35.43
	.05	13.54	13.99	14.39	14.75	15.08	15.38	15.65
	.10	9.41	9.73	10.01	10.26	10.49	10.70	10.89
3	.01	16.20	16.69	17.13	17.53	17.89	18.22	18.52
	.05	9.18	9.46	9.72	9.95	10.15	10.35	10.53
	.10	7.06	7.29	7.49	7.67	7.83	7.98	8.12
4	.01	11.93	12.27	12.57	12.84	13.09	13.32	13.53
	.05	7.60	7.83	8.03	8.21	8.37	8.53	8.66
	.10	6.14	6.33	6.50	6.65	6.78	6.91	7.03
5	.01	9.97	10.24	10.48	10.70	10.89	11.08	11.24
	.05	6.80	7.00	7.17	7.32	7.47	7.60	7.72
	.10	5.65	5.82	5.97	6.10	6.22	6.34	6.44
6	.01	8.87	9.10	9.30	9.48	9.65	9.81	9.95
	.05	6.32	6.49	6.65	6.79	6.92	7.03	7.14
	.10	5.34	5.50	5.64	5.76	5.88	5.98	6.08

TABLE D.10 (continued)

Error df	EF	\multicolumn{7}{c}{Number of ordered means}
		9	10	11	12	13	14	15
7	.01	8.17	8.37	8.55	8.71	8.86	9.00	9.12
	.05	6.00	6.16	6.30	6.43	6.55	6.66	6.76
	.10	5.14	5.28	5.41	5.53	5.64	5.74	5.83
8	.01	7.68	7.86	8.03	8.18	8.31	8.44	8.55
	.05	5.77	5.92	6.05	6.18	6.29	6.39	6.48
	.10	4.99	5.13	5.25	5.36	5.46	5.56	5.64
9	.01	7.33	7.50	7.65	7.78	7.91	8.03	8.13
	.05	5.60	5.74	5.87	5.98	6.09	6.19	6.28
	.10	4.87	5.01	5.13	5.23	5.33	5.42	5.51
10	.01	7.06	7.21	7.36	7.49	7.60	7.71	7.81
	.05	5.46	5.60	5.72	5.83	5.94	6.03	6.11
	.10	4.78	4.91	5.03	5.13	5.23	5.32	5.40
11	.01	6.84	6.99	7.13	7.25	7.36	7.47	7.56
	.05	5.35	5.49	5.61	5.71	5.81	5.90	5.98
	.10	4.71	4.84	4.95	5.05	5.15	5.23	5.31
12	.01	6.67	6.81	6.94	7.06	7.17	7.27	7.36
	.05	5.27	5.40	5.51	5.62	5.71	5.80	5.88
	.10	4.65	4.78	4.89	4.99	5.08	5.16	5.24
13	.01	6.53	6.67	6.79	6.90	7.01	7.10	7.19
	.05	5.19	5.32	5.43	5.53	5.63	5.71	5.79
	.10	4.60	4.72	4.83	4.93	5.02	5.10	5.18
14	.01	6.41	6.54	6.66	6.77	6.87	6.96	7.05
	.05	5.13	5.25	5.36	5.46	5.55	5.64	5.71
	.10	4.56	4.68	4.79	4.88	4.97	5.05	5.12
15	.01	6.31	6.44	6.56	6.66	6.76	6.85	6.93
	.05	5.08	5.20	5.31	5.40	5.49	5.57	5.65
	.10	4.52	4.64	4.75	4.84	4.93	5.01	5.08
16	.01	6.22	6.35	6.46	6.56	6.66	6.74	6.82
	.05	5.03	5.15	5.26	5.35	5.44	5.52	5.59
	.10	4.49	4.61	4.71	4.81	4.89	4.97	5.04
17	.01	6.15	6.27	6.38	6.48	6.57	6.66	6.73
	.05	4.99	5.11	5.21	5.31	5.39	5.47	5.54
	.10	4.46	4.58	4.68	4.77	4.86	4.94	5.01
18	.01	6.08	6.20	6.31	6.41	6.50	6.58	6.66
	.05	4.96	5.07	5.17	5.27	5.35	5.43	5.50
	.10	4.44	4.55	4.66	4.75	4.83	4.91	4.98
19	.01	6.02	6.14	6.25	6.34	6.43	6.51	6.59
	.05	4.92	5.04	5.14	5.23	5.32	5.39	5.46
	.10	4.42	4.53	4.63	4.72	4.80	4.88	4.95
20	.01	5.97	6.09	6.19	6.29	6.37	6.45	6.52
	.05	4.90	5.01	5.11	5.20	5.28	5.36	5.43
	.10	4.40	4.51	4.61	4.70	4.78	4.86	4.92
24	.01	5.81	5.92	6.02	6.11	6.19	6.26	6.33
	.05	4.81	4.92	5.01	5.10	5.18	5.25	5.32
	.10	4.34	4.45	4.54	4.62	4.71	4.78	4.85
30	.01	5.65	5.76	5.85	5.93	6.01	6.08	6.14
	.05	4.72	4.82	4.92	5.00	5.08	5.15	5.21
	.10	4.28	4.38	4.47	4.56	4.64	4.71	4.77

Error df	EF	\multicolumn{7}{c}{Number of ordered means}						
		9	10	11	12	13	14	15
40	.01	5.50	5.60	5.69	5.76	5.84	5.90	5.96
	.05	4.64	4.74	4.82	4.90	4.98	5.04	5.11
	.10	4.22	4.32	4.41	4.49	4.56	4.63	4.70
60	.01	5.36	5.45	5.53	5.60	5.67	5.73	5.79
	.05	4.55	4.65	4.73	4.81	4.88	4.94	5.00
	.10	4.16	4.25	4.34	4.42	4.49	4.56	4.62
120	.01	5.21	5.30	5.38	5.44	5.51	5.56	5.61
	.05	4.47	4.56	4.64	4.71	4.78	4.84	4.90
	.10	4.10	4.19	4.28	4.35	4.42	4.49	4.54
∞	.01	5.08	5.16	5.23	5.29	5.35	5.40	5.45
	.05	4.39	4.47	4.55	4.62	4.69	4.74	4.80
	.10	4.04	4.13	4.21	4.29	4.35	4.41	4.47

Error df	EF	\multicolumn{7}{c}{Number of ordered means}						
		16	17	18	19	20	30	40
2	.01	36.00	36.53	37.03	37.50	37.95	41.32	43.61
	.05	15.91	16.14	16.37	16.57	16.77	18.27	19.28
	.10	11.07	11.24	11.39	11.54	11.68	12.73	13.44
3	.01	18.81	19.07	19.32	19.55	19.77	21.44	22.59
	.05	10.69	10.84	10.98	11.11	11.24	12.21	12.87
	.10	8.25	8.37	8.48	8.58	8.68	9.44	9.95
4	.01	13.73	13.91	14.08	14.24	14.40	15.57	16.37
	.05	8.79	8.91	9.03	9.13	9.23	10.00	10.53
	.10	7.13	7.23	7.33	7.41	7.50	8.14	8.57
5	.01	11.40	11.55	11.68	11.81	11.93	12.87	13.52
	.05	7.83	7.93	8.03	8.12	8.21	8.88	9.33
	.10	6.54	6.63	6.71	6.79	6.86	7.44	7.83
6	.01	10.08	10.21	10.32	10.43	10.54	11.34	11.90
	.05	7.24	7.34	7.43	7.51	7.59	8.19	8.60
	.10	6.16	6.25	6.33	6.40	6.47	7.00	7.36
7	.01	9.24	9.35	9.46	9.55	9.65	10.36	10.85
	.05	6.85	6.94	7.02	7.10	7.17	7.73	8.11
	.10	5.91	5.99	6.06	6.13	6.20	6.70	7.04
8	.01	8.66	8.76	8.85	8.94	9.03	9.68	10.13
	.05	6.57	6.65	6.73	6.80	6.87	7.40	7.76
	.10	5.72	5.80	5.87	5.94	6.00	6.48	6.80
9	.01	8.23	8.33	8.41	8.49	8.57	9.18	9.59
	.05	6.36	6.44	6.51	6.58	6.64	7.15	7.49
	.10	5.58	5.66	5.72	5.79	5.85	6.31	6.62
10	.01	7.91	7.99	8.08	8.15	8.23	8.79	9.19
	.05	6.19	6.27	6.34	6.41	6.47	6.95	7.28
	.10	5.47	5.54	5.61	5.67	5.73	6.17	6.48
11	.01	7.65	7.73	7.81	7.88	7.95	8.49	8.86
	.05	6.06	6.13	6.20	6.27	6.33	6.79	7.11
	.10	5.38	5.45	5.51	5.57	5.63	6.07	6.36
12	.01	7.44	7.52	7.59	7.67	7.73	8.25	8.60
	.05	5.95	6.02	6.09	6.15	6.21	6.66	6.97
	.10	5.31	5.37	5.44	5.50	5.55	5.98	6.27

Error df	EF	Number of ordered means						
		16	17	18	19	20	30	40
13	.01	7.27	7.35	7.42	7.49	7.55	8.04	8.39
	.05	5.86	5.93	6.00	6.06	6.11	6.55	6.85
	.10	5.25	5.31	5.37	5.43	5.48	5.90	6.19
14	.01	7.13	7.20	7.27	7.33	7.40	7.87	8.20
	.05	5.79	5.85	5.92	5.97	6.03	6.46	6.75
	.10	5.19	5.26	5.32	5.37	5.43	5.84	6.12
15	.01	7.00	7.07	7.14	7.20	7.26	7.73	8.05
	.05	5.72	5.79	5.85	5.90	5.96	6.38	6.67
	.10	5.15	5.21	5.27	5.32	5.38	5.78	6.06
16	.01	6.90	6.97	7.03	7.09	7.15	7.60	7.92
	.05	5.66	5.73	5.79	5.84	5.90	6.31	6.59
	.10	5.11	5.17	5.23	5.28	5.33	5.73	6.00
17	.01	6.81	6.87	6.94	7.00	7.05	7.49	7.80
	.05	5.61	5.68	5.73	5.79	5.84	6.25	6.53
	.10	5.07	5.13	5.19	5.24	5.30	5.69	5.96
18	.01	6.73	6.79	6.85	6.91	6.97	7.40	7.70
	.05	5.57	5.63	5.69	5.74	5.79	6.20	6.47
	.10	5.04	5.10	5.16	5.21	5.26	5.65	5.92
19	.01	6.65	6.72	6.78	6.84	6.89	7.31	7.61
	.05	5.53	5.59	5.65	5.70	5.75	6.15	6.42
	.10	5.01	5.07	5.13	5.18	5.23	5.62	5.88
20	.01	6.59	6.65	6.71	6.77	6.82	7.24	7.52
	.05	5.49	5.55	5.61	5.66	5.71	6.10	6.37
	.10	4.99	5.05	5.10	5.16	5.21	5.59	5.85
24	.01	6.39	6.45	6.51	6.56	6.61	7.00	7.27
	.05	5.38	5.44	5.49	5.55	5.59	5.97	6.23
	.10	4.91	4.97	5.02	5.07	5.12	5.49	5.74
30	.01	6.20	6.26	6.31	6.36	6.41	6.77	7.02
	.05	5.27	5.33	5.38	5.43	5.48	5.83	6.08
	.10	4.83	4.89	4.94	4.99	5.03	5.39	5.64
40	.01	6.02	6.07	6.12	6.17	6.21	6.55	6.78
	.05	5.16	5.22	5.27	5.31	5.36	5.70	5.93
	.10	4.75	4.81	4.86	4.91	4.95	5.29	5.53
60	.01	5.84	5.89	5.93	5.97	6.02	6.33	6.55
	.05	5.06	5.11	5.15	5.20	5.24	5.57	5.79
	.10	4.68	4.73	4.78	4.82	4.86	5.20	5.42
120	.01	5.66	5.71	5.75	5.79	5.83	6.12	6.32
	.05	4.95	5.00	5.04	5.09	5.13	5.43	5.64
	.10	4.60	4.65	4.69	4.74	4.78	5.10	5.31
∞	.01	5.49	5.54	5.57	5.61	5.65	5.91	6.09
	.05	4.85	4.89	4.93	4.97	5.01	5.30	5.50
	.10	4.52	4.57	4.61	4.65	4.69	5.00	5.20

Source: Adapted from Table II.2 in *The Probability Integrals of the Range and of the Studentized Range,* prepared by H. L. Harter, D. S. Clemm, and E. H. Guthrie. The original tables are published in WADC Tech. Rep. 58–484, Vol. 2, 1959, Wright Air Development Center, and are reproduced with the permission of the authors.

TABLE D.11 CRITICAL VALUES FOR THE WILCOXON SIGNED-RANK TEST

One-tailed	Two-tailed	Number of pairs									
		5	6	7	8	9	10	11	12	13	14
.05	.10	0	2	3	5	8	10	13	17	21	25
.025	.05		0	2	3	5	8	10	13	17	21
.01	.02			0	1	3	5	7	9	12	15
.005	.01				0	1	3	5	7	9	12
		15	16	17	18	19	20	21	22	23	24
.05	.10	30	35	41	47	53	60	67	75	83	91
.025	.05	25	29	34	40	46	52	58	65	73	81
.01	.02	19	23	27	32	37	43	49	55	62	69
.005	.01	15	19	23	27	32	37	42	48	54	61
		25	26	27	28	29	30	31	32	33	34
.05	.10	100	110	119	130	140	151	163	175	187	200
.025	.05	89	98	107	116	126	137	147	159	170	182
.01	.02	76	84	92	101	110	120	130	140	151	162
.005	.01	68	75	83	91	100	109	118	128	138	148
		35	36	37	38	39	40	41	42	43	44
.05	.10	213	227	241	256	271	286	302	319	336	353
.025	.05	195	208	221	235	249	264	279	294	310	327
.01	.02	173	185	198	211	224	238	252	266	281	296
.005	.01	159	171	182	194	207	220	233	247	261	276
		45	46	47	48	49	50				
.05	.10	371	389	407	426	446	466				
.025	.05	343	361	378	396	415	434				
.01	.02	312	328	345	362	379	397				
.005	.01	291	307	322	339	355	373				

TABLE D.12   TRANSFORMATION OF *r* TO *Z*

r	Z	r	Z	r	Z	r	Z	r	Z
0.000	0.000	0.200	0.203	0.400	0.424	0.600	0.693	0.800	1.099
0.005	0.005	0.205	0.208	0.405	0.430	0.605	0.701	0.805	1.113
0.010	0.010	0.210	0.213	0.410	0.436	0.610	0.709	0.810	1.127
0.015	0.015	0.215	0.218	0.415	0.442	0.615	0.717	0.815	1.142
0.020	0.020	0.220	0.224	0.420	0.448	0.620	0.725	0.820	1.157
0.025	0.025	0.225	0.229	0.425	0.454	0.625	0.733	0.825	1.172
0.030	0.030	0.230	0.234	0.430	0.460	0.630	0.741	0.830	1.188
0.035	0.035	0.235	0.239	0.435	0.466	0.635	0.750	0.835	1.204
0.040	0.040	0.240	0.245	0.440	0.472	0.640	0.758	0.840	1.221
0.045	0.045	0.245	0.250	0.445	0.478	0.645	0.767	0.845	1.238
0.050	0.050	0.250	0.255	0.450	0.485	0.650	0.775	0.850	1.256
0.055	0.055	0.255	0.261	0.455	0.491	0.655	0.784	0.855	1.274
0.060	0.060	0.260	0.266	0.460	0.497	0.660	0.793	0.860	1.293
0.065	0.065	0.265	0.271	0.465	0.504	0.665	0.802	0.865	1.313
0.070	0.070	0.270	0.277	0.470	0.510	0.670	0.811	0.870	1.333
0.075	0.075	0.275	0.282	0.475	0.517	0.675	0.820	0.875	1.354
0.080	0.080	0.280	0.288	0.480	0.523	0.680	0.829	0.880	1.376
0.085	0.085	0.285	0.293	0.485	0.530	0.685	0.838	0.885	1.398
0.090	0.090	0.290	0.299	0.490	0.536	0.690	0.848	0.890	1.422
0.095	0.095	0.295	0.304	0.495	0.543	0.695	0.858	0.895	1.447
0.100	0.100	0.300	0.310	0.500	0.549	0.700	0.867	0.900	1.472
0.105	0.105	0.305	0.315	0.505	0.556	0.705	0.877	0.905	1.499
0.110	0.110	0.310	0.321	0.510	0.563	0.710	0.887	0.910	1.528
0.115	0.116	0.315	0.326	0.515	0.570	0.715	0.897	0.915	1.557
0.120	0.121	0.320	0.332	0.520	0.576	0.720	0.908	0.920	1.589
0.125	0.126	0.325	0.337	0.525	0.583	0.725	0.918	0.925	1.623
0.130	0.131	0.330	0.343	0.530	0.590	0.730	0.929	0.930	1.658
0.135	0.136	0.335	0.348	0.535	0.597	0.735	0.940	0.935	1.697
0.140	0.141	0.340	0.354	0.540	0.604	0.740	0.950	0.940	1.738
0.145	0.146	0.345	0.360	0.545	0.611	0.745	0.962	0.945	1.783
0.150	0.151	0.350	0.365	0.550	0.618	0.750	0.973	0.950	1.832
0.155	0.156	0.355	0.371	0.555	0.626	0.755	0.984	0.955	1.886
0.160	0.161	0.360	0.377	0.560	0.633	0.760	0.996	0.960	1.946
0.165	0.167	0.365	0.383	0.565	0.640	0.765	1.008	0.965	2.014
0.170	0.172	0.370	0.388	0.570	0.648	0.770	1.020	0.970	2.092
0.175	0.177	0.375	0.394	0.575	0.655	0.775	1.033	0.975	2.185
0.180	0.182	0.380	0.400	0.580	0.662	0.780	1.045	0.980	2.298
0.185	0.187	0.385	0.406	0.585	0.670	0.785	1.058	0.985	2.443
0.190	0.192	0.390	0.412	0.590	0.678	0.790	1.071	0.990	2.647
0.195	0.198	0.395	0.418	0.595	0.685	0.795	1.085	0.995	2.994

Appendix **E**

Control Information for Computer Programs for Statistical Analysis

At several points in this book, we presented output generated by three commonly used statistical packages: SYSTAT, BMDP, and SSPS[X]. In this appendix, we present the control information used to carry out the analyses presented in the book. By themselves, such examples of the use of statistical programs are not sufficient for you to immediately use the programs. However, in conjunction with a careful reading of the relevant manual, the material that follows should prove helpful. Another source of help will be your local computing center, either through consultants, or written instructions for using the programs.

E.1 SYSTAT

SYSTAT is a statistical package designed for use with microcomputers. It includes a broad range of applications, many of which have not been referenced in this book (e.g., factor analysis, multidimensional scaling). The version of SYSTAT current as we write this is Version 4.1; this consists of 14 modules, each containing one or more programs. Of particular interest are the EDIT and DATA modules. The EDIT module allows you to type in data and save it in a SYSTAT file; only SYSTAT files are analyzed by the system. The DATA module allows you to take data files typed using your word processor and convert them into SYSTAT files, or to convert SYSTAT files into a code (known as ASCII) that can be printed by your operating system or your word processor. In addition, the DATA module enables you to transform data and to carry out various arithmetic operations by using statements from a version of the BASIC programming language. The DATA module has many other functions including sorting data, converting raw scores to z scores, deleting cases or dropping variables, and merging files. Other modules relevant to the analyses in this book are GRAPH, STATS, NPAR, CORR, and MGLH. We will briefly describe the functions of each of these in what follows. (Note: We will use capital letters to refer to modules and to present SYSTAT commands. However, SYSTAT will respond to commands typed in lowercase. Also, we will type out commands in full. However, with a few exceptions, SYSTAT responds only to the first two characters of a command, so, usually, only the first two letters of a command need to be typed.)

Output appears on the screen unless you request that it also be printed or written to a file. The command OUTPUT@ sends output to the printer. OUTPUT followed by a file name between quotation marks (e.g., OUTPUT 'RESULTS.LST') writes to the file having that name. To stop sending output to the printer or to a file, type OUTPUT*.

Commands such as OUTPUT@ are typed in response to the SYSTAT prompt, the symbol >. You can also move between modules by typing the name of the next module in response to the prompt. Thus, if we are in the DATA module and now wish to plot the data, we type GRAPH in response to the prompt. To exit SYSTAT, just type QUIT in response to the prompt.

E.1.1 Creating a SYSTAT File

A SYSTAT file contains data and has a name of the form fn.SYS ("fn" represents any file name); note the ".SYS" extension. You cannot read or print such files unless you are in SYSTAT. Nor can you type such files with your ordinary word processor. There are two ways to create SYSTAT files. One way is to enter SYSTAT and use its editor to type in the data. Type EDIT to access the editor.

The SYSTAT editor is described in the SYSTAT manual and is very easy to use. A label is typed for each variable, and then the values of the variables for each case (a case is usually a subject) are typed in the columns headed by the appropriate variable labels. For example, suppose we wish to create a file named EXPER1.SYS. There are two between-subjects variables, gender and drug type, and three scores for each subject. The variable labels GENDER, DRUG, TRIAL1, TRIAL2, and TRIAL3 are entered in the top row of the data window, each preceded by an apostrophe or quote symbol. In any succeeding row, the first column will contain a 1 or 2 to indicate the subject's gender; the second column will contain a 1, 2, or 3 to indicate which of three drug conditions the subject was assigned to; and the next three columns will contain the scores for that subject on the three trials. When all the data have been typed in, press the ESCAPE key on your keyboard and then type SAVE EXPER1. Note that we do not type the file name extension ".SYS". The SAVE command automatically creates a file with that extension.

A second way to create a SYSTAT file is to type the file with your word processor. Save it under the name fn.DAT (remember fn stands for any file name you wish to use). The file must be in ASCII (American Standard Code for Information Interchange) code. For example, if you use WordStar, the file must be saved in nondocument mode; in WordPerfect, it would be saved as a DOS file. There should be no nonprinting characters, or special characters indicating such things as page breaks, or boldface, or underlining. ASCII files are converted to SYSTAT files by entering the DATA module (just type DATA) and typing four commands. For the example of the gender by drug experiment, we would have

```
SAVE EXPER1
GET EXPER1
INPUT GENDER, DRUG, TRIAL1, TRIAL2, TRIAL3
RUN
```

The first command opens a file to be called EXPER1.SYS. The second command "gets" a file labeled EXPER1.DAT. (The ASCII, or .DAT, and SYSTAT, or .SYS, files need not have the same file name.) The INPUT command labels the variables, and the RUN command completes execution. The result is a file labeled EXPER1.SYS that contains your data in the SYSTAT format.

E.1.2 The GRAPH Module

In Chapter 2, we presented stem-and-leaf, box, and normal probability plots of data sets (Figures 2.1, 2.2, 2.3). To obtain such plots for the EXPER1 data, we first type GRAPH to gain access to the files that provide these data plots. Once in the GRAPH module, type USE EXPER1. Again, note that the extension .SYS is not part of the command; USE assumes that extension. To obtain the desired plots, type

```
STEM TRIAL1, TRIAL2, TRIAL3
BOX TRIAL1, TRIAL2, TRIAL3
PPLOT TRIAL1, TRIAL2, TRIAL3
```

These yield a combined plot for the six experimental groups. To see a plot for each gender separately, precede the STEM (or BOX or PPLOT) command by

```
BY GENDER
```

To obtain six plots, one for each gender-drug combination, use the command

```
BY GENDER, DRUG
```

To obtain a plot only for females assigned to the third drug condition, type

```
SELECT GENDER=2, DRUG=3
```

The BY and SELECT commands are available in all data analysis modules. Note that you could just select a subset of variables (e.g., PPLOT TRIAL1) if you wished.

E.1.3 The STATS Module

Obtaining statistics such as the mean and standard deviation is also very easy in SYSTAT. Type STATS to enter the module. Then type

```
USE EXPER1
STATISTICS TRIAL1, TRIAL2, TRIAL3
```

This will print out the mean, standard deviation, N, and the minimum and maximum scores for each variable listed in the command. You can also calculate other statistics. For example, measures of the skewness and kurtosis can be obtained by the command

```
STATISTICS TRIAL1, TRIAL2, TRIAL3/SKEW,KURT
```

In fact, we can request various combinations of statistics following the slash (/). Note that, as with the GRAPH module commands, the STATISTICS command gives results for all subjects. If we simply used the command

```
STATISTICS TRIAL1
```

with the data set EXPER1, we would obtain the grand mean and the standard deviation of the scores of all $6n$ subjects on trial 1. Preceding the STATISTICS command by BY or SELECT, as in the previous section, enables us to obtain statistics for individual groups or sets of groups.

The STATS module also allows us to do t tests. To compare the means for males and females on trial 1 under the first drug condition, we would use the commands

```
SELECT DRUG = 1
TTEST TRIAL1*GENDER
```

To carry out the same contrast for each of the three trials, we type

```
TTEST TRIAL1, TRIAL2, TRIAL3*GENDER
```

This will perform three t tests contrasting males and females who were administered drug 1 (assuming the SELECT command) on each trial. Suppose we wished to compare the first and second trial means with each other, and we do this for each of the six groups (combinations of gender and drug). The command for this correlated-measures t test is

```
BY GENDER, DRUG
TTEST TRIAL1, TRIAL2
```

E.1.4 The CORR Module

In Chapter 8, we presented covariances for four subjects tested under three levels of a variable A (see Table 8.4). SYSTAT's CORR module easily generates the variance-covariance matrix. Staying with our example, EXPER1.SYS, we type CORR to access the module. Then just type COVARIANCE. The BY and SELECT commands, as usual, can be used to obtain covariance matrices for each of the six between-subject conditions or to select specific conditions. Also, if we want the covariances for a subset of dependent variables, we can specify that subset. For example, suppose we have scores for each of a battery of five tests, but we only want the covariances for tests 1, 2, and 5. Assume that the scores are labeled TEST1, TEST2, and so on. Then the appropriate command is COVARIANCE TEST1, TEST2, TEST5.

The CORR matrix module also calculates many other measures of relationships. For example, the commands PEARSON, SPEARMAN, and TAU yield the Pearson product-moment correlation coefficient, Spearman's rank-order correlation coefficient, and Kendall's rank-order correlation coefficient, respectively.

E.1.5 The NPAR Module

At several points in this text, we noted the possibility of using nonparametric tests. The Kruskal-Wallis H test is discussed in Chapter 4. In the example analyzed in Table 4.6, we had three groups of subjects, each run using a different method; there was one score for each subject. The SYSTAT file should have two columns, one representing the independent variable and the other the dependent variable. We label these METHOD and Y. METHOD has a value of 1, 2, or 3, depending on the condition to which the subject had been assigned, and Y is the original reading score. Type NPAR to access the module, then the USE command to designate the SYSTAT file, and then KRUSKAL to obtain the value of H and the p value reported in Table 4.6.

In Chapter 8, we considered several nonparametric tests for repeated measures. The original data of Table 8.13 can be analyzed by using Friedman's χ^2 by accessing the module, retrieving the data (the USE command), and then typing FRIEDMAN. In Chapter 8, we noted that the Wilcoxon signed-rank test was more powerful when there were only two measures for each subject. In this case, the appropriate command is WILCOXON.

E.1.6 The MGLH Module

The MGLH (multivariate general linear hypothesis) module carries out univariate and multivariate analyses of variance and covariance, tests of contrasts, and regression analyses. The MGLH program is not convenient to use for the types of univariate repeated-measures and mixed ANOVAs described in Chapters 8 and 9 if they have more than two within-subjects factors, and it lacks some of the assets of mainframe programs for these analyses, such as calculations of ϵ, the df adjustment, and the associated adjusted p values. For this reason, we have tended to employ other programs for such analyses. However, MGLH is readily applied to between-subjects ANOVAs and ANCOVAs. We have found MGLH to be particularly useful and convenient for performing regression analyses, and we have provided a number of examples of its output in this book.

Using MGLH for ANOVA

Before considering regression analysis, let's use the data set EXPER1.SYS to illustrate how MGLH can be used to perform ANOVAs. Suppose we wish to perform a one-factor between-subjects ANOVA in which drug is the independent variable and trial1 is the dependent variable. We get into the module by typing MGLH and access the data by typing USE EXPER1. The ANOVA is performed if we then type

```
CATEGORY DRUG=3
MODEL TRIAL1 = CONSTANT + DRUG
ESTIMATE
```

The CATEGORY statement informs the program that drug has three levels and is to be treated as a qualitative, categorical variable. If the CATEGORY statement is omitted, the program will regress trial1 on drug, treating drug as a quantitative variable that takes on the values 1, 2, and 3. Note that the MODEL statement reflects the one-factor between-subjects ANOVA model without the error term:

$$Y = \mu + \alpha_j$$

A two-factor between-subjects ANOVA using drug and gender as the factors and trial1 as the dependent variable can be obtained by typing

```
USE EXPER1
CATEGORY GENDER=2 DRUG=3
MODEL TRIAL1 = CONSTANT + GENDER + DRUG + GENDER*DRUG
ESTIMATE
```

Note that the MODEL statement again reflects the between-subjects ANOVA model

$$Y_{ijk} = \mu + \alpha_j + \beta_k + (\alpha\beta)_{jk}$$

in which the grand mean, gender and drug effects, and their interaction are all represented.

Finally, a mixed-design ANOVA in which trial is the within-subjects factor and drug and gender are between-subjects factors can be performed by replacing the previous MODEL statement by

```
MODEL TRIAL1 TRIAL2 TRIAL3 = CONSTANT + GENDER + DRUG + GENDER*DRUG/REPEAT
```

The levels of the within-subjects factor are included to the left of the equality in the MODEL statement and /REPEAT is included to the right of the between-subjects ANOVA model statement. If we had a pure repeated-measures $(S \times A)$ ANOVA in which there were no between-subjects factors, only CONSTANT/REPEAT would appear to the right of the equality in the MODEL statement.

Using MGLH for Regression

Panel a of Table 12.3 contains values for two quantitative variables, size and time. The regression of time on size can be conducted by typing, after the USE statement,

```
MODEL TIME = CONSTANT + SIZE
ESTIMATE
```

Panel b of Table 12.3 contains the resulting output. Note that no CATEGORY statement is included because we are performing a regression with quantitative variables. Note also that if we left out the CONSTANT in the MODEL statement, we would find the best regression line that passed through the origin.

It is easy to obtain the residuals and their diagnostics with MGLH. Had we preceded the MODEL statement by

```
SAVE RESID
```

not only would MGLH perform the regression, it would save the residual diagnostics into a file called RESID and include as part of the standard regression output warnings about outliers and influential data points and the results of a Durbin-Watson test. The form of this expanded output can be seen in panel b of Table 12.6. Also, the SYSTAT files containing the information about residuals can be listed by using the DATA module. This has been done in Table 12.5 and in panel d of Table 12.6.

Regression analyses can easily be extended to include additional predictor variables. Table 15.2 contains data for three variables: final, pretest, and SAT. The multiple regression of final on pretest and SAT was obtained by typing, after the USE statement,

```
MODEL FINAL = CONSTANT + PRETEST + SAT
ESTIMATE
```

If there are a large number of possible predictor variables, any subset of them may be included in the regression equation by including them in the MODEL statement and then typing ESTIMATE.

To perform a stepwise regression, include all the potential predictors in the MODEL statement along with the CONSTANT term, but now type STEP instead of ESTIMATE. If you wish to force the first two variables into the regression equation before stepping, use the FORCE option; that is, replace ESTIMATE by STEP/ FORCE = 2. For reasons mentioned in Chapter 15 and in the SYSTAT manual, MGLH does not provide the p values at each step.

In Section 15.8, we considered the regression approach to trend analysis. Using hierarchical regression, we can easily perform trend analyses even for data sets with unequal n and unequal spacing. Such a data set is presented in Table 15.8. In order to perform the trend analysis, we must find the squared multiple correlations $R^2_{Y.X}$, $R^2_{Y.X,X^2}$, and $R^2_{Y.X,X^2,X^3}$. These can be obtained by using the following MODEL statements:

```
MODEL Y = CONSTANT + X
MODEL Y = CONSTANT + X + X*X
MODEL Y = CONSTANT + X + X*X + X*X*X
```

E.2 BMDP

The BMDP (biomedical programs) package was designed as a package to be run on mainframe computers. It contains over 40 programs that perform a wide variety of analyses. Commands to access the programs and data files depend on the operating system in use and, therefore, are different at different computer installations; users should check with local consultants before attempting to run any of the programs. Users who have microcomputers with large hard disks may wish to investigate the relatively new PC version of BMDP. Most of the programs require at least 500,000 bytes of the disk (for example, the data manager, which functions in a manner similar to SYSTAT's DATA module, requires 757,712 bytes, and P2V requires 682,880 bytes). However, if there are a few programs you use frequently, the investment of both money and hard disk space may be worthwhile.

We have used BMDP to perform ANOVAs for this book, and we will present the control statements we used. Not all of these statements are required to obtain our output, and other statements could have been added to obtain additional information. Some of the statements could have been written differently. Also, many of the programs overlap in their functions (for example, P7D, P1V, P2V, P4V, and P8V are all capable of handling a one-factor between-subjects design), and it is wise to consider the choices carefully. All of this suggests the need to read the BMDP software manual carefully.

E.2.1 BMDP7D

Some of the output from this program was presented in Table 4.8. The program is applied to one- and two-factor between-subjects designs. It performs Levene's test for heterogeneity of variance and the Welch, Brown-Forsythe, and standard F tests of the null hypothesis of equality of the means. It also provides histograms for each group for each dependent variable. To analyze the data of Table 4.4, we constructed a data file consisting of two columns; the first contained a 1, 2, or 3 to indicate the condition (method of instruction), and the second contained the subject's score. The statements we used for the analysis of the data of Table 4.4 are

```
/PROBLEM      TITLE IS 'TABLE 4-4 ANOVA'.
/INPUT        VARIABLES ARE 2.
              FORMAT IS FREE.
/VARIABLE     NAMES ARE METHOD, Y.
/GROUP        NAMES(1) ARE METH1, METH2, METH3.
/HISTOGRAM    GROUPING IS METHOD.
/END
```

Note that the information consists of paragraphs; each paragraph begins with a slash (/) and, except for the END paragraph, consists of one or more commands. All commands end with a period.

The PROBLEM paragraph is optional. Because the control information is part of the output, we find it useful to include this paragraph as a reminder of which data set was being analyzed. The INPUT paragraph is required. The first sentence tells the computer that there are two variables (the method and the score). The FORMAT statement tells the computer just to read the two numbers for each case as they are presented. An option is to use FORMAT statements based on the programming language FORTRAN. Such statements permit us more flexibility when typing our data file. For example, decimal points may be omitted, and a subject's data may take up several lines; the FORTRAN FORMAT statement tells BMDP where the decimal point should go and how many lines constitute a single case. Interested readers should consult the BMDP manual.

The VARIABLE paragraph names the variables, and the GROUP paragraph names the levels of the first variable. The HISTOGRAM passage states that a histogram is to be plotted for each method. It is necessary to end any set of control statements with the END paragraph.

Many of these statements are not required. In fact, the same analysis can be run with the control information

```
/INPUT        VARIABLES ARE 2.
              FORMAT IS FREE.
/HISTOGRAM    GROUPING IS 1.
/END
```

Because this set of statements doesn't name the variables, BMDP labels them X(1) (Method) and X(2) (Y). Note the change in the HISTOGRAM paragraph; because we have no name for the independent ("grouping") variable, we indicate that it is the first variable. If we had typed the data in the order Y, Method, we would have used GROUPING IS 2.

There are many optional statements that could have been added. BMDP7D is capable of providing pairwise t tests and Bonferroni probabilities, Winsorized means based on trimming outliers, and plots of standard deviations against means. As we discussed in the book, this last function is useful for deciding on transformations to obtain homogeneous variances.

E.2.2 BMDP2V

Repeated Measures

Table 8.10 presents part of the output of BMDP2V's analysis of the data of Table 8.9. In that example, each of three subjects was tested under each of nine combinations of A and B. The data were typed in the order indicated by the control information here (VARIABLE paragraph), with one line for each subject. The control information used for this analysis was

```
/PROGRAM     TITLE IS 'ANOVA OF TABLE 8-9'.
/INPUT       VARIABLES ARE 9.
             FORMAT IS FREE.
/VARIABLE    NAMES ARE A1B1,A2B1,A3B1,A1B2,A2B2,A3B2,A1B3,A2B3,A3B3.
/DESIGN      DEPENDENT ARE 1 to 9.
             LEVELS ARE 3,3.
             NAMES ARE B,A.
/END
```

The PROGRAM, INPUT, and VARIABLE paragraphs are much the same as in our BMDP7D example. The DESIGN paragraph is new. It tells us that the nine measures are all levels of dependent variables, that they form a 3×3 design, and it names the two variables.

Reconsider the LEVELS and NAMES statements in the DESIGN paragraph. Assume that we had four levels of A and two levels of B and that the data for each subject were typed in the order

```
A1B1, A1B2, A2B1, A2B2, A3B1, A3B2, A4B1, A4B2
```

We refer to A as the "slowest-moving variable" because the levels of B change while those of A are held constant. In the LEVELS and NAMES statements, the variables are always indexed from slowest- to fastest-moving. Therefore, for this 4×2 example, we would have

```
/DESIGN      DEPENDENT ARE 1 to 8.
             LEVELS ARE 4,2.
             NAMES ARE A,B.
```

Note that in the original 3×3 example, A was the fastest-moving variable, so the variables were named in the order B, A.

Mixed Designs

BMDP2V also is used to analyze the designs of Chapter 9, in which we have between- and within-subject factors. Table 9.11 presented data for two groups of six subjects each, with each subject tested under all combinations of B (three levels) and C (two levels). The values for each subject are either a 1 or 2, depending on the level of A to which the subject had been assigned, and then the six scores in the order indicated by the NAMES sentence in the following VARIABLE paragraph. The control information that generated the output presented in Table 9.13 is

```
/PROBLEM      TITLE IS 'P2V ANOVA OF TABLE 9-11'.
/INPUT        VARIABLES ARE 7.
              FORMAT IS FREE.
/VARIABLE     NAMES ARE A,Y11,Y21,Y31,Y12,Y22,Y32.
/DESIGN       GROUPING IS A.
              DEPENDENT ARE 2 TO 7.
              LEVELS ARE 2,3.
              NAMES ARE C,B.
/GROUP        CODES(1) ARE 1,2.
              NAMES(1) ARE A1,A2.
/END
```

We have seven variables because we have the value of A (1 or 2) and the six repeated measures. We have added a GROUPING sentence to the DESIGN paragraph we used for the simple repeated-measures design. We have also added a GROUP paragraph. This tells BMDP2V what codes have been typed in the first column and what names they correspond to. Note that the order of LEVELS and NAMES in the DESIGN paragraph reflects the fact that C is the slowest-moving variable.

E.2.3 BMDP8V

We used BMDP8V to analyze the data of Table 10.6; the results of the analysis were presented in Table 10.7. In this design, there are five subjects within each of two levels of A. Each subject is tested under 12 combinations of B and C. This design differs from the last design discussed in the preceding section in that the levels of B are nested within the levels of C; a more detailed description of this common design is provided in Chapter 10.

The control information that generated the output of Table 10.7 is

```
/PROBLEM      TITLE IS 'ANOVA OF DATA OF TABLE 10.6'.
/INPUT        VARIABLES ARE 12.
              FORMAT IS FREE.
/DESIGN       LEVELS ARE 2,5,3,4.
              NAMES ARE A,S,C,B.
              FIXED ARE A,C.
              RANDOM ARE S,B.
              MODEL IS 'A,S(A),C,B(C)'.
/END
```

BMDP8V differs from most of the BMDP programs in that it does not make use of the code for the between-subjects variable. (You could include that code for your own use, but then a FORTRAN FORMAT statement must be used to instruct BMDP8V to skip over that entry.) For this reason, we state that there are 12 variables (four levels of B nested within three levels of C), not 13. The LEVELS sentence is very critical here because it must correspond to the order of all four variables. If you turn back to Table 10.6, you will note that the fastest-moving variable is B (four levels), then C (three levels), then subjects (S, five levels), and then A (two levels). The LEVELS and NAMES statements reflect this with variables indicated as usual from slowest- to fastest-moving.

BMDP8V also demands an explicit statement of the model, including designations of each variable as fixed or random. This enables it to generate expected mean squares and to choose proper error terms. In this example, we had assumed that there were four randomly sampled problems (levels of B) nested within each of three levels of difficulty (C). Thus B (and, as usual, S) is treated as random in its effects.

E.3 SPSS^X

The other mainframe package we used in providing output for this book was SPSS^X (Statistical Package for the Social Sciences). SPSS^X is a comprehensive data analysis system that operates only in batch mode; however, a sister package, SPSS/PC+, is available for microcomputers.

SPSS^X control language consists of lists of *commands*. Each command (which can consist of more than one word) may be followed by one or more *subcommands*. If there is more than one subcommand, they are separated from one another by slashes (/).

For example, we used SPSS^X with the class example data of Table 12.1 to produce the scatter diagram and regression statistics in Figure 12.1. In using SPSS^X, we must first define the data by using a DATA LIST command that informs the package about the location and format of the data and names the variables. For the data of Table 12.1, this was accomplished by the statements

```
DATA LIST    FILE = DATA121,   FREE/
             PRETEST, FINAL
```

These two lines indicate that the data are located in a file called DATA121, are in free format, and consist of values of two variables called pretest and final. Figure 12.1 was produced by the following PLOT command:

```
PLOT FORMAT=REGRESSION
/PLOT = FINAL WITH PRETEST
```

By default, the PLOT command produces only a bivariate scatter diagram. However, the FORMAT subcommand can be used to request that the regression statistics be printed. The regression line is not actually drawn in the scatter diagram; however, the R's marked on the left and right vertical axes indicate its location. The PLOT subcommand in the second line (note that the word PLOT may be used as a command and as a subcommand) indicates that values of FINAL and PRETEST will be the vertical and horizontal coordinates, respectively, in the plot. If we had a longer list of variables, we could specify a series of scatter diagrams by using additional PLOT subcommands. The more complete regression output for these data presented in Table 12.2 was produced by using the REGRESSION command

```
REGRESSION   VARIABLES = PRETEST, FINAL/
             DEPENDENT = FINAL/
             ENTER PRETEST
```

To use the REGRESSION command, you must provide a variable list for which a correlation matrix is calculated, specify a dependent (criterion) variable, and indicate how predictor variables are to be entered into the equation. Here we have only a single predictor, pretest. If we had a longer list of possible predictors, they could be entered into the equation one at a time by using a series of ENTER commands. If one wished to use one of the automated procedures for entering potential predictors into a regression equation (see Section 15.6), one could replace the ENTER subcommand by STEPWISE, FORWARD, or BACKWARD. Residuals and their diagnostics can be requested by using the RESIDUALS subcommand.

Although we did not use SPSS^X to produce ANOVA or ANCOVA outputs in this book, we should point out that the ANOVA command is easily used to analyze factorial designs and that the MANOVA command invokes a powerful and flexible program that can be used to perform univariate and multivariate ANOVAs and MANOVAs and that can handle repeated-measures, mixed, and hierarchical designs. The most recent version of SPSS^X as we write this (release 3.0) now provides the Greenhouse-Geisser and Huynh-Feldt epsilons (although still not the associated p values) for repeated-measures designs in the output of the MANOVA program, making it more useful for analyzing the types of designs discussed in Chapters 8 and 9 than previous versions.

To illustrate, if we wished to use the EXPER1 data set discussed earlier and perform a factorial ANOVA that had gender and drug as the factors and trial1 as the dependent variable, we could use

```
ANOVA TRIAL1 BY GENDER(1,2) DRUG(1,3)
```

The numbers in parentheses indicate the ranges of values taken on by the factors that are to be considered in the analysis. If data for a covariate, X, had been included in the file, the factorial ANCOVA could be requested by using the WITH keyword:

```
ANOVA TRIAL1 BY GENDER(1,2) DRUG(1,3) WITH X
```

It should be noted that for unbalanced (i.e., unequal-n) factorial designs, ANOVA uses the "classic experimental approach" (i.e., method 2 described in Chapter 16) by default. We indicated in Chapter 16 that we do not recommend this method. The method that would be appropriate if we wished to weight each cell mean equally (method 1 of Chapter 16, the "classic regression approach") can be requested by using the METHOD subcommand with the keyword UNIQUE:

```
ANOVA TRIAL1 BY GENDER(1,2) DRUG(1,3)
        /METHOD = UNIQUE
```

With the MANOVA program, the same factorial analysis would be produced by the command.

```
MANOVA TRIAL1 BY GENDER(1,2) DRUG(1,3)
```

The MANOVA program uses method 1 for unbalanced factorial designs by default. Although we shall not go into the MANOVA program further here, a careful reading of the users' manual will indicate how it can be used to perform a wide variety of analyses.

Answers to Selected Exercises

CHAPTER 2

2.2 (a) 31.938;
(b) 32.000;
(c) 261,121;
(d) 16,311;
(e) 43.304;
(f) 153.263;
(g) .367;
(h) $H_L = 23.5$; $H_U = 37.5$.

2.4 Outliers and the shape of the normal probability plot indicate that data set (a) is heavy-tailed. Both stem-and-leaf and normal probability plots indicate that data set (c) is skewed to the right. Data set (b) appears to be normally distributed.

2.6 The proportions are sufficient for these calculations.
(a) The average score is $(.4)(3) + (.3)(2) + (.2)(1) + (.1)(0) = 2.0$.
(b) The variance is the average squared deviation; therefore, $S^2 = (.4)(3 - 2)^2 + (.3)(2 - 2)^2 + (.2)(1 - 2)^2 + (.1)(0 - 2)^2 = .76$.

2.8 (a) $E(X) = .2$; $\text{Var}(X) = E(X^2) - [E(X)]^2 = .2 - .04 = .16$.
(b)

Y	$p(Y)$	$\bar{X}$	$S^2_{\bar{X}}$	$\hat{\sigma}^2_{\bar{X}}$
0	.512	0	0	0
1	.384	.333	.222	.333
2	.096	.667	.222	.333
3	.008	1	0	0

(c) $E(Y) = 6 = 3 \times E(X)$; $E(\bar{X}) = 2 = E(X)$; $E(S^2) = .107 = (2/3) \times \text{Var}(X)$; $E(\hat{\sigma}^2_X) = .16 = \text{Var}(X)$. S^2 is a biased estimator.

2.10 (a) .7;

(b) .21;

(c) $X_1 - X_2 = 0 \quad -1 \quad 1$

$p(X_1 - X_2) = .5 \quad .2 \quad .3$

$E(X_1 - X_2) = .1;\ \mathrm{Var}(X_1 - X_2) = .5 - .1^2 = .49$

(d) $\sigma_{12} = E(X_1 X_2) - E(X_1)E(X_2) = .4 - (.7)(.6) = -.02.$

Note that $\sigma_1^2 + \sigma_2^2 - 2\sigma_{12} = .21 + .24 - (2)(-.02) = .49$, the variance of the difference scores computed in part (c).

2.12 (a) $\sigma_d^2 = (\sigma_1^2 + \sigma_2^2)/n;$

(b) (i) $\sigma_d^2 = \sigma_1^2 + \sigma_2^2 - 2\sigma_1\sigma_2\rho_{12};$

(ii) $\sigma_d^2 = (\sigma_1^2 + \sigma_2^2 - 2\sigma_1\sigma_2\rho_{12})/n;$

(iii) The repeated-measures design should involve less variance in the sampling distribution of the mean and therefore more precise inferences about $E(d)$. The reason for the smaller standard error is that we are subtracting out a covariance term in part (b); presumably the covariance of two scores taken from the same subject will be positive.

2.14 (a) Let $\bar{X}$ be the mean of the four scores for a subject. Then $E(\bar{X}) = 70$ because $E(\Sigma X/4) = \Sigma E(X)/4 = \Sigma (70/4) = (4)(70/4)$.

(b) The variance will be less than 100/4 because the variance of the mean will equal the sum of the four variances of the mean (σ^2/n, or 100/4) minus 2 times the sums of the six possible covariances; the covariances usually will be greater than zero.

2.16 (a) $P(D) = (.5)(.9 + .3) = .6;$

(b) $\mathrm{Var}(P) = (.6)(.4)/50 = .0048;$

(c) (i) $\mathrm{Var}(P_L) = (.9)(.1)/25 = .0036;$

(ii) $\mathrm{Var}(P_C) = (.3)(.7)/25 = .0084;$

(iii) $\mathrm{Var}(P) = (1/2)^2[\mathrm{Var}(P_L) + \mathrm{Var}(P_C)] = .0030.$

(d) Stratification is preferred because it yields a smaller standard error for the proportion being estimated.

CHAPTER 3

3.2 (a) .023, .840, .977, .069;

(b) 80.8, 119.2;

(c) 110.125;

(d) 490;

(e) .159;

(f) essentially zero (.0008); the denominator of the z score is the S.E.M., which is $15/\sqrt{10} = 4.743$. Thus, we need the area beyond the z score of $(115 - 110)/4.743$, or 3.162.

3.4 (a) (i) .401; (ii) .067; (iii) .465;

(b) $\mu_W = 50$ and $\sigma_W = \sqrt{20^2 + 16^2} = 25.6125$; (i) .721; (ii) .332; (iii) .879 [note that this is just $p(W > 20)$].

(c) An X score at the 85th percentile $(X_{.85})$ corresponds to a z score of 1.03; that is, $1.03 = (X - 30)/20$. Solving, $X = 50.6$. Similarly, $Y = 11.68$. Then $W = 62.28$, its z transform is .479, and the percentile rank is 68 (specifically, 62.68 exceeds .684 of the W scores).

3.6 (a) $H_0: \pi \geqslant .4; H_1: \pi < .4$;

(b) $\sigma_p = \sqrt{(.6)(.4)/50} = .0693$. Then $z = (.24 - .4)/\sigma_p = -2.309$. Assuming $\alpha = .05$, $z_{crit} = -1.645$; the computed z is less than this, so we reject H_0.

(c) It is assumed that the sampling distribution of p is normal. Because p is a sum of 0's and 1's divided by n, and n is fairly large (50), the central limit theorem provides a justification for the normality assumption. With $n = 10$, or with a more skewed value of p such as .1, the sampling distribution of p would be much less like the normal distribution.

3.8 (a) $H_0: \mu_2 - \mu_1 = 0; H_1: \mu_2 - \mu_1 \neq 0$. On 9 df, with $\alpha = .05$ (two-tailed), reject if $|t| > 2.262$. To carry out the test, let $\bar{d} = \bar{Y}_2 - \bar{Y}_1$. Then $\bar{d} = .9$, $\hat{\sigma}_{\bar{d}} = .809$, and $t = 1.123$. We cannot reject H_0.

(b) The .95 CI is $.9 \pm (2.262)(.809) = -.93, 2.73$.

(c) $H_0: (.5)[(\mu_3 + \mu_4) - (\mu_1 + \mu_2)] = 0; H_1: (.5)[(\mu_3 + \mu_4) - (\mu_1 + \mu_2)] \neq 0$. The rejection region is the same as in part (a). To test H_0, let $\bar{d} = (.5)[(\bar{Y}_3 + \bar{Y}_4) - (\bar{Y}_1 + \bar{Y}_2)]$. Then $t = 1.6/.770 = 2.078$; this also fails to reach significance at the .05 level.

3.10 (a) (i) $H_0: \mu_{10,G} - \mu_{8,G} \leqslant 0; H_1: \mu_{10,G} - \mu_{8,G} > 0$. Assuming $\alpha = .05$, and with df = 18, reject if $t > 1.734$.

(ii) The standard error of the difference in the means is $\sqrt{(2.9^2 + 2.2^2)/10} = 1.15; t = (60 - 53)/1.15 = 6.087$. Therefore, reject H_0.

(b) (i) The .90 CI = $7 \pm (1.734)(1.15) = 5.006, 8.994$.

(ii) The lower limit of the CI is well above the null hypothesized value of zero. Therefore, we can reject H_0 at the .05 level.

(c) (i) $H_0: (\mu_{10,B} - \mu_{10,G}) - (\mu_{8,B} - \mu_{8,G}) \leqslant 0$,
$H_1: (\mu_{10,B} - \mu_{10,G}) - (\mu_{8,B} - \mu_{8,G}) > 0$.

(ii) The standard error of the linear combination of interest is $\sqrt{(2.7^2 + 2.1^2 + 2.9^2 + 2.2^2)/10} = 1.580$; therefore, $t = [(72 - 60) - (58 - 53)]/1.58 = 4.432$, which is clearly significant at the .05 level.

3.12 (a) $H_0: \mu_H - \mu_L = 0; H_1: \mu_H - \mu_L \neq 0$; for $\alpha = .05$, df = 34, reject if $|t| > 2.034$.

$$\hat{\sigma}_{\bar{Y}_H - \bar{Y}_L} = \sqrt{[14(6.102^2) + 20(6.128^2)]/34}\sqrt{1/15 + 1/21} = 2.068$$

and $t = (67.333 - 66.048)/2.068 = .621$, which is not significant.

(b) $H_0: \mu_M - (.5)(\mu_H + \mu_L) \leqslant 0, H_1: \mu_M - (.5)(\mu_H + \mu_L) > 0$, for $\alpha = .05$ and df = .51. Reject if $t > 1.676$. Let $\bar{C} = \bar{Y}_M - (.5)(\bar{Y}_H + \bar{Y}_L) = 1.921$; the numerator of our t test. The denominator estimates

$$\sigma_{\bar{C}} = \sqrt{\sigma_M^2/n_M + (.5^2)[\sigma_L^2/n_L + \sigma_H^2/n_H]}.$$

However, under the assumption of homogeneity of variance, the three variances can be replaced by a pooled variance estimate:

$$\hat{\sigma}^2_{pool} = [14(6.102^2) + 17(6.137^2) + 20(6.128^2)]/51 = 37.502.$$

Then $\hat{\sigma}_{\bar{C}} = \sqrt{(37.502)[1/18 + (.25)(1/15 + 1/21)]} = 1.776$ and $t = 1.931/1.776 = 1.081$, which is not significant.

3.14 (a) The mean increase is 4.133, and the SE of the difference is $4.340/\sqrt{15} = 1.121$. For $\alpha = .05$ and df $= .14$, the critical t value (two-tailed) is 2.145. Therefore, the .95 CI is $4.133 \pm (1.121)(2.145) = 1.729, 6.538$.

(b) H_0: $\mu_{1983} - \mu_{1982} = 0$; H_1: $\mu_{1983} - \mu_{1982} \neq 0$. It is unnecessary to calculate the t statistic. Because 0 is not in the .95 CI, we can reject H_0.

3.16 (a) (i) .90; (ii) .75;

(b) .5.

3.18 (a) (i) .10; (ii) .975;

(b) We assume that we have two independently and normally distributed populations with identical variances.

CHAPTER 4

4.2 (a)

SV	df	SS	MS	F
A	2	3886.867	1943.433	31.532
S/A	27	1664.100	61.633	

(b)

A	4	3541.869	885.467	1.299
S/A	26	17727.550	681.829	

4.4 Because a given experimental group contains only n of the infinite number of values of ϵ_{ij} in the population, the sum of these n values will not generally be zero. The α_j, on the other hand, are fixed-effect variables; all deviations of the μ_j about μ are viewed as being represented in the experiment, and the sum of these deviations therefore will be zero. If the α_j were a random sample from a larger population of such values, they would no longer sum to zero, although $E(\Sigma \alpha_j) = 0$.

4.6 (a) With 10 subjects in each group, $MS_A = 10\Sigma(\bar{Y}_{.j} - \bar{Y}_{..})^2/2$. Entering the three group means into a calculator with a variance key and multiplying by 10 is a simple way to calculate MS_A. Similarly, with equal group sizes, $MS_{S/A}$ is the average of the three group variances. Therefore we have

SV	df	MS	F
A	2	323.333	4.619
S/A	27	70.000	

Setting $\alpha = .05$, reject H_0 if $F > 3.35$. Therefore, we conclude that the three population means differ significantly.

(b) (i) $\text{Var}(\hat{\psi}) = (1/2)^2[\text{Var}(\bar{Y}_A) + \text{Var}(\bar{Y}_C)] + \text{Var}(\bar{Y}_B)$. Assuming homogeneity of variance, $\sigma^2_{\bar{Y}_A} = \sigma^2_{\bar{Y}_B} = \sigma^2_{\bar{Y}_C} = \text{MS}_{S/A}/10$. Therefore, the estimate of $\text{Var}(\hat{\psi})$ is $(\text{MS}_{S/A}/10)(1/4 + 1/4 + 1) = (6/40)(\text{MS}_{S/A}) = 10.5$.

(ii) $t = [(1/2)(16 + 13) - 24]/\sqrt{10.5} = -2.93$, which, on 27 df, is clearly significant at the .01 level.

4.8 Under the null hypothesis that the four samples of size 5 and the additional 15 observations are from the same population, we have two independent estimates of σ^2, the variance of that population. Based on the means of the four samples, we have $\text{MS}_A = (5)(84) = 420$. The additional 15 scores give us $\text{MS}_{error} = 384$. Under the null hypothesis, $\text{MS}_A/\text{MS}_{error}$ is distributed as F on 3 and 14 df. The ratio is not significant. There are two differences between this and the usual ANOVA problem: (1) the two estimates in this case are based on different data sets, and (2) we must consider the possibility that either $\text{MS}_A/\text{MS}_{error}$ or $\text{MS}_{error}/\text{MS}_A$ is significantly large, whereas in the usual ANOVA problem logic dictates that only the first ratio be evaluated.

4.10 The Kruskal-Wallis H statistic is 6.464 and is significant at the .05 level ($p = .039$). The value of the F statistic is 3.132 ($\text{MS}_A = 1532.067$; $\text{MS}_{S/A} = 489.216$) and fails to reach significance ($p = .054$). The data are clearly skewed to the right. Under these circumstances, the Kruskal-Wallis H test may have more power than the F test, accounting for the difference in results for the two tests.

4.12 (a)

SV	df	SS	MS	F
A	3	698.2	212.733	5.076
S/A	16	733.6	45.850	

The result is significant.

(b) $\hat{\phi} = \sqrt{(.16)(5)/(.84)} = .976$. For 3 and 16 df, power is approximately .30. If the variance of the population effects is small, we were very fortunate to detect that variance; in general, this is much too small a sample size.

4.14 (a) $n = 10$;

(b) no, the F ratio of 2 is far less than 3.35, the critical value.

(c) The four statistics are

$$\hat{\theta}^2_A = (\text{MS}_A - \text{MS}_{S/A})/n = 4.0$$

$$\hat{\delta}^2_A = [(a - 1)/a]\hat{\theta}^2_A = 2.667$$

$$\hat{\omega}^2_A = \hat{\delta}^2_A/(\hat{\delta}^2_A + \hat{\sigma}^2_e) = .0625$$

$$\hat{\eta}^2_A = \text{SS}_A/\text{SS}_{tot} = (2)(80)/[(2)(80) + (27)(40)] = .1290$$

$\hat{\theta}^2_A$ and $\hat{\delta}^2_A$ are estimates of the absolute variability of the treatment effects. $\hat{\omega}^2_A$ is preferable because it is an estimate of the proportion of population variance accounted for by the independent variable. $\hat{\eta}_A^2$ overestimates this proportion because of the contribution of error and is also sensitive to the value of n. [Part (c) of Exercise 4.13 illustrates the problems with $\hat{\eta}_A^2$.]

4.16 We begin by defining the term of interest:

$$E(MS_{S/A}) = \frac{1}{a(n-1)} E\sum_j\sum_i (Y_{ij} - \bar{Y}_{.j})^2$$

Assuming the usual structural model, $Y_{ij} = \mu + \alpha_j + \epsilon_{ij}$, and summing over i and dividing by n to obtain the structural equation for the group means, we have

$$\bar{Y}_{.j} = \mu + \alpha_j + \bar{\epsilon}_{.j}$$

where $\bar{\epsilon}_{.j} = \Sigma_i \epsilon_{ij}/n$. Then

$$Y_{ij} - \bar{Y}_{.j} = \epsilon_{ij} - \bar{\epsilon}_{.j}$$

and

$$\sum_j\sum_i (Y_{ij} - \bar{Y}_{.j})^2 = \sum_j\sum_i (\epsilon_{ij}^2 + \bar{\epsilon}_{.j}^2 - 2\epsilon_{ij}\bar{\epsilon}_{.j})$$

$$= \sum_j\sum_i \epsilon_{ij}^2 - n\sum_j \bar{\epsilon}_{.j}^2$$

Then

$$E\sum_j\sum_i (Y_{ij} - \bar{Y}_{.j})^2 = \sum_j\sum_i E(\epsilon_{ij}^2) - n\sum_j E(\bar{\epsilon}_{.j}^2)$$

If we assume independence, the average squared deviation about the mean is σ^2/n. Therefore,

$$\sum_j\sum_i E(\epsilon_{ij}^2) - n\sum_j E(\bar{\epsilon}_{.j}^2) = \sum_j (n)\sigma_j^2 - \frac{n\Sigma_j\sigma_j^2}{n}$$

Assuming homogeneity of variance, this last expression becomes

$$an\sigma_e^2 - a\sigma_e^2 = a(n-1)\sigma_e^2$$

Dividing by $a(n-1)$ to obtain the expression for a mean square completes the proof.

CHAPTER 5

5.2 **(a)** The cell and marginal means are

	A_1	A_2	A_3	A_4	Mean
B_1	17.333	26.667	35.667	19.333	24.750
B_2	28.000	27.000	20.000	38.333	28.333
Mean	22.667	26.833	27.833	28.833	26.542

There is a slight indication of A and B main effects, and a more pronounced indication that an interaction may be present; the function at B_1 is roughly an inverted U, whereas that at B_2 is somewhat U-shaped.

(b) As the following ANOVA indicates, only the interaction is significant.

SV	df	SS	MS	F
A	3	132.125	44.042	0.521
B	1	77.042	77.042	0.912
AB	3	1003.458	334.486	3.960
S/AB	16	1351.333	84.458	

5.4 **(a)** All effects are significant at the .05 level.

SV	df	SS	MS	F
A	2	2013.333	1,006.667	71.905
B	1	601.667	601.667	42.976
AB	2	93.333	46.667	3.960
S/AB	54	756.000	14.000	

(b) $SS_{B/A_3} = 405$. The average of the two cell variances is 14. Therefore, $F = 405/14 = 28.3$. On 1 and 18 df, this is clearly significant.

(c) $SS_{A/B_2} = 646.667$ and the average of the three B_2 cell variances is 10.333. Therefore $F = (646.667/2)/10.333 = 31.29$. On 2 and 27 df, this is clearly significant.

(d) If we could assume homogeneity of the six population variances, it would be appropriate to use $MS_{S/AB}$ as the error term, gaining df and, accordingly, power. However, the cell variances differ sufficiently that it seems safer to just use those corresponding to the means involved in the analysis.

5.6 Let $t_c = 40$, 80, and 120 for the three list lengths. Let $t_e = 100$ for the fuzzy probe and 50 for the clear probe. The "data" are

	L		
Q	2	3	4
Fuzzy	480	520	560
Clear	430	470	510

The L and Q terms should be significant, but not $L \times Q$.

5.8 The significant SV should be *therapy* and *therapy* $\times$ *socioeconomic level*. Means consistent with this would be

	S-E Level		
	Low	Middle	High
Psychotherapy	10	12	14
Behavior therapy	20	18	16

5.10 (a) Let A, I, and X represent the variables age, irrelevant information, and sex, respectively. Then the SV, df, and EMS are

SV	df	EMS
A*	2	$\sigma_e^2 + 60\theta_A^2$
I	2	$\sigma_e^2 + 60\theta_I^2$
X	1	$\sigma_e^2 + 90\theta_X^2$
AI*	4	$\sigma_e^2 + 20\theta_{AI}^2$
AX	2	$\sigma_e^2 + 30\theta_{AX}^2$
IX	2	$\sigma_e^2 + 30\theta_{IX}^2$
AIX*	4	$\sigma_e^2 + 10\theta_{AIX}^2$
S/AIX	162	σ_e^2

(b) SV followed by an asterisk (*) are predicted to be significant.

5.12 The terms A, D, AD, and CD should be significant.

5.14 (a) R: $.05 < p < .1$; P: $.05 < p < .1$; RP: $p \approx .2$

(b) $\hat{\delta}_R^2 = (3/4)(750 - 300)/18 = 18.75$; $\hat{\delta}_P^2 = (2/3)(900 - 300)/24 = 16.67$

5.16 First calculate the estimates of the δ^2's:

$$\hat{\delta}_A^2 = \left(\frac{2}{3}\right)\frac{512 - 62}{30} = 10.000$$

$$\hat{\delta}_B^2 = \left(\frac{4}{5}\right)\frac{512 - 62}{18} = 20.000$$

$$\hat{\delta}_{AB}^2 = \left(\frac{8}{15}\right)\frac{152 - 62}{6} = 8.00$$

Summing these plus the $MS_{S/AB}$ yields the denominator of our values of $\hat{\omega}^2$. We have

SV	$\hat{\omega}^2$
A	.10
B	.20
AB	.08
S/AB	.62

(b) To obtain the value of power, we first estimate ϕ:

$$\hat{\phi} = \frac{\sqrt{n\Sigma_j\Sigma_k\Sigma_m(\widehat{\alpha\beta\gamma})_{jkm}^2/(df_{AB} + 1)}}{\sqrt{MS_{S/AB}}} = \frac{\sqrt{(6)(15)(.08)/9}}{\sqrt{.62}} = 1.136$$

Turning to the power chart in Appendix D for df $= 8$, and interpolating between the df $= 60$ and the df $= \infty$ curves, we find power to be approximately .60 for the interaction effects estimated by this data set.

5.18 (a) There is some variability of the estimated effects of A, B, and AB; therefore, with errorless data, all would be significant.

(b)

SV	df	SS	MS	F
A	2	570.000	285.000	3.644
B	2	443.333	221.667	2.834
AB	4	726.667	181.667	2.322
S/AB	36	2816.000	78.222	
Total	44	4556.000		

(c) Only the A main effect is significant at the .05 level.

(d)

SV	df	SS	MS	F
A	2	570.000	285.000	3.003
Error	42	3986.00	94.905	
Total	44	4556.000		

Note that the A effect is no longer significant. Researchers often ignore some variable that is not of interest (e.g., B); these data suggest that if the variable contributes substantial (not necessarily significant) variability, tests of SV that are of interest may be affected.

(e) For the one-factor design, $\hat{\omega}^2 = .082$; for the two-factor design, $\hat{\omega}^2 = .089$.

(f) For the one-factor design, $\hat{\phi} = 1.15$ and, with df $= 2$ and 44, power is approximately .35. For the two-factor design, $\hat{\phi} = 1.33$, and, with df $= 2$ and 36, power is approximately .45.

5.20 (a) $SS_{cells} = (20^2/2) + \cdots + (4^2/2) - 80^2/16 = 240$.

(b) $SS_A = (60^2/6) + (20^2/10) - 80^2/16 = 240$; similarly, $SS_B = 52.267$. The variability due to A and B is greater than the variability among the four cell means. If we were to calculate SS_{AB} by the usual method, we would obtain a negative sum of squares. The problem is that the A and B effects are correlated.

(c) The row means are $\bar{Y}_{.1} = 60/6 = 10$ and $\bar{Y}_{.2} = 20/10 = 2$. The grand mean is $\bar{Y}_{..} = 80/16 = 5$. The $\hat{\alpha}_j$ are 5 and -3, respectively; note that $(6)(5) + (10)(-3) = 0$, where 6 and 10 are the n_j.

(d) Adjusting for the $\hat{\alpha}_j$, we have the following cell means and n_{jk}:

$$
\begin{array}{c}
\quad\quad B_1 \quad B_2 \\
\begin{array}{c} A_1 \\ A_2 \end{array}
\left[\begin{array}{cc} 5/2 & 5/4 \\ 5/8 & 5/2 \end{array} \right]
\end{array}
$$

The adjusted cell totals and n_{jk} are now

$$
\begin{array}{c}
\quad\quad B_1 \quad\quad B_2 \\
\begin{array}{c} A_1 \\ A_2 \end{array}
\left[\begin{array}{cc} 10/2 & 20/4 \\ 40/8 & 10/2 \end{array} \right]
\end{array}
$$

$SS_A = 0$ for this adjusted table because we have subtracted the A main effect for the original data. $SS_B = (50^2/10) + (30^2/6) - 80^2/16 = 0$. Similarly, the SS_{cells} (for this adjusted table) is zero, and therefore $SS_{AB} = 0$. In this data set, the A, B, and AB effects were perfectly correlated, and the removal of the A effects removed the other effects as well. In most data sets involving disproportionate n's, the correlation will not be perfect. Nevertheless, ordinary ANOVA methods will give misleading and even [as in part (b)] absurd results. Regression methods that adjust data for effects in a sequential order, much as we did here, are described in Chapter 16.

CHAPTER 6

6.2 **(a)** (i) H_0: $(1/2)(\mu_{.2} + \mu_{.3}) - \mu_{.1} = 0$.

(ii)

$$t = \frac{(1/2)(\bar{Y}_{.2} + \bar{Y}_{.3}) - \bar{Y}_{.1}}{\sqrt{MS_{S/AB}(\Sigma w_k^2)/an}} = \frac{7.667}{\sqrt{78.222(1.5)(30)}} = 3.877$$

With 81 df, this result is clearly significant.

(b) $SS_\psi = [(1/2)(\bar{Y}_{.2} + \bar{Y}_{.3}) - \bar{Y}_{.1}]^2/(\Sigma w_k^2)/an = 1175.555$.
Note that $SS_\psi/MS_{S/AB} = 15.028 = 3.877^2$; that is, $F = t^2$.

(c) Let $\psi_j = (1/2)(\mu_{j2} + \mu_{j3}) - \mu_{j1}$. Then H_0: $\psi_1 = \psi_2 = \psi_3$; that is, the contrast is the same at all levels of A. To test this null hypothesis, calculate $SS_{A \times \hat{\psi}(B)} = [(10^2 + 10^2 + 3^2)/(1.5/10)] - 1175.555 = 217.777$. Dividing by $MS_{S/AB}$, $F = 2.784$. On 2 and 81 df, a critical value of approximately 3.12 is required, so we cannot reject H_0.

6.4 **(a)** The A means are 13, 10, and 7. Using the weights dictated by the null hypotheses, $SS_{\hat{\psi}_1} = [13 - (1/2)(10 + 7)]^2/(1.5/20) = 270$, and $SS_{\hat{\psi}_2} = (10 - 7)^2/(2/20) = 90$.

(b) Checking the sum of the cross products of coefficients, we find that the contrasts are orthogonal.

(c) $SS_A = 360 = 270 + 90$, consistent with our conclusion that the contrasts are orthogonal.

(d)

$$SS_{B \times \hat{\psi}_1} = \frac{[20 - (1/2)(10 + 6)]^2}{1.5/10} + \frac{[6 - (1/2)(10 + 8)]^2}{1.5/10} - SS_{\hat{\psi}_1} = 750$$

$$SS_{B \times \hat{\psi}_2} = \frac{[(10 - 6)^2 + (10 - 8)^2]}{2/10} - SS_{\psi_2} = 10$$

The sum of the two SS values is 760. SS_{AB} also equals 760.

6.6 **(a)** The weights are

$$\begin{array}{cc} & C_1 & C_2 \\ & \begin{array}{cc} B_1 & B_2 \end{array} & \begin{array}{cc} B_1 & B_2 \end{array} \\ \begin{array}{c} A_1 \\ A_2 \end{array} & \begin{bmatrix} 1 & -1 \\ 1 & -1 \end{bmatrix} & \begin{bmatrix} 1 & -1 \\ 1 & -1 \end{bmatrix} \end{array}$$

Summing the weighted cell totals, squaring, and dividing by 80 yields $SS_B = .45$.

(b)

$$
\begin{array}{cc}
C_1 & C_2 \\
\end{array}
$$

$$
\begin{array}{c}
 \\
A_1 \\
A_2
\end{array}
\begin{array}{cc}
B_1 & B_2 \\
\begin{bmatrix} 1 & 1 \\ -1 & -1 \end{bmatrix}
\end{array}
\begin{array}{cc}
B_1 & B_2 \\
\begin{bmatrix} -1 & -1 \\ 1 & 1 \end{bmatrix}
\end{array}
$$

(c)

$$
\begin{array}{cc}
C_1 & C_2 \\
\end{array}
$$

$$
\begin{array}{c}
 \\
A_1 \\
A_2
\end{array}
\begin{array}{cc}
B_1 & B_2 \\
\begin{bmatrix} 1 & -1 \\ -1 & 1 \end{bmatrix}
\end{array}
\begin{array}{cc}
B_1 & B_2 \\
\begin{bmatrix} -1 & 1 \\ 1 & -1 \end{bmatrix}
\end{array}
$$

(d) Multiply the $B_1 C_1$ and $B_2 C_2$ cell totals by 1 and all others by -1. Square the sum of these weighted cell totals and divide by 120. Then $SS_{BC} = 1.875$.

6.8 (a)

$$
\begin{array}{c}
B_1
\end{array}
\qquad
\begin{array}{c}
B_2
\end{array}
$$

$$
\begin{array}{c}
 \\
A_1 \\
A_2 \\
A_3
\end{array}
\begin{array}{cccc}
C_1 & C_2 & C_3 & C_4 \\
\begin{bmatrix} 1 & 1 & 1 & 1 \\ 1 & 1 & 1 & 1 \\ 1 & 1 & 1 & 1 \end{bmatrix}
\end{array}
\begin{array}{cccc}
C_1 & C_2 & C_3 & C_4 \\
\begin{bmatrix} -1 & -1 & -1 & -1 \\ -1 & -1 & -1 & -1 \\ -1 & -1 & -1 & -1 \end{bmatrix}
\end{array}
$$

(b)

$$
\begin{array}{c}
B_1
\end{array}
\qquad
\begin{array}{c}
B_2
\end{array}
$$

$$
\begin{array}{c}
 \\
A_1 \\
A_2 \\
A_3
\end{array}
\begin{array}{cccc}
C_1 & C_2 & C_3 & C_4 \\
\begin{bmatrix} -1 & 3 & -1 & -1 \\ -1 & 3 & -1 & -1 \\ -1 & 3 & -1 & -1 \end{bmatrix}
\end{array}
\begin{array}{cccc}
C_1 & C_2 & C_3 & C_4 \\
\begin{bmatrix} -1 & 3 & -1 & -1 \\ -1 & 3 & -1 & -1 \\ -1 & 3 & -1 & -1 \end{bmatrix}
\end{array}
$$

(c)

$$
\begin{array}{c}
B_1
\end{array}
\qquad
\begin{array}{c}
B_2
\end{array}
$$

$$
\begin{array}{c}
 \\
A_1 \\
A_2 \\
A_3
\end{array}
\begin{array}{cccc}
C_1 & C_2 & C_3 & C_4 \\
\begin{bmatrix} -1 & 3 & -1 & -1 \\ -1 & 3 & -1 & -1 \\ -1 & 3 & -1 & -1 \end{bmatrix}
\end{array}
\begin{array}{cccc}
C_1 & C_2 & C_3 & C_4 \\
\begin{bmatrix} -1 & 3 & -1 & -1 \\ -1 & 3 & -1 & -1 \\ -1 & 3 & -1 & -1 \end{bmatrix}
\end{array}
$$

(d)

$$
\begin{array}{c}
B_1
\end{array}
\qquad
\begin{array}{c}
B_2
\end{array}
$$

$$
\begin{array}{c}
 \\
A_1 \\
A_2 \\
A_3
\end{array}
\begin{array}{cccc}
C_1 & C_2 & C_3 & C_4 \\
\begin{bmatrix} -2 & 6 & -2 & -2 \\ 1 & -3 & 1 & 1 \\ 1 & -3 & 1 & 1 \end{bmatrix}
\end{array}
\begin{array}{cccc}
C_1 & C_2 & C_3 & C_4 \\
\begin{bmatrix} 2 & -6 & 2 & 2 \\ -1 & 3 & -1 & -1 \\ -1 & 3 & -1 & -1 \end{bmatrix}
\end{array}
$$

6.10 (a) $t = 7/\sqrt{(4)(2/5)} = 5.534$. With df $= 20$ and four comparisons, the critical values for the Bonferroni and Dunnett tests are 2.74 and 2.65, respectively.

(b) Tukey's test should be performed. Compare the t calculated above ($= 5.534$) with $q/\sqrt{2}$, where q is the value needed for significance when there are five means, df $= 20$, and $\alpha = .05$. Then $q/\sqrt{2} = 4.23/\sqrt{2} = 2.99$. Again, the comparison is significant.

(c) Both tests in part (a) are more powerful than the Tukey test in part (b), as is evident by comparing the three critical values. Basically, we have sacrificed some power in part (b) in order to gain more information in the form of additional comparisons among means. The Dunnett test is more powerful than the Bonferroni, which tends to be conservative.

(d) Let $C =$ the critical values: 2.65, 2.74, and 2.99 for the Dunnett, Bonferroni ($k = 4$), and Tukey procedures. Then all three confidence intervals are of the form $(\bar{Y}_{.1} - \bar{Y}_{.2}) \pm C\sqrt{\text{MS}_{\text{error}}(2/n)}$. The resulting CI are

Dunnett	3.648,	10.352
Bonferroni	3.529,	10.471
Tukey	3.218,	10.782

The widths are consistent with our comments about power in part (c).

6.12 (a) With $k = 5$ and df $= 45$, the critical t value for the Bonferroni procedure is approximately 2.70. The calculated t is $1.575/\sqrt{(4)(5/4)(10)} = 2.227$; we cannot reject H_0.

(b) The critical t value for the Scheffé procedure is $\sqrt{(a-1)F_{.05,4,45}} = \sqrt{(4)(2.59)} = 3.219$; the result is not significant.

(c) The Bonferroni CI is $1.575 = \pm (2.70)\sqrt{(4)(5/4)10} = -.334, 3.483$; the Scheffé CI is $1.575 \pm (3.219)\sqrt{(4)(5/4)/10} = -.701, 3.851$. The Scheffé interval is wider (and the test is less powerful) because the family of contrasts is much larger than in part (a).

(d) With the Tukey procedure, the critical value of q is 4.03. The value of t we calculate will be compared with $4.03/\sqrt{2}$, or 2.850. We find $t = .9/\sqrt{(4)(2)/10} = 1.006$; the result is not significant.

(e) This is just the standard t test on 45 df; therefore, the criterion value is approximately 2.015. We find

$$t = [(1/2)(8.6 + 9.5) - (1/3)(9.2 + 8.0 + 10.2)]/\sqrt{(4)(5/6)/10} = .144,$$

which is clearly not significant.

CHAPTER 7

7.2 (a) $b_1 = -.864$.

(b) $\hat{Y}_j = \bar{Y}_{..} + b_1(X_j - \bar{X}) = 4.84 - (.864)(X_j - 2.5)$. The predicted times are 6.136, 5.272, 4.408, 3.544.

(c) By Equation 7.6, $8\Sigma(\hat{Y}_j - \bar{Y}_{..})^2 = 29.860$. The single df approach of Chapter 6 yields

$$\frac{[(-3)(6.49) + (-1)(4.82) + (1)(4.25) + (3)(3.80)]^2}{20/8} = 29.860$$

The F ratio is 21.028, signifying that the best-fitting straight line has a slope significantly different from zero.

(d) The potential advantage is that the error variance is based on more degrees of freedom than in the test of part (b). However, if the population means do not fall on a straight line, variability due to curvature [see part (e)] may contribute to the error term, negatively biasing the F test.

(e) $SS_A = 33.221$; $SS_{curv} = SS_A - SS_{lin} = 3.361$; $MS_{curv}/MS_{S/A} = 1.681/1.42 = 1.18$, which is not significant. Therefore, we conclude that the four group means do not depart significantly from the best-fitting straight line.

7.4

SV	df	SS	MS	F
T	2	8.133	4.067	2.222
P	3	35.000	11.333	6.193[a]
lin(P)	1	27.000	27.000	14.754[a]
quad(P)	1	6.667	6.667	3.643
cub(P)	1	1.133	1.133	.619
TP	6	16.400	2.733	1.493
$T \times$ lin(P)	2	10.640	5.320	2.907
$T \times$ quad(P)	2	5.733	2.867	1.567
$T \times$ cub(P)	2	.027	.013	.007
S/TP	48	87.840	1.830	

[a] $p < .01$.

(b) (i) H_0: $\beta_{11} = (1/2)(\beta_{12} + \beta_{13})$; (ii) $MS_{\hat{\psi}} = SS_{\hat{\psi}} = .960$; $F < 1$.

7.6 $SS_A = 35.160$; most of this variability is due to quadratic curvature, as evidenced by $SS_{quad(A)} = 32.593$. With respect to the analysis of the interaction variability, $SS_{AC} = 24.342$. The interaction variance is not significant [$F(8, 135) = (24.342/8)/(2.084) = 1.478$], nor are any of the polynomial components. There is some indication that the slopes at the three levels of C differ, but $SS_{C \times lin(A)} = 10.220$; then $F(2, 135) = 5.110/2.084 = 2.452$, which is less than the approximate critical value of 3.0 for $\alpha = .05$.

7.8 (a) The following is one of many set of values consistent with the four hypotheses:

Ability	Time 30	Time 45	Time 60	Time 30	Time 45	Time 60	Time 30	Time 45	Time 60
Low	50	73	82	48	71	80	31	44	53
Medium	54	75	82	52	73	80	37	48	55
High	58	80	82	56	78	80	43	52	57
		Linear			Branching			Reading	

(b) (i) Let μ_L, μ_B, and μ_R be the populations means for the three instructional methods. Then H_1: $(1/2)(\mu_L + \mu_B) - \mu_R = 0$. These weights are used to contrast the corresponding three means in the data set.

(ii) H_2 implies testing lin(time) and quad(time).

(iii) Let $p(P)$ designate the contrast defined in part (i). H_3 implies a test of $p(P) \times$ ability.

(iv) H_4 implies a test of lin(time) $\times$ ability.

7.10 If no A or B main effects are present, the following data set will meet the requirement:

$$
\begin{array}{c}
& A_1 \ A_2 \ A_3 \\
B_1 & \begin{bmatrix} 10 & 20 & 30 \\ B_2 & 30 & 20 & 10 \\ B_3 & 30 & 20 & 10 \\ B_4 & 10 & 20 & 30 \end{bmatrix}
\end{array}
$$

If the values of the A_j are 1, 2, and 3, the slopes $[\text{lin}(A)]$ at the levels of B are 10, -10, -10, and 10, a quadratic function of B.

CHAPTER 8

8.2(a, b)

SV	df	MS	F	EMS
Subjects	3	23.556		$\sigma_e^2 + 3\sigma_S^2$
A	2	16.583	14.559	$\sigma_e^2 + 4\theta_A^2$
SA	6	1.139		σ_e^2

The result is significant.

(c) $\hat{\delta}_A^2 = [(a-1)/a][\text{MS}_A - \text{MS}_{S/A})/n] = 3.861;$ $\hat{\sigma}_S^2 = (\text{MS}_S - \text{MS}_{S/A})/a = 7.472;$ $\hat{\omega}_A^2 = \hat{\delta}_A^2/(\hat{\delta}_A^2 + \hat{\sigma}_S^2 + \hat{\sigma}_e^2) = .310.$

(d) $\text{MS}_{S/A} = 7.253.$ (Note the reduction in error variance that results from removing subject variability through the use of a repeated-measures design.)

In both designs, we calculate $\hat{\phi} = \sqrt{n\hat{\delta}_A^2/\hat{\sigma}_e^2}$; the difference is in the estimate of error variance. Designate the repeated-measures and between-subjects designs by 1 and 2, respectively. Then

$$\hat{\phi}_1 = \sqrt{(4)(3.861)/1.139} = 3.68 \quad \text{and} \quad \hat{\phi}_2 = \sqrt{(4)(3.861)/7.472} = 1.44.$$

The estimated power for the repeated-measures design is essentially 1.0, whereas that for the between-subjects design is about .62.

8.4 **(a)** (i) The variance-covariance matrix is

$$
\begin{array}{c}
& A_1 \quad\ A_2 \quad\ A_2 \\
\begin{array}{c} A_1 \\ A_2 \\ A_3 \end{array} & \begin{bmatrix} 2.943 & & \\ 2.228 & 2.612 & \\ 3.646 & 3.330 & 6.360 \end{bmatrix}
\end{array}
$$

(ii) $\text{Var}(Y_{1i} - Y_{2i}) = \hat{\sigma}_1^2 + \hat{\sigma}_2^2 - 2\hat{\sigma}_{12} = 2.943 + 2.612 - (2)(2.228)$
$= 1.099$. Similarly,

$$\text{Var}(Y_{1i} - Y_{3i}) = 2.011 \text{ and } \text{Var}(Y_{2i} - Y_{3i}) = 2.312.$$

To check these results, you can find the difference score for each subject and calculate the variance of the six difference scores.

(iii)

SV	df	MS	F
A	2	10.002	11.070
SA	10	.904	

Note that $(1/2)(1/3)(1.099 + 2.011 + 2.312) = .904$.

(b) (i) $t_{10} = (5.2 - 4.367)/\sqrt{.904/6} = 2.146; .05 < p < .1$.

(ii) $t_5 = (5.2 - 4.367)/\sqrt{1.099/6} = 1.946; .1 < p < .2$. Because MS_{SA} is an average of variances for different contrasts, it is inappropriate as an error term for any particular contrast unless the variances of difference scores are homogeneous. Therefore, the t in part (ii) is the correct test statistic.

(c) The variance of the six contrast scores is

$$\hat{\sigma}_1^2 + \left[-\frac{1}{2}\right]^2 (\hat{\sigma}_2^2 + \hat{\sigma}_3^2) + 2 \left[(1)\left[-\frac{1}{2}\right](\hat{\sigma}_{12} + \hat{\sigma}_{13}) + \left[-\frac{1}{2}\right]^2 \hat{\sigma}_{23}\right] = .977$$

This can be checked by obtaining a contrast score for each subject and calculating the variance of the six contrast scores. Then

$$t = [(1/2)(5.2 + 6.9) - 4.367]/\sqrt{.977/6} = 4.172.$$

8.6 Compound symmetry is violated because the variances and covariances are heterogeneous. However, there is sphericity because $\text{Var}(Y_1 - Y_2) = \text{Var}(Y_1 - Y_3) = \text{Var}(Y_2 - Y_3) = 3.0$. As we would expect when the variances of difference scores are homogeneous, $\epsilon = 1.0$.

8.8 The variance of the eight contrast scores is

$$\left[\frac{1}{2}\right]^2 (\hat{\sigma}_1^2 + \hat{\sigma}_2^2) + \hat{\sigma}_4^2 + 2 \left[\left[\frac{1}{2}\right]^2 \hat{\sigma}_{12} + (-1)\left[\frac{1}{2}\right](\hat{\sigma}_{24} + \hat{\sigma}_{14})\right] = .684$$

Dividing by 8 and taking the square root gives the estimated standard error of the contrast: $\hat{\sigma}_{\hat{\psi}} = .242$. Then $t_7 = -.831/.242 = -3.437$.

8.10 On the basis of the clearly nonsignificant test of words, we can revise the model, obtaining MS_{res} as a weighted average of MS_W and MS_{SW}:

$$MS_{\text{res}} = (19/760)(739.141) + (741/760)(853.157) = 850.307.$$

Reliability is the estimated proportion of total variance due to subjects. Therefore, $r_{11} = \hat{\sigma}_S^2/(\hat{\sigma}_S^2 + \hat{\sigma}_e^2)$. We find $\hat{\sigma}_S^2 = (MS_S - MS_{\text{res}})/20 = 10,372.736$ and, letting $\hat{\sigma}_e^2 = MS_{\text{res}}$, $r_{11} = .924$.

8.12 (a) $\hat{X}_{12} = 147/6 = 24.5$.

(b) The estimates of the missing two scores are

	Cycles			
	1	2	3	4
Y_{12}	30.5000	24.6806	24.5189	24.5144
Y_{43}	37.9167	38.8866	38.9135	38.9143

8.14

(a)

SV	df	EMS
Subjects (S)	$n - 1$	$\sigma_e^2 + ot\sigma_S^2$
Occasions (O)	$o - 1$	$\sigma_e^2 + nt\sigma_O^2$
Tasks (T)	$t - 1$	$\sigma_e^2 + noo_T^2$
Residual	$not - (n + o + t - 2)$	σ_e^2

(b) $\hat{\sigma}_S^2 = (MS_S - MS_{res})/ot; \; \hat{\sigma}_O^2 = (MS_O - MS_{res})/nt;$
$\hat{\sigma}_T^2 = (MS_T - MS_{res})/no.$

(c)

SV	df	EMS
Subjects (S)	$n - 1$	$\sigma_e^2 + oo_{ST}^2 + ot\sigma_S^2$
Occasions (O)	$o - 1$	$\sigma_e^2 + nt\sigma_O^2$
Tasks (T)	$t - 1$	$\sigma_e^2 + oo_{ST}^2 + noo_T^2$
ST	$(n - 1)(t - 1)$	$\sigma_e^2 + oo_{ST}^2$
Residual	$(nt - 1)(o - 1)$	σ_e^2

$\hat{\sigma}_S^2 = (MS_S - MS_{ST})/ot; \; \hat{\sigma}_O^2 = (MS_O - MS_{res})/nt;$
$\hat{\sigma}_T^2 = (MS_T - MS_{ST})/no.$

8.16 (a)

SV	df	SS	MS	Error	F
S	4	34.467	8.617		
A	1	112.133	112.133	SA	30.170
B	2	9.800	4.900	SB	15.474
AB	2	4.067	2.033	SAB	.859
SA	4	14.867	3.717		
SB	8	2.533	.317		
SAB	8	18.933	2.367		

(b)

SV	df	SS	MS	F
S	4	11.489	2.872	
A	1	37.378	37.378	30.170
SA	4	4.956	1.239	

Note that the MS and SS are one-third of their original values because the "scores" in part (b) are an average of three scores in part (a). The F ratio for A is unchanged. Therefore, as long as B has fixed effects, averaging over its levels will not change the test of the effects of interest.

(c) The EMS are presented in the table. The terms in parentheses do not contribute to the variability in the data when B has fixed effects.

SV	EMS
S	$\sigma_e^2 + (a\sigma_{SB}^2) + ab\sigma_S^2$
A	$\sigma_e^2 + b\sigma_{SA}^2 + (n\sigma_{AB}^2) + (\sigma_{SAB}^2) + nb\theta_A^2$
B	$\sigma_e^2 + a\sigma_{SB}^2 + na\sigma_B^2$
SA	$\sigma_e^2 + \sigma_{SAB}^2 + b\sigma_{SA}^2$
SB	$\sigma_e^2 + a\sigma_{SB}^2$
AB	$\sigma_e^2 + \sigma_{SAB}^2 + n\sigma_{AB}^2$
SAB	$\sigma_e^2 + \sigma_{SAB}^2$

When effects of B are random, we require a quasi-F test of A. This can be done either by $F' = (MS_A + MS_{SAB})/(MS_{SA} + MS_{AB})$ or $F' = MS_A/(MS_{SA} + MS_{AB} - MS_{SAB})$. Note that averaging over the levels of B and treating the analysis as if there were only two scores for each subject is no longer correct because this process loses the variability due to AB and SAB that are required in the quasi-F test.

8.18 (a)

SV	df	SS	MS	Error	F
S	2	16.074	8.037		
A	2	288.074	144.037	SA	3.250
B	2	36.074	18.037	SB	3.746
AB	4	7.926	1.981	SAB	.148
SA	4	177.259	44.315		
SB	4	19.259	4.815		
SAB	8	107.407	13.426		

(b) The relevant EMS are

$$E(MS_A) = \sigma_e^2 + 3\sigma_{SA}^2 + 3\sigma_{AB}^2 + \sigma_{SAB}^2 + 9\theta_A^2$$

$$E(MS_{SA}) = \sigma_e^2 + \sigma_{SAB}^2 + 3\sigma_{SA}^2$$

$$E(MS_{AB}) = \sigma_e^2 + \sigma_{SAB}^2 + 3\sigma_{AB}^2$$

$$E(MS_{SAB}) = \sigma_e^2 + \sigma_{SAB}^2$$

We need a quasi-F ratio: $F_1' = (MS_A + MS_{SAB})/(MS_{SA} + MS_{AB}) = 3.401$, and the df (see Equation 8.29) are 2 (2.385) and 4 (4.357); $F_2' = MS_A/(MS_{SA} + MS_{AB} - MS_{SAB}) = 4.382$, and the df are 2 and 2 (2.100).

(c) The EMS are exactly as in part (b). The expected mean squares for B and SB are the only ones that would change, and they are not relevant to the test of A.

(d)

SV	df	MS	F
$\psi(A)$	1	76.056	.091
$S \times \psi(A)$	2	84.389	

(e)

SV	df	MS	F
$\psi(B)$	1	25.389	5.813
$S \times \psi(B)$	2	5.056	

(f)

SV	df	MS	F
$\text{lin}(B)$	1	.222	.308
$S \times \text{lin}(B)$	2	.722	

(g)

SV	df	MS	F
$A \times \text{lin}(B)$	2	.389	.049
$S \times A \times \text{lin}(B)$	4	7.889	
$A \times \text{quad}(B)$	2	3.574	.189
$S \times A \times \text{quad}(B)$	4	18.963	

8.20 (a) The only negative difference is that for S_8. Ranking the absolute differences, we find that the difference score for S_8 has a rank of 6. Therefore, the sum of the negative ranks is 6. The test is two-tailed because no directionality is implied for the alternative to the null hypothesis. For $n = 8$, the result is not significant.

(b) For each subject, multiply the scores from A_1 to A_4 by -3, -1, 1, and 3, respectively, and sum the cross products to obtain a measure of linearity. The only such contrast that is negative is that for S_7. Ranking absolute contrast values, we find that the value for S_7 is the smallest. Therefore, the sum of the negative ranks is 1. This is significant at the .05 level.

The exact two-tailed probabilities associated with parts (a) and (b) are .1194 and .0156. It is important to note that computer packages provide an approximate p value based on the standardized normal deviate (Equation 8.34). When n is small (less than 25), ignore the reported p value and enter Appendix Table D.11 with the value of T (reported by BMDP and SSPS^X).

CHAPTER 9

9.2 (a)

SV	df	EMS
A	1	$\sigma_e^2 + 3\sigma_{S/A}^2 + 9\sigma_A^2$
S/A	4	$\sigma_e^2 + 3\sigma_{S/A}^2$
B	2	$\sigma_e^2 + \sigma_{SB/A}^2 + 3\sigma_{AB}^2 + 9\theta_B^2$
AB	2	$\sigma_e^2 + \sigma_{SB/A}^2 + 3\sigma_{AB}^2$
SB/A	8	$\sigma_e^2 + \sigma_{SB/A}^2$

(b) B is now tested against AB rather than SB/A. The F ratio is $F = 42/168 = .25$, which clearly is not significant.

9.4 The correct approach is to drop the B_3 data and redo the analysis.
(a) The test of the contrast, $\psi(B)$, is $F = 150/130 = 1.153$.
(b) The test of the interaction of the contrast with A, $\psi(B) \times A$, is $F = 600/130 = 4.615$.

9.6 (a)

SV	df	SS	MS	Error term	F
A	1	154.083	154.083	S/A	17.61
S/A	6	52.500	8.750		
B	1	736.333	736.333	SB/A	63.42
AB	1	192.000	192.000	SB/A	16.54
SB/A	6	69.667	11.611		
C	2	26.542	13.271	SC/A	5.84
AC	2	19.542	9.771	SC/A	4.30
SC/A	12	27.250	2.271		
BC	2	13.042	6.521	SBC/A	5.76
ABC	2	3.375	1.688	SBC/A	1.49
SBC/A	12	13.583	1.132		

(b) The next three parts are best understood if we look at the EMS under the revised model. Rather than write out each expectation, we use letters to designate the variance components; for example, A represents $bnc\theta_A^2$.

SV	EMS
A	$A, S/A, AC, SC/A$
S/A	$S/A, SC/A$
B	$BC, SB/A, SBC/A, B$
AB	$SB/A, ABC, SBC/A, AB$
SB/A	$SBC/A, SB/A$
C	$SC/A, C$
AC	$SC/A, AC$
SC/A	SC/A
BC	$SBC/A, BC$
ABC	$SBC/A, ABC$
SBC/A	SBC/A

To test A, we construct the quasi-F test,

$$F = \text{MS}_A/(\text{MS}_{S/A} + \text{MS}_{AC} - \text{MS}_{SC/A})$$

with denominator

$$\text{df} = \frac{(\text{MS}_{S/A} + \text{MS}_{AC} - \text{MS}_{SC/A})^2}{\text{MS}_{S/A}^2/\text{df}_{S/A} + \text{MS}_{AC}^2/\text{df}_{AC} + \text{MS}_{SC/A}^2/\text{df}_{SC/A}}$$

Similarly, we have $\text{MS}_B/(\text{MS}_{BC} + \text{MS}_{SB/A} - \text{MS}_{SBC/A})$ and $\text{MS}_{AB}/(\text{MS}_{ABC} + \text{MS}_{SB/A} - \text{MS}_{SBC/A})$. The df are obtained just as in the test for A with the appropriate MS and df substituted.

(c) Suppose we were quite confident that $\sigma_{BC}^2 = 0$. Then $\text{MS}_{SB/A}$ is an appropriate error term against which to test B. Similar reasoning suggests preliminary tests of S/A against SC/A and ABC against SBC/A. If the preliminary test is not significant at the .25 level, we can assume that the tested variance component is zero throughout the EMS. Only the test of ABC meets this criterion. Therefore, AB can now be tested against SB/A, but we still require quasi-F tests for A and B.

(d) The SS and MS are exactly one-third of those obtained in part (a). Therefore, the F ratios remain the same as in part (a) provided that the tests are against S/A and SB/A. These tests are valid if C has fixed effects because, in that case, terms involving C do not influence the EMS for A, B, and AB. However, as we saw in part (c), when C has random effects, C and its interactions may contribute to the A, B, and AB sources of variance. In that event, quasi-F tests must be constructed that involve interactions of C with other variables. This is impossible when we average over the levels of C.

9.8 **(a)** The main and interaction effects of A, G, and C are all between-subject sources and are tested against S/ACG. T and its interactions with the between-subject sources are tested against ST/ACG. In order to answer the next part, we note that

$$E(\mathrm{MS}_{S/ACG}) = \sigma_e^2 + 2\sigma_{S/ACG}^2$$

and

$$E(\mathrm{MS}_{ST/ACG}) = \sigma_e^2 + \sigma_{ST/ACG}^2$$

(b) In order to obtain an estimate of ω_C^2, we require an estimate of each variance component. We lack an estimate of $\sigma_{S/ACG}^2$ unless we assume that

$$\sigma_{ST/ACG}^2 = 0.$$

CHAPTER 10

10.2

SV	df	EMS	Error term
T	1	$\sigma_e^2 + 6\sigma_{G/T}^2 + 36\theta_T^2$	G/T
G/T	10	$\sigma_e^2 + 6\sigma_{G/T}^2$	$S/GX/T$
X	1	$\sigma_e^2 + 3\sigma_{GX/T}^2 + 36\theta_X^2$	GX/T
TX	1	$\sigma_e^2 + 3\sigma_{GX/T}^2 + 18\theta_{TX}^2$	GX/T
GX/T	10	$\sigma_e^2 + 3\sigma_{GX/T}^2$	$S/GX/T$
$S/GX/T$	48	σ_e^2	

10.4 (a)

SV	df	EMS	Error term
P	1	$\sigma_e^2 + 2\sigma_{S/C/Sc/P}^2 + 20\sigma_{C/Sc/P}^2 + 40\sigma_{Sc/P}^2 + 120\theta_P^2$	Sc/P
Sc/P	4	$\sigma_e^2 + 2\sigma_{S/C/Sc/P}^2 + 20\sigma_{C/Sc/P}^2 + 40\sigma_{Sc/P}^2$	$C/Sc/P$
$C/Sc/P$	6	$\sigma_e^2 + 2\sigma_{S/C/Sc/P}^2 + 20\sigma_{C/Sc/P}^2$	$S/C/Sc/P$
$S/C/Sc/P$	108	$\sigma_e^2 + 2\sigma_{S/C/Sc/P}^2$	
T	1	$\sigma_e^2 + \sigma_{ST/C/Sc/P}^2 + 10\sigma_{CT/Sc/P}^2 + 20\sigma_{ScT/P}^2 + 120\theta_T^2$	ScT/P
PT	1	$\sigma_e^2 + \sigma_{St/C/Sc/P}^2 + 10\sigma_{CT/Sc/P}^2 + 20\sigma_{ScT/P}^2 + 60\theta_{PT}^2$	ScT/P
ScT/P	4	$\sigma_e^2 + \sigma_{ST/C/Sc/P}^2 + 10\sigma_{CT/Sc/P}^2 + 20\sigma_{ScT/P}^2$	$CT/Sc/P$
$CT/Sc/P$	6	$\sigma_e^2 + \sigma_{ST/C/Sc/P}^2 + 10\sigma_{CT/Sc/P}^2$	$ST/C/Sc/P$
$ST/C/Sc/P$	108	$\sigma_e^2 + \sigma_{ST/C/Sc/P}^2$	

(b) We have no measure of variability due to classes within schools in this design. Assuming such variability does not contribute, we have:

SV	df	EMS	Error term
P	1	$\sigma_e^2 + 2\sigma_{S/ScP}^2 + 20\sigma_{ScP}^2 + 120\theta_P^2$	ScP
Sc	5	$\sigma_e^2 + 2\sigma_{S/ScP}^2 + 40\sigma_{Sc}^2$	S/ScP
ScP	5	$\sigma_e^2 + 2\sigma_{S/ScP}^2 + 20\sigma_{ScP}^2$	S/ScP
S/ScP	108	$\sigma_e^2 + 2\sigma_{S/ScP}^2$	
T	1	$\sigma_e^2 + \sigma_{ST/ScP}^2 + 20\sigma_{ScT}^2 + 120\theta_T^2$	ScT
PT	1	$\sigma_e^2 + \sigma_{ST/ScP}^2 + 10\sigma_{ScPT}^2 + 60\theta_{PT}^2$	$ScPT$
ScT	5	$\sigma_e^2 + \sigma_{ST/ScP}^2 + 20\sigma_{ScT}^2$	ST/ScP
$ScPT$	5	$\sigma_e^2 + \sigma_{ST/ScP}^2 + 10\sigma_{ScPT}^2$	ST/ScP
ST/ScP	108	$\sigma_e^2 + \sigma_{ST/ScP}^2$	

10.6 (a)

SV	df	EMS
S	43	$\sigma_e^2 + \sigma_{SW/H}^2 + 50\sigma_S^2$
H	1	$\sigma_e^2 + \sigma_{SW/H}^2 + 44\sigma_{W/H}^2 + 25\sigma_{SH}^2 + 1100\theta_H^2$
W/H	48	$\sigma_e^2 + \sigma_{SW/H}^2 + 44\sigma_{W/H}^2$
SH	43	$\sigma_e^2 + \sigma_{SW/H}^2 + 25\sigma_{SH}^2$
SW/H	2064	$\sigma_e^2 + \sigma_{SW/H}^2$

To test the effects of H, a quasi-F ratio is required. We can either calculate $F_1' = (MS_H + MS_{SW/H})/(MS_{W/H} + MS_{SH})$ or $F_2' = MS_H/(MS_{W/H} + MS_{SH} - MS_{SW/H})$. The df follow from Equation 8.29.

(b) W/H and SH can be tested against SW/H. If either proves clearly nonsignificant ($p > .25$), then the other provides an error term for the test of H. This is unlikely in the present example because of the probable high power (many numerator and denominator df) of the preliminary tests.

10.8

SV	df	EMS
O	1	$\sigma_e^2 + \sigma_{GP/StO}^2 + 4\sigma_{OP/St}^2 + 6\sigma_{G/O}^2 + 24\theta_O^2$
G/O	6	$\sigma_e^2 + \sigma_{GP/StO}^2 + 6\sigma_{G/O}^2$
St	1	$\sigma_e^2 + \sigma_{GP/StO}^2 + 3\sigma_{GSt/O}^2 + 6\sigma_{P/St}^2 + 24\theta_{St}^2$
P/St	4	$\sigma_e^2 + \sigma_{GP/StO}^2 + 6\sigma_{P/St}^2$
StO	1	$\sigma_e^2 + \sigma_{GP/StO}^2 + 3\sigma_{GSt/O}^2 + 4\sigma_{OP/St}^2 + 12\theta_{OSt}^2$
OP/St	4	$\sigma_e^2 + \sigma_{GP/StO}^2 + 4\sigma_{OP/St}^2$
GSt/O	6	$\sigma_e^2 + \sigma_{GP/StO}^2 + 3\sigma_{GSt/O}^2$
GP/StO	24	$\sigma_e^2 + \sigma_{GP/StO}^2$

Quasi-F ratios must be formed. To test the O effects,

$$F_2' = \frac{MS_O}{MS_{OP/St} + MS_{G/O} - MS_{GP/StO}}$$

and the denominator df follow Equation 8.29. Similar F tests of St and StO follow from the EMS.

10.10 (a)

SV	df	EMS	Error term
A	2	$\sigma_e^2 + 10\sigma_{AE/V}^2 + 150\theta_A^2$	AE/V
V	2	$\sigma_e^2 + 30\sigma_{E/V}^2 + 150\theta_V^2$	E/V
AV	4	$\sigma_e^2 + 10\sigma_{AE/V}^2 + 150\theta_{AV}^2$	AE/V
E/V	12	$\sigma_e^2 + 30\sigma_{EV}^2$	$S/AE/V$
AE/V	24	$\sigma_e^2 + 10\sigma_{AE/V}^2$	$S/AE/V$
$S/AE/V$	405	σ_e^2	

(b) Preliminary tests of E/V and AE/V, if not significant at the .25 level, would permit revision of the model and the corresponding EMS. If both variance components could be assumed to be equal to zero, they could be pooled with $S/AE/V$.

10.12 (a)

SV	df	EMS	Error term
A	1	$\sigma_e^2 + 3\sigma_{G/AP/E}^2 + 15\sigma_{AP/E}^2 + 120\theta_A^2$	AP/E
E	1	$\sigma_e^2 + 3\sigma_{G/AP/E}^2 + 30\sigma_{P/E}^2 + 120\theta_E^2$	P/E
AE	1	$\sigma_e^2 + 3\sigma_{G/AP/E}^2 + 15\sigma_{AP/E}^2 + 60\theta_{AE}^2$	AP/E
P/E	6	$\sigma_e^2 + 3\sigma_{G/AP/E}^2 + 30\sigma_{P/E}^2$	$G/AP/E$
AP/E	6	$\sigma_e^2 + 3\sigma_{G/AP/E}^2 + 15\sigma_{AP/E}^2$	$G/AP/E$
$G/AP/E$	64	$\sigma_e^2 + 3\sigma_{G/AP/E}^2$	$S/G/AP/E$
$S/G/AP/E$	160	σ_e^2	

(b) This analysis is likely to involve less error variance and provide more powerful, and certainly simpler, tests of A, E, and AE. However, monkeys are expensive to purchase and maintain, and the design of Exercise 10.11 involves far fewer subjects.

CHAPTER 11

11.2 **(a)** From Equation 11.2, the estimate of $MS_{S/A}$ is 5.5.
(b) Substituting the estimate from (a) into Equation 11.3, the relative efficiency of the Latin square to the repeated-measures design is 1.062.
(c) If the effect of C is increased, the Latin square's relative efficiency increases. This is because such variability would contribute to σ_e^2 in the repeated-measures design, whereas in the Latin square the effects are systematic and can be removed.

11.4

SV	df	SS	MS	F
S	3	.685	.228	34.350
P	3	.245	.087	12.250
E	1	1.210	1.210	181.500
I	1	1.000	1.000	150.000
EI	1	.010	.010	1.500
Res	6	.040	.007	

11.6 Following Equation 11.9 (also, see Table 11.7), we have the following ANOVA:

SV	df	Error term
Squares (Q)	8	S/Q
S/Q	27	
Days (D)	3	DQ
Payoff (P)	3	PQ
DQ	24	Res
PQ	24	Res
Residual	54	

11.8

SV	df	Error term
$R(PD')$	3	S/R
S/R	32	
D	3	W cells res
P	3	W cells res
B cells res	6	W cells res
W cells res	96	

11.10 (a)

SV	df
Drug type (T)	1
R	3
TR	3
S/TR	24
Occasions (O)	3
Dosages (D)	3
TO	3
TD	3
B cells res	6
B cells res $\times T$	6
W cells res	72

T, R, and TR are tested against S/TR, and all other terms are tested against W cells res.

(b)

SV	df
R	7
S/R	8
O	7
T	1
D	3
TD	3
B cells res	42
W cells res	56

R is tested against S/R, and all other terms are tested against W cells res.

(c) Design (b) requires fewer subjects and involves a simpler analysis. On the other hand, it may be impractical to run each subject on eight different occasions [as opposed to four in design (a)]. Of course, there is also some question as to whether either drug types or dosages should be within-subjects variables. The answer to that depends on how quickly the effect of a drug wears off and what recovery period is allowed between sessions.

CHAPTER 12

12.2 **(b)** $r = .620$;

  (c) $\hat{Y}_i = 3.00 + 2.00X_i$;

  (d) the proportion of variability accounted for is $r^2 = .385$; the amount of variability accounted for is $r^2 SS_Y = 70.00$;

  (e) $\hat{\sigma}_e = 5.29$;

  (f) $\hat{X}_i = 1.58 + 0.192Y_i$.

12.4 **(a)** The proportion of variance of W accounted for by X, $r_{XW}^2 = .64$, is greater than the proportion of variance of Y accounted for by X, $r_{XY}^2 = .49$. On the other hand, the standard error of estimate for the regression of Y on X is less than the standard error of estimate for the regression of W on X.

  (b) Yes. $r_{XQ} = b_{XQ}\hat{\sigma}_Q/\hat{\sigma}_X = 2\hat{\sigma}_Q/\sqrt{20} = .477\hat{\sigma}_Q$ and $r_{YQ} = b_{YQ}\hat{\sigma}_Q/\hat{\sigma}_Y = \hat{\sigma}_Q/\sqrt{10} = .316\hat{\sigma}_Q$.

  (c) Yes. Because neither $\hat{\sigma}_Y$ nor $\hat{\sigma}_X$ can be negative, and $b_{YX} = r_{XY}\hat{\sigma}_Y/\hat{\sigma}_X$, r_{XY} and b_{YX} must have the same sign.

12.6 **(a)** The correlation and slope are both zero, indicating that there is linear independence (i.e., no linear relation) between X and Y.

  (b) On the other hand, stochastic independence does not occur; for example, $p(Y = 1) = 4/10 = .4$ but $p(Y = 1 | X = 4) = 0$.

12.8 **(a)** The standard error of prediction is given by Equation 12.23 and increases as a function of the distance between the value of X for the new case and the mean of the X's.

(b) Given no information about the students other than that they have signed up for the course, the best prediction for each of their final exam scores is 78.29, the mean of the Y scores in Table 12.1. The standard error of the mean is $\hat{\sigma}_Y/\sqrt{N} = 3.352$ and therefore, the 90% confidence interval for the mean final exam score for new students is $78.39 \pm (1.740)(3.532)$, or 78.39 ± 6.15.

(c) (i) The regression equation of Y on X is $\hat{Y}_i = -36.08 + 3.546X_i$. Substituting, the predicted Y scores for $X = 33$ and $X = 26$ are 81 and 56, respectively (rounding to the nearest integer).

(ii) The $\hat{SE}$'s for the conditional means are given by Equation 12.23 and substituting yields $\hat{SE} = 2.59$ and 5.85, for $X = 33$ and $X = 26$, respectively. The estimates of the conditional means obtained from the regression equation are 80.94 and 56.12 for $X = 33$ and $X = 26$, respectively, and the two 90% confidence intervals are $80.94 \pm (1.746)(2.59) = 80.94 \pm 4.52$ and 56.12 ± 10.23.

(iii) If we are interested in confidence intervals around the actual scores, we use the $\hat{SE}$'s given by Equation 12.25. The $\hat{SE}$'s are 10.95 and 12.14 for the students with $X = 33$ and $X = 26$, respectively, and the confidence intervals are 80.94 ± 19.12 and 56.12 ± 21.20.

12.10 (a) No, they are testing different hypotheses. Anne is testing whether there are *any* effects of array size; her null hypothesis is $\mu_1 = \mu_2 = \mu_3 = \mu_4$. On the other hand, Reg is testing the hypothesis $H_0: \beta_1 = 0$; he is concerned with whether response time varies *linearly* with array size. If Reg's H_0 is false, then so is Anne's, but the reverse is not necessarily true.

(b) An ANOVA can be performed using $F = MS_A/MS_{S/A}$, where

$$SS_A = n\sum_j (\bar{Y}_{.j} - \bar{Y}_{..})^2 = 10[(-40)^2 + 0^2 + 20^2 + 20^2] = 24,000$$

and

$$MS_A = 24,000/3 = 8000.$$

Also,

$$MS_{S/A} = \frac{\sum_j \hat{\sigma}_j^2}{a} = \frac{360 + 315 + 324 + 333}{4} = 333$$

so $SS_{S/A} = (333)(36) = 11,988$. $F = 8000/333 = 24.02$ with 3 and 36 df. This is highly significant.

To test $H_0: \beta_1 = 0$, use $t = b_1/\hat{SE}(b_1)$, where $SE(b_1) = \hat{\sigma}_e/\sqrt{SS_X}$. $b_1 = SP_{XY}/SS_X$, where

$$SP_{XY} = \sum_i \sum_j (X_j - \bar{X}_.)(\bar{Y}_{.j} - \bar{Y}_{..}) = (10)(200) = 2000$$

and

$$SS_X = \sum_i \sum_j (X_j - \bar{X})^2 = (10)(20) = 200$$

Therefore, $b_1 = 2000/200 = 10$. To find $\hat{\sigma}_e$, we must first find SS_{error}.

$$SS_{error} = SS_Y - SS_{reg}$$

$$SS_{reg} = b_1^2 SS_X = (100)(200) = 20,000$$

$$SS_Y = SS_{tot} = SS_A + SS_{S/A} = 24,000 + 11,988 = 35,988$$

Therefore,

$$SS_{error} = 35,988 - 20,000 = 15,988$$

and

$$\hat{\sigma}_e = \sqrt{SS_{error}/(N-2)} = \sqrt{15,988/38} = 20.51.$$

$\hat{SE}(b_1) = 20.51/\sqrt{200} = 1.45$. So the hypothesis is tested by

$$t = b_1/\hat{SE}(b_1) = 10/1.45 = 6.90,$$

which is also highly significant.

(c) SS_A must always be at least as large as SS_{reg} and will be larger unless the means for each of the conditions fall exactly on the regression line (see Chapter 16 for a more complete explanation).

(d) $SS_{error} = SS_{pure\ error} + SS_{nonlinearlity}$. From part (b), $SS_{pure\ error} = SS_{S/A} = 11,988$ and $SS_{error} = 15,988$. Therefore, $SS_{nonlinearity} = 4000$. The hypothesis that there is no nonlinearity is tested by

$$F = MS_{nonlinearity}/MS_{pure\ error} = (4000/2)/333 = 6.01$$

with 2 and 36 df. The hypothesis can be rejected at $p < .01$.

12.12 Here, equality of slopes can be tested using an independent-groups t test with individual slopes as scores. For males, mean $= 30.00$ and $\hat{\sigma} = 8.569$. For females, mean $= 20.00$ and $\hat{\sigma} = 5.345$.

$$t = \frac{30 - 20}{\sqrt{\left[\dfrac{7(8.569)^2 + 7(5.345)^2}{14}\right]\left[\dfrac{1}{8} + \dfrac{1}{8}\right]}} = 2.80 \text{ with } 14 \text{ df}$$

The hypothesis of equality of slopes can be rejected at $p < .025$.

CHAPTER 13

13.2

SV	df	SS	MS	F	p
(a) A	2	172.111	86.056	1.936	.179
S/A	15	666.833	44.456		
(b) A	2	0.778	0.389	.032	.969
S/A	15	184.167	12.278		
(c) Het	2	2.554	1.277	.354	.709
Error	12	43.339	3.612		
(d) A(adj)	2	182.822	91.411	27.886	.000
S/A(adj)	14	45.892	3.278		

(e) Because here subjects are assigned randomly to groups, the hypotheses tested by performing an ANOVA on the Y scores and by performing an ANCOVA on the Y scores using X as the covariate are the same.

13.4

SV	df	SS	MS	F	p
A	2	182.755	91.377	29.821	.000
S/A	15	45.960	3.064		

The ANOVA on the residuals is *not* equivalent to an ANCOVA, although the difference in the significance levels is small for this data set. ANCOVA involves testing the full model

$$Y_{ij} = \mu + \alpha_j + \beta(X_{ij} - \bar{X}_{..}) + \epsilon_{ij}$$

against the restricted model

$$Y_{ij} = \mu + \beta(X_{ij} - \bar{X}_{..}) + \epsilon_{ij}$$

where the β in the restricted model is the slope of the overall regression estimated by b_{tot}, and the β in the full model is the common slope that is estimated by $b_{S/A}$.

Using the residuals from the overall regression in an ANOVA is equivalent to testing

$$Y_{ij} = \mu + \alpha_j + \beta(X_{ij} - \bar{X}_{..}) + \epsilon_{ij}$$

against

$$Y_{ij} = \mu + \beta(X_{ij} - \bar{X}_{..}) + \epsilon_{ij}$$

where in both models β is the slope of the overall regression.

13.6   We first test the assumption of homogeneity of slope. Using a statistical package such as SYSTAT, the results are

SV	df	SS	MS	F	p
X	1	0.618	0.618	0.037	.850
A	1	34.341	34.341	2.059	.171
B	1	0.111	0.111	0.007	.936
A × B	1	34.887	34.887	2.092	.187
A × X	1	56.287	56.287	3.375	.085
B × X	1	3.991	3.991	0.239	.631
A × B × X	1	13.624	13.624	0.817	.380
Error	16	266.883	16.680		

Because there are no significant interactions with the covariate X, we can go ahead with the ANCOVA. The results are

SV	df	SS	MS	F	p
X	1	120.762	120.762	4.925	.039
A	1	434.748	434.748	17.730	.000
B	1	129.889	129.889	5.297	.033
A × B	1	145.560	145.560	5.936	.025
S/AB(adj)	19	465.895	24.521		

13.8   **(a)** No, it is not appropriate to use ANCOVA here. We have a nonequivalent-groups design (the workers for which we have job satisfaction scores have not been randomly assigned to the four departments), and an ANOVA using X as the dependent variable yields a highly significant effect of department, $F(3, 28) = 5.602$, $p = .004$.

   (b) No, it is not appropriate to use ANCOVA here. The data violate the assumption of homogeneity of regression slope. The test for heterogeneity of slope yields a highly significant interaction between A and X, $F(2, 24) = 7.137$, $p = .004$. Students who did well on the pretest also tend to do well on the achievement test in programs 1 and 3; however, in program 2, students who did poorly on the pretest perform about as well on the achievement test as students who did well on the pretest.

CHAPTER 14

14.2   **(a)** The reasoning of the committee member is that because there is a high correlation between pre- and posttest scores, no change in IQ has occurred. This reasoning is silly; r is sensitive to the relative standing on the two tests but not the absolute scores. For example, if *all* students had their scores increased by about 20 points, the correlation would be very high.

(b) The patients in the hospital probably constitute a restricted sample that should not be used to make inferences about the general population.

(c) The discrepancy scores must correlate with one another because the sociologist's scores are part of both measures. If $r_{XW} = r_{XY} = r_{YW} = 0$, it can be shown that the correlation between $z_X - z_Y$ and $z_W - z_Y$ must be .5.

(d) This is a classic case of fallaciously inferring causation from correlation. People who graduate from college may indeed make more money, but it is not obvious how much of their financial success can be attributed directly to graduating from college. Graduates may also be smarter and more motivated and organized than nongraduates and therefore more successful; they will also tend to come from wealthier and more stable families, who may be able to help them.

14.4 (a) We can show that $\text{Corr}(z_X, z_X - z_Y) = \sqrt{(1 - r_{XY})/2} = .8$. Solving, we have $r_{XY} = -.28$, so r_{XY}^2, the proportion of variance in Y accounted for by X, is .078. Therefore, X accounts for about 8% of the variance in Y.

14.6 If $r_{\text{pre,post}} = .70$ then $r_{\text{pre,change}} = -.39$. Notice that correlations involving change scores are not free of the influence of the pretest scores.

14.7 The correlation between the two halves is .28.

14.8 The estimate can be obtained from Equation 14.4, which yields $\hat{r} = .61$.

14.10 (a) $\text{Corr}(Y, \hat{Y}) = r_{XY} = .6$; this follows directly from the fact that $\hat{Y}$ is a linear function of X.

(b) Substituting into $\text{Corr}(Y, Y - \hat{Y}) = \Sigma z_Y z_{Y - \hat{Y}}/(N - 1)$ and simplifying, we obtain $\text{Corr}(Y, Y - \hat{Y}) = \sqrt{1 - r_{XY}^2}$. Therefore the correlation between Y and $Y - \hat{Y}$ is .8.

(c) Substituting into $\text{Corr}(\hat{Y}, Y - \hat{Y}) = \Sigma z_{\hat{Y}} z_{Y - \hat{Y}}$, we find that the correlation between $\hat{Y}$ and $Y - \hat{Y}$ is 0.

14.12 (a) Use the test statistic $\chi^2 = (N - 3)\Sigma\Sigma_{i<j}Z_{ij}^2$ with 6 df.

$$\chi^2 = 17(.203^2 + .310^2 + .424^2 + .549^2 + .693^2 + .867^2) = 31.45$$

$\chi_{\text{crit},6,.05}^2 = 12.59$, so we can reject H_0: $\mathbf{R} = \mathbf{I}$ at $\alpha = .05$.

(b) $r_{23|4} = .140$; H_0: $\rho_{23|4} = 0$ is tested using $t = .14\sqrt{17(1 - .14^2)} = 0.57$ with $N - 3 = 17$ df. We can't reject H_0.

(c) $r_{23|41} = (r_{23|4} - r_{21|4}r_{31|4})\sqrt{(1 - r_{21|4}^2)(1 - r_{31|4}^2)}$; $r_{23|4} = .140$, $r_{21|4} = -.054$, and $r_{31|4} = .031$; therefore, $r_{23|41} = .142$. To test H_0: $\rho_{23|41} = .20$, use $z = (Z_r - Z_{\text{hyp}})\sqrt{N - 5}$; $z = (.143 - .203)\sqrt{15} = -.232$, so we can't reject H_0.

(d) To test H_0: $\rho_{12} = \rho_{14}$, we use the test statistic given in Equation 14.12. Substituting, we have $t = -1.005$, with 17 df, so we can't reject H_0.

14.14 In this case, $r_{XY|W}$ must be greater than r_{XY}.

14.16 (a) For both the young and older people taken separately, there is no tendency for high verbal ability to be matched with large size. However, the older people tend to have high V and big S, and the young people tend to have low V and small S. Therefore, if the groups are combined, there will be a tendency for high V to be paired with big S and low V to be paired with small S. We can make an analogy between the situation described here and the discussion in Section 14.2.3 about correlations of combined groups.

(b) Using Equation 14.14, the partial correlation is

$$\phi = [.64 - (.8)(.8)]/\sqrt{(1 - .8^2)(1 - .8^2)} = 0.$$

(c)

	Department A Male-dominated High acceptance rate		Department B Female-dominated Low acceptance rate		Overall	
	Accept	Reject	Accept	Reject	Accept	Reject
Female	8	2	45	45	53	47
Male	60	30	3	7	63	37

Here, the proportion of acceptances is higher for females than for males in both departments, but the overall proportion is higher for males. This occurs because the acceptance rate is higher for the department with the large number of male applicants. If we code females and accepts as 1's and males and rejects as 0's, the phi coefficients are positive for both departments, but the overall phi is negative. This can be seen as another example of the changes that can occur in correlations when groups are combined.

14.18 The recommendations are the same for 82% of the patients. However, the phi coefficient for the 2 × 2 table is 0—the recommendations of the two clinicians are independent of one another. The high percentage of agreements occurs because of the high base rates for the "don't hospitalize" judgments.

CHAPTER 15

15.2 (a)

	POP	TAX	BELT	DEATHS	DHTRATE	PRLIC	PCFUEL
Mean	4670.42	13.60	.58	954.73	22.03	.72	.61
$\hat{\sigma}$	4744.30	2.93	.50	957.62	6.59	.07	.10

	POP	TAX	BELT	DEATHS	DHTRATE	PRLIC
TAX	−.364					
BELT	.345	.051				
DEATHS	.928	−.374	.309			
DHTRATE	−.242	−.196	−.105	.024		
PRLIC	−.208	−.032	.024	−.023	.438	
PCFUEL	−.349	−.187	−.082	−.122	.711	.392

(b) The three regression equations are

(i) $\widehat{\text{DTHRATE}} = -9.213 + 43.696\text{PRLIC}$

(ii) $\widehat{\text{DTHRATE}} = -6.323 + 46.438\text{PCFUEL}$

(iii) $\widehat{\text{DTHRATE}} = -16.818 + 18.802\text{PRLIC} + 41.608\text{PCFUEL}$

The coefficients of PRLIC and PCFUEL in the third equation are different from these in the first two. For example, when we regress DTHRATE only on PRLIC, part of the rate of change of $\widehat{\text{DTHRATE}}$ with PRLIC (and hence the regression coefficient) comes from the fact that, as PRLIC changes from state to state, both DTHRATE and PCFUEL tend to change in the same way. Therefore, the coefficient of PRLIC reflects changes in DTHRATE that are associated with changes in *both* PRLIC and PCFUEL. When DTHRATE is regressed on both PRLIC and PCFUEL, the coefficient of PRLIC represents the change of $\widehat{\text{DTHRATE}}$ with PRLIC, adjusting both variables for the effect of PCFUEL. That is, when we add PRLIC to a regression equation that already contains PCFUEL as a predictor, we, in effect, try to account for the part of DTHRATE that is not accounted for by PCFUEL by the part of PRLIC that is not accounted for by PCFUEL. That is why we get the same coefficient when we regress DTHRATE on both PRLIC and PCFUEL as if we were to regress RDF on RLF in part (e).

(d) When the residuals for the three regressions were saved, the MGLH module of SYSTAT identified several cases as possibly having undue influence because they had large leverages. Case 48 (Wyoming) had large leverages in both regressions ii and iii (.288 and .405, respectively) because of its large value of PCFUEL. Cases 2 (Arizona) and 26 (Nevada) had high leverages in regression i (.155 and .171, respectively) because of their large values of PRLIC. However, these cases do not have Cook's distances that exceed 1.

(e) See part (a).

(f) Not much can be said about the effectiveness of seat belt laws from these data, even though the states that had seat belt laws also tended to have lower rates of traffic fatalities. When DTHRATE is regressed on BELT, the coefficient of BELT is -1.396. However, the states with belt laws tend to be the larger states that also have lower values of PCFUEL and PRLIC. When DTHRATE is regressed on BELT, PCFUEL, and PRLIC, the coefficient of BELT is reduced to -0.770. In addition, factors such as quality of roads and enforcement of speed limits and other regulations may vary from state to state, making it very difficult to make any causal statements.

15.4 **(a)** For the data set, a standard ANOVA yields

SV	df	SS	MS	F	p
Dosage (D)	3	132.31	44.10	7.29	.003
Error (S/D)	15	90.74	6.05		

(b) A trend analysis yields

SV	df	SS	MS	F	p
Dosage (D)	3	132.31	44.10	7.29	.003
lin(D)	1	62.11	62.11	10.27	< .01
quad(D)	1	65.38	65.38	10.81	< .01
cubic(D)	1	4.82	4.82	< 1	ns
Error	15	90.74	6.05		

(c) The best-fitting quadratic equation is $\hat{Y} = -4.71 + 1.12X - 0.02X^2$.

15.6 (a) The best-fitting equation that contains both X_1 and X_2 and the corresponding overall ANOVA table are

$$\hat{Y} = -16.294 + 9.196X_1 + 9.941X_2$$

and

SV	df	SS	MS	F	p
Regression	2	2044.196	1022.098	13.883	.000
Error	14	1030.745	73.625		

(b) $SS_{\text{pure error}} = 264.50 + 0 + 92.667 + 226.75 + 293.00 + 60.50 + 0 = 937.417$ with $df_{\text{pure error}} = 1 + 0 + 2 + 3 + 3 + 1 + 0 = 10$. From (a), $SS_{\text{error}} = 1030.745$ with 14 df, so $SS_{\text{nonlinearity}} = 93.328$ with 4 df. $F = MS_{\text{nonlinearity}}/MS_{\text{pure error}} = .249$. There is no significant departure from linearity.

(c) The coefficients of X_1 and X_2 both differ significantly from 0, $t(14) = 2.42$, $p = .03$ and $t(14) = 3.38$, $p < .005$, respectively. Therefore, both variables should be included.

15.8 (a) Regressing final exam score (Y) on pretest score (X_1) and SAT (X_2), we obtain the equation $\hat{Y} = -57.423 + 1.719X_1 + 0.120X_2$. Using the equation, the best prediction for a student with $X_1 = 33$ and $X_2 = 700$ is 83.295.

(b) The appropriate estimated standard error is given by Equation 15.13, $\hat{SE}(\hat{Y}) = \hat{\sigma}_e\sqrt{1 + x'(X'X)^{-1}x}$, where $x' = [1 \quad 33 \quad 700]$. We have $\hat{\sigma}_e = 12.282$ and $(X'X)^{-1}$ from the regression output. Substituting, we obtain $\hat{SE}(\hat{Y}) = 12.282\sqrt{1 + .1564} = 13.2076$. The 95% confidence interval is therefore $83.295 \pm (2.074)(13.2076) = 83.30 \pm 27.39$.

(c) To find the confidence interval for the conditional mean of Y at $X_1 = 33$ and $X_2 = 700$, we use the $\hat{SE}$ given by Equation 15.12 with $x' = [1 \quad 33 \quad 700]$. $\hat{SE}(\hat{Y}) = 12.282\sqrt{.1564} = 4.8572$, so that the 95% confidence interval is given by 83.30 ± 10.08.

15.10 (a) If we use three predictors and have only four cases, the resulting equation must predict Y perfectly for the four cases, so $R_{Y.123} = 1$.

(b) and **(c)** The predicted scores are obtained using the regression equation from part (a), $\hat{Y} = 0.770 + 1.182X_1 - 1.731X_2$. The predicted scores have a correlation of 1.000 with the first 4 cases, but a correlation of 0.244 with the remaining 11 cases. As indicated in Section 15.4.4, any four data points must be fit perfectly by a regression equation with three predictors; even if the regression equation obtained fits the population poorly. R capitalizes on chance and is not a good measure of how well the regression fits the population if the N/p ratio is small.

15.12 (a) Both externalization and criticism contribute significantly to the predictability of attachment when the other is held constant.

(b) Adding the new variable to the regression equation is an appropriate way of determining whether the joint effect of externalization and criticism contributes to the predictability of attachment over and above the contributions made by the two measures. Because of the necessary correlation between the externalization (X_1) and criticism (X_2) measures and the variable formed by multiplying them ($X_1 * X_2$), it is important that the joint effect be evaluated by a hierarchical regression—that is, by determining whether $X_1 * X_2$ makes a significant contribution to the predictability of Y when added to an equation that already contains X_1 and X_2. Also, because of this correlation, it is likely that in the regression equation that contains X_1, X_2 and $X_1 * X_2$, the coefficients of X_1 and X_2 will not differ significantly from zero. However, they are of no interest. On the basis of the two regressions, we conclude that externalization and criticism both contribute significantly to the predictability of attachment, but that their joint effect does not add significantly to this predictability.

CHAPTER 16

16.2 (a) The t test is significant, $t(18) = -2.216$, $p < .05$, where

$$t = \frac{-15.80}{\sqrt{[(17.36 + 14.38)/2][2/10]}} = -2.216$$

(b) and **(c)** One dummy variable is needed. The dummy variables for effect (X_E) and dummy (X_D) coding are as follows:

Y	X_E	X_D	Y	X_E	X_D
53	1	1	79	−1	0
51	1	1	50	−1	0
88	1	1	79	−1	0
63	1	1	99	−1	0
56	1	1	80	−1	0
42	1	1	92	−1	0
55	1	1	69	−1	0
78	1	1	94	−1	0
48	1	1	70	−1	0
92	1	1	72	−1	0

(d) When Y is regressed on X_E, $b_1 = -7.900$ and $SE(b_1) = 3.565$; for the regression on X_D, $b_1 = -15.800$ and $SE(b_1) = 7.130$. In both cases, the t equals -2.216, the same value obtained in part (a).

(e) For effect coding, $b_0 = (\bar{Y}_{.1} + \bar{Y}_{.2})/2$ and $b_1 = (\bar{Y}_{.1} - \bar{Y}_{.2})/2$; for dummy coding, $b_0 = \bar{Y}_{.2}$ and $b_1 = \bar{Y}_{.1} - \bar{Y}_{.2}$.

16.4 **(a)** Three dummy variables are needed.

(b), (c), and **(d)**

	Y	Dummy (1, 0) coding			Effect coding			Contrast weights		
		X_{D1}	X_{D2}	X_{D3}	X_{E1}	X_{E2}	X_{E3}	X_{C1}	X_{C2}	X_{C3}
A_1	53	1	0	0	1	0	0	.5	1	0
	51	1	0	0	1	0	0	.5	1	0
	88	1	0	0	1	0	0	.5	1	0
	63	1	0	0	1	0	0	.5	1	0
	56	1	0	0	1	0	0	.5	1	0
A_2	42	0	1	0	0	1	0	.5	-1	0
	55	0	1	0	0	1	0	.5	-1	0
	78	0	1	0	0	1	0	.5	-1	0
A_3	48	0	0	1	0	0	1	-.5	0	1
	42	0	0	1	0	0	1	-.5	0	1
	79	0	0	1	0	0	1	-.5	0	1
	50	0	0	1	0	0	1	-.5	0	1
A_4	79	0	0	0	-1	-1	-1	-.5	0	-1
	99	0	0	0	-1	-1	-1	-.5	0	-1
	80	0	0	0	-1	-1	-1	-.5	0	-1
	92	0	0	0	-1	-1	-1	-.5	0	-1
	69	0	0	0	-1	-1	-1	-.5	0	-1
	94	0	0	0	-1	-1	-1	-.5	0	-1
	70	0	0	0	-1	-1	-1	-.5	0	-1

(e)

Type of coding	R^2	SS_{reg}	$F = MS_{reg}/MS_{error}$
Dummy	.462	2785.51	4.291
Effect	.462	2785.51	4.291
Contrast weights	.462	2785.51	4.291

(f)

Type of coding	Quantity	Interpretation	Value	SE	t
	Intercept	$\bar{Y}_{.4}$	83.286	5.560	14.980
Dummy	b_{D1}	$\bar{Y}_{.1} - \bar{Y}_{.4}$	−21.086	8.613	−2.448
	b_{D2}	$\bar{Y}_{.2} - \bar{Y}_{.4}$	−24.952	10.151	−2.458
	b_{D3}	$\bar{Y}_{.3} - \bar{Y}_{.4}$	−28.536	9.220	−3.095
	Intercept	$\bar{Y}_U = \dfrac{\bar{Y}_{.1} + \bar{Y}_{.2} + \bar{Y}_{.3} + \bar{Y}_{.4}}{4}$	64.642	3.539	18.265
Effect	b_{E1}	$\bar{Y}_{.1} - \bar{Y}_U = \dfrac{3\bar{Y}_{.1} - \bar{Y}_{.2} - \bar{Y}_{.3} - \bar{Y}_{.4}}{4}$	−2.442	5.845	−0.418
	b_{E2}	$\bar{Y}_{.2} - \bar{Y}_U$	−6.309	6.971	−0.905
	b_{E3}	$\bar{Y}_{.3} - \bar{Y}_U$	−9.892	6.291	−1.573
	Intercept	$\bar{Y}_U$	64.642	3.539	18.265
Contrast	b_{C1}	$\dfrac{\bar{Y}_{.1} + \bar{Y}_{.2}}{2} - \dfrac{\bar{Y}_{.3} + \bar{Y}_{.4}}{2}$	−8.751	7.078	−1.236
Weights	b_{C2}	$\dfrac{\bar{Y}_{.1} - \bar{Y}_{.2}}{2}$	1.933	5.371	0.360
	b_{C3}	$\dfrac{\bar{Y}_{.3} - \bar{Y}_{.4}}{2}$	−14.268	4.610	−3.095

16.6 **(a)** The coding of the A, C, and AC effects is as follows:

		A	C	AC
	Y	X_1	X_2	X_3
$A_1 C_1$	53	1	1	1
	51	1	1	1
	88	1	1	1
	63	1	1	1
	56	1	1	1
$A_1 C_2$	42	1	−1	−1
	55	1	−1	−1
	78	1	−1	−1
$A_2 C_1$	48	−1	1	−1
	42	−1	1	−1
	79	−1	1	−1
	50	−1	1	−1

$A_2 C_2$	79	-1	-1	1
	99	-1	-1	1
	80	-1	-1	1
	92	-1	-1	1
	69	-1	-1	1
	94	-1	-1	1
	70	-1	-1	1

(b) Yes, because of the unequal cell frenquencies.

(c–g) Some statistics from regressions of Y on different combinations of X_1, X_2, and X_3 are as follows:

SV	df	SS	MS	F
Between cells	3	$R^2_{Y.1,2,3} SS_Y = 2785.51$	928.50	4.291
X_1	1	$r^2_{Y1} SS_Y = 684.75$	684.75	3.165
X_2	1	$r^2_{Y2} SS_Y = 1354.67$	1354.67	6.261
X_3	1	$r^2_{Y3} SS_Y = 1466.73$	1466.73	6.779
$X_1 \mid X_2, X_3$	1	$(R^2_{Y.1,2,3} - R^2_{Y.2,3}) SS_Y = 330.75$	330.75	1.529
$X_2 \mid X_1, X_3$	1	$(R^2_{Y.1,2,3} - R^2_{Y.1,3}) SS_Y = 657.06$	657.06	3.037
$X_3 \mid X_1, X_2$	1	$(R^2_{Y.1,2,3} - R^2_{Y.1,2}) SS_Y = 1133.58$	1133.58	5.239
Residual	15	$(1 - R^2_{Y.1,2,3}) SS_Y = 3245.65$	216.38	

Interpretation of regression coefficients for simultaneous and individual regressions of Y on X_1, X_2, and X_3:

Type of regression	Coefficient	Interpretation
Simultaneous on X_1, X_2, and X_3	b_1	$\dfrac{\bar{Y}_{.11} + \bar{Y}_{.12}}{4} - \dfrac{\bar{Y}_{.21} + \bar{Y}_{.22}}{4}$
	b_2	$\dfrac{\bar{Y}_{.11} + \bar{Y}_{.21}}{4} - \dfrac{\bar{Y}_{.12} + \bar{Y}_{.22}}{4}$
	b_3	$\dfrac{\bar{Y}_{.11} + \bar{Y}_{.22}}{4} - \dfrac{\bar{Y}_{.12} + \bar{Y}_{.21}}{4}$
Individual on X_1	b_1	$\dfrac{1}{2}\left[\dfrac{n_{11}\bar{Y}_{.11} + n_{12}\bar{Y}_{.12}}{n_{11} + n_{12}} - \dfrac{n_{21}\bar{Y}_{.21} + n_{22}\bar{Y}_{.22}}{n_{21} + n_{22}} \right]$
Individual on X_2	b_2	$\dfrac{1}{2}\left[\dfrac{n_{11}\bar{Y}_{.11} + n_{21}\bar{Y}_{.21}}{n_{11} + n_{21}} - \dfrac{n_{12}\bar{Y}_{.12} + n_{22}\bar{Y}_{.22}}{n_{12} + n_{22}} \right]$

When a simultaneous regression on X_1, X_2, and X_3 is performed, the tests of b_1 and b_2 correspond to tests of the hypotheses that (i) the unweighted average of the population means at A_1 equals the unweighted average of the population means at A_2, and (ii) the unweighted average at C_1 equals the unweighted average at C_2, respectively. The test of b_3 corresponds to the test of the AC interaction. When Y is regressed only on X_1, the test of b_1 corresponds to a test of whether the weighted average of the population means at A_1 and the weighted average at A_2 are equal. When Y is regressed only on X_2, the test of the regression coefficient corresponds to a test of whether the weighted means at C_1 and C_2 are equal.

16.8 **(a)** The effect coding for the design is as follows:

	A	B		AB	
Y	X_1	X_2	X_3	X_4	X_5
17	1	1	0	1	0
33	1	1	0	1	0
26	1	1	0	1	0
27	1	1	0	1	0
21	1	1	0	1	0
11	1	0	1	0	1
18	1	0	1	0	1
14	1	0	1	0	1
9	1	−1	−1	−1	−1
12	1	−1	−1	−1	−1
10	1	−1	−1	−1	−1
17	−1	1	0	−1	0
8	−1	1	0	−1	0
22	−1	1	0	−1	0
12	−1	1	0	−1	0
9	−1	1	0	−1	0
9	−1	0	1	0	−1
7	−1	0	1	0	−1
4	−1	0	1	0	−1
3	−1	0	1	0	−1
13	−1	−1	−1	1	1
10	−1	−1	−1	1	1
6	−1	−1	−1	1	1
6	−1	−1	−1	1	1
8	−1	−1	−1	1	1

(b) and **(c)**

SV	df	SS	MS	F
A	1	$R_{Y.1}^2 SS_Y = 437.611$	437.611	22.27
B	2	$R_{Y.2}^2 SS_Y = 584.226$	292.113	14.87
$A \mid B,AB$	1	$(R_{Y.1,2,3,4,5}^2 - R_{Y.2,3,4,5}^2)SS_Y = 305.253$	305.253	15.54
$B \mid A,AB$	2	$(R_{Y.1,2,3,4,5}^2 - R_{Y.1,4,5}^2)SS_Y = 528.542$	264.271	13.45
$AB \mid A,B$	2	$(R_{Y.1,2,3,4,5}^2 - R_{Y.1,2,3}^2)SS_Y = 98.575$	49.287	2.51
Error	19	$(1 - R_{Y.1,2,3,4,5}^2)SS_Y = 373.289$	19.646	

In both cases, the $A \times B$ interaction is tested using the $AB \mid A,B$ SV. If we believe that the cell populations are of equal size and therefore wish to test hypotheses about unweighted means, we test the A and B main effects using the $A \mid B,AB$ and $B \mid A,AB$ SV's, respectively. If we believe that the population sizes are proportional to the cell frequencies and therefore wish to test hypotheses about the weighted means, we test the A and B main effects using the A and B SV's.

References

Achen, C.H. (1982). *Interpreting and using regression.* Beverly Hills: SAGE.

Alf, E.F., and Abrahams, N.M. (1975). The use of extreme groups in assessing relationships. *Psychometrika, 40*, 563–572.

Anderson, L.R., and Ager, J.W. (1978). Analysis of variance in small group research. *Personality and Social Psychology Bulletin, 4*, 341–345.

Anscombe, F.J. (1973). Graphs in statistical analysis. *American Statistician, 27*, 17–21.

Appelbaum, M.I., and Cramer, E.M. (1974). Some problems in the nonorthogonal analysis of variance. *Psychological Bulletin, 81*, 335–343.

Atiqullah, M. (1964). The robustness of the covariance analysis of a one-way classification. *Biometrika, 51*, 365–373.

Belsley, D.A., Kuh, E., and Welsch, R.E. (1980). *Regression diagnostics.* New York: Wiley.

Bentler, P.M. (1980). Multivariate analysis with latent variables: Casual modeling. *Annual Review of Psychology, 31*, 419–456.

Bevan, M.F., Denton, J.Q., and Myers, J.L. (1974). The robustness of the *F* test to violations of continuity and form of treatment populations. *British Journal of Mathematical and Statistical Psychology, 27*, 199–204.

Blair, R.C., and Higgins, J.J. (1980). A comparison of the power of Wilcoxon's rank-sum statistic to that of student's *t* statistic under various non-normal distributions. *Journal of Educational Statistics, 5*, 309–335.

Blair, R.C., and Higgins, J.J. (1985). A comparison of the paired samples *t* test to that of Wilcoxon's signed-rank test under various population shapes. *Psychological Bulletin, 97*, 119–128.

Boik, R.J. (1981). A priori tests in repeated measures designs: Effects of nonsphericity. *Psychometrika, 46*, 241–255.

Box, G.E.P. (1954). Some theorems in quadratic forms applied in the study of analysis of variance problems: II. Effects of inequality of variance and covariance between errors in the two-way classification. *Annals of Mathematical Statistics, 35*, 484–498.

Box, G.E.P., Hunter, W.G., and Hunter, J.S. (1978). *Statistics for experimenters: An introduction to design, data analysis, and model building.* New York: Wiley.

Bozivich, H., Bancroft, T.A., and Hartley, H.O. (1956). Power of analysis of variance test procedures for certain incompletely specified models. *Annals of Mathematical Statistics, 27,* 1017–1043.

Brown, M.B., and Forsythe, A.B. (1974a). The small sample behavior of some statistics which test the equality of several means. *Technometrics, 16,* 129–132.

Brown, M.B., and Forsythe, A.B. (1974b). The ANOVA and multiple comparisons for data with heterogeneous variances. *Biometrics, 30,* 719–724.

Bryant, J.L., and Paulson, A.S. (1976). An extension of Tukey's method of multiple comparisons to experimental designs with random concomitant variables. *Biometrika, 63,* 631–638.

Carlson, J.E., and Timm, N.H. (1974). Analysis of nonorthogonal fixed-effect designs. *Psychological Bulletin, 81,* 563–570.

Clark, H.H. (1973). The language-as-fixed-effect fallacy: A critique of language statistics in psychological research. *Journal of Verbal Learning and Verbal Behavior, 12,* 335–359.

Clinch, J.J., and Keselman, H.J. (1982). Parametric alternatives to the analysis of variance. *Journal of Educational Statistics, 7,* 207–214.

Cochran, W.G. (1950). The comparison of percentages in matched samples. *Biometrika, 37,* 256–266.

Cohen, J. (1977). *Statistical power analysis for the behavioral sciences.* New York: Academic Press.

Cohen, J., and Cohen, P. (1983). *Applied multiple regression/correlation analysis for the behavioral sciences,* 2nd ed. Hillsdale, NJ: Erlbaum.

Cook, R.D. (1977) Detection of influential observations in linear regression. *Technometrics, 19,* 15–18.

Cook, R.D., and Weisberg, S. (1982). *Residuals and influence in regression.* New York: Chapman and Hall.

Cook, T.D., and Campbell, D.T. (1979). *Quasi experimentation: Design & analysis issues for field settings.* Chicago: Rand McNally.

Cox, D.R. (1957). The use of a concomitant variable in selecting an experiemental design. *Biometrika, 44,* 150–158.

Cramer, E.M., and Appelbaum, M.I. (1980). Nonorthogonal analysis of variance—once again. *Psychological Bulletin, 87,* 51–57.

Cronbach, L.J., and Furby, L. (1970). How should we measure "change"—or should we? *Psychological Bulletin, 74,* 68–80.

Dalton, S., and Overall, J.C. (1977). Nonrandom assignment in ANCOVA: The alternative ranks design. *The Journal of Experimental Education, 46,* 58–62.

Davenport, J.M., and Webster, J.T. (1973). A comparison of some approximate F tests. *Technometrics, 15,* 779–789.

Dawes, R.M. (1971). A case study of graduate admissions: Application of three principles of human decision making. *American Psychologist, 26,* 180–188.

Delucchi, K.L. (1983). The use and misuse of chi-square: Lewis and Burke revisited. *Psychological Bulletin, 94,* 166–176.

Dijkstra, J.B., and Werter, P.S. (1981). Testing the equality of several means when the population variances are unequal. *Communications in Statistics, B. Simulation and Computation, 10,* 557–569.

Dillon, W.R., and Goldstein, M. (1984). *Multivariate analysis: Methods and applications*. New York: Wiley.

Donaldson, T.S. (1968). Robustness of the F test to errors of both kinds and the correlation between the numerator and denominator of the F ratio. *Journal of the American Statistical Association, 63,* 660–676.

Draper, N.R., and Smith, H. (1981). *Applied regression analysis*, 2nd ed. New York: Wiley.

Dunn, O.J. (1961). Multiple comparisons among means. *Journal of the American Statistical Association, 56,* 52–64.

Dunnett, C.W. (1955). A multiple comparison procedure for combining several treatments with a control. *Journal of the American Statistical Association, 50,* 1096–1121.

Dunnett, C.W. (1964). New tables for multiple comparisons with a control. *Biometrics, 20,* 482–491.

Edgell, S.E., and Noon, S.M. (1984). Effect of violation of normality on the t test of the correlation coefficient. *Psychological Bulletin, 95,* 576–583.

Edwards, A.L. (1979). *Multiple regression and the analysis of variance and covariance*. New York: Freeman.

Einhorn, H.J., and Hogarth, R.M. (1986). Judging probable cause. *Psychological Bulletin, 99,* 3–19.

Feldt, L.S. (1958). A comparison of the precision of three experimental designs employing a concomitant variable. *Psychometrika, 23,* 335–353.

Fisher, R.A. (1921). On the probable error of a coefficient of correlation deduced from a small sample. *Metron, 1,* 3–32.

Fisher, R.A. (1952). *Statistical methods for research workers,* 12th ed. London: Oliver and Boyd.

Fisher, R.A., and Yates, F. (1963). *Statistical tables for biological, agricultural, and medical research.* Edinburgh: Oliver and Boyd.

Forster, K.I., and Dickinson, R.G. (1976). More on the language-as-fixed-effect fallacy: Monte Carlo estimates of error rates for F_1, F_2, F', and $min F$. *Journal of Verbal Learning and Verbal Behavior, 15,* 135–142.

Friedman, M. (1937). The use of ranks to avoid the assumption of normality implicit in the analysis of variance. *Journal of the American Statistical Association, 217,* 929–932.

Games, P.A., Keselman, H.J., and Clinch, J.J. (1979). Tests for homogeneity of variance in factorial designs. *Psychological Bulletin, 86,* 978–984.

Gideon, R.A., and Hollister, R.A. (1987). A rank correlation coefficient resistant to outliers. *Journal of the American Statistical Association, 82,* 656–666.

Graybill, R.A. (1961). *An introduction to linear statistical methods,* Vol. I. New York: McGraw-Hill.

Greenhouse, S.W., and Geisser, S. (1959). On methods in the analysis of profile data. *Psychometrika, 55,* 431–433.

Hanushek, E.A., and Jackson, J.E. (1977). *Statistical methods for social scientists.* New York: Academic Press.

Harris, R.J. (1985). *A primer of multivariate statistics*, 2nd ed. New York: Academic Press.

Hays, W.L. (1988). *Statistics*, 4th ed. New York: Holt, Rinehart, and Winston.

Hildebrand, D.K. (1986). *Statistical thinking for behavioral scientists.* Boston: Duxbury Press.

Hoaglin, D.C., Mosteller, F., and Tukey, J.W. (1983). *Understanding robust and exploratory data analysis.* New York: Wiley.

Hoaglin, D.C., and Welsch, R. (1978). The hat matrix in regression and ANOVA. *American Statistician, 32,* 17–22.

Hocking, R.R. (1983). Developments in linear regression methodology: 1959–1982. *Technometrics, 25,* 219–245.

Hodges, J., and Lehmann, E. (1956). The efficiency of some nonparametric competitors of the *t* test. *Annals of Mathematical Statistics, 27,* 324–335.

Hogg, R.V., Fisher, D.M., and Randles, R.K. (1975). A two-sample adaptive distribution-free test. *Journal of the American Statistical Association, 70,* 656–671.

Holland, B.S., and Copenhaver, M.D. (1988). Improved Bonferroni-type multiple testing procedures. *Psychological Bulletin, 104,* 145–149.

Hora, S.C., and Iman, R.L. (1988). Asymptotic relative efficiencies of the rank-transformation procedure in randomized complete-block designs. *Journal of the American Statistical Association, 83,* 462–470.

Horst, P., and Edwards, A.L. (1982). Analysis of nonorthogonal designs: The 2^k factorial experiment. *Psychological Bulletin, 91,* 190–192.

Hsu, T.C., and Feldt, L.S. (1969). The effect of limitations on the number of criterion score values on the significance of the *F* test. *American Educational Research Journal, 6,* 515–527.

Huck, S.W., and McLean, R.A. (1975). Using a repeated measures ANOVA to analyze the data from a pretest-posttest design: A potentially confusing task. *Psychological Bulletin, 82,* 511–518.

Hudson, J.D., and Krutchkoff, R.C. (1968). A Monte Carlo investigation of the size and power of tests employing Satterthwaite's synthetic mean squares. *Biometrika, 55,* 431–433.

Huitema, B.E. (1980). *The analysis of covariance and alternatives.* New York: Wiley.

Huynh, H., and Feldt, L.S. (1970). Conditions under which mean square ratios in repeated measurements designs have exact *F* distributions. *Journal of the American Statistical Association, 65,* 1582–1589.

Huynh, H., and Feldt, L.S. (1976). Estimation of the Box correction for degrees of freedom from sample data in randomized block and split-plot designs. *Journal of Educational Statistics, 1,* 69–82.

Iman, R.L., Hora, S.C., and Conover, W.J. (1984). Comparison of asymptotically distribution-free procedures for the analysis of complete blocks. *Journal of the American Statistical Association, 79,* 674–685.

Jaccard, J., Becker, M.A., and Wood, G. (1984). Pairwise multiple comparison procedures: A review. *Psychological Bulletin, 96,* 589–596.

Jennings, E. (1988). Models for pretest-posttest data: Repeated measures ANOVA revisited. *Journal of Educational Statistics, 13,* 273–280.

Johnson, P.O., and Neyman, J. (1936). Tests of certain linear hypotheses and their application to some educational problems. *Statistical Research Memoirs, 1,* 57–93.

Joreskog, K.G., and Sorbom, D. (1986). *LISREL: Analysis of linear structural relationships by the method of maximum likelihood* (Version VI). Mooresville, IN: Scientific Software, Inc.

Kendall, M.G. (1963). *Rank correlation methods*, 3rd ed. London: Griffin.

Kenny, D.A. (1979). *Correlation and causality.* New York: Wiley.

Kepner, J.L., and Robinson, D.H. (1988). Nonparametric methods for detecting treatment effects in repeated-measures designs. *Journal of the American Statistical Association, 83,* 456–461.

Keselman, H.J. (1974). The statistic with the smaller critical value. *Psychological Bulletin, 81,* 130–131.

Keselman, H.J., Games, P.A., and Rogan, J.C. (1979). An addendum to "A comparison of modified-Tukey and Scheffé methods of multiple comparisons for pairwise contrasts." *Journal of the American Statistical Association, 62,* 626–633.

Kirk, R.E. (1982). *Experimental design: Procedures for the behavioral sciences,* 2nd ed. Belmont, CA: Brooks/Cole.

Kohr, R.L., and Games, P.A. (1977). Testing complex a priori contrasts in means from independent samples. *Journal of Educational Statistics, 1,* 207–216.

Lehmann, E.L. (1975). *Nonparametrics: Statistical methods based on ranks.* San Francisco: Holden-Day.

Levene, H. (1960). Robust tests for equality of variances. In I. Olkins (Ed.), *Contributions to probability and statistics.* Stanford: Stanford University Press.

Levine, D.W., and Dunlap, W.P. (1982). Power of the F test with skewed data: Should one transform or not? *Psychological Bulletin, 92,* 272–280.

Lindeman, R.H., Merenda, P.F., and Gold, R.Z. (1980). *Introduction to bivariate and multivariate analysis.* Glenview, IL: Scott Foresman.

Lindquist, E.F. (1953). *Design and analysis of experiments in education and psychology.* Boston: Houghton-Mifflin.

Linn, R.L., and Slinde, J.A. (1977). The determination of the significance of change between pre- and posttesting periods. *Review of Educational Research, 47,* 121–150.

Lorch, R.F., and Myers, J.L. (1990). Regression analyses of repeated measures data in cognitive research. *Journal of Experimental Psychology: Learning, Memory, and Cognition, 16,* 149–157.

Lunney, G.H. (1970). Using analysis of variance with a dichotomous variable: An empirical study. *Journal of Educational Measurement, 7,* 263–269.

Marston, A.R. (1971). It is time to reconsider the Graduate Record Examination. *American Psychologist, 26,* 653–655.

Mauchly, J.W. (1940). Significance test for sphericity of a normal n-variate distribution. *Annals of Mathematical Statistics, 11,* 204–209.

Maxwell, S.E. (1980). Pairwise multiple comparisons in repeated measures designs. *Journal of Educational Statistics, 5,* 269–287.

Maxwell, S.E., and Bray, J.H. (1986). Robustness of the quasi F statistic to violations of sphericity. *Psychological Bulletin, 99,* 416–421.

Maxwell, S.E., Delaney, H.D., and Dill, C.A. (1984). Another look at ANCOVA versus blocking. *Psychological Bulletin, 95,* 136–147.

Maxwell, S.E., Delaney, H.D., and Sternitzke, M.E. (1983). Complex comparisons in repeated measures designs. *Unpublished manuscript.*

Maxwell, S.E., and Howard, G.S. (1981). Change scores—necessarily anathema? *Educational and Psychological Measurement, 41,* 747–756.

Mitzel, H.C., and Games, P.A. (1981). Circularity and multiple comparisons in repeated measures designs. *British Journal of Mathematical and Statistical Psychology, 34,* 253–259.

Morrison, D.F. (1976). *Multivariate statistical methods,* 2nd ed. New York: McGraw-Hill.

Myers, J.L. (1979). *Fundamentals of experimental design,* 3rd ed. Boston: Allyn and Bacon.

Myers, J.L., DiCecco, J.V., and Lorch, R.F. (1981). Group dynamics and individual performances: Pseudogroup and quasi-*F* analyses. *Journal of Personality and Social Psychology, 40,* 86–98.

Myers, J.L., DiCecco, J.V., White, J.B., and Borden, V.M. (1982). Repeated measurements on dichotomous variables: *Q* and *F* tests. *Psychological Bulletin, 92,* 517–525.

Nesselroade, J.R., Stigler, S.M., and Baltes, P.B. (1980). Regression toward the mean and the study of change. *Psychological Bulletin, 88,* 622–637.

Nunnally, J.C. (1978). *Psychometric theory,* 2nd ed. New York: McGraw-Hill.

O'Brien, R.G., and Kaiser, M.K. (1985). MANOVA method for analyzing repeated measures designs: An extensive primer. *Psychological Bulletin, 97,* 316–333.

Odeh, R.E. (1977). Extended tables of the distribution of Friedman's *S*-statistic in the two-way layout. *Communications in Statistics, Part B. Simulation and Computation, 6,* 29–48.

Overall, J.E., Lee, D.M., and Hornick, C.W. (1981). Comparison of two strategies for analysis of variance in nonorthogonal designs. *Psychological Bulletin, 90,* 367–375.

Overall, J.E., and Spiegel, D.K. (1969). Concerning least squares analysis of experimental data. *Psychological Bulletin, 72,* 311–322.

Pedhazur, E.J. (1977). Coding subjects in repeated measures designs. *Psychological Bulletin, 84,* 298–305.

Pedhazur, E.J. (1982). *Multiple regression in behavioral research,* 2nd ed. New York: Holt, Rinehart, and Winston.

Perlmutter, J., and Myers, J.L. (1973). A comparison of two procedures for testing multiple contrasts. *Psychological Bulletin, 79,* 181–184.

Pollatsek, A., Lima, S.D., and Well, A.D. (1981). Concept or computation: Students' understanding of the mean. *Educational Studies in Mathematics, 12,* 191–204.

Price, B. (1977). Ridge regression: An application to nonexperimental data. *Psychological Bulletin, 84,* 759–766.

Reichardt, C.S. (1979). The statistical analysis of data from non-equivalent group designs. In T.D. Cook and D.T. Campbell (Eds.), *Quasi-experimentation: Design and analysis issues for field settings.* Chicago: Rand McNally.

Rencher, A.C., and Pun, F.C. (1980). Inflation of *R*-squared in best subset regression. *Technometrics, 22,* 49–54.

Rogan, J.C., Keselman, H.J., and Mendoza, J.L. (1979). Analysis of repeated measurements. *British Journal of Mathematical and Statistical Psychology, 32,* 269–286.

Rouanet, H., and Lepine, D. (1970). Comparisons between treatments in a repeated-measurement design: ANOVA and multivariate methods. *British Journal of Mathematical and Statistical Psychology, 23,* 147–163.

Rousseeuw, J.R., and Leroy, A.M. (1987). *Robust regression and outlier detection.* New York: Wiley.

Rozeboom, W.W. (1979). Ridge regression: Bonanza or beguilement? *Psychological Bulletin, 86,* 242–249.

Satterthwaite, F.E. (1946). An approximate distribution of variance components. *Biometrics Bulletin, 2,* 110–114.

Scheffé, H. (1959). *The analysis of variance.* New York: Wiley.

Sidak, Z. (1967). Rectangular confidence regions for the means of multivariate normal distributions. *Journal of the American Statistical Association, 62,* 626–633.

Siegel, S., and Castellan, N.J. (1988). *Nonparametric statistics for the behavioral sciences,* 2nd ed. New York: McGraw-Hill.

Smith, H.F. (1957). Interpretation of adjusted treatment means and regressions in analysis of covariance. *Biometrics, 13,* 282–308.

Snedecor, G.W., and Cochran, W.G. (1967). *Statistical methods,* 6th ed. Ames: Iowa State University Press.

Srivastava, S.R., and Bozivich, H. (1961). Power of certain analysis of variance test procedures involving preliminary tests. *Bulletin de l'Institut International Statistique,* 33rd Session.

Steiger, J.H. (1980). Tests for comparing elements of a correlation matrix. *Psychological Bulletin, 87,* 245–251.

Stevens, J.P. (1984). Outliers and influential data points in regression analysis. *Psychological Bulletin, 95,* 334–344.

Stevens, J.P. (1986). *Applied multivariate statistics for the social sciences.* Hillsdale, N.J.: Erlbaum.

Tamhane, A.C. (1979). A comparison of procedures for multiple comparisons of means with unequal variances. *Journal of the American Statistical Association, 74,* 471–480.

Tomarken, A.J., and Serlin, R.C. (1986). Comparison of ANOVA alternatives under variance heterogeneity and specific noncentrality structures. *Psychological Bulletin, 99,* 90–99.

Tukey, J.W. (1949). One degree of freedom for nonadditivity. *Biometrics, 5,* 232–242.

Tukey, J.W. (1953). The problem of multiple comparisons. Unpublished manuscript, Princeton University.

Tukey, J.W. (1965). A test for nonadditivity in the Latin square. *Biometrics, 21,* 111–113.

Tukey, J.W. (1969). Analyzing data: Sanctification or detective work? *American Psychologist, 24,* 83–91.

Tukey, J.W. (1977). *Exploratory data analysis.* Reading, MA: Addison-Wesley.

Velleman, P.F., and Hoaglin, D.C. (1981). *Applications, basics, and computing of exploratory data analysis.* Boston: Duxbury Press.

Velleman, P., and Welsch, R. (1981). Efficient computing of regression diagnostics. *American Statistician, 35,* 234–242.

Wainer, H., and Thissen, D. (1976). Three steps toward robust regression. *Psychometrika, 41,* 9–34.

Weisberg, S. (1980). *Applied regression analysis.* New York: Wiley.

Welch, B.L. (1938). The significance of the difference between two means when the population variances are unequal. *Biometrika, 29,* 350–362.

Welch, B.L. (1947). The generalization of Student's problem when several different population variances are involved. *Biometrika, 34,* 28–35.

Wherry, R.J. (1931). A new formula for predicting the shrinkage of the coefficient of multiple correlation. *Annals of Mathematical Statistics, 2,* 440–457.

Wilk, M.B., and Kempthorne, O. (1957). Nonadditivities in a Latin square. *Journal of the American Statistical Association, 52,* 218–236.

Wilkinson, L. (1979). Tests of significance in stepwise regression. *Psychological Bulletin, 86,* 168–174.

Wilkinson, L., and Dallal, G.E. (1982). Tests of significance in forward selection regression with an *F*-to-enter stopping rule. *Technometrics, 24,* 25–28.

Yates, F. (1934). The analysis of multiple classifications with unequal numbers in the different classes. *Journal of the American Statistical Association, 29,* 57–66.

Yuen, K. (1974). The two-sample trimmed *t* for unequal population variances. *Biometrika, 61,* 165–170.

Yuen, K., and Dixon, W. (1973). The approximate behavior and performance of the two-sample trimmed *t*. *Biometrika, 60,* 369–374.

Index

ISBN 0-673-46414-8

90000